Baruch S. Blumberg

HEPATITIS B AND THE PREVENTION OF PRIMARY CANCER OF THE LIVER

Selected Publications
of Baruch S Blumberg

World Scientific Series in 20th Century Biology – Vol. 4

HEPATITIS B AND THE PREVENTION OF PRIMARY CANCER OF THE LIVER

Selected Publications of Baruch S Blumberg

Edited by

Baruch S Blumberg

Fox Chase Cancer Center, USA

World Scientific

Singapore • New Jersey • London • Hong Kong

Published by

World Scientific Publishing Co. Pte. Ltd.

P O Box 128, Farrer Road, Singapore 912805

USA office: Suite 1B, 1060 Main Street, River Edge, NJ 07661

UK office: 57 Shelton Street, Covent Garden, London WC2H 9HE

British Library Cataloguing-in-Publication Data
A catalogue record for this book is available from the British Library.

HEPATITIS B AND THE PREVENTION OF PRIMARY CANCER OF THE LIVER: SELECTED PUBLICATIONS OF BARUCH S BLUMBERG

ISBN 981-02-3217-9

Printed in Singapore by Uto-Print

PREFACE

Contemporary biological and medical science is characterized by team work. The research which is described in this volume is the product of the efforts of many of my colleagues and would not have been possible without their contributions. The research of our group was built on, and guided by, the work of many physicians and scientists who preceded us and would not have been possible had we not been able to stand "on the shoulders of giants."

I was asked by the publishers to prepare this volume and to include not only a series of publications for reproduction, but also an essay to clarify and provide a guide to the corpus of research represented by the reprints. It has been written from a personal perspective and therefore represents my own perception of the events surrounding the projects. They start with my first scientific paper published in 1951, based on studies of tropical diseases in Suriname, South America, and end with the most recent studies in virology published in 1997. At the request of the publishers, also included are several papers, both published and unpublished, which are not directly based on the main areas of my scientific research.

I have already acknowledged my debt to colleagues both past and present; they will be discussed in greater detail in the introductory essay and the published papers are a testament to their scientific accomplishments. I especially wish to acknowledge the help of my secretaries and administrative assistants, Joyce Codispoti and Ann Dortort, who have made it possible to prepare this book. I am also indebted to Joyce and her husband, Dr. Joseph Codispoti, and my daughter, Jane Blumberg, for the excellent editorial help they have provided. Any errors that still remain are my doing.

The research that will be described here took many hours, days, weeks, and months of my life. A good part of it was accomplished on field trips, often to remote places, that kept me away from home for long periods. My wife, Jean, managed to keep our growing family thriving during those difficult times and to raise our four wonderful children. She has always encouraged and supported my labors during the 43 years of our married life, even though it often placed a great burden on her and took time away from her own career as an artist. My love and admiration for Jean are unbounded.

My four children, Anne, George, Jane, and Noah have been a joy and their interest in my work has been sustaining. It has been a great privilege to be able to watch them grow and develop; they have always maintained a strong sense of family love and loyalty. I acknowledge my great thanks to them.

Baruch S. Blumberg
1998

To Jean

CONTENTS

I INTRODUCTORY ESSAY

1. INTRODUCTION

1.1. General Comments

The guiding theme of my research has been the study of human genetic variation and its relation to susceptibility to disease. Although the research commenced with a focus on human population genetics, the major outcome was in the field of virology and preventive medicine. The research extended over a wide variety of subjects, for most of which I did not have any formal training. I have always had the sense of being an amateur, which has disadvantages, but it does have merit in an arena where new information is always flowing in.

My first formal research was a series of field studies on tropical diseases conducted in Suriname, South America, in 1949. I have continued, nearly without interruption, for close to fifty years. This book contains published and unpublished papers selected to represent the work my colleagues and I have done over this period. This introductory essay is an attempt to provide a synthesis of the research program as well as the background for the studies presented in the papers.

It would be useful to start with a few general comments on the scientific process. Science can be considered as *inductive* or *deductive*, although the division is often difficult to establish. I use the term *inductive* to mean the collection of data *first*, followed by the generation of hypotheses. *Deductive* implies that the hypothesis is stated first, and then data is collected to test it. In science there is a nearly seamless alteration between these approaches. It is often difficult to obtain support for inductive sciences. There is a demand for a well formulated hypothesis so that the direction of the research, and even its conclusion, can be seen at its start.

But inductive research has the advantage of producing new ideas from new data. Hypothesis testing provides the solid support required for scientific rigor and of course, also merges into an inductive phase when another wave of new ideas can be introduced.

There is a freshness to data newly collected. Data collected by the scientist, particularly field data, carries with it a scent of possessiveness, of uniqueness, of knowing about the circumstances and environment in which they were collected, that cannot be articulated in the published description of the data. But this extra knowledge, even if not consciously perceived, can inform hypothesis-formulation and testing and give it a greater imaginative content. We enjoyed collecting our own descriptive data, particularly in the field, and enjoyed the inductive and categorizing aspects of science. This did not diminish the allure of hypothesis creation and testing which is an essential part of the cycle of the scientific endeavor.

There was a seasonal character to the research. Field trips were usually in the summer, partly because travel was easier but, to a large extent, because in the summer my mother-in-law could spend her vacations with the family and help with the care of our children. The autumn and winter were spent assembling the data, doing the laboratory studies that were essential to the program, and analyzing the data. Then, with the coming of spring, we were like the Vikings after the seeds had been planted in the ground, off to wandering and collecting.

1.2. A Brief Chronology

My undergraduate degree in Physics was from Union College, Schenectady, New York. I attended graduate school in Physics and Mathematics at Columbia University from 1946 to 1947 and then entered medical school at the College of Physicians and Surgeons of Columbia University, graduating in 1951. I completed my clinical house staff training in Internal Medicine on the First Division at Bellevue Hospital and the Columbia-Presbyterian Medical Center between 1951 and 1955. In 1955, I went to Oxford University and received a D. Phil. (Ph.D.) in Biochemistry. I was a senior investigator at the National Institutes of Health from 1957 to 1964. In 1964, I was appointed Associate Director for Clinical Research at the Institute for Cancer Research of the Fox Chase Cancer Center in Philadelphia. Between 1989 and 1994, I was Master of Balliol College, Oxford University, and then returned to a continuing appointment at Fox Chase. I have been on the faculty of the University of Pennsylvania for many years with an appointment as Professor of Medicine and Anthropology.

2. EARLY TRAINING AND RESEARCH

2.1. Inherited Differences in Susceptibility to Disease, 1950–51

I first became aware of the power of inherited differences in disease susceptibility during my third year at medical school when, in 1950, I went to the then Dutch colony of Suriname in northern South America. I worked in the hospital in

Moengo, a remote community on the Cottica River in the coastal swamps of the north. It was the site of a large aluminum bauxite mine run by a Suriname subsidiary of the Aluminum Company of America. In addition to clinical responsibilities, we undertook several research projects, primarily surveys of tropical diseases. These included intestinal parasites, malaria, and filariasis.[1,2,3]

Suriname was (and is) a remarkably interesting country, nearly entirely undeveloped, with vast rain forests covering most of the land. There were only a few miles of automobile road in the northern part of the country, very few, if any, airstrips in the interior, and helicopters were not available. The jungle was so dense that walking trails were rare, and the main method of travel and communication was the mighty rivers which drained the interior. The population was heterogeneous. There were descendants of African slaves brought to the country during the period of the sugar plantations. Many lived in the few towns along the coast, but there were also sizable groups of *Djukas,* the descendants of slaves who, in the mid seventeen hundreds, successfully rebelled against the colonial Dutch government. They established self-governing communities in the interior, often on the isolated southern side of the escarpment that ran parallel to the coast. This high ground turned the rivers which served as the major method of communication into powerful rapids as they flowed northward toward the Caribbean Sea. Only the Djukas could negotiate these rapids, and to travel south of them, into the interior, required their help. In order to conduct a malaria survey and provide medical treatment for the Djukas, we organized a trip up the Marowyne River to the island of Langatabatje, the seat of the King of the Paramacanners. We had a *corrial* built for us — a dugout canoe — about 35 feet long. We contracted with *Moiman*, a famous guide to take us through the rapids and to the island headquarters of the Paramacanners. (*Moiman* means "Goodman" in the local language, *Talkee-talkee*, a brand of circum-Caribbean pidgin which contained many African words and included contributions from several European languages including English, Dutch, French and Spanish.) It was exhilarating maneuvering the heavy canoe with a full load of supplies, food and medicine, trade goods and presents, myself and two other members of the expedition, Moiman and his three paddlers, and the younger brother of one of them who simply came along for the ride. It was a demanding task to manhandle the *corrial* through the rapids where we often had to drag the heavily laden boat across the rocks as we went up-river to gain the escarpment.

The other inhabitants of Suriname included East Indians (mostly Javanese) brought into the country during the period when the Netherlands was the colonial power in what is now Indonesia, Hindus and Moslems from sub-continent India, Chinese, and a few Europeans. In Moengo, the employees lived mostly under uniform conditions in company-built housing although they maintained much of their indigenous culture.

There was a striking difference among the various populations in the prevalence of microfilaria, a phase in the reproductive cycle of *Wuchereria bancrofti,* a causative agent of filariasis (elephantiasis) in South America. The residents of African descent had five times the prevalence of the Javanese. There was no obvious cause for this difference and we did not have the facilities at the time to carry this observation forward. But it did bring home the concept of population and individual differences in susceptibility to an infectious health hazard, a concept that became a guiding influence in my later research.

2.2. Clinical Training and Research, 1951–55

The early part of my career did not allow me the opportunity of undertaking research in the fascinating field of inherited human variation. My undergraduate training and a year of graduate training were in Physics and Mathematics. My insensitivity to my own character was such that it took me all those years of intense education to realize that I had no talent or the appropriate intelligence for these demanding subjects. At the recommendation of my father and the strong support of my Professor of Physics at Union College, who recommended that I should *not* continue in his subject, I changed my course to medicine. Following my graduation in 1951, I had four years of postgraduate clinical training at Bellevue and Presbyterian Hospitals in New York City as a House Staff officer. I was totally immersed in the care of patients, an experience that I thoroughly enjoyed.

The concept of difference in susceptibility to diseases and their effects was enhanced during my clinical experience as an intern and resident at Bellevue Hospital, a large New York City municipal hospital caring for the needy ill. The heterogeneous population of the City provided many examples of differences in disease susceptibility, particularly among the patients on the teeming tuberculosis wards. Tuberculosis was still a major problem; public health measures were not fully effective, and antibiotic treatment was in its early stages. There were striking differences in the outcome of the disease in different populations. Our clinical impressions were that Scandinavians and the Irish had a much poorer prognosis than others, and a variety of studies appeared to bear out these impressions.

Bellevue was a surrealistic experience. As young physicians we were given immense responsibility. We were charged with the care of the city's poor. Funds, equipment, and resources were inadequate. We did not even have enough needles to draw bloods on all of our patients in the morning rounds and they were reused after superficial cleaning. This was in the days before disposable needles and other blood-contact instruments were in wide use. Ironically, our later work on the blood-borne transmission of hepatitis B virus, and the recognition that it could be transmitted by re-used needles impelled the adoption of disposable needles and other devices. The staff were too few in number and, as interns and residents, we often had to perform routine tasks. For example, taking patients to and from X-ray and surgery and selecting and typing donor blood before we administered it to our patients. But the flip side of these inadequate arrangements, if they did not result in complete frustration, was the instilling of a strong sense of responsibility. If we did not care for our patients, then there often was no one else who could do so. It also accustomed us to the notion of hard work. There was no fixed length to the workday; one stayed until the patients were cared for. The definition of a "half-day" meant twelve hours! I'm sure that my recollections are clouded by the conceits of the young, but they were rich and intense days and nights. Surrounded as we were day and night by illness, the misery imposed on the poor of a big city, the Bellevue staff had, I would like to think, a sense of compassion and concern for our patients who often had no other source of comfort.

We also developed a gallows humor, a protection against the sorrow of losing so many of our patients, and a happy ability to recreate excessively during the little free time we had. Mortality among the poor for most diseases is greater than among the rich; and this was reflected in the sad outcome for many of those for whom we were responsible. Having had this experience with the terrible social and personal problems generated by inadequate medical care, I have retained ever since a concern for the provision of health care, particularly preventive care, for those who aren't adequately provided for in our society. This applies even more so in the underdeveloped parts of the world where poor health often dominates the prospects of the population.

Medicine is an excellent preparation for biological research. The clinical investigator, by which I mean a scientist who studies humans, develops as a consequence of clinical experience, an intimate understanding of his/her subject of study. The knowledge approaches the intuitive. Obviously intuition may be appropriate for generating hypothesis about a natural phenomenon but objective experiment and observation, not intuitive hunches, are required to test the hypothesis.

Physicians receive a very broad biological education. I understand that in recent years the amount of time spent in the laboratory has decreased but during the period I received my training, we spent most afternoons of our first few years at the laboratory bench actually doing the experiments and making natural observations. These included anatomy (we had half a year of 5 days a week of anatomy dissection), biochemistry, pharmacology, physiology, parasitology, and microbiology. We did not become expert in these basic science subjects but we knew what they were. This has the advantage of providing an interdisciplinary background which can be called on in the research years to address a problem that could not be dealt with by only one discipline.

A further advantage is that scientists with clinical experience are aware of the nature of medical problems because they have seen them. For example, when we discovered "Australia antigen" we were aware of the post-transfusion hepatitis problem and this hastened our identification of the mysterious antigen with the hepatitis virus. Clinical experience also helps in the process of "technology transfer", the current term for application of basic science discoveries to practical (and commercial) uses.

Perhaps the most important feature of medical training and experience is that it accustoms its practitioners to arrive at a decision with incomplete but, hopefully, adequate knowledge. As discomforting as this may be to putative patients, it is often essential to arrive at a plan of action before the complete story unfolds and this may happen in research as well.

As is often the case, the direction of my research did not immediately take an unambiguous direction. Research is *Shandian,* in the sense of the novel "Tristram Shandy" (1759–67) by the Irish-born English novelist Laurence Sterne (1713–1768). In this comic novel, any sense of narrative order is subordinated to the free associations of characters and the narrator. Events ramble from one apparent irrelevancy to another but, nevertheless, a strange sense of order emerges. There is also an interesting exploration of the quality of time. A description of an event takes far longer to relate than the event itself. For example, he tells of his own conception but gets so embroiled in the tale of his parents' ineptitude in initiating the process that it takes three volumes to get himself born! Research often is ordered and planned (for example, Arthur Miles' X-ray research described by C. P. Snow in his excellent novel of science, *The Search*), but it is also circuitous and driven by the immediate event rather than a pattern determined at its beginning. It can be argued that much of original research fits the circuitous pattern. However, when reconstructing the sequence of events the scientist may attempt to present it in a form to suggest that there was an *a priori* plan of rigorous order, whereas the actual sequence was less obviously logical. I will attempt to provide an actual rather than imagined order, not only in the interests of verity, but because it is more interesting.

My clinical training was in internal medicine with a specialization in arthritis and rheumatism. I completed a study of the natural history of ankylosing spondylitis,[4] a relatively uncommon arthritis of the spine which often ends in a grossly bent back. It was primarily descriptive but included the first estimation of the life expectancy of individuals with this disease compared to those without it. The disease did not appear to decrease their survival. I used long term outcome in evaluating medical phenomenon extensively in my later research and I shall refer to it as the narrative continues.

2.3. Studies on Hyaluronic Acid, 1953–57

While completing my Fellowship in the Arthritis group in the Department of Medicine at Columbia-Presbyterian Medical Center, it was suggested that I undertake a project in physical biochemistry. It was assumed that because of my training in physics and mathematics, this would be a suitable topic for me. Little did they or I realize that I had reached the limits of my mathematical capabilities and this was not an ideal subject for me. My thinking was much more biological than I had imagined it would be. In any case, they recommended that I work on hyaluronic acid, a natural biochemical which was of interest to basic scientists in the arthritis field since it is a major component of synovial fluid which lubricates the joints. Ever the dutiful student, I complied.

Hyaluronic acid, one of a large group of natural polyelectrolytes, is a long chain carbohydrate polymer with a discaccharide, beta-glucuronido-1-3 N-acetylglucosamine constituting the repeating unit. Each of the monomers contains an ionizable carboxyl group. Hyaluronic acid is a constituent of the ground substance which permeates the connective tissue and is involved in ion transport, support, and storage. It is also found in the umbilical cord, the vitreous humor and, as noted, in joint fluid, where it is thought to impart unique lubricating properties because of its non-Newtonian viscosity. At Columbia-Presbyterian Medical Center I worked (1953–4) in the laboratory of Professor Karl Meyer who had first isolated the chemical. I also worked with Gerald Oster, then at the Institute of Polymer Research of the Polytechnic Institute of Brooklyn. We did a series of physical chemical studies on the molecule[5,6] and learned that size and shape were very dependent on sodium chloride concentration. We suggested that hyaluronate played a role in the regulation of salt and water under physiological conditions and in pathological states such as inflammation, edema and dehydration.

However, most of my work on the physical characteristics of hyaluronate was done in the laboratory of Alexander ("Sandy") G. Ogston at the Department of Biochemistry at Oxford University (1955–57). This came about as a consequence of my interest in continuing the studies on the physical characteristics of hyaluronate after completing my clinical training. There were few people working in this field, and Ogston was one of the leaders. But there were other reasons. In 1954 I had married Jean Liebesman, whom I had met when I was an intern at Bellevue Hospital. We wanted to live and work abroad for a period and England was an exciting possibility. Before I started, Sandy suggested that I enroll as a candidate for the D. Phil. (Ph.D.) degree and that my research would constitute my thesis topic. I was matriculated as a student of Balliol College, where Sandy was a Fellow and the Tutor in Biochemistry and this began a lifelong association with the College. But more of this later.

In its native state in joint fluid, hyaluronic acid is associated with protein. We wanted to determine if the presence of the protein is required to provide the unique physical characteristics of the molecule, in particular its non-Newtonian viscosity which, it was thought, imparted helpful lubricating characteristics. The plan was to study the native hyaluronate-protein complex and then remove the protein gradually with a protease. In the mid-1950s science was do-it-yourself, with little equipment and few experimental chemicals available from the science supply industry. The project started with the collection of several papaya from which I purified the potent protease papain and double crystallized it to insure purity. My measurements were made on a Coue viscometer which I believe had been fabricated in the departmental machine shop. The ultracentrifuge was a Svedberg instrument modeled after the original machine designed in Sweden. It occupied an entire room and the machinery associated with it reminded me of the engine room of the Landing Ship I had served on while I was in the Navy, the only difference being that it was substantially more finicky than the diesels which propelled our ship. We determined in a series of studies[7,8] that the non-Newtonian viscosity was significantly diminished when the protein was removed by proteolysis.

Although I enjoyed this bit of research, I never returned to it after the completion of my thesis. I wonder how often this occurs in science? However, the years I spent in Oxford were a turning point in my scientific training. Sandy was a wonderful mentor. We shared a laboratory and I saw him nearly daily. He had a great curiosity about nature and his research was designed to satisfy this curiosity. Although I later learned the value and excitement of practical application, Sandy taught me to follow a problem for its own intrinsic value. We remained friends for more than forty years. I was particularly pleased when many years later Sandy's son, Walter Ogston, came to work in my laboratory in Philadelphia.

Sandy Ogston died in 1996. I was very honored and moved when his family asked me to deliver the eulogy address at his memorial service held in Oxford in November, 1996.[9]

3. GENETIC POLYMORPHISM

3.1. The Concept of Polymorphism

During my second year at the Department of Biochemistry my research took an unplanned turn which eventually brought me back to my original interest in inherited variation. There has been a long-standing interest, one could say preoccupation, in Darwinian Evolution in Oxford science and particularly in the Department of Zoology. Oxford was the scene, at the 1860 meeting of the British Association for the Advancement of Science, of the great debate between Thomas Henry Huxley, Darwin's colleague and spokesperson ("bulldog"), and Samuel Wilberforce, Bishop of Oxford. Wilberforce, a strong defender of religious orthodoxy, challenged the notion that humans were descended from lower primates and in the heat of argument asked Huxley if he were descended from an ape. Exercising his considerable rhetorical talents Huxley replied,

> "If the question is put to me, would I rather have a miserable ape for a grandfather or a man highly endowed by nature and possessed of great means of influence, and yet who employs these faculties and that influence for the mere purpose of introducing ridicule into a grave scientific discussion — I unhesitatingly affirm my preference for the ape."

This devastating reply won the day and the cause of Darwinism secured a strong hold in the intellectual life of the country.

The concept of a genetic polymorphism was developed, among others, by E. B. Ford, Professor of Ecological Genetics at Oxford. The definition which he used, and which is still useful is:

> "... the occurrence together in the same habitat of two or more discontinuous forms or "phases" of a species in such proportions that the rarest of them cannot be maintained merely by recurrent mutation."

Sickle cell hemoglobin is often used to explain this concept. Most individuals have normal hemoglobin A in double dose (Hb^A / Hb^A). In some regions of the world, including many parts of Africa, southern Europe, and portions of Asia, an alternative allele, sickle cell hemoglobin may be common. Individuals with two of these genes, Hb^S / Hb^S , are likely to develop sickle cell hemoglobin disease, which often decreases survival because of the limited ability of sickle hemoglobin to transport oxygen and the detrimental effects it produces by clogging blood vessels. However, young people who are heterozygotes, Hb^A / Hb^S, are protected against death from *falciparum* malaria, a common infection in regions where sickle hemoglobin is common. Each time a person who is homozygous for Hb^S fails to contribute his or her genes to the next generation, because of death or a less-than-average fertility, two Hb^S genes are removed from the population. In the absence of any selective factors, the prevalence of Hb^S would drop to a level that could be maintained only by recurrent mutation, that is, at a very small number. However, because of the advantage conferred on the heterozygote he or she is able to contribute more genes to the next generation than bearers of the other genotypes, and the Hb^S gene is maintained at relatively high levels in the population.

The original concept of polymorphism implied that there are survival benefits conferred by some combinations of alleles compared to others, and that there is heterozygote advantage. However it has been difficult to demonstrate such an advantage for most of the polymorphic traits; it has been proposed that many polymorphisms are selectively neutral and occur as a consequence of chance. This may be the case for some polymorphisms, but it is much more interesting to hypothesize a selective advantage, since that leads to further investigation. It is likely that in most cases the selective advantage of a single polymorphic system may be too small to detect. But, if it is considered along with other polymorphic systems, or with environmental factors with which it interacts, then the selective advantage may be detectable.

Independent of the question of selection, the existence of polymorphic systems provided an excellent mechanism for studying biochemical and immunologic variation in human populations. In the mid-1950s when our work in this area began, the red blood cell antigens were the most studied human polymorphisms. These included the ABO and Rh systems which were required in order to provide compatible blood transfusions. If a transfused patient received blood from individuals whose ABO and/or Rh systems were incompatible, then a serious and sometimes life-threatening antigenic reaction could occur. There were other polymorphic red blood cell antigenic systems known, but these were used primarily in theoretical studies and by physical anthropologists who wished to characterize a population by the distribution of these genes. The technique used was to collect blood specimens from a sample of the population in question and determine the frequency of the antigens in the various systems, and from this infer the frequency of the genes which were known to control the antigens. The population could then be described by the distribution of these genes and then compared to other populations with which they might have affinities. There was already a considerable literature on the population distributions of the red cell antigens often with conjectures on differences in disease susceptibility associated with different alleles. The notion was that the frequencies of the alternate alleles would differ from population to population depending on the disease and evolutionary history of the populations both remote and recent.

In addition to the red cell polymorphic systems, several other human polymorphisms were known. Some were rather bizarre; for example, the ability to taste the chemical phenylthiocarbamide,[10] tongue twisting, and the presence of an odiforous material in the urine after eating asparagus. For the most part, the chemical basis of these traits were not known and one can imagine that their esoteric nature did not convince people that they would be of survival value. Although there was little interest in this field, those who were involved included a wide spectrum of scientists including the highly quantitative and theoretical population biologists, blood bank scientists, anthropologists, geneticists, and others. It was not in the mainstream of what was then deemed to be an important scientific and medical subject. A major step forward was taken in 1955 when Oliver Smithies (who had also been a student of Sandy Ogston) introduced the technique of electrophoresis in gels. This allowed the separation of proteins and other chemicals not only as a function of their charge differences but also their mass and shape. Smithies found a polymorphism of the serum protein haptoglobin (which binds hemoglobin) and there were indications that other polymorphic serum proteins could be identified. This appeared to be an appropriate technique for studying population distributions of human biochemical traits and identifying interesting and useful disease associations.

I was introduced to the concepts of population biology and biochemical polymorphisms by Anthony C. Allison, who was a member of the Department of Biochemistry at Oxford when I was pursuing my D. Phil degree. He had done a considerable amount of field work in East Africa, in part during an expedition of the Oxford University Exploration Club, and had been responsible in large part for demonstrating the association between the sickle cell trait and protection against *falciparum* malaria referred to above. He induced me to read E.B. Ford, J.B.S. Haldane, R.A. Fisher, Sewell Wright, and others who had pioneered the field of population biology and polymorphisms. It began to dawn on me that the study of human polymorphisms would be the conceptual framework with which to study inherited human variation and its connections with disease and survival. In addition, the style of the research was very appealing. It would require travel, working with populations outside of western culture, and the prospect of active questing and discovery in the fashion of the explorer-scientists of previous centuries.

3.2. Polymorphisms in Spain and Nigeria, 1956, 1957

Beginning in 1956, while I was still completing my thesis, we started on a series of collecting trips in different environments which would be expected to result in significantly different selective forces and hence different distributions of the polymorphic alleles and possibly, unique polymorphic systems. The first studies were in Spanish Basques. They were selected because their language is considered to be unique and this implied that they might have different genetic origins from other Europeans. At the time we did the studies, there was little evidence of their ancient affinities, although there were suggestions of a connection with populations of the Caucasus and, possibly, North Africa. Anthony Allison and I collected blood specimens in several mountain villages in the Pyranees near San Sebastain in northern Spain, and confirmed the blood group differences (high *Rh* negative, low gene *B,* and low *Duffy*) that had been reported in earlier studies.[11]

The next field trip (July 19 to August 27, 1957) was to Nigeria, West Africa, which, at the time, was still a British colony. B.G.T. Elmes, Professor of Pathology at the medical school of the University of Ibadan, which had only recently been opened, was my host and arranged introductions to local physicians and scientists. I stayed at the Medical School for a few weeks where I collected biological specimens from the Yoruba, the dominant population in the western region of Nigeria.

I then embarked on an extensive trip to the North to collect blood specimens from other populations. The University provided me with a long-wheel-base Land Rover and an experienced driver, a former sergeant in the West African Frontier Force, who had served with his regiment in Burma during the Second World War. As the time for the trip grew close, the traveling party increased in size. I had been assigned a medical student to assist me and he was an indispensable member of the field party. Just as we were about to leave, the young man who looked after my rooms at the medical school asked if he could come along in order to visit his family in the north, and I welcomed him to join us. He insisted on acting as my valet during the trip but, in practice, assisted in the blood collecting logistics. He was joined by two cousins just as we left Ibadan. When we arrived at a collection point we were usually joined by friends or family of the permanent party, and, in the field, we were a formidable group. On one occasion, when we were trekking through the bush to a remote village, I looked back to see a line of about 20 people snaking up the trail behind me, each carrying a bit of supplies and equipment, chatting amiably with each other. I thought we formed a coherent social group!

It was a fascinating trip through the heart of Nigeria going nearly the entire length of The Great North Road, unpaved, but well maintained. We traveled through Illorin, Bida, the home of the Nupe, Kaduna, Jos, Vom and ended in Kano, capital of the Northern Region. Our immediate goal was to obtain bloods from the nomadic Fulani, a population of obscure origin who range across Nigeria, Mali, Guinea, Cameroon, Senegal, and Niger. The Fulani language (*Fulfulde* or *Fula)* is classified within the West Atlantic branch of the Niger-Congo family and is related to some of the languages of Senegal. They swept into sub-Saharan Africa in the early 19th century and in a series of holy wars established a reformist Moslem empire. Many have become urban dwellers and intermixed with the predominant Hausa culture. A portion have remained pastoral and nomadic and their economy is based on the cattle they graze over large areas of their former domains. I planned to obtain specimens from the nomadic group. It required a certain amount of intelligence work to find out where they were, and a fair amount of trekking over the countryside trying to find their semi-permanent villages.

We published several papers from this trip, including a description of the distribution of the red blood cell groups in the Yoruba and the Fulanis,[12] and studies on several serum protein polymorphisms including the hemoglobin binding haptoglobins,[13] the iron binding transferrin proteins,[14] and the *Gm* gamma globulin groups.[15] These studies provided data on the distribution of the genes which determine these red cell antigens and inherited serum proteins and have been used in subsequent studies of the population genetics and anthropology of these regions.

Although the studies were primarily descriptive, we also made hypotheses related to disease susceptibility. For example, we found that haptoglobin appeared to be absent in many of the Nigerian samples. Since haptoglobin binds free hemoglobin, we conjectured that in Africa, where breakdown of red cells due to malarial infection is common, the absence of haptoglobin would be disadvantageous because a great deal of iron would be lost through renal excretion and the African diet is low in iron. This has a resonance in the later studies which we did on the relation of ferritin, the iron binding protein present in cells and blood, to the presence of carrier state and the subsequent development of cancer of the liver in carriers of HBV. We also suggested that there might be some compensatory selective advantage related to the unusual distribution of haptoglobin genes in the African populations.

As discussed in the Introduction, the field trips and polymorphism studies were primarily inductive investigations. We had a broad hypothesis, i.e., that the genes determining polymorphic alleles would have different distributions in populations living in different environments and at risk for different diseases. But we did not have a more specific hypothesis. The main thrust was to make observations in the field and in the laboratory in the expectation, or to put it more strongly, the faith, that we would observe relationships with health and disease that would generate hypotheses that could be tested more directly in subsequent studies. In testing a hypothesis, the scientist seeks to be economical, that is to collect just that amount of data required to provide a rigorous test. However, there is an advantage to collecting additional data in the same venue — both geographic and intellectual — to allow the testing of additional hypotheses which may or may not be related to the original studies. In many of our field trips we also collected biological material from other organisms to determine the distribution of polymorphisms in other species. This was analogous, on the phenotype level, to the current practice in the genome project of simultaneously determining the genomes of species in addition to humans. The parallel body of data is extraordinarily useful in identifying the location of genes and their functions.

In Nigeria, we found a previously undetected polymorphism in the alpha-lactalbumin milk whey proteins of *Bos bovis*, which has been of interest in subsequent studies in other laboratories.[16] We also studied the hemoglobin polymorphism in cattle in collaboration with A.D. Bangham, then at the Agricultural Research Station in Babraham, near Cambridge, England.[17] Most northern European Cattle (and most American cattle), have a single hemoglobin designated *A*. However, Jersey, Guernsey, Brown Swiss and a few other breeds of cattle have a high frequency of hemoglobin B which is also found in much lower frequency in several breeds in central and southern France. On the basis of archeological and other evidence, it had been suggested that Jersey cattle originated in the Indus Valley civilization in India, and migrated or were brought by nomadic herdsman through Asia and Africa to Europe. We found that the Nigerian White Fulani breed, humped animals similar to the Zebu cattle of India, also had a high frequency of hemoglobin B. We suggested that this might be construed as collateral evidence of the similarities of the Jersey, Indian and African cattle. We also attempted to correlate the hemoglobin distribution with protection against trypanosomiasis, a common and damaging disease of African cattle. Two breeds of Nigerian cattle, the N'Dama and Muturu ("Coast shorthorn") are resistant to trypanosomiasis and do not have hemoglobin B, while the Fulani cattle are not resistant and do have hemoglobin B. It was a rather feeble association which did not, to my knowledge, engender very much interest. The program to find new polymorphisms, describe the population distribution of newly discovered and previously known polymorphisms, and to search for disease associations, was proceeding at a reasonable pace.

3.3. Continuing Studies. The National Institutes of Health, 1957–1964

In 1957 after completing my D. Phil at Oxford, I accepted an appointment at the National Institute of Arthritis and Metabolic Diseases (NIAMD) of the National Institutes of Health in Bethesda, Maryland. The NIH was in a period of rapid and rich growth. The mood in the United States after the successful completion of the Second World War was triumphant. The country had the sense that they had accomplished this great victory and anything was possible. We were to take the major role in the reconstruction and growth of the world. The Cold War with the Soviet Union intervened almost immediately and the country was preoccupied with this threat until the ultimate dissolution of the Soviet Empire in 1992. Funding for scientific research had increased greatly after the end of World War II, and it was even further expanded after the Soviets, on Oct. 4, 1957, launched Sputnik, the first orbiting satellite. My wife and I were at sea returning to the New World from our years in Oxford, and saw Sputnik orbiting through the night sky, a demanding beacon urging forward the scientific development of our country.

I joined the Division of Clinical Research and continued to see patients, primarily those with arthritic complaints. There was an expectation that I would continue with my laboratory studies in hyaluronic acid, but I found this research less exciting than the studies on polymorphisms. Trials of various steroid hormones for the treatment of rheumatoid arthritis and other diseases was a primary mission of the Clinical Research group, and I found these to be unrewarding scientifically and clinically. Taking advantage of the freedom that was encouraged at NIH, I continued my field and laboratory studies in polymorphisms. However, I do not think that my work was entirely understood. It was difficult to classify it by scientific discipline. Because of the extensive field work it was thought to be epidemiology which was not regarded as a very sophisticated branch of medical science. But there was also a considerable amount of laboratory research involved. The Director of Research of NIAMD at the time was DeWitt ("Hans") Stetten, a brilliant scientist who had made significant contributions to the biochemistry of intermediate metabolism. I had the impression that he mainly valued reductionist science — that is, that biology could be explained entirely in terms of chemistry and physics — and that population studies involving whole humans was imprecise. Years later, when I had been awarded the Nobel Prize, Hans called to congratulate me . He said, candidly, that while at NIAMD he had not really understood what I was doing, but that I had been funded just in case I knew something they did not. This came as a surprise to me as I had always thought that my work was well received in the Division.

3.4. Arctic Alaska, 1958

Anthony Allison and I had decided to continue our field investigations in a population very different from the West Africans I had studied the previous summer. We selected the arctic and sub-arctic populations of Alaska for the next trip. On June 30, 1958, I departed Seattle for Alaska, joining a U.S. Air Force military research group going to the village of Wainwright on the Bering Sea littoral to study nutrition in relation to heart disease. The research party included a large party of Air Force and academic scientists, the latter primarily from the University of Washington. Dr. Charles T. Dotter, Professor of Radiology at the University of Washington, was the scientific leader. In return for transportation and support, I agreed to help with the clinical examinations they planned to undertake on the Inuit (Eskimo) populations. On July 3, 1958, in Fairbanks, we boarded a C123, a large military cargo plane, and with four tons of cargo and nine military and academic members of the expedition we lumbered into the air, bundled in survival clothing against the chill of an unheated and unpressurized cabin. We climbed over the Brooks range sucking on the oxygen provided and descended over the tundra on the approach to Barrow. The airfield was at near zero visibility and the pilot contemplated returning to Fairbanks. We finally landed with less than 100 feet of ceiling. We were fogged in for several days and were lodged at the Arctic Medical Research Laboratory housed in a series of Quonset huts near the large Inuit village. The laboratory was run by the United States Navy and had been active in arctic research for many years. I took the opportunity to become acquainted with several of the scientists and arranged to obtain bloods from local species of small mammals which I subsequently used in a variety of studies.

A few days later, I was out on the tundra on board a "Weasel," a small tracked amphibious vehicle useful for tundra transport, helping the permafrost team, when we heard the engines of our C-123 warming up. We took this as a signal that we were about to move and hastened back, quickly packed our field equipment and were soon aloft flying south and east. After a flight of about an hour we set down at LIZ-3, a "Dew Line" site. This was one of a series of radar stations set out along northern Alaska and Canada to provide early warning of Soviet overflights. Unfortunately, these sites were not effective in warning against missile attack, nor was the radio communication with the airfields at Fairbanks very effective. Nevertheless, they were manned for many years during the period of the cold war. They were beautifully built and supplied at great expense. I marveled then, as I have often since, how ready society is to spend money on military needs while other activities, such as the funding of science and the arts, always seems to be at the end of the feeding hierarchy.

We were several miles from the village of Wainwright, and I volunteered to walk across the tundra and along the beach to tell the villagers we had arrived and ask them to send a large Umiak skin boat to ferry the men and equipment across a lagoon (actually, a large bay) which separated the mainland from the beach-side location of the village. It was a beautiful bright and sunny morning, the tundra was alive with the flowers of early summer, the insects humming and buzzing. It was a wonderful walk, until I arrived at the edge of the village to be greeted by packs of snarling and belligerent sled dogs, fortunately securely chained. Fred Milan, an anthropologist from Fairbanks and an advance man for the party, was already there and we arranged to transport the rest of the research team to the village.

Wainwright was established during the period of the whaling trade. It consisted of a cluster of frame houses without streets or the ordinary amenities of an American village. There were no roads to the outside and no airstrip. At low tide, light aircraft could land on the beach or when the weather was favorable, the nearby Dew Line site provided access. They still lived a traditional life in the sense that a large part of their diet was obtained by hunting, fishing and pelagic capture of sea mammals, including the occasional whale.

Over the next few days we performed medical examinations and collected blood from about half of the 220 or so people who were resident in the village at the time. We determined the phenotype frequencies of a large number of polymorphisms, including red cell antigens,[18] haptoglobins and hemoglobins,[19] the ability to taste phenythiocarbamide,[10] and the excretion of beta amino-isobutyric acid.[20] The sera were stored and in later years were used for other studies on polymorphisms and viruses, which were not known at the time the sera were collected.

After a week of collecting, I was anxious to get the bloods back to Boston where the red cell antigens which can deteriorate with time were to be tested. Fortunately, a Cessna landed on the beach to refuel. The pilot, a man in his 70s, who had been a fighter pilot in World War I, offered to take me back to Barrow if I could get my things together in time. I quickly assembled the bloods and my survival gear, and after a bumpy take off we were airborne *en route* to Barrow and a commercial airline that could speed the bloods to their testing place. We circled the cairn marking the spot where Will Rogers and his pilot, the famous aviator Wiley Post, had crashed and died while on a trip to the Orient and soon landed in Barrow. After all night processing of the bloods, they were ready to wing their way to Boston.

During the next few weeks I (accompanied part of the time by Anthony Allison), scurried about above the Arctic Circle, in perpetual daylight, collecting blood specimens and assembling pedigree information in small Inuit and Athabascan Indian communities. I traveled mainly by bush plane and walking, sometimes over considerable distances. As a government employee, I could charter aircraft or hitch rides on military aircraft. Our trip was facilitated by Alaskan hospitality; Alaskans had a frontier attitude which emphasized helping others.

The communities I visited included Fort Yukon, a large Indian town on a big bend of the Yukon River; Arctic Village, a small community that had broken away from Fort Yukon to facilitate a more peaceful existence; and Anaktuvak Pass, high up in the Brooks Range with one of the few Inuit settlements in the north which was distant from the coast. We also did some studies on disease, including one of the first epidemiological surveys in any population of rheumatoid arthritis and osteoarthritis.[21] It was during this trip that I began to conjecture about the relation between American indigenous populations and the Asian population from which they emerged. This became an important part of our research later when Liisa Melartin discovered Albumin Naskapi, a variant which appeared to be unique in American Indians (see below).

I have gone into some detail describing this trip, first because it was an exciting and productive venture, and also to capture the flavor of field work; rapid and rough travel, quick decision making, the sense of adventure and discovery. Alaska is a wonderful place to encourage the imagination and I was fortunate to be there on a purposeful mission in the high summer of my (relative) youth.

3.5. Rongelap Atoll and the Central Pacific, 1954

In 1954 the United States Government exploded a large nuclear weapon (the second in the series of hydrogen bomb tests) in the test site in the Marshall Islands in the Central Pacific. The fallout from this device, which included radioactive dust generated from the islet on which the bomb was exploded and the tower from which it was suspended, was carried downwind for more than 120 miles. In its path was the atoll of Rongelap where about 64 Micronesians were exposed. They were evacuated within a few days and resettled on another atoll until five years after the event when they were repatriated to their home island. The crew of a Japanese fishing vessel, the Lucky Dragon, which was within the fallout area, were also irradiated and their exposure became known when they returned to Japan. One member of the crew died, ironically, as a consequence of hepatitis which he incurred following transfusion for the treatment of his radiation associated illness.

Following the initial evaluation of the population, a medical team organized by the Brookhaven National Laboratory visited the island regularly to continue the surveillance. I took part in the fifth year (February and March 1959) follow-up. In addition to assisting with the medical examinations, I undertook to measure several polymorphic traits[22,23,14] to determine if there were any associations of the traits with outcome following radiation exposure. The results of the fifth survey, including our data, have been published.[24,25] There was an additional observation which, I believe, had an impact on further studies on radiation. Prior to my departure for the Pacific, I had spoken to Dr. F.R. Fosberg, a botanist specializing in Atoll biology. The uninhabited northern islets of Rongelap Atoll had received far more radiation than the inhabited islets in the southeast of the Atoll. Whereas it had been estimated that the human populations in the southeast had received about 175 roentgens of whole body gamma radiation (an estimate later revised upward), the northern islets had received more than 2000. Dr. Fosberg had reported observing damage to several of the rather limited number of species of plants which grow on the Atoll, and he asked me if I could confirm his observations. Fortunately, there was a helicopter on the LST (Landing Ship Tank) which brought us to Rongelap and I, along with two other members of the expedition, were flown to the islets of Naen and Gegen in the far northern part of the Atoll. We skimmed low over the lagoon, disturbing *en route* the nesting of Frigate birds and viewing very clearly from our height several enormous sharks in the shallow waters surrounding the islets. After landing, we did a compass-course traverse of the islets and found evidence of damage similar to Fosberg's report. This was published in an annotation to the Brookhaven Medical Report.[23] I was told that the results did have an impact on scientists at the Brookhaven National Laboratory, who were concerned with the possibility of damage to vegetation, particularly food crops, that might result from a nuclear bombing. This was my first foray into botany, a field that I returned to on several occasions later in my studies.

Over the years there was considerable discussion concerning the exposed population of Rongelap. I had attempted, when I returned to Washington in 1959, to assist with the efforts to obtain reparations for them. They eventually were compensated, but their lives have been disturbed and they have been moved to other islands on several occasions. There is current agitation in the now-independent Republic of the Marshall Islands for additional compensation.

3.6. The Albumin Polymorphism, 1966

In 1966, while studying sera of the Naskapi and Montagnais Indians from the Labrador border area of northern Quebec, Liisa Melartin Prehn found an electrophoretic variant of albumin.[26] This variant, termed albumin Naskapi after the population in which it was discovered, was shown to be inherited. It had an interesting distribution.[27,28,29,30] It was found only in northern North American Indians including the Algonkians, Plains Indians, and northern Athabascans and the southern relatives of the Athabascans, the Apache and the Navaho. Later, we found a slow-moving electrophoretic variant, termed albumin Mexico, found primarily in southwestern US Indians, Mexican Indians, and several Indian groups in Central America.[31,32,33] The variants were not found in sera from persons of European or African origin nor in Aleuts or Inuits. They appeared to be restricted to American Indians.

At a meeting held in Austria in 1975 dedicated to the study of the origins of indigenous Americans, we postulated that the albumin variants could be used to identify contemporary Asian populations that had common ancestors with contemporary American Indians.[34] This prediction was soon fulfilled. Electrophoretic variants similar to Albumin Naskapi were identified in Turkish[35] and Punjabi[36] populations. The Turkish variant was found to have the same amino acid substitution as the variant found in Americans. It was difficult to obtain funding for these anthropological projects and we were unable to continue with them. The current availability of many genetic polymorphic markers would now make it possible to compare these populations to see if the connections implied by the albumin association could be validated.

The discovery of an extensive polymorphic system raises the question of what selective forces might be operating to maintain the polymorphism. Albumin binds to physiological and exogenous materials, including many drugs. We studied the differential binding of a plant-derived drug coumarin and found that the albumin variants bound less well than the more common albumin. We postulated that the polymorphism arose as a consequence of interactions of humans with plant substances over the course of many generations. It also raised the question of differential effects of these drugs in individuals with different albumin variants, but this has not been tested directly. Dr. Richard Raichelson, one of my post-doctoral students (in folklore), did a survey of plants traditionally used by southwestern Indians and found that a large number contained coumarins. It would be interesting to pursue this possible connection between a particular culture and a specific polymorphism.

4. THE STUDY OF TRANSFUSED PATIENTS, 1960

4.1. Ag, An Isoantibody Polymorphism of Low density Lipoproteins, 1962

The research was going well. We were beginning to understand the distribution of polymorphisms in different human and animal populations, and identifying several not previously recognized. We had also made conjectures on disease associations but it is fair to say that nothing spectacular had emerged from the research. However, these were much less stressful times in science than at present, and financial support was not as competitive as it has become. I was in a particularly fortunate situation because I was funded for seven years by the National Cancer Institute of the National Institutes of Health through a grant to Dr. Timothy Talbot, the Director of Fox Chase Cancer Center. I will have more to say about this later.

In the summer of 1960 the research took an interesting and unexpected turn, based primarily on the introduction of a new method and a new concept. Anthony Allison, with whom I had worked in Oxford and Alaska, came to spend the summer in my laboratory in Bethesda and suggested that we change our method for the discovery of new serum protein polymorphisms. We reasoned that if there were unknown polymorphisms some of them could be antigenic, and that patients who received many transfusions might develop an immune reaction, in particular, precipitating antibodies against antigenic specificities that they themselves had not inherited or acquired. We used the method of double diffusion in agar gel. A pattern of six wells is cut into a thin layer of agar. The serum of a person who had received many transfusions is placed in the center well. In the surrounding peripheral wells we placed the sera of other individuals who might contain the putative antigenic polymorphic variants. If an antibody had developed in the transfused persons blood, then it would diffuse into the agar gel and form a precipitin band with the specific antigenic protein that might be present in one or more of the surrounding sera.

We obtained blood specimens from patients who had received a large number of transfusions for a variety of reasons. In the panel against which these were to be tested, we placed sera from the populations we had used to determine the distribution of polymorphisms and which we had retained frozen in our "Blood Collection." We used sera from many different populations since the frequency of polymorphic alleles can vary greatly from population to population. In order to increase the probability of finding a new polymorphism, we randomized the populations in the test panel.

The gel was cast in small glass Petri dishes, six tests to a dish, and a rack of these dishes were filled and stored overnight at room temperature to allow time for the migration to take place and the precipitin, if any, to form. One morning we examined the plates and found that in one series, containing the serum of patient "C. de B.," there were a series of beautiful parabolic precipitin bands between the transfused and panel sera; about half of the sera in the test panel had reacted and the others had not.[37] It was a wonderful moment of discovery, for we knew that our hypothesis had been supported by this observation and that it could lead to many new experiments. We gave the system the unimaginative title of Ag, an arbitrary choice, but with the vague notion that it involved an antigen. The designation Ag(a+) was used to designate the presence of the protein that reacted with the serum of "C. de B.", and non-reactors were designated Ag(a–).

Our immediate task was clear. We wanted to know the nature of the protein, if Ag(a+) was inherited and if there were any disease associations. In our first paper we had reported Mendellian segregation of Ag(a+) in American and Micronesian families.[37] Individuals who were homozygous or heterozygous for a gene termed Ag^A carried a protein that reacted with C. de B. Those who were homozygous for the alternate silent allele *Ag* were non-reactors. This was confirmed in subsequent family and population studies.[38] Furthermore, there were striking differences in the distribution of the gene in different populations, in accordance with our expectations of a polymorphism.

To determine the nature of the protein we determined its electrophoretic characteristics, approximate size, and other attributes. To expedite discovery, I took my experimental data, including the electrophoretic and immunodiffusion experiments, and showed them to colleagues at the National Institutes of Health who were working on proteins. They tended to guess that it was related to the protein that they themselves were working on. I showed it to Dr. Martin Goodman, later the Director of the Immunology program at the World Health Organization (WHO), who immediately suggested that it was a lipoprotein, one of the family of proteins in which he was interested. We stained the precipitin band with a fat-stain, azocarmine, and determined that it had a large component of lipid. Then in a series of electrophoretic, immunologic, and chemical studies, we established that it was a low-density lipoprotein.[39]

We continued our search for additional antibodies in the sera from the transfused persons. This was both to enrich our knowledge of the Ag system, which we had inferred from the start to include many specificities, alleles and loci, and also to identify whole new systems of iso-antibodies that might reveal new polymorphisms.

The second antibody we studied was found in the serum of a transfused patient from New York City and was designated as the New York phenotype. However, it reacted with a very high percentage of the local population which made it difficult to study family segregation since non-reactors were rare. We decided to try to identify another population with a higher prevalence of non-reactors which would make it easier to ascertain families in which segregation could be studied. In a preliminary survey we determined that the non-reactors were common in American Indians.

I had been collaborating with an old friend and colleague, Dr. John Martin, a physician and scientist at McGill University in Montreal. His University had a medical and scientific interest in the Indians in the northern part of the province of Quebec, and in August 1962, he arranged a field trip to the community of Schefferville situated on Knob Lake in the northern section of Quebec (variously called New Quebec or Ungava.) It was adjacent to the remote Labrador-Quebec border on the pre-Cambrian Labrador Shield and was the site of an immense open pit mine owned and operated by the Iron Ore Company of Canada. There were two Indian communities, the Montagnais and the Naskapi living in and near the town. Both belonged to the Algonkian language group. Before European contact they had roamed over the vast interior of Quebec and Labrador, subsisting largely on the caribou herds that ranged over this sub-arctic region. With the coming of the Europeans, the northern group, the Naskapi, had focused on the British trading posts and Anglican missions in the northern part of the peninsular. The southern group, the Montagnais, were oriented to the French-speaking and Catholic communities of the northern bank of the St. Lawrence River and Bay. We were able to establish that the New York specificity of the lipoprotein was also inherited, probably controlled by a gene allelic to Ag^A or at a locus linked to it.[40] As usual, we preserved the sera we had collected, and these were of much use in studies of a considerably different nature that I will describe later. We also studied the relations of the polymorphisms to disease and found that there was an association between the Ag(a+) polymorphism and diabetes.[41]

We found several new inherited antigenic specificities[42,40] and, in due course, our work and that of other scientists in the field, revealed multiple specificities that were under a complex genetic control analogous to the gene complex controlling the Rh antigens on human red cells. The investigators in the field met at a seminar sponsored by the World Health Organization in Geneva in 1965 in order to standardize antisera and bring order into what was becoming a very confused system. Unfortunately, it was difficult to arrive at a method of standardization and subsequent work in the study of lipoprotein inherited antigenic specificities determined by human antisera did not progress very rapidly. In the meanwhile, Kare Berg in Norway had developed antisera in rabbits which determined a series of other loci and alleles related to the lipoprotein polymorphisms. This area of work has flourished and led to fascinating and important medical and genetic findings.

Our hypothesis that new polymorphic systems could be discovered by the use of blood from transfused persons had proven to be productive. We continued the search for additional antibodies and antigens and this soon resulted in the discovery of the antigen that was on the hepatitis B virus.

On November 18, 1963, I gave the Fifth Bernadine Becker Memorial Lecture at the New York Academy of Medicine. On the last page of the published version of the lecture,[41] I reported that we had found an iso-precipitin that reacted with an antigen that appeared to be different from the Ag system we had initially described. It had less lipid and the antigen had a much different distribution than the Ag antigens. Whereas the Ag antigens had been common in most populations, the new antigen was very rare in Western populations. However, it was common in Micronesians, Vietnamese, Taiwanese, and Australian Aborigines. It was also common in patients with leukemia. This "new" antigen antibody system had been initially seen by Harvey Alter who was working in my laboratory at the National Institutes of Health and has subsequently become a leading investigator in the hepatitis field.

I will continue with this narrative after a brief section of my move from the National Institutes of Health to the Fox Chase Cancer Center in 1964.

5. THE INSTITUTE FOR CANCER RESEARCH, FOX CHASE CANCER CENTER, 1964–1997

In 1964 I moved to The Institute for Cancer Research (later incorporated into the Fox Chase Cancer Center) located on the outskirts of Philadelphia, Pennsylvania where I was appointed Associate Director for Clinical Research. I stayed at Fox Chase for more than thirty years. Despite the title, I did not have a great deal of administrative responsibility since the Division of Clinical Research was essentially my own research group and the administrative tasks were assumed by the staff of the Institute. As noted above, an additional advantage of the position was that it was fully funded for seven years. Even in the 1960s, when research funds were more available than they subsequently became, this was unusual. It meant, in effect, that I could work on any project I thought suitable within the general area of clinical research which we had any reason to believe was related to cancer. An unfortunate characteristic of contemporary research funding in the United Sates and elsewhere is that it is necessary to tell the funding committees the anticipated results before the project is actually accomplished. This may limit the free exercise of the imagination which is an essential constituent of original research. We were freed from this burden and, although we had to report yearly on our results, and publish completed work regularly, we did not have the burden of budgetary restrictions. I subsequently obtained a grant of funds specifically directed to our project, but for many years our research was funded in large part by the program project granted to Tim Talbot. Tim was a remarkable administrator and director of a research institute. It was in large part because of his vision and understanding of the needs of scientists that the Fox Chase Cancer Center flourished. Because of Tim's intelligent and consistent support and encouragement, my colleagues and I were able to pursue our research unburdened by administrative responsibilities and financial anxieties.

During the next few years we assembled a group of workers in the Division of Clinical Research who, in most cases, remained for many years. They are responsible for the studies I have and will be describing and for the application of their outcomes. They included: W. Thomas London, Alton I. Sutnick, Irving Millman, Edward D. Lustbader, Hei-wan L. Hann, Anna O'Connell, Betty

Gerstley, Liisa Melartin Prehn, J. Bruce Smith, Barbara Werner, Veronica Coyne, Scott Mazzur, Terry Halbherr, Cyril Levene, William Wills, Pinayur S. Venkateswaran, Susan Wolf, Mark Feitelson, Jean S. Drew, Richard G. Stevens, Robert L. Yarrish, Samuel G. Franklin, and others. We also had collaborators within Fox Chase Cancer Center including Manfred Bayer and Lawrence Loeb. There were pre- and post-doctoral students, primarily in Anthropology, including Jana Hesser, Betsy Dickey, Lawrence Schell, and others who contributed the enthusiasm and questions that only students can provide. Names of co-workers are cited in the references and elsewhere in the text.

The research I will be describing was very much a team effort, a joint endeavor of a large number of scientists working on the same problem and dedicated to its solution and application to the improvement of health. We were very aware of the deadly effects of hepatitis B virus, particularly in the tropics and in Asia, and this fired our drive to move forward rapidly.

Several of the investigators were essential to the project nearly from its inception and were part of the team during the period when the most important results were obtained. Tom London was a member of the group for the longest time. He came to Fox Chase Cancer Center shortly after I arrived and was central to its activities from then on. When I went to England on an extended leave of absence in 1989, he assumed the direction of the group and continued in this role after my return to Philadelphia in 1994. Tom went to medical school at Cornell University and was certified in Internal Medicine. His expertise was in epidemiology although, as was the case with all of us, he participated in a wide range of the laboratory activities. Alton Sutnick was a graduate of the Medical School of the University of Pennsylvania. He was also certified in Internal Medicine and was, initially, responsible for the patients on our research ward. Irving Millman received a Ph.D. in Microbiology from Northwestern University in Chicago. He was skilled in immunochemistry and immunology. Prior to coming to Fox Chase, he was employed as an investigator at Merck Pharmaceuticals and was responsible for the invention of an effective Rubella vaccine. His experience was invaluable during the period we were inventing the hepatitis B vaccine. Edward Lustbader was a statistician who obtained his Ph.D. degree at the University of Pennsylvania. He was a very innovative and original scientist and made many contributions to study design and data analysis. Anna O'Connell served for many years as the head of the laboratory technical activities and probably has the longest and best institutional memory of our history. Barbara Werner, who started as a technician with us, returned to

University to obtain a Ph.D., returned to our laboratory briefly, and then went on to a successful career elsewhere.

There were foreign colleagues who participated in the field and laboratory work and also, in some cases, came to our laboratory for research training and experience. These included Alberto Vierucci from Italy and many of his students, Liisa Melartin Prehn and other colleagues from Finland including Veijo K. Raunio and Eino Kaarsalo. There were a large group of workers from France and Senegal who took part in our field work in West Africa, including Bernard Larouze, Veronique Barrois, Elisabeth Feret, Elisabeth Marinier, Gerard Saimot, Phillipe Maupas and others. The names of these and other collaborators and colleagues are cited in the individual papers.

Over the years we created an unusual research group. In contemporary university and institutional research, a program is usually constituted within the confines of a disciplinary department, for example, biochemistry, immunology, virology, or pharmacology. The scientists within the group have mostly had their training within the discipline represented by the department. Research groups within a clinical department, say medicine or pediatrics, may have a broader representation but they usually are focused on the area for which their department is responsible. In these University and medical school departments, the researchers have a "loyalty" to their own discipline, for it is within the confines of these disciplines that their careers are determined.

At Fox Chase Cancer Center we did not (at that time) have discipline-oriented departments, but rather independent laboratories working in specific research areas. My laboratory was dedicated to the study of inherited biochemical variation and susceptibility to disease and later the hepatitis virus. The disciplinary training of the staff, which at its maximum reached about forty people including scientists with advanced degrees, technicians, secretaries, physicians, nurses, field workers, etc., was very diverse. During the 25 years or so that it flourished it included physicians, surgeons, nurses, biochemists, natural product chemists, immunologists, anthropologists, botanists, molecular biologists, virologists and even a post-doctoral scientist with a degree in Folklore. Such multidisciplinary groups were not common then but they are likely to become more so as scientists begin to deal with the complex problems that will be addressed in the next millennium. It was my responsibility to provide a certain measure of leadership and direction but, I like to think, initiative was encouraged. Although we often decided jointly on the direction the research would take, the individual scientist had control of his or her own project. With rare exceptions, we were a well functioning crew without much acrimony.

6. AUSTRALIA ANTIGEN AND THE HEPATITIS B VIRUS, 1964

In our first definitive publication concerning the "new" precipitin we had identified,[43] we referred to it as "Australia antigen" because our earliest studies were done on the antigen found in the serum of an Australian aborigine. As already noted, the antigen was rare in the U.S.A and western European populations but common in Australian aborigines and populations from Taiwan, Korea, the Pacific Islands, the Philippines, Vietnam and elsewhere. We also studied the distribution of the antigen in many diseases and found a high prevalence in patients with leukemia. This was an observation which lent itself to testable hypotheses. Several were selected:

1. Individuals with Australia antigen (abbreviated Au) had an increased susceptibility to leukemia. This susceptibility was under genetic control and constituted a polymorphism similar to the Ag lipoprotein polymorphism we had already described.
2. Leukemia causes Au.
3. Au is related to the virus that had been postulated to be the cause of leukemia.

We continued to test the polymorphism model for the Australia antigen, but also considered the possibility that it was a virus. During the course of the next few months, a series of observations inclined our thinking to the possibility that Au was related to a hepatitis virus. These observations were made in the context of many other observations that did not yet form a discernible pattern. I will describe a few of these that established a mind-set that eventually led to the identification of Au with hepatitis.

1. We did a survey of healthy employees at a large research institution. Only two of these had Au. We interviewed one of the individuals with Au. She was not ill, but some years prior to the blood drawing she had a blood transfusion. There were no symptoms of hepatitis.
2. We had collected blood from a Pacific Island population over the course of several years. Serial studies identified one individual who, in one year had Au and in a subsequent year had antibody against it. This was suggestive of an infection followed by the development of antibody against the infectious virus.
3. A colleague in a rural community had provided us with sera from non-hospitalized people from the population that he served. We tested these for Au and found one serum that scored positive. When he identified this

individual, he learned that it had been drawn not from a normal person but from a patient suffering from clinical hepatitis.

These observations stimulated us to collect sera from patients with hepatitis to test the hypothesis that Au was associated with this disease. However, the event that triggered a much more intensive testing of the Au/hepatitis hypothesis occurred in one of our own patients.

We had embarked on the testing of the first of the hypotheses listed above. A corollary of this hypothesis is that individuals who are at increased risk of leukemia would have a greater probability of also having Au. Children with Down's syndrome (mental retardation associated with trisomy of chromosome 21) are at much greater risk than other children of developing a form of childhood leukemia. We tested patients with Down's Syndrome in a large public institution for the mentally retarded and found that our hypothesis was amply supported. About 30% had Au compared to about 5% in other mentally retarded patients in the same institution. The support of our hypothesis was very encouraging and it also allowed us to observe persons with Au over an extended time. We found that individuals who were negative when first tested remained negative on subsequent testing, and those who were positive remained so. On one occasion, a patient who had previously been negative became positive on a subsequent testing. He was found to have developed a form of anicteric hepatitis coinciding with the transition from negative to positive. This observation accelerated our interest in the hepatitis hypothesis and we soon completed a series of studies which confirmed the association.

Patients with acute hepatitis had a much higher frequency of Au than normal people.[44,45] It was much more common in patients with post-transfusion hepatitis than in acute hepatitis patients without known exposure to blood or serum. That is, its strongest association was with what had been designated as serum hepatitis or hepatitis B. There was also an association of the Au with chronic hepatitis, and carriers of Au had significantly higher levels of liver chemistry abnormalities than normal individuals who did not have the antigen.[44,46] All the data were consistent with an association of acute and chronic hepatitis (particularly hepatitis B) with Au. In the first paper emanating from this project,[100] we suggested that Au was associated with a hepatitis virus, and a variety of studies were done to confirm this.

Particles with the appearance of a virus were detected by electron microscopy in the serum of individuals with Au.[47] These were subsequently shown to be the surface antigen particles, different from the whole virus particles which were subsequently identified by British investigators. I will return to this later in the discussion of the hepatitis vaccine.

My colleagues Millman and Coyne were able to localize Au in the liver cells of patients with hepatitis, which would be expected if the Au was part of the hepatitis virus. They were also able to sustain the growth of the virus for weeks and months in liver cells obtained from patients with liver disease associated with Au.[48,49]

We published our paper showing the hepatitis association and stating the virus hypothesis in 1967.[44] There were soon confirmations of our findings in Japan, Italy and the United States. We arranged to distribute samples of serum containing the antibody or the antigen to other investigators. The method we had used for the detection of Au — double diffusion in gel — was very easy to duplicate and the equipment to do the experiments consisted only of some glass slides and agar gel. Hence it was easy for others to repeat and extend our observations even in laboratories with little equipment. There were numerous confirmations of our findings. In retrospect, our decision to distribute the reagents widely was one of the most important factors in gaining acceptance for our claims since many others were able to confirm and extend our findings.

Due to the efforts over the years of a large group of scientists interested in hepatitis, it was known that there were at least two hepatitis viruses, hepatitis A, transmitted by the fecal-oral route, and hepatitis B or serum hepatitis, transmitted by injections of blood and serum. As our initial findings had indicated, the Australia antigen was part of hepatitis B and the designation HBV became accepted. Australia antigen itself was seen to be the surface antigen of HBV and designated as HBsAg. This system of nomenclature has been adapted for the hepatitis viruses discovered after HBV and they have been designated as HAV, HCV, HDV, HEV, HGV. Undoubtedly there will be others.

7. CONTROL OF POST-TRANSFUSION HEPATITIS B, 1969

We quickly realized the clinical implications of our findings. Although the immunodiffusion technique, which had been used in most of our experiments was insensitive, it is highly specific. As clinicians, we were well aware of the problem of post transfusion hepatitis. In the 1960s in the United States, a very large percentage of blood donor units were obtained from paid donors, although this was not the case in many countries in Europe, Canada and elsewhere. The incidence of post transfusion hepatitis, both clinical and occult, was enormous.

In some studies, post-transfusion hepatitis occurred in up to 50% of patients who had received large numbers of transfusions for extensive surgical treatments such as open-heart surgery. Even if insensitive, the immunodiffusion method could detect a significant number of carriers of the hepatitis virus.

There were several observations that recipients of donor blood containing Au developed hepatitis, but there were no systematic observations on the extra risk imposed. Our colleague, Kazuo Okochi in Tokyo, undertook such a study in the early 1970s, which clearly showed a significantly increased risk of hepatitis in recipients of the positive donor units. We conducted a study to determine the effect of a blood screening program at the Philadelphia General Hospital, where the incidence of post-transfusion hepatitis was extremely high.[50] The results were very convincing; prior to the screening program, the incidence of post-transfusion hepatitis had been 17.9% and after, it had been reduced by two thirds to 5.9%. In retrospect, the residual cases were probably a consequence of the insensitivity of the immunodiffusion method, and the transmission of non-B hepatitis, most of which we now know is hepatitis C virus. The results indicate that in this early period most of the post transfusion hepatitis was a consequence of hepatitis B.

The possibility of drastically reducing post-transfusion hepatitis galvanized the commercial medical and health care communities. (These events have been described in a paper I presented at the British Blood Transfusion Society in 1989.[51]) There were frequent meetings of government and professional societies, for example, the American Association of Blood Banks, and eventually professional and government regulations required the testing of blood donor units for Au. There was a scramble by commercial companies to produce kits of increasing sensitivity to tap the enormous market for testing the more than 10 million units of donor blood collected each year. In addition, there was a vast market for the clinical diagnosis of hepatitis B. We had invented, and Fox Chase Cancer Center had patented, a sensitive radioimmunoassay for HBsAg[52] and this technique and similar ones were commercialized and realized enormous profits to several companies. (Fox Chase did not profit from this development since the government, who had partially supported the research, retained the patent rights.) I gave samples of the Australia antigen and the antibody against it to scientists from Abbott Pharmaceuticals. I was later told that the diagnostic division of Abbott Pharmaceuticals whose business is in billions of dollars was established as a direct result of their commercialization of the radioimmunoassay application. With the improvements in the tests and the nearly universal requirement of donor blood screening, post transfusion hepatitis due to HBV became a rarity, although cases due to other viruses continued to be a problem until detection methods for HCV became available in the 1990s. There are still other causes of post transfusion hepatitis, but the problem is far less grievous than it was before the introduction of the screening tests.

8. INVENTION OF HEPATITIS B VACCINE, 1969

The most important outcome of all our research on Australia antigen and the hepatitis B virus has been the invention of the vaccine against HBV. This came about as a consequence of a series of events that coalesced in 1969. The data confirming the hypothesis that Australia antigen was part of HBV was accumulating rapidly and the observations were repeatedly validated by other scientists. None of our original group of investigators was trained as a virologist but from our clinical experience we were knowledgeable about vaccines and their effects. Since I had experience in preventive medicine, primarily because of my work in Suriname and elsewhere in the tropics, the advantages of a vaccine were obvious. My experiences in the South American jungles where large percentages of the population were devastated by infectious diseases — malaria, tuberculosis, yaws, diarrhea-causing bacteria, elephantiasis, and many others — whose prevention or control was at least in part understood, was an inspiration to find something that would prevent large scale infection.

Perhaps the most important factor was the arrival of Irving Millman in our laboratory in June of 1967. Irving's doctoral degree was in microbiology and immunology and he was familiar with the developments in virology. Even more pertinent was the experience he had working for Merck Pharmaceuticals at their Pennsylvania facility. He had invented a purified pertusis vaccine which, although superior to existing vaccines, had not reached the production stage. When he had initially visited our laboratory we told him of our studies on protein polymorphisms and he was looking forward to applying his knowledge of protein chemistry to population molecular genetics. By the time he started working, we had already brought forward the virus hypothesis and Irving realized that, once again, he would be working on infectious diseases.

His reaction to the data was influenced by his experience with other infectious agents. We had calculated the amount of Australia antigen in the serum of carriers and estimated that in some it amounted to about 1% of the serum proteins! His immediate response was that if this were all virus it would be incompatible with the life of the carrier. This became of considerable importance later when we devised the vaccine.

Sometime in 1969, Timothy Talbot informed us that he had been advised by the U.S. Government, which funded most of the basic research in medical science, that they would expect scientists to generate their own funds to replace, in part, government grants. The inference was that commercial exploitation and patenting of discoveries would be encouraged. On receiving this information I spoke to Irving and told him that it was now time for us to attempt to obtain a patent for the vaccine. From our collaboration with Bayer and Loeb[53] we knew that the Australia antigen particles which we had isolated and which had been visualized in the electron microscope by Bayer, Werner, and Blumberg[47] did not contain nucleic acid. We had also observed that the surface antigen (HBsAg, Australia antigen) which marked the presence of the virus, and the antibody against the surface antigen (anti-HBs), were never detected in the same individual. This was consistent with the explanation that anti-HBs was protective. Further, Professor Okochi in Tokyo had shown that transfused patients who had anti-HBs, or acquired it after transfusion, were less likely to develop post-transfusion hepatitis than transfused patients who had not developed the antibody. This also suggested that anti-HBs was protective.

The significance of this finding is emphasized by the present status of vaccine research for HIV, the AIDS-causing virus. Although the molecular biology is well understood and recombinant virus proteins can be relatively easily produced, it is not clear which of these antigens generate a protective antibody, or protective cellular immunity. (The development of an HIV vaccine is further complicated by the successful assault on the immune cells by the virus, and the extreme mutability of the virus.) There are also difficulties in identifying protective antigens for the HCV virus.

Based on the knowledge that the small particles visualized in the electron microscope did not contain nucleic acid, we postulated that these were a non-replicating, incomplete form of the virus. We reasoned that there must be additional particles which contained nucleic acid and were replicating, infectious and pathogenic. The small particles were highly resistant to a variety of proteinases, which allowed the purification of the particles by using proteolytic enzymes to remove the serum proteins. In a series of animal experiments, we showed that *partially* purified Australia antigen particles, which presumably also contained the whole virus particles which had not yet been visualized, could be transmitted by inoculation to vervet monkeys, while the fully purified particles were not transmitted. The implication was that we could separate the non-infectious particles containing only the surface antigen from the whole viruses which were pathogenic.

Based on these findings we submitted a patent application on October 8, 1969 for the extraction of the surface antigen particles from the serum of human carriers of the virus.[54] The particles were purified by centrifugation, enzyme digestion, column gel filtration, differential density centrifugation in sucrose and cesium chloride, and dialysis. Standard adjuvants and preservatives were added to complete the formulation of the vaccine. This was a unique method for producing a vaccine which had never previously been used. It took some time before the concept was accepted by virologists and vaccine manufacturers who were more accustomed to dealing with vaccines produced by attenuation of viruses, or the use of killed viruses produced in tissue culture, or related viruses which were non-pathogenic but protective (i.e., smallpox). However, by 1971 we were able to interest the Merck Pharmaceutical Corporation which had considerable experience with vaccines and they, through the agency of Maurice Hilleman, offered to collaborate with us.

During the next few years a series of human and primate observations by scientists including Hilleman, Krugman, Purcell, Maupas, and others provided additional support for the validity of our claims. In 1975 at a meeting in Bethesda, Maryland, the most recent reports on the animal studies were presented. We recognized that this was now the time to encourage commercial development and after discussions with several foreign and domestic companies, we granted a license to Merck for the vaccine. In 1980 the results of the first extensive field trial were published by Szmuness and his colleagues. They showed that the vaccine was highly effective (over 90% appeared to be protected) and that, within the limits of their observations, there were no untoward side effects. The appropriate FDA approvals were obtained and, by 1982, the vaccine was available for general use. Soon afterwards, a recombinant method for producing hepatitis B vaccine was developed in several laboratories and enormous quantities of the vaccine have been produced by this method. HBV vaccine is the first vaccine to be produced commercially by recombinant methods. However, vaccine is still produced from the blood of carriers and, I understand, is at present the major source of vaccine in the world.

The major methods for the transmission of HBV are:

1. From a carrier mother to a child, at, or soon after birth and, in some cases, probably before birth.
2. From a carrier mother and/or carrier sibs in the first few years of life.
3. Venereal.
4. By transfusion or injection of blood from a carrier to another person.

A large percentage of transmission can be decreased by the vaccination of children soon after birth, preferably within the first few months. Universal childhood vaccination has been adapted in over 70 countries. There is still a great deal to be done to introduce programs in other countries, particularly in Africa and sub-continent India where the need is great but the resources are wanting.

Reports are now available from several places that demonstrate that the vaccine is highly effective in decreasing the prevalence of carriers. The results in some locations are dramatic (Blumberg 1997). Major programs involving millions of people are now in place in the People's Republic of China. In Native Americans in Alaska, acute hepatitis B infection dropped from 215 cases/ 100 000 to 7 to 14 cases in 1993. In 1995, zero cases were reported. In Taiwan, where universal childhood vaccination has been in use for more than ten years, the prevalence in the 1–2 year age group decreased from 10.7% in 1984 to 1.5% in 1989, and the decrease since that time has been even greater (for review, see Refs. 55, 56).

A vaccination program was initiated in 1983 in Afragola, a community in southern Italy selected because of its very high rates of HBV carriers and of morbidity and mortality from liver disease. There were major decreases in the prevalence of carriers in the vaccinees; for example, in males aged 5–10 the prevalence of HBsAg dropped from 11.9% in 1978 to 1.6% in 1989 after the vaccination program was in place for about five years. There were similar decreases in other age groups. Of particular interest was the drop in the prevalence of HBsAg and anti-HBc (an indicator of viral infection) in the *unvaccinated* population. HBsAg dropped from 13.4% in 1978 to 7.3% in 1989. This implies that the reduction of the carrier prevalence in the vaccinated group has an indirect effect on unvaccinated carriers and susceptibles, and suggests that there may an amplification characteristic to the vaccination program.

There are other examples in Korea, The Gambia, and elsewhere that demonstrate the considerable success of the vaccination programs. They have raised the question of the eradication of HBV particularly if it becomes possible to treat carriers as well as extend the vaccination programs. I will come back to this question after discussing our publications and the work of others in the field on the prevention of primary cancer of the liver.

9. HBV AND PRIMARY CANCER OF THE LIVER, 1969

In 1975 we published an article entitled, *The relation of infection with the hepatitis B agent to primary hepatic carcinoma.*[57] In the introduction we stated,

> "During recent years, there have been parallel developments in understanding, on the one hand, the pathogenesis of primary hepatic carcinoma (PHC, hepatoma) and, on the other, the biology of Australia antigen (hepatitis B surface antigen) and the infectious agent, hepatitis B virus (HBV), to which it is intimately related. Recently, the paths of these developments have begun to converge, and from this it is possible to design a preliminary strategy which could, if the interpretations of these data are correct, result in the prevention of many, and perhaps most, cases of one of the most widespread and deadly cancers of humans."

This prediction appears to be accurate. The etiologic relation between HBV and hepatocellular carcinoma is now generally accepted and it has been stated that about 85% of HCC is caused by HBV. HCV is also causally related to HCC and these two hepatitis viruses can account for most cases of HCC in the world. It also now appears that the vaccination program is decreasing the incidence of HCC, but more of this later.

We became interested in the possible relation of HBV and HCC as early as 1969, only two years after the discovery of the virus and the same year in which the vaccine was invented. Bruce Smith and myself,[58] using the insensitive immunodiffusion method did not find an association between HBV and HCC. We suggested that the levels of HBV were too low to be detected and recommended the use of more sensitive methods. We now know that the levels of HBV drop significantly late in the course of the pathogenesis of HCC, and more sensitive methods such as radioimmunoassay, introduced several years after our first study, have shown a strong correlation between HBV and HCC. This observation later became an important feature of the model for the pathogenesis of HCC formulated by W. Thomas London and myself[59] and rendered into a mathematical model by Payne and Nowack some years later.[60,61]

In the 1950s, pathologists working in West and East Africa (Payette, Steiner, Davies, and others) had suggested, based on pathological and epidemiological evidence, that hepatitis might be the cause of HCC, but they did not have any methods for testing the hypothesis directly. In the summer of 1971, I attended a meeting in Kampala, Uganda, on Cancer in Africa sponsored by the National Cancer Institute of the National Institutes of Health, USA. The results of several preliminary surveys of HBV prevalence in HCC and controls were presented by investigators working in Asia and Africa. These showed that there was a highly significant association between HCC and HBV, and the high prevalence indicated that HBV could account for a large percentage of HCC in these areas of the world. This was a very exciting finding and when I returned to my laboratory in Philadelphia we resolved to direct our research to HBV and cancer of the liver.

This association with cancer was for a variety of reasons extremely important in the work of our group. The Fox Chase Cancer Center had always dedicated itself to the study of basic scientific problems, independent of their apparent association with cancer. The guiding principle was that a fundamental understanding of normal and pathological processes would in due course help to unravel problems more directly associated with cancer. However, the funding climate had begun to thicken and there was a certain imperative to demonstrate a direct oncology connection. The reports from Uganda more than fulfilled this need.

The meeting in Kampala was exciting for other reasons. A secondary purpose for my trip was to collect mosquitoes in areas highly indigenous for HBV to see if they carried the virus and could be considered as possible vectors for the spread of the disease. In order to collect samples in different parts of the country, I joined a group visiting the national game parks and, while the others were viewing animals, I would set up my traps in and near villages to collect night and day flying mosquitoes. The traps consisted of a series of net chambers to which the insects were attracted by a small battery powered light, and solid CO_2 (dry ice). (Some species of mosquitoes are attracted to the carbon dioxide expired by mammals.) Since the batteries had a limited lifetime, I had to place the traps just before sunset and retrieve them just after dawn. The collection areas were teeming with animals, some of them quite vicious, and I was cautioned not to roam around by myself during the dark. Hence, in the evening I would hurriedly rush out to set the traps and return to the compound before nightfall. In the morning, after a fitful sleep, I would awake at first light to collect the traps and retrieve the live insects retained in the traps by a stream of air from the battery powered fan. It was a frenetic ballet, but I managed to collect a large number of mosquitoes. I will discuss the results of this in greater detail later.

To return to the hepatocellular carcinoma story. In 1972, while I was on sabbatical leave in the sub-Department of Genetics in the Department of Biochemistry at Oxford University, I was approached by Dr. Bernard Larouze, a junior colleague of Professor M. Payette, Dean of the School of Medicine at the University of Dakar, Senegal, who, as already noted, was one of the pioneers who had worked on the association of hepatitis and HCC. He suggested that we collaborate in a study of the problem in West Africa. This started a long, profitable and happy association with a group of colleagues in France and West Africa which contributed to the understanding of HCC and its control. My first trip to Senegal and Mali was in June 1974, and there were return trips in 1976 and

1977. Tom London joined the 1976 trip. In the late 1980s, he and colleagues in the U.S.A and Senegal organized a major epidemiological and interventional study involving the armed forces of Senegal which as of 1997, is still continuing. The project will, it is hoped, eventually lead to the establishment of universal HBV vaccination.

In this and other studies we showed that there was a significant association between HBV and HCC in Senegal, Bamako (in Mali), Hong Kong, and in Philadelphia.[62,63] This confirmed studies by many others, and was similar to our own studies done later in Korea.[64] We found that patients in Dakar with HCC were much more likely to have mothers who were also carriers of HBV, confirming similar reports from Asia.[65,66] This was consistent with the notion that individuals who acquired HBV at an early age from their carrier mothers, and themselves became carriers, were at high risk of developing HCC. This hypothesis was supported in many other studies in the following years. The vital statistical analysis for this study was done by Ed Lustbader. He estimated the risk ratio for HCC for individuals with carrier mothers at about 15. This compares to about 10 for cancer of the lung for one-pack-a-day cigarette smokers. Studies in Taiwan have shown a risk ratio of over 100 for carriers of HBV, while lower risk have been found in other locations (e.g., Alaska). These calculations emphasize the importance of HBV control for the prevention of HCC.

The finding that mothers could transmit HBV to their children at a very young age raised some perplexing psychological problems. Mothers are very concerned about the heath of their offspring and the knowledge that they had themselves might have transmitted an agent that could cause cancer in their children is a difficult burden to bear. Advising the parents of the pathogenesis has to be done with sensitivity and appropriate counseling when necessary.

There were additional interesting results of the African studies which had important consequences for the design of vaccination strategies.[67,68] A large study was initiated in Thies, a major provincial city in the interior of Senegal some 80 km east of Dakar. Blood was collected from pregnant women, and from their newborns at the time of birth. Additional observations were made on the children in successive field studies up to three years after the initial collections. Infants did not become carriers until 5 months of age or older, therefore vaccine could be administered in a convenient window of time after birth. This was coupled with data from East Asia. Although the children of carrier mothers had a high likelihood of becoming carriers, they did not do so until about three months of age.

When the vaccine was first developed, we did not know if it would be effective in new born children. Based on our genetic studies, we inferred that some individuals, when infected, were more likely to become carriers of the virus, while others, presumably with a different genetic endowment, would more likely develop anti-HBs. There was a possibility that those who were likely to become carriers would not respond to the vaccine. There was no experimental or theoretical manner in which this important question could be answered; only actual use of the vaccine could provide the information required.

It was therefore very encouraging when we learned of the results of the first studies on newborns whose mothers were carriers. They did respond to the vaccination, developed satisfactory levels of antibodies and were protected against subsequent infection. These results made it highly likely that large scale programs of infant vaccination would be successful and whole communities and nations could be protected against the morbid consequences of long-term HBV infection.

The validity of our prediction (that vaccination programs would decrease and possibly eliminate most of the HCC) quoted at the beginning of this section, has been strengthened within the past year. At the international hepatitis meeting held in Rome early in 1996, a study from Taiwan has shown that the incidence of HCC has decreased by two-thirds in the vaccinated population. In another study from Korea, the results indicate that, not only the vaccinated group, but also several unvaccinated cohorts have experienced a decrease in the incidence of HCC. This has happened after only ten years of the vaccination program. There is good reason to be hopeful that cancer of the liver will decrease significantly in many parts of the world where HBV is the main cause of the cancer and could eventually become a rare disease. Our laboratory has published several papers discussing the possibilities of control[69,70] and in 1989 I co-organized the International Conference on Prospects for Eradication of HBV.[71] Representatives from many countries with a high prevalence of HBV carriers attended this meeting, and several of these countries established universal vaccination programs after the meeting. I believe it had an important effect on pushing forward the international vaccination program. In collaboration with Dr. A. McLean of the biomathematics group at the Department of Zoology at Oxford University, we published a mathematical model of the vaccination program which could help to provide a theoretical framework for the design of control and eradication programs.[72]

In a remarkably short time after the introduction of the vaccine, there is a striking decrease in the incidence of hepatitis carriers, and there is a measurable effect on the incidence of HCC, one of the most common and deadly cancers in the world. Mark Kane, the scientist responsible for the hepatitis programs at the WHO has said that, second to the smoking cessation campaign, control of

HBV-caused HCC is the largest interventional cancer prevention program in the world.

10. IRON AND IRON-BINDING PROTEINS, 1974

The etiological role of HBV in cancer began to be appreciated in the early 1970s. I was frequently asked about the role of nutritional factors in the origin of liver cancer and, in a review of one of our grant applications, it was recommended that we add a nutritional study to the already complex project we had proposed. I had some experience with nutritional surveys in 1960 when I was asked to act as an observer on a large survey in Chile organized by the National Institutes of Health. This was sufficient experience to discourage me from attempting a comprehensive survey. The actual collection process was very costly in time and money, and the results were often difficult to analyze. We decided to focus our studies on iron. Iron is a nutritional material and it is relatively easy to measure. In addition there was experimental data that linked it with infection and susceptibility to cancer.[73] Iron binding proteins, such as ferritin, provide a measure of the body load of iron, and may also be measured. A further impetus to the study of ferritin was the research conducted in my laboratory by my colleague Hei-Wan Hann. She had studied the relation of increased ferritin levels to poor prognosis in childhood neuroblastoma. Also, the amount of iron in an individual can be manipulated by regulating intake or by removing iron from the body with chelating agents. We reasoned that if a biologically significant relation were found, it might be possible to develop therapeutic and preventive interventions for altering iron and iron binding.

We found that iron is increased in HBV carriers[74,75] and that this appears to be a consequence of infection and does not precede it.[76] Ed Lustbader showed that individuals with elevated ferritin, when exposed to infection with HBV, were more likely to become carriers than individuals who had lower levels, and the latter were more likely to develop anti-HBs.[77] The elevation of ferritin preceded the infection.

We then used an epidemiological approach made possible by the large numbers of sera collected in our field trips and stored in deep freezes in Philadelphia. Sera had been collected in the Solomon Islands by Damon and his colleagues between 1966 and 1970 and we had participated in some of the blood studies. The alive or dead status of members of this population had been determined in 1978 and, in some cases, in 1980. In the 1980s Richard Stevens, then a doctoral student of mine, went to the Solomon Islands and collected additional information. The sera samples that had been collected earlier were tested to determine the ferritin and transferrin levels. As predicted, the individuals who had died had higher levels of ferritin and lower levels of transferrin, that is, higher body iron stores, than those who had survived.[78] These studies are called "historic prospective" in that the specimens are from the past, but they are utilized in a prospective fashion.

A similar technique was used in a cancer study in Taiwan. In collaboration with Beasley,[79] we found that increased ferritin levels conferred an increased probability of cancer of the liver and other cancers developing later in life. In 1994, Stevens and others[80] used data collected during the first National Health and Nutrition Examination Survey conducted between 1971 and 1975. This again confirmed that even moderate elevations of body iron levels increased the risk of cancer. This outcome has been observed by other investigators.

These and similar studies by ourselves and others (primarily Stevens) may lead to a reconsideration of the practice common in many countries of fortifying foodstuffs, particularly bread, with iron. We have recommended that iron should be prescribed only for those who need it since increased iron intake in the general population might increase total body iron stores and possibly increase the probability of decreased survival and cancer. This may have had an effect in Sweden which, traditionally, had fortified all bread with added iron. Stevens and myself were invited to speak about our work at a series of meetings which received a considerable amount of media attention. The Swedish government subsequently changed its iron fortification program.

All these findings, in addition to ongoing experimental studies, indicate that increased iron may hasten viral replication and the development of cancer. We are currently considering therapeutic trials to test this notion.

11. HBV, GENETICS AND RELATED TOPICS, 1967

11.1. Gender

There is an interesting relation between infection with HBV and the gender of the human host. In our earliest papers[81,46] and others[82] we observed that there was a higher prevalence of carriers among males than females, and this has been confirmed in many studies from our laboratory and others. Male carriers are more likely to remain carriers than are females. Males when infected are more likely to develop the carrier state, while females when infected are more likely to develop anti-HBs. HCC is much more common among males than females and male carriers are more likely to develop HCC than are female carriers.

Perhaps the most curious observation concerning gender was the observation we reported in an epidemiological study in the community of Plati in northern Greece[83,84] undertaken by Jana Hesser, then a graduate student. We collected blood from most of the inhabitants and classified them as: a) carriers (HBsAg); b) individuals who had developed anti-HBs; or, c) those with no signs of infection. The results were arranged according to these family classifications. There were smaller numbers of offspring and the ratio of males to females (sex ratio) was significantly higher in families in which one or the other parent was a carrier than in the families in which the parents (mainly mothers) had developed anti-HBs. We repeated these studies in populations in other parts of the world; East Greenland,[85] Papua New Guinea[86] and the Philippines.[87] In each case the sex ratio was higher in the offspring of carriers than in the offspring of families in which the parents had anti-HBs.

In my view, this is one of the most interesting aspects of HBV. The implications (and the hypotheses that could be tested in subsequent studies) are that the virus has an effect on sex determination, and that there is a polymorphic balance between the host response to HBV and the gender of their offspring. Further, this could account for the high sex ratio (more males than females) that has been observed in regions such as South China where there is a high prevalence of HBV carriers. However, sex ratio/HBV studies have not, to my knowledge, been done in China. There are important and fascinating social and psychological problems related to the larger than expected number of male offspring. In many traditional societies males are preferred to females. The Government of China has advocated small families. Parents faced with the prospect of a single child family could have a strong desire that the child should be a male. It is important to know if the carrier state actually influences gender determination in China and if so, the effect of the extensive childhood vaccination program that is now in place.

Sex ratio is one of the crucial biological factors in demography and mammalian evolution. Is HBV a major force in the evolutionary process, and if it is, how should it be studied? To my knowledge there has been little if any interest in this problem and few attempts have been made to confirm or reject our observations and inferences on sex ratio. This may be because of our failure to provide a theoretical model for our observations. Professor H. Sharat Chandra of the Indian Institute of Science, Bangalore, and I formulated a tentative model to explain these observations, but it has not been published. Briefly, it postulates that female conceptus are more likely to die *in utero* as a consequence of chronic HBV infection during pregnancy. We postulated that there was a region on the male Y chromosome that had an affinity for HBV, and that the equivalent region on the X chromosome was much smaller. When the male conceptus was infected from the mother (or, indirectly, the father), HBV would be bound or in some way rendered less replicative and/or pathogenic. The male fetus would be more likely to survive and the likelihood of chronic infection would be higher. In the female conceptus, to the contrary, HBV would not be bound (or otherwise compromised) and could replicate and eventually the fetus would die.

An inference of this is that there would be DNA homologies between HBV and portions of the Y chromosome. Using the genome data base, we found significant homologies between portion of the HBV genome and the male determining gene on the Y chromosome (the SRY gene), which was consistent with our prediction. However, these findings have not been published and we are seeking further validation of the model.

These observations on the gender relations may prove to be the most interesting biological consequences of our research program.

11.2. Genetics, Maternal Transmission, and Early Childhood Infection

The discovery of the hepatitis B virus had originated with our studies of genetic polymorphisms and the genetics of the carrier state. (See, for example, the pedigree of an extended Samaritan family published in the first definitive paper in the Australia antigen investigation[43]). In 1966, even before it was realized that Australia antigen was the surface antigen of HBV, we did an extensive analysis of the segregation of Au in families from Cebu in the Philippines; Rongelap Atoll, in the then Marshall Islands District of the U.S. Trust Territories of the Pacific Islands (now Republic of the Marshall Islands); and from the Cashinahua Indians of Peru and Brazil.[81] I had collected the specimens of all but the last of these populations. The Cashinahua were a remote group of Amazonian Indians who, at the time of the collection had only recently come into contact with Europeans. They were collected by K. Kesinger, a graduate student in Anthropology and Frank Johnston, later the Chairman of the Department of Anthropology at the University of Pennsylvania, the Department with which I was associated for nearly thirty years. A diverse group of populations, but the magic of segregation analysis, admittedly a very unsophisticated one, allowed us to test and support the hypothesis of Mendellian segregation for the trait of persistent Au, i.e., the carrier state of HBV.

We collaborated with the anthropologist Jonathan S. Friedlaender, then at Harvard and later at Temple University in Philadelphia, on the segregation analysis of an extensive group of families that he had collected on the island of Bougainville, Papua New Guinea.[88] The results of these studies, combined with the families studied on Cebu, supported the autosomal recessive hypothesis for the genetic control of the carrier state of HBV. We termed this gene Au^1, and suggested that the immune response of the putative host to infection with HBV was under control of a polymorphic genetic system of which Au^1 was one of the alleles. We had combined two major hypotheses, namely, that Australia antigen was part of an infectious agent, HBV, and that it also constituted a polymorphic system. It was both genetic and environmental.

We did not do much more for several years on the formal genetics. There was little interest in our results, and several other investigators cast doubts on our conclusions. Recently, using modern methods Ed Lustbader did a segregation analysis of family data selected from a Chinese population with a high prevalence of HBV carriers. His model confirmed the genetic hypothesis and also accounted for the maternal effect (see below). Sadly, Ed's death has left this work uncompleted. The development of molecular biology completely transformed the field. Whereas in the past we had studied the protein phenotype produced by the gene, it was now possible to look directly at the genetic material. We are planning a family study in West Africa, in conjunction with a large public health research program to attempt to localize, identify, and sequence the *Au* gene or genes that we have postulated to control the immune response to HBV infection.

The study of families also revealed a maternal effect. These are summarized in part in a paper presented at the Princess Takamatsu Cancer Research Conference held in Tokyo in November 1972,[89] one of the many occasions when we presented our research results in Japan. We looked at families in several communities where HBV was common. Families in which the mothers were carriers were compared to families in which the fathers were carriers in respect to the ratio of children who were carriers to those who were not. In all the populations the ratio was higher when the mother was a carrier. This implied that transmission from mothers to children was relatively common and that this would be an important factor in public health control.

We found one possible exception to this general rule. The virologist, Scott Mazzur worked in our laboratory for several years. After leaving she obtained a senior appointment in the American Red Cross Laboratory in Washington, D.C. This allowed her to continue her work on HBV and also to indulge her interests in anthropology. She completed several field trips to Garciosa Bay in the Santa Cruz Islands in the Central Pacific and sampled a large part of the community including many family groups. The traditional culture included the institution of Male Houses where men and young boys would live separately from the conventional biological family. Institutionalized male homosexuality was widely practiced. In this community the parental effect was different. When fathers were carriers they had a higher percentage of carrier offspring than the families where the mothers were carriers. Sadly, Scott died of a rapidly malignant cancer diagnosed when she was in the field. She asked Tom London and myself to be the "scientific executors" of her data, and her work was completed by members of our laboratory and published after Scott's death.[90]

There was another interesting and unexpected observation made by one of our co-workers, Kay Kukowski, working in India. Kay was a member of The Medical Mission Sisters, a Catholic order, whose United States headquarters were located near the Fox Chase Cancer Center. The Order was made up of women religious health care workers, doctors, nurses, laboratory personnel, social workers, and others who staffed mission hospitals in developing countries. We had created a training program for the Sisters and several of them spent a period working in our laboratory gaining competence in HBV testing and other laboratory skills. Kay's project in India was to examine pregnant woman who were carriers and to compare them to mothers who were not.[91] She also tested the cord blood of the offspring, and blood collected from infants up to three months after birth but none were found to contain HBsAg. This is consistent with later studies in which it was found that the carrier state does not develop in newborns until about three months of age.

There were, however, interesting differences between the carrier and non-carrier mothers. Carrier mothers had a longer gestation period and were younger than the non-carrier mothers. The lower age was consistent with previous observations that the carrier state peaks in young adults. The shorter gestation time does not appear to be a consequence of the younger age of the mothers since, when correction was made for age, the difference persisted. This implies that carriage of HBV has a significant effect on the biology of parturition and might have an evolutionary effect in making carrier women less fertile. This is consistent with our other studies on gender and sex reviewed above.

11.3. HBV and Human Evolution

Does HBV influence human evolution? It does appear to have significant, although sometimes, subtle effects on important biological processes. It affects morbidity and mortality, but most of this would occur after the child bearing age. Death from cancer of the liver and chronic liver disease, the major consequences of chronic HBV infection usually occur after the 5th or 6th decade, past the most active period of reproduction. In pre-modern times when life expectancy was much shorter than at present, the long term consequences of HBV infection would not have been realized. However, if our data are valid, HBV infection is related to and may have an effect on the reproductive process even before any symptoms of liver disease are apparent to the carrier. As already discussed, HBV infection may affect the number of children ever born, and increase the ratio of males to females. The postulated HBV carrier polymorphism appears to interact with the human gender polymorphism, one of the most critical forces in mammalian evolution. The increase in gestation time, described above, would also have an effect on reproductive capability.

If *Au* is a polymorphic system, what are the advantages? One possible advantage is the increased iron storage of HBV carriers. In areas where iron is in short supply in the diet (as is the case in Africa and other regions where HBV carriers are common) then carriers might be at an advantage, particularly at periods of their life when iron is required for metabolism. There are probably other advantages to the carrier state which will be uncovered as the biology of the virus is better understood. It is unusual to wish to attribute advantages to the carrier state for a virus which can lead to death-causing illness late in life, but it is also intriguing to seek them out.

It is essential to understand the biology of HBV and how it interacts with human reproductive biology and human physiology as we begin to consider the eradication of HBV.

12. METHODS OF TRANSMISSION, 1967

12.1. Perinatal, Venereal, Needles and Other Devices

In a previous section, maternal transmission has been discussed. Transmission may occasionally occur before birth, at the time of birth (which appears to be common in parts of Asia) or in the weeks, months and years after birth when mothers and their children are in close contact. It is one of the major mechanisms for the transmission of HBV, and the early childhood vaccination programs are directed toward preventing this form of transmission.

Sexual activity is a common mechanism of HBV transmission.[92] Often the sexual partner does not develop chronic infection but does develop acute hepatitis with the subsequent production of anti-HBs which provides protection against future infection. The goal of adolescent vaccination programs is to provide protection before young men and women start their sexual careers. These are now being introduced in countries of relatively low prevalence where childhood transmission is less common.

Transmission by transfusion and its prevention has already been discussed. This form of transmission is now controlled in most countries. By 1970 we recognized that intravenous injection of medications or illegal drugs using contaminated needles carried a high risk of HBV transmission.[93] Any equipment which is exposed to human blood and reused can be a source of transmission. One of the most important measures for the prevention of HBV and other blood borne viruses (for example HIV), has been the use of disposable needles and other equipment.

In the 1970s, we collaborated with U.S. Army medical officers in a study of Armored Force troopers at Fort Knox, Kentucky.[94,95] The units preparing to go to Vietnam had a much lower prevalence of HBV carriers than a group of troopers who had recently returned from active service there. The military authorities had instituted a voluntary drug addiction treatment program for returnees. When the individuals who had volunteered for the drug treatment program were removed from the sample of returnees, the prevalence of carriers dropped to that seen in the soldiers before their year in Vietnam. The presumption was that many of the carriers were infected as a consequence of drug use. We were told that hepatitis, much of it due to HBV, was the second most common cause of morbidity during the Vietnam campaigns.

12.2. Insects, 1973

Many viruses and other infectious agents are spread by mosquitoes and other insects and an understanding of the role, if any, of these vectors is of great importance in public health control. We wished to determine if insects in the wild carried HBV. I have already mentioned that I collected mosquitoes on my trip to Uganda in 1971. I continued the collection in Ethiopia on my way back to Europe. We detected HBsAg in individual mosquitoes in high prevalence. More extensive studies were then done in Senegal and elsewhere by William Wills, the Medical Entomologist for the Commonwealth of Pennsylvania who subsequently joined our staff. He found a field infection of about 1 in 100, an extremely high prevalence. We also showed that mosquitoes experimentally fed on human blood containing HBV carried the virus. However, there was no evidence that HBV could replicate in the mosquitoes.[96,97,98]

We did not continue experimental studies on mosquito borne transmission because of our concerns about maintaining the security of infected mosquitoes. Mosquitoes seemed capable of exiting an insectorium even when there were multiple cages designed to contain them. We decided to switch our studies to bedbugs, the genus *Cimex,* because they are common in regions of the world where HBV is common, and they are much easier to contain.[99] Many of these studies were undertaken by Will Wills and later, Walter Ogston (the son of Sandy Ogston.) We eventually developed a thriving colony of tropical bedbugs, *Cimex hemipterus,* which Ogston used for many of his studies. I understand that it was one of the few such colonies in the United States.

In the bedbugs collected in Senegal, we found an extraordinarily high rate of infection in the bugs that had been collected from beds whose main occupant was a HBV carrier; in some of the collections the prevalence of HBV in bedbugs was over 70%. It was also shown that the virus persisted for weeks in the insects after they had imbibed a blood meal containing HBV and that it could be detected in their waste for weeks after infection. This raised the possibility that bedbugs could be a cause of transmission from an infected to a non-infected person who occupied the same bed, i.e., sexual partners and/or mothers and their newborn infants. Despite all this circumstantial evidence there has not been any direct data or experiments to show that insects are a major mechanism for transmission. It became increasingly difficult to obtain funding for these studies — in part because medical entomology altogether is not well funded — and we were unable to continue. One of the few possible sources of funding was the U.S. military research groups who were probably interested in these mechanisms for the protection of troops operating in insect infested areas, and also possibly for the development of anti-biological warfare measures.

There was an unexpected application of our research. Soon after the discovery of AIDS and the Human Immunodeficiency Virus (HIV), there was a great concern about mosquito transmission based, in part, on an alarming report from a small community in south central Florida of a very high prevalence of AIDS. Several local physicians and others raised the possibility that this was a consequence of mosquito transmission. This was a matter of public health and political importance for, if HIV was transmitted by mosquitoes, there could be calls for restricting the mobility and civil rights of those infected with HIV. The Office of Science and Technology of the U.S. Congress called an "emergency" meeting of scientists and others working in this field. The general conclusion was that there was *not* any evidence of HIV transmission by mosquitoes. The insect studies with HBV were cited as an example of how an infectious agent similar to HIV *might* be transmitted by insects. By contrast, there was no similar data on HIV. It was a very spirited and emotionally charged meeting followed closely by the media.

There are some interesting conjectures that arise from the possibility that HBV and its genes can be transported from one individual to another by a flying or crawling insect. Using a flight of fancy (please excuse the pun) and imaginative assumptions concerning the molecular biology of HBV, some interesting conjectures can be made. When the HBV genome, or portions of it, integrate into the host genome it may acquire segments of the fringing or nearby sequences of the host. If the integrated virus sequence is packaged and transmitted via an insect bite it could carry with it portions of the genome of the previous host. (In following this conjecture one must keep in mind that the integrated HBV sequences are not thought to enter into the replication process.) Thus, the insect could transmit genetic material from one individual to the next, circumventing the sexual mechanism for transferring genes.

Enough of speculation for the moment. At present it is not clear what, if any, is the role of insects in HBV transmission.

13. ANTI-VIRALS

13.1. Initial Studies, 1984

By the late 1980s there was a certain measure of satisfaction in the community of scientists and clinicians working with HBV based on the accomplishments of the two decades since the discovery of the virus. They included the following:

1. The virus had been identified.
2. Sensitive and specific clinical diagnostic methods were available.
3. Donor blood was routinely screened and post-transfusion hepatitis due to HBV was essentially controlled.
4. The vaccine we had introduced, prepared either from blood or by the recombinant method, was highly effective and safe. Several national universal vaccination programs for children were in effect. The price of the vaccine, originally very high, was beginning to decrease in cost sufficiently to allow its use in public health programs.
5. The etiologic relation between HBV and primary cancer of the liver was fairly well understood and ongoing molecular biological studies were providing insights into how viruses cause cancer.

6. The prevention of infection with HBV was recognized as the major interventional cancer prevention program advocated by the WHO, with the exception of smoking cessation.
7. The discovery of HBV had facilitated the subsequent discovery of HAV, HDV, HCV, HEV, HGV, etc.

All this had been accomplished without the media attention that has been given to HIV where, for the most part, the practical problems were still not resolved despite the expenditure of vast amounts of research funds. HBV infects far more people than HIV and, probably, *in toto,* has a greater mortality and morbidity, but the attention of the public and that of politicians had not been as focused. This is in many ways curious, since the affected populations are similar. This could be the subject in a study of the public perception of disease and how it reacts to it.

There was, however, one area in which there was only limited success; the treatment of chronic HBV infection. Interferon in several forms had been used for some years, but it was not a very effective treatment. Only a portion of the patients responded and the virus replication re-started after the cessation of treatment. In addition, the side effects were profound and many patients did not wish to continue the treatment. There were few if any other potentially useful anti-virals and there was little to offer to the approximately three hundred million carriers in the world, many of whom would go on to develop chronic liver disease and HCC. We decided to attempt to develop an anti-viral.

Many of the drugs in the current therapeutic list are derived from natural substances, particularly plants. In a surprising number of cases these were introduced into modern medicine from various indigenous medical systems. At present, there is a growing reliance on the use of rational drug design based on the remarkable progress in determining molecular structure and the advances in the Human Genome Project. Despite this, we decided to proceed with the traditional method of looking for drugs in plants.

Initially we established a screen to determine if extracts of the plant in question inhibited the interaction between the surface antigen of HBV (HBsAg) and the antibody against it (anti-HBs). I collected about 1200 species in various parts of the world (nearly always when I was traveling for other purposes) and found that about one-third inhibited the interaction. It was hard to know what to do with this data.

I must admit that a major attraction of this project was the opportunity to do botanical collections. I had always envied scientists who undertake extended visits to exotic places to search for new species of plants and animals. It was a particular pleasure to tramp through the woods and mountains collecting specimens for these studies. I had always enjoyed field work; but my previous field expeditions had involved collecting biological specimens from humans (and occasionally domestic animals) which usually was done indoors. Plant collecting was another matter; it was conducted in the open air with fewer problems associated with obtaining permission for collections and the many complication of clinical research. I had some wonderful collecting trips in California, Florida, Hawaii, Maryland, South Carolina, India, Nepal, Australia, England, Scotland, Korea, and other places.

After this initial survey, we decided to make more systematic collections. We assembled a list of about 2000 plants that had been used for the treatment of jaundice in folk pharmacopoeia. Jaundice is seen in several diseases, but worldwide, hepatitis viruses are probably the major causes. The species on the list were then sorted alphabetically and species belonging to genera that had been used in three or more continents or geographical locations were selected for further investigations. The notion was that plants that had been used very widely would have a higher likelihood of containing effective and active components against jaundice. This process shortened the list to about one hundred species.

While this process was going on (I was at the time on sabbatical at Balliol College in Oxford), my colleague in Philadelphia, Dr. P.S. Venkateswaran, who had joined our group as a natural products chemist, began to work on a plant *Phyllanthus niruri* (subsequently classified as *P. amarus*) that he was familiar with from his previous work in India. It was one of the species on our short list of widely used plants and has been used as a remedy for jaundice (as well as many other conditions) in India, China, in the Pacific, South America, the Caribbean, West and East Africa, and elsewhere.[100] The initial results were very promising. Extracts of the plants inhibited the DNA polymerase of HBV and also inhibited the HBsAg/anti-HBs interaction. Woodchuck (*Marmota monax)* carriers of the woodchuck hepatitis virus (WHV) appeared to have decreased amounts of virus during the course of and following treatment with extracts and whole plant. We prepared for the next step in the project.

13.2. Studies in India, 1986

My wife and I spent the first three months of 1986 in Bangalore, a large city in the state of Karnataka in south India, where I served as the Raman Visiting

Professor at the Indian Institute of Science. While there, I lectured at the Institute and many other places in India and also completed several fascinating plant collecting expeditions in the western Ghats, in central Karnataka, and in Nepal. The collecting trip in Nepal was particularly gratifying. We had prepared a list of the species of plants which had been used for the treatment of jaundice, as discussed above and which were known to occur in Nepal. I had engaged a local botanist to join us on the trek (he was the father of the *surdar,* the expedition leader) to identify and locate the desired plants. We spent nine days proceeding north, and later east from the provincial town of Ghorka, the roadhead from which the footpaths started. Through a happy error in the arrangements we had about 12 porters, cooks, and others to support us on the trek. It was the most luxurious hike I have ever done. I thought of it often when in subsequent years I was humping a thirty pound pack up high passes in the Trinity Alps of California. The trip was successful and we returned with most of the plants on our "wanted" list.

I was also asked to consult on the hepatitis B problem in India. I spoke with senior officials in the Departments of Science and Technology and agencies related to health and research. The Prime Minister, Rajit Ghandi, spoke with me at some length about the problem and exhibited a surprising insight and knowledge of the current situation relevant to hepatitis and public health in general. It was not long after our meeting that he was tragically assassinated. I had the impression that the health and science officials were primarily interested in the manufacture of the hepatitis B vaccine, preferably by the recombinant method. They wanted the country to enter the new field of Biotechnology, and this seemed an appropriate mechanism by which to do so.

In my report to the Indian Government, I did not emphasize the high tech manufacture of vaccine as the first priority. I had learned that in all of India, where there were several hundred medical schools, there was only one school of public health. I suggested the establishment of at least five such schools in order to enhance the prospects of public health research and application. In many parts of India the public water supply is contaminated with human waste. This is probably the single largest cause of disease in the sub-continent. In urban areas the problem is particularly difficult because the waste pipes and the water supply pipes are often in the same ditch. Since some of these systems have not been maintained or replaced for decades, there are many breaks, and leakage from the waste pipes can contaminate the water supply. This is most likely to happen during periods of drought when there is a partial vacuum in the water pipes and the probability of contamination is increased. Replacing these water pipes would probably cost billions of dollars, but it would be one of the most important health measures that could be taken·

I advocated the development of an industry to supply disposable needles and other medical supplies that have contact with human blood. My final recommendation included a survey of the extent of the hepatitis B problem and the manufacture of vaccine, first from blood and later, when the infrastructure for biotechnology was further advanced, by the recombinant method. I received an acknowledgment of the receipt of the report, but it is, of course, difficult to know what effect, if any, it had. I understand that some of the measures I had suggested have been accomplished.

Another consequence of this visit, more pertinent to the *Phyllanthus* project, was a collaboration with Dr. S.P. Thyagarajan of the Department of Microbiology at the University of Madras. He had worked with *Phyllanthus* and was proposing to do a clinical trial. We helped in the design of a controlled study and also confirmed the testing of the sera for HBsAg. The clinical studies themselves were carried out by Thyagarajan in India. The results were striking; 59% of the carriers who received the *Phyllanthus* lost HBsAg and it did not return for up to 9 months in the patients that could be followed. Only a few of the placebo group lost HBsAg. This was a very dramatic result, far better than with any other medication that had been used for HBV. We cautioned in our written report[101] that these results were preliminary and subject to error and encouraged further controlled trials. Under the direction of David Unander, a plant scientist who joined our group at Fox Chase Cancer Center, we completed several studies on the botany and folk uses of the plants[102,103,104,105,106] and methods for domestic cultivation.[107] Several species of *Phyllanthus* were raised as an experimental crop at the Fairchild Gardens in tropical Florida south of Miami.

During the course of the investigation, Unander and I collected species of the genus *Phyllanthus* in the United States and in many locations in the tropics and sub-tropics. These have been deposited in the Herbarium of the Academy of Natural Sciences in Philadelphia. We established a Herbarium at the Fox Chase Cancer Center which contains the voucher specimens for all the plants we collected.

13.3. Anti-Virals from Plants. Further Studies

There was a brisk response to our publications. I was contacted by patients and their relatives requesting the drug. The natural food and alternative medicine advocates saw this as an important support of their mission. However, I cautioned all who asked that the results were preliminary, that the drug had not been approved for use and that the experiments should continue. Unhappily, the clinical findings from Madras were not confirmed in other studies from Thailand and elsewhere which are reviewed in a paper I presented at a meeting on plant derived drugs.[108] Drs. Millman and Venkateswaran and Terry Halbherr did a series of very difficult studies on woodchuck WHV carriers. Although the results were in the same direction we had seen in our previous studies, they were far less impressive. Venkateswaran and Millman attempted to isolate the active principle but this was unsuccessful. In essence, our original findings could not be confirmed. In the world of science unless an observation can be confirmed it has little validity. We had to decide whether to continue with this project or recognize the uncertainty of using natural products and await the isolation of an active principle before proceeding. We decided to close the project for these reasons, but also because of the break-up of the research team at Fox Chase.

Strangely, despite these setbacks there was continuing interest in *Phyllanthus* and several laboratories have continued laboratory and clinical studies. Recently, several companies have expressed an interest in the project and may continue in attempts to develop a useable medication. On balance, the plant project could be considered a failure. However, the research is ongoing and may still result in a wholesome and useable agent.

13.4. Glycoproteins. Balliol College, Oxford University, 1989–1994

In 1989, I was elected Master of Balliol College, one of the thirty or more colleges that make up Oxford University. It was an unusual appointment; I was the first American and the first scientist (with the possible exception of an alchemist in the fourteenth century) and I enjoyed my five years in the post immensely. Balliol is an old (established 1263!) and respected institution which has trained many leaders of British society, and in particular, many of its politicians and statesman. We left Philadelphia and I devoted most of my energies to my College responsibilities, although I retained a part-time position at Fox Chase Cancer Center and maintained my laboratory for several years. I did, however, develop collaborations with scientific colleagues in Oxford and this resulted in some exciting new projects.

Oxford is a major center for population biology. I started a seminar on this subject which met regularly at the College and which was run by Professor Adrian Hill, who we had appointed as a Huxley-Royal Society Research Fellow at Balliol. We are now collaborating on a project on the genetics of the carrier state of HBV. The interest in mathematical models and evolutionary biology has continued to thrive in the Department of Zoology. I have already referred to the mathematical analysis Drs. Nowak and Payne[60,61] devised for the selection-developmental model for the pathogenesis of primary hepatocellular carcinoma that Thomas London and I had published.[59] McLean modeled the impact of a mass vaccination program directed against HBV.[72]

Professor Raymond Dwek graciously appointed me as a visiting scientist in the Glycobiology Institute, a unit of the Department of Biochemistry. Timothy Block, a colleague from Philadelphia, came to work in his department for a year and we initiated studies on the glycoproteins of the surface antigen of HBV. *N*-Butyl-DNJ, an imino sugar compound developed at the Glycobiology Institute, inhibits the terminal glycosylation of HBsAg.[109] This prevents the export of HBV from liver cells and has a profound effect on the distribution of the virus in the liver cells that it infects. This may have therapeutic implications and studies on this process are continuing in Oxford and in Philadelphia.[110]

14. MISCELLANEOUS, EPHEMERAL AND UNPUBLISHED ARTICLES

During the course of our research, biological samples we collected were usually frozen and stored. The anonymity of the donors was preserved and the specimens were identified by a code number. It has been possible to use these for historic studies. For example, in collaboration with Drs. Dube and Poiesz and others, we studied sera from Equateur Province, Zaire that had been collected by our Belgian colleague Professor Michel Lechat in 1969 to determine the distribution of HBV in patients with leprosy.[111] HIV is now very common in this region but was absent in the archive collection. This represents one of the earliest "negative" dates for HIV infection and helps to establish the timing of the entry of the virus into humans in Africa. We did find a very high prevalence of HTLV-I, another retrovirus which is common in the region now. Sequencing of the viruses indicated that they represent a unique phylogenetic cluster. The single HTLV-II isolate that could be sequenced showed a very high homology with sequences previously found in several New World Indian groups.

HTLV-I is associated with a variety of diseases, including Adult T Cell Leukemia and Tropic Spastic Paralysis. However, these diseases are rare and, with the exception of a study from Japan, there was little information on the effect on overall mortality of persistent infection with HTLV-I. Using the "historic prospective" method, we determined who was dead and who alive in 1991 among the persons whose sera had been collected in 1969. We found a 40% increased mortality in the people who were HTLV-I carriers during the period of observation.[112] This finding, combined with the Japanese study, requires first, additional studies to determine if this increased mortality in the HTLV-I carriers is found in other populations, and secondly, the identification of the causes of the early deaths. It also illustrates the value of these stored specimens to determine the long term effects of factors that can be measured in stored sera but were not known when the collections were made. The historic prospective study is very powerful; it can permit the completion of a "prospective" study of 20 or more years in a few months.

I have written several articles on the history of medicine. These include an essay on Bernard Connor[113] an Irish physician, who, in 1693, had written the first description of the pathology of ankylosing spondylitis, a form of arthritis of the spine, of which I had conducted a clinical and epidemiological study.[4] Connor was an interesting polyglot who had been accused of being a French spy, written on the "natural oeconomy" and traveled widely. In a philosophic treatise he described what must be one of the earliest expressions of the "biophysical" approach to an understanding of biology and medicine, that all of biology can ultimately be expressed in physical or chemical terms. It was a remarkably prescient statement of the reductionist approach to biology which has resulted in such extraordinary advances. Connor also claimed to have been the physician to John Sobieski, King of Poland, who became a central figure in the history of 17th century Europe when he successfully defended Vienna against the advancing Islamic force. During his stay at the court of Sobieski, Connor collected data later incorporated into the first history of Poland in English which remained a standard for many decades. I was sufficiently impressed with the special qualities of Connor that I included his name among the several bestowed on my eldest son.

While Master of Balliol, at the invitation of the chaplain, Douglas Dupre — a fellow American — I gave two sermons in the College chapel.[114,115] Both of these drew on my experience in science and although they were directed primarily to my students, they may be of more general interest. While in Oxford, the President of Fox Chase Cancer Center, Robert Young, asked me to write articles for the annual scientific report of the Center. These included an essay on the concept of independent evidence,[116] and a lament on the anti-scientistic character of the contemporary cinema including a critical review of the very popular *Jurassic Park* .[117] I had been an active cyclist before I came to Oxford, but while there became even more involved with the sport. At a lunch party with the then Editor of *The Daily Telegraph*, one of the major broadsheets in Britain, we discussed a variety of topics concerning science and government funding. I also mentioned my cycling interests. He later wrote and commissioned me to write an article, not on the scientific and political issues we had discussed, but on bicycling.[118]

During this period I became fascinated with the Frankenstein story as a metaphor for the public perception of science. I learned a great deal about this fascinating novel from the thesis and other publications of my daughter Jane whose academic work has been on Mary Shelley. Another metaphor for science that caught my imagination was Daedalus, the Greek mythic figure who was the inventor of sculpture and a pioneer in the invention of bio-mechanical prostheses, (i.e., the wings he devised to escape from Crete along with his son Icarus.) However, in an essay written by René Fox and myself,[119] we develop a different theme. Daedalus was a problem-solver, but each time he solved a problem he created another. He seemed to cherish the problems as much as, if not more, than the solutions. Scientists also solve problems and answer questions; but a main outcome of their effort is to reveal new mysteries. This theme is illustrated with our experience in the application of the HBV research.

My wife, Jean, was a student at the Pennsylvania Academy of the Fine Arts. I was asked to address her graduation class and presented a paper attempting to show the similarities between the scientific process and artistic creativity.[120]

In 1996 I spoke at the commencement on the 200th anniversary of the founding of Union College, which I attended as an undergraduate.[121] Alexander Ogston, died in 1997. I was very honored when his widow, Elizabeth, asked me to speak at his memorial service at the University Church in Oxford.[9] Our families had been friends for more than 35 years and I attempted to express my feelings of respect and gratitude towards a teacher who had a profound influence on my life and career and that of many others.

15. HEPATITIS B VIRUS; A SUBJECTIVE VIEW

In a humorous mood, Sandy Ogston told me many years ago that a goal of his research was to know what it felt like to *be* a hyaluronic acid molecule. This was such a striking idea that I remembered it for forty years. What would it be like to be a hepatitis B virus? I have worked with this virus for a large part of my adult life, spent days and nights in its company, and thought about it in the reveries of my resting moments. We are cautioned not to think teleologically about biology but it can be the best way to understand or explain to others the complex behavior of even so simple an organism as a virus with only 3200 base pairs. Consciousness, the ability to be aware of oneself, is said to be an attribute of humans which distinguishes them from other organisms. One of the qualities of consciousness is the ability to make plans; if you are aware that you exist then plan-making is essential to a continuing existence. Can a virus make plans? The answer has to be "No". But it *seems* to make plans and imagining that it does simplifies explanation and the synthesis of the idea of the virus. HBV has "devised" clever schemes for making its way in a hostile environment and ensuring its survival over the eons. In doing so, it has "arranged" accommodation with its host species to guarantee that the host will also survive in order to play its role in the perpetuation of HBV. This is reminiscent of the definition of a chicken as the mechanism that eggs have devised to produce more eggs.

It is likely that hepatitis viruses have infected humans for many generations and in earlier evolutionary forms may have preceded the development of *Homo sapiens* and pre-human species. This long experience with its host has given it ample time to develop evolutionary accommodation. Viruses with high levels of homology to human HBV have been found in woodchucks, ducks, ground squirrels, tree squirrels,[122] herons, and other species, and human HBV can readily infect chimpanzees and other higher apes. It is likely that HBV has been transmitted from other species to humans, and, probably *vice versa.*

HBV has adapted methods of transmission that link its perpetuation with functions that are essential to the survival of humans, namely, sex and childbirth. Before modern times, when blood transmission by needles was uncommon, sexual and mother-to-child were the main methods of transmission; they probably still are in most parts of the world. HBV has even adopted the subtle, but very clever, maneuver of increasing the ratio of males to females in offspring of carrier mothers, if our interpretation of the gender studies is correct. Males are more likely to become carriers than are females and, on balance, this is favorable to the virus.

HBV is very hardy. It can survive drying, appears to remain infectious in a cell free environment, and can be transmitted even in very low doses. It is a very small and simple genome — four reading frames — may contribute to this survivability. However, compactness may have its disadvantages. The four reading frames overlap in order to crowd into the small genome. This may limit the possibilities of mutation and therefore the possibilities of evading the host immune system. If, say, there is a mutation in the surface antigen gene it might also mutate the polymerase gene and this could render the virus incapable of replication. Since HBV is transmitted via the blood and other body fluids, it is advantageous to HBV to maintain high titers in its host organism. Titers of virus particles are higher during the reproductive age when sexual and mother to child transmission is more common and decreases with advanced age when transmission is less likely. Although the virus is usually benign early in the life of its host when it is in the "carrier" mode, it can become nasty later when the patient develops chronic liver disease and cancer of the liver. It is a great killer; an estimated 1.3 million deaths a year which places it in the top five causes of deaths due to infectious agents. In the pre-modern age when life expectancy was in the 40s and 50s, very few carriers would have lived long enough to develop the deadly phases of chronic infection, i.e., chronic liver disease and cancer of the liver. HBV would have contributed less than it does now to worldwide mortality and morbidity.

HBV has "designed" excellent strategies to use the immune system of the host to its own advantage and maintain high titers of virus particles in the blood. The large number of small particles containing only the surface antigen which are present in the peripheral blood may act as a "decoy" to attract anti-HBs and spare the less common whole virus particles. Ironically, these surface antigen particles have become disadvantageous to the virus in that they are used in the preparation of the vaccine which controls its spread and may lead to its eradication.

There is an elaborate mechanism whereby the C reading frame can produce both HBcAg and HBeAg. The former remains in the cells and is one of the antigens which the host immune systems attack in order to destroy infected cells. The HBeAg can escape the cell and render the host's immune cells "tolerant", probably by acting on the cytokine systems, and mute the host's destruction of the HBV-containing liver cells. The assault by HBV on the host immune system seems altogether more subtle than that of the HIV virus which invades and wantonly destroys important elements of the host immune system. The host infected with HBV is usually spared to survive the reproductive period. The incubation time of HBV is often four or five decades, far longer than that of HIV which can be less than a decade.

HBV is involved with other organisms and agents in its environment such as malaria and the other hepatitis viruses, and carcinogens such as aflatoxin.[123] It may even use insects to help its transmission, although there is little data to support this contention. It interacts directly and indirectly with the genetics of the host; the MHC class II genes, the postulated *Au*^{1} gene, the genes that act to detoxify aflatoxin, the polymorphisms of the iron binding proteins, and undoubtedly many others. We have stressed that the carrier state of HBV has the characteristics of a polymorphism or, to put it more explicitly, there are one or more polymorphic loci which control susceptibility to become a carrier when infected by HBV. If that is the case, there may be advantages to the heterozygote or homozygotes for these polymorphic traits. We have noted one possible advantage to the carrier. Carriers retain more dietary iron than non-carriers, and this may be of physiological importance in regions where there is iron deficiency. There are probably other advantages that have not yet been observed but should be known as we face the possibility of the marked decrease or eradication of the virus. Even though the virus may disappear, the genetic susceptibility polymorphisms may still remain in a population since they are probably associated with environmental hazards in addition to HBV.

HBV leads a complex, dynamic and busy "life". It is a remarkable, if deadly organism, and attempting to understand how it behaves and fits into its environment has given me insights into the grand pattern that pervades the seeming disorder of existence.

II SELECTED PUBLICATIONS: BIBLIOGRAPHY

This Bibliography contains 123 publications cited in the Introductory Essay. All but the papers with an asterisk are reproduced in this book.

*1. Blumberg B S, McGiff J and Guicherit I. Filariasis in Moengo (Surinam) in 1950. *Documenta Neerlandica Indonesica Morbis Tropicus*. **3**: 368–372, 1951.

*2. Blumberg B S, McGiff J and Guicherit I. Malaria survey among the Bush Negroes of Marowyne District, Surinam, S.A., in 1950. *Doc Med Geograph Trop* **4**: 2–4, 1952.

*3. Blumberg B S, McGiff J and Guicherit I. A survey of intestinal parasites in the school children of Moengo, Surinam, 1950. *Doc Med Geograph Trop* **5**: 137–140, 1953.

4. Blumberg B S and Ragan C. The natural history of rheumatoid spondylitis. *Medicine* **35**: 1–31, 1956.

5. Blumberg B S and Oster G. Light-scattering studies on hyaluronic acid. *Science* **120**: 432–433, 1954.

*6. Blumberg B S, Oster G and Meyer K. Changes in the physical characteristics of the hyaluronate of ground substance with alterations in sodium chloride concentration. *J Clin Invest* **34**: 1454–1461, 1955.

*7. Blumberg B S and Ogston A G. The effects of proteolytic enzymes on the hyaluronic acid complex of ox synovial fluid. *Biochem J* **66**: 342–346, 1957.

*8. Blumberg B S and Ogston A G. Further evidence on the protein complexes of some hyaluronic acids. *Biochem J* **68**: 183–188, 1957.

9. Blumberg B S. A memorial oration, The University Church of St. Mary the Virgin, Oxford, England, November 2, 1996. Alexander George Ogston, MA (Oxon), D. Phil, FRS, FAA. January 30, 1911 — June 29, 1996.

*10. Allison A C and Blumberg B S. Ability to taste phenylthiocarbamide among Alaskan Eskimos and other populations. *Human Biol* **31**: 352–359, 1959.

11. Alberdi F, Allison A C, Blumberg B S, Ikin E W and Mourant A E. The blood groups of the Spanish Basques. *J Roy Anthrop Inst* **87**: 217–221, 1957.

12. Blumberg B S, Ikin E W and Mourant A E. The blood groups of the pastoral Fulani of northern Nigeria and the Yoruba of western Nigeria. *Am J Phys Anthropol* **19**: 195–201, 1961.

*13. Allison A C, Blumberg B S and ap Rees. Haptoglobin types in British, Spanish Basque and Nigerian African populations. *Nature* **181**: 824–825, 1958.

*14. Blumberg B S and Gentile Z. Haptoglobins and transferrins of two tropical populations. *Nature* **189**: 897–899, 1961.

*15. Steinberg A G, Stauffer R, Blumberg B S and Fudenberg H. Gm phenotypes and genotypes in U.S. Whites and Negroes; in American Indians and Eskimos; in Africans; and in Micronesians. *Am J Hum Genet* **13**: 205–213, 1961.

*16. Blumberg B S and Tombs M T. Possible polymorphism of bovine alpha-lactalbumin. *Nature* **181**: 683–684, 1958.

17. Bangham A D and Blumberg B S. Distribution of electrophoretically different haemoglobins among some cattle breeds of Europe and Africa. *Nature* **181**: 1551–1552, 1958.

18. Corcoran P A, Allen F H Jr, Allison A C and Blumberg B S. Blood groups of Alaskan Eskimos and Indians. *Am J Phys Anthropol* **17**: 187–193, 1959.

*19. Blumberg B S, Allison A C and Garry B. The haptoglobins and hemoglobins of Alaskan Eskimos and Indians. *Ann Hum Genet* **23**: 349–356, 1959.

*20. Allison A C, Blumberg B S and Gartler S M. Urinary excretion of beta-amino-isobutyric acid in Eskimo and Indian populations of Alaska. *Nature* **183**: 118–119, 1959.

21. Blumberg B S, Bloch K J, Black R L and Dotter C. A study of the prevalence of arthritis in Alaskan Eskimos. *Arthritis and Rheumat* **4**: 325–341, 1961.

*22. Blumberg B S and Gartler S M. High prevalence of high level beta-amino-isobutyric acid excretors in Micronesians. *Nature* **184**: 1990–1992, 1959.

*23. Blumberg B S and Conard R A. A note on the vegetation of the northern islets of Rongelap Atoll, Marshall Islands, March 1959. Brookhaven National Laboratory, September 1960, pp. 85–86, 1960.

*24. Conard R A, Meyer L M, Sutow W W, Blumberg B S, Lowery A, Cohn S H, Lewis W H Jr, Hollingsworth J W and Lyon H W. Medical survey of Rongelap people five and six years after exposure to fallout. Brookhaven National Laboratory, September 1960, pp. 25–27, 1960.

*25. Conard R A, Meyer L M, Sutow W W, Blumberg B S, Lowery A, Cohn S H, Lewis W H Jr, Hollingsworth J W and Lyon H W. Medical status of Marshall Islanders in 1959, five years after exposure to fallout radiation. *Nuclear Medicine* **1**: 314–330, 1960.

26. Melartin L and Blumberg B S. Albumin Naskapi: A new variant of serum albumin. *Science* **153**: 1664–1666, 1966.

*27. Blumberg B S, Martin J R and Melartin L. Alloalbuminemia. Albumin Naskapi in Indians of the Ungava. *J Am Med Assoc* **203**: 180–185, 1968.

*28. Melartin L, Blumberg B S and Martin J R. Albumin polymorphism (albumin Naskapi) in Eskimos and Navajos. *Nature* **218**: 787–789, 1968.

*29. Johnston F E, Blumberg B S, Agarwal S S, Melartin L and Burch T A. Alloalbuminemia in southwestern U.S. Indians. Polymorphism of albumin Naskapi and albumin Mexico. *Human Biology* **41**: 263–270, 1969.

*30. Schell L M, Agarwal S S, Blumberg B S, Levy H, Bennett P H, Laughlin W S and Martin J P. Distribution of albumin variants Naskapi and Mexico among Aleuts, Frobisher Bay Eskimos, and Micmac, Naskapi, Mohawk, Omaha and Apache Indians. *Am J Phys Anthropol* **49**: 111–117, 1978.

*31. Melartin L, Blumberg B S and Lisker R. Albumin Mexico. A new variant of serum albumin. *Nature* **215**: 1288–1289, 1967.

32. Schell, L M and Blumberg B S. The genetics of human serum albumin. In *Albumin Structure, Function and Uses*, Rosenoer V M, Oratz M and Rothschild M A (eds.). Pergamon Press, Oxford and New York pp. 113–141, 1977.

*33. Franklin S G, Wolf S I, Zweidler A and Blumberg B S. Localization of the amino acid substitution site in a new variant of human serum albumin, albumin Mexico-2. *Proc Natl Acad Sci USA*. **77**: 2505–2509, 1980.

34. Lampl M and Blumberg B S. Blood polymorphisms and the origins of New World populations. In *The First Americans: Origins, Affinities, and Adaptations*, Laughlin W S and Harper A B (eds.). Gustav Fischer, New York, Stuttgart pp. 107–123, 1979.

35. Franklin S G, Wolf S I, Ozdemir Y, Yuregir G T and Isbir T. Albumin Naskapi varian in North American Indians and Eti Turks. *Proc Natl Acad Sci USA* **77**: 5480–5482, 1980.

*36. Kaur H, Franklin S G, Shrivastava P K and Blumberg B S. Alloalbuminemia in North India. *Am J Human Genet* **34**: 972–979, 1982.

37. Allison A C and Blumberg B S. An isoprecipitation reaction distinguishing human serum protein types. *Lancet* **1**: 634–637, 1961.

38. Blumberg B S, Bernanke D and Allison A C. A human lipoprotein polymorphism. *J Clin Invest* **41**: 1936–1944, 1962.

*39. Blumberg B S, Dray S and Robinson J C. Antigen polymorphism of a low-density beta-lipoprotein. Allotypy in human serum. *Nature* **194**: 656–658, 1962.

*40. Blumberg B S and Riddell N M. Inherited antigenic differences in human serum beta-lipoproteins. A second antiserum. *J Clin Invest* **42**: 867–875, 1963.

41. Blumberg B S. Polymorphisms of serum proteins and the development of iso-precipitins in transfused patients. *Bull N Y Acad Med* **40**: 377–386, 1964.

42. Blumberg B S. Iso-antibodies in humans against inherited serum low density beta-lipoproteins, the Ag system. *Ann N Y Acad Sci* **103**: 1052–1057, 1963.

43. Blumberg B S, Alter H J and Visnich S. A "new" antigen in leukemia sera. *J Am Med Assoc* **191**: 541–546, 1965.

44. Blumberg B S, Gerstley B J S, Hungerford D A, London W T and Sutnick A I. A serum antigen (Australia antigen) in Down's syndrome leukemia and hepatitis. *Ann Int Med* **66**: 924–931, 1967.

45. London W T, Sutnick A I and Blumberg B S. Australia antigen and acute viral hepatitis. *Ann Int Med* **70**: 55–59, 1969.

46. Sutnick A I, London W T, Gerstley B J S, Cronlund M M and Blumberg B S. Anicteric hepatitis associated with Australia antigen. Occurrence in patients with Down's syndrome. *J Am Med Assoc* **205**: 670–674, 1968.

*47. Bayer M E, Blumberg B S and Werner B. Particles associated with Australia antigen in the sera of patients with leukemia, Down's syndrome and hepatitis. *Nature* **218**: 1057–1059, 1968.

48. Coyne (Zavatone) V, Millman I, Cerda J, Gerstley B J S, London W T, Sutnick A I and Blumberg B S. The localization of Australia antigen by immunofluorescence. *J Exp Med* **131**: 307–320, 1970.

*49. Coyne V E, Blumberg B S and Millman I. Detection of Australia antigen in human tissue culture preparations. *Proc Soc Exp Biol Med* **138**: 1051–1057, 1971.

50. Senior J R, Sutnick A I, Goeser E, London W T, Dahlke M B and Blumberg B S. Reduction of post-transfusion hepatitis by exclusion of Australia antigen from donor blood in an urban public hospital. *Am J Med Sci* **267**: 171–177, 1974.

51. Blumberg B S. Blood transfusion and the discovery of viruses. James Blundell Award Lecture: 1989. British Blood Transfusion Society Newsletter No. 19, January 1990, pp. 1–3, 1990.

52. Coller J A, Millman I, Halbherr T C and Blumberg B S. Radioimmunoprecipitation assay for Australia antigen, antibody, and antigen-antibody complexes. *Proc Soc Exp Biol Med* **138**: 249–257, 1971.

53. Millman I, Loeb L A, Bayer M E and Blumberg B S. Australia antigen (a hepatitis-associated antigen). Purification and physical properties. *J Exp Med* **131**: 1190–1199, 1970.

*54. Blumberg B S. Viral hepatitis, Au antigen, and hope for a vaccine. *Gastroenterology* (*Med World News*): pp. 14–18, 1972.

*55. Chen D-S, Hsu H-M, Bennett C L, Pajeau T S, Blumberg B, Chang P-Y, Nishioka K, Huang A and Sung J-L. A program for eradication of hepatitis B from Taiwan by a 10 year, four-dose vaccination program. *Cancer Causes and Control* **7**: 305–311, 1996.

56. Blumberg B S. The current state of the prevention of HBV infection, the carrier state and hepatocellular carcinoma. *Res Virol* **148**: 91–94, 1997.

57. Blumberg B S, Larouze B, London W T, Werner B, Hesser J E, Millman I, Saimot G and Payet M. The relation of infection with the hepatitis B agent to primary hepatic carcinoma. *Am J Pathol* **81**: 669–682, 1975.

*58. Smith J B and Blumberg B S. Viral hepatitis, post-necrotic cirrhosis and hepatocellular carcinoma. *Lancet* **2**: 953, 1969.

59. London W T and Blumberg B S. A cellular model of the role of hepatitis B virus in the pathogenesis of primary hepatocellular carcinoma. *Hepatology* **2**: 10s–14s, 1982.

60. Payne R J H, Nowak M A and Blumberg B S. Analysis of a cellular model to account for the natural history of infection by the hepatitis B virus and its role in the development of primary hepatocellular carcinoma. *J Theor Biol* **159**: 215–240, 1992.

61. Payne R J H, Nowak M and Blumberg B S. The dynamics of hepatitis B virus. *Proc Natl Acad Sci USA* **92**: 6542–6546, 1996.

62. Maupas Ph, Larouze B, London W T, Werner B, Millman I, O'Connell A, Blumberg B S, Saimot G and Payet M. Antibody to hepatitis-B core antigen in patients with primary hepatic carcinoma. *Lancet* **2**: 9–11, 1975.

*63. Yarrish R L, Werner B G and Blumberg B S. Association of hepatitis B virus infection with hepatocellular carcinoma in American patients. *Int J Cancer* **26**: 711–715, 1980.

*64. Kim C-Y, Bae S-K, Hann H L, London W T and Blumberg B S. Prevalence of HBeAg and anti-HBe in chronic active hepatitis, cirrhosis and primary hepatocellular carcinoma in Korea. *Hepatology* **5**: 54–56, 1985.

65. Larouze B, London W T, Saimot G, Werner B G, Lustbader E D, Payet M and Blumberg B S. Host responses to hepatitis B infection in patients with primary hepatic carcinoma and their families. A case/contol study in Senegal, West Africa. *Lancet* **2**: 534–538, 1976.

66. Larouze B, Blumberg B S, London W T, Lustbader E D, Sankale M and Payet M. Forecasting the development of primary hepatocellular carcinoma by the use of risk factors: Studies in West Africa. *J Natl Cancer Inst* **58**: 1557–1561, 1977.

67. Marinier E, Barrois V, Larouze B, London W T, Cofer A, Diakhate L and Blumberg B S. Lack of perinatal transmission of hepatitis B virus infection in Senegal, West Africa. *J Pediatrics* **106**: 843–849, 1985.

68. Feret E, Larouze B, Diop B, Sow M, London W T and Blumberg B S. Epidemiology of hepatitis B virus infection in the rural community of Tip, Senegal. *Am J Epidemiol* **125**: 140–149, 1987.

69. Blumberg B S and London W T. Hepatitis B virus and the prevention of primary cancer of the liver. Guest Editorial. *J Natl Cancer Inst* **74**: 267–273, 1985.

*70. Blumberg B S. Feasibility of controlling or eradicating the hepatitis B virus. *Am J Med* **87**: 3A–2S–4S, 1989.

71. Blumberg B S. Proc. International Conference on Prospects for Eradication of Hepatitis B Virus. *Vaccine* Vol. 8, Introduction: pS5; Conclusion pS139, 1990.

*72. McLean A and Blumberg B S. Modelling the impact of mass vaccination against hepatitis B. I. Model formulation and parameter estimation. *Proc R Soc London B* **256**: 7–15, 1994.

73. Blumberg B S. Hepatitis B virus, iron and iron-binding proteins. In *Viruses, Immunity and Immunodeficiency*. Szentivanyi A and Friedman H (eds.). Plenum Publishing Corp. pp. 81–99, 1986.

*74. Sutnick A I, Blumberg B S and Lustbader E D. Elevated serum iron levels and persistent Australia antigen (HBsAg). *Ann Internal Med* **81**: 855–856, 1974.

75. Felton C, Lustbader E D, Merten C and Blumberg B S. Serum iron levels and response to hepatitis B virus. *Proc Natl Acad Sci USA* **76**: 2438–2441, 1979.

*76. Blumberg B S, Lustbader E D and Whitford P L. Changes in serum iron levels due to infection with hepatitis B virus. *Proc Natl Acad Sci USA* **78**: 3222–3224, 1981.

77. Lustbader E D, Hann H L and Blumberg B S. Serum ferritin as a predictor of host response to hepatitis B virus infection. *Science* **220**: 423–425, 1983.

*78. Stevens R G, Kuvibidila S, Kapps M, Friedlaender J and Blumberg B S. Iron-binding proteins, hepatitis B virus, and mortality in the Solomon Islands. *Am J Epidemiol* **118**: 550–561, 1983.

79. Stevens R G, Beasley R P, Blumberg B S. Iron-binding proteins and risk of cancer in Taiwan. *J Natl Cancer Inst* **76**: 605–610, 1986.

80. Stevens R G, Graubard B I, Micozzi M S, Nerushi K and Blumberg B S. Moderate elevation of body iron level and increased risk of cancer occurrence and death. *Int J Cancer* **56**: 364–369, 1994.

81. Blumberg B S, Melartin L, Guinto R A and Werner B. Family studies of a human serum isoantigen system (Australia antigen). *Am J Human Genet* **18**: 594–608, 1966.

82. Blumberg B S, Sutnick A I, London W T and Melartin L. Sex distribution of Australia antigen. *Arch Intern Med* **130**: 227–231, 1972.

*83. Hesser J E, Economidou J and Blumberg B S. Hepatitis B surface antigen (Australia antigen) in parents and sex ratio of offspring in a Greek population. *Human Biol* **47**: 415–425, 1975.

84. Drew J S, London W T, Lustbader E D, Hesser J E and Blumberg B S. Hepatitis B virus and sex ratio of offspring. *Science* **201**: 687–692, 1978.

*85. Drew J S, Blumberg B S and Robert-Lamblin J. Hepatitis B virus and sex ratio of offspring in East Greenland. *Human Biology* **58**: 115–120, 1986.

86. Drew J, London W T, Blumberg B S and Serjeantson S. Hepatitis B virus and sex ratio on Kar Kar Island. *Human Biol* **54**: 123–135, 1982.

*87. London W T, Stevens R G, Shofer F S, Drew J S, Brunhofer J E and Blumberg B S. Effects of hepatitis B virus on the mortality, fertility, and sex ratio of human populations. In *Viral Hepatitis*, Szmuness W, Alter H J and Maynard J E (eds.). Franklin Institute Press, Philadelphia, pp. 195–202, 1982.

88. Blumberg B S, Friedlaender J S, Woodside A, Sutnick A I and London W T. Hepatitis and Australia antigen: Autosomal recessive inheritance of susceptibility to infection in humans. *Proc Nat Acad Sci* **62**: 1108–1115, 1969.

89. Blumberg B S. The natural history of Australia antigen. In *Analytic and Experimental Epidemiology of Cancer. Proceedings of the 3rd International Symposium of the Princess Takamatsu Cancer Research Fund.* Nakahara W, Hirayama T, Nishioka K and Sugano H (eds.). University of Tokyo Press, pp. 109–151, 1973.

90. Langendorfer A, Davenport W, London W T, Blumberg B S and Mazzur S. Sex-related differences in transmission of hepatitis B infection in a Melanesian population. *Am J Phys Anthropol* **64**: 243–254, 1984.

*91. Kukowski K, London W T, Sutnick A I, Kahn M and Blumberg B S. Comparison of progeny of mothers with and without Australia antigen. *Human Biology* **44**: 489–500, 1972.

92. Blumberg B S. Sex-related aspects of hepatitis B infection and its consequences. In *Hepatitis B: A sexually transmitted disease in heterosexuals*, Piot P and Andre F E (eds.). Elsevier Science Publishers B.V. (Biomedical Division), pp. 3–7, 1990.

93. Sutnick A I, Cerda J J, Toskes P P, London W T and Blumberg B S. Australia antigen and viral hepatitis in drug abusers. *Arch Int Med* **127**: 939–941, 1971.

94. Blumberg B S, London W T, Sutnick A I, Camp F R Jr, Luzzio A J and Conte N F. Hepatitis carriers among soldiers who have returned from Vietnam. Australia antigen studies. *Transfusion* **14**: 63–66, 1974.

*95. Luzzio A J, Camp Col F R Jr, Blumberg B S, Conte Col N F and Coley Maj V R. Prevalence of Australia antigen in military populations. *Military Med* **140**: 41–43, 1975.

*96. Blumberg B S, Wills W, Millman I and London W T. Australia antigen in mosquitoes. Feeding experiments and field studies. *Res Commun Chem Path Pharm* **6**: 719–732, 1973.

*97. Metselaar D, Blumberg B S, Millman I, Parker A M and Bagshawe A F. Hepatitis B antigen in colony mosquitoes. *Lancet* **2**: 758–760, 1973.

*98. Wills W, Saimot G, Brochard C, Blumberg B S, London W T, Dechene R and Millman I. Hepatitis B surface antigen (Australia antigen) in mosquitoes collected in Senegal, West Africa. *Am J Trop Med Hyg* **25**: 186–190, 1976.

*99. Wills W, Larouze B, London W T, Millman I, Werner B G, Ogston W, Pourtaghva M, Diallo S and Blumberg B S. Hepatitis-B virus in bedbugs (Cimex hemipterus) from Senegal. *Lancet* **2**: 217–219, 1977.

100. Venkateswaran P S, Millman I and Blumberg B S. Effects of an extract from *Phyllanthus niruri* on hepatitis B and woodchuck hepatitis B viruses: *In vitro* and *in vivo* studies. *Proc Natl Acad Sci USA* **84**: 274–278, 1987.

101. Thyagarajan S P, Subramanian S, Thirunalasundari T, Venkateswaran P S and Blumberg B S. Effect of *Phyllanthus amarus* on chronic carriers of hepatitis B virus. *Lancet* **2**: 764–766, 1988.

102. Unander D W, Venkateswaran P S, Millman I, Bryan H H and Blumberg B S. Phyllanthus species: Sources of new antiviral compounds. In *Advances in New Crops*, Janick J and Simon J E (eds.). Timber Press Inc., Portland, OR pp. 518–521, 1990.

*103. Unander D W, Webster G L and Blumberg B S. Records of usage or assays in Phyllanthus (Euphorbiaceae) I. Subgenera Isocladus, Kirganelia, Cicca and Emblica. *J Ethnopharmacol* **30**: 233–264, 1990.

*104. Unander D W, Webster G L and Blumberg B S. Uses and bioassays in Phyllanthus (Euphorbiaceae): A compilation. II. The subgenus Phyllanthus. *J Ethnopharmacol* **34**: 97–133, 1991.

*105. Unander D W, Webster G L and Blumberg B S. Usage and bioassays in Phyllanthus (Euphorbiaceae): A compilation III. The subgenera Eriococcus, Conami, Gomphidium, Botryanthus, Xylophylla and Phyllanthodendron, and a complete list of the species cited in the three-part series. *J Ethnopharmacol* **36**: 103–112, 1992.

*106. Unander D W, Webster G L and Blumberg B S. Usage and bioassays in Phyllanthus (Euphorbiaceae). IV. Clustering of antiviral uses and other effects. *J Ethnopharmacology* **45**: 1–18, 1995.

*107. Unander D W and Blumberg B S. *In vitro* activity of Phyllanthus (Euphorbiaceae) species against the DNA polymerase of hepatitis viruses: Effects of growing environment and inter- and intra-specific differences. *Economic Botany* **45**: 225–242, 1991.

*108. Blumberg B S. A case history of plant-derived drug research — *Phyllanthus* and hepatitis B virus. *Proceedings of the International Symposium from the Utilization of Medicinal Plants*. (In press, 1997)

109. Block T M, Lu X, Platt F M, Foster G R, Gerlich W H, Blumberg B S and Dwek R A Secretion of human hepatitis B virus is inhibited by the imino sugar N-butyldeoxynojirimycin. *Proc Natl Acad Sci USA* **91**: 2235–2239, 1994.

*110. Lu X, Mehta A, Dadmarz M, Dwek R, Blumberg B S and Block T M. Aberrant trafficking of hepatitis B virus glycoproteins in cells in which N-glycan processing is inhibited. *Proc Natl Acad Sci USA* **94**: 2380–2385, 1997.

*111. Lechat M, Prehn L M, Blumberg B S and Moris R. Australia antigen in Zaire. Studies on leprosy. *Ann Soc Belg Med Trop* **53**: 173–178, 1973.

112. Lechat M F, Shrager D I, Declercq E, Bertrand F, Blattner W A and Blumberg B S. Decreased survival of HTLV-1 carriers among leprosy patients from Zaire, a historic prospective study. *JAID&HR* (In press, 1997.

113. Blumberg B S and Blumberg J L. Bernard Connor (1666–1698) and his contribution to the pathology of ankylosing spondylitis. *J Hist Med and Allied Sci* **13**: 349–366, 1958.

114. Blumberg B S. Creativity. A sermon given at Balliol College Chapel on Rogation Sunday. Oxford, England. May 5, 1991.

115. Blumberg B S. A Sermon. Balliol College Chapel, Oxford University, Oxford, England. June 12, 1994.

*116. Blumberg B S. A Letter from Oxford. Scientific interaction and the generation of new ideas. In Fox Chase Cancer Center Scientific Report 1988–1989. Fox Chase Cancer Center, Philadelphia, PA.

117. Blumberg B S. A Letter from Oxford. Science in Contemporary Movies. In Fox Chase Cancer Center Scientific Report 1992–1993. Fox Chase Cancer Center, Philadelphia, PA USA.

118. Blumberg B S. Reflections on a two-wheel way of life. London Telegraph, Saturday, October 8, 1994. Lonfon, October

119. Blumberg B S and Fox R C. The Daedalus effect: Changes in ethical questions relating to hepatitis B virus. *Ann Int Med* **102**: 390–394, 1985.

120. Blumberg B S. Scientific and artistic process. Awards Ceremony, Pennsylvania Academy of the Fine Arts, Philadelphia, PA USA, May 14, 1980.

121. Blumberg B S. An Address. Perceptions of Science. Union College Bicentennial Celebration, Schenectady, New York, USA, February 25, 1995.

122. Feitelson M A, Millman I, Halbherr T, Simmons H and Blumberg B S. A newly identified hepatits B type virus in tree squirrels. *Proc Natl Acad Sci USA* **83**: 2233–2237, 1986.

123. Blumberg B S. Complex interactions of hepatitis B virus with its host and environment. *J Royal College Physicians London* **29**: 31–40, 1995.

III FACSIMILES OF SELECTED PUBLICATIONS

This collection contains 67 publications
(The papers are numbered according to
the sequence given in Section II.)

Reprinted from MEDICINE
Vol. 35, No. 1, February, 1956
Printed in U.S.A.

THE NATURAL HISTORY OF RHEUMATOID SPONDYLITIS

BARUCH BLUMBERG, M.D.* AND CHARLES RAGAN, M.D.

INTRODUCTION

Rheumatoid spondylitis (ankylosing spondylitis, pelvo-spondylitis ossificans, spondylitis rhizomelique, Marie-Strumpell disease, morbus Bechterew, etc.) is an affliction primarily of young men, characterized by back pain and deformity resulting from involvement of the sacroiliac and small joints of the spine, paravertebral calcification and, in advanced disease, bony ankylosis of these regions. The hips, knees, shoulders, and occasionally other peripheral joints may also be involved. It is a chronic malady with its onset in the prime of life, whose stigmata remain until death and, indeed, after death, being preserved in the interred bones. It is important to consider such a disease (as one must consider all chronic diseases) over an extended period of time in order to evaluate clinical findings, complications, and results of treatment. Short-term studies, particularly at a point in time, when the patient has symptoms, would tend to emphasize aspects of acute exacerbations out of proportion to the entire temporal extent of the disease, during a major portion of which the patient may be asymptomatic. The purpose of the present study is to give a picture of the natural history of spondylitis over a period of time thereby providing useful prognostic material.

History: The late Dr. C. W. Buckley (18, 19, 30) among others (47, 56, 64, 80, 85, 103) has called attention to the primacy of the contributions to the pathology of rheumatoid spondylitis of the Irish physician and traveler, Bernard Connor. Connor (or O'Connor) was born in County Kerry in 1666. He received his medical education at Paris, Montpelier and Rheims, and in 1691 published in French at Paris a volume describing a skeleton with spondylitis (31). In 1695, his observations were abstracted and commented upon in the Transactions of the Royal Society (32). He described and diagramed (Fig. I) a skeleton "found in some churchyard or charnel house" which conforms to our present understanding of the gross pathology of rheumatoid spondylitis (10).

In the succeeding years, several descriptions of skeletal and clinical material, which were probably rheumatoid spondylitis, were published. These are discussed by Dunham and Kautz (36), Polley (80), and Jones (64). The differentiation from what is now called degenerative joint disease of the spine, however, was not clearly made. In 1893, the Russian neurologist, Bechterew published a description of what he believed was a new neurological disease, characterized by kyphosis of the upper spine, spinal stiffness and nerve root symptoms (5). On the basis of

From the Department of Medicine, Columbia University College of Physicians and Surgeons, and the Edward Daniels Faulkner Arthritis Clinic of the Presbyterian Hospital, New York.

Supported in part by grant from the Masonic Foundation for Medical Research and Human Welfare.

*Fellow, Arthritis and Rheumatism Foundation.

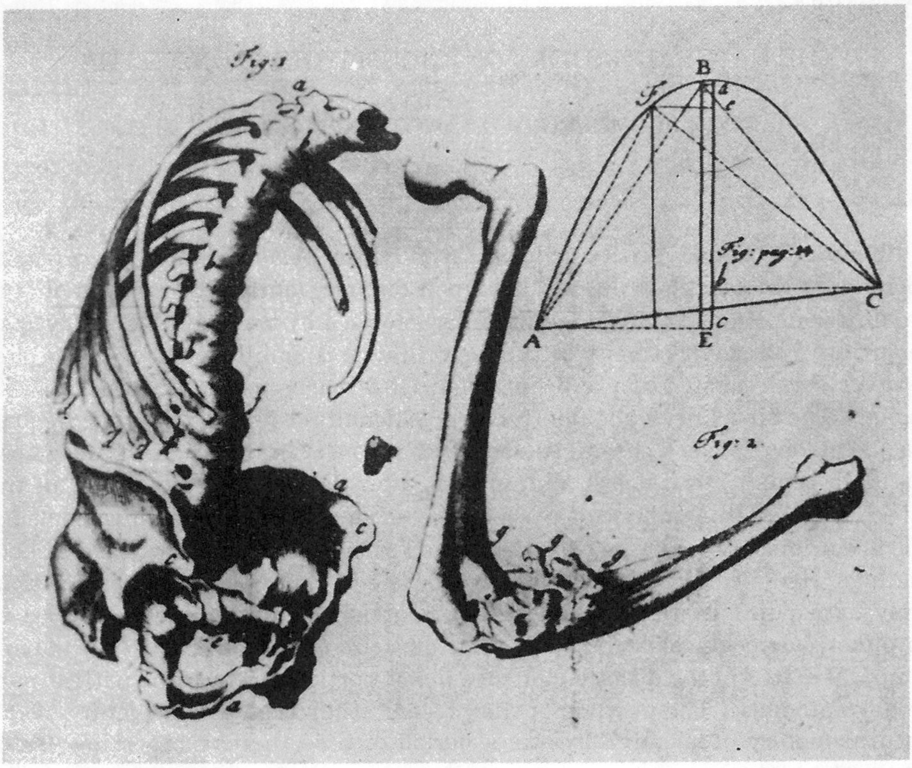

FIG. I. The illustration from Connor's description of a pathological specimen of rheumatoid spondylitis. Reproduction of plate from Connor's article describing the skeleton of a person with rheumatoid spondylitis. Spine, pelvis and ribs are shown in Fig. 1. Fig. 2 is an ankylosed knee from another skeleton which, from the description of the text, might be due to rheumatoid arthritis or infectious arthritis.

this original paper and subsequent description of other cases, it appears that he was confusing kyphosis associated with advanced degenerative joint disease of the spine and rheumatoid spondylitis (36). In 1897, Strümpell (99) gave an accurate clinical description of the disease as it is now understood. In 1897, Marie, with Astie, had published a description of a case similar to Bechterew's (36). In the following year, however, he described three cases of apparently true rheumatoid spondylitis and for the first time correlated the clinical with pathological findings, using a skeleton from Dupuytren's collection (73). The major difficulty in the definition of the disease as a separate entity was in differentiating it from degenerative joint disease of the spine with senile kyphosis ("spondylitis deformans"). Finally in 1904, Simmonds (91) and Fraenkel (42), correlating clinical studies with autopsy material, established the classical clinico-pathological description of the disease emphasizing ankylosis of the diarthrodial joints of the spine with sparing of the vertebral bodies and intervertebral discs. Interestingly enough, the controversy has by no means subsided. Van Swaay (103), Engfeldt and his associates (37), and others have raised the question again of primary involvement of the intervertebral disc and vertebral body (see below under Pathology). The resolution of the problem must await further understanding of the etiology and pathogenesis of this intriguing disease.

Frequency: It is difficult to determine the "true" frequency of rheumatoid spondylitis since all studies have been made on highly selected groups. In this clinic, of 1,200 patients, the ratio of rheumatoid spondylitis to peripheral rheumatoid arthritis was 1:7 (34). The ratio of rheumatoid spondylitis to rheumatoid arthritis at the Mayo Clinic during a 3-year period was approximately 1:6 (80). The ratio in civilian practice is said to be 1:15 (12). Schmorl, in his classic necropsy study of 10,000 spines, found only eight examples (quoted in 36, 60). Holst and Iversen (63) studied the incidence in a Norwegian county by computing the admission rate to the local hospitals. The incidence was 1.73 per thousand admissions for males and 0.38 per thousand for females. Bachman, in a large series of x-rays of the spine taken for various purposes, found it appeared in 1.5 per cent (quoted in 36). West estimated an incidence of 1 case in 2,000 of the general population "at risk" (106). In susceptible groups, the incidence is far higher. In a hospital admitting mainly young male servicemen, Boland (14) found that 18 per cent of patients with chronic back complaints had this disease. He also found that 18.1 per cent of 6,000 consecutive admissions for all types of rheumatic diseases had rheumatoid spondylitis. Almost as many patients were admitted with spondylitis as were admitted with peripheral rheumatoid arthritis (11).

Since many patients with mild disease do not come to a hospital, the frequency is probably understated by these studies. Interviews, and physical and x-ray examinations of a large group may be necessary before the true frequency is determined.

Pathology: There were no autopsies at this hospital on the patients in this series. There was one at another hospital where the patient had died as a result of a head injury and the spine was not examined. Indeed, a search of the autopsy files of the Department of Pathology of the College of Physicians and Surgeons from 1906 to the present uncovered no cases in which rheumatoid spondylitis (or variants of this name) had been indexed either as a pathological or clinical diagnosis. During this period, over 10,000 autopsies had been performed. One patient in the series with definite clinical and x-ray evidence of rheumatoid spondylitis had a surgical biopsy of one sacroiliac joint apparently involved by x-ray evidence. This was reported as normal. There were no other examples of surgical biopsies of the spine or pelvis of patients with rheumatoid spondylitis in over 160,000 biopsies included in the files of the Department of Surgical Pathology of the Presbyterian Hospital.

There has been surprisingly little study of pathological material in this disease entity. Cruickshank (33) points out that prior to the publication of his paper in 1951, only three others had appeared in which the histopathology of the diarthrodial joints was described (27, 43, 52) and that these were primarily patients with late and inactive disease. Van Swaay (103) states that prior to his report in 1950 on three autopsied cases, the pathological-anatomical data had been based on museum preparations and twenty-one descriptions of postmortem findings, in 8 cases with histologic studies.

The present view of the pathology (27) holds that the diarthrodial joints of the spine and pelvis, the costovertebral joints, the costotransverse joints, and in some instances the symphysis pubis (51), the sternoclavicular manubriosternal and other peripheral joints are involved in an inflammatory process similar to that seen in peripheral rheumatoid arthritis. Destructive changes have been demonstrated in the pelvic bones roentgenographically (51). Cruickshank (33) has found that these inflammatory changes are by no means specific and may be seen in such widely different diseases as systemic lupus erythematosus, "non-rheumatic arthritis", bursitis and nonspecific tenosynovitis. Since the joint can react only in a limited manner to exciting agents, the similarity of histologic appearance cannot of necessity imply a common etiology.

A different view of the histopathology is proposed by Van Swaay (103). On the basis of study of three autopsied cases (who had died as a result of leukemia 12, 21 and 25 years after the onset of spondylitis), he believes that the process in the sacroiliac joints is a proliferation of new cartilage replacing the old without evidence of inflammation. The joint space is destroyed by this new growth and subsequently ossified.

The result in advanced disease is well described by Connor in the title of a paper ". . . An extraordinary humane skeleton, whose vertebrae, the ribs, and several bones down to the os sacrum, were all firmly united into one solid bone, without joynting or cartilage" (32). The drawing from Connor's article (Fig. I) shows this striking pathological specimen.

The importance of the ankylosis of the sacroiliac joint in this disease has been emphasized. It should be recalled, however, that fusion of the sacroiliac joints, apparently due to aging processes, may be found frequently in elderly males and is not uncommon after the sixth decade. Sashin (88) stated that bony ankylosis of the sacroiliac was present in 82 per cent of 23 skeletons of males who had died after the age of 60. Macdonald and Hunt (72) found that 6 of 21 patients over 70 years of age had fused joints. One of these patients had rheumatoid spondylitis and one had rheumatoid arthritis, all the others apparently represented degenerative joint disease. The latter authors believe, however, that there is a complete bony ankylosis in rheumatoid spondylitis and that it can be distinguished from degenerative joint disease. Through the courtesy of Dr. H. Shapiro of the Department of Anthropology of the New York Museum of Natural History, we were permitted to examine the collection of contemporary skeletal remains. Four examples of fusion of the sacroiliac joints without other evidence of rheumatoid spondylitis were seen in an examination of seventy specimens of elderly males.

Collins (27) and Van Swaay (103) have observed that the bamboo appearance of the spine is due to calcification in the annulus fibrosis of the intervertebral disc and the latter author believes that calcification of ligaments is secondary to this. Connor (32), in his early account, also noted this: "The cartilagineous edge of the vertebrae themselves were turned to perfect bone. In short, they were as entire as a skeleton cut out of the same piece of wood by a carver would be. Being willing to see if these vertebrae were united throughout their whole diameter, or at the edges only, I sawed two of them asunder at the commissure, and found this uniting did not enter above two lines deep and that afterwards their middles were separated as they usually are, and touched each other only at the edges which was raised up a little above the middle part". Although it has been generally believed that the vertebral body itself is unaffected by the disease, more recently Engfeldt, Romanus, and Yden (37, 87) on the basis of x-ray evidence and some histological studies have claimed that changes do occur at this site. This is evidenced by "squaring" of the vertebral bodies as seen by x-ray early in the disease. They believe that the intervertebral disc is also primarily involved.

Although muscle wasting and weight loss may be seen occasionally with far-advanced disease, and muscle spasm is of great importance in early disease, there has been little, if any, pathological evidence of generalized systemic involvement which, contrariwise, has been reported frequently in peripheral rheumatoid arthritis. There have been reports of heart valve involvement and aortitis (4, 6, 7, 45, 95) but no definite relationship with rheumatoid spondylitis has been demonstrated. There is good reason to believe that most of these cases of valvular damage were due to a pre-existing rheumatic fever (6, 7). (vide infra). Peripheral nodules of the juxta-articular type common in peripheral rheumatoid arthritis and thought by many to be the lesion located in various viscera in the disseminated or malignant form of the disease (8) are rarely if ever seen in spondylitis.

It is apparent, therefore, that the pathology of this disease is incompletely understood, due primarily to the scarcity of carefully studied pathological material representing early disease.

Etiology: The cause of rheumatoid spondylitis is unknown. Various agents have been implicated, i.e., tuberculosis, brucellosis and the gonococcus. Although these organisms may produce lesions in the sacroiliac joints which in some ways resemble those of rheumatoid spondylitis, they have not been generally accepted as the etiologic agent of the disease. Parathyroid dyscrasias, genitourinary pathology, allergy, metabolic and hormonal factors have also been suggested but not proven.

There are, however, several facts which may serve as a point of departure in approaching the problem of etiology. Hersh, et al. in a definitive paper, decided on the basis of a study of pooled data on fifty families that rheumatoid spondylitis was due to "... an autosomal dominant factor which has about 70 per cent penetrance in males and about 10 per cent or slightly greater penetrance in females." (61). In the present study, 11 per cent of 93 patients gave a history of rheumatoid spondylitis in the immediate family, and an additional 10 per cent gave a history of rheumatoid arthritis. One female had 2 brothers with documented disease and 1 male patient knew of the disease in his father, grandfather and great grandfather. Hence, we may assume that, whatever the agent or agents may be, they act in genetically susceptible individuals.

There are several morbid conditions which mimic some of the changes of rheumatoid spondylitis. It is not suggested that their causative agent is the one implicated in rheumatoid spondylitis but that study of these diseases may lead to a better understanding of pathogenesis.

Crippling fluorosis was first described by Flemming Møller and Gudjonsson (38) in workers chronically exposed to cryolite dust. (Cryolite is a fluoride of sodium and aluminum which is essential to the manufacture of aluminum by the electrolytic process.) Roholm has provided an excellent study of the disease (84). The clinical picture of crippling fluorosis is difficult to distinguish from rheumatoid spondylitis with stiffness of the back and neck, forward flexion, and eventually complete stiffness and ankylosis of the spine. Osteosclerosis characterizes the pathological picture, with complete effacement of the osseous structures of the pelvis, extensive calcification of the ligaments and fibrocartilaginous muscle attachments, obliteration of the sacroiliac joints and the small joints of the vertebrae. This is in contrast to the normal or at times decreased radiolucency of the bones seen in spondylitis. Hence, although the clinical pictures are often similar, radiological differentiation is usually readily made. Pathologically, fluorosis is usually confined to the bone and adjacent structures and joints are involved only secondarily and often spared.

The disease also occurs in areas where there are very high concentrations of fluorine in the ground water. A description of the problem in Madras State, India, is given in a series of interesting papers (78, 90). It has also been described in South Africa (65).

Steinberg et al. (97) have called attention to the relationship of these two diseases and have stressed the importance of the problem in connection with the impending water fluoridation program for the prophylaxis of dental caries. They report that the fluorine level in the bones of 1 patient with rheumatoid spondylitis was within normal limits.

There is a high incidence of fusion of the sacroiliacs and spine changes by x-ray in paraplegics with spinal cord injuries. Abramson and Kamberg (2) reported that 85 per cent of 35 patients studied had involvement of that joint to some degree and 8 had complete fusion. Fifty per cent of them also had fusion of the spinal diarthrodial joints. Paravertebral calcification was less common. None of the lesions were located above the lumbar region regardless of the level of the spinal cord lesion. Abel (1) reported that 70 of 98 patients showed severe sacroiliac changes and 44 showed complete ankylosis. All but 3 of the patients studied were in persons who had suffered flexion injuries of the spine as a result of mine cave-ins. He could find no correlation between the presence and extent of the bony lesions and the site of injury, level of cord transection, duration of bed rest or age. It was even found in 5 persons with minimal paraplegia of the sacral cauda equina variety. The latter author considered that since the lesions were so similar to those of rheumatoid spondylitis, the same etiological factors were operating on the paraplegics. The strikingly high incidence of rheumatoid spondylitis-like lesions associated with this disease suggests that common factors in their etiology may provide a method for experimental study.

The presence of sacroiliac fusion, ligamentous calcification and occasionally spinal diarthrodial joint changes in old persons which is probably the result of degenerative joint disease has been discussed above.

Sacroiliac involvement, detected on x-ray, appears frequently in juvenile rheumatoid arthritis (50). Cervical spine fusion is also commonly seen. Symptoms and findings of pain, immobility and stiffness are associated with the findings in the cervical region but not in the lumbar and sacral region.

There has been considerable discussion in previous studies concerning the relationship of rheumatoid spondylitis to peripheral rheumatoid arthritis. To say that two diseases are "the same" implies that they have a common etiology. Since the etiology of neither of these diseases is known, it is impossible to decide definitely on their relationship. Another measure of "sameness" is similarity of clinicopathological findings. Although peripheral joints are involved in rheumatoid spondylitis, the distribution of the joint involvement is only occasionally the same as that of peripheral rheumatoid arthritis. Transient swelling, not uncommon in many inflammatory diseases, is found in about one-half of the cases of rheumatoid spondylitis (see below and 82) whereas objective changes in the joints occur in only approximately 30 per cent of the cases. As we shall see below (Fig. IX), when it does occur, the small joints of the wrist and hand are involved only rarely in contrast to the very frequent involvement of these areas in peripheral rheumatoid arthritis. On the other hand, the sacroiliacs and back are only occasionally significantly involved in peripheral rheumatoid arthritis. Juxta-articular subcutaneous nodules, a frequent finding in peripheral rheumatoid arthritis, are only rarely found in rheumatoid spondylitis (29). None was seen in the present study.

Although the pathology of the joint lesions is said to be the same in both diseases (27), it should be recalled that this information is based on a small number of cases, as noted above. Cruickshank (33) has pointed out that the changes are by no means specific for rheumatoid arthritis and do not imply a common etiology. Van Swaay's (103) findings of a completely different pathology require further confirmation especially in early disease. The other clinical differences between the diseases, i.e., sex ratio, age of onset, differences in motivation and functional ability, etc., will be discussed below and have been pointed out in previous studies.

The agent which affects the genetically susceptible individual to cause rheumatoid spondylitis is still to be discovered.

Selection of Cases: Every series of patients collected to study a disease, barring a complete survey of the world's population, is unrepresentative and reflects some selection on the part of the collector. No two series, therefore, are exactly comparable in all respects. Information on clinical findings, complications, results of therapy, etc., apply to the studied group and perhaps to those similar to it but no exact extrapolations to other studies should be attempted. Our series was collected in the arthritis clinic of a university hospital in a large city and applies to patients seen here and at similar clinics elsewhere. A series collected from other sources, i.e., chronic disease hospital in-patients, orthopedic hospitals, private patients, routine low-back x-ray surveys, etc., would, no doubt, give different results. For this study, we considered all patients seen at the Edward Daniels Faulkner Arthritis Clinic from 1928 to 1954 on whom the diagnosis of rheumatoid spondylitis (or variants of this name) was made. This is a referral clinic and the patients come from several sources: (1) the clinics and wards of the Presbyterian Hospital, (2) area doctors and smaller hospitals, and (3) The New York Orthopedic Hospital.

In general, a higher percentage of refractory cases might be expected in such a selection, and the patients of the present study probably represent a more severely affected group than a "true" sample of this disease would show. In addition, there are some persons who have the disease but never have sufficiently severe symptoms to require their seeking medical help. These are, of course, excluded from all hospital series and would be noted only on surveys of the entire population of an area regardless of the presence or absence of symptoms. Cobb, in Pittsburgh, is at present conducting such studies on rheumatoid arthritis (24, 25, 26).

Although the symptoms on the initial visits may have been very severe, when these patients were requested to return to the hospital for the study, they were very often asymptomatic. Therefore, our experience on this follow-up would be different than that gained by general clinic experience where the patients come in only for the treatment of complaints.

The diagnosis was made on 311 persons during this period. Some were excluded from the series when re-examination made it apparent that the original diagnosis was incorrect. In the final selection of cases, only those in whom sacroiliac joint involvement was present were included. A letter requesting that the patient come to the clinic was sent to the last address given on the chart. One hundred and two came in for examination, x-rays and blood tests; 40 answered questions either by letter or telephone; 14 were reported as dead; 26 did not respond to multiple letters and telephone contacts; 129 letters were returned by the Post Office Department as undeliverable. We have no information about this latter group other than that they have moved and left no forwarding address. The follow-up of this group as respects clinical findings, mortality, functional status, etc., is, of course, not known. On the basis of available information on these patients, we have no reason to suspect that these cases were more severe than those adequately followed.

The majority of these patients had received treatment of some kind, including aspirin, radiotherapy, exercise, casts and jackets, and in a few instances gold, steroid hormones, butazolidin or surgery. No attempt has been made to divide the series into separate treatment groups.

Composition of Group: Eighty per cent were males and 20 per cent females, a ratio of 4:1. This is a larger proportion of females than has been recorded in most other studies (3, 6, 29, 36, 49, 58, 60, 82, 92, 108) but not all (39, 79). The age-at-onset distribution curve is shown in Fig. II. It is contrasted with similar data for patients with peripheral rheumatoid arthritis studied at this clinic. The onset is dated from the time at which the patient could first recall symptoms of low back pain or discomfort, sciatic pain or other symptoms. The patients could usually date this within 6 months, an error of small magnitude in a disease of 20 or more years' duration. The maximum incidence of onset is in the third decade and the average age of onset is 27 years. There is a striking difference between

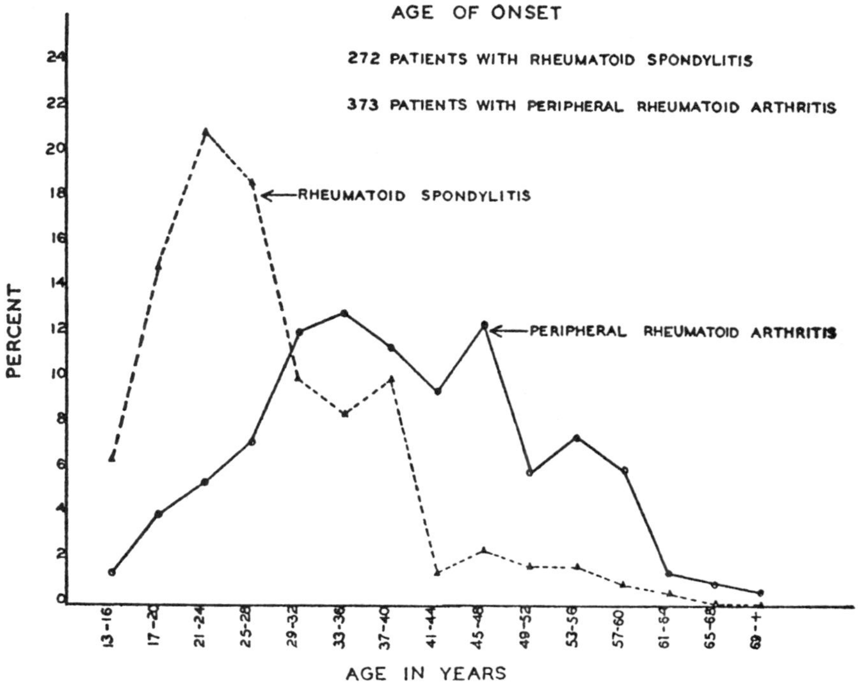

FIG. II. Age-of-onset distribution curve. Age of onset in a group of patients with rheumatoid spondylitis and a group with peripheral rheumatoid arthritis.

the two diseases, another evidence of their clinical dissimilarities. Twelve dated the onset of symptoms after the age of 50 years. Although there is no definite reason to believe that the disease cannot start at this age or have a protracted asymptomatic phase, it could easily be confused with osteoarthritis of the spine with x-ray evidence of osteophyte bridging mirroring paravertebral calcification and even sacroiliac joint fusion.

Eight patients noted symptoms prior to the age of 15 years. Differentiation from Still's disease may be necessary in this age group, since sacroiliac fusion and cervical spine involvement are not uncommon in juvenile rheumatoid arthritis (50). Luchesi and Luchesi have reported on 17 children under the age of 10 on whom the diagnosis was made (70). Of the patients who responded to our letters, 40 per cent had a follow-up of better than 10 years. Another 30 per cent were followed for more than 5 years. Sixty per cent of 150 had the disease for more than 10 years and 30 per cent of them for more than 20 years.

Results: In evaluating the status of the disease through the years, the following parameters have been studied: (1) functional ability, (2) pain, (3) physical deformity, (4) x-ray changes in the spine and pelvis, and (5) laboratory tests. Each of these will be discussed in detail and in addition certain clinical findings will be mentioned.

The overall functional ability of the patient is perhaps most important since it represents the net result of all the stigmata of the disease. The patient is most interested to know if his illness will enable him to live an independent life and continue to care for himself and his family.

We have been content to evaluate the gross functional condition, i.e., the patient's ability to work at regular employment. No attempt has been made to quantitate specific functional abilities (i.e., dexterity, bending, range of motion, etc.) nor to measure downgrading of jobs, increase in the amount of rest time necessary, decrease of time spent at social activities and recreation, motivation and other complex factors, whose quantitation is as yet poorly worked out. Such a study is at present being conducted by Levin and his associates (68). Functional class was determined by a modification of the A.R.A. criteria (98): (1) working regularly, supporting self and family, (2) working irregularly, only partially self-supporting, (3) care for self but no other activities, and (4) cannot care for self; bed and chair existence. Fig. III shows the functional class of 121 patients, at a point in time, namely the time that they were interviewed. These patients had the disease for varying lengths of time. Seventy-six per cent of the group were in Class 1, leading essentially normal occupational lives. Nineteen per cent were in Class 2 and 3 and only 5 were reduced to bed or chair existence. The patients were then divided into seven 5-year duration-of-disease groups and the percentage in each functional class determined in each group (Fig. IV). The

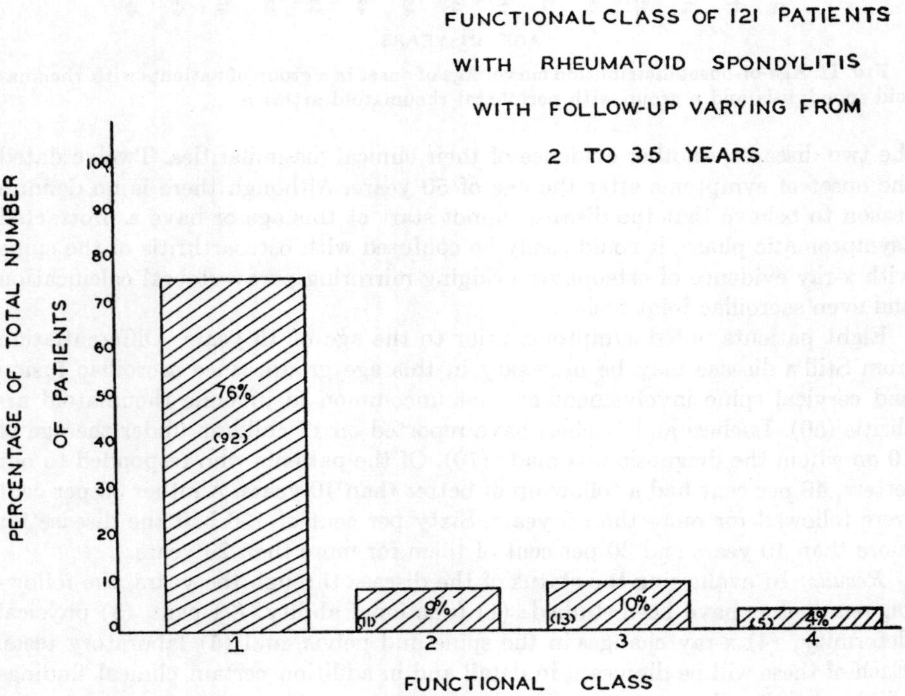

Fig. III. Functional class of patients at the time they were interviewed for the study.

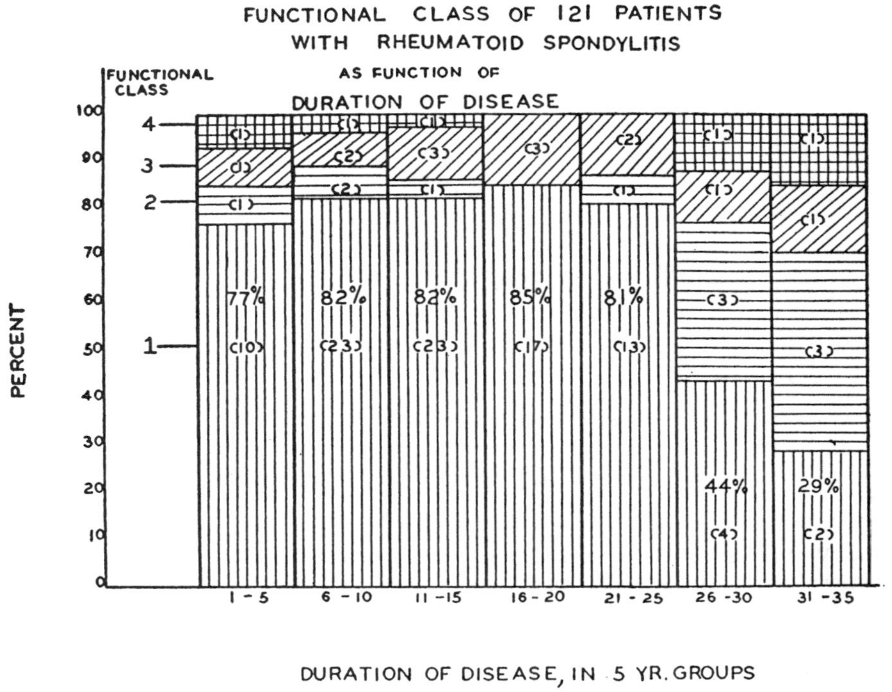

FIG. IV. Functional class during different periods of the disease.

patients in each group were then arranged according to their class, an average determined and then plotted as a function of the duration of the disease. This is shown in Figure V where this information is arranged with other data as a function of duration of disease. The ability to work is remarkably well maintained until the time the patients, on the average, have exceeded the age of 55. In Fig. VI, a comparison is made between the patient's functional class at the time they first sought medical attention and at the time they were recalled to the hospital for this study. It will be seen that only 7 per cent had regressed. The remainder had either remained in the same class, in all cases Class 1, or had improved. Hence, it would appear from these data that the patients do remarkably well functionally and, indeed, tend to improve after the acute stage of the disease has passed.

We were able to gain some information concerning the causes of poor functional ability. Physical deformity, so striking a feature of the disease, is not the sine qua non of poor functional ability. It was estimated roughly on the following quantitative scale: (1) no deformity, (2) loss of lumbar lordosis, slight straightening of spine, (3) stiffness and restriction of motion of a portion of the back with slight forward flexion, or (4) stiffness of entire back with forward flexion deformity. In Fig. VII, functional class, as described above, is shown as a function of physical deformity. If these two parameters were related in a 1:1 manner

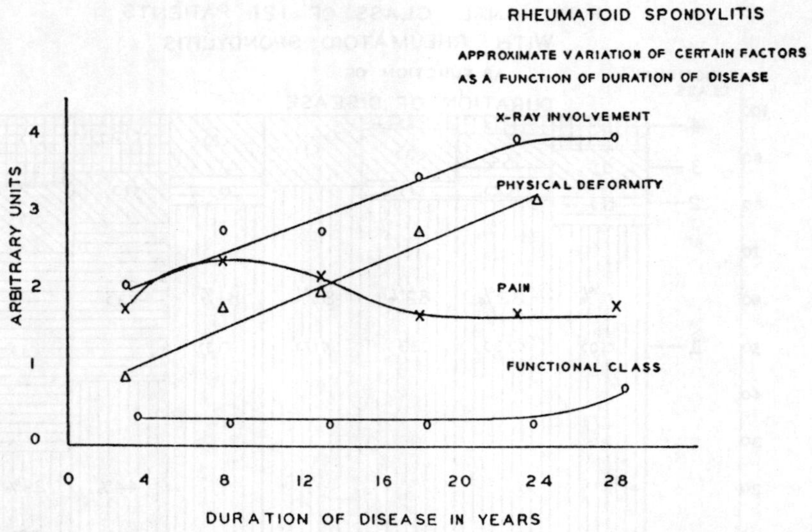

Fig. V. Averages of various parameters as a function of duration of disease. Averages of functional class, pain, physical deformity and x-ray involvement of the sacroiliac joint in a group of patients with rheumatoid spondylitis, arranged as a function of duration of disease. (For clarity, the curve for functional class has been lowered one unit.)

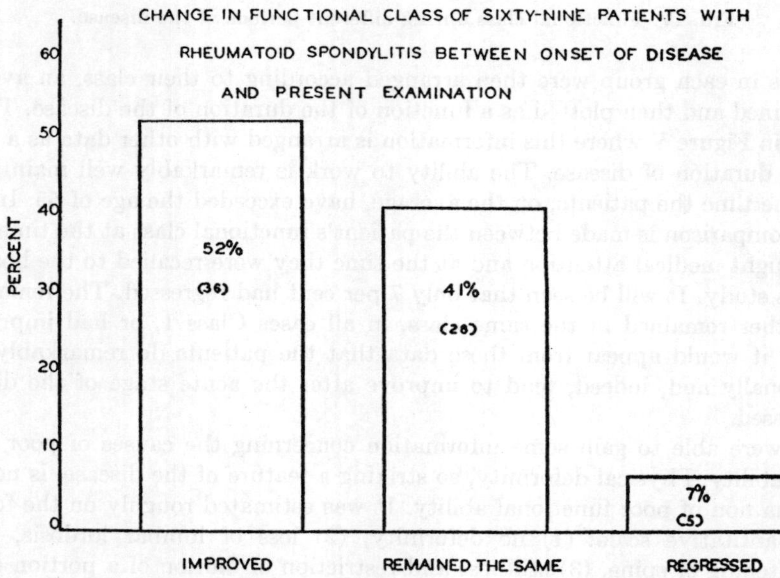

Fig. VI. Change in functional class in 69 patients with rheumatoid spondylitis between onset of disease and present examination.

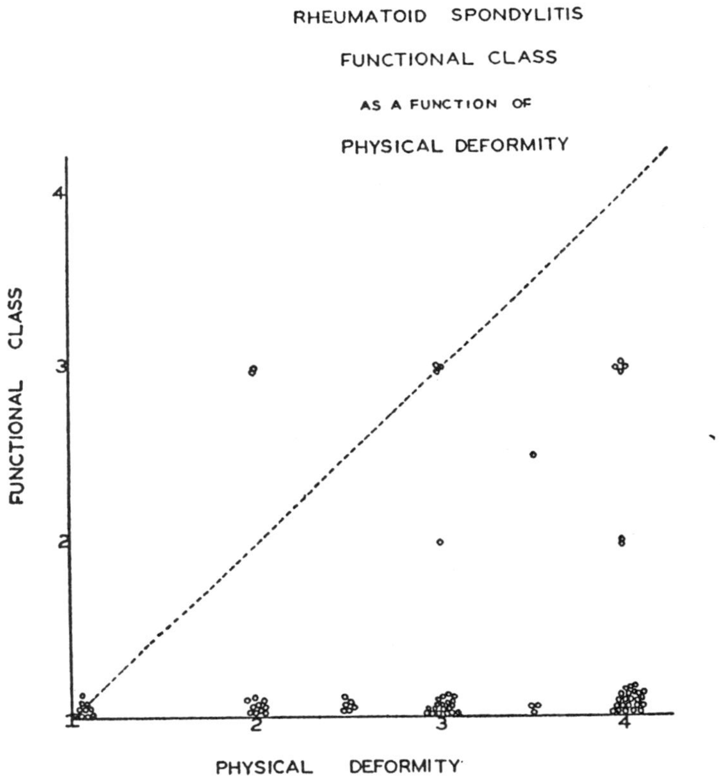

FIG. VII. Functional class vs. physical deformity. The dotted line indicates the 1:1 coincidence curve.

and an increase in deformity necessarily led to an increase in poor functional ability, then the points would cluster about the dotted 45° line. They do not, indeed, most of them fall below this line suggesting that the advanced skeletal changes do not necessarily decrease the patient's ability to be gainfully employed.

Twenty-four of the 121 patients fell into functional Class 2, 3 or 4. The sex distribution for these 24 was the same as that of the general group. They had the disease for an average of 20 years, longer than the average for the general group and they tended to be somewhat older. Twelve of them, or 50 per cent, had severe involvement of the hip(s) with or without other peripheral joint involvement. In 3, heart disease was disabling and interfered with normal activity. One patient had disabling multiple sclerosis and 1 was blind as a result of repeated attacks of iritis. The activity of 4 was limited because of emotional problems manifested by anxiety states of varying degree and they appeared to be poorly motivated. In 3 patients, the cause of disability could not be determined. There is a clinical impression that these persons are more effective individuals than those with peripheral arthritis.

In summary, severe hip and other peripheral joint involvement, heart disease with failure, and psychiatric problems were the major causes of functional embarrassment. Spinal involvement alone is frequently a remarkably benign condition insofar as the patient's functional ability is concerned.

Pain: Pain is a frequent accompaniment of this disease. However, if one considers this symptom over the 30 to 40 year span of the life history of rheumatoid spondylitis, it is found to be present during only a portion of this time. It is a difficult symptom to evaluate. Many factors enter into the final sensation which the patient experiences and then expresses to the physician; the painful stimulus; the nature and location of the endorgan stimulated; the pathways to the central nervous system; the subjective interpretation of the individual; the method by which he conveys his impression and finally, the interpretation which the observer places upon it. Hence, in an attempt to evaluate pain, we are approaching a pathogenetically complex matter. However, it is the end result in which the patient and clinician is interested and it is this which we have attempted to measure by rough quantitation on a scale of 1 to 4 plus, where 1 indicated little or no pain and 4 considerable, constant pain. Fig. VIII shows the situation at a point in time, i.e., the time the patients were interviewed. Only 30 per cent of these patients had severe pain and the remainder were relatively comfortable. These estimations were then arranged in 5-year duration-of-disease groups and averaged. This represents an approximation of the pain in a hypothetical "average" patient during the course of his disease. The resultant curve is shown in Fig. V, where it is compared with other factors discussed elsewhere. The pain appears to be greatest in the early acute phase of the disease and to decrease,

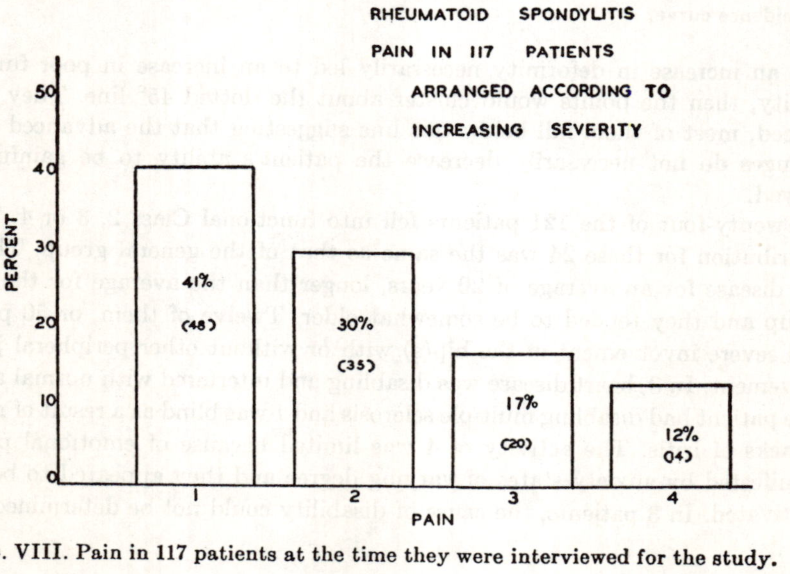

FIG. VIII. Pain in 117 patients at the time they were interviewed for the study.

TABLE 1

Approximate physical deformity in 91 patients with rheumatoid spondylitis, arranged in order of increasing severity

Duration of Disease (yrs.)	Degree of Severity							
	1+		2+		3+		4+	
	No.	%	No.	%	No.	%	No.	%
0–5	4	40	3	30	3	30	0	0
6–10	3	14	5	24	8	38	5	24
11–15	1	5	1	5	15	68	5	22
16–20	0	0	2	13	5	31	9	56
21 and over	0	0	1	4	4	18	17	78

and in many, disappear, in the later years. This has been noted previously (66). It must not be construed that all patients will act in this manner. Back pain in advanced disease is not uncommon and it may be present throughout life. In addition, some patients have an episodic type of disease with exacerbations and remissions. On the average, however, the curve appears to indicate the trend of symptoms.

Physical Deformity: The degree of physical deformity was evaluated by one observer on the basis of inspection and physical examination of the spine, using the grading scale described above. The number and per cent of 91 patients in each of these groups is shown in Table 1. The deformity as a function of duration of disease was computed in the same manner as for pain and is shown in Fig. V. From these data, it may be inferred that the physical deformity progresses relentlessly in a large number of patients although some do not go on to marked deformity. The small number of patients in the last duration-of-disease group makes this figure less accurate.

Peripheral Arthritis: Arthritis of the peripheral joints is frequently present. It is important to differentiate between a history of peripheral joint involvement and actual objective evidence of deformity. Sixty-seven per cent of the patients examined had peripheral joint pain and stiffness in the past which often was accompanied by objective but transient swelling, tenderness and heat. Only 30 per cent of them had actual objective evidence of peripheral joint deformity at the time seen. This phenomenon has been noted in other studies (29, 82). Indeed, some of the patients had a history of joint involvement preceding the disease or early in its course which was difficult to differentiate clinically from rheumatic fever (36, 57). When objective involvement does occur, the joint involved as a rule, differs from those affected in classical peripheral joint rheumatoid arthritis. A comparison from data collected in this clinic between peripheral rheumatoid arthritis and rheumatoid spondylitis for involvement of peripheral joints (arranged in order of their proximity to the trunk) is shown in Fig. IX. In the former, the incidence of involvement decreases as we proceed from more peripheral to more centrally located "root" joints. The reverse is true for patients with rheumatoid spondylitis. Indeed, only about 7 per cent of patients with

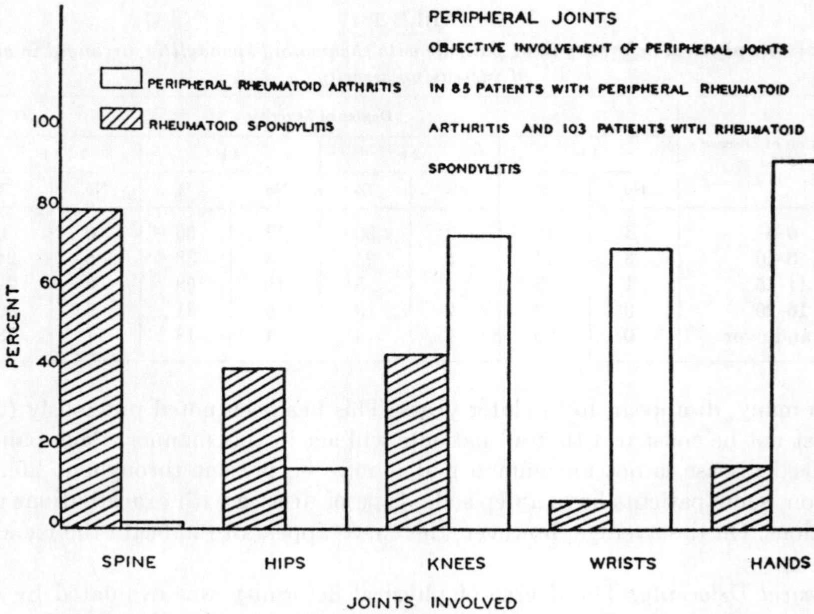

Fig. IX. Peripheral arthritis. A comparison of the involvement of peripheral joints, arranged in order of their proximity to the spine, in patients with rheumatoid arthritis and in patients with rheumatoid spondylitis with peripheral involvement.

spondylitis had the typical wrist changes so common in peripheral rheumatoid arthritis. Simpson and Stevenson (92) reported no hand involvement in 200 cases. Bùckley (20) and Dunham and Kautz (36) confirmed these findings. Swaim found small joint involvement in only 5 of 106 patients seen (100). Hence, the nature of the peripheral joint involvement changes during the course of the disease and in its final stages appears to have a different distribution than in peripheral rheumatoid arthritis.

Chest Expansion: In the present series, chest expansion was nearly always markedly limited. Sixty per cent of those examined had a chest expansion of less than ½ cm. and all less than the normal of about 7 cm. (23). The figures in general agree with those of other studies (56, 92). The nature of the selection of our series excluded very early cases in which good chest expansion might have been retained. However, along with Hart, et al. (56), we may still conclude that the decrease in chest expansion appears to be a relatively early finding. In spite of this limitation, there were very few complaints referable to the chest other than occasional pain. These people apparently had adjusted adequately to their thoracic embarrassement. D'Silva et al. (35) in a study of maximal breathing capacity of 14 spondylitics have noted that none of them had cardiorespiratory symptoms. They pointed out that the patients could use the air volumes available to them as well as normals even though this was considerably reduced by the chest restriction. On the other hand, Hamilton (54) and Hart et al. (56) have called attention to the frequency of infectious pulmonary complications. The former states that prior to the introduction of sulfonamides, deaths due to respiratory disease were not infrequent.

Involvement of Other Organ Systems: Twenty-five per cent of patients examined had one or more attacks of iritis during the course of the disease. This is a somewhat higher figure than has been reported in other studies (9, 11, 21, 29, 58, 79, 82, 92) although Traut (101) stated that the incidence was 50 per cent and Dunham and Kautz (36) 25 per cent. These differences are probably not of great significance and are a consequence of the methods of case selection.

Iritis may be the presenting symptom while the back is still asymptomatic (9). Very often the only reason for hospital admission and in some cases treatment with hormones was the iritis. The uncomplicated spondylitic required inpatient treatment in only a small percentage of cases.

Abdominal hernias, predominantly indirect inguinal, were found in 23 per cent of the spondylitics specifically questioned for this finding. This is significantly higher than the incidence in the general population (28) and to the best of our knowledge has not been commented on before. The reason for this high incidence is unknown. Several explanations may be offered for further investigation: (1) the chronic deformity of the back may lead to increased intra-abdominal pressure which, over many years, can produce herniation, (2) the predisposition to abdominal hernias is considered to be an hereditary trait, probably a dominant factor (107). About 25 per cent of patients with herniae gave a history of the disease in their parents or grandparents (105). The hereditary aspects of rheumatoid spondylitis are discussed elsewhere in this paper. It is conceivable that both of these characteristics are transmitted together in some of the cases. If the physical deformity leads to increase in the intra-abdominal pressure for a sufficiently long period, then the hernia will develop in the genetically susceptible individual. This problem cannot be resolved by the present study and might be a matter for subsequent investigations.

Heart Disease: Clinical reviews of this nature are of only limited value in elucidating the role of heart disease in rheumatoid spondylitis particularly in the present study where routine electrocardiograms were not performed and no autopsy material is available. Clinical evidence of heart disease was noted in 25 of 128 patients and is recorded in Table 2. Contrary to the findings in other studies (6, 7, 45), heart involvement was not always benign. Four of these patients had died and, in 3, the doctors in attendance attributed the demise to the heart disease. In one, subacute bacterial endocarditis with subsequent heart failure had developed in a patient with pre-existing mitral and aortic stenosis and insufficiency. A second developed aortic insufficiency three years prior to his death and succumbed to intractable heart failure. The third died as a consequence of arteriosclerotic heart disease. A rough, loud systolic murmur had developed several years before his death.

The nature of valvular heart disease associated with rheumatoid spondylitis is as yet uncertain and the present study has done nothing to make it clearer. It is not known whether the valvular lesions found in this disease are due to rheumatic fever or are specific for spondylitis (6, 7, 45). The problem is further

TABLE 2

Heart disease in patients with rheumatoid spondylitis

Clinical observations of 128 patients

1. Some evidence of cardiac disease		25 (20%)
2. Valvular lesions detected clinically		6
a. AI, MS, MI	History of A.R.F.	
b. MS, MI	History of A.R.F.	
c. MS, MI, AI, AF	History of A.R.F.	
d. MS, MI, AI	History of A.R.F.	
e. AI	? History of A.R.F.	
f. MS, MI	No history of A.R.F.	
3. Apical systolic murmur		9
4. ASCVD and HCVD		5
5. Various findings:		5
a. enlarged heart		
b. frequent extrasystoles		
c. myocarditis, ? pericarditis		
d. history of valvular disease		
e. first degree heart block		

complicated by the fact that patients with rheumatoid spondylitis often give a history of acute peripheral joint swelling and pain which cannot clinically be differentiated from rheumatic fever. Forty per cent of the 25 patients who had heart symptoms and 70 per cent of those who had valvular lesions gave a history of acute rheumatic fever. Of the group without heart involvement, only 8 per cent had acute rheumatic fever. The elucidation of this problem awaits further pathological studies and, of course, a better understanding of etiology.

*Mortality**: There has been little study of the mortality in this disease or of the causes of death. It would be of interest to know the number of deaths in this group compared to an equivalent unaffected group in the general population. The present clinical view, supported indirectly by the paucity of autopsy material in hospitals, is that people do not die as a result of rheumatoid spondylitis (86). We can approach this problem statistically from the data of the present study. The patient-years of experience in each age group of 274 patients was determined and is shown in Column 2 of Table 3. (37 of the original 311 patients were not included because of incorrect diagnosis, loss of charts, etc.) One hundred thirty-nine of these were seen at the time of the study and our experience with them extends from the time of their first visit to the present. The remainder have not been seen or heard from for varying lengths of time and could not be located by the Post Office Department. Their experience was calculated from the time of the first visit to the time they were last heard from. This, of course, introduces an unreliability as we do not know their fate after the last visit. We can only assume, perhaps wrongly so, that the incidence of deaths in this group is not greatly different than in those we were able to follow.

* We are indebted to Dr. John W. Fertig, Professor of Biostatistics of the School of Public Health, Columbia University, for his help with the following study.

TABLE 3

Deaths in 287 persons with rheumatoid spondylitis compared to "normals"

1	2	3	4	5
Age Group	Patient years of experience	Deaths in R.S. group	Age specific death rate per 1000 New York City, 1940	Expected deaths in "normals" Col. 4 × Col.2 / 1,000
15–19	25	0	1.3	.03
20–24	85	1	1.7	.15
25–29	210	0	2.2	.45
30–34	330	1	3.0	.98
35–39	380	2	4.2	1.61
40–44	330	2	6.5	2.17
45–49	250	1	10.5	2.61
50–54	185	0	16.1	2.97
55–59	100	2	24.2	2.42
60–64	58	2	35.8	2.08
65–69	23	1	52.2	1.20
70–74	8	2	75.6	0.56
Total		14		17.5

The patient years experience is determined on the basis of when patients were first seen at the clinic (as suggested by Merrell and Shulman (74)) and not from the time of onset of the disease. Using the age specific death rates per 1,000 population in the City of New York for the year 1940 (104), a table for deaths per 1,000 population in each age group with a mixture of 4:1 males to females was computed. This is shown in Column 4 of Table 3. Deaths to be expected in these age groups in the general population for the computed number of patient-years is determined and is shown in Column 5. The sum is 17.5. For the equivalent number of patient-years, there were 14 deaths in the group of the spondylitics studied (Column 3). Hence, we may say (bearing in mind the uncertainty in our data due to the incomplete follow-up discussed above) that there does not appear to be a greater number of deaths in the spondylitics studied than one would expect in a similar group of the general population of New York.

We were able to determine the cause of death in only 7 of the 14. These are shown in Table 4. It is interesting that none of these deaths could be attributed to "arthritis" per se and most of them were apparently asymptomatic as far as their back pain was concerned, at the time of their demise.

Laboratory: With the possible exception of the erythrocyte sedimentation rate, no laboratory tests of practical value are currently in use for the diagnosis of rheumatoid spondylitis.

As noted in other studies, (14, 36, 40, 82), the sedimentation rate is not always elevated. It was normal in 20 per cent of the present group at the time of examination. The average value was 29. From the data, we were able to determine an "average" value as a function of the duration of disease. The ESRs were arranged in 5-year duration-of-disease groups, with the exception of duration of

TABLE 4

Cause of death in patients with rheumatoid spondylitis

Pt.	Sex	Age at death	Cause of death
G. B.	M	71	ASHD with failure. Arthritis asymptomatic at time of death.
J. G.	M	38	Head injury with skull fracture and subdural hematoma. Knee and sternoclavicular joint normal at autopsy.
S. R.	M	73	Acute appendicitis. Also had pernicious anemia and HCVD.
A. R.	M	56	Subacute bacterial endocarditis with heart failure. AS, AI, MS, MI. Gastric ulcer
S. R.	M	22	Diagnosis at a psychiatric hospital "exhaustion manic-depressive psychosis. Manic type". Ulcerative colitis.
S. W.	F	67	Carcinoma of lung with metastasis.
R. W.	M	57	Valvular heart disease (A.I.) with failure.
N. B.	M	30?	Not known. Had AI and MS.
M. H.	M	60	Not known.
C. H.	M	36	Not known.
K. M.	M	60	Not known.
M. O.	M	42	Not known.
A. R.	M	41	Not known.
R. W.	M	46?	Not known.

disease less than 5 years which were arranged in 1-year groups. These were then averaged and the resultant values plotted at the median for the duration-of-disease group (Fig. X). The ESR tended to be elevated during the early acute phase and returned towards normal values during the later quiescent stages of the disease, as has been noted in previous studies (58, 67, 92). The reciprocal relationship between skeletal deformity and x-ray changes, on one hand, and elevated sedimentation rate on the other, illustrates the diagnostic role of the latter. In the early stages of the disease when the objective findings are equivocal, the elevated ESR increases the index of diagnostic suspicion. When, in the later stages of the disease, the physical deformity and x-rays are more characteristic and diagnosis is evident, the ESR may be normal, but it is no longer necessary for diagnosis. As noted before, this curve applies to the hypothetical "average" value and atypical cases are not infrequent.

The streptococcus agglutination test and the sensitized sheep cell test were done on all the patients who came to the hospital including those with peripheral joint involvement. In only one case was the former positive and in no case was the latter positive. This is in striking contrast to peripheral rheumatoid arthritis where the streptococcus agglutination and sensitized sheep cell agglutination are positive in more than 50 per cent of cases. Only 14 of the 84 patients on whom data was available showed any depression of their hemoglobin. In all patients (except one female who had a reading of 9.0 Gm.), the fall was only 1 or at most 2 Gm. below the normal of 12.5 for females and 13.5 for males.

The cerebrospinal fluid in rheumatoid spondylitis has been studied by Boland, et al., Ludwig, et al., and Polley (13, 71, 80). In the present study, cerebrospinal fluid was examined in 20 patients who had been admitted to the Neurological

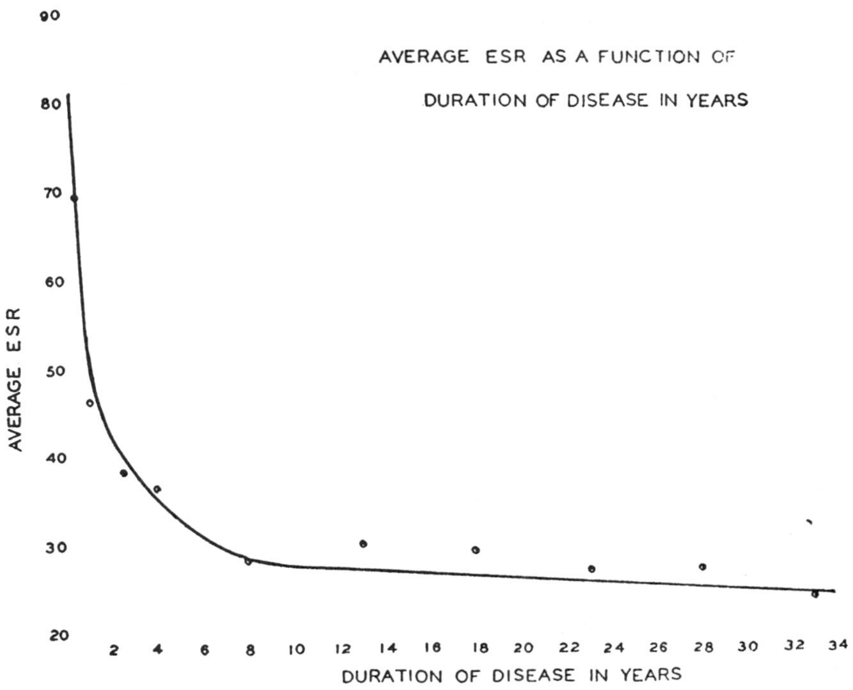

FIG. X. Sedimentation rate averaged in 5-year groups of duration of disease. (One-year groups for years 1-5.)

Institute, Columbia-Presbyterian Medical Center, for the diagnosis of what were thought to be neurological symptoms. The only abnormality found was a slight to moderate rise (in the range of 46–60 mgm. %) in protein in 5 cases. There was no abnormality in the colloidal curve as reported by Ludwig, et al. (71).

There was no evidence of abnormalities in the white blood count and differential examination of the peripheral blood of the patients who came in. Seventy per cent of this group had received radiotherapy, averaging about 1500 r sometime during the course of their disease. Deaths due to leukemia in persons with rheumatoid spondylitis who had been treated with radiation have been reported. Van Swaay (103) reported 6 cases. Graham in discussing Robinson and Lampe's paper (48) reported 2 cases; and Hernaman-Johnson and Law (59) saw a case of aplastic anemia after x-ray and thorium therapy. The Medical Research Council of Great Britain has published an inquiry (69) in an effort to determine if there is an increase in incidence of leukemia in persons who have had radiotherapy for spondylitis. Apparently none of those on whom we have a definite follow-up has developed leukemia and none of the known deaths was due to it. Of course, we do not know the fate of those who did not respond to our letters.

Radiology: Since the early radiological studies (75, 102), many discussions of x-ray findings have appeared (41, 15, 37, 80). In general, the changes first occur in the sacroiliac joints then spread to the diarthrodial joints of the spine with ankylosis of these joints and paravertebral calcification in more or less ascending progression from the lumbar regions cephaled. Oppenheimer (77) and Borak (16) believe that the primary changes are in the diarthrodial joints of the spine and that the sacroiliac changes are secondary to them. Scrupulous care in positioning the patients must be used in roentgenography of the small joints of the spine. Although it has been assumed that the intervertebral discs and vertebra are not primarily involved in this disease, recent work (referred to in the discussion of pathology above) suggests that this is not necessarily true. It is generally agreed, however, that involvement of the sacroiliac joints will eventually be present in nearly all the well-authenticated cases (14).

In the present study, AP and lateral views of the cervical spine, AP and oblique views of the thoracic and lumbar spine, a 35° film of the pelvis and an PA view of the chest were taken on each of the patients. Peripheral joint x-rays were made when indicated. All the films were reviewed by Dr. Josephine Wells of the Department of Radiology. X-ray evidence of generalized demineralization of bone is not a necessary concomitant of this disease. Only 22 (30 per cent) of 89 patients with adequate x-rays showed this finding and in 17, it was of a mild degree. Oppenheimer (76, 77) has pointed out that rarification of the vertebral bodies was absent when early changes only were seen in the spinal diarthrodial joints (at the onset of disease), present with increased diarthrodial joint involvement and again decreased when these joints were fused, i.e., late in the disease. In the present study of patients who had the disease for less than 5 years, 25 per cent showed demineralization; from 6 to 10 years, 25 per cent and from 11 to 15 years only 8 per cent. In the older age groups, as may be expected, the incidence of demineralization increased somewhat again. The incidence of ligamentous calcification on the other hand rose with passage of time. For the same year intervals specified above, the per cent of patients showing ligamentous calcification in the thoracic region of moderate or greater degree is 12, 24 and 35 per cent, respectively.

The degree of involvement of the sacroiliac joints was established by a crude quantitative method, as follows: 1 plus indicated slight irregularity in the joint; 2 plus parajoint bony condensation and irregularity; 3 partial joint obliteration; 4 complete joint obliteration. The assigned values were then arranged in order of duration of disease and averaged in 5-year groups. They were then plotted on a scale with the other parameters to show the extent of changes with increasing duration of disease (Fig. V). For the hypothetical "average" patient, it appears that sacroiliac involvement by x-ray proceeds relentlessly independent of the changes in pain, ESR and functional ability.

In Table 5, the incidence of involvement of various portions of the spine at different stages of disease is shown. The incidences vary greatly during different stages of the disease and may account for differences in incidence noted in other studies. A selection containing primarily early cases would show a much lower incidence of involvement of the spine diarthrodial joints and ligamentous calcification than would a compilation weighted with late cases. The relentless nature of the x-ray involvement is again obvious although some cases will be spared and arrested at a stage of minor involvement.

TABLE 5

X-ray evidence of involvement of different portions of the spine and pelvis in patients with rheumatoid spondylitis

Region of spine or pelvis	Duration of disease (yrs.)	Number of patients	% with no involvement	% with moderate involvement	% with advanced involvement
Sacroiliac joint	1–5	35	3	60	37
	6–15	65	2	28	70
	16 and over	35	0	0	100
Lumbar diarthrodial fusion	1–5	29	80	10	10
	6–15	63	56	12	32
	16 and over	35	23	3	74
Lumbar paravertebral calcification	1–5	28	60	32	8
	6–15	64	44	30	26
	16 and over	27	37	26	37
Thoracic paravertebral calcification	1–5	26	65	35	0
	6–15	60	55	30	15
	16 and over	33	12	58	30
Cervical diarthrodial fusion	1–5	18	89	11	0
	6–15	54	63	7	30
	16 and over	33	18	6	76
Cervical paravertebral calcification	1–5	21	95	5	0
	6–15	54	50	28	22
	16 and over	24	21	29	50

Treatment: There are four major objectives of treatment in rheumatoid spondylitis: (1) relief of pain, (2) prevention of back deformity, and hip and knee ankylosis, (3) maintenance of functional ability adequate to permit the patient to care for himself, earn a living and to partake in normal social and recreational activities, and (4) provide information to the patient concerning the nature of the affliction and its effect on his mode of living. Retrospective analysis such as the present one does not lend itself to a long-term study since the treated and untreated groups are not randomly selected and therefore not comparable. We can, however, make some comments on the basis of our experience and of information in the literature.

(1) Effective means are available to control acute and chronic pain. Large doses of aspirin, preferably enteric coated, are effective in some of the milder cases. X-ray therapy has been a standard mode of treatment. Of 57 patients in the present study who had received radiotherapy to the back, one-quarter reported complete immediate improvement, one-half improvement ranging from moderate to considerable and the remaining one-quarter no improvement. Some of those who improved subsequently relapsed and required further therapy, to which they usually responded. Our findings are in general agreement with those of other series (44, 46, 83, 89, 92, 94, 96).

Butazolidin has also proven to be effective in the treatment of pain. Holbrook (62) states that 89 per cent of patients responded favorably to the drug initially and that improvement was well maintained in a high percentage of these. This confirms our experience with the drug. He believes it to be the drug of choice when simpler methods have been ineffective. Cortisone, hydrocortisone, corticotropin and their analogues are usually not necessary since less toxic medications are available, but they may be of value in the patient with refractory symptoms.

Medications to relieve muscle spasm associated with the back pain have been developed (curare, mephanesin, etc.) but have generally been unsatisfactory primarily because of the brief duration of action. Longer-acting preparations are at present being tried.

(2) There are at present no therapeutic modalities known which can be shown to prevent back deformity in the end-stage.

Surgery has only a limited usefulness. The osteotomy procedure first introduced by Smith-Peterson (93) should be used only in the rare case where the mechanical deformity prevents or seriously limits functional activity. Occasionally it may be of use for cosmetic reasons, particularly in young females. This is a hazardous operation (93) and should be attempted only after the careful consideration of the indications.

Vitallium mold arthroplasties of the hip are not often successful (53) and other prosthetic procedures involving the hip have been used too briefly to permit an adequate evaluation. The two patients who had hip surgery in our group did not appear to derive benefit from it. It may be of value in an occasional carefully selected case.

Long bedrest with casting is rarely advisable (55, 100). It is more important to keep the patient active and at his work. Supporting jackets are also not as popular as they were several years ago (22, 55) and simple procedures, i.e., use of bed board and small pillow, physical exercises for back stretching, chest expansion and extension (see below) appear to be more efficacious.

(3) Functional ability is remarkably well maintained (vide supra) even in the untreated case. Procedures advocated to minimize the development of physical deformity which interfere with maintenance of function should be used cautiously if at all. The objective of all the modalities of therapy should be directed towards maintaining the maximum possible function.

(4) It is extremely important to apprise the patient of the nature of his disease and its prognosis. There are several optimistic features which should be emphasized. People do not die from rheumatoid spondylitis. The oft-repeated homily, concerning the long life to be expected in a person who has a chronic disease and cares for it, is probably applicable in this disease. Functional ability is well maintained. Since the disease affects young people during the height of their sexual activity, it is important to reassure the patient that he (or she) can have a more or less normal sexual life. Pain is not a constant accompaniment of the disease; it can be controlled by several forms of therapy, and it may disappear completely in the late stages. It is equally important to tell the patient frankly that further deformity may occur and to attempt to have him accept his disease

and make the best of it. This is often the first step in rehabilitating the patient and can be of greater importance than any form of medical, surgical or physical treatment.

To summarize briefly these suggestions for treatment:

(a) Acute exacerbations of pain can be treated with measures of increasing potency: 1. large doses of aspirin, 2. x-ray, 3. butazolidin, and 4. cortisone, hydrocortisone and their analogs.

(b) A daily series of exercises can be performed by the patient at home. Appendix A contains suggestions for physical therapy compiled with the help of Dr. Edward Gordon of the Institute for Crippled and Disabled, New York City. Surgery of back and hips is only rarely of help.

(c) Inform the patient of the nature of his disease and wherever possible point out the optimistic features of it.

DISCUSSION

From the vantage point of a long-term follow-up, certain features of the disease become apparent. The progress of five of the major parameters are shown in Fig. V and X. These grossly represent the history of the disease in the hypothetical "average" patient but individuals may, of course, vary considerably from this "average".

While the physical deformity and the confirmatory x-ray changes appear to proceed relentlessly with the passage of time, the manifestations of acute disease, i.e., the pain and ESR, are at their highest during the early phases and then decrease. The period of acute symptoms and signs is short, relative to the entire length of the disease. "Systemic" illness is not as common in this disease as it is in peripheral rheumatoid arthritis and the relationship of valvular heart disease to rheumatoid spondylitis remains obscure. Aside from the pain and stiffness, the patients do not feel particularly debilitated as evidenced by their excellent occupational history. They are rarely admitted to hospital and when so usually for the concomitant iritis. Although exacerbations may occur late in the disease, particularly in patients who develop objective evidence of peripheral joint involvement, the period of acute illness is usually restricted to the early years. Acute attacks of iritis, on the other hand, may occur at any time but bear little relationship to the arthritic complaints.

The insulting agent, whatever it may be, does not appear to have a sustained effect whereas the ankylosis of the joints, presumably a reaction to the original insult, may take years to achieve completion. The patient is not placed in constant jeopardy of repeated devastating attacks which may lead to further acute injury to the joints (and viscera) as is commonly seen in peripheral rheumatoid arthritis. The clinical picture rather suggests that the etiologic agent acts subacutely for a length of time (which may be one, two or more years, but a period short relative to the entire length of the disease) and then stops or slows down its activity. The sclerosing process may start soon after the initiation of the original excitant and continue when the latter has become quiescent, eventually leading to ankylosis and calcification. For further understanding of this disease, it is important to discover the nature of this agent, its reasons for selecting specific joints in genetically susceptible individuals and the manner in which it causes aberrant ossification, calcification and ankylosis.

SUMMARY

1. The diagnosis of rheumatoid spondylitis or variants of this name was made on 311 patients at this clinic from 1928 to 1954. One hundred and two came in for examination, 40 were contacted by phone or letter and information from the chart was used for the remainder.

2. Seventy-six per cent of the group contacted were in functional Class 1. Only 5 were leading a bed and chair existence. The functional status in the average patient was well maintained during most of his life. Pain was greatest during the early phase of the disease but in most patients decreased in the latter years. X-ray changes and physical deformity appeared to progress relentlessly in most of the patients.

3. Treatment should be directed towards maintaining functional ability. Body restricting procedures should be avoided.

4. In the majority of patients, the acute phase is present during the early years for a period short with respect to the entire life span of the disease. Sclerosis and ankylosis may take years to proceed to completion.

5. Etiology, pathology and certain clinical features are discussed.

APPENDIX A

The role of physical therapy in preventing deformity, or insuring ankylosis of the spine in good position is difficult to evaluate. Because of the nature of this chronic disease, it would be extremely difficult to select two groups of patients with equal severity of involvement, subject one group to intensive physical therapy for several years and then compare the results in them to the untreated group. However, it seems logical to assume that these exercises could be of value and it has been the impression of clinicians of long experience in this field that they are of help. In addition, aside from any direct value they may have, by causing the patient to perform the exercises at home and at work, they serve to interest him personally in managing his disability. This aspect, however, should not be over-emphasized since these patients tend to be overly body-centered and this additional fixation may in some cases increase this morbid attention.

In general, physical treatment should be directed toward the following objectives:

1. Maintaining good posture.
2. Prevention of flexion deformity of the back by means of extension exercises.
3. Maintaining range of motion of the costovertebral joints. When and if ankylosis of the ribs occurs, which appears to happen in most instances in spite of exercises, then effective diaphragmatic breathing should be taught. In many cases the patients appear to learn this without specific instruction.

Once the decision has been made to use physical therapy, it should be applied with enthusiasm. The patient should have several sessions of instruction, preferably at a Department of Physical Medicine, and then encouraged to perform the exercises at home. *Pain control must be achieved before the patient will do the exercises.* They should not be conducive to pain nor cause excessive fatigue. The following instruction sheet is given to the patient as an aid in performing the procedures.

1. Maintain good posture. Look at yourself in the mirror from time to time and check on your posture. Do not exaggerate the backward pull of the shoulders but attempt to maintain a good "military" posture.

2. Use a firm mattress with no pillow or, if necessary, a small one under the head and neck. A hair mattress, firmly supported, is the best. If this is not procurable, then a felt or cotton stuffed mattress, or a thin foam rubber mattress with a stiff bedboard that will not sag may be substituted. The bedboard should be supported on the frame of the bed and *not* by the springs. A spring mattress is unsatisfactory irrespective of how it is supported.

3. There are several simple exercises which can help in the maintenance of good posture. It is often of value to take a warm bath prior to doing the exercises. This helps to relax the muscles, relieve pain, and make the process more enjoyable. If you are unable to take a bath, the application of towels soaked in hot water, rung out and then applied along the spine and back is a good substitute.

(a) Lie on back with knees drawn up. Keep the arms at your side. Arch the back placing the weight on the forearms, shoulders, and soles of the feet.

(b) Lie on stomach, placing a pillow under the hips. Arch the back raising the upper portion of the body off the table. Then allow the upper portion of the body to remain on the table and raise the lower extremities by arching the lower portion of the body.

Do both these exercises for short periods two to three times daily.

4. Practice deep breathing three times daily for 5 minute intervals. Lie on the back. Take a deep breath while raising the arms slowly over the head simultaneously. Then let the breath out slowly with a hissing noise, bringing the arms back to your sides.

5. If the chest cavity has become fixed and there is some difficulty in breathing, it may be necessary to practice diaphragmatic breathing. Lying on the back, hold the hands on the abdomen. On inspiration, the abdomen should swell out. On expiration, the abdomen should flatten. This should then be repeated in sitting and then standing position.

With practice, this method of breathing will become automatic.

Remember: GOOD POSTURE AT ALL TIMES.

REFERENCES

1. Abel, M. S.: Sacroiliac joint changes in traumatic paraplegics. Radiology **55:** 235, 1950.
2. Abramson, D. J. and Kamberg, S.: Spondylitis, pathological ossification and calcification associated with spinal cord injury. J. Bone and Joint Surg. **31A:** 275, 1949.
3. Baker, L. D., Coonrad, R. W., Reeves, R. J. and Hoyt, W. A.: Marie-Strümpell Arthritis. Follow-up study of roentgenographic, physical and orthopedic therapy. J. Bone and Joint. Surg. **32A:** 848, 1950.

4. BAUER, W., CLARK, W. S. AND KULKA, J. P.: Aortitis and aortic endocarditis, an unrecognized manifestation of rheumatoid arthritis. Ann. Rheum. Dis. **10:** 470, 1951.
5. BECHTEREW, W.: Steifigkeit, der wirbelsaule und ihre verkrummung als besondere erkrankungsform. Neurologisches Centralblatt **12:** 426, 1893.
6. BERNSTEIN, L.: Cardiac complications in spondylarthritis ankylopoietica. Rheumatism **7:** 18, 1951.
7. BERNSTEIN, L. AND BROCH, O. J.: Cardiac complication in spondylarthritis ankylopoietica. Acta med. Scand. **135:** 185, 1949.
8. BEVANS, M., NADELL, J., DEMARTINI, F. AND RAGAN, C.: The systemic lesions of malignant rheumatoid arthritis. Am. J. Med. **16:** 197, 1954.
9. BIRKBECK, M. Q., BUCKLER, W. ST. J., MASON, R. M. AND TEGNER, W. S.: Iritis as the presenting symptom in ankylosing spondylitis. Lancet **2:** 802, 1951.
10. BLUMBERG, B. S. AND BLUMBERG, J. L.: Bernard Connor and his contribution to the pathology of rheumatoid arthritis. In Preparation.
11. BOLAND, E. W.: Rheumatoid Spondylitis: its general features and management. California Med. **65:** 1, 1946.
12. BOLAND, E. W.: Rheumatoid Spondylitis: clinical features and current methods of treatment. Bull. Rheum. Dis. **2:** 11, 1952.
13. BOLAND, E. W., HEADLEY, N. E. AND HENCH, P. S.: The cerebrospinal fluid in rheumatoid spondylitis. Ann. Rheum. Dis. **7:** 195, 1948.
14. BOLAND, E. W. AND PRESENT, A. J.: Rheumatoid Spondylitis. J.A.M.A. **129:** 843, 1945.
15. BOLAND, E. W. AND SHEBESTA, E. M.: Rheumatoid Spondylitis: correlation of clinical and roentgenographic features. Radiology **47:** 551, 1946.
16. BORAK, J.: Significance of the sacroiliac findings in Marie Strümpell spondylitis. Radiology **47:** 128, 1946.
17. BUCHMANN, E.: Spondylitis anchylopoetica. Nord med. **5:** 12, 1940.
18. BUCKLEY, C. W.: Spondylitis deformans. Brit. M. J. **1:** 1108, 1931.
19. BUCKLEY, C. W.: Spondylitis deformans: its differential diagnosis and treatment by physical methods. Proc. Roy. Soc. Med. **26:** 105, 1932–33.
20. BUCKLEY, C. W.: The differential diagnosis of spinal arthritis in young subjects. Brit. M. J. **1:** 4, 1943.
21. BUCKLEY, C. W.: Ankylosing Spondylitis. Its aetiology and pathology. Ann. Rheum. Dis. **5:** 49, 1945.
22. CAPENER, N.: Ankylosing Spondylitis. Ann. Rheum. Dis. **7:** 19, 1948.
23. Causes of rejection and incidence of defects. Medical Statistics Bulletin #2. National Headquarters, Selective Service System, Washington, D. C. August 1, 1943.
24. COBB, S., WARREN, J., THOMPSON, D., AND CIOCCO, A.: The epidemiology of rheumatoid arthritis with particular reference to the importance of morning stiffness. Pennsylvania Med. J. **57:** 37, 1954.
25. COBB, S., MERCHANT, W. R., AND WARREN, J.: An epidemiologic look at the problem of classification in the field of arthritis. Proc. Ann. Meet., Amer. Rheum. Assoc., Ann. Rheum. Dis. **13:** 361, 1954.
26. COBB, S., MERCHANT, W. R. AND WARREN, J. E.: An epidemiologic look at the problem of classification in the field of arthritis. J. Chronic Dis. **2:** 50, 1955.
27. COLLINS, D. H. The pathology of articular and spinal diseases. E. Arnold & Co., London, page 316, 1949.
28. COLLINS, S. D., Trantham, K. S. AND LEHMANN, J. L.: Sickness experience in selected areas of the United States. Public Health Monograph #25, U. S. Govt. Printing Office, Washington, D. C., 1955.
29. COMROE, B. I.: Arthritis and Allied Diseases. Philadelphia, Lea & Febiger, 1949, page 548.
30. COPEMAN, W. S. C.: Textbook of the Rheumatic Diseases. E & S Livingstone, Ltd., London. 2nd Edition, 1955.
31. CONNOR (or O'Connor), B.: in Letters écrit à monsieur le chevalier . . . , Paris, 1691. In Dictionary of National Biography, Stephen, L. and Lee, S.

32. Connor, B.: An extract of a letter from Bernard Connor, M.D. to Sir Charles Walgrave. . . . Philosophical Transactions **19:** 21, 1695.

33. Cruickshank, B.: Histopathology of diarthrodial joints in ankylosing spondylitis. Ann. Rheum. Dis. **10:** 393, 1951.

34. Dawson, M. H. and Ragan, C. Chronic Arthritis. Chap. II, Nelson's Loose-Leaf System of Medicine, New York.

35. D'Silva, J. L., Freeland, D. E. and Kazantzis, G.: The performance of patients with ankylosing spondylitis in the maximum ventilatory capacity test. Thorax **8:** 303, 1953.

36. Dunham, C. L. and Kautz, F. G.: Spondylitis ankylopoietica. Review and report of twenty cases. Am. J. Med. Sci. **201:** 232, 1941.

37. Engfeldt, B., Romanus, R. and Yden, S.: Histological studies of pelvospondylitis ossificans (ankylosing spondylitis) correlated with clinical and radiological findings. Ann. Rheum. Dis. **13:** 219, 1954.

38. Flemming Møller, P. and Gudjonsson, S. V.: Massive fluorosis of bones and ligaments. Acta radiol. **13:** 269, 1932.

39. Fletcher, E.: Ankylosing spondylitis. Lancet **1:** 754, 1944.

40. Fletcher, E.: The diagnosis and modern treatment of early ankylosing spondylitis. M. Press **221:** 451, 1949.

41. Forestier, J.: The importance of sacro-iliac changes in the early diagnosis of ankylosing spondylarthritis. Radiology **33:** 389, 1939.

42. Fraenkel, E.: Über chronische ankylosierende wirbel säulenversteifung. Fortsch. a.d. Geb. d. Röntgenstrahlen **7:** 62, 1904.

43. Freund, E.: Contribution to pathogenesis of spondylitis ankylopoietica. Edinburgh M. J. **49:** 91, 1942.

44. Freyberg, R. H.: Roentgen therapy for rheumatic diseases. M. Clin. N. A. **30:** 603, 1946.

45. Gleckler, W. J.: Rheumatoid spondylitis with carditis. Am. J. Med. **16:** 284, 1954.

46. Godfrey, M. F.: X-ray diagnosis and treatment in arthritis. California Med. **69:** 16, 1948.

47. Golding, F. C.: Spondylitis ankylopoietica (spondylitis ossificans ligamentosa). Brit. J. Surg. **23:** 484, 1936.

48. Graham, W. in Robinson, W. D. and Lampe, I.: Long range evaluation of radiotherapy in rheumatoid spondylitis. Proc. Ann. Meet., Amer. Rheum. Assoc., Ann. Rheum. Dis. **7:** 245, 1948.

49. Graham, W. and Ogryzlo, M. A.: Ankylosing (Marie-Strümpell) spondylitis (analysis of 100 cases). Canad. M. A. J. **57:** 16, 1947.

50. Grokoest, A. W., Demartini, F., Snyder, A. I. and Ragan, C. Personal communication.

51. Guest, C. M. and Jacobson, A. G.: Pelvic and extrapelvic osteopathy in rheumatoid spondylitis. J. Bone and Joint Surg. **65:** 650, 1951.

52. Guntz, E.: Beitrag zur pathologischen Anatomie der spondylarthritis ankylopoetica. Fortsch. a.d. Geb. d. Röntgenstrahlen **47:** 683, 1933.

53. Hammond, G., Crawford, H. R. and Haggart, G. E.: Vitallium mold arthroplasty of hip. J.A.M.A. **158:** 161, 1955.

54. Hamilton, K. A.: Pulmonary disease manifestations of ankylosing spondylarthritis. Ann. Int. Med. **31:** 216, 1949.

55. Hart, F. D.: Ankylosing spondylitis: a survey. Ann. Rheum. Dis. **13:** 186, 1943.

56. Hart, F. D., Bogdanovitch, A., and Nichol, W. D.: The thorax in ankylosing spondylitis. Ann. Rheum. Dis. **9:** 116, 1950.

57. Hart, F. D. and Maclagan, N. F.: Ankylosing spondylitis, a review of 184 cases. Ann. Rheum. Dis. **14:** 77, 1955.

58. Hart, F. D., Robinson, K. C., Allchin, F. M. and Maclagan, N. F.: Ankylosing spondylitis. Quart. J. Med. **18:** 217, 1949.

59. Hernaman-Johnson, F. and Law, W. A.: Ankylosing spondylitis. Butterworth & Co., Ltd., London. 1949.
60. Herrick, W. W. and Tyson, T. L.: The medical aspect of ankylosis spondylitis (Marie-Strümpell). Ann. Int. Med. **15:** 994, 1941.
61. Hersh, A. H., Stecher, R. M., Solomon, W. M., Wolpaw, R. and Hauser, H.: Heredity in ankylosing spondylitis. A study of fifty families. Am. J. Human Genetics **2:** 391, 1950.
62. Holbrook, W. P.: Cortisone, ACTH and phenylbutazone in long-term therapy of rheumatoid arthritis. Med. Clin. N.A. **39:** 405, 1955.
63. Holst, H. and Iversen, P. F.: On the incidence of spondylarthritis ankylopoetica in a Norwegian county. Acta med. Scand. **142:** 333, 1952.
64. Jones, L.: Arthritis deformans. William Wood & Co., New York, page 278, 1949.
65. Kemp, F. M., Murray, M. M. and Wilson, D. C.: Spondylosis deformans in relation to fluorine and general nutrition. Lancet **2:** 93, 1942.
66. Kersley, G. D. The rheumatic diseases. Wm. Heinemann, London, 1945, page 52.
67. Lennon, W. and Chalmers, I. S.: Ankylosing spondylitis. Lancet **1:** 12, 1948.
68. Levin, M. H.: Personal communication.
69. Leukemia and ankylosing spondylitis. J.A.M.A. **158:** 214, 1955.
70. Lucchesi, M. and Lucchesi, O.: Rheumatoid spondylitis, a pediatric problem. Ann. Rheum. Dis. **9:** 372, 1950.
71. Ludwig, A. O., Short, C. L. and Bauer, W.: Rheumatoid arthritis as a cause of increased cerebrospinal fluid protein. New England J. Med. **228:** 306, 1943.
72. MacDonald, G. R. and Hunt, T. E.: Sacroiliac joints. Observations on the gross and histological changes in the various age groups. Canad. Med. Assoc. J. **6:** 157, 1952.
73. Marie, P.: sur la Spondylose Rhizomelique. Revue de Med. **18:** 285, 1898.
74. Merrell, M. and Shulman, L. E.: Determination of prognosis in chronic disease illustrated by systemic lupus erythematosus. J. Chronic Dis. **1:** 12, 1955.
75. Mery, H. quoted in Dunham, C. L. and Kautz, F. G. Spondylitis ankylopoietica, review and report of 20 cases. Am. J. Med. Sci. **201:** 233, 1941.
76. Oppenheimer, A.: Diseases of the apophyseal (intervertebral) articulations. J. Bone and Joint Surg. **20:** 285, 1938.
77. Oppenheimer, A. Development, clinical manifestations and treatment of rheumatoid arthritis of the apophyseal intervertebral joint. Am. J. Roent. **49:** 49, 1943.
78. Pandit, C. H., Raghavachari, T. N. S., Rao, D. S. and Krishnamurti, V.: Endemic fluorosis in South India. Indian J. Med. Res. **28:** 533, 1940.
79. Parr, L. J. A., White, P. and Shipton, E.: Some observations on 100 cases of ankylosing spondylitis. Med. J. Aus. **1:** 544, 1951.
80. Polley, H. F.: A study of 1035 cases of rheumatoid spondylitis. Thesis. Graduate School, University of Minnesota, March 1945.
81. Polley, H. F.: The diagnosis and treatment of rheumatoid spondylitis. Med. Clin. N.A. **39:** 509, 1955.
82. Polley, H. F. and Slocumb, C. H.: Rheumatoid spondylitis, a study of 1035 cases. Ann. Int. Med. **26:** 240, 1947 and Ann. Rheum. Dis. **6:** 95, 1947.
83. Rees, S. E. and Murphy, W.: Roentgenotherapy of spondylitis adolescens. Northwest Med. **41:** 164, 1942.
84. Roholm, K.: Fluorine Intoxication. H. K. Lewis & Co., Ltd. London, 1937.
85. Rolleston, H.: The history of chronic rheumatism in A Survey of Chronic Rheumatic Diseases. R. G. Gordon, ed. London, Oxford Univ. Press, 1938, page 3.
86. Romanus, R.: Pelvospondylitis ossificans in the male and genito-urinary infection. Acta med. Scand. **145:** 1953, suppl. 280.
87. Romanus, R. and Yden, S.: Destructive and ossifying spondylitic changes in ankylosing spondylitis. Acta orthoped. Scand. **22:** 88, 1952.
88. Sashin, D.: A critical analysis of the anatomy and the pathological changes of the sacroiliac joints. J. Bone and Joint Surg. **12:** 891, 1930.

89. SHARP, J. AND EASSON, E. C.: Deep x-ray therapy in spondylitis. Brit. Med. J. **1:** 619, 1954.
90. SHORTT, H. E., MCROBERT, G. R., BARNARD, T. W. AND NAYAR, A. S. M.: Endemic fluorosis in the Madras Presidency. Indian J. Med. Res. **25:** 553, 1937.
91. SIMMONDS, M.: Über spondylitis deformans und ankylosierende spondylitis. Fortschr. a.d. Geb. d. Röntgenstrahlen **7:** 51, 1904.
92. SIMPSON, N. R. W. AND STEVENSON, C. J.: An analysis of 200 cases of ankylosing spondylitis. Brit. Med. J. **1:** 214, 1949.
93. SMITH-PETERSEN, M. N., LARSON, C. B. AND AUFRANC, O. E.: Osteotomy of the spine for correction of flexion deformity in rheumatoid arthritis. J. Bone and Joint Surg. **27:** 1, 1945.
94. SMYTH, C. J., FREYBERG, R. H. AND LAMPE, I.: Roentgen therapy for rheumatoid arthritis of the spine. J.A.M.A. **117:** 826, 1941.
95. SOKOLOFF, L.: The heart in rheumatoid arthritis. Amer. Heart J. **45:** 635, 1953.
96. SPISHAKOFF, N. M. AND LOW-BEER, B. V. A.: Roentgen therapy of rheumatoid spondylitis. California Med. **70:** 124, 1949.
97. STEINBERG, C. L., GARDNER, D. S., SMITH, F. A. AND HODGE, H. C.: A comparison of rheumatoid spondylitis and crippling fluorosis. Ann. Rheum. Dis. In press.
98. STEINBROCKER, O., TRAEGER, C. H. AND BATTERMAN, R. C.: Therapeutic criteria in rheumatoid arthritis. J.A.M.A. **140:** 659, 1949.
99. STRÜMPELL, A.: Bemerkung über die chronische ankylosirende entzündung der wirbelsäule und der hüftgelenke. Deut. Ztsch. f. Nerven Heilkunde **11:** 338, 1897.
100. SWAIM, L. T.: The orthopedic treatment of Strümpell-Marie arthritis. J. Bone and Joint Surg. **21:** 983, 1939.
101. TRAUT, E. F.: Rheumatic diseases, diagnosis and treatment. C. V. Mosby Co., 1952, St. Louis, page 294.
102. VALENTINI, quoted in DUNHAM, C. L. AND KAUTZ, F. G.: Spondylitis ankylopoietica. A review and report of 20 cases. Am. J. Med. Sci. **201:** 232, 1941.
103. VAN SWAAY, H.: Spondylosis Ankylopoëtica. Een pathogenetische studie. Thesis, Leiden, 1950, page 182.
104. Vital Statistics, 1950, Summary of. Department of Health, City of New York, John F. Mahoney, M.D., Comm.
105. WATSON, L. F.: Hernia, C. V. Mosby Co., St. Louis, 1948, page 37.
106. WEST, H. F.: The aetiology of ankylosing spondylitis. Ann. Rheum. Dis. **8:** 143, 1949.
107. WEST, L. S.: Two pedigrees showing predisposition to hernia. J. Hered. **27:** 449, 1936.
108. YOUNG, J. H.: Rheumatoid (ankylosing) spondylitis: its early manisfestations and early diagnosis. Med. J. Aus. **3:** 761, 1951.

Reprinted from SCIENCE, September 10, 1954, Vol. 120, No. 3115, pages 432–433.

Light-Scattering Studies on Hyaluronic Acid

Baruch S. Blumberg*

Department of Medicine, Columbia University, College of Physicians and Surgeons, and *the Edward Daniels Faulkner Arthritis Clinic, Presbyterian Hospital, New York*

Gerald Oster

Institute of Polymer Research, Polytechnic Institute of Brooklyn, Brooklyn, New York

Hyaluronic acid is an important constituent of many body tissues and occurs in synovial fluid, vitreous humor, and umbilical cord. This acidic polysaccharide is also present in some connective tissue and, as such, may take part in the physiological functions of connective tissue—that is, transport, storage, repair, and resistance to infection (*1*). In addition, it appears that there are physicochemical changes of the hyaluronic acid of joint fluids in rheumatoid arthritis (*2*). It is felt that a study of the size and shape of the molecule may contribute to an understanding of the role of this substance in its various physical functions. In the present work we are concerned with the results of light-scattering determinations. Other physicochemical studies are in progress.

Ogston and Stanier (*3*) have studied raw ox-synovial fluid and its ultrafiltrate containing hyaluronic acid. The material in the ultrafiltrates contains approximately 30 percent protein with the hyaluronic acid. By means of flow birefringence and viscosity studies, Ogston and Stainer concluded that the particles are highly hydrated spheres and that under the influence of a shear gradient the particles are deformed.

Light scattering affords a means whereby the size and shape of the particles in solution are determined without subjecting the particles to any external stress and where the theoretical interpretation of the experimental data is explicit.

Hyaluronic acid was prepared in the laboratory of Karl Meyer from umbilical cord by alcohol fractionation after peptic and tryptic digestion (*4*). Chemical analysis of the material gave the following results in percentages: nitrogen, 3.03; hexosamine, 38.0; sulfate, less than 0.4; hexosamine/nitrogen, 0.98; uronic acid (Dische method), 45.5. Our starting solutions were 0.5 percent by weight of hyaluronic acid in a 0.10*M* acetate buffer medium at *p*H 5.0 containing 0.15 molar sodium chloride. The solutions were clarified by centrifugation in a clinical centrifuge at 5000 rev/min for 20 min. The buffer solution used for dilution was clarified by filtering through a fine-grade sintered glass filter and by centrifugation.

Measurements were made in the Aminco Light Scattering apparatus (*5*) using the green line of mercury (λ, 546 mμ) throughout. A micro light-scattering cell was used that required approximately 3 ml of solution (*5*). This cell was surrounded by a larger cylindrical cell containing the solvent to decrease the reflectivity.

The refractive index increment of the solutions $(n - n_0)/c$, where n_0 is the refractive index of the solvent and n is the refractive index of the solution at concentration c (g/ml), needed in the molecular-weight determinations was measured in a Rayleigh interferometer. It was found to equal 0.180 independent of the concentration.

Information from light scattering concerning the size, shape, and molecular weight is gained from two sources, the dissymmetry and the turbidity (*6*). The observed values of the angular scattering given in terms of the dissymmetries (intensity at angle θ divided by the intensity at angle $180° - \theta$, where θ is measured from the transmitted beam) were 5.65, 5.32, 4.77, 3.40, 2.39, and 1.53 for θ equal to 40°, 45°, 50°, 60°, 70°, and 80°, respectively. The dissymmetries were found to be practically independent of concentration from 0.125 to 0.5 percent. The scattering for the solution is greater in the forward directions ($\theta < 90°$) than in the backward directions ($\theta > 90°$). Theory indicates (*6*) that the particles are not rigid rods, since $I_{45°}/I_{135°}$ cannot exceed 2.4, no matter what the length of the rod may be. From theoretical considerations for a random coil (in the high-polymer sense) and a sphere, graphs have been constructed of the dissymmetry for various pairs of supplementary angles, $I_\theta/I_{180°-\theta}$, versus L/λ', where L is the largest dimension of the particle and λ' is the wavelength of the light in the medium; $\lambda' = \lambda/n_0$, where n_0 is the refractive index of the solvent and λ the wavelength of the incident light. The observed values in the present study are incompatible with the theoretical values calculated for a random coil but are all in close agreement with those expected for a sphere where $L/\lambda' = 0.51$. Hence we conclude from the dissymmetry measurements that the particles are spheres of diameter $L = 546 \times 0.51/1.33 = 210$ mμ.

From turbidity measurements, we obtain the molecular weight of the material having a different refractive index from that of the solvent—that is, the "dry" hyaluronic acid. To determine the turbidity, the observed intensity of scattering at 90° was related to the absolute turbidity by the use of Ludox whose absolute turbidity is measured by transmission measurements (*5*). The turbidity at 90° must be multiplied by a factor to account for the lack of symmetry of the scattering envelope (*6, 7*), and this factor is calculated from our observed dissymmetry values to be 2.90. The turbidity divided by the concentration was found to be independent of concentration over the range studied (0.125 to 0.500 percent), although there is evidence for interaction at concentrations higher than these. The solutions that we studied may therefore be regarded as thermodynamically ideal.

For ideal solutions the molecular weight M is given (*6*) by

$$M = \frac{\tau}{Hc}, \text{ where } H = \frac{32\pi^3 n_0^2}{3N\lambda^4}\left(\frac{n - n_0}{c}\right)^2;$$

N is the Avogadro number and τ is the turbidity corrected for the dissymmetry. Since $\lambda = 546$ mμ and for hyaluronic acid $(n - n_0)/c$ is 0.180, then $H = 3.54 \times$

2

10^{-6}. In our determinations the average value of Hc/τ over the concentration range studied was 1.26×10^{-7} and therefore $M = 8 \times 10^6$.

The dissymmetry measures the outline of a molecule and gives a length corresponding to its greatest dimension. The turbidity gives a value for the molecular weight of the "dry" material present. The molecular weight that we obtained from the corrected turbidity is considerably less than that expected for a sphere of diameter 210 mμ determined from the dissymmetry. It appears, therefore, that the particles are highly swollen spheres of diameter 210 mμ and that the hyaluronic acid contained within the sphere has a molecular weight of 8.0×10^6. Simple calculations show that such spheres contain 0.27 percent by weight of anhydrous hyaluronic acid, and the remainder is presumably water or solvent. Hence the "swelling" is considerable.

It is interesting that the molecular model arrived at by our methods for an essentially protein-free hyaluronate resembles to a great extent the model of Ogston and Stanier (*3*) who used a hyaluronic-protein complex. This finding might indicate that the protein does not contribute to the general shape and physical properties of the molecule.

We wish to thank Karl Meyer for his advice and assistance with this problem and for the preparation of the hyaluronic acid. We further wish to thank Alfred Linker for carrying out the analyses of the material.

References and Notes

* Fellow, Arthritis and Rheumatism Foundation. Supported in part by USPHS grant A-570.

1. C. Ragan, *Ann. Rev. Physiol.* **14**, 51 (1952).
2. ———, *Proc. Soc. Exptl. Biol. Med.* **63**, 572 (1946).
3. A. G. Ogston and J. E. Stanier, *Biochem. J.* **49**, 585 (1951).
4. B. Weissman *et al., J. Biol. Chem.* **205**, 205 (1953).
5. G. Oster, *Anal. Chem.* **25**, 1165 (1953).
6. ———, *Chem. Revs.* **43**, 319 (1948).
7. P. Doty and R. F. Steiner, *J. Chem. Phys.* **18**, 1211 (1950).

9 June 1954

MS Ogston Memorial 11.2.96

Alexander George Ogston, MA (Oxon), DPhil, FRS, FAA

January 30, 1911 - June 29, 1996

A memorial oration

The University Church of St. Mary the Virgin

Oxford

November 2, 1996

Baruch S. Blumberg, MA (Oxon), M.D., DPhil

11.4.96

2

We are here today to remember and honor Alexander George Ogston a friend, teacher and devoted family member. He loved Oxford, and particularly this Church, of which he was a comunicant for many years and which he remembered fondly, in the memoir he prepared for his family.

The day I joined Sandy's laboratory at the Department of Biochemistry in Oxford in the autumn of 1955, nearly his first words to me were "Call me Sandy." All who knew him at all well called him by his first name, a reflection of his lack of pomposity, his warmth and his modesty. I shall continue to call him Sandy, as I have for more than forty years.

Sandy was born in Bombay in 1911, the son of a successful businessman, whose family originated in Aberdeen. In 1914, at the age of 3 ½ years, when the family returned to England on leave, he was taken on deck, by his mother, wrapped in a rug, to see the sands stretching away from the banks of the Suez Canal, shining like snow, bathed in desert moonlight. He recalled this memorable incident many years later in 1955 when he went by sea to Australia, and came on deck at night to revive his childhood memory.

The family returned to England during the first World War. Sandy won a scholarship to Eton and then, in 1929, to Balliol where he was a Brackenbury Scholar reading Chemistry. His training was rigorous. This included tutorials with the renowned Harold Hartley with a strong emphasis in Physical Chemistry, a particularly quantitative and demanding part of the curriculum. His laboratory exercises were in the combined Balliol-Trinity laboratories, a premonition of his later connection with both of these Colleges. He rowed enthusiastically and, in his fourth year was elected President of the Junior Common Room, an experience, I am sure which proved very valuable in his later role as the President of Trinity. He also took to climbing, joined the Oxford University Mountaineering Club and as an initial exercise climbed the classic "Idwal Slabs by the ordinary route" in north Wales, another experience I share with Sandy. However, whereas he described it as a "very easy climb" taken to improve his confidence, it was, in my estimation, formidable.

His undergraduate academic career was eminently successful and, after its' completion and a brief period at the London Hospital, he returned to Balliol in 1937 when he was appointed a Fellow. He had heavy teaching responsibilities, but at the same time continued his research which in time lead to the granting of an Oxford DPhil. His interest had shifted to biological problems and his teaching and research where henceforth in Physiology and later in the Honour School in Biochemistry when that became a separate subject, in no small part due to Sandy's influence.

In 1934 he married Elizabeth Wicksteed. They have had three daughters and a son. Sandy and Elisabeth were devoted parents and, particularly in their later years, focused on their children and now many grandchildren. Sandy and Elisabeth were loving and extraordinarily compatible. In our later years I hardly ever remember seeing one without the other. In addition to their offspring they enjoyed the out-of-doors and vigorous walking together.

Sandy's science was mathematically based, but with an insightful understanding of the complexity of biological processes. He favored the identification of an interesting problem, providing an original and exquisite solution, and then going on to another project. As his brilliant student, Oliver Smithies has pointed out, his scientific bent was for sensible solutions. This I believe, was associated with his joy in solving mechanical problems well illustrated by Sandy's account of a motoring trip in Germany in 1930. Their Lagonda automobile suffered a seizure of the gearbox in a small town in Germany. Parts were not available, and the local garageman did not have the requisite experience with this rare foreign car. Sandy and his friends, disassembled the gear box, removed the broken brass bits of the gear, welded them together, and then hand-shaped the assembled part to fit its' rightful place. They spent no less than <u>eight days</u> in the town and then motored back to England. The Lagonda company subsequently included the repaired gear as an exhibit in their museum. Sandy describes this event as the "real fun" part of this expedition to continental Europe.

During the Second World War, Sandy was assigned to the Research Team of the Ministry of Supply and to the Inter-Service Research Bureau. His experience in physical Chemistry was useful in attempts to develop methods for the inactivation of some of the poisonous gasses which were known to be available to the enemy, but which, fortunately, were never used. His research interests for the military were undoubtedly broader than that. In his writing, he once referred to a colleague "...with whom I had collaborated over the ventilation of a submarine during the war."

In the 1940's Hans Krebs (later, Sir Hans, Professor of Biochemistry at Oxford) published his classical studies that have helped to unravel the metabolic pathways of living organisms. An apparent paradox arose when it appeared that a symmetrical precursor resulted in only one of a possible two asymmetric products. Several attempts at explanation were offered, but none were satisfactory. Sandy was intrigued by this problem and I would like to quote directly from his recollection of this event as an example of his lively and elegant approach to problem solving.

> "In the course of casual reading of a recent volume of J.Biol.Chem. I had a sudden flash of insight into the steric relationship of asymmetric enzyme and symmetrical substrate, which I expressed in terms of "three-point attachment" and which

> became known as "the Ogston concept". I sent off a letter to Nature, but did not at the time regard the discovery as being of great interest until shortly after its publication, I received an excited letter from Hans Krebs (then of Sheffield). He gave my idea great weight in his Harvey Lecture the following year; and my surmise with regard to the reaction of citrate in the Tricarboxcylic Acid Cycle was shortly afterwards strikingly confirmed experimentally by Potter and Heidelberger. It was, I think, Potter who remarked to Rudolph Peters: "We just do the experiments, Dr. Ogston does the thinking for us." The "concept" has been regarded as the beginning of the theory of Pro-chirality. Except for two later notes, I did not myself follow the matter up. It was not within my regular field, and I had plenty else to occupy me."

Modestly, and characteristically he later uncovered an earlier reference which had a bearing on his paper, but this did nothing to detract from the general admiration for his solution.

He covered a wide range of subjects, each of which he associated with one or more students who he had supervised. He was meticulous in assigning credit to the students themselves. For example one of his most interesting solutions that has had very wide application is the "Johnston-Ogston" effect. It derived from observations he made while at the London Hospital where he worked on the centrifugation of serum proteins. It predicted the changes in concentration in moving particles when they are slowed down by other objects in their path. As Oliver Smithies has pointed out this model has broad applications beyond serum proteins in solution; for example it can predict the changes in the spacing of automobiles as they travel through different speed zones.

Several of his projects related to hyaluronic acid, a long chain polysaccharide - sugar, associated with a small amount of protein. It has novel physical characteristics and Ogston and his pupils showed how these might coincide with the demands on mammalian joint fluid, of which it is a major component. At one point, the conclusion was that it didn't He was the world's expert on the physical biochemistry of this unusual molecule and in particular contributed to the problem of the interaction of these long chain sugars with protein particles.

In 1960 Sandy was elected Professor of Physical Biochemistry at the Australian National University in Canberra. This was during what could be called the early modern period of a subject which now dominates biochemistry and, indeed, all of biological science, namely molecular biology. Sandy's quantitative approach to biology has been vindicated by the incredible success of these methods. He thoroughly enjoyed his ten years in Australia. He and Elisabeth became ardent bush-walkers. His description of some of their cross country walks, dead-reckoning with map and compass leaves one green with envy. While in Australia he was actively involved in the student Christian movement.

In 1970 he returned to Oxford as President of Trinity College.

Sandy was delighted with this post. He told me on one occasion that he greatest interest was in his students and their growth. In addition, he was an expert on the administrative workings of the University and of its relations with the Colleges. I recall that even in the mid-1950's when he was a young Fellow at Balliol, College and University faculty and staff would come to Sandy for advice on University rules and procedures. In retrospect I find this odd. Sandy's mind was guided by reason and sensibility. The governance of the University has no relation at all to these mind-sets. For example, during my stay at Balliol I had recommended that we inscribe, in stone, over the entrance to our College, "Abandon reason all you who enter here" but others recommended that we not do so. Sandy must have accepted the University administration as a challenge and a form of game playing which he enjoyed and learned to execute extremely well.

Sandy was a great scientist and educational leader. But his most extraordinary attributes relate to his guidance of young scientists. In looking through his reminisces one is struck by how often he writes of his former students, and how often they visit him and he they. Many of his students did not carry on in the field in which they were supervised. But there was something of greater importance that transcended the subject matter itself. He taught that science was an effective method for problem solving but that the problems it could solve were limited. That it had a value in its' own right and that application, although important and even essential, should not necessarily lead research, but follow from it. The need to know about nature, to understand the beauty of its organization was not only an appropriate avocation of humans, but a responsibility and a duty. And, I conjecture, this view flowed from his appreciation of the message of Scripture with its story of a Creator who gave us an organized cosmos. He also saw the world of science as a social organism whose members had broad areas of shared interest not only in the subject matter itself, but in the cooperative style and process which is essential for the prosecution of our endeavor. All of us who were taught and influenced by Sandy are united by a bond of appreciation and memory which we share with the members of his family.

The dead live on in their children, in their works, and in those whose lives they have influenced. On that basis Sandy is assured of a long and wide remembrance. I was one of those who he influenced. His scientific style had an enormous effect on mine. He brought my family and I to Oxford, and a major part of our lives has been spent here. Two of our children and four grandchildren live here now. His son Walter worked in my laboratory in

Philadelphia for many years and provided a wonderful opportunity to continue our connections with the Ogston family. When I think of Sandy my mind turns to the passage of the Hebrew liturgy that is read when the scrolls of the Bible are returned to rest in the Ark after the Sabbath reading.

It is a tree of life to those who grasp it, and happy is everyone who upholds it.
Its ways are ways of pleasantness and all its paths are peace.

NOT USED IN THE SPOKEN ADDRESS

["For the whole earth is the sepulchre of famous men; and their story is not graven only on stone over their native earth, but lives on, far away, without visible symbol, woven into the stuff of other men's lives."]

NOTES FROM "REMINISCENCES 1911-1988, A.G.OGSTON".

p.20 Repairing gearbox of Lagonda in Germany
p.36 "We were a happy, shifting party, interested in each other's work.
p.38 The "three-point attachment " model or "the Ogston concept" "I had a sudden flash of insight into the steric relationship of asymmetric enzyme and symmetrical substrate ..."We just do the experiments, Dr. Ogston does the thinking for us." Beginning of the theory of pro-chirality
p.42 The moon-lit desert going through the Suez Canal. Reminded him of the same passage when he was three and a half years old
p. 45 Keen knowledge of University admin. and politics
Retirement unnumbered. "...with whom I had collaborated over the ventilation of a submarine during the war."

Reprinted from *J Roy Anthrop Institute*
87: 217–221, 1957 (Jul–Dec 1957)

The Blood Groups of The Spanish Basques

F. ALBERDI, A. C. ALLISON, B. S. BLUMBERG, ELIZABETH W. IKIN & A. E. MOURANT

From the Medical Service, Red Cross Hospital, San Sebastian (F.A.); the Department of Biochemistry, Oxford University (A.C.A., B.S.B.); and the Blood Group Reference Laboratory (Medical Research Council), London, (E.W.I., and A.E.M.)

THE BASQUES ARE A GROUP OF PEOPLE living in northern Spain and south-western France who differ from their neighbours in many physical and cultural features. The most conspicuous point of distinction is their language, which appears to be unrelated to any other European tongue. However, comparative philologists have shown a resemblance between the Basque language on the one hand and languages spoken by some populations in the north-west Caucasus on the other (Lafon 1933; Bouda 1938; Holmer 1947; Meillet & Cohen 1952). Points of similarity between Basque and certain pre-Hamitic features of Berber languages have also been emphasized (Gèze 1883; Gabelentz 1894; Schuchardt 1913; Meillet & Cohen 1952).

In their blood groups also, the Basques differ considerably from all other European populations. They were first shown by Boyd & Boyd (1937) to have, like many other populations on the periphery of Europe, a rather high frequency of group *O*, but also to have an exceptionally low frequency of group *B*; and Etcheverry (1945) showed them to have a frequency of *Rh*-negatives higher than any previously recorded. These findings have been repeatedly confirmed. The most extensive blood group survey undertaken with a wide range of antisera is that carried out by Chalmers, Ikin & Mourant (1948) who examined 383 persons of the purest available Basque ancestry, some born and resident in France, and others born in Spain but resident in France and England. The authors not only confirmed the exceptionally high frequency of *Rh*-negatives (or of the gene *d*) but in addition showed the frequency of the *CDe* chromosome to be high and that of *cDE* to be the lowest recorded in Europe. No systematic tests were done for the D^u antigen. The C^w antigen appeared to be absent, but with only a weak antiserum available and in the absence of a comparable positive control, this result remained uncertain. Van der Heide, Magnee, Van Loghem & Souchard (1951) subsequently found one C^wc heterozygote among 181 Basques tested, indicating a C^w frequency below that recorded for any other European community.

In the discussion of their results Chalmers *et al.* pointed out that on genetical grounds it would appear that deaths from haemolytic disease were slowly lowering the frequency of the *d* gene in Europe. Such a lack of equilibrium must be transitory, but might be explained if the present population was the result of relatively recent mixing. It was suggested that the modern European population was the result of hybridization between the ancestors of the Basques, who had long been in Europe and had contributed the *d* gene, and more recent immigrants who had brought with them the *D* gene. In the light of present knowledge of the distribution of blood groups in Europe and south-western Asia few workers would now subscribe to this hypothesis in its simplest form; the difficulties which it presents have recently been discussed by Allison (1955). It probably, however, contains an element of truth since it has now been shown that the Basques have the lowest, or almost the lowest, European frequencies not only of the genes *B* and *D* but also of the Duffy gene, Fy^a (Nijenhuis, 1956); all three genes have higher frequencies in Asia than in Europe.

In the region surrounding the Basque country we have evidence from place-names that the Basque language was once much more widespread than it is today, while the low *B* and *D* frequencies found over a large part of south-western France suggest a considerable Basque physical contribution to the population.

The discovery in Africa and the Mediterranean region of high frequencies of the *Rh* gene D^u, which in the absence of special tests is easily confused with the *Rh*-negative gene *d*, has raised doubts regarding the genuineness of the high frequency of the latter gene among the Basques, especially since the Berbers, an African population, appear to resemble the Basques in their *Rh* blood group frequencies.

Thus, while the main blood group picture of the Basques has for some time been firmly established, a number of points have remained to be cleared up by further investigation. The opportunity for such an investigation arose when two of us (A.C.A. and B.S.B.) were planning a survey of the incidence among the Basques of the genetically determined serum protein types identifiable by the starch-gel technique of electrophoresis (Smithies 1955); and it was arranged that blood grouping tests should be performed in addition on the specimens collected for electrophoresis. The observations on the serum types will be reported elsewhere (Allison, Blumberg & ap Rees 1957).

MATERIALS AND METHODS

One hundred and sixty-one bloods were collected in the widely separated farming villages of Goizueta and Echalar in Navarre, and Ataun in Guipúzcoa, and from workers at the factory of Barrio de Oria, Guipúzcoa. Not more than two members of any one family were included in the sample. In every village the local physician was at hand to assist in the selection of the candidates for bleeding, and only persons known to be of 'pure Basque descent' were included. An additional precaution was the insistence that each of the four names such as are used by most Basques was of Basque origin. The blood was collected in 'vacunes', taken to Biarritz by car, and then forwarded by air to London for testing at the Blood Group Reference Laboratory.

The blood samples were tested for the antigens A_1, A_2, *B*, *C*, C^w, *c*, *D*, D^u, *E*, *e*, *V*, Lu^a (Lutheran) and Fy^a (Duffy). In order to detect the D^u antigen all samples which gave negative results with a saline-agglutinating anti-*D* serum were further examined by means of anti-human-globulin tests with several selected incomplete anti-*D* sera. This is the most sensitive test available for detecting low grades of D^u. Owing to an oversight it was thought that Chalmers, Ikin & Mourant (1949) had included tests with anti-*S* in their investigation, and an examination for the antigens of the *MNS* system was therefore unfortunately omitted from the present survey. The only published results of tests on the Basques with anti-*S* are those of Nijenhuis (1957).

DISCUSSION OF RESULTS

The results of our tests and of the genetical analysis are shown in Tables 1–3. The most important result was the finding of only one example of the D^u antigen among 161 persons tested, giving a low D^u frequency approximately the same as that found in Europe generally; but it is of interest that the one person with this antigen carried the cD^ue chromosome which is characteristically African and Mediterranean rather than North European. The typically African *cDe* chromosome was not found, nor was any example of the *V* antigen, a component of the *Rh* complex which is found almost exclusively among persons of African ancestry (De Natale, Cahan, Jack, Race & Sanger 1955). No example of the C^w antigen was found; this antigen has been found, although with a low frequency, in nearly every European population tested for it, but shows a tendency to increase in frequency towards the east and north. The rather low frequency of the Fy^a antigen previously reported was confirmed; that of the Lu^a antigen was found to be about the same as in other parts of Europe. In general, our findings for the *ABO* and *Rh* systems confirm the previously published results. The frequencies of the *B* and *D* genes are both somewhat higher than those reported previously for Basques; it is possible that the relatively high frequency of the *B* gene was mainly due to persons from the village of Ataun, where it is possible that by chance the sample included rather a large number of persons who were distantly related to one another.

Our results are consistent with the Basques being the relatively unmixed descendants of an ancient 'autochthonous' European stock, though perhaps but one of many such stocks. Thus, the population of Spakenburg and Bunschoten, a pair of isolated villages in Holland, has rather low frequences of *B* and *D* (Van der Heide *et al.* 1951). These people may be derived from the same stock, though in the case of such a very small population the possibility of random gene frequency fluctuation cannot entirely be excluded.

With regard to the remote origin of the Basques, little reliable information is to be found. As mentioned above, comparative philologists have demonstrated relationships between the Basque language on the one hand and languages in the Caucasus and North Africa on the other. It is therefore of interest that the limited blood group data available should suggest similar relationships. Bru and Messerlin (1955) describe a number of Berber tribes in the Great Atlas Mountains with *Rh* and *ABO* frequencies comparable to those of the Basques but with much higher *N* frequencies. Similar *Rh* frequencies were found by Jacquemin (1952) in a small group of Touaregs. Data quoted by Boyd (1939) show that the peoples of the Caucasus have much higher *O* and lower *B* frequencies than the surrounding populations. The South Caucasians, listed by Boyd as 'Western Group', have the most extreme values, with an average of 73 per cent of *O* and only 6 per cent of *B* genes. Forty Georgians (South Caucasians) were, however, found by Kherumian, Moullec & Desabie (1954) to include only 12½ per cent of *Rh* negatives, a frequency which, even on this small sample, is highly significantly less than that among the Basques. On a very small sample of 18 individuals the frequency of Duffy (Fy^a) positives is higher than in either Basques or in other western Europeans. The parallelism between the data of philology and of serology is nevertheless striking and makes it most desirable that further blood grouping should be undertaken both in North Africa and in the Caucasus.

SUMMARY

Blood samples from 161 Spanish Basques have been tested for the blood group antigens A_1, A_2, *B*, *C*, C^w, *c*, *D*, D^u, *E*, *e*, *V*, Lu^a and Fy^a. In confirmation of previous studies, the frequency of gene *B* was found to be low and that of *d* (*Rh*-negative) to be high. The typically African *V* component of the *Rh* complex was absent. The frequency of Fy^a (Duffy) was lower than that recorded for any other European population, thus confirming the observation of Nijenhuis (1956). The frequency of the Lu^a (Lutheran) gene was within the narrow western European range. The findings are discussed in relation to the origin of the Basques and their relation to other European populations.

ACKNOWLEDGEMENTS

The vacunes were supplied by Winthrop Products Ltd. Valuable help in collecting blood specimens was given by the following physicians: Dr Gil Rudrigo, Ataun; Dr Arroyo, Goizueta; and Dr Bayona, Echalar. Dr J. C. de Oya, Madrid, helped in making the original arrangements for this study. We are grateful to Dr Ada C. Kopeć for performing the calculations. B. S. Blumberg is a Fellow of the Arthritis and Rheumatism Foundation, and A. C. Allison a member of the Scientific Staff of the Medical Research Council.

TABLE I. *The* ABO *Blood Groups of the Spanish Basques*

Blood Group	Number Observed	Frequency Observed	Frequency Expected	Number Expected	Gene Frequencies
O	89	0.5528	0.5501	88.57	p^1 0.1961
A_1	54	0.3354	0.3401	54.75	p^2 0.0275
A_2	7	0.0435	0.0415	6.69	q 0.0347
B	8	0.0497	0.0527	8.49	r 0.7417
A_1B	3	0.0186	0.0136	2.19	
A_2B	0	0.0000	0.0019	0.31	
TOTAL	161	1.0000	0.9999	161.00	TOTAL 1.0000

TABLE II. *The* Rh *Blood Groups of the Spanish Basques*

Phenotype	Number Observed	Frequency Observed	Frequency Expected	Number Expected	Chromosome Frequencies
CCDee	33	0.2050	0.2080	33.48	*CDe* 0.4357
CCddee	0	0.0000	0.0004	0.07	*Cde* 0.0209
CcDEe	13	0.0807	0.0822	13.24	*cDE* 0.0901
CcDee	65	0.4037	0.3951	63.61	cD^ue 0.0067
CcD^uee	0	0.0000	0.0003	0.05	*cde* 0.4467
Ccddee	3	0.0186	0.0186	3.00	
ccDEE	3	0.0186	0.0081	1.31	
ccDEe	10	0.0621	0.0817	13.15	
ccD^uee	1	0.0062	0.0060	0.97	
ccddee	33	0.2050	0.1995	32.13	
TOTAL	161	0.9999	0.9999	161.01	TOTAL 1.0000

TABLE III. *The* Lutheran *and* Duffy *Blood Groups of the Spanish Basques*

Group	Number Observed	Frequency Observed	Gene Frequencies
Lu(a+)	13	0.0807	Lu^a 0.0412
Lu(a–)	148	0.9193	Lu^b 0.9588
TOTAL	161	1.0000	
Fy(a)	81	0.5031	Fy^a 0.2951
Fy(a–)	80	0.4969	Fy^b 0.7049
TOTAL	161	1.0000	

REFERENCES

ALLISON, A. C. 1955. Aspects of polymorphism in man. *Cold. Spr. Harb. Symp. quant. Biol.* **20**, pp. 239–255.

ALLISON, A. C., BLUMBERG, B. S., & ap REES, W. 1957. Haptoglobulin types of the Spanish Basques. (In preparation.)

BOUDA, K. 1938. Die Beziehungen des Sumerischen zum Baskischen, Westkaukasischen und Tibetischen. *Mitt. altorient. Gesell.* **12**, no. 3.
BOYD, W. C. 1939. Blood groups. *Tabul. biol.* (*Hague*). **17**, pp. 113–240.
BOYD, W. C., & BOYD, LYLE, G. 1937. New data on blood groups and other inherited factors in Europe and Egypt. *Amer. J. Phys. Anthrop.* **23**, pp. 49–70.
BRU, P., & MESSERLIN, A. 1955. Transfusion sanguine et groupes sanguins. *Maroc. méd.* **34**, pp. 1551–1562.
CHALMERS, J. N. M., IKIN, ELIZABETH W., & MOURANT, A. E. 1949. The *ABO*, *MN* and *Rh* blood groups of the Basque people. *Amer. J. Phys. Anthrop.* n.s.**7**, pp. 529–544.
DE NATALE, A., CAHAN, A., JACK, J. A., RACE, R. R., & SANGER, RUTH 1955. V, a 'new' *Rh* antigen, common in Negroes, rare in white people. *J. Amer. med. Ass.* **159**, pp. 247–250.
ETCHEVERRY, M. A. 1945. El factor rhesus, su genética e importancia clínica. *Día méd.* **17**, pp. 1237–1259.
GABELENTZ, G. VON DER. 1894. *Die Verwandschaft des Baskischen mit den Berbersprachen Nord-Africas.* Braunschweig.
GÈZE, L. 1883. *De quelques rapports entre les langues berbères et le basque.* Cited by Meillet, A. & Cohen, M. (see below), p. 258.
HEIDE, H. M. VAN DER, MAGNEE, WILLY, LOGHEM, J. J. VAN & SOUCHARD, L. 1951. Blood group distribution in Basques. *Amer. J. hum. Genet.* **3**, pp. 356–361.
HOLMER, N. M. 1947. Ibero-caucasian as a linguistic type. *Studia Linguistica.* **1**, pp. 11–44.
JACQUEMIN, P. 1952. Personal communication, quoted by Mourant, A. E. 1954, in *The distribution of the human blood groups.* Oxford, p. 276.
KHERUMIAN, R., MOULLEC, J. & DESABIE, J. 1955. Les groupes sanguins A_1 A_2, *BO*, *MN*, *Rh* et *P* chez les Arméniens et les Géorgiens. 5th *Int. Congr. Blood Transfus.* Paris, pp. 264–270.
LAFON, R. 1933. Basque et langues kartvèles. *Rev. Int. Et. Basques.* **24**, pp. 150–172.
MEILLET, A. & COHEN, M. (eds.) 1952. *Les langues du monde.* Paris. pp. 228–230, 257–258.
NIJENHUIS, L. E. 1957. Blood group frequencies in French Basques. *1st Int. Congr. hum. Genet* (Copenhagen), pp. 375–379.
SCHUCHARDT, H. 1913. Baskisch-hamitische Wortvergleichungen. *Rev. Int. Et. Basques*, **7**, pp. 289–340.
SMITHIES, O. 1955. Zone electrophoresis in starch gels: group variations in the serum protein of normal human adults. *Biochem. J.* **61**, pp. 629–641.

Reprinted from THE AMERICAN JOURNAL OF PHYSICAL ANTHROPOLOGY
N. S. Vol. 19, No. 2, June 1961

The Blood Groups of the Pastoral Fulani of Northern Nigeria and the Yoruba of Western Nigeria

B. S. BLUMBERG, ELIZABETH W. IKIN AND A. E. MOURANT
National Institute of Arthritis and Metabolic Diseases, National Institutes of Health, Bethesda, Maryland and The Blood Group Reference Laboratory, The Lister Institute, London, England

The Fulani are one of the largest tribal groups of Africa, spread over nearly half its width, from Senegal in the west to Wadai, east of Lake Chad. In part they are settled town dwellers who, as a result of vigorous military and political action before and shortly after the arrival of the Europeans, became the ruling class in several areas of West Africa. The town dwellers have, in general, intermarried with the groups they governed and, to a large extent, forgotten or abandoned many of their ancient customs and their language. The seminomadic "cow" Fulani, who still follow their ancient pastoral life, have remained more aloof from their neighbors and, to a large extent, retained the Fulani language and customs (de St. Croix, '45).

There has been a great deal of conjecture as to the affinities of the Fulani, and several theories of Caucasoid origin have been advanced. Delafosse (with very little evidence) considers them a group of Aramaic-speaking Judeo-Syrians who entered Negro Africa from Cyrenaica about 200 A.D. (quoted in Greenberg, '49). However, Greenberg ('49) has shown that their language does not belong to the Hamito-Semitic group but closely resembles Serer-Sin (a language spoken in Senegal) and has a more remote connection with Wolof—a language spoken in the same region and elsewhere in West Africa. This places it in the northern sub-group of Westermann's West Atlantic section of the West Sudanic family, or in the westernmost branch of Greenberg's Niger-Congo group, neither of which is a Hamitic language group (Greenberg, '49). The Fulani are known by other names in other regions of West Africa, Foulah, Peul, Foul-be, etc., and anthropological and other differences may exist between these populations. Blood group studies have been reported on some of these populations but none, to our knowledge, on the nomadic Fulani of Nigeria (Koerber, '47; Koerber and Linhard, '51; Kossovitch, '34; de Lestrange, '49; Moullec and Linhard, '55; and Trincão, '53).

The Yoruba are the dominant tribal group in western Nigeria, and enclaves of Africans of Yoruba descent who still maintain some of their customs are found in Brazil and Cuba (Verger, '55). Blood specimens were collected from the Yoruba for studies reported elsewhere (Allison, Blumberg and ap Rees, '58; Steinberg et al., '61), and blood group tests were done on these specimens for comparison with previously published data. Previous blood group tests on the Yoruba have been reported by Muller ('27), Garlick ('57), Garlick and Barnicot ('57) and Chalmers, Ikin and Mourant ('53), most of whose South-west Nigerians were Yoruba.

MATERIALS AND METHODS

Blood was collected in "venules" (Winthrop Products, Ltd.), allowed to clot, refrigerated at approximately 4°C, and sent by air to London for blood typing. The tests were completed within three to 7 days of collection, and all bloods were deemed satisfactory for typing. Tests were carried out for the following antigens: A_1, A_2, B, M, N, S, Henshaw, P_1, C, c, C^w (all negative), D, D^u, E, e, V, K. All specimens were also tested for Lu^a, all with negative results, and for Fy^a, with only one positive result among the Fulani. In view of the known difficulty in testing specimens for the last two antigens when they are several days old, and in the absence of adequate positive controls, these results have not been included in the tables. Fifty-nine Yoruba specimens were tested for the Diego (Di^a) antigen, all with negative results.

Fulani. The Fulani bloods were collected in two groups. The first group consisted of 89 specimens collected by one of the authors (BSB, collection 1). These included 74 from Rimi and Kakio villages, both near Kaduna, Northern Nigeria, and 15 from visiting herdsmen in the market place at Gasawaba, Northern Nigeria. The second group consisted of 59 bloods for which we are indebted to Dr. Hayhurst, Medical Officer at Pankshin, (collection 2). These were collected from Pastoral Fulani from the region near Pankshin, Northern Nigeria. The samples were unselected as to sex and age. It was at first intended to treat these groups together, but it was found that they differed markedly in their MNS phenotype frequencies. After considering treating them as completely separate populations, it was finally decided to perform separate calculations upon their MNS groups but not to separate them otherwise. One reason for this decision was the large number of Rh phenotypes detected by means of the antisera used, including anti-V: as a result separate treatment would have meant that the numbers of each phenotype within each population would have been unduly small for statistical treatment.

Yoruba. One hundred and thirty-five bloods were collected from unrelated individuals in the village of Ilora, near Ibadan, in the western region of Nigeria. The majority (approximately 85%) of the subjects were school children, but some adults were included and the sexes were about equally represented. Gene frequency calculations were performed by the methods described by Mourant ('54). From the gene frequencies, the expected frequency of each phenotype was calculated, and from this the expected absolute number. Comparison with the number originally observed gave a value of χ^2 for that phenotype. Expected numbers less than 5 were combined for this purpose. The total value of χ^2 taken in conjunction with the number of degrees of freedom gave the probability of such a distribution in a random sample. χ^2 values are not given for tables 3 and 9 because of the small numbers in some of the groups.

RESULTS AND DISCUSSION

The results of our tests, and gene frequencies calculated from them, are shown in tables 1 to 11. In tables 4 and 9 the method used for representing the Rh phenotypes is that proposed by Mourant ('49), the presence or absence of the V antigen being expressed on exactly the same lines as for the other antigens, by V or vv. Most of the comparative data discussed below will be found tabulated, set out in maps, and discussed in Mourant ('54), Mourant, Kopec and Domaniewska-Sobczak ('58) and Roberts, Ikin and Mourant ('55).

Fulani. As already stated, the two batches of Fulani are treated separately as far as their MNS blood groups are concerned. Those of collection 2, though small in number, show a particularly high frequency of the N gene, which suggests a relationship with the Berber and Tuareg populations to the north (Messerlin and Lohro, '51; Jacquemin, '52; Barnicot, Ikin and Mourant, '54). The high O frequency in the same batch supports this suggestion, as do the Rh frequencies in both batches, for CDe is considerably higher and cDe lower than in any other African populations so far tested from south of the Sahara other than the Tuaregs. However, the frequency of the d gene (all as cde) is about that usually found in Africans and nowhere near the high figure found in Berbers and Tuaregs. This is compatible with a Mediterranean rather than a Berber admixture. The frequency of 45% of cDe chromosomes is, however, consistent with the thesis that these populations are in the main of Negroid origin.

The only other available data on the MNS and Rh groups of the Fulani (apart from 11 individuals included among the "Northern Nigerians" by Chalmers, Ikin and Mourant, '53) are an unpublished series by Garlick, Ikin and Roberts. These have a cDe chromosome frequency of 55% and thus appear to be more negroid than either of the present series. A higher cDE frequency of 10% suggests connections with other Nigerian tribes and with the

TABLE 1

The A_1A_2BO groups of the Fulani

Group	Number observed	Frequency observed	Frequency expected	Number expected	χ^2
O	82	0.5541	0.5381	79.64	0.0699
A_1	24	0.1622	0.1582	23.41	0.0147
A_2	8	0.0541	0.0753	11.14	0.8869
B	26	0.1757	0.1932	28.59	0.2345
A_1B	3	0.0203	0.0232	3.43	1.4897
A_2B	5	0.0338	0.0121	1.79	
Total	148	1.0002	1.0001	148.00	2.6957

Gene frequencies

P_1	0.0952
P_2	0.0496
q	0.1216
r	0.7336
Total	1.0000

TABLE 2

MNS groups in Fulani

Collection 1; 89 specimens sent by Dr. Blumberg

Phenotype	Number observed	Frequency observed	Frequency expected	Number expected	χ^2
MMS	11	0.1235	0.1223	10.89	0.0011
MsMs	12	0.1348	0.1333	11.86	0.0015
MNS	13	0.1461	0.1473	13.11	0.0009
MsNs	31	0.3483	0.3526	31.38	0.0047
NNS	1	0.0112	0.0112	1.00	0.0028
NsNs	21	0.2360	0.2332	20.75	
Total	89	0.9999	0.9999	88.99	0.0110

Chromosome frequencies

MS	0.1405
Ms	0.3651
NS	0.0115
Ns	0.4829
Total	1.0000

TABLE 3

MNS groups in Fulani

Collection 2; 59 specimens sent by Dr. Hayhurst

Phenotype	Number observed	Frequency observed	Frequency expected	Number expected
MMS	3	0.0508	0.0197	1.16
MsMs	8	0.1356	0.0538	3.18
MNS	2	0.0339	0.0888	5.24
MsNs	8	0.1356	0.3065	18.08
NNS	7	0.1186	0.0949	5.60
NsNs	31	0.5254	0.4362	25.74
Total	59	0.9999	0.9999	59.00

Chromosome frequencies

MS	0.0391
Ms	0.2320
NS	0.0683
Ns	0.6605
Total	0.9999

TABLE 4

The Rh groups of the Fulani

Phenotype	Number observed	Frequency observed	Frequency expected	Number expected	χ^2
ccDeeV	36	0.2432	0.2607	38.59	0.1733
ccDeevv	24	0.1622	0.1481	21.92	0.1969
CcDeeV	26	0.1757	0.1530	22.65	0.4965
CcDeevv	35	0.2365	0.2350	34.79	0.0013
ccDEeV	3	0.0203	0.0198	2.93	0.8027
ccDEevv	2	0.0135	0.0304	4.50	
CCDeevv	10	0.0676	0.0825	12.20	0.3982
ccD^{u}eeV	2	0.0135	0.0136	2.02	
CcDEevv	4	0.0270	0.0213	3.16	0.4877
ccDEEvv	1	0.0068	0.0014	0.20	
ccddeeV	2	0.0135	0.0147	2.18	0.0004
ccddeevv	3	0.0203	0.0194	2.86	
Total	148	1.0001	0.9999	148.00	2.5570

Chromosome frequencies

CDev	0.2872
cDEv	0.0372
cDeV	0.1872
cDev	0.2701
cD^{u}eV	0.0338
cdeV	0.0454
cdev	0.1391
Total	1.0000

TABLE 5

The P and Kell groups of the Fulani

P_1^+	141	K+	1
P_1^-	7	K−	147

TABLE 6

The Henshaw groups of the Fulani

MNS phenotype	He+
MMS	0
MsMs	0
MNS	0
MsNs	1
NNS	1
NsNs	0
Total[1]	2

[1] Out of 148 tested, and listed in tables 2 and 3.

Belgian Congo. The MNS frequencies are intermediate between the two present series but the rather high N supports the theory of Berber admixture in the Fulani. The ABO frequencies also are intermediate between those of our two series in most respects. A fuller comparison will be made with the present data when the new material is published.

Results of previous tests on the ABO groups of the Yoruba show a high degree of consistency. The present results show a higher frequency of B, but the excess (of B+AB) is not quite significant statistically ($\chi^2 = 3.69$, $P > 0.05$).

The only published results of Rh and MNS tests on Yoruba are those of Chalmers, Ikin and Mourant ('53) whose "South-western Nigerians" were mainly Yoruba (102 out of 145). The present results differ little from these for the MNS system, and for the Rh system the main difference consists in a lower frequency of cDE in the present series, with a chromosome frequency of 1.1% as compared with 8.3% for the series of Chalmers et al. ('53). Dr. Ada Kopeć has kindly extracted the figures for Yoruba alone from the combined data of Chalmers et al. ('53), and calculated chromosome frequencies: that of cDE is 7.35%. Chalmers et al. ('53) found higher frequencies of cDE than this in all the other Nigerian populations examined and the frequency is high also in the Belgian Congo.

The frequencies of the Henshaw antigen, 1.4% in the Fulani and 2.2% in the Yoruba, are near to those found previously in Nigeria. The present data, taken by themselves, give little information as to

TABLE 7

A_1A_2BO groups of Yoruba

Group	Number observed	Frequency observed	Frequency expected	Number expected	χ^2
O	60	0.4444	0.4414	59.60	0.0027
A_1	21	0.1556	0.1426	19.25	0.1595
A_2	8	0.0593	0.0756	10.21	0.4793
B	38	0.2815	0.2848	38.45	0.0053
A_1B	3	0.0222	0.0350	4.72	0.0342
A_2B	5	0.0370	0.0205	2.77	
Total	135	1.0000	0.9999	135.00	0.6810

Gene frequencies

P_1	0.0931
P_2	0.0547
q	0.1878
r	0.6644
Total	1.0000

TABLE 8

MNS groups of Yoruba

Phenotype	Number observed	Frequency observed	Frequency expected	Number expected	χ^2
MMS	7	0.0519	0.0456	6.16	0.1156
MsMs	27	0.2000	0.1687	22.78	0.7823
MNS	12	0.0889	0.0999	13.49	0.1642
MsNs	45	0.3333	0.3973	53.64	1.3920
NNS	8	0.0593	0.0545	7.36	0.0565
NsNs	36	0.2667	0.2339	31.58	0.6187
Total	135	1.0001	0.9999	135.01	3.1293

Chromosome frequencies

MS	0.0522
Ms	0.4108
NS	0.0534
Ns	0.4837
Total	1.0001

which of the MNSs chromosomes carry the Henshaw gene. They are consistent with its going only with N, but if so then in at least two out of the 5 positives it accompanied Ns, not NS as previous tests showed to be the rule in Nigeria.

The frequency of the V antigen also has proved to be very high both in Fulani (46.6%) and in Yoruba (40%). In agreement with previous findings, the V gene appears to accompany only cDe, cD^{u}e or cde, and mainly cDe, in these populations.

The Yoruba have been found to have a high incidence of the haptoglobin type 1-1, and a large number of individuals with very little or no haptoglobin (Allison, Blumberg and ap Rees, '58). Preliminary studies on the Fulani reveal a similar pattern. The Fulani also have a high frequency of the transferring D gene (Tf^D). (Blumberg and Gentile, '61).

SUMMARY

The blood groups of Nigerian Yoruba and Fulani have been studied. There is a suggestion of some Mediterranean admixture in the Fulani. The results in the Yoruba are consistent with previous studies.

ACKNOWLEDGMENTS

We would like to acknowledge the assistance and encouragement given to us by Professor B.G.T. Elmes, University College, Ibadan. Mr. O. Oyefesu assisted in the collection of blood specimens. A part of this work was carried out during an expedition supported by the Wenner-Gren Foundation, Inc.

TABLE 9

The Rh groups of the Yoruba

Phenotype	Number observed	Frequency observed	Frequency expected	Number expected
ccDeeV	47	0.3481	0.3439	46.43
ccDeevv	56	0.4148	0.4201	56.72
CcDeeV	4	0.0296	0.0347	4.68
CcDeevv	16	0.1185	0.0974	13.14
ccDEeV	—	—	0.0050	0.68
ccDEevv	1	0.0074	0.0151	2.05
ccD^{u}eeV	1	0.0074	0.0074	1.00
CCDeevv	—	—	0.0056	0.75
CcDEevv	—	—	0.0018	0.24
ccDEEvv	1	0.0074	0.0001	0.02
CcddeeV	1	0.0074	0.0011	0.15
Ccddeevv	1	0.0074	0.0137	1.85
CCddeevv	—	—	0.0011	0.14
CcD^{u}eeV	—	—	0.0010	0.14
ccddeeV	1	0.0074	0.0073	0.99
ccddeevv	6	0.0444	0.0446	6.02
Total	135	0.9998	0.9999	135.00

Chromosome frequencies

CDev	0.0490
cDEv	0.0111
Cdev	0.0325
cDeV	0.1933
cDev	0.4705
cD^{u}eV	0.0157
cdeV	0.0166
cdev	0.2112
Total	0.9999

TABLE 10

The P and Kell groups of the Yoruba

P_1^+	122	K+	4
P_1^-	13	K−	131

TABLE 11

The Henshaw groups of the Yoruba

MNS phenotype	He+
MNS	0
MsMs	0
MNS	1
MsNs	0
NNS	1
NsNs	1
Total	3

LITERATURE CITED

Allison, A. C., B. S. Blumberg and W. ap Rees 1958 Haptoglobin types in British, Spanish Basque and Nigerian populations. Nature, *181:* 824–825.

Barnicot, N. A., E. W. Ikin and A. E. Mourant 1954 Les groupes sanguins ABO, MNS et Rh des Touareg de l'Air. Anthropologie, *58:* 231–240.

Blumberg, B. S., and Z. Gentile 1961 The haptoglobins and transferrins of two tropical populations. Nature, *189:* 897–899.

Chalmers, J. N. M., E. W. Ikin and A. E. Mourant 1953 The ABO, MNS, and Rh blood groups of the Nigerians. Ann. Eugenics, *17:* 168–176.

Garlick, J. P. 1957 Personal communication quoted by A. E. Mourant, A. C. Kopec and K. Domaniewska-Sobczak 1958 The ABO Blood Groups: Comprehensive Tables and Maps of World Distribution. Blackwell Scientific Publications, Oxford.

Garlick, J. P., and N. A. Barnicot 1957 Blood groups and hemoglobin variants in Nigerian (Yoruba) schoolchildren. Ann. Hum. Gen., *21:* 420–425.

Greenberg, J. H. 1949 Studies in African linguistic classification. II. The classification of Fulani. Southwestern J. Anthrop., *5:* 190–198.

Jacquemin, P. 1952 Personal communication quoted by A. E. Mourant, 1954. The Distribution of the Human Blood Groups. Blackwell Scientific Publications, Oxford.

Koerber, R. 1947 Contribution à l'étude de la répartition des groupes sanguins chez quelques races de l'Afrique Occidentale Française. C. R. Soc. Biol., *141:* 1013–1015.

Koerber, R., and J. Linhard 1951 Note sur la répartition des groupes sanguins AB, A, B, et O, dans divers groupes ethniques de l'Afrique Occidentale Francaise. Bull. Soc. Anthrop., Paris, *2:* 158–160.

Kossovitch, N. 1934 Recherches Séro-Anthropologiques chez quelques peuples du Sahara Français. C. R. Biol., *116:* 759–761.

Lestrange, M. de 1949 Contribution à l'étude de l'Anthropologie des Noirs d'A.O.F.; 1) Physiologie et pathologie des Coniaqui, Bassari, Badyaranké, et Peuls de la Région de Youkounkoun (Guinée Française). Bull. Soc. Anthrop., Paris, *10:* 156–165.

Messerlin, A., and Y. Lohro 1951 Groupes sanguins et facteur rhésus au Maroc. Bull. Inst. Hyg. Maroc., n.s. *11:* 97–119.

Moullec, J., and J. Linhard 1955 2nd Congres Nat. Transfus. Sang, Bordeaux, pp. 114–118.

Mourant, A. E. 1949 Rh phenotypes and Fisher's CDE notation. Nature, *163:* 913.

Muller, H. R. 1926–1927 Blood groups among the Yoruba Tribe of West African Negroes. Proc. Soc. Exp. Biol., *24:* 437–438.

St. Croix, F. W. de 1945 The Fulani of Northern Nigeria. The Govt. Printer, Lagos, 74 pp.

Steinberg, A. G., R. Stauffer, B. S. Blumberg and H. Fudenberg 1961 Gm phenotypes and genotypes in U. S. Whites and Negroes; in American Indians and Eskimos; in Africans; and in Micronesians. Am. J. Human Gen., *13:* 205–213.

Trincão, C. 1953 Grupos sanguineos dos indigenas da Guiné Portuguesa. Prim. Coloquio Hemat. Afric., Lisbon, pp. 65–67.

Verger, P. 1955 Yoruba influences in Brazil. ODU, J. of Yoruba and related studies, *1:* 3–11.

(*Reprinted from Nature, Vol.* 181, *pp.* 1551–1552, *May* 31, 1958)

Distribution of Electrophoretically Different Hæmoglobins among Some Cattle Breeds of Europe and Africa

Two hæmoglobins, *A* and *B*, can be distinguished in the blood of cattle by means of paper electrophoresis[1,2]. It has been shown that the presence or absence of the hæmoglobins is determined by two allelic autosomal genes which are fully expressed in the heterozygote[1]. The three phenotypes, bovine *A* (slow moving in electrophoresis), bovine *B* (fast moving component in electrophoresis), and bovine *AB* (both hæmoglobins present), have been found. In a previous study[1], it was shown that among British breeds, only Jerseys, Guernseys and South Devons possessed the *B* hæmoglobin.

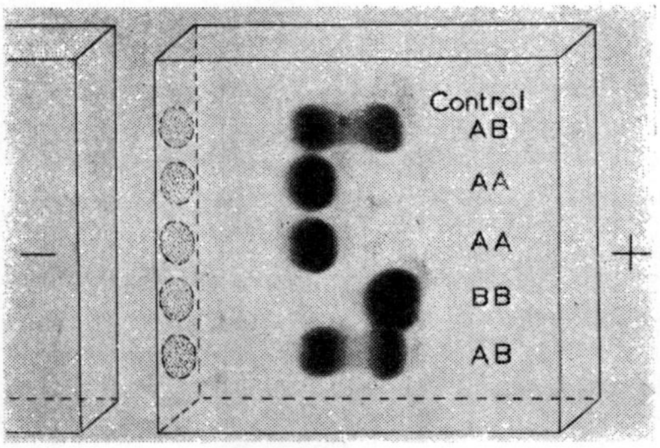

Fig. 1

One of the suggested ancestral lines of the Jersey breed was outlined by Boston[3] on the basis of archæological and other evidence, and led from the ancient Indus Valley civilization through Africa to Europe. Using the presence of bovine *B* hæmoglobin in a breed of cattle as a marker of common ancestry with Jersey cattle, as many breeds as possible were examined which were living on or near this geographical line.

Ford[4] has defined polymorphism as ". . . the occurrence together in the same habitat of two or more discontinuous forms of a species in such proportions that the rarest of them cannot be maintained merely by recurrent mutation". The existence in certain cattle breeds of these two forms of hæmoglobin fits within this definition and suggests that

Table 1. The Distribution of Hæmoglobin Genotypes in the Polymorphic European and African Cattle Breeds

Location	Breed	Genotype		
		AA	*AB*	*BB*
Jersey	Jersey	32	77	73
N. France	Brittany Shorthorn	0	2	0
N. France	Flamande	8	1	0
Mid France	Limousine	13	6	0
Mid France	Charolais	9	1	0
Mid France	Parthenaise	10	2	0
S. France	Salers and Aubrac	10	5	0
S. France	Garronaise	2	1	0
S.E. France	Tarentaise	9	1	0
S.E. France Switzerland	Brown Swiss Brown Swiss	24	9	0
W. Africa : Nigeria	Zebu (White Fulani)*	30	17	2
E. Africa : Sudan	Zebu*	13	29	10

* Low tolerance to trypanosomiasis infection.

some selective factors and especially factors favouring the heterozygote may be operating under natural conditions or under those of man-controlled breeding. No such selective factors have yet been confirmed, although in the present study a possible association with lack of tolerance to trypanosomiasis in some African breeds has been noted.

The hæmoglobins were detected by a modification of the starch gel electrophoresis method of Smithies[5]. A drop of blood, collected from the jugular vein or by ear puncture, was placed on a 2-in. circle of Whatman filter paper, No. 1 or 3, and, when dry, forwarded by air to the laboratory in England. On arrival, 5 mm. circles of the dried blood were punched out of the filter paper and inserted into the starting slit of the starch gel. Borate buffer *p*H 8·0, ionic strength 0·16, was used for the preparation of the gel and borate buffer *p*H 8·6 was used in the electrode bath solution. Readings were made 4 hr. later, and provided that a control sample of polymorphic hæmoglobin was included in all runs, excellent identification of hæmoglobin was possible on dried samples up to one month old (Fig. 1). Identifications using this method were checked against determinations on whole blood using paper electrophoresis[1] and were found to be consistent.

In Table 1, it will be seen that bovine *B* hæmoglobin is present in many European herds, but its frequency is low compared with that found in Jersey Island Cattle. The high hæmoglobin *B* frequency again appears in breeds of Zebu stock.

The possible presence of selective factors may affect the use of the hæmoglobins as simple genetic markers of breed inter-relationships. However, the presence of the *B* gene in breeds located geographically on a

Table 2. Numbers of Animals examined in Breeds exclusively Bovine Hæmoglobin *A*

Location	Breed	No.
Europe		
Denmark	Danish Red	21
Holland	Friesians	35*
N. France	Normande	29
Africa		
Nigeria	Muturu†	50
Nigeria	N'Dama†	50

* From the table published by Bangham (ref. 1).
† High tolerance to trypanosomiasis infection.

line running southwards through France, and again in breeds in West and East Africa, is consistent with at least one of Boston's suggested[5] geneological lines of the Jersey breed. As it is probable that hæmoglobins bovine *A* and *B* are similar to the two described by Cabannes and Serain[2], there is yet further evidence of the *B* gene in Africa, although in this case it may have been imported into Algeria by Europeans through the medium of French breeds. On the other hand, and in view of the apparent mode of inheritance of the hæmoglobin genes, it is improbable that the cattle breeds of north-east France, Holland and Denmark have in recent times received any genes from the cattle of Jersey, central and southern France, or of cattle of Zebu stock (Table 2).

Although little is known of the nature or tolerance of cattle breeds to trypanosomal infection, there is evidence that N'Dama and Muturu[6] have a greater tolerance to the disease than do the White Fulani (Zebu) cattle[7]. This receives further support from the quantitative studies of Desowitz[8], on the immune reactions of N'Dama and Zebu breeds to trypanosomal infections.

The present study shows a *prima facie* correlation between the absence of bovine *B* hæmoglobin and relatively high tolerance to trypanosomiasis (see Tables 1 and 2) and assumes some interest when related to the work of Allison on the presence of sickle cell hæmoglobin and resistance to falciparum malaria in man[9]. In order to determine if this is more than coincidence, it will be necessary to study, first, other breeds in the same environment; second, the relationship between the hæmoglobin phenotypes and tolerance to disease in members of the African breed; and third, breeds presumably related to the Zebu in areas remote from trypanosomiasis exposure. It has recently been shown that Nigerian White Fulani (Zebu) cattle are polymorphic for the genetically determined milk proteins *B*-lactoglobulin *A* and *B*, as are some European breeds, and that they are the only breed so far studied to have two types of α-lactalbumin, which may also be genetically determined[10]. A parallel study of the distribution of these proteins may help in the problems outlined.

It is with pleasure that we acknowledge the help given by Mr. N. le Q. Blampied, Mr. C. D. Cooke, Dr. M. Theret, A. A. Karib, H. K. Littlewood, A. G. Wilder and Dr. Hans Fey. We would also like to thank those breeders and owners who so readily offered material for this survey ; for the loyal and patient technical help of A. Bernard ; and for the invaluable help of O. Oyefeso in the field work in Nigeria. Prof. B. G. T. Elmes, University College, Ibadan, was instrumental in arranging for the collections in Nigeria. The investigations in Nigeria were supported in part by a research grant from the Wenner–Gren Foundation for Anthropological Research and by a Research Grant, *H*–3161, from the National Heart Institute, U.S. Public Health Service. A grant was made to one of us (A. D. B.) by the Agricultural Research Council for visits to France.

A. D. BANGHAM

Agricultural Research Council
Institute of Animal Physiology,
Babraham Hall, Babraham, Cambridge.

B. S. BLUMBERG

National Institute of Arthritis and Metabolic Diseases,
National Institutes of Health,
U.S. Public Health Service,
Bethesda, Maryland.

1 Bangham, A. D., *Nature*, **179**, 467 (1957).
2 Cabannes, R., and Serain, C., *C.R. Soc. Biol., Paris*, **149**, 7 (1955).
3 Boston, E. J., "Jersey Cattle", 32 (Faber and Faber, London, 1954).
4 Ford, E. B., "Genetics for Medical Students", 101 (Methuen, London, 1956).
5 Smithies, O., *Biochem. J.*, **61**, 629 (1955).
6 Gates, G. M., *Farm and Forest*, **11**, 19 (1952).
7 Chandler, R. L., *Ann. Trop. Med. Parasit.*, **46**, 127 (1952).
8 Nash, T. A. M., Ann. Rep. West African Institute for Trypanosomiasis Research, 20 (Gasakiya Corp., Zaria (Nigeria), 1955).
9 Allison, A. C., *Brit. Med. J.*, i, 290 (1954).
10 Blumberg, B. S., and Tombs, M. P., *Nature*, **181**, 683 (1958).

Printed in Great Britain by Fisher, Knight & Co., Ltd., St. Albans.

Reprinted from THE AMERICAN JOURNAL OF PHYSICAL ANTHROPOLOGY
N.S. Vol. 17, No. 3, September 1959

Blood Groups of Alaskan Eskimos and Indians

PATRICIA A. CORCORAN, F. H. ALLEN, JR., A. C. ALLISON AND B. S. BLUMBERG
The Blood Grouping Laboratory of Boston,[1] National Institute for Medical Research, London, and National Institute of Arthritis and Metabolic Diseases, National Institutes of Health, Bethesda, Md.

During the course of recent field work in Alaska, blood specimens were collected from Eskimos and Indians for blood grouping and other investigations. This paper records the results of blood group tests and provides a basis for comparison with findings in other groups of indigenous Americans. Other investigations of genetically determined metabolic traits carried out on the same populations included hemoglobin and haptoglobin analyses (Blumberg et al., '59), urinary amino acid excretion patterns (Allison et al., '59), and taste-testing with phenylthiocarbamide (Allison and Blumberg, '59). A full discussion of the anthropological significance of these findings will be presented later.

Populations from which samples were taken

The Eskimos were all northern dialect (Inupik) speakers (fig. 1). The first group consisted of 111 people out of 220 living in Wainwright; the second of 64 not-closely-related inhabitants of the large town of Barrow; the third of 55 people out of a total of approximately 70 now living near Summit Lake in Anaktuvuk Pass in the Brooks Range. Of the 70, approximately 10 were three years or younger from whom no attempt was made to remove blood. The fourth group consisted of 11 full-blooded Eskimos derived from the

[1] Supported in part by grant G-3524 from the National Institutes of Health.

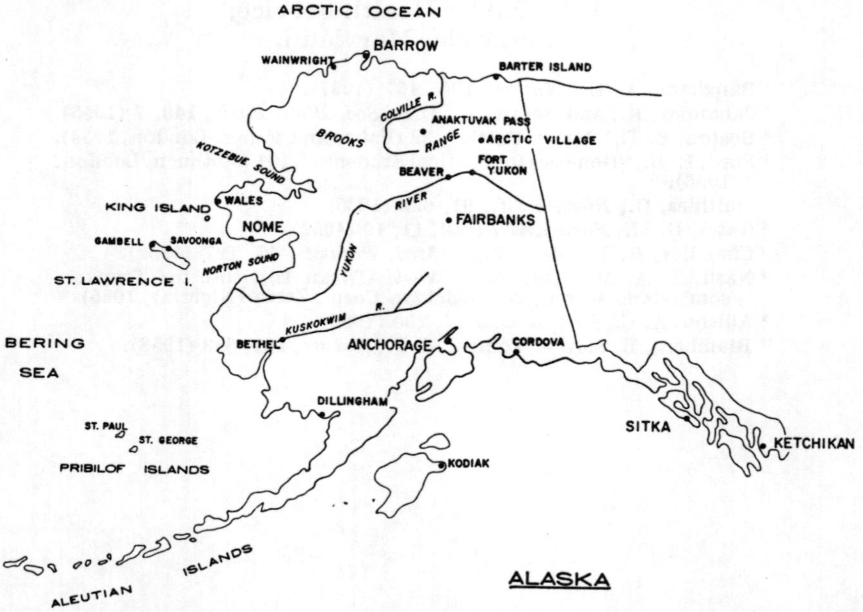

Fig. 1 Outline map of Alaska showing the locations at which blood collections were made.

Arctic slope now living in the mixed Eskimo-Indian community of Beaver on the Yukon River.

The northern Athabascan speaking Indians included 78 out of a total of approximately 110 from Arctic Village, Brooks Range, 110 from Fort Yukon and 18 from Beaver. The majority of the Aborigines tested were recorded as full blooded in the census records, and were regarded as being so in their communities. However, in Wainwright and Fort Yukon approximately 15% were listed as more than half, but less than full blooded. Tlingits were obtained from Mt. Edgecombe and Sitka, all but 5 of them being recorded as full blooded and the remainder as 3/4 or 7/8 Indian.

In some of the communities (those of Wainwright, Anaktuvuk Pass, Arctic Village and Beaver) a large proportion of the population was sampled, and the results include several members of the same sibship. Exogamy is common, and only one instance of a first-cousin marriage appeared in kinships from Wainwright, Anaktuvuk, Arctic Village, and Beaver. The findings in these various villages are sufficiently different one from another that it was considered advisable to list them separately (tables 1–4).

METHODS OF STUDY

Blood specimens were collected by venepuncture using Vacutainers without anticoagulant (Becton, Dickinson and Company, Rutherford, N. J.). Specimens were kept at 4°C and dispatched to the Blood Grouping Laboratory, Boston, within 80 hours of collection. Travelling time from Alaska to Boston was approximately 48 hours for each lot except for the first 49 samples from the Tlingits which were delayed in transit. All samples except those of 49 of the Tlingits arrived in Boston in excellent condition. Those 49 Tlingit bloods were in relatively poor condition, and gave unsatisfactory results with the P and Lewis serums.

Reagents and the methods of their use are outlined in table 5. Anti-A_1 was used only on A and AB bloods. Anti-D plus D^u and anti-G were used only on the D-negative. Anti-rh_1 was used only with C-positives, anti-k with K-positives. Tests were deemed satisfactory with the exception of the tests for P and Lewis in the first 49 Tlingits. The tests for P and Lewis on the Beavers were omitted because of lack of time. Anti-Vel was tested only with O's since the serum was from a type O donor.

TABLE 1

Phenotype and gene frequencies—A-B-O groups of Alaskans

	Alaskan Eskimos				Alaskan Indians			
	Wainwright	Barrow	Anaktuvuk Pass	Beaver	Arctic Village	Fort Yukon	Beaver	Tlingit
Phenotypes								
O	33 *.30*	17 *.27*	9 *.16*	6 .55	74 *.95*	92 *.84*	15 *.83*	63 *.80*
A_1	58 *.52*	42 *.66*	29 *.53*	5 *.45*	4 *.05*	16 *.14*	3 *.17*	11 *.14*
A_2	1 *.01*	1 *.02*	0	0	0	2 *.02*	0	0
B	13 *.12*	4 *.06*	12 *.22*	0	0	0	0	4 *.05*
A_1B	6 *.05*	0	5 *.09*	0	0	0	0	1 *.01*
Total	111	64	55	11	78	110	18	79
Genes								
O	*.55*	*.54*	*.44*	—	*.97*	*.915*	—	*.89*
A_1	*.35*	*.41*	*.39*	—	*.03*	*.075*	—	*.08*
A_2	*.005*	*.02*	*.00*	—	*.00*	*.01*	—	*.00*
B	*.09*	*.03*	*.17*	—	*.00*	*.00*	—	*.03*

TABLE 2

Phenotype and gene frequencies—Rh groups of Alaskans

	Alaskan Eskimos				Alaskan Indians			
	Wainwright	Barrow	Anaktuvuk Pass	Beaver	Arctic Village	Fort Yukon	Beaver	Tlingit
Phenotypes								
Rh_2rh	1 *.01*	3 *.05*	0	1 *.09*	7 *.09*	11 *.10*	0	6 *.08*
Rh_2Rh_2	24 *.22*	19 *.30*	15 *.27*	4 *.36*	23 *.295*	44 *.40*	6 *.33*	37 *.47*
Rh_1Rh_2	65 *.59*	28 *.44*	30 *.55*	4 *.36*	34 *.44*	38 *.345*	10 *.56*	23 *.29*
Rh_1Rh_1	20 *.18*	11 *.17*	10 *.18*	2 *.18*	7 *.09*	9 *.08*	1 *.06*	7 *.09*
Rh_1rh	1 *.01*	3 *.05*	0	0	1 *.01*	6 *.055*	0	3 *.04*
Rh_2Rh_z	0	0	0	0	3 *.04*	0	1 *.06*	2 *.03*
Rh_1Rh_z	0	0	0	0	2 *.03*	1 *.01*	0	0
rh	0	0	0	0	1 *.01*	0	0	0
rh″rh	0	0	0	0	0	0	0	1 *.01*
$Rh_1(D^u)rh$	0	0	0	0	0	1 *.01*	0	0
Totals	111	64	55	11	78	110	18	79
Genes								
R^2	.51	.54	.55	—	.58	.62	—	.60
R^1	.48	.41	.45	—	.33	.29[1]	—	.25
r	.01	.05	.00	—	.06	.08	—	.063
R^z	.00	.00	.00	—	.03	.0045	—	.013
r''	.00	.00	.00	—	.00	.00	—	.066

[1] Includes the one known D^u variant.

TABLE 3

Phenotype and gene frequencies—M-N-S groups of Alaskans

	Alaskan Eskimos				Alaskan Indians			
	Wainwright	Barrow	Anaktuvuk Pass	Beaver	Arctic Village	Fort Yukon	Beaver	Tlingit
Phenotypes								
M, S−	49 *.44*	25 *.39*	42 *.76*	5 *.45*	56 *.72*	47 *.43*	6 *.33*	42 *.53*
M, S+	24 *.22*	14 *.22*	1 *.02*	1 *.09*	16 *.205*	23 *.21*	10 *.56*	15 *.19*
MN, S−	29 *.26*	19 *.30*	10 *.18*	3 *.27*	5 *.06*	18 *.16*	0	15 *.19*
MN, S+	8 *.07*	5 *.08*	2 *.04*	1 *.09*	1 *.01*	22 *.20*	1 *.06*	7 *.09*
N, S−	1 *.01*	1 *.02*	0	0	0	0	0	0
N, S+	0	0	0	1 *.09*	0	0	1 *.06*	0
Totals	111	64	55	11	78	110	18	79
Genes								
M	.82	.80	.89	—	.96	.82	—	.86
N	.18	.20	.11	—	.038	.18	—	.14
S	.16	.16	.03	—	.12	.23	—	.15
Ms	.66	.63	.87	—	.85	.65	—	.73
MS	.16	.17	.02	—	.115	.16	—	.13
Ns	.18	.20	.10	—	.037	.115	—	.12
NS	.00	.00	.01	Present	.001	.067	Present	.02

Calculations of gene frequencies

O was obtained from the square root of the frequency of type O. Similarly, the frequencies for *Ms* were obtained by taking the square root of the frequency of phenotype M,S−. Frequencies for all the Rh genes, and for *M*, *N*, and the genes of the Kell system were obtained by direct count. For this purpose individuals of phenotype Rh_2Rh_2, for example, were considered to be of genotype R^2R^2, etc. Since no R^0, r', or r^y genes were identified, it is impossible to guess how many (if any) of the supposed *r*, R^1, and R^z genes were really R^0, r', or r^y. It is certain that none of these genes are common in this population.

Frequencies for *A*, A_1, *B*, *S*, Fy^a, Jk^a, *P*, and Di^a were obtained from the formula:

$$\text{Frequency X} = 1 - \sqrt{\text{frequency of X-negative.}}$$

The frequency of *MS* was obtained by subtracting *Ms* from *M*. The frequency of *NS*

TABLE 4

Phenotype and gene frequencies—other blood groups in Alaskans

	Alaskan Eskimos				Alaskan Indians			
	Wainwright	Barrow	Anaktuvuk Pass	Beaver	Arctic Village	Fort Yukon	Beaver	Tlingit
Phenotypes								
Fy(a+)	106 .955	64 1.00	55 1.00	11 1.00	77 .99	110 1.00	18 1.00	79 1.00
Fy(a−)	5 .045	0	0	0	1 .013	0	0	0
Jk(a+)	78 .70	43 .67	39 .71	10 .91	60 .77	79 .72	14 .78	62 .78
Jk(a−)	33 .30	21 .33	16 .29	1 .09	18 .23	31 .28	4 .22	17 .22
P+	19 .17	24 .38	20 .36	*	14 .18	33 .33	*	9 .30
P−	92 .83	40 .62	35 .64	*	63 .82	66 .67	*	21 .70
Le(a+b−)	0	0	0	*	0	0	*	1 .03
Le(a−b+)	109 .98	61 .95	55 1.00	*	73 .95	94 .95	*	25 .83
Le(a−b−)	2 .02	3 .05	0	*	4 .052	5 .05	*	4 .13
K+Kp(a−)	0	0	0	0	0	·1 .009	0	1 .01
K−Kp(a+)	1 .01	0	0	0	0	0	0	0
K−Kp(a−)	110 .99	64 1.00	55 1.00	11 1.00	78 1.00	109 .99	18 1.00	78 .99
Di(a+)	2 .02	0	0	0	0	1 .009	0	0
Di(a−)	109 .98	64 . 1.00	55 1.00	11 1.00	78 1.00	109 .99	18 1.00	79 1.00
Genes								
Fy^a	.79	1.00	1.00	—	.89	1.00	—	1.00
Jk^a	.46	.43	.46	—	.52	.47	—	.54
P	.09	.21	.20	—	.10	.18	—	.16
K^b	.00	.00	.00	—	.00	.004	—	.01
k^a	.004	.00	.00	—	.00	.00	—	.00
k^b	.996	1.00	1.00	—	1.00	.996	—	.99
Di^a	.01	.00	.00	—	.00	.004	—	.00

* Not done

TABLE 5
Reagents used in testing Alaskans

Reagent	Lot no.	Source	Method of use
anti-A	SA 127	BGL	Saline tube test
anti-B	SB 116	BGL	Saline tube test
anti A_1	SAA 56	BGL	Saline tube test
anti-H	Ulex 38	BGL	Saline tube test
anti-C	SRC 77	BGL	Saline tube test
anti-C^w	ARCW 67	BGL	Albumin tube test
anti-D	ARD 38	BGL	Albumin tube test
anti-E	ARE 117	BGL	Albumin tube test
anti-E^w		T. J. Greenwalt	Indirect Coombs test
anti-c	AHC 67	BGL	Albumin tube test
anti-e	AHE 67	BGL	Albumin tube test
anti-D + D^u	CRDU 68	BGL	Indirect Coombs test
anti-G	ARDC 68	BGL	Albumin tube test
anti-f	Cla.	R. E. Rosenfield	Ficin technique
anti-rhi[1]	Bau	R. E. Rosenfield	Ficin technique
anti-M	SM 117	BGL	Saline tube test
anti-N	N15	Ortho	Saline tube test
anti-S	SFA 237	R. E. Rosenfield	Saline tube test
anti-M^g	SMG 97	BGL	Saline tube test
anti-Vw	SVW 58	BGL	Saline tube test
anti-Mi^a	UB 4	J. F. Mohn	Saline tube test
anti-K	CK 117	BGL	Indirect Coombs test
anti-k	M.N.	P. J. Schmidt	Indirect Coombs test
anti-Kp^a	CKPA 97B	BGL	Indirect Coombs test
anti-Kp^b	CKPB 116	BGL	Indirect Coombs test
anti-Fy^a	CFYA 97B	BGL	Indirect Coombs test
anti-Jk^a	CJKA 48	BGL	Indirect Coombs test
anti-Jk^a	Mc C.	BGL	Indirect Coombs test
anti-P	Daw.	BGL	Saline (cold)
anti-P	SP 27	BGL	Saline (cold)
anti-Le^a	SLEA 47	BGL	Saline (cold)
anti-Le^b	Mc G. 56	BGL	Saline (cold)
anti-Lu^a	Gin.	BGL	Saline tube test
anti-Di^a	Woj.	J. F. Mohn	Indirect Coombs test
anti-Di^a	—	M. Layrisse	Indirect Coombs test
anti-Wr^a	SWRA 18B	BGL	Saline tube test
anti-Be^a	—	K. Stern	Indirect Coombs test
anti-Vel	—	J. J. van Loghem	Saline tube test
anti-Js[2]	—	R. E. Rosenfield	Indirect Coombs test

[1] See Rosenfield, R. E., '58. [2] See Giblett, E. R., '58.

was obtained by subtracting *MS* from *S*. Similarly, $Ns = N - NS$. Expected phenotype frequencies, calculated from the gene frequencies as shown in the tables, fitted closely the observed frequencies. Gene frequencies for *A*, *B*, and *O* were corrected when necessary, using Bernstein's simple correction. In the Tlingit group, the frequency of r'' was derived from the formula

$$r'' = \sqrt{r^2 \text{ plus observed frequency type rh}''} - r,$$

which resulted in a calculated frequency of r'' of 0.066, indicating that about 9 of the 105 supposed R^2 genes in this population may actually be r''. The frequency of R^2 as counted was therefore reduced correspondingly.

RESULTS

Most of the data are contained in tables 1–4. In addition 239 Eskimos and 255 Indians were tested with anti-Mg, all with negative results. 241 Eskimos and 255 Indians were all negative with anti-Lu^a. 241 Eskimos and 223 Indians were all positive with anti-Kp^b. 241 Eskimos and 255 Indians were all negative with anti-C^w, anti-Wr^a, anti-Be^a, anti-Vw, anti-E^w, anti-Mi^a, and anti-Js, and all 309 type O's were Vel-positive.

COMMENT

The very low incidence of Di^a in the Alaskan Indians was the most interesting finding. This is the only American Indian population reported to date that has not had at least five per cent Di^a-positive. The northern Athabascans are thought to have been among the last of the Indians to arrive in America from Asia, some authorities dating their arrival after that of some of the Eskimo groups (Jenness, '41). The highest incidence of Di^a so far reported from America has been from South America Indians (Levine et al., '56) with lower frequencies in North American Indians (Lewis, Chown and Kaita, '56; Lewis et al., '56; Chown et al., '58; Allen et al., '58). Diego-positives are found also, though not in high frequency, in Chinese (Layrisse and Arends, '56) and Japanese (Lewis, Ayukawa, Chown and Levine, '56). The gene is also found in Burmese and Dyaks of Sarawak (Colbourne et al., '58), but not in Micronesians (Sussman et al., '58), Polynesians (Simmons, '57), New Zealand Macris (Lehmann, '58) or Australian Aborigines (Simmons, '57). It was also not found in Europeans (Levine et al., '56) nor Negroes (Junqueira and Wishart, '57). The present finding is consistent with the suggestion that the incidence is highest in Indians who have been separated from Asia for the longest time (Colbourne et al., '58). Further understanding of this problem must await studies on South American Indians and in particular on contemporary Asian groups with whom American Indians may have affinities. The low incidence in Alaskan Eskimos shown in the present study and the complete ababsence in Eastern Canadian Eskimos (Lewis, Chown and Kaita, '56) is also consistent with this explanation. The observed cline in Diego blood group incidence, if substantiated by subsequent studies, can be explained by one of the following: (a) separate origins for each of the groups, (b) varying degrees of intermixture with other populations, and (c) the operation of selection at different intensity in the different populations. There is not sufficient information to decide which of these explanations, if any, is applicable.

The absence of Js-positives indicates at least no large admixture of Negro genes which could conceivably have been introduced by the crews of New England whaling ships.

The results on the Anaktuvuk Pass Eskimos agree with those reported for ABO, MN, and Rh by Laughlin ('57) who did his blood group studies in the field. With the exception of the MN system, the gene frequencies of the Anaktuvuk group are similar to those of the coastal Eskimos suggesting closer genetic relation to them than to the Indian groups.

The results for both the Eskimo and Indian groups suggest that admixture with Europeans has not been extensive. This is indicated by the low frequency of B and Fy^b and the high incidence of Ms in both the Indians and Eskimos. Further comment should be postponed until publication of other studies from North America, particularly those of Steinberg and of Chown and Lewis.

ACKNOWLEDGMENTS

Collections of specimens in Alaska were facilitated by many individuals and agencies to whom we are indebted. In particular, Col. H. F. Currie, C. Murray and F. Milan of the Arctic Aeromedical Laboratory, Ladd Field; M. C. Brewer of the Arctic Research Laboratory, Barrow; Dr. Burns Jones of the Hudson Stuck Memorial Clinic, Fort Yukon; Drs. C. Dotter and C. Hopkins of the University of Oregon Medical School and Mr. Patrick O'Connell of the O'Connell Trading Post, Anaktuvuk Pass, Alaska. The participation of one of us (A.C.A.) in the field work in Alaska was made possible by a grant from the Wenner-Gren Foundation for Anthropological Research.

We are very grateful to Drs. Greenwalt, Layrisse, Mohn, Rosenfield, Schmidt, Stern, and van Loghem for gifts of various reagents as listed in table 5.

LITERATURE CITED

Allen, F. H. Jr., P. A. Corcoran and L. K. Diamond 1958 Blood Groups of Eastern American Indians. Prox. X Int. Cong. of Genetics, Vol. 2, p. 5.

Allison, A. C., and B. S. Blumberg 1959 The taste test in Alaskan Eskimos. Human Biology, in press.

Allison, A. C., B. S. Blumberg and S. M. Gartler 1959 Urinary β-aminoisobutyric acid excretion in Eskimo and Indian populations of Alaska. Nature, *183:* 118.

Blumberg, B. S., A. C. Allison and B. Garry 1959 The haptoglobins of Alaskan Eskimos and Indians. Annals of Human Genetics, in press.

Chown, B., M. Lewis and H. Kaita 1958 The Diego blood group system. Nature, *181:* 268.

Colbourne, M. J., E. W. Ikin, A. E. Mourant, H. Lehmann and H. Thein 1958 Haemoglobin E and the Diego blood group antigen in Sarawak and Burma. Ibid., *181:* 119–120.

Giblett, E. R. 1958 Js, a "new" blood group antigen found in Negroes. Ibid., *181:* 1221–1222.

Jenness, D. 1940 Prehistoric culture waves from Asia to America. Annual Report of the Board of Regents of the Smithsonian Institution. p. 383–396. U.S. Government Printing Office, Washington, 512 pp.

Junqueria, P. C., and P. J. Wishart 1957 Diego blood group in Negro. Nature, *180:* 341.

Laughlin, W. S. 1957 Blood groups of the Anaktuvuk Eskimos, Alaska. Anthropological Papers of the University of Alaska, *6:* 5–16.

Layrisse, M., and T. Arends 1956 The Diego blood factor in Chinese and Japanese. Nature, *177:* 1083–1084.

Lehmann, H., A. North and J. M. Staveley 1958 Absence of the Diego blood group and abnormal hemoglobins in 92 Maoris. Ibid., *181:* 791–792.

Levine, P., E. A. Robinson, M. Layrisse, T. Arends and R. Dominques Sisco 1956 The Diego blood factor. Ibid., *177:* 40–41.

Lewis, M., H. Ayukawa, B. Chown and P. Levine 1956 The blood group antigen Diego in North American Indians and in Japanese. Ibid., *177:* 1084.

Lewis, M., B. Chown and H. Kaita 1956 Further observations on the blood factor Di^a. Ibid., *178:* 1125.

Rosenfield, R. E. 1958 An Rh blood factor (Ce) and its relationship to hr (ce). Am. J. Human Genetics, *10:* 473–480.

Simmons, R. T. 1957 The Diego (Di^a) blood group: tests in some Pacific peoples. Nature, *179:* 970–971.

Sussman, L. N., L. H. Meyer and R. A. Conard 1959 Blood grouping in Marshallese. Science, *129:* 644–645.

Reprinted from
ARTHRITIS AND RHEUMATISM
Vol. 4, No. 4, **August, 1961**
Printed in U.S.A.

Arthritis and Rheumatism

Official Journal of the American Rheumatism Association

VOL. IV, NO. 4 AUGUST, 1961

A Study of the Prevalence of Arthritis in Alaskan Eskimos

By BARUCH S. BLUMBERG, KURT J. BLOCH, ROGER L. BLACK AND CHARLES DOTTER

Rheumatoid arthritis is present in an Alaskan Eskimo population. The prevalence of X-ray evidence of osteoarthritis of the hands and wrists has been determined in an Eskimo population. Comparison with available U. S. white data suggests that the prevalence may be lower in the Eskimo. There was a high prevalence of individuals in the village of Wainwright whose sera yielded a positive bentonite flocculation test.

Arthritis rheumatoide es presente in un population de eschimos alaskan. Le prevalentia de manifestationes roentgenologic de osteoarthritis del manos e carpos esseva determinate in un population de eschimos. Comparation con disponibile datos pro le population statounitese de racia blanc suggere que le prevalentia es possibilemente plus basse in le eschimos. Esseva trovate in le village de Wainwright un alte prevalentia de individuos con seros que reageva positivemente in tests de flocculation a bentonite.

THERE IS AN increasing interest in determining the prevalence of the arthritides in different populations, both as a measure of the disease load these afflictions impose on the population and as a clue to their etiology and pathogenesis. The value of "geographical" studies in arthritis has been discussed in a recent publication.[1] Attempts at epidemiological investigations prior to 1950 were hampered by the absence of accepted diagnostic criteria, and many of these early studies are of limited value because their nomenclature of diseases does not correspond to modern usage. In the past few years several developments have made possible an estimation of the prevalence and incidence of arthritic diseases. These developments include: (1) establishment by the American Rheumatism Association of criteria for the diagnosis of rheumatoid arthritis,[2] (2) use of serologic tests for rheumatoid factor as an aid in the diagnosis of rheumatoid arthritis (as a result of the present and other investigations, it is known that these tests cannot be used alone as an indication of the prevalence of the disease); and (3) use of the methods for population studies of rheumatoid arthritis that were established by Cobb and his co-workers,[3] Kellgren and Lawrence,[4] Miall,[5] and others.[6]

During a recent field trip to Northern Alaska in which two of us took part (B. S. B. and C. D.), opportunity arose to collect data that could be used for

From the National Institute of Arthritis and Metabolic Diseases, Public Health Service, Bethesda, Md.

a preliminary study of the prevalence of rheumatoid arthritis and for an estimation of the prevalence of osteoarthritis in an Eskimo population. Studies on the distribution of positive reactions in serological tests used in rheumatoid arthritis were also made.

The results reported in the present paper are arranged under three major headings: (1) the prevalence of osteoarthritis determined from radiographs of hands and wrists in North Alaskan Eskimos and Southwest Alaskan Eskimos; (2) the prevalence of individuals with positive bentonite flocculation tests in Alaskan communities; and (3) the prevalence of rheumatoid arthritis in the North Alaskan Eskimo community of Wainwright obtained from history, physical examinations, x-rays, and serology. Reports on the prevalence of cardiovascular disease[7] and the distribution of some inherited biochemical traits[8-11] have been prepared separately.

Populations Studied

There are three major Eskimo groups in the Arctic which are distinguished from each other by linguistic, cultural, and anthropological differences (fig. 1). North of Norton Sound and extending along the north coast of Alaska, northern Canada, the Labrador, and Greenland are the Northern (Inupik) Eskimos. South of Norton Sound, and along the western edge of Alaska are the Southern (Yupik) Eskimos who have affinities with the Eskimos of St. Lawrence Island and Eastern Siberia. West of approximately the 160th meridian west longitude

PRINCIPAL ETHNOGRAPHIC DIVISIONS OF ALASKA

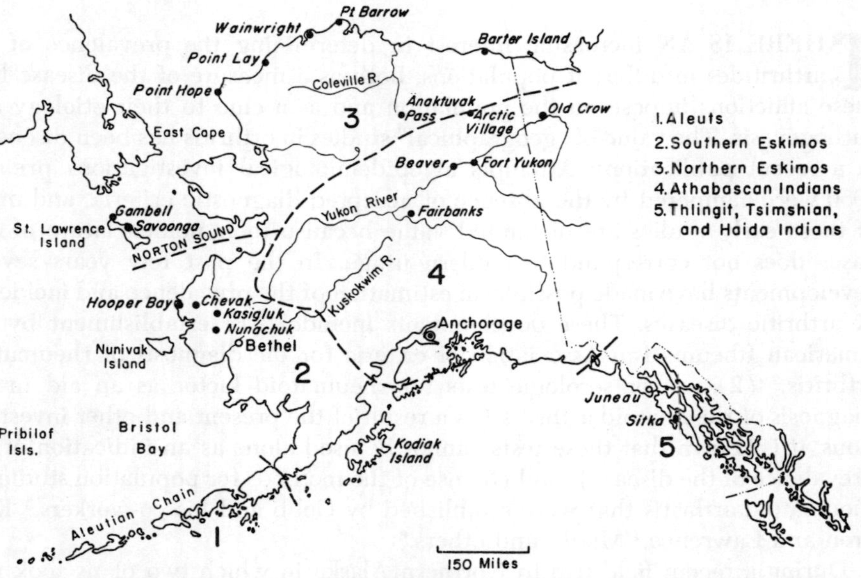

Fig. 1.—An ethnographic map of Alaska showing the location of communities mentioned in this article.

Table 1.—*Studies Completed in Wainwright, Alaska*

Population of Village at Time of Visit	219
Joint Examination	208
History and Physical Examination	219
Bentonite Flocculation Test	125
Radiograms of Hands, Wrists	167

is the ancient home of the Aleutians, although many of them now live elsewhere in Alaska and in the U.S.S.R.[12] In addition to the Eskimos there are Indians of various tribal affinities living in the east and southeast of the country. These include the Kuchin and kindred Athabascan-speaking people, the Haida, Tlingit, and others. The majority of the studies reported in the present paper were done on the Northern Alaskan and Southern Alaskan Eskimos, although sera from Indian groups was also used.

Northern Eskimos

The most complete survey was done in the North Eskimo village of Wainwright, a community of approximately 225 inhabitants situated on bluffs overlooking the Arctic Ocean (table 1).[13] The housing consists of frame structures, native-built homes made of driftwood, whalebone, boxwood and other manufactured building materials, and several turf igloos. The rooms in all the habitations are small and crowded. It has been stated that the Eskimos live in a warm microclimate, since they dress in warm clothing, leaving only their face and hands exposed, and keep their homes adequately heated when fuel is available. The Wainwright people eat about 70 per cent of a native diet consisting primarily of the meat of caribou, whale, seal, walrus and fish. This is supplemented with food bought at the Eskimo cooperative store and includes tea, coffee, tinned milk, fruit, and flour for making bread, sourdough pancakes, and biscuits. An extensive study of the nutrition of Alaskan Eskimos has recently been completed.[14] The major occupations of the men are hunting and fishing; the women are occupied in caring for the children, preparing the food and skins, and general housework.*

Southwestern Eskimos

Only radiograms and sera were available from this group (see below). The villages, (mainly in the Kuskokwim and Bethel regions) in which x-rays were taken, are shown in table 2 (Group B) and in figure 1. Serum was collected from members of the Eskimo Scout Batallion of the Alaska National Guard during their 1958 encampment (table 3) and from villagers on St. Lawrence Island.†

U. S. White Population

We were unable to locate any published data for comparison with the x-ray results. Through the kindness of Dr. Alice Waterhouse of the Division of

*We are indebted to Mr. F. Milan of the Arctic Acromedical Laboratory, Fairbanks, for the anthropological information on Wainwright.[13]

†We are indebted to Drs. E. N. Scott and K. E. Reinhard of the Arctic Health Research Center for the sera and to Dr. Scott for providing the radiographs.

Table 2.—*X-rays of Hands and Wrists Used in Osteoarthritis Survey*

Population Group	Village	No.
A. Eskimo, North Alaska	Wainwright	167
"	Pt. Lay and Pt. Hope	144
Total Eskimo, North Alaska		311
B. Eskimo, Southwest Alaska	Ninapitchuk	41
"	Savoonga, St. Lawrence Is.	40
"	Kasilgluk	16
"	Nunachuk	2
"	Chevak	12
"	Hooper Bay	35
Total Eskimo, Southwest Alaska		146
Total Eskimo, Alaska		457
C. Whites, U.S.A.	Washington, D. C.	77
"	Fort Wayne, Indiana	189
"	Cresco, Iowa	170
Total Whites, U.S.A.		436
Total, All X-rays		893

Table 3.—*Sera Used for the Bentonite Flocculation and Other Tests (See Text). All Individuals Were Not Hospitalized*

Population Group	Village	No.
A. Eskimo, North Alaska	Wainwright	125
"	Barrow	61
"	Anaktuvuk Pass	56
"	Beaver	14
B. Eskimo, Southwest Alaska	St. Lawrence Is.	86
Eskimo, Southwest and North Alaska	Alaska National Guard	102
Total Eskimo		444
Total Eskimo Excluding Wainwright		319
B'. Indian, Athabascan, Alaska	Arctic Village	78
"	Fort Yukon	91
"	Beaver	13
Total Indian		182
Total Alaska		626
Total Alaska Excluding Wainwright		501
C. Total U. S. White		545
Total Sera Studied		1171

Public Health Methods, United States Public Health Service, radiographs of hands and wrists obtained from a U. S. white population were made available. In two communities (Fort Wayne, Indiana, and Cresco, Idaho) the subjects were selected randomly by methods to be described elsewhere.[20] In a third community (Washington, D. C.) casual selection was used. Since the U. S. white and Eskimo samples are not comparable because of differences in method of selection and other factors, no direct statistical comparison can be made. However, the results of both surveys arranged by age and sex are given in tables 7 and 8 for approximate comparison. Kellgren and Lawrence,[22] who surveyed a British population and used methods not directly comparable to those of the present study, found the prevalence of osteoarthritis to be slightly higher (males 83 per cent, females 87 per cent) than in the U. S. white population cited in the present study. The studies of Kellgren and Lawrence were restricted to the 55–64 year age group. The sera from the U. S. white population used for the bentonite flocculation test were obtained from laboratory personnel, blood bank donors in Baltimore and Bethesda, Maryland, employees of the National Institutes of Health, and the parental members of the families participating in the "Family Health Maintenance Demonstration" Program at Montefiore Hospital, New York.* All these individuals were not hospitalized and were presumably healthy.

Methods

Interviews and Physical Examinations

These were done only in the village of Wainwright (table 1). A standard medical history was taken of each individual with the aid of a local interpreter. The histories were deemed satisfactory for the detection of gross illness, but could not be relied upon to detect arthritis which was not accompanied by physical or X-ray evidence of deformity. Every subject was given a physical examination by one of the physician members of the survey team, and 208 of them were given an additional joint examination by a physician experienced in the diagnosis of rheumatic diseases.

Serologic Studies

Collection of blood.—Blood was collected in dry, sterile "Vacutainers" (Becton, Dickinson and Company, Rutherford, New Jersey), kept at room temperature for several hours to permit clotting, and stored at approximately 4 C. The blood collected in Wainwright were hand-carried to Barrow where the serum and blood clots were separated, using laboratory facilities of the Arctic Research Laboratory.† The sera were frozen and sent by air express to Bethesda, where they were kept frozen until ready for use. Blood collected in the other Alaskan communities was handled in approximately the same manner (table 3). The bentonite flocculation test was performed on all sera listed in table 3. The sensitized sheep cell agglutination test and serological test for syphilis were performed on all sera collected in Wainwright which yielded a positive bentonite flocculation test.

Bentonite Flocculation Test (BFT).—The BFT was performed according to the method of Bozicevich et al.[15] The same gamma globulin preparation or preparations with equivalent reactivity were used for all tests reported in this paper.

Sensitized sheep cell agglutination test (SSCA).—The SSCA was performed according to the modification developed by Williams.[16] The BFT and SSCA on all serum samples in this study were done by one of the authors (K. J. B.).

Venereal disease research laboratory slide flocculation test.—This test was carried out in the serology laboratory of the Clinical Pathology Department, NIH, according to a standard method.[17]

Effect of freezing and thawing.—Previous experiments had indicated that freezing and thawing of serum did not interfere with the BFT[18] or SSCA.[19] However, since several of the Wainwright sera were positive in low titer in the BFT and negative in the SSCA test, the effect of freezing and thawing on sera with low titers was again tested. Results of this experiment are presented in table 4 and indicate that repeated (five times) freezing and thawing does not destroy sensitized sheep cell agglutinins when present in low titer. The

---

*We are indebted to Dr. George A. Silver for his cooperation in securing the last group of sera.

†We are indebted to the Office of Naval Research and to the University of Alaska for permission to use the facilities of the Arctic Research Laboratory, and to Mr. M. C. Brewer, director, for his help and cordial hospitality.

Table 4.—*The Effect on the Agglutination Titer of Freezing and Thawing (Five Times) Sera Which Were Positive for the BFT and SSCA Tests. Five Sera With No Agglutination Titers Remained Negative after the Same Treatment*

	Before Freezing		After Freezing and Thawing Five Times	
Serum	BFT	SCA	BFT	SCA
A	1:128	1:32	1:256	1:64
B	1:512	1:512	1:8192	1:1024
C	1:128	1:32	1:256	1:64
D	1:512	qns	1:2048	1:128
E	1:512	1:16	1:512	1:32

table also indicates that the BFT titers generally tend to be higher than the corresponding SSCA titers.

Radiographs

Sources.—The various population groups that were examined radiologically are listed in table 2. The groups in Wainwright and Point Lay-Point Hope represent 76 per cent and 50 per cent respectively of the population of the villages at the time of the visit.

Methods of taking and reading radiograms.—The radiograms from Wainwright, Point Hope and Point Lay were taken with a Picker Field Portable unit rated at 30 milliamperes on Eastman Blue Brand film. Electrical generators available at the school house were used for power. Some of the processing was done in the field and some at the University of Oregon Medical School. It was found that sea water was adequate for washing the radiograms (although a final rinsing in fresh water was necessary) and had the advantage of being usable in freezing weather. The radiograms of the Southwestern Eskimos were taken with similar equipment. The radiograms obtained in the United States (non-Alaskan) survey were made with a General Electric mobile unit rated at 200 milliamperes on Ansco film in cardboard film holders. All the radiograms of the hands and wrists were examined by one observer (R.L.B.). The degree of osteoarthritis, when found, was graded according to the method of Kellgren and Lawrence[21] on a scale of 1 to 4. The features listed by these observers as evidence of osteoarthritis ("osteo-arthrosis") are as follows: (a) formation of osteophytes on joint margins; (b) periarticular ossicles; (c) narrowing of cartilage associated with sclerosis of subchondral bone; (d) small pseudocystic areas with sclerotic walls, usually in the subchondral areas; and (e) alteration of the bone ends. The paper[21] contains excellent detailed photographs illustrating the features of each grade of disease for the small joints of the hands and also for the wrists. For the grading of our films, only changes of grade 2 or greater were considered osteoarthritis and only those with osteoarthritis of 2 or more joints were considered positive. The score of the most severely involved joint was utilized as the score for that patient. Radiographs of hands and wrists permit a survey of but 30 joints in each patient. Thus, there is the possibility of failing to discover disease in the spine and other peripheral joints. Kellgren and Lawrence[22] have shown in a survey of an English population (age 55–64) that there is a definite association between osteoarthritis ("osteo-arthrosis") of the distal interphalangeal joints (Heberden's nodes) and radiographic evidence of multiple osteoarthritis in the other small joints of the hands and feet, interfacetal joints of the spine, and knees. Since the distal interphalangeal joints of the hands are included in the present survey, there is reason to believe that this survey may provide an "index" of the prevalence of osteoarthritis elsewhere in the body. The observations of Kellgren and Lawrence have not been tested, however, in other age groups or in other populations, and we cannot be certain of the application of their results to our data. Hence, the present material must be considered as a comparative study of osteoarthritis of the hands and wrists only.

Criteria for radiologic evidence for rheumatoid arthritis included juxta-articular osteoporosis, cartilage destruction, bone destruction, and ankylosis.

Results

Serology

With the exception of the Wainwright group there was no significant difference in the prevalence of positive BFT's in the different Alaskan groups, so they were combined. In table 5 the Wainwright, other Alaskan, and U. S. populations are classified according to age. There is a slightly higher prevalence in the Alaska-minus-Wainwright group than in the U. S. group, but this is not significant. There is, however, a considerably higher prevalence in the Wainwright group in each decade, and in total, than in the U. S. white population. Since the methods of selection of the U. S. white and Eskimo sample were different, an exact comparison is not possible. It is unlikely, however, that the large difference in prevalence between Wainwright Eskimos and the U. S. white controls is due to differences in sample selection.

The prevalence of positive BFT's decreases in the over 60 age group in the three populations but in no case is this drop statistically significant.

Of the 124 sera from Wainwright, twelve yielded positive BFT's. These included eleven apparently normal individuals and a female with probable rheumatoid arthritis (see below). Sera obtained from five men and six women had BFT titers of 1:32 or 1:64 and one man had a titer of 1:128. All others had titers below 1:32, which is considered negative in the method used. The modified sensitized sheep cell agglutination test was negative in all the males and four of the females with a positive BFT. Two women had positive SSCA and BFT tests; one of these had probable rheumatoid arthritis (case 2, I. A., see below) and the other was apparently normal. (As noted below, blood was not obtained from M. A., case 1, who had definite rheumatoid arthritis).

Positive reactions in serologic tests for rheumatoid arthritis occur in diseases other than rheumatoid arthritis. In several diseases, including syphilis,[23] such positive reactions are obtained only with serologic tests employing pooled normal human gamma globulin (FII) but not with tests employing sheep cells sensitized with rabbit amboceptor. The VDRL slide flocculation test was negative in all of the 11 individuals with a positive BFT.

Table 5.—*The Prevalence of Positive BFT Sera in Wainwright, among other Alaskans, and in a United States White Population*

	Wainwright			Alaskans minus Wainwright			U.S. Whites		
Age	No.	No. BFT+	% BFT+	No.	No. BFT+	% BFT+	No.	No. BFT+	% BFT+
0–9	4	0	0	54	0	0	36	0	0
10–19	46	1	2.2	107	2	1.9	73	0	0
20–29	26	3	11.5	146	3	2.1	131	3	2.3
30–39	15	1	6.7	83	3	3.6	149	3	2.0
40–49	13	3	23.1	50	2	4.0	127	4	3.1
50–59	8	2	25.0	22	2	9.1	35	0	0
60+	12	1	8.3	39	1	2.6	3	0	0
Total	124	11	8.9	501	13	2.6	554	10	1.8

The subject with possible rheumatoid arthritis (case 2, I. A.) is omitted from the Wainwright column. There is a higher number of positive BFT sera in Wainwright compared to the U. S. White Population ($p < .001$).

Table 6.—*Wainwright Subjects with Positive BFT. BFT, Bentonite Flocculation Test; SSCA, Sensitized Sheep Cell Agglutination Test; STS, Serologic Test for Syphilis; TBC, Radiographic Evidence of Tuberculosis, Past or Present*

No.	Subject No.	Age	Sex	BFT	SSCA	STS	TBC
1.	301*	55	F	1:64	1:256	$\bar{0}$	+
2.	313	16	M	1:32	–	–	0
3.	343	54	M	1:128	$\bar{0}$	$\bar{0}$	+
4.	365	23	F	1:32	$\bar{0}$	$\bar{0}$	+
5.	386	52	M	1:32	$\bar{0}$	$\bar{0}$	0
6.	401	47	F	1:64	$\bar{0}$	$\bar{0}$	0
7.	406	46	F	1:32	$\bar{0}$	$\bar{0}$	+
8.	416	60	M	1:32	$\bar{0}$	$\bar{0}$	+
9.	801	46	M	1:64	$\bar{0}$	$\bar{0}$	0
10.	817	29	F	1:64	$\bar{0}$	$\bar{0}$	+
11.	826	33	M	1:32	$\bar{0}$	$\bar{0}$	+
12.	827	26	F	1:32	1:128	$\bar{0}$	+

*Subject I. A. (Case 2) with probable rheumatoid arthritis.

There is a high prevalence of tuberculosis among the Wainwright and other Eskimos.[24,25] Positive serologic tests for rheumatoid arthritis employing human gamma globulin have been reported in 5 of 100[26] and 4 of 18[27] patients with tuberculosis. Data from a chest radiographic survey of 240 Wainwright people is available (this includes some individuals not in the village in 1958). Seventy-nine or 33 per cent of these inhabitants had X-ray evidence of tuberculosis. Eight of the 12 positive BFT individuals had X-ray evidence of tuberculosis, although none had active disease when examined in 1958. This is significantly different from the prevalence in the group without positive BFT's.

Radiography

The radiographic studies were of most value in the osteoarthritis survey. Two of the radiographs contributed to the study of the prevalence of rheumatoid arthritis and these will be discussed below.

Osteoarthritis.—Although clinical information was available on the subjects in Wainwright, only the X-rays of the hands and wrists were used in estimating the prevalence of osteoarthritis in the populations studied.

Since there was no significant difference in the prevalence in the several communities making up the Northern Eskimo group, Southwestern Eskimo group, and U.S.A. white group, they were combined to make up the three groups shown in table 7. Grades 0 and 1 were deemed to represent no arthritis and Grades 2, 3, and 4, presence of arthritis. The North and Southwest Eskimo groups do not differ significantly from each other and they are combined in the table to form the total Eskimo group. Since the sex differences were not significant in the Eskimos, the male and female groups were also combined.

For the reasons stated above, statistical comparisons between the U. S. white and Eskimo populations were not made. However, there appears to be less osteoarthritis in the Eskimos in each decade than in the U. S. whites. If the samples are divided into two groups of individuals 40 and above and 39 and

Table 7.—***Radiographic Evidence of Osteoarthritis (Grades 2, 3, or 4) in Eskimo and White Populations Arranged by Age Groups and Sex. "Number" Indicates the Total Number of Individuals in Each Age Group, on Whom Radiographs Were Taken and "Per Cent Positive X-ray," the Per Cent with Grades 2, 3, or 4 Osteoarthritis***

	Males		Females		Males and Females	
Age	Number	Per cent Positive X-ray	Number	Per cent Positive X-ray	Number	Per cent Positive X-ray
			Eskimos, North			
0–9	20	0	20	0	40	0
10–19	37	0	26	0	63	0
20–29	27	0	29	0	56	0
30–39	21	0	19	5.3	40	2.5
40–49	21	0	25	12.0	46	6.5
50–59	23	13.0	8	37.5	31	19.4
60+	20	55.0	16	62.5	36	58.3
Total	169	8.3	143	11.9	312	9.9
40+	64	21.9	49	32.7	113	26.5
			Eskimos, Southwest			
0–9	0	0	0	0	0	0
10–19	0	0	0	0	0	0
20–29	1	0	1	0	2	0
30–39	6	0	4	0	10	0
40–49	36	11.1	32	9.4	68	10.3
50–59	23	26.1	17	23.5	40	25.0
60+	14	42.9	12	50.0	26	46.0
Total	80	20.0	66	19.7	146	19.9
40+	73	21.9	61	21.3	134	21.6
			Eskimo, Total			
0–9	20	0	20	0	40	0
10–19	37	0	26	0	63	0
20–29	28	0	30	0	58	0
30–39	27	0	23	4.3	50	2.0
40–49	57	7.0	57	10.5	114	8.8
50–59	46	19.6	25	28.0	71	22.5
60+	34	50.0	28	57.1	62	53.2
Total	249	12.0	209	14.4	458	13.1
40+	137	21.9	110	23.6	247	23.9
			Whites, U.S.A.			
0–9	0	0	0	0	0	0
10–19	10	0	12	0	22	0
20–29	41	4.9	48	2.1	89	3.4
30–39	47	12.8	58	5.2	105	8.6
40–49	46	23.9	61	26.2	107	25.2
50–59	35	57.1	44	56.8	79	57.0
60+	13	69.2	13	84.6	26	76.9
Total	192	25.0	236	23.7	428	24.3
40+	94	42.6	118	44.1	212	43.4

below, there again appears to be less osteoarthritis in the Eskimo than the U. S. whites (table 8).

Rheumatoid arthritis.—Of the 211 people examined in Wainwright, one was found to have definite rheumatoid arthritis and a second to have probable

Table 8.—*Radiographic Evidence of Osteoarthritis of Hands (2 or Greater) in Total U.S. Whites and Total Eskimos, by Age*

	Age Group	U.S. White Total		Eskimo Total	
		No.	% with Osteo	No.	% with Osteo
1	0–9	0	0	40	0
2	10–19	22	0	63	0
3	20–29	89	3.4	58	0
4	30–39	105	8.6	50	2.0
5	40–49	107	25.2	114	8.8
6	50–59	79	57.0	71	22.5
7	60+	26	76.9	62	53.2
	Total	428		458	

rheumatoid arthritis according to the criteria of the American Rheumatism Association.[2]

Case Histories

Case 1

M. A. (No. 460) was a 61-year-old woman who complained of intermittent pain in fingers, wrists, elbows, feet and knees for more than 20 years. During the early phase of her disease, the pain was accompanied by periarticular swelling and, in the last 5–10 years, permanent deformities became noticeable. Physical Examination: the changes in the hands and wrists were characteristic of rheumatoid arthritis (fig. 2). The wrists were enlarged and fixed; flexion, extension, abduction and adduction were markedly limited. There was enlargement and a hyperextension deformity of all the proximal interphalangeal joints; the metacarpophalangeal joints were enlarged and deformed. The hand was in ulnar deviation. There was a flexion deformity of the left elbow which prohibited extension beyond 90 degrees, while the right elbow was not prominently involved. The knees were enlarged and deformed, there was pain on flexion and extension with marked crepitation bilaterally, and flexion was limited to approximately 90 degrees bilaterally. Bilateral hallux valgus and rheumatoid deformity of her toes were present. X-rays of her wrists, knees and elbows showed changes characteristic of rheumatoid arthritis (figs. 3 and 4). Blood for serologic studies was not available.

Case 2

I. A. (No. 301), a 55-year-old woman had a history of morning stiffness and intermittent swelling and pain in her left knee and interphalangeal joints of both hands for ten years. Physical Examination: Heberden's nodes were present on the fingers of both hands. There was moderate swelling of the left knee and flexion was limited to 120 degrees. There were no obvious changes in the other joints. Radiograms showed only mild osteoarthritis of the hands and wrists, but no changes characteristic of rheumatoid arthritis. The bentonite flocculation titer was 1:64 and the sheep cell agglutination titer was 1:256.

There were a total of 14 females, 45 years or over, in Wainwright (table 9) and two of these had either definite or probable rheumatoid arthritis. The prevalence of rheumatoid arthritis in the females in this age group is similar to that in other populations.[6]

Other Findings

There were other findings pertaining to the joints which are of some clinical interest. Trauma is very common among Eskimos and is considered to be one

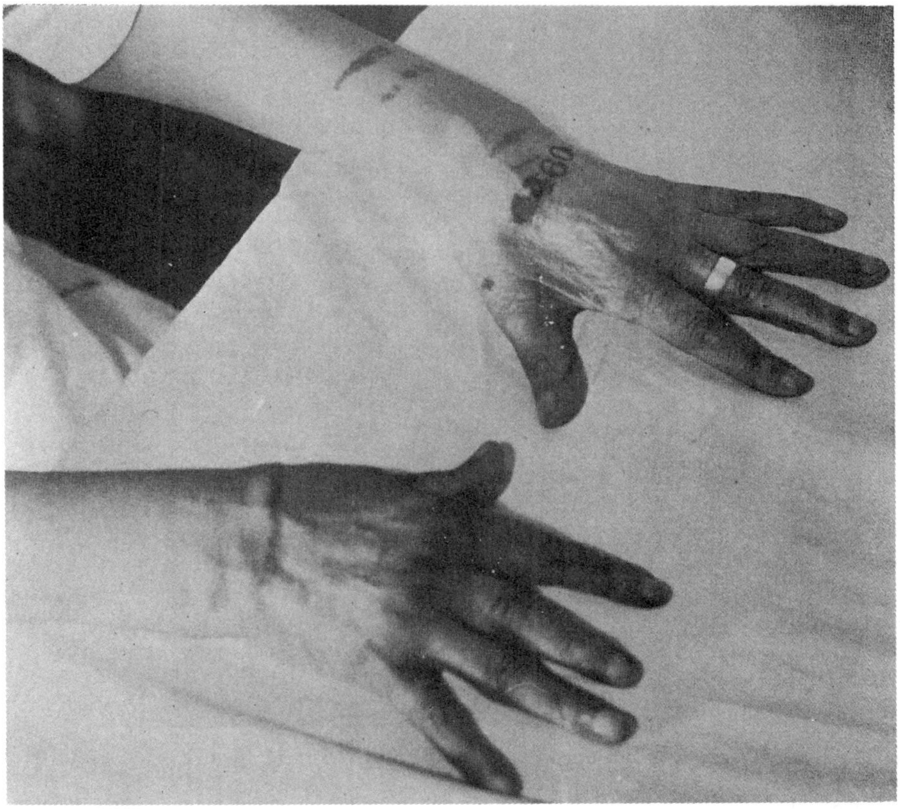

Fig. 2.—Photograph showing deformities of rheumatoid arthritis in hands and wrists of Case 1.

of the major causes of death. There were seven cases of probable traumatic arthritis of fingers, wrists or toes seen in the 167 Wainwright X-rays. Trauma not directly involving the joints was seen in an equal number. Six additional cases of joint trauma and two cases of trauma elsewhere on the extremities which were not detected by the X-rays, were found on clinical examination. Many of the other subjects gave histories of trauma, but had no obvious injuries when examined.

Many of the Eskimos examined in Wainwright had hypermobility of the joints, particularly noticeable in the hands and wrists (fig. 5). The prevalence of the trait was not determined exactly, but on the basis of clinical experience in the United States, and observations among laboratory personnel in Bethesda, it appears to be very much more common in the Eskimos. The significance of this is not apparent. There was one case of rheumatic heart disease with mitral insufficiency. There were no cases of acute rheumatic fever in Wainwright at the time of our visit, but two individuals gave a history of acute rheumatic fever in the past. Studies conducted in 1949 provided serologic evidence that streptococcal infections were probably common in this Eskimo population.[28]

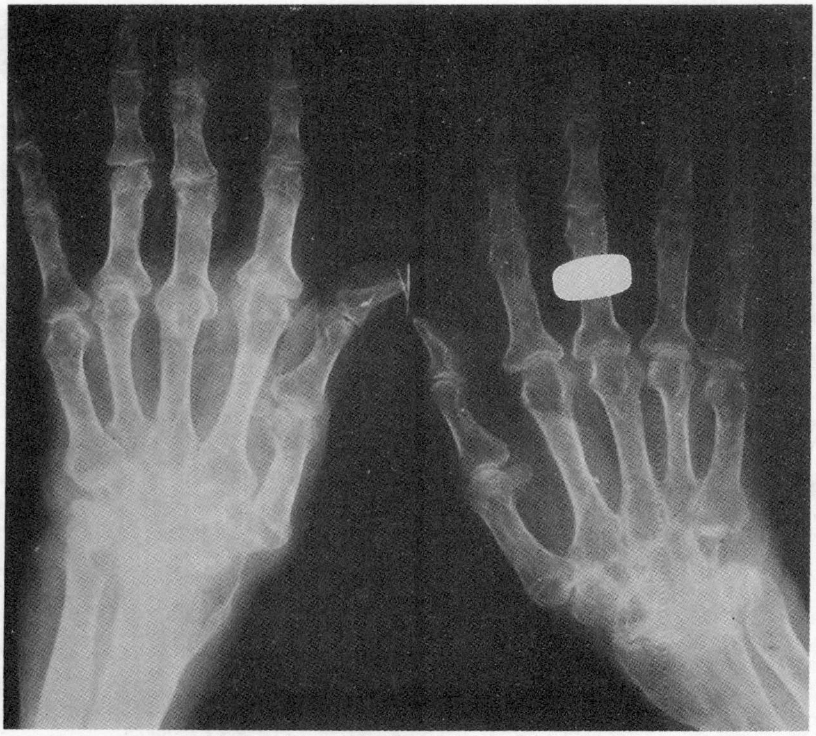

Fig. 3.—Radiographs of hands and wrists of Case 1.

No clinical gout was seen, and uric acid studies on the blood or urine were not done.

Discussion

Serology

With the exception of the village of Wainwright, the prevalence of individuals with positive BFT's (that is, with a titer of 1:32 or higher) was not significantly higher in the Alaskan population than in the control population. The prevalence in Wainwright, however, was higher than in any of the other communities. It is unlikely that this is due to technical error, since all the sera were collected and tested in the same manner. The positive sera were not due to syphilis. There were significantly more individuals with chest X-ray evidence of tuberculosis in the BFT positive group than in those with negative BFT's. However, none of these had active tuberculosis. Furthermore, the prevalence of tuberculosis is also very high in the other Alaskan communities where the BFT positive prevalence was not elevated. There is no good explanation for the high incidence of positive BFT's in this community.

Of the 12 individuals with positive BFT's, 11 were tested with the sheep cell agglutination test. Only two were found to be positive. One of these was patient I. A. with probable rheumatoid arthritis, and the other was apparently asymptomatic and free of joint disease. The difference in the results obtained

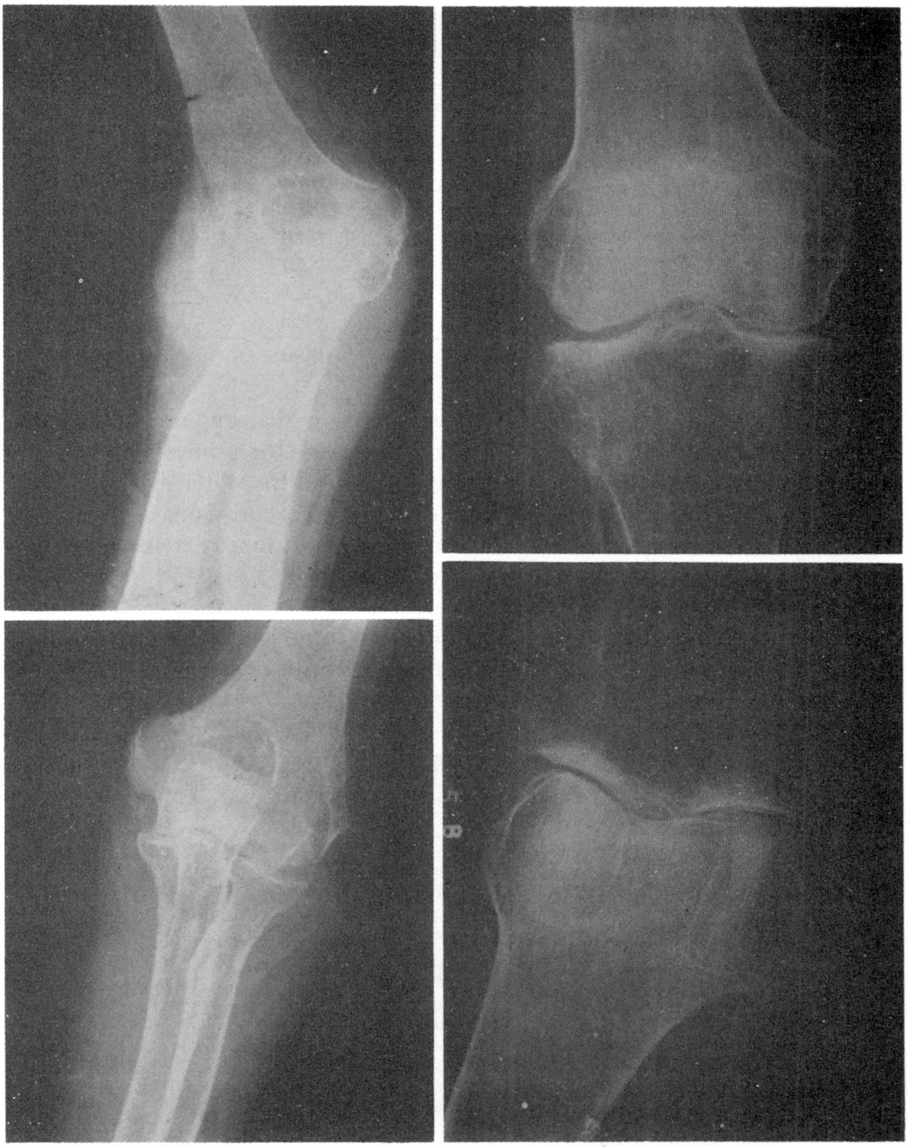

Fig. 4.—Radiographs of elbows (left) and knees (right) of Case 1.

with BFT and SSCA may be based on differences in sensitivity of the methods or may be due to the presence of two factors, one reacting with human gamma globulin and the other with rabbit amboceptor.[29]

The apparent drop in the prevalence of BFT positive individuals in the 60-plus age group was seen in each of the three populations, but as noted previously, this drop is not significant and could be due to chance. If, however, this decrease in the older age groups is confirmed in subsequent studies on the prevalence of BFT reactors in apparently normal populations, there are several

Table 9.—*Age and Sex Distribution of the Population of Wainwright, Alaska, in July, 1958*

	Males		Females		Total Males & Females	
Age Group	No.	%	No.	%	No.	%
0–9	37	29.6	36	38.3	73	33.3
10–19	35	28.0	19	20.2	54	24.7
20–29	16	12.8	17	18.1	33	15.1
30–39	13	10.4	4	4.3	17	7.8
40–49	5	4.0	9	9.5	14	6.4
50–59	11	8.8	3	3.2	14	6.4
60–69	5	4.0	2	2.1	7	3.2
70+	3	2.4	4	4.3	7	3.2
Total	125	100.0	94	100.0	219	100.0

possible explanations. One interesting possibility is that apparently normal individuals with positive BFT tests die at an earlier age than those with a negative test and, therefore, do not survive to be counted in the older age groups; that is, there may be selection against apparently normal individuals with positive tests. Any further conjecture on this point must await more extensive studies.

Osteoarthritis

A determination of the "true" prevalence of osteoarthritis is not possible at present, for no adequate method for an osteoarthritis survey is available short of a complete radiographic and clinical examination of a population. It would

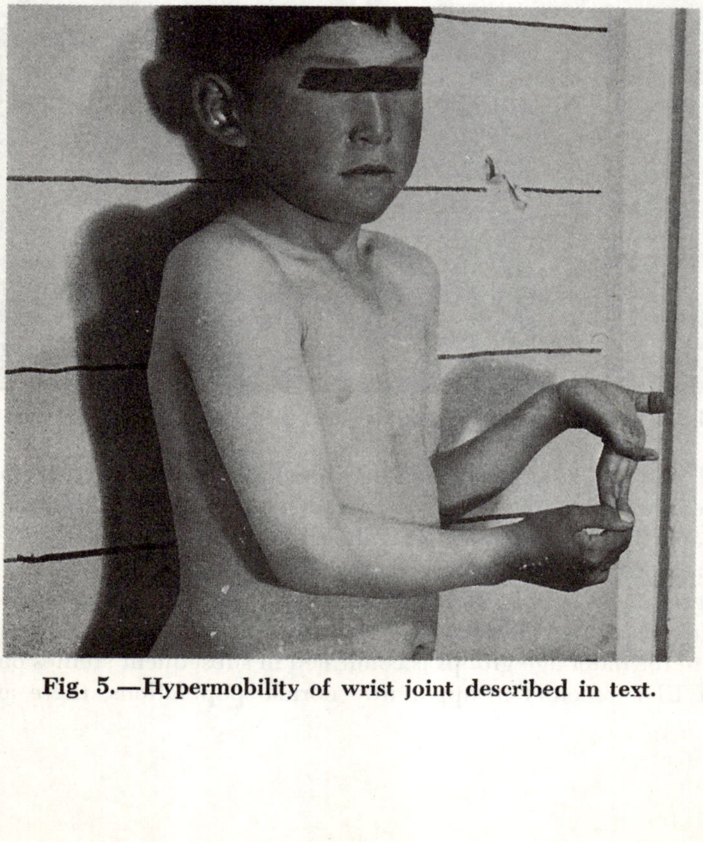

Fig. 5.—Hypermobility of wrist joint described in text.

be desirable to have a "short-cut" method to obtain an "index" of the prevalence of osteoarthritis by utilizing radiographic examinations of only a few joints. The study of Kellgren and Lawrence[22] already discussed gives hope that such an "index" may be determined. If this "index" were studied in two different communities, using the same criteria, a valid comparison within its definition would be possible. The present survey yields information only on the prevalence of radiographic evidence of osteoarthritis in the hands and wrists.

As noted above, there was no appropriate sample with which the Eskimo X-ray data could be accurately compared. Comparison by age and sex with the available U. S. white X-ray data suggests a lower prevalence of X-ray evidence of osteoarthritis of hands and wrists in the Eskimo. The prevalence of osteoarthritis in the British group studied by Kellgren and Lawrence[22] appared to be slightly higher than in the U. S. white population of the present study. The methods of study, however, were not directly comparable.

It is interesting to speculate that what appears to be a low prevalence of osteoarthritis in the Eskimo may have arisen as a result of selection over many generations against individuals prone to develop osteoarthritis, since such individuals would be less likely to survive the rigours of Arctic life. This may imply that some trait related to the development of osteoarthritis is inherited, which is consistent with some previous studies. Any further conjecture must await more complete investigations on this interesting population.

Summary

Rheumatoid arthritis is present in an Alaskan Eskimo population. The prevalence of X-ray evidence of osteoarthritis of the hands and wrists has been determined in an Eskimo population. Comparison with available U. S. white data suggests that the prevalence may be lower in the Eskimo. There was a high prevalence of individuals in the village of Wainwright whose sera yielded a positive bentonite flocculation test.

REFERENCES

1. Doll, R.: Methods of Geographical Pathology. Springfield, Ill., Charles C Thomas, 1959, p. 72.
2. Ropes, M. W., Bennett, G. A., Cobb, S., Jacox, R., and Jessar, R. A.: 1958 Revision of proposed diagnostic criteria for rheumatoid arthritis. Arth. and Rheumat. *2:*16, 1959.
3. Cobb, S., Warren, J. E., Merchant, W. R., and Thompson, D. J.: An estimate of the prevalence of rheumatoid arthritis. J. Chron. Dis. *5:*636–643, 1957.
4. Kellgren, J. H., and Lawrence, J. S.: Rheumatoid arthritis in a population sample. Ann. Rheumat. Dis. *15:*1–11, (March) 1956.
5. Miall, W. E.: Rheumatoid arthritis in males; An epidemiological study of a welsh mining community. Ann. Rheum. Dis. *14:*150, 1955.
6. Kellgren, J. H.: Population Studies in Rheumatoid Arthritis. Washington, D. C. McGregor and Werner, 1958, p. 97.
7. Griswold, H. E., Dotter, C. T., Rennie, D., and Hopkins, C. E.: The prevalence of cardiovascular disease in northwestern coastal alaskan eskimos. In preparation.
8. Corcoran, P., Allen, F. H., Allison, A. C., and Blumberg, B. S.: Blood groups of alaskan eskimos and indians. Am. J. Phys. Anthropol. *17:*187, 1959.
9. Allison, A. C., Blumberg, B. S., and Gartler, S. M.: Urinary excretion of β-amino-isobutyric acid in eskimo and indian populations. Nature *183:* 118–119, 1959.

10. Blumberg, B. S., Allison, A. C., and Garry, Barbara: The haptoglobins and hemoglobins of alaskan eskimos and indians. Ann. Human Genet. *23:* 349, 1959.
11. Allison, A. C., and Blumberg, B. S.: Ability to taste phenylthiocarbamide among alaskan eskimos and other populations. Human. Biol. *31:*352, 1959.
12. Laughlin, W. S.: Contemporary problems in the anthropology of southern alaska, in Collins, H. B., Ed., Science in Alaska, Washington, D. C., The Arctic Institute of North America, 1952, pp. 66–84.
13. Milan, F.: Observations on the Contemporary Eskimo of Wainwright, Alaska, Technical Report (No. 57–14). Alaska. Arctic Acromedical Laboratory, Ladd Air Force Base, 1958.
14. An Appraisal of the Health and Nutritional Status of the Alaskan Eskimo. Report of the Interdepartmental Committee on Nutrition for National Defense, 1958.
15. Bozicevich, J., Bunim, J. J., Freund, J., and Ward, S. B.: Bentonite flocculation test for rheumatoid arthritis. Proc. Soc. Exper. Biol. & Med. *97:* 180–183, 1958.
16. Lerner, E. M., Williams, N. R., and Jenkins, J. C.: Sensitized sheep cell hemagglutination reaction in rats with experimental infection of bone and joint. Proc. Soc. Exper. Biol. & Med. *99:*249–252, 1958.
17. Serologic Tests for Syphilis: 1959 Manual; U.S. Department of Health, Education and Welfare, Public Health Service Publication.
18. Bloch, K. J., and Bunim, J. J.: A simple, rapid diagnostic test for rheumatoid arthritis—The bentonite flocculation test. J. A. M. A. *169:*307–314, 1959.
19. Pike, R. M., Sulkin, S. E., and Coggeshall, N. C.: Serological reactions in rheumatoid arthritis. I. Factors affecting the agglutination of sensitized sheep erythrocytes in rheumatoid arthritis serum. J. Immunol. *63:*441, 1949.
20. Waterhouse, A.: Personal communication.
21. Kellgren, J. H., and Lawrence, J. S.: Radiological assessment of osteo-arthrosis. Ann. Rheumat. Dis. *16:*494, 1957.
22. —, and —: Osteo-arthrosis and disk degeneration in an urban population. Ann. Rheumat. Dis. *17:*388, 1958.
23. Peltier, A., and Christian, C. L.: The presence of the "rheumatoid factor" in sera from patients with syphilis. Arth. and Rheumat. *2:*1–7, 1959.
24. Paul, J. R., and Riordan, J. R.: Observations on serological epidemiology antibodies to the lansing strain of poliomyelitis virus in sera from alaskan eskimos. Am. J. Hyg. *52:*202–212, 1950.
25. Indian Health Highlights, Division of Indian Health, U.S. Department of Health, Education and Welfare, December, 1957.
26. Rothermich, N. O., and Philips, V. K.: The serological diagnosis of rheumatoid arthritis. J. A. M. A. *164:*1999–2004, 1957.
27. Rheins, M. S., McCoy, F. W., Burrell, R. G., and Buehler, E. V.: A modification of the latex-fixation test for the study of rheumatoid arthritis. J. Lab. and Clin. Med. *50:*113–119, 1957.
28. Paul, J. R., Riordam, J. T., and Kraft, L. H.: Serological epidemiology; antibody patterns in north alaskan eskimos. J. Immunol. *66:*695–713, 1951.
29. Lospalluto, J., and Ziff, M.: Chromatographic studies of the rheumatoid factor. J. Exper. Med. *110:*169–186, 1959.

Baruch S. Blumberg, M.D., Geographic Medicine and Genetics Section, Epidemiology and Biometry Branch, National Institute of Arthritis and Metabolic Diseases, Bethesda, Md.

Kurt J. Bloch, M.D., Senior Investigator, National Institute of Arthritis and Metabolic Diseases, Bethesda, Md.

Roger L. Black, M.D., Senior Investigator, National Institute of Arthritis and Metabolic Diseases, Bethesda, Md., and Clinical Assistant Professor of Medicine, Georgetown University, Washington, D. C.

Charles Dotter, M.D., Department of Radiology, University of Oregon Medical School, Portland, Ore.

Reprinted from *Science*
September 30, 1966, Vol. 153, No. 3744,
pp. 1664–1666

Albumin Naskapi: A New Variant of Serum Albumin

Liisa Melartin
B. S. Blumberg

Abstract. *An apparently new variant of human serum albumin, albumin Naskapi, has been found in high frequency in the Naskapi Indians of Quebec and, in lower frequency, in other North American Indians. The family and population data of the albumin are consistent with its inheritance as a simple autosomal trait controlled by a gene designated* Al Naskapi. *This gene is allelic with the gene* Al^A *which controls the common albumin. Both homozygotes and heterozygotes have been distinguished. This is the first report of a homozygote for an albumin variant.*

Rare electrophoretic variants of serum albumin have been reported several times since the original description of bisalbumin by Scheurlen (*1*). In nearly all of the reports, the rare variant has an electrophoretic mobility less than that of the common form of albumin (*1–4*), but two investigators reported fast-moving forms (*5, 6*). Family studies are consistent with simple autosomal inheritance. There have been no reports of individuals homozygous for the rare variants, and the bis forms are apparently heterozygotes.

All of the albumin variants are rare in the populations tested (for example, one per 1015 in Norway) (*2*), and apparently disease is not associated with the trait. However, increase in concentration of cholesterol is associated with bisalbumin, but this elevation is not statistically significant (*3, 5*).

We now report an inherited, fast-moving, electrophoretic variant of albumin which is different from at least one of the other fast-moving variants (*6*). The trait is common in several North American Indian tribes, but has not been seen in other populations tested. Several homozygotes for the variant have also been identified.

The original finding of the albumin variant, and the subsequent screening, was made with whole serum with Ashton's discontinuous buffer system at *p*H 8.6 (*7*). The gels were prepared from hydrolyzed starch (Connaught) at a concentration 25 percent higher than that recommended by the manufacturers. The electrophoresis was performed in a horizontal system with a constant voltage of 9 volt/cm. The discontinuous buffer system (*p*H 8.6) described by Poulik (*8*) and Ashton's acid buffer system (*p*H 5.6) (*9*) with a constant voltage of 16 volt/cm were also used.

Cellulose acetate electrophoresis was performed with a microzone electrophoresis system in barbital buffer at *p*H 8.6. Immunoelectrophoresis was done in 1 percent special agar-noble gel with a modified barbital buffer (*10*) and horse antiserum to human albumin. Barbital buffer (*p*H 8.6; ionic strength, 0.75) was used in the paper electrophoresis experiments.

The populations studied are shown in Table 1. In several cases, the blood specimens were used in other studies, and the populations are described in the indicated references. Blood was collected by venepuncture, and the serum or plasma was separated and stored at −20°C until tested. The dates of collection vary from 1958 for the Athabascans to 1962 for the Naskapi and Montagnais Indians.

The fast-moving variant can be distinguished by electrophoresis in starch gel with Ashton's discontinuous buffer, Ashton's acid buffer, or cellulose acetate. It can also be seen with Poulik's buffer, but only when the serum is diluted. Good separations were not obtained with barbiturate buffer with paper as a supporting medium. The three phenotypes can be easily distinguished (Fig. 1). In immunoelectrophoresis the heterozygote has a slightly elongated albumin band, and the new fast homozygote has a greater mobility than the common slow homozygote. However, antigenic differences between the two albumins are not seen. In Fig. 1, the new variant is compared to that described by Wieme (*6*); it is clearly different. The fast-moving variant described by Tarnoky and Lestas (*5*) was not available for comparison.

The distribution of the trait in several populations and the expected frequencies as calculated by the Hardy-Weinberg formula are shown in Table 1. Serums from 365 Haida Indians from Canada (*12*), 100 Quechua and 92 Cashinahua Indians from Peru (*13*), 443 Eskimos from Alaska (*12*), and 114 Americans of European descent were tested. None of these had albumin Naskapi. The results correlate well with the expected distribution. The gene frequency of the new variant in the Indian populations ranges from 0.13 in the Naskapi to 0 in the Haida and South American groups. None was seen

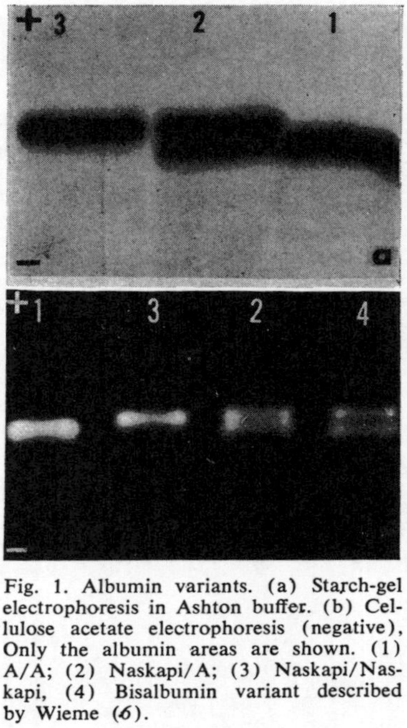

Fig. 1. Albumin variants. (a) Starch-gel electrophoresis in Ashton buffer. (b) Cellulose acetate electrophoresis (negative), Only the albumin areas are shown. (1) A/A; (2) Naskapi/A; (3) Naskapi/Naskapi, (4) Bisalbumin variant described by Wieme (*6*).

in the Eskimos or Caucasian Americans tested. Many of the smaller populations of Indians included related individuals, a fact which may affect the calculation of the gene frequency. Therefore, the frequencies given in Table 1 are only approximate. Three pedigrees in which the trait is segregating are shown in Fig. 2. The first (Naskapi) shows the segregation of the trait in three generations; the albumin types of all parents are known. In the second (Montagnais), a mother with both types of albumin has three children, of which one has only the common albumin, one has only the new variant, and the third has both forms. The third (Naskapi) shows a large family where the variant apparently has segregated to five of seven children in the oldest generation. A man in the second generation has only the new type, while his three children from two marriages have both types. An analysis of eleven Naskapi and Montagnais families (including some not in the pedigrees of Fig. 1) is shown in Table 2.

The albumin is different from the slow-moving variant and from the fast-moving albumin described by Wieme (*6*). It is proposed that this new variant be called albumin Naskapi, after the Indian population in which it was first discovered, and that the gene determining the trait be designated *Al Naskapi* (abbreviated Al^{Na}). The common albumin variant is designated albumin A (gene notation Al^{A}) and the originally described slow-moving variant, albumin B (gene notation Al^{B}). As new variants are discovered, it would be convenient to give them geographic, population, or other specific designations.

The segregation of *Al Naskapi* in the families is consistent with simple autosomal inheritance, with full expression of *Al Naskapi* and the alternate gene Al^{A} in the heterozygote. On the basis of the populations tested, albumin Naskapi appears to be restricted to Northern American Indians. The high frequency in the Naskapi could be associated with the inclusion of many family members in the sample tested. However, the Montagnais and other groups included mainly nonrelated individuals. This gene may serve as a useful marker for Northern Indian groups, as does the red blood cell Diego factor for other Indians.

None of the albumin variants described previously occurred with such high frequency in the general population. The high frequency of albumin Naskapi could be due to the operation of selective forces in the environment of these populations, or to genetic drift.

The Naskapi and Montagnais have a high frequency of two transferrin variants which are rare in most other populations (*14*). The transferrins are not found in any of the other Northern Indian groups, with the exception of the Tlingit, where they are found in very low frequency. If the high frequencies of albumin Naskapi in the Indians were due to simple gene mixture, and if selection were not operating on either of these protein polymorphic traits, then it would be expected that the transferrin variants would occur in the same Indian groups in which the albumin variant occurs. This was not found to be the case.

There appears to be a broad north-south cline for the frequency of albumin Naskapi which cuts across known ethnic relations of the Indians. This is the reverse of the cline for the genes which determine the Diego factor and haptoglobin-1, both of which occur with low frequency in North American tribes but in high frequency in Central and South American tribes.

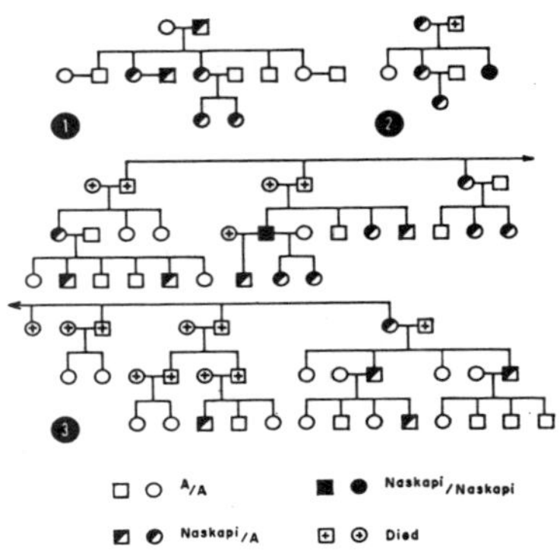

Fig. 2. Segregation of albumin Naskapi in two Naskapi families (genealogies 1 and 3) and one Montagnais family (genealogy 2.

In addition to the albumin Naskapi, we have found an albumin variant which occurs with high frequency in the Eskimo population of a small Indian-Eskimo village in Alaska. This variant may be different from albumin Naskapi, but additional specimens will be needed to complete the comparison. A slow-moving variant occurs with

Table 1. Distribution of albumin types in different populations.

Occurrence of phenotypes	Naskapi/ Naskapi (No.)	A/Naskapi (No.)	A/A (No.)	Gene frequency Al^{Na}	Gene frequency Al^{A}
		Naskapi, Canada (11)			
Observed	1	37	113	0.130	0.870
Expected	2.6	34.1	114.3		
		Montagnais, Canada (11)			
Observed	2	14	96	.080	.920
Expected	0.6	16.5	94.8		
		Sioux, United States			
Observed	0	2	158	.007	.993
Expected	0	2.2	157.8		
		Athabascan, Alaska and Canada (12)			
Observed	1	11	218	.028	.972
Expected	0.2	12.4	217.4		
		Tlingit, Alaska (12)			
Observed	0	1	99	.005	.995
Expected	0	1	99		

Table 2. Segregation of albumin types in families.

Occurrence of phenotype	Albumin type of children: A/A (No.)	A/Naskapi (No.)	Naskapi/ Naskapi (No.)
		One family, A/A – Naskapi/Naskapi	
Observed	0	2	0
Expected	0	2	0
		Ten families, A/A – A/Naskapi	
Observed	17	15	0
Expected	16	16	0

relatively high frequency in Mexican Indians. Therefore, our findings indicate that albumin variants occur with greater frequency than previous evidence had indicated.

LIISA MELARTIN*

Department of Medical Microbiology, University of Turku, Turku, Finland

B. S. BLUMBERG

Institute for Cancer Research, 7701 Burholme Avenue, Fox Chase, Philadelphia, Pennsylvania

References and Notes

1. P. G. Scheurlen, *Klin. Wochschr.* **33**, 198 (1955); F. Wuhrman, *Schweiz. Med. Wochschr.* **89**, 150 (1959).
2. G. Efremov and M. Braend, *Science* **146**, 1679 (1964).
3. D. P. Earle, M. P. Hutt, K. Schmid, D. Gitlin, *J. Clin. Invest.* **38**, 1412 (1959).
4. V. M. Knedel, *Blut* **3**, 129 (1957); H. J. Nennstiel and T. Becht, *Klin. Wochschr.* **35**, 689 (1957); H. Bennhold, H. Ott, G. Scheurlen, *Verhandl. Deut. Ges. Inn. Med.* **64**, 279 (1958); G. Franglen, N. H. Martin, T. Hargraves, M. J. H. Smith, D. I. Williams, *Lancet* **1960-I**, 307 (1960); F. Miescher, *Schweiz. Med. Wochschr.* **90**, 1273 (1960); K. B. Cooke, T. E. Cleghorn, E. Lockey, *Biochem. J.* **81**, 39 (1961); P. L. Adner and A. Redfors, *Nord. Med.* **65**, 623 (1961); E. J. Sarcione and C. W. Aungst, *Blood* **20**, 156 (1962); J. L. Robbins, G. A. Hill, S. Marcus, J. H. Carlquist, *J. Lab. Clin. Med.* **62**, 753 (1963); O. Drachmann, N. M. G. Harboe, P. J. Svendsen, T. S. Johnsen, *Danish Med. Bull.* **12**, 74 (1965); S. Ungari, and V. Lopez, *Minerva Pediat.* **17**, 288 (1965); G. Sandor, L. Martin, M. Porsin, A. Rousseau, R. Martin, *Nature* **208**, 1222 (1965).
5. A. L. Tarnoky and A. N. Lestas, *Clin. Chim. Acta* **9**, 551 (1964).
6. R. J. Wieme, *ibid.* **5**, 443 (1960).
7. G. C. Ashton and A. E. Braden, *Australian J. Biol. Sci.* **14**, 284 (1961).
8. M. D. Poulik, *Nature* **180**, 144 (1957).
9. G. C. Ashton, *Genetics* **50**, 1421 (1964).
10. J. Hirschfeld, *Sci. Tools* **7**, 19 (1960).
11. B. S. Blumberg, R. J. Martin, F. H. Allen, Jr., J. Weiner, E. M. Vitagliano, A. Cooke, *Human Biol.* **36**, 263 (1964).
12. P. A. Corcoran, F. H. Allen, Jr., A. C. Allison, B. S. Blumberg, *Am. J. Phys. Anthropol.* **17**, 187 (1959).
13. K. Kensinger, *Expedition* **7**, No. 4, 5 (1965).
14. L. Melartin and B. S. Blumberg, in preparation.
15. We thank R. J. Martin, A. Cooke, J. Cooke, A. C. Allison, M. T. Newman, C. Leyshon, K. Kensinger, R. A. Donato, Early Sites Foundation, Arctic Aeromedical Laboratory, Arctic Research Laboratory, Alaskan Native Hospital, and the Wenner-Gren Foundation for serums. We thank R. J. Wieme for the bis-albumin sample. Supported in part by PHS grants CA-06551-03 and CA 08069-02.
* Present address: The Institute for Cancer Research, 7701 Burholme Avenue, Fox Chase, Philadelphia, Pennsylvania.

5 July 1966 ■

Reprinted from *Albumin Structure, Function and Uses*
Eds. V. M. Rosenoer, M. Oratz and M. A. Rothschild
Pergamon Press, Oxford and New York, 1977

THE GENETICS OF HUMAN SERUM ALBUMIN†

LAWRENCE M. SCHELL AND BARUCH S. BLUMBERG

The Institute for Cancer Research, The Fox Chase Cancer Center, 7711 Burholme Avenue, Philadelphia
and
Department of Anthropology, University of Pennsylvania, Philadelphia, Pennsylvania

CONTENTS

I. INTRODUCTION

A. Historical Review

The study of the genetics of human serum albumin began with Scheurlen's report in 1955 of a diabetic female whose serum on paper electrophoresis had a slow-moving albumin band in addition to the common one.[1] Subsequently, Wuhrmann[2] reported that the two albumin bands were present in the patient irrespective of her clinical status and that two albumins, one slow, the other of the usual mobility, were present in the sera of her father, brother and son. This suggested that the double albumin was an inherited trait.[1] The distribution of the variants in all the families that were tested was consistent with a codominant mode of inheritance, each allele controlling the synthesis of each albumin fraction.[3,4]

†This work was supported by USPHS Grants CA-06551, RR-05539 and CA-06927 from the National Institutes of Health, and by an appropriation from the Commonwealth of Pennsylvania.

The genetic hypothesis gained support as additional cases of uncommon albumin bands were reported. Slow-moving variants were found in Europeans from several countries including Germany,[5-8] the United States,[9-11] France,[12] Italy,[13,14] the United Kingdom,[15,16] Sweden,[17] Norway[18,19] and Denmark.[20] These slow albumins all appeared to have the same electrophoretic mobility and were termed albumin B. The common more rapidly moving band was designated albumin A. Two fast-moving albumin bands were reported in families of European origin, albumin Gent from a Belgian family[3] and albumin Reading from a Welsh/English family.[21] These variants were rare in the populations in which they were found (in the order of one in several thousand) and seem to be examples of mutations at the albumin locus which had not increased to high frequencies in any of the populations surveyed.

In 1966 Melartin and Blumberg[22] discovered a fast albumin variant among American Indians. It differed significantly from those previously reported not only in its mobility but because it occurred relatively commonly (several per hundred) in the populations in which it had been detected. It was found originally in the Naskapi and Montagnais Indians of Quebec Province, Canada, and was termed albumin Naskapi. Some Naskapi and Montagnais Indians lacked the common albumin A and had only the fast-moving albumin band in their sera. This was the first report of individuals apparently homozygous for an albumin other than the common one. In 1967 another case of a single fast-moving albumin band was reported in sera from an Indian family living in Alberta, Canada,[23] and this was subsequently shown to have the same electrophoretic mobility as albumin Naskapi.[24] In the same year a new slow-moving variant, albumin Mexico, was discovered among Indians of Mexico.[25] This also was relatively common among the Indian populations in which it occurred.

By 1967 the numerous family studies, electrophoretic comparisons and the reports of apparent homozygotes confirmed the earlier suggestions of genetic control of electrophoretic variation of albumin. Cases of transient double albumins have also been reported.[26-29] There is no evidence that these are under genetic control.

When the first review of what had then become a sizable literature on albumin variation was published by Melartin in 1967[30] there were only a half-dozen albumin variants. The number has since increased about fourfold. The rapid increase is primarily due to the application of relatively simple screening techniques to sera from South America and Asia. New variants have been found among Japanese,[31] Malaysians,[32,33] Indonesians,[32,34] Indo-Dravidians,[35-39] New Guinea aborigines[40,41] and South American Indians.[42-48]

The distribution of albumin variation appears to be worldwide. The only exception is sub-Saharan Africa, although this may be due to the paucity of observations. To date only two cases of alloalbuminemia in individuals of African ancestry have been reported.[49-51]

B. Terminology

There are now over two dozen albumin variants. Originally, the term bisalbuminemia (from the Greek, meaning two albumins in the blood) was used to describe the presence of two types of albumin in the sera (i.e. the heterozygote). When homozygotes were discovered,[22] the term bisalbuminemia became inaccurate (there was only one albumin) and alloalbuminemia (from the Greek, meaning different albumins in the blood) was suggested to cover both the heterozygote and the homozygote variant conditions,[52] and this term has since gained some acceptance.[43,46,52–58]

The condition, alloalbuminemia, is recognized by the presence of an albumin band which has an electrophoretic mobility different from normal albumin. When a new variant is identified, it is often conventionally named for the population in which it was first discovered.

Melartin[30] suggested that a nomenclature system analogous to that used for transferrins, hemoglobin and other human polymorphic systems be adopted for the albumins. The phenotype of the common albumin is designated Al A (gene symbol Al^{A}), and the common genotype is Al^{A}/Al^{A}. Variant phenotypes and genotypes are similarly identified but with the variant name or abbreviation in the appropriate script. For example, Al Na is the phenotype for an albumin Naskapi homozygote and Al^{Na}/Al^{Na} for the genotype.

II. LABORATORY STUDIES

A. The Relative Percentages of Albumin Variants in the Serum

In alloalbuminemic individuals the total albumin fraction of the serum, varian and normal components combined is within normal limits.[10,13,30,59–61]

The variant type usually comprises about half of the total albumin whether the variant is fast or slow.[3,5,10–13,15,21,23,30,35,39,59–62] There appears to be one exception to this. A very low quantity of the variant fraction (albumin Maku) was reported by Weitkamp and Chagnon.[42]

B. The Antigenic Specificity of Albumin Variants

The antigenic specificity of albumin variants has been tested by Ouchterlony techniques, immunoelectrophoresis against anti-whole human serum, and immunoelectrophoresis against horse, goat, sheep and rabbit anti-human albumin serum. In general, these tests indicate that albumin variants are antigenically identical to normal albumin.[3,10,13,15,20,21,30,59,61–67]

There are, however, two exceptions. Robbins *et al.*[11] noted in a report of an unnamed variant from Utah that during immunoelectrophoresis with undiluted serum one precipitin arc was formed, but when the variant serum was diluted (1:30) prior to immunoelectrophoresis, two overlapping precipitin arcs were formed. Margni *et al.*[68] has also noted (using absorption studies) slightly different antigenic properties between sera from two alloalbuminemic families.

C. Electrophoretic Mobilities of Albumin Variants

Variants of human serum albumin have been recognized by differences in their electrophoretic mobility. Such differences would be found only if the amino acids which distinguish the variants carry an electric charge. However, an albumin molecule with a variant amino acid composition may not be detected if the distinguishing amino

TABLE 1. LIST OF VARIANTS

Name of variant	Notes	Reference	Provenance/ethnic origin
A_2	Slow	Knedel[5]	European
Afghanistan	Same mobility as "Kashmir" according to Weitkamp *et al.*[50] "Kashmir" is formally precedent	Weitkamp & Buck[38]	India
Albuquerque	May be same as "Mexico" of Melartin & Blumberg[25]	Melartin *et al.*[69]	Native Americans, Mexico
Algonkian	Subsequently found identical to "Naskapi".[24] "Naskapi" is formally precedent	Polesky & Rokala[70]	Native Americans (Algonkian tribes)
Al^r		Emannelli *et al.*[71]	European
Ann Arbor	First identified by Adams,[59] termed "Jensen" in Weitkamp *et al.*,[24] termed "B" in Melartin[30]	Weitkamp *et al.*[72]	European
B		Earle *et al.*[10]	European
B		Fine[67]	European
Belem I		Salzano *et al.*[46]	Brazil (Trihybrid descent: European, Black, Native American)
Belem II	Same mobility as "Mexico"	Salzano *et al.*[46]	Brazil (Trihybrid descent: European, Black, Native American)
Belem III	Same mobility as "Maku" & "Makiritare-2"[50]	Salzano *et al.*[46]	Brazil (Trihybrid descent: European, Black, Native American)
Birmingham	Same mobility as "Kashmir", different binding properties	Bradwell *et al.*[39]	India
Boal		Izquierdo *et al.*[73]	Spain
Caracas		Arends *et al.*[43]	Not stated
Carib		Arends *et al.*[43]	Carib Indian, Venezualian descent
Cartago		Lau *et al.*[57]	Costa Rican (Spanish descent)
Cayemite		Weitkamp *et al.*[49]	Haitian Negro
Fast	Same as "Reading";[74] "fast" Weitkamp *et al.*;[75] same as "New Guinea" of Weitkamp *et al.*[50]	Tarnoky & Lestas[21]	European
Gainesville		Sunderman *et al.*[54]	European descent (U.S.A.)
Gent	"Very fast" Weitkamp *et al.*;[100] MI/FAST source from Petrini, cited in Porta *et al.*, 1972 (from Weitkamp *et al.*)[76]	Wieme[3]	.European
Gombak	Same as "Paris".[50] "Paris" is formally precedent	Lie-Injo *et al.*[32]	Malaysian Indigenes
Jensen	See "Ann Arbor"	Weitkamp *et al.*[24]	European
Kashmir	Same mobility as "Afghanistan".[50] "Kashmir" is formally precedent	Tarnoky & Dowding[35]	India

Kuala Lumpur		Welch & Lie-Injo[33]	Chinese Malaysians
Kyoto		Weitkamp[77]	Not stated
Larson	Slow	Weitkamp *et al.*[72]	Unknown
Luarca		Izquierdo *et al.*[62]	Spain
Makiritare	Dimer; same as "Makiritare-1"[50] and "Piaroa"[45]	Arends *et al.*[44]	South American Indigenes
Makiritare-1	Dimer; same as "Makiritare",[50] "Piaroa"[45]	Geerdink *et al.*[47]	South American Indigenes
Makiritare-2	Same as "Maku", "Belem III". [46,50] "Maku" is formally precedent	Tanis *et al.*[45]	South American Indigenes
Makiritare-3	Tanis & Neel (unpublished)	Weitkamp *et al.*[50]	South American Indigenes
Maku	See "Makiritare-2", "Belem III"; same as "Wapishana-1" from Tanis *et al.*[45]	Weitkamp & Chagnon[42]	South American Indigenes
Maprik	Probably same as "New Guinea"[41]	McDermid[41]	New Guinea
Medan		Lie-Injo *et al.*[32]	Chinese Malaysians
Mexico	A polymorphic variant, may be same "Belem II"[46] or "P" of Polesky *et al.*[78]	Melartin & Blumberg[25]	Native Americans, several tribes; see Table 7
MI/TN	May be "SO/BS"	Petrini *et al.*[79]	European
Naskapi	A polymorphic variant: same as "Algonkian" of Polesky & Rokala.[70] "Naskapi" has formal precedence	Melartin & Blumberg[22]	Native Americans, several tribes; see Table 7
New Guinea	Also called "Greek" from Franglen (unpublished) cited in Weitkamp *et al.*;[100] same as "Reading" according to Weitkamp;[77] and "Maprik" of McDermid[41]	Weitkamp *et al.*[40]	New Guinea
Oliphant	Slow	Weitkamp *et al.*[72]	European
Otsu		Weitkamp[77]	Not stated
Paris	"Slow" Weitkamp *et al.*;[75] may be same as "Gombak"	Sandor *et al.*[12]	European
Piaroa	Same as "Makiritare-1"	Tanis *et al.*[45]	South American Indigenes
Pollibauer	Author refers to Geerdink (unpublished for source of sample)	Weitkamp *et al.*[50]	European
Pushtoon		Weitkamp & Buck[38]	India
Reading	Same as "New Guinea" and "Maprik"[77]	Tarnoky & Lestas[21]	European
Roma	Author cites Porta *et al.*[58] for source of sample	Weitkamp *et al.*[50]	European
RS-I	Author cites Ortalli for source of sample	Weitkamp *et al.*[50]	Not stated
RS-II	Author cites Ortalli for source of sample	Weitkamp *et al.*[50]	Not stated
Santa Ana		Kueppers *et al.*[64]	Mexican
Sentul		Welch & Lie-Injo[33]	Malaysians
SO/BS	Author cites Porta for source of sample	Weitkamp *et al.*[50]	Not stated
Syracuse	Called "Reading" in Weitkamp;[77] "Reading" is formally precedent	Schneiderman *et al.*[80]	European
Uinba		Weitkamp *et al.*[40]	New Guinea

(*continued overleaf*)

TABLE 1 (*cont.*)

Name of variant	Notes	Reference	Provenance/ethnic origin
Wapishana	Same as "Maku".[45] "Maku" is formally precedent	Tanis *et al.*[45]	South American Indigenes
Warao	May be "Makiritare" of Arends *et al.*[44]	Arends *et al.*[43]	Warao tribe, South American Indigenes
Westcott	Probably "New Guinea"	Weitkamp *et al.*[81], cites Peters (unpublished)	English/American
Xavante	Weitkamp & Neel (unpublished)	Weitkamp *et al.*[81]	South American Indigenes
Yanomama	Dimer	Weitkamp & Neel[82]	South American Indigenes
Yanomama-2	Compared in Weitkamp *et al.*[81]	Tanis *et al.*[45,48]	South American Indigenes
Unnamed (B?)	Antigenic differences	Margni *et al.*[68]	Argentine families of European backgrounds
Unnamed (B?)	Referred to as Z.N. and indistinguishable from SO/CZ Weitkamp *et al.*[50]	Bonazzi[65]	Not stated (probably European)
Unnamed (B?)	"Ann Arbor" of Weitkamp *et al.*[72]	Adams[59]	Not stated
Unnamed (B?)	"Very slow" Weitkamp *et al.*[75]	Drachmann *et al.*[20]	European
Unnamed (B?)	"Very slow" Weitkamp *et al.*[75]	Robbins *et al.*[11]	United States (no ethnic affiliation stated)
Unnamed (B?)	"Very slow" Weitkamp *et al.*[75]	Wuhrmann[2]	European
Unnamed (B?)	"Very slow" Weitkamp *et al.*[75]	Cooke *et al.*[16]	Not stated
Unnamed (B?)	"Very slow" Weitkamp *et al.*[75] Same as "slow" of Efremov & Braend[18]	Braend *et al.*[19]	European
Unnamed (B?)	"Very slow" Weitkamp *et al.*[75]	Adner & Redfors[17]	
Unnamed (B?)	"Very slow" Weitkamp *et al.*[75]	Adner (unpublished) first appeared in Weitkamp[75]	Not stated
Unnamed (B?)	"Very slow" Weitkamp *et al.*[75]	Weitkamp *et al.*[72]	European
Unnamed (B?)		Sarcione & Aungst[13]	European
Unnamed (B?)		Franglen *et al.*[15]	Not stated (probably European)
Unnamed	Dimer?	Fraser *et al.*[83]	Probably European
Unnamed	May be "Gainesville"	McDermid & Vos[84]	Indian residents of Natal, So. Africa
Unnamed		Atal *et al.*[80]	India
Unnamed		Fukunaga & Glober[31]	Hawaii, Japanese ancestry
Unnamed	Slow, dimeric?	Laurell & Nilehn[63]	Not stated (probably European)
Unnamed	"Faster" Weitkamp *et al.*[75]	Robbins (unpublished) first mentioned in Weitkamp *et al.*[75]	Not stated
Unnamed	"Faster" Weitkamp *et al.*[75]	Weitkamp & Rucknagel (unpublished, first appeared in Weitkamp *et al.*[75]	Not stated
Unnamed	"Faster" Weitkamp *et al.*[75]	Weitkamp *et al.*[72]	Not stated
Unnamed	"Faster" Weitkamp *et al.*[75]	Drachmann (unpublished) first appeared in Weitkamp *et al.*[75]	Not stated
Unnamed	"Very fast" Weitkamp *et al.*[75]	Wieme (unpublished) appeared in Weitkamp *et al.*[75]	Not stated

acids carry no charge. In this case the variant molecules will behave identically in an electric field and could not be distinguished. It is possible, therefore, that albumin variants not detectable through electrophoresis will be discovered as the techniques for mapping the amino acid sequence of albumin, or of detecting non-charge amino acid differences, are more widely applied.

The first variants were discovered by paper electrophoresis and were characterized by electrophoretic mobilities markedly different from that of normal albumin. As new combinations of support medium, buffer and conductors were used in screening serum proteins new albumin variants were discovered. Many of these new variants have electrophoretic mobilities quite close to normal albumin and cannot be detected by paper or cellulose acetate electrophoretic systems used in routine screening but can be detected by the use of other support media (i.e. cellulose acetate or acrylamide gel; see, for example, Fig. 1). At present no single system is able to distinguish all the albumin variants from each other and from normal albumin.

A list of albumin variants that have appeared in the literature is presented here (Table 1). Some of these variants, however, have not yet been distinguished from other known variants by electrophoresis. It is possible that there are more names than there are actually different variants.

Recently Weitkamp *et al.* [50,75,77] have compared samples of many of the variants using three starch gel systems. The first employs the sodium acetate-ethylenediaminetetraacetic acid system (pH 5.0) described by Kueppers and Bearn. [85]

TABLE 2. RELATIVE MOBILITY OF TWENTY-FIVE ALBUMIN VARIANTS IN THREE STARCH GEL ELECTROPHORETIC SYSTEMS (FROM WEITKAMP [77])

Variant	Buffer system		
	pH 5.0	pH 6.0	pH 6.9
1. RS-I	1	2	6
2. Pollibauer	1	2	4
3. Belem I	2	1	4
4. B	3	3	1
5. Roma	3	1	3
6. Gainesville	3	3	4
7. Paris (Gombak)	4	3	5
8. Kashmir (Afghanistan)	5	2	1
9. Otsu	6	4	2
10. Santa Ana	7	4	3
11. SO/BS	7	5	3
12. Cartago	7	3	5
13. Xavante	8	6	7
14. Pushtoon	8	4	6
15. Cayemite	9	5	6
16. Mexico	9	4	5
17. Uinba	9	6	7
18. Yanomama-2	10	6	6
Normal albumin	10	6	7
19. Medan	11	6	8
20. Maku	12	7	10
21. Reading (New Guinea)	13	9	9
22. Makiritare-3	13	8	9
23. Naskapi	14	10	10
24. Gent	15	11	10
25. Kyoto	16	11	10

The second employs a tris-lithium-succinate-citrate gel buffer (pH 6.0) and a vessel buffer of the same composition but diluted 3:1.[43,50] The third system[50] has a tray buffer of 0.03 M tris (hydroxymethyl) amino-methane, 0.043 M ethylenediaminetetra-acetic acid, 0.27 M boric acid (pH 6.4) and a gel buffer of the same but diluted 1:6.2 (pH 6.9). With these three systems twenty-five variants were distinguished. These results are presented in Table 2 (adapted from Weitkamp[77]).

Some variants (Table 2) which appear to have the same mobility in one electrophoretic system may be distinguished from each other by another combination of buffers and media. The most acidic buffer (pH 5.0) is best for comparing the fast variants but is relatively ineffective for distinguishing the slower ones. Both of the more alkaline systems are necessary to distinguish all of the slow variants. Screening for albumin variants with only one or two of these systems does not assure the investigator that all known variants will be detected.

Since Weitkamp's most recent comparison in 1973, there have been additional reports of albumin variants. Determining whether these variants are new or not requires the extensive comparisons just described. Since most investigators do not have a bank of variants for comparison, some of these "new" variants may be identical to those already discovered. These new variants include albumin Kuala Lumpur,[34] albumin Birmingham,[39] albumin Boal,[73] albumin Luarca[62] and an unnamed variant discovered among Japanese émigrés to Hawaii.[31]

The recognition of these and future discoveries as "new" variants rests entirely upon comparisons with those variants already known. At present there is no centralized system for the extensive electrophoretic testing required to determine the status of "new" variants. Ultimately, the recognition of new species of albumin molecules will depend upon amino acid mapping, but until then the status of new variants will be somewhat ambiguous.

III. FAMILY STUDIES

A. Mode of Inheritance

When albumin variants were discovered in the families of propositi the hypothesis was advanced that the albumin phenotype was under genetic control.[2,5,9,13,15] Since then numerous pedigrees have been published (examples are given in Fig. 2) all of which support the hypothesis that two autosomal codominant alleles, each having complete penetrance and segregating at one locus, control the albumin phenotype.[3,4,11,12,22,30,43,49,57,63,79,86] Segregation analysis (Table 3) also supports this hypothesis.

B. Linkage to Other Loci

Pedigree data from families with alloalbuminemic members have also been analyzed to test the possibility of linkage between structural loci for albumin and other genetic systems. Table 4 presents a list of the genetic loci systems tested for linkage with the albumin locus. Several statistical techniques have been employed; the most common of these are those of Finney,[87] Haldane and Smith,[88] Morton[89] and Smith.[90]

Fig. 1. Electrophoretic patterns of serum albumin; vertical electrophoresis using a 5% acrylamide gel at pH 8.4. O = origin; + = direction of anode; (1) Al^{A}/Al^{Me}; (2) Al^{A}/Al^{Na}; (3) Al^{Na}/Al^{Na}; (4) Al^{A}/Al^{A}. (From Johnston *et al.* (53)).

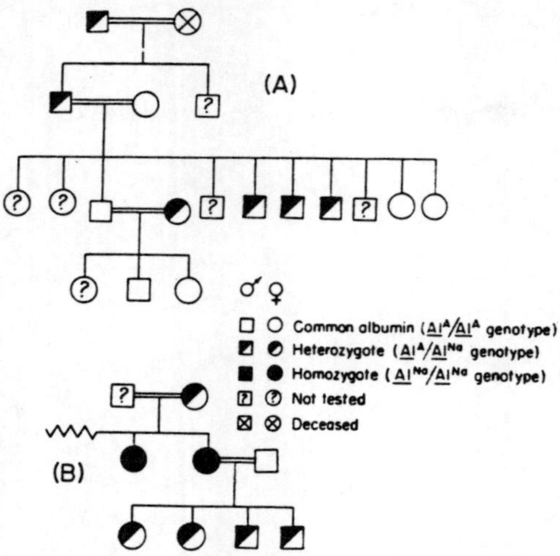

Fig. 2. Pedigrees of Montagnais families with heterozygotes (A) and homozygotes (B). (From Melartin.[30])

TABLE 3. SEGREGATION OF ALBUMIN TYPES IN FAMILIES (FROM MELARTIN [30])

Albumin types of parents	No. of families	Albumin type of children						No. of children
		A/A		A/Na		Na/Na		
		obs.	exp.	obs.	exp.	obs.	exp.	
A/A × Na/Na	2	0	0	11	11	0	0	11
A/A × A/Na	14	24	26	28	26	0	0	52
		A/A		A/Me		Me/Me		
A/A × A/Me	1	0	2	4	2	0	0	4

Weitkamp *et al.* [72] and Kaarsalo *et al.* [91] initially recognized linkage between the albumin and Gc loci. Since these first analyses more pedigrees have become available in the literature. From a study of twenty-one pedigrees, fifteen from the literature and six new ones, Weitkamp *et al.* [92] concluded that the most likely value for the recombination fraction of albumin and the Gc genotypes is 0.023 (with 95% probability limits of 0.008 and 0.047). Furthermore, recombination values computed for each pedigree were about the same although six different albumin variants were segregating in the pedigrees (probably albumins B, Santa Ana, Gainesville, Syracuse, Naskapi and a dimeric variant). This suggests that all variants are allelic at the same locus. Bowen *et al.* [93] using the twenty-one pedigrees from Weitkamp *et al.* [92] and three new ones, computed a map distance for the albumin and Gc loci of 2.5 centimorgans with 95% probability limits of 0.9 to 5.0 centimorgans.

TABLE 4. LINKAGE BETWEEN ALBUMIN AND OTHER LOCI

Genetic system	Linkage	Statistical methods [a]	Reference
Group specific component (Gc)	+	?	Arends *et al.*[43]
	+	a	Bowen *et al.*[93]
	+	b, c, d, e	Kaarsalo *et al.*[91]
	+	?	Kueppers *et al.*[64]
	+	a, c, d, e	Sunderman *et al.*[54]
	+	a, c, d	Weitkamp *et al.*[72]
	+	a	Weitkamp *et al.*[76]
	+	c	Weitkamp *et al.*[92]
ABO	−	a	Bowen *et al.*[93]
	−	b, c, d, e	Kaarsalo *et al.*[91]
	−	a, b, c, d, e	Sunderman *et al.*[54]
	−	a, c, d	Weitkamp *et al.*[72]
Rh	−	a	Bowen *et al.*[93]
	−	b, c, d, e	Kaarsalo *et al.*[91]
	−	a, c, d	Weitkamp *et al.*[72]
MNS	−	a	Bowen *et al.*[93]
	−	b, c, d, e	Kaarsalo *et al.*[91]
	−	a, c, d	Weitkamp *et al.*[72]
Duffy	−	b, c, d, e	Kaarsalo *et al.*[91]
	−	a, c, d	Weitkamp *et al.*[72]
Kidd	−	a	Bowen *et al.*[93]
	−	b, c, d, e	Kaarsalo *et al.*[91]
	−	a, c, d	Weitkamp *et al.*[72]
P	−	a	Bowen *et al.*[93]
	−	b, c, d, e	Kaarsalo *et al.*[91]
Haptoglobin	−	b, c, d, e	Kaarsalo *et al.*[91]
Transferrin	−	b, c, d, e	Kaarsalo *et al.*[91]
Methemoglobinemia	−	d	Scott and Wright[94]
Alpha chain haptoglobin		?	Kueppers *et al.*[64]

[a] a, Haldane and Smith;[88] b, Finney;[87] c, Smith;[90] d, Morton;[89] e, Mohr.[95]

There is a very low probability of linkage between the albumin locus and the other loci which have been tested. Scott and Wright[94] reported that the probability of linkage between hereditary methemoglobinemia and albumin was less than 0.08. The probability of linkage between the albumin locus and the ABO, MNS, Duffy, Kidd, P, haptoglobin and transferrin loci is also small.[57,72,91,93] Kueppers *et al.*[64] have tentatively suggested that there may be linkage between the loci for albumin and the alpha chain of haptoglobin, but this hypothesis has not been tested by any other investigators.

IV. THE DISTRIBUTION OF ALBUMIN VARIANTS

A. Classification

At present it is difficult to determine the exact number of albumin variants. There are approximately twenty-five variants which have been distinguished from each other using one or more electrophoretic systems (see Table 2), [50,77,81] but this figure represents less than half the number of albumin variants that have been reported. Undoubtedly the discrepancy is due to the use of several names to apply to a smaller number of variants. Several practices are responsible for the introduction of synonyms into the literature on albumin variation. Occasionally, a new variant is not named by its discoverer and is later given several names by investigators working independently. At other times a variant which had been given a provisional but formally precedent name is renamed. In some instances a variant which was thought to be new is later recognized to be identical with a previously described variant. The result of these practices is a proliferation of variant names, many of which are not commonly used or recognized.

A list of variants is presented in Table 1. This table includes variants which were not named by their discoverers, variants which have been renamed, variants which may be rediscoveries of previously identified variants and variants which appear in comparative studies but which have not been described in the literature individually. In Table 1 the variant name is listed (when given by the discoverer), notes on similarities to other variants, the geographical provenance (when given by the discoverer), the reference for the first mention of the variant, and when there are synonyms for a variant, the formally precedent name is indicated.

This extensive list serves several purposes. It serves as a reference for investigators interested in the original source of any albumin variant discovered. It may help to resolve the history of those variants which have been lost from the literature by multiple naming or by other means. Furthermore, it serves to document the history of the field, and is a testimony to the large amount of albumin variation and the difficulty of settling on a uniform system of reporting and naming albumin variants.

The variants listed in Table 1 can be classified according to their frequency and distribution. Variants may be either rare or polymorphic depending on whether they occur at a frequency of 1% or more. This distinction is based on E. B. Ford's classic definition of genetic polymorphism, "the occurrence together in the same habitat of two or more discontinuous forms or 'phases', of a species in such proportions that the rarest of them cannot be maintained merely by recurrent mutation" (ref. 96, page 11). Although Ford's definition and his comments elsewhere [96] indicate that genetic polymorphisms are maintained by natural selection alone rather than by genetic drift or recurrent mutation, we cannot be certain that natural selection alone is responsible for the polymorphic albumin variants. The role of genetic drift, particularly drift that is caused by cultural practices and historical accident, cannot be accurately estimated in the American Indian populations where the few albumin variants that reach high frequencies occur. In the classification presented here, the term polymorphic means simply that the variant alleles occur at a frequency above 1%, and by implication at too high a frequency to be maintained by recurrent mutation.

The classes, rare and polymorphic, may be subdivided on the basis of distribution. There are three classes of rare variants: (1) unique variants which have been seen only in the family in which they were discovered; (2) restricted rare variants which occur in a small number of unrelated individuals of one ethnic group; (3) unrestricted rare variants which occur in a small number of unrelated individuals of different ethnic groups and presumably different genomes. The polymorphic variants may be

TABLE 5. CLASSIFICATION OF ALBUMIN VARIANTS BY FREQUENCY AND DISTRIBUTION [a]

Rare			Polymorphic	
Unique[b]	Restricted	Unrestricted	Restricted	Dispersed
Belem I	B (Ann Arbor, Jensen	Paris (Gombak)	Warao (Makiritare,	Mexico (Belem II, P)
Birmingham	SO/CZ)	Reading (Maprik,	Makiritare-I)	Naskapi (Algonkian)
Boal	Gainesville	New Guinea,	Yanomamo-2	
Caracas	Gent (MI/FAST)	Syracuse,		
Carib	Kashmir (Afghanistan)	Westcott)		
Cayemite	Maku (Belem III,			
Luarca	Makiritare-2,			
Makiritare-3	Wapishana)			
Pollibauer	Medan			
Santa Ana	Yanomamo			
Sentul				
Uinba				

[a] The following variants from Table I cannot be classified since detailed information on the genetic relatedness or distribution of the individuals having the variants is not available: A_2; Albuquerque; Al^r; Kyoto; Larson; Oliphant; Otsu; Pushtoon; RS-I; RS-II; SO/BS; Xavante.

[b] As noted in the text, not all variants have been compared directly and so some may not be different.

either restricted to one ethnic group and locality, or dispersed among several ethnic groups and localities. A classification of variants according to these criteria is presented in Table 5.

B. Rare Variants

As can be seen from Table 5 most variants are unique. The number of unique variants may be inflated somewhat since not all variants have been compared directly to all others and so some may be identical to already described and named variants.

Of the restricted rare variants, albumins B, Gainesville and Gent have been found only in families of European origin. Albumins Kuala Lumpur and Medan are restricted to families of Far Eastern origin. Albumins Maku and Yanomama are found only in South American Indian families. Although these variants are not unique they are very rare.

Albumin Kashmir is restricted to families of Indo-Dravidian descent, but it occurs more often than any of the other restricted rare variants. Data on albumin Kashmir and other variants which also occur among Indo-Dravidian populations are presented in Table 6. As can be seen from the table, several different variants appear to be involved but this impression is misleading. Albumin Kashmir, originally described by

TABLE 6. ALBUMIN VARIANTS FROM INDIAN POPULATIONS

Location and population	Variant name	Number tested	Number variant	Reference
India				
New Delhi	unnamed (Kashmir) [a]	518	2	37
Madras	—	275	0	37
Bombay	—	485	0	37
Bombay (Parsis)	—	363	0	97
Bombay (Iranis)	—	33	0	97
West Bengal (Koaras)	—	202	0	98
Kerala (Kadars)	—	213	0	99
South Indians	—	50	0	51
England				
Kashmiri family (Pakistan)	Kashmir	?	2	35
Birmingham émigré (from India)	Birmingham	hospital screening	1	39
Afghanistan				
Gawargia (Pushtoon)	Afghanistan	251		38
	(Kashmir)[a]		10	
	Pushtoon		7	
Malaysia				
Indians	unnamed	36	0	37
Indians	—	532		33
	Gombak		1	33
	Sentul		2	33
South Africa				
Natal		193	1	37

[a] Albumins "Kashmir" and "Afghanistan" are electrophoretically identical.[81] Albumin "Kashmir" is the formally precedent name.

Tarnoky and Dowding,[35] albumin Afghanistan[38] and the unnamed variant reported by McDermid[37] are indistinguishable when compared directly.[50,81] (McDermid's original suggestion of similarity between the unnamed Indian variant and albumins Paris and Gainesville may have been based on arithmetic descriptions of migration rates and not on direct comparisons.) Albumin Pushtoon is electrophoretically distinguishable from Kashmir[50] and albumin Birmingham shows different binding properties, but not different electrophoretic mobility from albumin Kashmir.[39] There are then three variants involved, Pushtoon, Birmingham and Kashmir. Weitkamp *et al.*[50,81] have suggested that albumin Kashmir may be a low-frequency polymorphism in Indo-Dravidian populations. This possibility is based on the presence of the variant in geographically diverse Indo-Dravidian populations from New Delhi (India), Natal (South Africa), Kashmir (Pakistan) and Gawargia (Afghanistan). The variant does not reach polymorphic frequencies, although calculation of the gene frequency is an estimate since alloalbuminemic individuals from one village may be related or may not have been ascertained through a random sampling of the population.

The unrestricted rare variants, albumins Paris and Reading, occur in both European and Asian populations. Albumin Paris was first discovered in a French family.[12] Several years later a variant termed "Gombak" was discovered among Malays and Indians of Malaysia.[33] In a direct comparison of the variants they displayed identical electrophoretic mobility,[50] consequently the name "Paris" should be adopted to refer to this variant and the name "Gombak" dropped.

Similarly albumin Reading was first found in a European family[21] and other variants, "Maprik"[41] and "New Guinea"[40,100] from New Guinea, and "Syracuse"[80] and "Westcott"[81] from families of European origin, were subsequently found to be identical.[50] Albumin Reading is the formally precedent name. Weitkamp *et al.* have suggested that the distribution of albumin Reading in New Guinea may not be accounted for solely by random genetic forces.[81] The data from New Guinea on albumin Reading and albumin Uinba, a rare variant restricted to one village in New Guinea, are presented in Table 7. As can be seen from the table, albumin Reading just reaches the frequency of 1% which denotes a polymorphism when widely separated villages are combined to form one population.

C. Polymorphic Variants

Polymorphic variants, both restricted and dispersed ones, occur only among Native American populations. Furthermore, the dispersed polymorphic variants, albumins Naskapi and Mexico, have been detected only among North and Central American Indian and mestizo populations while the restricted polymorphic variants, albumins Yanomama-2 and Makiritare, are found only among South American Indian populations.

1. *Restricted polymorphic variants*

Two variants, albumins Yanomama-2 (Yan-2) and Makiritare, reach polymorphic frequencies within the ethnic groups for which they were named. Data on albumin variants in South America are presented in Table 8.

TABLE 7. ALBUMIN VARIANTS FROM NEW GUINEA POPULATIONS

Population and location	Variant name	Number tested	Number variant	Gene frequency	Reference
New Guinea Highlanders	—	50	0	0.0	51
New Guinea natives					
Marksham Valley	New Guinea (Reading) [a]	35	2		40
Papondetta	,, ,,	45	1		40
Rossel Island	,, ,,	31	4		40
Other villages	—	613	0		40
Total:	New Guinea (Reading) [a]	724	7	0.001 [b]	40
relatives of Rossel Is. Propositi	New Guinea (Reading) [a]	—	12		40
Uniba Village	Uniba	82	4		40
New Guinea natives Maprik region	Maprik (Reading) [a]	595	1	0.002 [b]	41
New Guinea natives Manus Islands	—	183	0	0.0	101
New Guinea natives Eastern Highlands and Islands	New Guinea (Reading) [a]	1307	7	0.0053 [b]	81, McDermid (unpublished)
New Guinea natives Tuade speakers, Goilala	—	145	0	0.0	102

[a] These gene frequencies were not published but were calculated for this paper from the published numbers of normal and heterozygotic individuals. They are estimates based on the application of the direct count method. In some cases individuals may be related.

[b] Albumins "Reading", "New Guinea" and "Maprik" are electrophoretically identical.[92] Albumin "New Guinea" is the formally precedent name.

TABLE 8. ALBUMIN VARIANTS IN SOUTH AMERICAN POPULATIONS

Population and location	Variant name	Number tested	Number variant	Gene frequency	Reference
Venezuela & Brazil					
Yanomama	Yanomama-2 (tested	2943	461 (30)[a]	0.076	48
Piraoa	(for Yan-2 only)		0		48
Macushi	,,		0		48
Wapishana	,,		0		48
Makiritare	,,		0		48
Xavante	,,				48
Total (Non-Yanomama)	,,		0	0.0	48
Makiritare					
Juduwaduna Village	—	71	0		44
Santa Maris Village	—	154	0		44
Wasana Village	Makiritare(-1)'	42	3		44
Chajorana Village	,,	74	5		44
Sharamana Village	—	47	0		44
Belem Village	Makiritare(-1)	72	3		44
Acanana Village	,,	74	2		44
Total	Makiritare(-1)	534	13	0.01	44
Warao	Warao (Makiritare)	222	5		43
Yanomama	Unnamed	984	49		45
Makiritare	,,	186	2		45
Piaroa	,,	146	2		45
Macushi	,,	188	0		45
Wapishana	,,	62	15		45
Cayapo	—	89	0		45
Yanomama					
Borabuk Village	Maku (Makiritare-2)	75	4		42
Other villages	—	1036	0		42
Yanomama		2256			103
	Maku (Makiritare-2)		4		103
	Yanomama		·1		103

Brazil					
Belem (Trihybrid)		1360			46
	Belem I	family	5		46
	Belem II (Mexico)	,,	6		46
	Belem III	. ,,	1		46
Porto Alegre (Trihybrid)	—	1628	0		46
Xavante					
San Marcos Village	Xavante	214	2		81
Sao Dimingos and Sinoes Lopes villages	—	245	0		81
Cayapo					
Xikrin Village	—	95	0		104
Grorotire Village	—	201	0		104
Kuben-kran-kegn Village	—	164	0		104
Mekranoti Village	—	92	0		104
Txukahamae Village	—	156	0		104
Surinam					
Trio					
Pelelu Tepu Village	Makiritare-1	158	9		47
Alalaparu Village	,,	255	2		47
Wajana					
Puleowime Village	Makiritare-1	81	7		47
Kawemkakan Village	,,	198	10		47
Colombia					
Noanama	—	155	0		105
Cofan	—	68	0		105
Ingana (Quechua)	—	41	0		105
Peru					
Quechua	—	100	0		30
Cashinahua	—	92	0		30

(a) Figure in parentheses refers to number of homozygotes.

Yanomama-2 is found at a frequency of 7.6% among sixty-four Yanomama villages from southern Venezuela and northern Brazil.[48] The frequency tends to be highest in the northern area; the highest value for any village is 40%. Screening specifically for Yanomama-2, which requires a novel electrophoretic system,[48] has been performed on 2562 sera from several tribes of North and South America. In only one case, and that from the nearby Makiritare tribe, was Yanomama-2 detected. Hence, Yanomama-2 appears to be polymorphic within a single ethnically defined group.[48]

Albumin Makiritare (also called Warao) has been found among the Trio and Wajana tribes of Surinam,[47] the Warao of Venezuela[43] and the Makiritare.[44] It reaches a frequency of 4% in some Makiritare villages[44] and may be even more frequent in some Wajana villages.[47] The average frequency for the variant in seven Makiritare villages is 1%.[44] Albumin Makiritare reaches polymorphic frequencies among several tribal groups, but its distribution appears to be limited to the northeastern area of South America only.

2. *Dispersed polymorphic variants*

(a) *Distribution.* The only variants that have been found at polymorphic frequencies in samples from several widely separated geographic areas are albumins Naskapi and Mexico. These variants are found only among American Indian populations where they often reach frequencies above 1%.[53,78,93,94,106] The distribution of these variants reaches from Alaska to Guatemala. Albumin Naskapi is found predominantly among northern populations while albumin Mexico is found among Middle American populations. Both variants are found among some populations in the greater southwest. In this survey the populations are grouped by linguistic affiliation to facilitate description. In the subsequent sections linguistic groupings form only a part of the interpretation.

Table 9 shows the distribution of albumin variants among Native Americans of North and Middle America. (Stated in this table are the linguistic affiliation, tribal name, location, total number in the sample, number of albumin Naskapi heterozygotes and, in parentheses, homozygotes, the number of albumin Mexico heterozygotes, and homozygotes in parentheses, and the gene frequencies of each variant. Gene frequencies are direct count estimations from the frequency of heterozygotes and homozygotes. Some authors have calculated gene frequencies with a correction factor for including related individuals in the populations: these data are noted in parentheses.)

Albumin Naskapi is found at high frequencies among the Algonkian-speaking Naskapi and Montagnais Indians of Quebec,[22] among Athabascan-speaking Slave and Beaver Indians of Alberta[93] and among Athabascan Indians of Alaska.[22,94] Lower frequencies, from 0.5% to 3.4%, are found among the Sioux,[22] Assiniboine,[93] Ojibwa,[48,70,107] Cree,[93] Chippewa[48,76,77] and Blackfoot[70] who are found in the plains and forested regions of the United States and Canada.

Very low frequencies of albumin Naskapi had been reported for a Tlingit Indian sample[22] and for two Eskimo samples.[69] Some of the alloalbuminemic individuals within each sample were later found to be of mixed ancestry, or of a different ethnic group than had originally been supposed. (All the samples in question were collected near Indian populations where the Naskapi gene is present.) Albumin Naskapi now appears to be absent from the Tlingit [30] and Haida Indians of the Canadian Pacific coast [22] and from numerous Eskimo samples from Alaska to Greenland. [22,30,56,108] Albumin Naskapi is also absent from samples of Omaha Indians of the Macy reservation [109] and from Floridian Indians. [110] Albumin Mexico has not been found in any of the northern Native Americans who have been tested.

The distribution of albumin Mexico is limited to more southern regions. Among Middle American Indian and Mestizo populations albumin Mexico is occasionally found at polymorphic frequencies. [25,106,111,112] It is present from Tamulipas, Mexico, on the Gulf Coast in the north [106] to Guatemala City, Guatemala, in the south. [112] The highest frequencies of albumin Mexico in Middle America have been reported for two Mestizo samples, both in Tlaxcala and Hidalgo, Mexico. [106,111]

Albumin Mexico also reaches polymorphic frequencies in the southwest United States. The highest frequencies have been found among the Yuman River-speaking Maricopa and Cocopah [77] and among the Uto-Aztecan-speaking Pima [53,78] and Zuni. [113] Lower frequencies have been noted for Papago samples. [30,53,113] Several other samples from southwest tribes have been tested, [53] but these samples are too small for an accurate estimate of gene frequency.

Occasionally, both albumin Mexico and Naskapi are found in the same population. These occurrences are limited to several southwestern groups, the Navajo and Apache [53] and the Mohave and Maricopa. [53] Within each group, the proportion of the variants to each other and to common albumin (Al^A) is different (see Table 9).

(b) *Anthropological interpretation.* These differences suggest that albumin variants may be a useful indicator of contacts and movements among Native American populations. There are features of the distribution of albumin variants which support anthropologists' theories regarding relationships among Native American peoples.

Johnston *et al.* [112] noted that the distribution of albumin Mexico in Guatemala paralleled the archaeological evidence of past cultural relationships there. Archaeological evidence suggests that the cultures of the Guatemalan highlands were greatly influenced by a northern culture, perhaps one centered near the Valley of Mexico which surrounds the modern Mexico City. [114] The distribution of albumin Mexico reflects this relationship, being present in the Guatemalan highlands [112] but absent from the lowland Yucatan [106] where influence from the Valley of Mexico was presumably less intense.

The distribution of albumin Mexico to the north of the Valley of Mexico may parallel cultural relationships there also. Archaeological research suggests that there may have been contact between the Valley of Mexico cultures and peoples in the southwest United States, in particular, the Hohokam culture [115] and peoples from the area around Tamulipas, Mexico. [116] Albumin Mexico has been found at polymorphic frequencies in a sample from Tamulipas, [106] in Pima samples [53,78] and in one Papago sample. [53] The Pima and Papago are thought to be descendants of the Hohokam. [117] The presence of albumin Mexico among them may reflect earlier contact with Valley of Mexico peoples, and if so, the far higher frequency of albumin Mexico among the Pima may further indicate much more contact with the Pima than with the Papago. The distribution of albumin Mexico appears to parallel the area of Valley of Mexico influence in the American southwest as well as in Guatemala although the direction of gene flow and cultural contact cannot be presumed from distributional data.

TABLE 9. THE DISTRIBUTION OF ALBUMIN MEXICO AND ALBUMIN NASKAPI IN NORTH AND CENTRAL AMERICA

Linguistic affiliation, tribal name (location)	Number tested	Number Naskapi (homozygotes in parentheses)	Number Mexico	Naskapi gene frequency	Mexico gene frequency	Reference
Algonkian						
Ojibwa (So. Ontario)	117	6	0	0.026	0.0	107
Ojibwa (No. Ontario)	94	1	0	0.005	0.0	107
Ojibwa (?)	120	8	0	0.033	0.0	48
Ojibwa/Anglo (No. Minnesota)	250	11 + (1)	0	0.026	0.0	70
Chippewa/Cree (Saskatchewan)	610	36 + (1)	0	0.031	0.0	77
Chippewa/Cree (Saskatchewan)	102	2	0	0.009	0.0	76
Chippewa (?)	87	2	0	0.011	0.0	48
Cree, Plains (Alberta)	605	25 + (1)	0	0.022 (0.023) [a]	0.0	93
Cree, Northern (Alberta)	187	10	0	0.026 (0.034) [a]	0.0	93
Blackfoot (Montana)	97	4	0	0.020	0.0	70
Naskapi (Schefferville, Quebec)	151	37 + (1)	0	0.130	0.0	22
Montagnais (Schefferville, Quebec)	112	14 + (2)	0	0.080	0.0	22
Siouian						
Sioux (?)	160	2	0	0.007	0.0	22
Assinboine (Alberta)	100	1	0	0.005 (0.007) [a]	0.0	93
Omaha (Macy Reservation, Nebraska)	96	0	0	0.0	0.0	109
Eskimoan						
Eskimo (Igloolik, Canada)	356	0	0	0.0	0.0	108
Eskimo (E. Greenland)	78	0	0	0.0	0.0	56
Eskimo (Thule, Greenland)	297	0	0	0.0	0.0	56
Eskimo (W. Greenland)	116	0	0	0.0	0.0	56
Eskimo (Wainwright, Alaska)	111	0	0	0.0	0.0	22
Eskimo (Anatavuk)	55	0	0	0.0	0.0	22
Eskimo (Ungava Bay, Canada)	124	3 [b]	0	0.012 [b]	0.0	69

Eskimo (Barrow, Alaska)	88	0	0	0.0	0.0	69
Eskimo (Beaver, Alaska)	14	0 [b]	0	— [b]	—	69
Eskimo (St. Lawrence Is., Alaska)	82	0	0	0.0	0.0	69
Eskimo (Nat'l Guard, Alaska)	107	0	0	0.0	0.0	69
Aleut (Sitka & Nat'l Guard, Alaska)	15	0 [b]	0	—	—	69
Na-Dene (non-Athabascan branch)						
Tlingit (Sitka & Mt. Edgecomb, Alaska)	91	0 [b]	0	0.0 [b]	0.0	30
Haida (Queen Charlotte Is., British Columbia)	365	0	0	0.0	0.0	22
Na-Dene (Athabascan branch)						
Alaskan Athabascans (Alaska)	230	11 + (1)	0	0.028	0.0	22
Alaskan Athabascans (Alaska)	137	14	0	0.051	0.0	94
Slave & Beaver (Alberta)	143	22 + (2)	0	0.090 (0.098) [a]	0.0	93
Navajo (Arizona & New Mexico)	95	12	0	0.063	0.005	69
Navajo (Sherman School, Ca.)	468	30	6	0.032	0.006	53
Navajo (?)	192	11	4	0.03	0.010	48
Apache (Sherman School, Ca.)	103	13	7	0.015	0.034	53
Yuman-River						
Cocopah (Arizona)	164	0	13	0.0	0.039	78
Maricopa (Arizona)	101	0	9 + (1)	0.0	0.054	78
Maricopa (Sherman School, Ca.)	9	1	0	—	—	53
Mohave (Sherman School, Ca.)	19	1	1	—	—	53
Yuman-Upland						
Havasupi (Sherman School, Ca.)	6	0	0	—	—	53
Walapai (Sherman School, Ca.)	8	0	0	—	—	53
Yavapai (Sherman School, Ca.)	3	0	0	—	—	53
Uto-Aztecan						
Pima (Arizona)	1528	0	100 + (7)	0.0	0.037	78
Pima (Sherman School, Ca.)	113	0	11	0.0	0.049	53

(For table notes [a] and [b] see page 135.)

(continued overleaf)

TABLE 9 (*cont.*)

Linguistic affiliation, tribal name (location)	Number tested	Number Naskapi (homozygotes in parentheses)	Number Mexico	Naskapi gene frequency	Mexico gene frequency	Reference
Uto-Aztecan (*cont.*)						
Papago (Sherman School, Ca.)	115	0	3	0.0	0.013	53
Papago (Arizona)	179	0	2	0.0	0.006	30
Papago (Arizona)	546	0	5	0.0	0.005	113
Hopi (Sherman School, Ca.)	82	1	0	0.006	0.0	53
Ute (Sherman School, Ca.)	17	0	1	—	—	53
Paiute (Sherman School, Ca.)	4	0	0	—	—	53
Chemehueve (Sherman School, Ca.)	2	0	0	—	—	53
Mission (Sherman School, Ca.)	1	0	0	—	—	53
Washoan						
Washoe (Sherman School, Ca.)	1	0	0	—	—	53
Zunian						
Zuni (New Mexico)	655	0	16	0.0	0.012	113
Eastern Muskogean						
Tribal affiliation indeterminate:						
Dania (Fla.)	150	0	0	0.0	0.0	110
Big Cypress (Fla.)	135	0	0	0.0	0.0	110
Brighton (Fla.)	128	0	0	0.0	0.0	110
Macro-Nahua						
Nahua (Puebla, Hidalgo, Veracruz, Mexico)	440	0	6	0.0	0.006	106
Otomi (Hidalgo, Mexico)	100	0	5	0.0	0.025	106
Cora (Nayrati, Mexico)	83	0	2	0.0	0.012	106
Macro-Maya						
Maya (Yucatan, Mexico)	263	0	0	0.0	0.0	111
Huasteco (Veracruz, Tamulipas, Mexico)	188	0	4	0.0	0.01	106
Tzeltal-Tzotzil (Chiapas, Mexico)	144	0	0	0.0	0.0	106

Chol (Chiapas, Mexico)	138	0	0	0.0	0.0	106
Dialect not reported						
(Guatemala City & San Marcos, Guatemala)	204	0	1	0.0	0.002	112
(San Antonio Ilotenango, Guatemala)	186	0	3	0.0	0.008	112
Macro-Mixteco						
Zapoteco (Oaxaca, Mexico)	255	0	3	0.0	0.005	106
Zapotec (Guelatao, Mexico)	123	0	1	0.0	0.004	111
Mixteco (Oaxaca, Mexico)	48	0	0	—	—	106
Mazateco (Oacaca, Mexico)	22	0	0	—	—	106
Mazatec (Huautla de Jimenez, Mexico)	20	0	0	—	—	111
Mixe (Oaxaca, Mexico)	21	0	0	—	—	106
Tarasco						
Tarasco (Michoacan, Mexico)	167	0	5	0.0	0.014	106
Mixed linguistic groups						
Mestizo-Zapotec (Pochutla, Mexico)	36	0	1	—	—	111
Mestizo-Mixtec						
(San Pedro Mixtec, Mexico)	20	0	0	—	—	111
(Ometepec, Mexico)	20	0	0	—	—	111
(Cauajinicuilapa, Mexico)	20	0	0	—	—	111
Mestizo (Tlaxcala, Hidalgo & Mexico, D.F., Mexico)	185	0	10	0.0	0.027	111
Mestizo (Tlaxcala–Hidalgo, Mexico)	257	0	12	0.0	0.023	106
Mestizo (Veracruz, Mexico)	109	0	1	0.0	0.004	106
Mestizo (Campeche, Mexico)	109	0	0	0.0	0.0	106
Mestizo (West Coast, Mexico)	99	0	1	0.0	0.005	106
Mestizo (Mexico City, Mexico)	1313	0	27	0.0	0.010	106

[a] Published gene frequency, calculated with correction for related individuals.
[b] A very low frequency was reported originally; this value has since been corrected to the value shown here.

The distribution of albumin Naskapi also parallels some cultural relationships among Indian groups of the American southwest. The Navajo and Apache for whom polymorphic frequencies of albumin Naskapi have already been noted are closely related linguistically to Canadian and Alaskan Athabascan-speaking Indians. Archaeological research suggests that the Navajo and Apache gradually moved into the southwest from 500 to 1000 years ago.[118,119] Glottochronological analyses support these dates.[120,121] The presence of albumin Naskapi among them may therefore be due to their past relationship with northern peoples who maintain albumin Naskapi at polymorphic frequencies. The presence of albumin Naskapi among indigenous peoples of the southwest may be due to contact with the Apache and Navajo. Indeed, there is evidence for such contact within the last 300 years.[115]

The distribution of albumin Naskapi in the far north substantiates theories regarding the relationship between Eskimos and Indians. Although both groups are considered Native Americans, a distinction between them has been made on biologic and cultural grounds (see, for example, references 117, 122 and 123). The absence of albumin Naskapi from most Eskimo samples and its presence among Indians lends support to this distinction.

Although the distribution of albumin Naskapi may reflect movements and contacts among Native Americans, the presence of the gene at polymorphic frequencies in some areas may have been produced by other factors. The nature of these evolutionary forces is not known, nor is it likely to be detected solely from distributional studies. However, the distribution of the gene might indicate environments where albumin Naskapi has been maintained at polymorphic frequencies by natural selection. For example, research into the functional differences between albumins, particularly into the different drug-binding properties of albumin Naskapi and common albumin (Al^A), might consider their activity with regard to substances common to the ecologies of the Naskapi, Montagnais, Slave, Beaver and Alaskan Indians. If selective forces operate on the albumin locus, they may be more apparent among those groups where albumin Naskapi is more frequent.

V. SUMMARY

The human serum albumin phenotype is controlled by two autosomal codominant alleles which segregate at a single locus. The albumin and Group Specific component loci are closely linked.

Variants of human serum albumin are indistinguishable from normal albumin by immunologic and ultracentrifugation techniques. The variants are distinguishable by their electrophoretic mobilities. There are approximately twenty-five electrophoretically distinguishable variants.

Most of these variants are unique although a few variants from different families have identical electrophoretic mobilities. Other variants, such as "Kashmir" and "New Guinea", may be more common, but do not reach a frequency of 1% and so may not be polymorphic. Albumin polymorphisms are known from Native American populations only. In South America there are two variants, albumin Makiritare and Yanomama-2, which reach polymorphic frequencies in limited areas. In North America albumin Naskapi and Mexico reach polymorphic frequencies in several widely separated geographic regions. The distribution of albumin Naskapi and Mexico may parallel relationships among Native American peoples and reflect their past movements and contacts. The evolutionary forces responsible for the high frequency of albumin variants in some areas are not known.

REFERENCES

1. Scheurlen, P. G. (1955) Über Serumeiweissveränderungen beim Diabetes Mellitus. *Klin. Wschr.* **33,** 198–205.
2. Wuhrmann, F. (1959) Albumindoppelzacken als vererbbare Bluteiweissanomalie. *Schweiz. Med. Wschr.* **89,** 150–152.
3. Wieme, R. J. (1960) On the presence of two albumins in certain normal human sera and its genetic determination. *Clin. Chem. Acta* **5,** 443.
4. Wieme, R. J. (1962) On the occurrence of different types in genetically determined bisalbuminemia. In: *Protides of the Biological Fluids. Proceedings of the Ninth Colloquium, Bruges, 1961,* H. Peeters, editor, pp. 221–274. Elsevier, N.Y.
5. Knedel, M. (1957) Die Doppel-Aibuminämie: eine neue erbliche Proteinanomalie. *Blut* **3,** 129–134.
6. Knedel, M. (1958) Über eine neue verebte Protein Anomalie. *Clin. Chem. Acta* **3,** 72–75.
7. Nennstiel, H. J. and T. Becht (1957) Über das erbliche Auftreten einer Albuminspaltung im Electrophoresediagram. *Klin. Wschr.* **35,** 689.
8. Nennstiel, H. J., T. Becht and J. Schwick (1969) Untersuchungen uber die hereditare Doppelalbuminämie. *Folia haemat.* **4,** 385–402.
9. Earle, D. P., M. P. Hutt, K. Schmid and D. Gitlin (1958) A unique human serum albumin transmitted genetically. *Trans. Ass. Am. Phycns* **71,** 69–76.
10. Earle, E. P., M. P. Hutt, K. Schmid and D. Gitlin (1959) Observations on double albumin: a genetically transmitted serum protein anomaly. *J. Clin. Invest.* **38,** 1412–1420.
11. Robbins, J. L., G. A. Hill, S. Marcus and J. H. Carlquist (1963) Paraalbumin: Paper and cellulose acetate electrophoresis and preliminary immunoelectrophoretic analysis. *J. Lab. Clin. Med.* **62,** 753.
12. Sandor, G., L. Martin, M. Porsin, A. Rousseau and R. Martin (1965) A new bisalbuminaemic family. *Nature* **208,** 1222.
13. Sarcione, E. J. and C. W. Aungst (1962) Studies on bisalbuminemia: Binding properties of the two albumins. *Blood* **20,** 156–164.
14. Ungari, S. and V. Lopez (1965) Doppia albuminaemia. *Minerva Pediat.* **17,** 288–291.
15. Franglen, G., N. H. Martin, T. Hargreaves, M. J. Smith and D. Williams (1960) A hereditary albumin abnormality. *Lancet* **I,** 307–308.
16. Cooke, K. B., T. E. Clegham and E. Lockey (1961) Two new families with bisalbuminemia: an examination of possible links with other genetically controlled variants. *Biochem. J.* **81,** 39.
17. Adner, P. L. and A. Redfors (1961) En slakt med tra albumin fraktioner i serumelektiofovesen (bisalbuminemia). *Nord. Med.* **65,** 623–624.
18. Efremov, G. and M. Braend (1964) Serum albumin: polymorphism in man. *Science* **146,** 1679–1680.
19. Braend, M., G. Efremov, M. K. Fagerhol and O. Hartman (1965) Albumin and transferrin variants in Norwegians. *Hereditas* **53,** 137–142.
20. Drachmann, O., N. M. G. Harboe, P. J. Svendsen and T. S. Johnsen (1965) Bis- or paraalbuminaemia. A genetic alteration in plasma albumin. *Dan. Med. Bull.* **12,** 74–79.
21. Tarnoky, A. L. and A. N. Lestas (1964) A new type of bisalbuminaemia. *Clin. Chim. Acta* **9,** 551–558.
22. Melartin, L. and B. Blumberg (1966) Albumin Naskapi: a new variant of serum albumin. *Science* **153,** 1664–1666.
23. Bell, H. E., S. F. Nicholson and Z. R. Thompson (1967) Bisalbuminemia of the fast type with a homozygote. *Clin. Chim. Acta* **15,** 247–252.
24. Weitkamp, L. R., G. Granglen, D. Rokala, H. Polesky, N. Simpson, F. Sunderman, H. Bell, J. Saave, R. Lisker and S. Bohls (1969) An electrophoretic comparison of human serum albumin variants. Eight distinguishable types. *Human Hered.* **19,** 159–169.
25. Melartin, L. and B. S. Blumberg (1966) Inherited variants of human serum albumin. *Clin. Res.* **14,** 482.
26. Gable, F. and E. G. Huber (1964) Passagere, nicht hereditare Doppelalbuminämie. *Ann. Paediat.* **202,** 81–91.

27. Arvan, D. A., B. S. Blumberg and L. Melartin (1968) Transient "Bisalbuminemia" induced by drugs. *Clin. Chim. Acta* **22,** 211–218.
28. Shashaty, G. G. and M. A. Atamer (1972) Acquired bisalbuminemia with hyperamlasemia. *Am. J. Digest. Dis.* **17,** 59–67.
29. Tarnoky, A. L. (1966) Varieties of bisalbuminemia. *Proc. Ass. Clin. Biochem.* **4,** 12–13.
30. Melartin, Liisa (1967) Albumin polymorphism in man. *Acta Pathologica et Microbiologica Scand.*, Suppl. 191.
31. Fukunaga, F. H. and G. A. Glober (1973) Serum albumin polymorphism: bisalbuminemia in Hawaii–Japanese. *A.J.C.P.* **60,** 867–869.
32. Lie-Injo, L. E., L. R. Weitkamp, E. N. Kosasih, J. M. Bolton and C. L. Moore (1971) Unusual albumin variants in Indonesians and Malaysian aborigines. *Human Heredity* **21,** 376–383.
33. Welch, Q. B. and L. E. Lie-Injo (1972) Serum variants in three Malaysian racial groups. *Human Heredity* **22,** 503–507.
34. Lie-Injo, L. E., E. N. Kosasih and G. Tann (1974) Variation of several erythrocyte enzymes and serum proteins of Indonesians from North Sumatra. *Humangenetik* **22,** 331–334.
35. Tarnoky, A. L. and B. Dowding (1969) Albumin Kashmir. A new variant and its behavior on routine analysis. *Clin. Chim. Acta* **26,** 455–458.
36. Blake, N. M., R. L. Kink, E. M. McDermid, K. Omoto and Y. R. Ahuja (1971) The distribution of serum protein and enzyme group systems among north Indians. *Human Heredity* **21,** 440–457.
37. McDermid, E. M. (1971) Serum albumin variation in Indian populations. *Vox Sang.* **21,** 462–464.
38. Weitkamp, L. R. and A. A. Buck (1972) Phenotype frequencies for four serum proteins in Afghanistan: two "new" variants. *Humangenetik* **15,** 335–340.
39. Bradwell, A., I. Deverill and R. Jefferis (1975) Bisalbuminaemia Birmingham—a new variant in an Indian family. *Vox Sang.* **28,** 383–388.
40. Weitkamp, L. R., D. C. Shreffler and J. Saave (1969) Serum albumin variants in New Guinea indigenes. *Vox Sang.* **17,** 237–240.
41. McDermid, E. M. (1971) Variants in human serum albumin and caeruloplasmin in populations from Australia, New Guinea, South Africa and India. *Austr. J. Exp. Biol. Med. Sci.* **49,** 308–312.
42. Weitkamp, L. R. and N. A. Chagnon (1968) Albumin Maku: a new variant of human serum albumin. *Nature* **207,** 759–760.
43. Arends, T., M. L. Gallango, M. Layrisse, J. Wilberg and H. D. Heinen (1969) Albumin Warao: new type of human alloalbuminemia. *Blood* **33,** 414–420.
44. Arends, T., L. Weitkamp, M. Gallango, J. Neel and J. Schultz (1970) Gene frequencies and microdifferentiation among the Makiritare Indians. II. Seven serum protein systems. *Human Genet.* **22,** 526–532.
45. Tanis, R. J., J. V. Neel, H. Dovey and M. Morrow (1973) The genetic structure of a tribal population, the Yanomama Indians. IX. Gene frequencies for 18 serum protein and erythrocyte enzyme systems in the Yanomama and five neighboring tribes: nine new variants. *Am. J. Human Genet.* **25,** 566–576.
46. Salzano, F. M., M. Helena, L. P. Franco and M. Ayres (1974) Alloalbuminemia in two Brazilian populations: a possible new variant. *Am. J. Human Genet.* **26,** 54–58.
47. Geerdink, R. A., H. A. Bartstra and J. M. Schillhorn van Veen (1974) Serum proteins and red cell enzymes in Trio and Wajana Indians from Surinam. *Am. J. Human Genet.* **26,** 581–587.
48. Tanis, R., R. E. Ferrel, J. V. Neel and M. Morrow (1974) Albumin Yanomama-2, a "private" polymorphism of serum albumin. *Am. Human Genet.* **38,** 179–190.
49. Weitkamp, L. R., A. Basu, J. C. Gall and W. Brown (1969) Albumin Cayemite: a Negro plasma albumin variant. *Humangenetik.* **7,** 180–182.
50. Weitkamp, L. R., F. M. Salzano, J. V. Neel, F. Porta, R. A. Geerdink and A. L. Tarnoky (1973) Human serum albumin: twenty-three genetic variants and their population distribution. *Ann. Human Genet.* **36,** 381–392.
51. Cohen, B. S. (1965) Paucity of albumin variants in man. *Nature* **207,** 1109–1110.
52. Blumberg, B. S., J. R. Martin and L. Melartin (1968) Alloalbuminemia. *J.A.M.A.* **203,** 180–185.
53. Johnston, F. E., B. S. Blumberg, S. S. Agarwal, L. Melartin, and T. A. Burch (1969) Alloalbuminemia in southwestern U.S. Indians: polymorphism of albumin Naskapi and Albumin Mexico. *Human Biology* **41,** 263–270.
54. Sunderman, F. W., S. S. Agarwal, A. I. Sutnick and B. S. Blumberg (1969) Genetics of albumin Gainesville, a new variant of human serum albumin. *Nature* **221,** 66–68.
55. Blumberg, B. S. (1969) The epidemiology of alloalbuminemia. *Arch. Environ. Health* **18,** 1–3.
56. Persson, I., L. Melartin and A. Gilberg (1971) Alloalbuminemia. A search for variants in Greenland Eskimos. *Human Heredity* **21,** 57–59.

57. Lau, T., F. Sunderman, L. Weitkamp, S. S. Agarwal, A. I. Sutnick, B. S. Blumberg and R. Carrillo de Jimenez (1972) Letters to the Editor: Albumin Cartago: a "new" slow moving alloalbumin. *Am. J. Clin. Path.* **57,** 247–251.
58. Porta, F., G. Ruffini, V. Ortalli and F. Fisauli (1972) Alloalbuminemia. Analytical separation by electrophoretic procedure. In: *Proc. Int. Symp. on Standardization in Haematology and Clinical Pathology. Sept. 1971, Foggio (Italy)*: S. Giovanni Rotondo, pp. 241–246.
59. Adams, M. S. (1966) Genetic diversity in serum albumin. *J. Med. Genet.* **3,** 198–202.
60. Atal, P. R., U. P. Mital and R. C. Kulshrestha (1970) Heterogeneity of human serum albumin. Report of a case of bisalbuminaemia. *Indian J. Med. Sci.* **12,** 797–798.
61. Baisden, R., R. B. Conn, Jr. and V. Amido (1969) Heterogeneity of human serum albumin. *Am. J. Clin. Path.* **51,** 760–764.
62. Izquierdo, J. M., P. Sotorrio, B. Dowding and A. L. Tarnoky (1971) Bisalbuminaemia in a Spanish family: characterization of the albumin variant. *J. Clin. Path.* **24,** 515–517.
63. Laurell, C. B. and J. E. Nilehn (1966) A new type of inherited serum albumin anomaly. *J. Clin. Invest.* **45,** 1935–1945.
64. Kueppers, F., P. V. Holland and L. R. Weitkamp (1969) Albumin Santa Ana: a new inherited variant. *Human Heredity* **19,** 378–384.
65. Bonazzi, L. (1968) On a rare genetic variation of plasma albumin: bisalbuminemia. *Clin. Chim. Acta* **20,** 362–363.
66. Payne, R. B. and J. P. Dickinson (1967) Immunochemical studies of bisalbuminemia. *Nature* **215,** 530–537.
67. Fine, J. M. (1970) Les allotypes de l'albumine humaine. Etude de 8 cas de bisalbuminémie observés en France. *Rev Europ. Etudes Clin. et. Biol.* **15]** 113–118.
68. Margni, R. A., E. F. Heer, E. Acerbo, S. Hajos, M. Beliuea and M. Bobbi (1970) Immunochemical and genetic studies in two bisalbuminaemic families. *Clin. Chim. Acta* **29,** 219–225.
69. Melartin, L., B. S. Blumberg and J. R. Martin (1968) Albumin polymorphism (Albumin Naskapi) in Eskimos and Navajos. *Nature* **218,** 787–789.
70. Polesky, H. F. and D. A. Rokala (1967) Serum albumin polymorphism in North American Indians. *Natnre* **216,** 184–185.
71. Emannelli, G., A. Monni, P. Congiu, V. Palomba and G. Pilleri (1970) Bisalbuminemia of the fast type. *Acta Haemat.* **44,** 246–250.
72. Weitkamp, L. R., D. L. Rucknagel and H. Gershowitz (1966) Genetic linkage between structur al loci for albumin and group specific component (Gc). *Am. J. Human Genet.* **18,** 559–571.
73. Izquierdo, J. M., D. Romeo, P. Sotorrio, B. Dowding and A. L. Tarnoky (1973) Bisalbuminaemia. A new Spanish variant. *I.R.C.S. Human Genet.* 40–6–2.
74. Tarnoky, A. L., B. Dowding and A. L. Lakin (1970) Eight types of bisalbuminemia. *Nature* **225,** 742–743.
75. Weitkamp, L. R., D. C. Schreffler, J. L. Robbins, O. Drachmann, P. L. Ardner, R. J. Wieme, N. M. Simon, K. B. Cooke, G. Sandor, F. Wuhrmann, M. Braend and A. L. Tarnoky (1967) An electrophoretic comparison of serum albumin variants from nineteen unrelated families. *Acta Genet.* **17,** 339–405.
76. Weitkamp, L. R., E. B. Robson, D. C. Shreffler and G. Corney (1968) An unusual serum albumin variant: further data on genetic linkage between loci for human serum albumin and group specific component (Gc). *Am. J. Human Genet.* **20,** 392–397.
77. Weitkamp, L. R. (1973) The contribution of variations in serum albumin to the characterization of human populations. *Israel J. Med. Sci.* **9,** 1238–1248.
78. Polesky, H. F., D. A. Rokala and T. A. Burch (1968) Serum albumin polymorphism in Indians of southwestern United States. *Nature* **220,** 175–176.
79. Petrini, C., F. Giovcelli, F. Porta and M. Fraccaro (1975) A homozygote for a serum albumin variant of the slow type. *Humangenetik* **26,** 245–248.
80. Schneiderman, H., J. Beyer and A. Krieg (1968) Albumin Syracuse: a variant demonstrated at acid pH. *Nature* **218,** 1159–1160.
81. Weitkamp, L. R., E. M. McDermid, J. V. Neel, J. M. Fine, C. Petrini, L. Bonazzi, V. Ortalli, F. Porta, R. Tanis, J. Harris, T. Peters, G. Ruffini and E. Johnston (1973) Additional data on the population distribution of human serum albumin: three new variants. *Ann. Human Genet.* **37,** 219–226.
82. Weitkamp, L. and J. V. Neel (1972) The genetic structure of a tribal population, the Yanomama Indians. IV. Eleven erythrocyte enzymes and summary of protein variants. *Ann. Human Genet.* **35,** 433–444.
83. Fraser, G. R., H. Harris and E. B. Robson (1959) A new genetically determined plasma protein in man. *Lancet*, **1** 1023–1024.

84. McDermid, E. M. and G. H. Vos (1971) Serum protein groups of South African Indians. *South African J. Med. Sci.* **36,** 57–62.
85. Kueppers, F. and A. G. Bearn (1966) Inherited variations of human serum alpha-antitrypsin. *Science* **154,** 407–408.
86. Martin, J. P., L. C. Legneult, C. Ropartz, J. Manze, A. Lescaroux, A. Richard and D. Chauvat (1971) Another example of slow albumin allotype. *Humangenetik* **13,** 320–327.
87. Finney, D. J. (1940) The detection of linkage. *Ann. Eugen.* **10,** 171–214.
88. Haldane, J. B. S. and C. A. B. Smith (1947) A new estimate of the linkage between the genes for color-blindness and hemophilia in man. *Ann. Eugen.* **14,** 10–31.
89. Morton, N. E. (1955) Sequential test for the detection of linkage. *Am. J. Human Genet.* **7,** 277–318.
90. Smith, C. A. B. (1959) Some comments on the statistical methods used in linkage investigations. *Am. J. Human Genet.* **11,** 289–304.
91. Kaarsalo, E., L. Melartin and B. S. Blumberg (1967) Autosomal linkage between the albumin and Gc loci in humans. *Science* **158,** 123–125.
92. Weitkamp, L. R., J. H. Renwick, J. Berger, D. C. Shreffler, O. Drachmann, F. Wuhrmann, M. Braend and G. Franglen (1970) Additional data and summary for albumin Gc linkage in man. *Human Heredity* **20,** 1–7.
93. Bowen, P., F. O'Callaghan and C. S. N. Lee (1971) Serum protein polymorphism in Indians of Western Canada. *Human Heredity* **21,** 242–253.
94. Scott, E. M. and R. C. Wright (1969) The absence of close linkage of methemoglobinemia and other loci. *Am. J. Human Genet.* **21,** 194–195.
95. Mohr, J. (1954) *A Study of Linkage in Man.* Munksgaard, Copenhagen.
96. Ford, E. B. (1965) *Genetic Polymorphisms.* M.I.T. Press, Cambridge, Mass.
97. Undevia, J. V., R. L. Kirk and E. M. McDermid (1973) Serum protein systems among Parsis and Iranis in Bombay. *Human Heredity* **23,** 492–498.
98. Das, S. R., B. N. Mukherjee, S. K. Das, R. Monami and S. S. Chhatui (1974) Blood groups, serum proteins, haemoglobin and some serum and red cell enzymes among the Koaras of 24-Paraganes in West Bengal (India). *Human Heredity* **24,** 24–31.
99. Saha, N., R. L. Kirk, S. Shanbhag, S. H. Joshi and H. M. Bhatia (1974) Genetic studies among the Kadar of Kerala. *Human Heredity* **24,** 198–218.
100. Weitkamp, L., N. Chagnon, J. Saave, F. Salzano and J. Gall (1968) Serum albumin variants in American and New Guinea Indigenes. *Clin. Res.* **16,** 298.
101. Malcolm, L. A., D. G. Woodfield, N. M. Blake, R. L. Kirk and E. M. McDermid (1972) The distribution of blood, serum protein and enzyme groups on Manus Island (Admiralty Islands, New Guinea). *Human Heredity* **22,** 305–322.
102. Woodfield, D. G., R. F. R. Scragg, N. M. Blake, R. L. Kirk and E. M. McDermid (1974) Distribution of blood, serum protein and enzyme groups among the Fuyuge speakers of the Boilala sub-district. *Human Heredity* **24,** 507–519.
103. Weitkamp, L. R., T. Arends, M. Gallango, J. V. Neel, J. Schultz and D. C. Shreffler (1972) The genetic structure of a tribal population, the Yanomama Indians. III. Seven serum protein systems. *Ann. Human Genet.* **35,** 271–279.
104. Salzano, F. M., J. V. Neel, L. R. Weitkamp and J. P. Woodall (1972) Serum proteins, hemoglobins and erythrocyte enzymes of Brazilian Cayapo Indians. *Human Biol.* **44,** 443–458.
105. Kirk, R. L., E. M. McDermid, N. M. Blake, D. C. Gajdusek, W. C. Leyshon and R. MacLennan (1974) Blood group, serum protein and red cell enzyme groups of Amerindian populations in Colombia. *Am. J. Phys. Anthrop.* **41,** 301–316.
106. Lisker, R., L. Cobo and G. Mora (1971) Distribution of albumin variants in Indians and non-Indians of Mexico. *Am. J. Phys. Anthrop.* **35,** 119–124.
107. Szathmary, E. J. E., D. W. Cox, H. Gershowitz, D. L. Rucknagel and M. S. Schanfield (1974) The northern and southeastern Ojibwa: serum proteins and red cell enzyme systems. *Am. J. Phys. Anthrop.* **40,** 49–66.
108. McAlpine, P. J., S. H. Chen, D. W. Cox, J. B. Dossetor, E. Giblett, A. G. Steinberg and N. E. Simpson (1974) Genetic markers in blood in a Canadian Eskimo population with a comparison of allele frequencies in circumpolar populations. *Human Heredity* **24,** 114–142.
109. Agarwal, S. S. (1975) Unpublished.
110. Pollitzer, W. S., D. Rucknagel, R. Tashian, D. Shreffler, W. Leyshon, K. Namboodiri and R. Elston (1970) The Seminole Indians of Florida: morphology and serology. *Am. J. Phys. Anthrop.* **32,** 65–81.
111. Melartin, L., B. S. Blumberg and R. Lisker (1967) Albumin Mexico, a new variant of serum albumin. *Nature* **215,** 1288–1289.

112. Johnston, F. E., O. Alarcon, F. Benedict, M. Dary, M. Galbraith and P. S. Gindhart (1973) Albumin Mexico (Al^{Me}) in the Guatemalan Highlands. *Am. J. Phys. Anthrop.* **38,** 27–30.
113. Brown, K. S. and R. S. Johnson (1970) Populations studies on southwestern Indian tribes. III. Serum protein variations of Zuni and Papago Indians. *Human Heredity* **20,** 281–286.
114. Wauchope, R. (1964) Southern Meso America. In: *Prehistoric Man in the New World,* J. D. Jennings and E. Norbeck, editors, pp. 331–386. University of Chicago Press, Chicago.
115. Wormington, H. M. (1970) *Prehistoric Indians of the Southwest.* Denver Museum of Natural History, Denver.
116. Weaver, M. P. (1972) *The Aztecs, Maya, and their Predecessors.* Seminar Press, New York.
117. Spencer, R. F. and J. D. Jennings (1965) *The Native Americans.* Harper & Row, New York.
118. Aikens, C. M. (1966) *Fremont–Promontory–Plains Relationships in Northern Utah.* Anthropological Papers, No. 82. Department of Anthropology, University of Utah.
119. Aikens, C. M. (1970) *Hogup Cave.* Anthropological Papers, No. 93. Department of Anthropology, University of Utah.
120. Hoijer, H. (1956) The chronology of the Athabascan languages. *Int. J. Am. Ling.* **22,** 219–232.
121. Hymes, D. (1957) A note on Athabascan glottochronology. *Int. J. Am. Ling.* **23,** 291–297.
122. Neumann, G. K. (1952) Archaeology and race in the American Indians. In: *Archaeology of Eastern United States,* J. B. Griffin, editor, pp. 13–36. University of Chicago Press, Chicago.
123. Allison, A., B. S. Blumberg and S. Gartler (1959) Urinary excretion of β-aminoisobutyric acid in Eskimo and Indian populations of Alaska. *Nature* **183,** 118–119.

ADDENDUM

A report by D. Bradley and C. Hornbeck (*Biochem. Med.* **8**, 92–97, 1973) noted the presence of albumins Naskapi and Mexico among a White River Apache sample. M. H. Crawford, W. C. Leyshon, K. Brown, Frances Lees and L. Taylor (*Am. J. Phys. Anthrop.* **41**, 251–268, 1974) noted the presence of albumin Mexico among samples from Tlaxcala and San Pablo del Monte, Tlaxcala, Mexico.

In: The First Americans: Origins, Affinities, and Adaptations. Eds. William S. Laughlin and Albert B. Harper. Gustav Fischer, New York, Stuttgart, pp. 107-123, 1979.

Chapter 5

Blood Polymorphisms and the Origins of New World Populations

Michelle Lampl and Baruch S. Blumberg

Introduction

An objective of this volume is to discuss methods that may be used to identify contemporary Asian populations that have affinities with contemporary native American populations by virtue of their descent from the same ancestral population. If such populations exist, their similarity to American populations can be demonstrated if they have retained genes present in the ancestral populations and in their related populations in the Americas. These would be particularly useful if they included genes of polymorphic loci found only in native Americans and their Asian counterparts. In this chapter we will describe genes that may be useful in such comparisons. It is obvious that if gene frequencies have altered very differently in the two continents by virtue of drift, gene admixture, or differences in selection in the two environments, then a comparison using this technique would not be possible. Hence, this discussion is based on the assumption that it is possible to identify similarities between North American and Asian populations by comparisons using the appropriate genes.

There are two systems (serum albumins and white blood cell HL-A antigens) for which a relatively large amount of data is available and these are discussed in greatest detail. In addition, other systems that may be useful are mentioned.

Albumin Variants

The albumin polymorphisms may be one of the most useful markers for the identification of American Indian affinities. Two genes of relatively wide distribution, albumin Naskapì and albumin Mexico, have been identified in North

The First Americans: Origins, Affinities, and Adaptations.

and Central American Indian populations. Two other genes of relatively restricted distribution, albumin Makiritare and albumin Yanomama-2, have been found in South American populations. Although rare variants (i.e., fewer than one per 100) have been found in European, Asian, and Oceanic populations (i.e., albumin B, albumin Kashmir), in no case have they reached the 1 percent figure characteristic of genetic polymorphisms. Hence, if a polymorphism for albumin is identified in an Asian population whose affinities with American Indians is suspected, and in particular if the polymorphic variant has the same electrophoretic mobility and other characteristics as Naskapi, Mexico, Makiritare, or Yanomama, then this would provide a strong argument in favor of the association of that population with American Indians (Schell and Blumberg 1977).

Human serum albumin phenotypes are determined by autosomal codominant alleles which segregate at a single locus. Albumin variants are distinguished by their electrophoretic mobilities (often using several techniques for the separations) and approximately 25 such variants have so far been identified (Table 5.1). Most of these variants occur rarely (one in several thousand) and presumably are maintained in the population primarily as a consequence of recurrent mutation.

Table 5.1. *Relative Mobilities of 25 Albumin Variants in Three Starch Gel Electrophoretic Systems*[a]

Variant	Buffer system pH 5.0	pH 6.0	pH 6.9
1. RS-I	1	2	6
2. Pollibauer	1	2	4
3. Belem I	2	1	4
4. B	3	3	1
5. Roma	3	1	3
6. Gainesville	3	3	4
7. Paris (Gombak)	4	3	5
8. Kashmir (Afghanistan)	5	2	1
9. Otsu	6	4	2
10. Santa Ana	7	4	3
11. SO/BS	7	5	3
12. Cartago	7	3	5
13. Xavante	8	6	7
14. Pushtoon	8	4	6
15. Cayemite	9	5	6
16. Mexico	9	4	5
17. Uinba	9	6	7
18. Yanomama-2	10	6	6
Normal albumin	10	6	7
19. Medan	11	6	8
20. Maku	12	7	10
21. Reading (New Guinea)	13	9	9
22. Makiritare-3	13	8	9
23. Naskapi	14	10	10
24. Gent	15	11	10
25. Kyoto	16	11	10

[a] From Weitkamp (1973).

The present chapter considers the four variants that reach polymorphic frequencies: Two of these have been observed in widely separated geographic regions (Naskapi and Mexico), whereas the other two have been primarily confined to the populations in which they were first discovered (Makiritare and Yanomama-2).

The first albumin variant to be found at polymorphic frequency was reported by Melartin and Blumberg (1966). It has a mobility faster than the common albumin A. It was found originally in the relatively isolated Naskapi and Montagnais Indians of Quebec Province, Canada and was termed albumin Naskapi. This study also contained the first report of individuals homozygous for an albumin other than the common albumin A. In 1967, a slow moving variant (albumin Mexico) was found in polymorphic frequency among Indians of the American Southwest and Mexico (Melartin *et al.* 1967). These variants (Naskapi and Mexico) have not been reported in populations other than American Indians, even though many tens of thousands of sera have been examined by the appropriate electrophoretic methods.

Albumin Naskapi is found predominantly among northern populations, whereas albumin Mexico is found among southeastern United States and middle American populations. Populations in the southwest United States have both Naskapi and Mexico (Table 5.2, Figure 5.1). Table 5.2 is adopted from Schell and Blumberg (1977) and Figure 5.1 is taken from Schell *et al.* (1978).

The highest frequency of the albumin Naskapi gene (13 percent) is found in the eastern regions of North America, among the Algonkian-speaking Naskapi of Schefferville, Quebec. High gene frequencies (8 percent) are also found among the Montagnais Indians from the same region. Naskapi living at Northwest River, however, have a lower frequency of Al^{Na} (3 percent). Albumin Naskapi was not found in samples of Mohawk and Micmac Indians from the St. Lawrence River area, and this river may represent the southern limit of the distribution of this variant in the east (Schell *et al.* 1978).

In the Ungava region of northern Quebec a few cases of albumin Naskapi have been found among Eskimos living close to Indian groups with a high frequency of the Naskapi variant. These appear to result from admixture, as no variant has been found among the more remote Eskimo populations from Alaska, Canada, and Greenland (McAlpine *et al.* 1974, Melartin 1967, Melartin and Blumberg 1966, Persson *et al.* 1971). Albumin Naskapi is also apparently absent from the Haida and Tlingit Indians of the Canada Pacific coast (Melartin 1967, Melartin and Blumberg 1966).

Albumin Naskapi, however, is found at high gene frequencies among Athabascan Indians of Alaska (2.8–5.1 percent) (Melartin and Blumberg 1966, Scott and Wright 1969) and Athabascan-speaking Slave and Beaver Indians of Canada (9.8 percent) (Bowen *et al.* 1971). Lower frequencies (less than 3.4 percent) are found among a number of other central and western Indian groups: Sioux (Melartin and Blumberg 1966), Assiniboine (Bowen *et al.* 1971), Ojibwa (Polesky and Rokala 1967, Szathmary *et al.* 1974, Tanis *et al.* 1974), Cree (Bowen *et al.* 1971), Chippewa (Tanis *et al.* 1974, Weitkamp 1973, Weitkamp *et al.* 1968), and

Table 5.2. ***The Distributions of Albumin Mexico and Albumin Naskapi in North and Central America***

Linguistic affiliation, Tribal name (location)	Number tested	Number Naskapi[b]	Number Mexico[b]	Naskapi gene frequency	Mexico gene frequency	Reference
Algonkian						
Ojibwa (No. Ontario)	117	6	0	0.026	0.0	Szathmary *et al.* (1974)
Ojibwa (So. Ontario)	94	1	0	0.005	0.0	Szathmary *et al.* (1974)
Ojibwa (?)	120	8	0	0.033	0.0	Tanis *et al.* (1974)
Ojibwa/Anglo (No. Minnesota)	250	11 + (1)	0	0.026	0.0	Polesky and Robala (1967)
Chippewa/Cree (Saskatchewan)	610	36 + (1)	0	0.031	0.0	Weitkamp (1973)
Chippewa/Cree (Saskatchewan)	102	2	0	0.009	0.0	Weitkamp *et al.* (1968)
Chippewa (?)	87	2	0	0.011	0.0	Tanis *et al.* (1974)
Cree, Plains (Alberta)	605	25 + (1)	0	0.022 (0.023)[c]	0.0	Bowen *et al.* (1971)
Cree, Northern (Alberta)	187	10	0	0.026 (0.034)[c]	0.0	Bowen *et al.* (1971)
Blackfoot (Montana)	97	4	0	0.020	0.0	Polesky and Rokala (1967)
Naskapi (Schefferville, Quebec)	151	37 + (1)	0	0.130	0.0	Melartin and Blumberg (1966)
Montagnais (Schefferville, Quebec)	112	14 + (2)	0	0.080	0.0	Melartin and Blumberg (1966)
Naskapi (N.W. River, Labrador)	66	2	0	0.03	0.0	Schell *et al.* (1978)
Micmac (Restigouche, Quebec)	102	0	0	0.0	0.0	Schell *et al.* (1978)
Iroquois						
Mohawk (Caugh Nawaga, Quebec)	112	0	0	0.0	0.0	Schell *et al.* (1978)
Siouian						
Sioux (?)	160	2	0	0.007	0.0	Melartin and Blumberg (1966)
Assinboine (Alberta)	100	1	0	0.005 (0.007)[c]	0.0	Bowen *et al.* (1971)
Omaha (Macy Reservation, Nebraska)	96	0	0	0.0	0.0	Schell *et al.* (1978)
Eskimoan						
Eskimo (Igloolik, Canada)	356	0	0	0.0	0.0	McAlpine *et al.* (1974)
Eskimo (E. Greenland)	78	0	0	0.0	0.0	Persson *et al.* (1971)
Eskimo (Thule, Greenland)	297	0	0	0.0	0.0	Persson *et al.* (1971)
Eskimo (W. Greenland)	116	0	0	0.0	0.0	Persson *et al.* (1971)
Eskimo (Wainwright, Alaska)	111	0	0	0.0	0.0	Melartin and Blumberg (1966)
Eskimo (Anaktuvuk)	55	0	0	0.0	0.0	Melartin and Blumberg (1966)
Eskimo (Ungava Bay, Canada)	124	3	0	0.012	0.0	Melartin *et al.* (1968)
Eskimo (Barrow, Alaska)	88	0	0	0.0	0.0	Melartin *et al.* (1968)
Eskimo (Beaver, Alaska)	14	0[d]	0	0.0	0.0	Melartin *et al.* (1968)

Eskimo (St. Lawrence Is., Alaska)	82	0	0	0.0	0.0	Melartin *et al.* (1968)
Eskimo (Nat'l. Guard, Alaska)	107	0	0	0.0	0.0	Melartin *et al.* (1968)
Aleut (Sitka and Nat'l. Guard, Alaska)	15	0[d]	0	—	—	Melartin *et al.* (1968)
Eskimo (Frobisher Bay Baffin Island N.W. Terr.)	101	0	0	0.0	0.0	Melartin *et al.* (1968)
Na-Dené (non-Athabascan branch)						
Tlingit (Sitka and Mt. Edgecomb, Alaska)	91	0[d]	0	0.0[d]	0.0	Melartin (1967)
Haida (Queen Charlotte Is., British Columbia)	365	0	0	0.0	0.0	Melartin and Blumberg (1966)
Na-Dené (Athabascan branch)						
Alaskan Athabascans (Alaska)	230	11 + (1)	0	0.028	0.0	Melartin and Blumberg (1966)
Alaskan Athabascans (Alaska)	137	14	0	0.051	0.0	Scott and Wright (1969)
Slave and Beaver (Alberta)	143	22 + (2)	0	0.090 (0.098)[e]	0.0	Bowen *et al.* (1971)
Navajo (Arizona and New Mexico)	95	12	0	0.063	0.005	Melartin *et al.* (1968)
Navajo (Sherman School, Ca.)	468	30	6	0.032	0.006	Johnston *et al.* (1969)
Navajo (?)	192	11	4	0.03	0.010	Tanis *et al.* (1974)
Apache (Sherman School, Ca.)	103	3	7	0.015	0.034	Johnston *et al.* (1969)
W. Apache (San Carlos, Arizona)	538	9	19	0.016	0.036	Schell *et al.* (1978)
Yuman River						
Cocopah (Arizona)	164	0	13	0.0	0.039	Polesky *et al.* (1968)
Maricopa (Arizona)	101	0	9 + (1)	0.0	0.054	Polesky *et al.* (1968)
Maricopa (Sherman School, Ca.)	9	1	0	—	—	Johnston *et al.* (1969)
Mohave (Sherman School, Ca.)	19	1	1	—	—	Johnston *et al.* (1969)
Yuman Upland						
Havasupi (Sherman School, Ca.)	6	0	0	—	—	Johnston *et al.* (1969)
Walapai (Sherman School, Ca.)	8	0	0	—	—	Johnston *et al.* (1969)
Yavapai (Sherman School, Ca.)	3	0	0	—	—	Johnston *et al.* (1969)
Uto-Aztecan						
Pima (Arizona)	1528	0	100 + (7)	0.0	0.037	Polesky *et al.* (1968)
Pima (Sherman School, Ca.)	113	0	11	0.0	0.049	Johnston *et al.* (1969)
Papago (Sherman School, Ca).	115	0	3	0.0	0.013	Johnston *et al.* (1969)
Papago (Arizona)	179	0	2	0.0	0.006	Melartin (1967)
Papago (Arizona)	546	0	5	0.0	0.005	Brown and Johnson (1970)
Hopi (Sherman School, Ca.)	82	1	0	0.006	0.0	Johnston *et al.* (1969)
Ute (Sherman School, Ca.)	17	0	1	—	—	Johnston *et al.* (1969)
Paiute (Sherman School, Ca.)	4	0	0	—	—	Johnston *et al.* (1969)
Chemehueve (Sherman School, Ca.)	2	0	0	—	—	Johnston *et al.* (1969)
Mission (Sherman School, Ca.)	1	0	0	—	—	Johnston *et al.* (1969)

Table Continued

Table 5.2.—Continued

Linguistic affiliation, Tribal name (location)	Number tested	Number Naskapi[b]	Number Mexico[b]	Naskapi gene frequency	Mexico gene frequency	Reference
Washoan						
Washoe (Sherman School, Ca.)	1	0	0		—	Johnston *et al.* (1969)
Zunian						
Zuni (New Mexico)	655	0	16	0.0	0.012	Melartin (1967)
Eastern Muskogean (tribal affiliation indeterminate)						
Dania (Fla.)	150	0	0	0.0	0.0	Pollitzer *et al.* (1970)
Big Cypress (Fla.)	135	0	0	0.0	0.0	Pollitzer *et al.* (1970)
Brighton (Fla.)	128	0	0	0.0	0.0	Pollitzer *et al.* (1970)
Macro-Nahua						
Nahua (Puebla, Hidalgo, Veracruz, Mexico)	440	0	6	0.0	0.006	Lisker *et al.* (1971)
Otomi (Hidalgo, Mexico)	100	0	5	0.0	0.025	Lisker *et al.* (1971)
Cora (Nayrati, Mexico)	83	0	2	0.0	0.012	Lisker *et al.* (1971)
Macro-Maya						
Maya (Yucatan, Mexico)	263	0	0	0.0	0.0	Melartin *et al.* (1967)
Huasteco (Veracruz, Tamulipas, Mexico)	188	0	4	0.0	0.01	Lisker *et al.*(1971)
Tzeltal-Tzotzil (Chiapas, Mexico)	144	0	0	0.0	0.0	Lisker *et al.* (1971)
Chol (Chiapas, Mexico)	138	0	0	0.0	0.0	Lisker *et al.* (1971)
Dialect not reported						
(Guatemala City and San Marcos, Guatemala)	204	0	1	0.0	0.002	Johnston *et al.* (1973)
(San Antonio Ilotenango, Guatemala)	186	0	3	0.0	0.008	Johnston *et al.* (1973)

Macro–Mixteco						
Zapoteco (Oaxaca, Mexico)	255	0	3	0.0	0.005	Lisker *et al.* (1971)
Zapotec (Guelatao, Mexico)	123	0	1	0.0	0.004	Melartin *et al.* (1967)
Mixteco (Oaxaca, Mexico)	48	0	0	—	—	Lisker *et al.* (1971)
Mazateco (Oaxaca, Mexico)	22	0	0	—	—	Lisker *et al.* (1971)
Mazatec (Huautla de Jimenez, Mexico)	20	0	0	—	—	Melartin *et al.* (1967)
Mixe (Oaxaca, Mexico)	21	0	0	—	—	Lisker *et al.* (1971)
Tarasco						
Tarasco (Michoacan, Mexico)	167	0	5	0.0	0.014	Lisker *et al.* (1971)
Mixed Linguistic Groups						
Mestizo–Zapotec (Pochutla, Mexico)	36	0	1	—	—	Melartin *et al.* (1967)
Mestizo–Mixtec						
(San Pedro Mixtec, Mexico)	20	0	0	—	—	Melartin *et al.* (1967)
(Ometepec, Mexico)	20	0	0	—	—	Melartin *et al.* (1967)
(Cauajinicuilapa, Mexico)	20	0	0	—	—	Melartin *et al.* (1967)
Mestizo (Tlaxcala, Hidalgo, and Mexico, D. F., Mexico)	185	0	10	0.0	0.027	Melartin *et al.* (1967)
Mestizo (Tlaxcala–Hidalgo, Mexico)	257	0	12	0.0	0.023	Lisker *et al.* (1971)
Mestizo (Veracruz, Mexico)	109	0	1	0.0	0.004	Lisker *et al.* (1971)
Mestizo (Campeche, Mexico)	109	0	0	0.0	0.0	Lisker *et al.* (1971)
Mestizo (West Coast, Mexico)	99	0	1	0.0	0.005	Lisker *et al.* (1971)
Mestizo (Mexico City, Mexico)	1313	0	27	0.0	0.010	Lisker *et al.* (1971)

[a] Adapted from Schell and Blumberg (1977).
[b] Homozygotes in parentheses.
[c] Published gene frequency, calculated with correction for related individuals.
[d] A very low frequency was reported originally; this value has since been corrected to the value shown here.

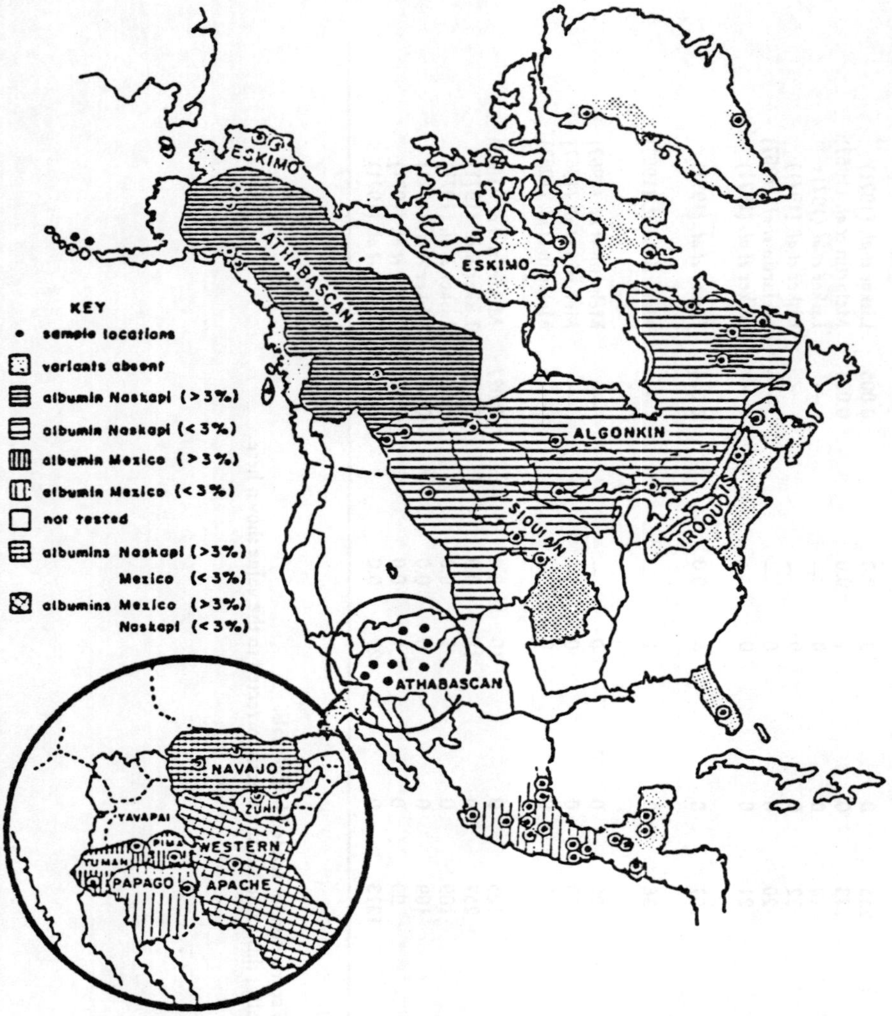

Figure 5.1. Distribution of albumins Naskapi and Mexico in native American populations. (From Schell *et al.* 1978.)

Blackfoot (Polesky and Rokala 1967), although the variant is absent in others, i.e., Omaha (Schell *et al.* 1978). It also occurs in relatively high frequency in the Athabascan-speaking Navajo and Apache (Schell *et al.* 1978).

This wide geographic dispersion is consistent with some linguistic and cultural–historical data. From archaeologic evidence it is postulated that the movement of the Navajo and Apache into the Southwest from the North took place about 500–1000 years ago (Aikens 1970). The Navajo and Apache, furthermore, are linguistically related to Canadian and Alaskan Athabascan-speaking Indians, and glottochronologic analyses support these dates (Hoijer 1956, Hymes 1957).

In the southern regions of North America and in Central America the slow albumin variant, albumin Mexico, has been found in a distribution extending from the southwest United States to Guatemala. Albumin Mexico has not (at this time) been found in any other native northern North American population. Polymorphic frequencies of albumin Mexico have been found among the Uto-Aztecan-speaking Pima (Johnston *et al.* 1969, Polesky *et al.* 1968) and Papago (Brown and Johnson 1970, Johnston *et al.* 1969, and Melartin 1967), the Zuni (Brown and Johnson 1970), and the Yuman River-speaking Maricopa and Cocopah (Weitkamp 1973).

The highest frequencies of albumin Mexico in middle America are reported for two Mestizo samples (Lisker *et al.* 1971, Melartin *et al.* 1967) and the variant is present at lower frequencies in populations from Tamulipas, Mexico on the Gulf coast (Lisker *et al.* 1971) to Guatemala City, Guatemala (Johnston *et al.* 1973).

The distribution of albumin Mexico parallels archaeologic data for past cultural contacts (Weaver 1972, and Wormington 1970). The evidence suggests that a culture centered near the Valley of Mexico (area of modern Mexico City) had contacts with the peoples of the Guatemalan highlands, to the north with the Hohokam culture in southwestern United States and with people from the area of Tamulipas, Mexico. The Pima and Papago are thought to be descendants of the Hohokam (Spencer and Jennings 1965).

Among several southwestern United States groups (i.e., Navajo, Apache, Mohave, and Maricopa), both albumins Mexico and Naskapi are found concurrently at varying frequencies (Johnston *et al.* 1969).

It is interesting that albumin Mexico is found among the Navajo and, to a greater extent, among the Apache, but albumin Naskapi is only rarely found among the Uto-Aztecan tribes. This is consistent with ethnohistorical evidence that, as a result of warfare and kidnapping, Pima and Papago and others were assimilated into the Navajo and Apache groups but the reverse happened less often.

In South America the two geographically restricted variants (albumins Yanomama-2 and Makiritare) are found in polymorphic frequencies in the ethnic groups after which they are named and in some nearby groups. Yanomama-2 has been found at a frequency of 7.6 percent among 64 Yanomama villages in southern Venezuela and Northern Brazil. The highest frequency of Yanomama-2 (40 percent) has been found in a northern village (Tanis *et al.* 1974).

Albumin Makiritare (which is identical with albumin Warao) has been found among the Makiritare, the Warao of Venezuela, and the Trio and Wajana tribes of Surinam (Arends *et al.* 1969, 1970, Geerdink *et al.* 1974). The variant reaches polymorphic frequencies among several tribal groups, but its distribution is confined to the northeastern area of South America. Although the average frequency for the albumin in the Makiritare is about 1 percent it reaches a frequency of 4 percent in several individual villages.

Hence, polymorphic variants of albumin have been found only in American Indians (but not Eskimos). The identification of similar variants in Asian populations could contribute to our understanding of affinities with Native Americans.

The HL-A System (Human Major Histocompatibility Complex)

The HL-A system is thought to be the major human histocompatibility system (apart from ABO) in humans and is present as an antigenic polymorphism on many human tissues. The HL-A antigens were first found as isoagglutinogens on platelets and white blood cells, and routine determinations utilize these cells. Subsequent research has revealed their medical importance in determining whether and to what extent the tissues of a potential transplant donor are compatible with the putative recipient. Histocompatibility antigens show a high degree of polymorphism, and the great variation of antigen frequencies in different populations makes the HL-A system a potentially powerful genetic marker for population studies. Research into the histocompatibility system is complicated by the large number of antigenic specificities and the difficulties in standardizing reagents. Despite this, these systems may prove to be among the most important genetic systems for studying relations between human populations. We use it here as an example of how such data may be applied to the study of Asian populations that may be related to Americans. A revised system of nomenclature for the HL-A system has recently been introduced (Amos *et al.* 1976). To maintain consistency with the published population data the earlier nomenclature is used here.

It is now generally assumed that the HL-A system consists of four autosomal linked loci, now termed *A*, *B*, *C*, and *D*, each with a complex series of allelic genes giving rise to distinct antigens. (The actual sequence on the chromosome is *D*, *B*, *C*, *A*.) Of these, the antigens defined by the *A*, *B*, and *C* loci are serologically determined and the *D* locus antigens are responsible for the mixed lymphocyte response (*MLR* locus). An HL-A haplotype generally refers to a section of a chromosome containing the *A* (or *LA*) locus and the *B* (second or fourth locus) (i.e., a pair of antigens). The reagents for the *C* locus were not generally available until recently. The use of haplotype designations, as we shall see later, can lead to very specific identifications of populations. There are some population-related antigens at each of the loci. HL-A1 is characteristic of Caucasians, whereas HL-A3 and HL-A8 characterize non-Caucasian populations. HL-A9 is high among Orientals, New Guineans, New Zealand Maoris, Eskimos, and American Indians. W10 is high among populations of Asians and North and South American Indians; and W15 is commonly found in Asia and South America (Bodmer and Bodmer 1974, Colombani and Degas 1973, Dausset and Colombani 1973).

There is a striking amount of homogeneity with respect to the HL-A system in American Indians; only two or three alleles of each allele series account for most of the variation. Summarizing from numerous sources (Bodmer 1973, Bowen *et al.* 1971, Cann *et al.* 1973a, b, Colombani and Degas 1973, Corley *et al.* 1973, Daveau *et al.* 1975, Dossetor *et al.* 1973, Kissmeyer-Nielson *et al.* 1973, Kostyu *et al.* 1975, Layrisse *et al.* 1973a, b, Perkins *et al.* 1973, Rubenstein *et al.* 1967, Spees *et al.* 1973, Tittor *et al.* 1973, Troup *et al.* 1973, Van der Does *et al.* 1973, Zaretskaya and Fedrunova 1973) it can be said that at the first locus HL-A2 is the most frequent gene, found at frequencies higher among American Indian populations than any other populations yet studied. HL-A9, W28, and W31 are the other common genes. W10, W16, and one of the antigens of the cross-reacting set of HL-A5, W5, and W15, account for the majority of the second series variability.

The HL-A antigens vary geographically among the New World populations. W31 of the first series and W15 of the second series are found at higher frequencies among South American populations, whereas W28, HL-A9, and W5 are found at highest frequencies among North and Central American Indian populations. There appears to be a north–south cline for HL-A9; the highest frequency for any population has been observed in Eskimos. Conversely, HL-A2 is observed at high frequencies in central and southern populations tested but is found at much lower frequencies among Eskimos.

Bodmer and Bodmer (1974) suggest that HL-A1, -3, -10, -11, W29, and 19.6 are absent in American Indians and their presence in a population designated as Indian represents admixture. Studies on Mexican Indians, Chilean Mapuchi, and Maya support this (Amos *et al.* 1970, Rubenstein *et al.* 1967).

The data available on HL-A antigens in Mongoloid populations other than American Indians indicate that the alleles HL-A2, -9, -10, -11, and W31 of the

Table 5.3. *Population Distribution of HL-A Haplotypes*

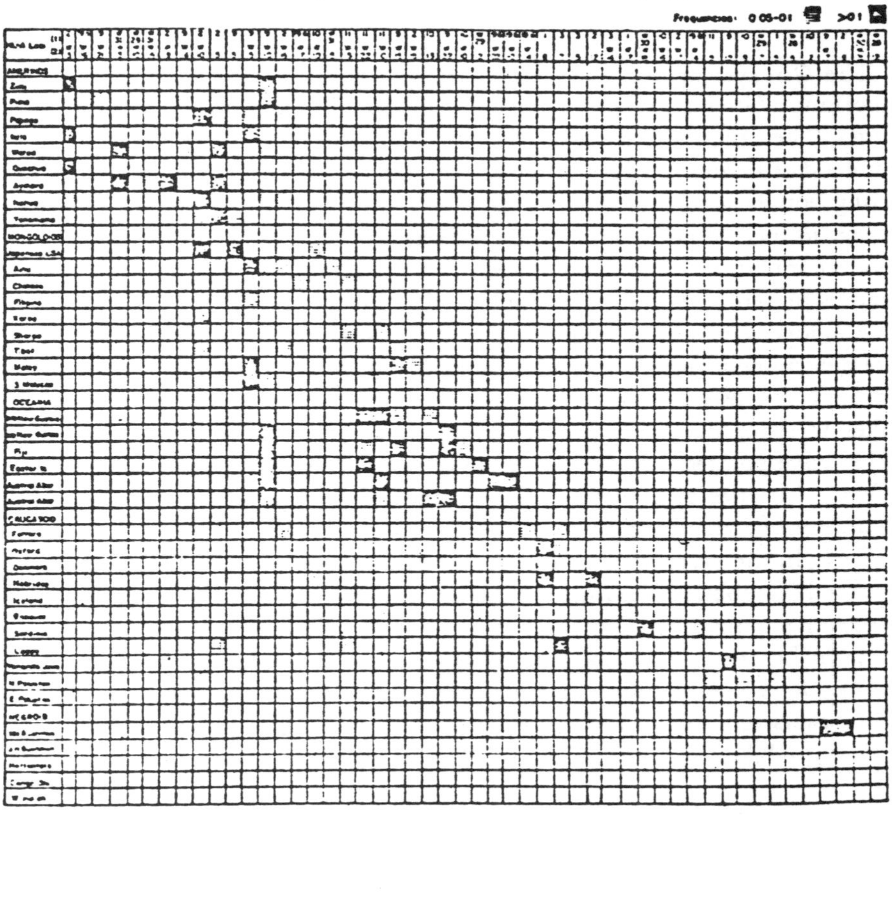

first locus and HL-A5, -12, W5, W10, W15, and W22 of the second locus characterize this group (Dausset and Colombani 1973, Albert *et al.* 1970, Neel *et al.* 1972).

At the International Workshop and Conference on Histocompatibility Testing held at Evian, France in 1972, papers were presented in which a large number of populations had been tested using the same or appropriately compared antisera. Frequencies of haplotypes for each of these populations were estimated and the most common haplotypes shown in tabular form (Bodmer and Bodmer 1974, Dausset and Colombani 1973). We have rearranged these data in Table 5.3 to show how the native American populations tested could be readily distinguished from most other populations, including the closely related Mongoloid and Oceanic groups. In this table high (>0.1) and intermediate (0.05–0.01) frequencies are indicated. From this, it can be seen that there are about 10 haplotypes that occur at these levels only in native Americans or at most in one other population. Furthermore, there are many haplotypes that occur commonly in one or more other population(s) but not at all (or very rarely) in native Americans. From this, the discriminating power of the HL-A haplotypes is apparent; it would be a very useful tool for comparing eastern Asian populations with native Americans.

The Gm System

The Gm system resembles HL-A in genetic complexity and immunologic importance. Gm factors contribute to the inherited variations of the heavy chains of IgG. The genetics of the Gm system is complex and similar to that of the Rh red blood cell antigens. The frequencies of some of the Gm factors vary strikingly among populations, and some of the Gm haplotypes (also called allotypes) are distinct to specific populations and characteristically either present or absent in certain populations. This makes the Gm system a most useful population marker. There is presently, however, a number of problems that make it difficult to use these factors to their full potential. As more antisera were developed, more factors were determined. In earlier studies only one or two factors may have been identified, and in subsequent studies four or five factors may have been tested. This makes interpopulation comparisons difficult. Furthermore, Szathmary *et al.* (1974) have pointed out the difficulties of comparing data because of the use of different varieties of antisera ("Ragg" and "Snag") which may have caused differences of classifications.

Early in the investigations the population associations of single factors were described (Grubb 1970). Gm(1) is found at nearly 100 percent frequency in non-European populations; Gm(6) is rare except in Negroes and Gm(16) is rare except in Japanese (Van Loghem 1970).

Haplotype combinations have also been used to identify populations. Mongoloid markers reported in the literature are Gm(1,17,21), Gm(1,2,17,21), Gm(1,13,17), Gm(1,11,16), Gm(1,3,5,13,14), Gm(1,3,5,11), Gm(1,11,13,15,16,17,1,5),

Gm(1,13), Gm(1,13,15,16,17), Gm(1,10,11,17,25), Gm(1,4,5,10,11,14,23,25), and Gm(1,4,22). Caucasoid markers include Gm(1,17,21), Gm(1,2,21), Gm(3,5, 13,14), Gm(4,5,13,14), and Gm(4,5,8,10,11) (Daveau *et al.* 1975, Giblett 1969, Grubb 1970, Steinberg 1969, Steinberg *et al.* 1974, Szathmary *et al.* 1974).

There are certain Gm factors thought to be characteristic of American Indian populations. Factors 1; 1,2; 1,2,21; and 1,21 are the most frequently and ubiquitously found. The Mongoloid marker 1,11,16, however, has been identified among populations of Ojibwa, and Szathmary *et al.* (1974) reports that Schanfield identified Gm(1,11,16) in an Indian group in the Valley of Mexico and in the Papago of the American Southwest.

The Gm system has promising potential for distinguishing native American populations from others, if comparison and "control" populations are studied with the same reagents in the same laboratory.

Red Blood Cell Antigens

The red blood cell antigens are the most extensively and intensively tested of the human polymorphisms. Specific gene frequency distributions for American Indians have been known since the earliest investigations of the red blood cell antigens. These have been thoroughly reviewed recently by Mourant *et al.* (1976) and will be presented only briefly here. In the ABO system, O occurs in extremely high frequencies. In fact, it is assumed that a great majority of prehistoric Indians were of the O blood type and that the presence of A and B usually indicates genetic admixture with nonnative Americans. Native American populations are characterized by a high frequency of M of the MNSs system and, in particular, by the alleles *MS* and *Ms*. The frequency of the R^2 gene of the Rh system is higher than anywhere else in the world and the frequency of R^or is extremely low or absent. The Kell antigen was probably absent in precontact Indians and occurs in extremely low frequency or is absent in most contemporary American Indian populations. There is an extremely high frequency of Fy^a of the Duffy system and the Diego gene (Di^a) occurs only in native American, Asian, and some Oceanic populations and, in general, is absent from other populations. There is in general a cline of increasing frequency of the Di^a gene from north to south in the Americas. The Lewis A and Lutheran A are thought to occur in American Indian populations only as results of genetic admixture.

Serum Protein Polymorphisms

A large number of serum protein polymorphisms is now known and in some cases alleles found only (or rarely) in Mongoloid populations and in native Americans have been identified (Mourant *et al.* 1976). The transferrin variant "D Chinese" is found in people of east Asia and also in many American Indian populations but rarely in Europeans or other populations. Transferrin B_{0-1} has been reported only in people of Asiatic origin (including Finns, Lapps, and Hungarians) and in native Americans.

The rare Gc variant, Gc Chippewa, has been found so far only in Chippewa and Ojibwa Indians (Szathmary *et al.* 1974).

The serum lipoprotein systems (Ag, Lp) have been used only occasionally in native American population studies, and no generalization can be made about their distribution at present. However, the Ag system is determined by alleles segregating at four or more closely linked locuses and complex combinations of "haplotypes" are known. If populations can be tested with the same antisera using the same techniques, then it is likely that interesting and useful distinctions between populations can be made.

Conclusions

On the basis of the currently available data, there are several polymorphic systems and/or alleles that appear to be unique to native American populations. These include the polymorphisms at the albumin locus, several of the haplotypes of the HL-A system, probably some of the haplotypes of the Gm system, and certain fairly rare alleles (such as Gc Chippewa) of other systems. There is also a large number of new polymorphic traits the distributions of which are not known but which may in due course be shown to have characteristics that make them useful for distinguishing native Americans from other populations.

Using these traits, and others with distinctive frequencies in native Americans, a study could be designed to identify contemporary Asian populations that might have ancient affinities with native Americans.

Acknowledgments

This work was supported by USPHS grants CA-06551, RR-05539, and CA-06927 from the National Institutes of Health, and by an appropriation from the Commonwealth of Pennsylvania.

References

Aikens, C.M. (1970). Hogup Cave. Anthropological Papers, No. 93. Dept. of Anthropol., University of Utah.

Albert, E.D., M.R. Mickey, A.C. McNichols, and P. Terasaki. (1970). Seven New HL-A Specificities and Their Distribution in Three Races. P. Terasaki, Ed., *Histocompatibility Testing 1970*. Copenhagen: Munksgaard, pp. 221–230.

Amos, D., G. Cabresa, W.B. Bias, J.M. McQueen, S.L. Lancaster, J.G. Southworth, and R.E. Ward. (1970). The Inheritance of Human Leukocyte Antigens. III. The Organization of Specificities. *In* P. Terasaki, Ed., *Histocompatibility Testing 1970*. Copenhagen: Munksgaard, pp. 259–275.

Amos, D.B., R. Batchelor, W.F. Bodmer, R. Ceppellini, and J. Dausset. (1976). WHO–IUS Terminology Committee. Nomenclature for Factors of the HL-A System. *Transplantation* 21:353–358.

Arends, T., M.L. Gallango, M. Layrisse, J. Wilberg, and H.D. Heinen. (1969). Albumin Warao: New Type of Human Alloalbuminemia. *Blood* 33:414–420.

Arends, T., L. Weitkamp, M. Gallango, J. Neel, and J. Schultz. (1970). Gene Frequencies and Microdifferentiation among the Makiritare Indians. II. Seven Serum Protein Systems. *Hum. Genet.* 22:526–532.

Bodmer, W.F. (1973). Population Genetics of the HL-A System: Retrospect and Prospect. *In* J. Dausset and J. Colombani, Eds., *Histocompatibility Testing 1972.* Copenhagen: Munksgaard, pp. 611–620.

Bodmer, J., and W.F. Bodmer. (1974). Population Genetics of the HL-A System. *In.* B. Ramot, Ed., *Genetic Polymorphisms and Disease in Man.* New York: Academic Press.

Bowen, P., F. O'Callaghan, and C.S.N. Lee. (1971). Serum Protein Polymorphism in Indians of Western Canada. *Hum. Hered.* 21:242–253.

Brown, K.S., and R.S. Johnson. (1970). Population Studies on Southwestern Indian Tribes. III. Serum Protein Variation of Zuni and Papago Indians. *Hum. Hered.* 20: 281–286.

Cann, H.M., J.G. Bodmer, and W.F. Bodmer. (1973a). The HL-A Polymorphism in Mayan Indians of San Juan La Laguna, Guatemala. *In* J. Dausset and J. Colombani, Eds., *Histocompatibility Testing 1972.* Copenhagen: Munksgaard, pp. 367–376.

Cann, H.M., K.M. Kidd, R. Lisker, R. Rodvany, and R. Payne. (1973b). Genetic Structure of the HL-A System in a Nahua Indian Population in Mexico. *Tiss. Antigens* 3:364–372.

Colombani, G., and L. Degas. (1973). Variations of HL-A Antigens in Populations. *Rev. Eur. Etud. Clin. Biol.* 17:551–563.

Corley, R.B., E.K. Spees, G. Cabrera M., J.L., Swanson, and D.B. Amos. (1973). HL-A Antigens of the Guatemalan Ixils. *In* J. Dausset and J. Colombani, Eds., *Histocompatibility Testing 1972.* Copenhagen: Munksgaard, pp. 351–358.

Dausett, J., and J. Colombani (Eds.) (1973). *Histocompatibility Testing 1972.* Copenhagen: Munksgaard.

Daveau, M., L. Rivat, A. Langaney, N. Afifi, E. Bois, and C. Ropartz. (1975). Gm and Inv. Allotypes in French Guiana Indians. *Hum. Hered.* 35:88–92.

Dossetor, J.B., W.T. Howson, J. Schlaut, P.R. McConnachie, J.D.M. Alton, B. Lockwood, and L. Olson. (1973). Study of the HL-A System in Two Canadian Eskimo Populations. *In* J. Dausset and J. Colombani, Eds., *Histocompatibility Testing 1972.* Copenhagen: Munksgaard, pp. 325–332.

Geerdink, R.A., H.A. Bartstra, and J.M. Schillhorn Van Veen. (1974). Serum Proteins and Red Cell Enzymes in Trio and Wajana Indians from Surinam. *Am. J. Hum. Genet.* 26:581–587.

Giblett, E. (1969). *Genetic Markers in Human Blood.* Philadelphia: F.A. Davis.

Grubb, R. (1970). *The Genetic Markers of Human Immunoglobulins.* Berlin, New York: Springer-Verlag.

Hoijer, H. (1956). The Chronology of the Athabascan Languages. *Intl. J. Am. Ling.* 22: 219–232.

Hymes, D. (1957). A Note on Athabascan Glottochronology. *Intl. J. Am. Ling.* 23:290–297.

Johnston, F.E., O. Alareon, F. Benedict, M. Dary, M. Galbraith, and P.S. Gindhart. (1973). Albumin Mexico (Al^{Me}) in the Guatemalan Highlands. *Am. J. Phys. Anthropol.* 38:27–30.

Johnston, F.E., B.S. Blumberg, S.S. Agarwal. L. Melartin, and T.A. Burch. (1969). Alloalbuminemia in Southwestern U.S. Indians: Polymorphism of Albumin Naskapi and Albumin Mexico. *Hum. Biol.* 41:263–270.

Kissmeyer-Nielson, F., K.E. Kjerbye, L.U. Lamm, J. Jorgensen, G. Bruun Petersen, and H. Gurtler. (1973). Study of the HL-A System in Eskimos. *In* J. Dausset and J. Colombani, Eds., *Histocompatibility Testing 1972.* Copenhagen: Munksgaard, pp. 317–324.

Kostyu, D.D., D.B. Amos, and S. Hinostroza. (1975). An Analysis of the 4c Complex of HL-A Based on Indian Populations. *Tiss. Antigens* 5:420–430.

Lisker, R., L. Cobo, and G. Mora. (1971). Distribution of Albumin Variants in Indians and non-Indians of Mexico. *Am. J. Phys. Anthropol.* 35:119–124.

Layrisse, Z., M. Layrisse, I. Malave, P. Teraski, R.H. Ward, and J.V. Neel. (1973a). Histocompatibility Antigens in a Genetically-Isolated American Indian Tribe. *Am. J. Hum. Genet.* 25:493–509.

Layrisse, Z., P. Terasaki, J. Wilbert, H.D. Heinen, B. Rodriquez, A. Soyano, K. Mittal, and M. Layrisse. (1973b). Study of the HL-A System in the Warao Population. *In* J. Dausset and J. Colombani, Eds., *Histocompatibility Testing 1972.* Copenhagen: Munksgaard, pp. 377–386.

McAlpine, P.J., S.H. Chen, D.W. Cox, J.B. Dossetor, E. Giblett, A.G. Steinberg, and N.E. Simpson. (1974). Genetic Markers in Blood in a Canadian Eskimo Population with a Comparison of Allele Frequencies in Circumpolar Populations. *Hum. Hered.* 24:114–142.

Melartin, L. (1967). Albumin Polymorphism in Man. *Acta Pathol. Microbiol. Scand.*, Suppl. 191.

Melartin, L., and B.S. Blumberg. (1966). Albumin Naskapi: A New Variant of Serum Albumin. *Science* 153:1664–1666.

Melartin, L., B.S. Blumberg, and R. Lisker. (1967). Albumin Mexico, a New Variant of Serum Albumin. *Nature (London)*, 215:1288–1289.

Melartin, K., B.S. Blumberg, and J.R. Martin. (1968). Albumin Polymorphism (Albumin Naskapi) in Eskimos and Navajos. *Nature (London)* 218:787–789.

Mourant, A.E., A.C. Kopec, and K. Domaniewska-Sobczak. (1976). *The Distribution of the Human Blood Groups and Other Polymorphisms.* London: Oxford Univ. Press.

Neel, J.V., T. Arends, C. Brewer, N. Chagnon, H. Gershowitz, M. Layrisse, Z. Layrisse, J. MacClever, E. Migliazza, W. Oliver, F. Salzano, R. Spielman, R. Ward, and L. Weitkamp. (1972). Studies on the Yanomama Indians. Proc. IV Int. Congr., Hum. Genet. Paris. *Excerpta Medica* 1972:96–111.

Perkins, H.A., R.O. Payne, K.K. Kidd, and D.W. Huestis. (1973). HL-A and Gm Typing of Papago Indians. *In* J. Dausset and J. Colombani, Eds., *Histocompatibility Testing 1972.* Copenhagen: Munksgaard, pp. 317–324.

Persson, I., L. Melartin, and A. Gilberg. (1971). Alloalbuminemia: A Search for Variants in Greenland Eskimos. *Hum. Hered*, 21:57–59.

Polesky, H.F., and D.A. Rokala. (1967). Serum Albumin Polymorphism in North American Indians. *Nature (London)* 216:184–185.

Polesky, H.F., D.A. Rokala, and R.A. Burch. (1968). Serum Albumin Polymorphism in Indians of Southwestern United States. *Nature (London)* 220:175–176.

Pollitzer, W.S., D. Rucknagel, R. Tankian, D. Shreffler, W. Leyshon, K. Namboodiri, and R. Elston. (1970). The Seminole Indians of Florida: Morphology and Serology. *Am. J. Phys. Anthropol.* 32:65–81.

Rubenstein, P., R. Costa, A. van Leeuwen, and J.J. van Rood. (1967). The Leukocyte Antigens of Mapuchi Indians. *In* E.S. Curtoni, P.L. Mattiuz, and R.M. Tosi, Eds., *Histocompatibility Testing 1967.* Baltimore: Williams and Wilkins, pp. 251–255.

Schell, L.M., and B.S. Blumberg. (1977). The Genetics of Human Serum Albumin. *In* V.M. Rosenoer, M.A. Rothschild, and M. Oratz, Eds., *Albumin—Structure, Function and Uses in Man*. London, New York: Pergamon Press, pp. 113–141.

Schell, L.M., S.S. Agarwal, B.S. Blumberg, H. Levy, P.H. Bennett, W.S. Laughlin, and J.P. Martin. (1978). Distribution of Albumin Variants Naskapi and Mexico among Aleuts, Frobisher Bay Eskimos, and Micmac, Naskapi, Mohawk, Omaha and Apache Indians. *Am. J. Phys. Anthrop.* 49: 111–118.

Scott, E.M., and R.C. Wright. (1969). The Absence of Close Linkage of Methemoglobinemia at Other Loci. *Am. J. Hum. Genet.* 21: 194–195.

Spees, E.K., D.D. Kostyu, R.C. Elston, and D.B. Amos. (1973). HL-A Profiles of the Pima Indians of Arizona. *In* J. Dausset and J. Colombani, Eds., *Histocompatibility Testing 1972*. Copenhagen: Munksgaard, pp. 345–350.

Spencer, R.F., and J.D. Jennings. (1965). *The Native American*. New York: Harper and Row.

Steinberg, A.S. (1969). Globulin Polymorphisms in Man. *Ann. Rev. Genet.* 3: 25–52.

Steinberg, A.G., A. Tiilikainen, M.R. Eskola, and A.W. Eriksson. (1974). Gammaglobulin Allotypes in Finnish Lapps, Finns, Aland Islanders, Maris (Cheremis) and Greenland Eskimos. *Am. J. Hum. Genet.* 26: 223–243.

Szathmary, E.J.E., D.W. Cox, H. Gershowitz, D.L. Ruchnagel, and M.S. Schanfield. (1974). The Northern and Southeastern Ojibwa: Serum Proteins and Red Cell Enzyme Systems. *Am. J. Phys. Anthropol.* 40: 49–66.

Tanis, R., R.E. Ferrel, J.V. Neel, and M. Morrow. (1974). Albumin Yanomama-2 a "Private" Polymorphism of Serum Albumin. *Am. J. Hum. Genet.* 38: 179–190.

Tittor, W., J. Sobenes, G.S. Smith, P. Sturgeon, E. Zeller, and R.L. Walford. (1973). Distribution of HL-A Antigens, Blood Group Antigens, and Serum Protein Groups in Quechua Indians of Peru. *In* J. Dausset and J. Colombani, Eds., *Histocompatibility Testing 1972*. Copenhagen: Munksgaard, pp. 387–390.

Troup, B.M., R.L. Harvey, R.L. Walford, G.S. Smith, and P. Sturgeon. (1973). Analysis of the HL-A, Erythrocyte and Gammaglobulin Antigen Systems in the Zuni Indians of New Mexico. *In* J. Dausset and J. Colombani, Eds., *Histocompatibility Testing 1972*. Copenhagen: Munksgaard, pp. 339–344.

Van der Does, J.A., J. D'Amaro, A. Van Leeuwen, P. Meera Khan, L.F. Bernibi, E. Van Loghem, L. Nijenhuis, G. Van der Steen, J.J. Van Rood, and P. Rubenstein. (1973). HL-A Typing in Chilean Aymara Indians. *In* J. Dausset and J. Colombani, Eds., *Histocompatibility Testing 1972*. Copenhagen: Munksgaard, pp. 391–396.

Van Loghem, E. (1970). Stability of Gm Polymorphism. *In* R. Grubb and G. Samuelson, Eds., *Human Anti-Human Gammaglobulins*. Oxford: Pergamon Press, pp. 29–37.

Weaver, M.P. (1972). *The Aztecs, Maya and Their Predecessors*. New York: Seminar Press.

Weitkamp, L.R. (1973). The Contribution of Variations in Serum Albumin to the Characterization of Human Populations. *Israel J. Med. Sci.* 9: 1238–1248.

Weitkamp, L.R., E.B. Robson, D.C. Shreffler, and G. Corney. (1968). An Unusual Serum Albumin Variant: Further Data on Genetic Linkage Between Loci for Human Serum Albumin and Group Specific Component (Gc). *Am. J. Hum. Genet.* 20: 392–397.

Wormington, H.M. (1970). *Prehistoric Indians of the Southwest*. Denver, Colo.: Denver Museum of Natural History.

Zaretskaya, Yu., and V. Fedrunova. (1973). Some General Characteristic of the HL-A System in the Russian Population. *Tiss. Antigens* 3: 218–221.

Proc. Natl. Acad. Sci. USA
Vol. 77, No. 9, pp. 5480–5482, September 1980
Medical Sciences

Albumin Naskapi variant in North American Indians and Eti Turks

(albumin B/albumin Mersin/albumin Adana/Central Asia)

SAMUEL G. FRANKLIN*, SUSAN I. WOLF*, YÜKSEL ÖZDEMIR†, GÜNES T. YÜREGIR†, TURGAY İSBIR†, AND BARUCH S. BLUMBERG*

*The Institute for Cancer Research, The Fox Chase Cancer Center, Philadelphia, Pennsylvania 19111; and †Biochemistry Department, Cukurova University Medical Faculty Adana, Turkey

Contributed by Baruch S. Blumberg, June 28, 1980

ABSTRACT **Both conventional polyacrylamide gel electrophoresis and a new type of electrophoretic screening procedure indicate that the polymorphic albumin variants Naskapi, found chiefly in the Naskapi Indians of Quebec, and Mersin, found in the Eti Turks of southeastern Turkey, are molecularly identical or very similar and that the amino acid substitution site in these variants is located between residues 330 and 446. This discovery is consistent with a genetic relationship between the Eti Turks and American Indians. We also report a new variant found in the Eti Turks, albumin Adana, which migrates similarly to albumin B on conventional gels but which our new system shows to differ from the common albumin A and albumin B by a substitution between residues 549 and 585.**

Of the approximately two dozen known electrophoretic variants of human serum albumin, albumin Naskapi is one of the few that occur in polymorphic frequency (1, 2). It has been found, until now, exclusively in North American Indians, with the highest reported frequencies being in the Naskapi Indians of Quebec (1, 2). Albumin Mersin was recently discovered in the Eti Turks of southeastern Turkey.‡ We have found that it migrates identically to albumin Naskapi on polyacrylamide gel electrophoresis at pH 8.6 and occurs with high frequency in the Eti Turks; 8.8% of 397 Eti Turks in Mersin and 9.7% of 93 Eti Turks in Tarsus were heterozygotes for albumin Mersin. However, in two districts of Adana where Eti Turks are numerous in the population, albumin Mersin was not detected in the 442 subjects tested. All of these locations are in southeastern Turkey. By using a newly developed, more discriminating electrophoretic technique (3, 4), we have confirmed (within the limits of this technique) that albumins Naskapi and Mersin are identical. We also report a new variant (proposed name, albumin Adana) which migrates like albumin B on conventional polyacrylamide gel electrophoresis but which our new method has revealed to be different. The cyanogen bromide peptides containing the amino acid substitution sites have been identified for these variants.

MATERIALS AND METHODS

Sera. The albumin A and Naskapi heterozygote standards used in this study were from the Pima Indians of the Gila River Community in Arizona. The Naskapi homozygote standard was from the Naskapi Indians of the Ungava region of Quebec collected in Schefferville, Quebec. These samples had been previously classified by cellulose acetate electrophoresis (Tris citrate buffer, pH 5.4). The Eti Turk samples were from the sources mentioned in the text.

Variant Screening. Cyanogen bromide cleavage and iodoacetamide alkylation of the albumin samples were done as described (4). Polyacrylamide gel electrophoresis of native sera and of cleaved, alkylated sera was done as described earlier (4) and in the legends to Figs. 1 and 2.

RESULTS

Fig. 1 shows the electrophoretic mobilities of albumins A/A, A/Naskapi, Naskapi/Naskapi, A/Mersin, A/Adana, and A/B on a nondenaturing polyacrylamide gel at pH 8.6. Albumin A/Naskapi exhibited two bands of about equal intensity, one band comigrating with albumin A and the other migrating more rapidly towards the anode, whereas homozygotic albumin Naskapi had only the more rapidly migrating band. (The diffuse quality of the latter band is probably due to deamidations resulting from several years of storage.) Albumin A/Mersin migrated identically to A/Naskapi under these conditions, and albumin A/Adana migrated essentially the same as albumin A/B.

It is possible that electrophoretic variants arising from different substitutions or substitutions in different regions of the molecule may comigrate in conventional nondenaturing gels. Thus, it was necessary to devise a more discriminating screening system. The strategy we used was first to cleave the molecule at each of its six methionine residues with cyanogen bromide and then to alkylate the resulting fragments with iodoacetamide to prevent disulfide formation. The fragments were then analyzed on polyacrylamide gels containing acetic acid, urea, and the nonionic detergent Triton X-100 (5, 6). Amino acid substitutions can be localized to a specific cyanogen bromide fragment by this method. Deamidation or aggregation artifacts are not seen in this acidic, denaturing system. In addition, we have shown that the Triton gel system is capable of resolving proteins differing in neutral as well as charged amino acid substitutions (5, 6).

Fig. 2 illustrates the electrophoretic mobilities of the same samples as those of Fig. 1 after they were cleaved and alkylated as described above. When compared to albumin A, the albumin Naskapi heterozygote possesses an additional band, CNBr VN, migrating more slowly than the usual CNBr V, whereas the Naskapi homozygote possesses only the variant band. (A minor band, probably arising from cysteine oxidation or lysine alkylation of CNBr V, comigrated with CNBr VN. Note also that only those samples possessing the Naskapi-like fragment exhibited a modified counterpart of this band, VNm.) These results strongly suggest that fragment CNBr V (residues 330–446) is the region containing the Naskapi substitution. Albumin Mersin also possesses a variant CNBr VN fragment identical to that of albumin Naskapi whereas the remainder of the fragments migrate identically to those of albumin A. This

The publication costs of this article were defrayed in part by page charge payment. This article must therefore be hereby marked "*advertisement*" in accordance with 18 U. S. C. §1734 solely to indicate this fact.

‡ Bingöl, G. & Aydin, G. (1975) Bulletin of Abstracts Fifth Scientific Congress of the Scientific and Technical Organization of Turkey, Istanbul, Turkey.

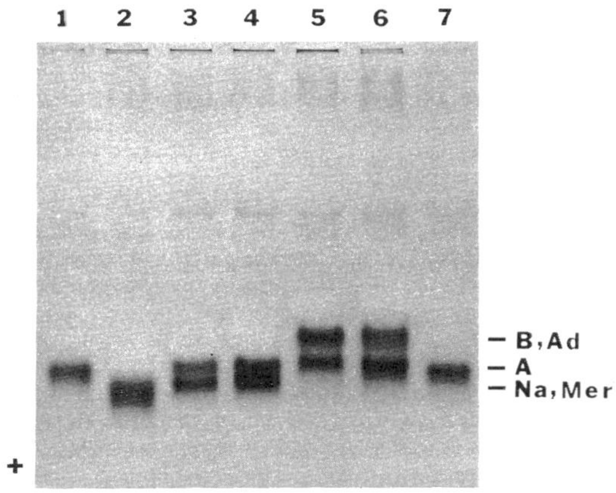

FIG. 1. Electrophoretic mobilities of native albumin variants. Tracks 1 and 7, A/A; track 2, Naskapi/Naskapi; track 3, A/Naskapi; track 4, A/Mersin; track 5, A/Adana; track 6, A/B. Gel was 8% polyacrylamide containing 50 mM Tris·HCl (pH 8.6). Ad, Adana; Na, Naskapi; Mer, Mersin. Stain was Coomassie brilliant blue R-250.

similarity greatly enhances the possibility that albumins Naskapi and Mersin are molecularly identical. Thus, we suggest that the precedent name Naskapi be used. Albumin Adana exhibited an additional band migrating ahead of fragment CNBr VI (CNBr VIAd), which suggests that this fragment (residues 447–548) contains the substitution site in this variant. Albumin A/B contains a very fast migrating counterpart of CNBr VII (residues 549–585), CNBr VIIB, which is consistent with its known lysine-glutamic acid substitution at position 570 (7). It is thus quite different from albumin Adana. Peptide mapping of CNBr VN from albumin Naskapi has shown the substitution site to be between residues 373 and 389.

DISCUSSION

Up to now, albumin variants reaching polymorphic frequencies have been found primarily among certain groups of North, Central, and South American Indians (2). Now two local populations of Eti Turks have also been shown to be polymorphic for a fast-migrating albumin variant which appears to be the same as albumin Naskapi. It is restricted to Eti Turks living in Mersin and Tarsus and is not found in Adana. This discontinuous distribution could result from ethnic diversity among the Eti or from inbreeding or genetic drift within local communities. The slow variant, albumin Adana, was discovered in a patient undergoing examination in a hospital, and its frequency among Eti Turks or the general Turkish population is unknown. It is possible that this variant also exhibits wider distribution but has been erroneously classified as albumin B.

The origins and affinities of American Indians have been questions of long-standing interest in population phylogeny. Skeletal (8), morphological (8), and genetic (9, 10) evidence clearly suggests that American Indians are descendants of Mongoloid populations that moved into the New World from Central and (or) Eastern Asia, most probably from the interior of Siberia. However, it is not known from what part of Asia the migrants came. An ancestral lineage can be constructed by identifying which contemporary Asian populations are most genetically similar to living American Indians. We have suggested that the restricted distribution of the Naskapi allele makes it particularly valuable in determining population affinities between Indian and Asian populations (11). Until now,

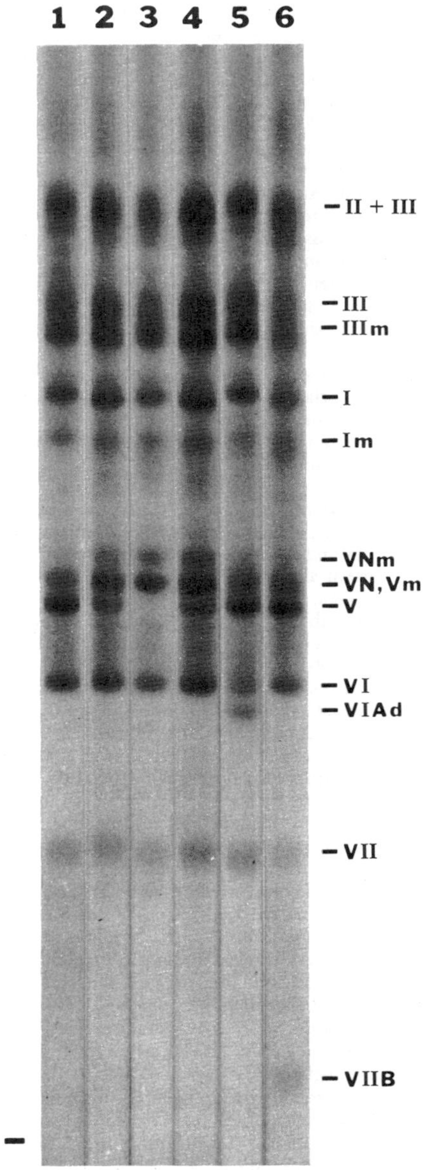

FIG. 2. Resolution of alkylated CNBr fragments of albumin variants. Gel 1, A/A; gel 2, A/Naskapi; gel 3, Naskapi/Naskapi; gel 4, A/Mersin; gel 5, A/Adana; gel 6, A/B. Gels were 12% polyacrylamide containing 5 M urea, 6 mM Triton X-100, and 5% (vol/vol) acetic acid. Stain was Amido Black 10B. CNBr fragments are numbered in accordance with their order in the known sequence of albumin A. N, Ad, and B refer to variant peptides found only in albumins Naskapi (Mersin), Adana, and B, respectively; m refers to modifications probably resulting from cysteine oxidation or lysine alkylation by iodoacetamide. The bands were originally identified by isolation on a preparative gel, $NaDodSO_4$ gel electrophoresis, amino acid analysis, and, in some cases, peptide mapping.

albumin Naskapi has been a genetic marker found exclusively among certain groups of North American Indians. Our discovery that albumin Naskapi may be molecularly identical to an albumin variant (Mersin) in an Asian group is consistent with a genetic relationship between that specific population and American Indians. The two variants could have arisen from the same mutation that either occurred in an ancient population ancestral to both American Indians and Eti Turks or was introduced into the Eti by subsequent admixture from descendants of this ancestral population. Alternatively, the two variants may have arisen from independent mutations, and thus no direct relationship exists between the populations. On historical grounds population admixture is a reasonable explanation. Modern Turkey was in antiquity part of the Eastern (Byzantine) Roman Empire, and historians of the period have provided extensive documentation of nearly continuous invasion of the area by nomadic peoples of Central and East Asian stock. During the sixth through ninth centuries, the Byzantines alternately allied with and fought with a number of groups, including the Avars and the Khazars (12). In the 11th, 13th, and 15th centuries, Asia Minor was successively conquered by the Seljuks (13), Mongols (14), and Ottomans (15), all of whom were ultimately of Asian origin. Most of these conquerors were relatively few in number and formed a ruling elite that was eventually absorbed into the indigenous population. Coon (15) points out that the modern Turks of Turkey show little evidence of their Asiatic ancestry in their physical appearance, body measurements, and ABO blood group frequencies. The presence of a variant albumin may be one of the few detectable indicators of gene flow from Asia. However, to distinguish between the possible origins of a Naskapi-like albumin in the Eti Turks requires collection of more data on the albumins of geographically intervening populations in Siberia and other parts of Central Asia.

Stewart (16) estimates that the ancestors of the American Indians probably entered the New World across the Bering Land Bridge between 23,000 and 28,000 years ago. If albumins Naskapi and Mersin arose from the same mutation, the Naskapi allele must have already been present in the gene pool of this ancestral population at the time of their migration, and thus the mutation would be at least 23,000–28,000 years old.

The results presented here demonstrate that electrophoresis of cyanogen bromide-cleaved albumins on Triton gels is a more sensitive and informative method for the detection and comparison of albumin variants than electrophoresis on conventional nondenaturing gels. Albumins Mersin and Naskapi migrate identically on nondenaturing gels, suggesting that they contain the same substitution. This conclusion is supported by Triton electrophoresis, which shows identical shifts in the CNBr V fragment for both variants. Localization of the difference to a much smaller region of the molecule increases the probability that the variants are molecularly identical. In contrast, albumin Adana migrates much like albumin B on conventional gels, but Triton electrophoresis shows shifts in CNBr fragments VI and VII, respectively, thus indicating that these two variants have amino acid substitutions at different sites.

The ability of our screening procedure to discriminate between variants that possess similar mobilities in conventional systems but that arise from substitutions in different regions in the molecule provides a new tool for genetic and anthropological applications. Additional genetic studies, for example, of loci located on chromosome number 4 (the location of the albumin locus), can now be designed to test the hypothesis raised in this paper that the Eti Turk populations of Mersin and Tarsus are genetically related to American Indians. Other genes and gene combinations unique to American Indian populations (i.e., some of the HL-A haplotypes) could also be studied in the Eti Turks.

The region of variance between albumin A and albumins Mersin and Naskapi (fragment CNBr V) constitutes part of a primary and a secondary long-chain fatty acid-binding site (17). We have previously shown that albumin Naskapi possesses a diminished affinity for the drug Warfarin (18) compared to albumin A. It is thus possible that these variants have functional differences of physiological importance.

We thank Ms. Judy Mamas for technical assistance. This research was supported in part by Grant RR-05539 awarded by the Biochemical Research Support Grant Program, Division of Research Resources, National Institutes of Health, by Grants CA-06551 and CA-06927 from the National Institutes of Health, and by an appropriation from the Commonwealth of Pennsylvania.

1. Johnston, F. E., Blumberg, B. S., Agarwal, S. S., Melartin, L. & Burch, T. (1969) *Human Biology* **41**, 263–270.
2. Schell, L. M. & Blumberg, B. S. (1977) in *Albumin Structure, Function and Uses.* eds. Rosenoer, V. M., Oratz, M. & Rothschild, M. A. (Pergamon, New York), pp. 113–141.
3. Zweidler, A. (1978) in *Methods in Cell Biology,* eds. Stein, G., Stein, A. & Kleinsmith, L. J. (Academic, New York), Vol. 17, pp. 223–233.
4. Franklin, S. G., Wolf, S. I., Zweidler, A. & Blumberg, B. S. (1980) *Proc. Natl. Acad. Sci. USA* **77**, 2505–2509.
5. Franklin, S. G. & Zweidler, A. (1977) *Nature (London)* **266**, 273–275.
6. Urban, M. K., Franklin, S. G. & Zweidler, A. (1979) *Biochemistry* **18**, 3952–3960.
7. Winter, W. P., Weitkamp, L. R. & Rucknagel, D. L. (1972) *Biochemistry* **11**, 889–896.
8. Wormington, H. M. (1957) *Ancient Man in North America* (The Denver Museum of Natural History, Denver, CO), pp. 225–260.
9. Szathmary, E. J. E. (1979) in *The First Americans: Origins, Affinities and Adaptations,* eds. Laughlin, W. S. & Harper, A. B. (Fischer, New York), pp. 185–209.
10. Mourant, A. E., Kopec, A. C. & Domaniewska-Sobczak, K. (1976) in *The Distribution of the Human Blood Groups and Other Polymorphisms* (Oxford Univ. Press, London), pp. 117–122.
11. Lampl, M. & Blumberg, B. S. (1979) in *The First Americans: Origins, Affinities and Adaptations,* eds. Laughlin, W. S. & Harper, A. B. (Fischer, New York), pp. 107–123.
12. Grousset, R. (1970) *The Empire of the Steppes* (Rutgers Univ. Press, New Brunswick, NJ), pp. 171–178.
13. Grousset, R. (1970) *The Empire of the Steppes* (Rutgers Univ. Press, New Brunswick, NJ), pp. 148–159.
14. Grousset, R. (1970) *The Empire of the Steppes* (Rutgers Univ. Press, New Brunswick, NJ), p. 263.
15. Coon, C. S. (1965) *The Living Races of Man* (Knopf, New York), p. 39.
16. Stewart, T. D. (1973) *The People of America* (Charles Scribner's Sons, New York), p. 18.
17. Peters, T., Jr. & Reed, R. G. (1977) in *Albumin Structure, Biosynthesis, Function,* eds. Peters, T., Jr. & Sjoholm, I. (Pergamon, New York), pp. 11–20.
18. Wilding, G., Blumberg, B. S. & Vesell, E. S. (1977) *Science* **195**, 991–994.

Reprinted from THE LANCET, *March* 25, 1961, *pp.* 634–637

AN ISOPRECIPITATION REACTION DISTINGUISHING HUMAN SERUM-PROTEIN TYPES

A. C. ALLISON
M.A., B.M., D.Phil. Oxon.
OF THE NATIONAL INSTITUTE FOR MEDICAL RESEARCH, LONDON, N.W.7

B. S. BLUMBERG
M.D. Columbia, D.Phil. Oxon.
OF THE NATIONAL INSTITUTES OF HEALTH, BETHESDA, MARYLAND

DURING the past decade it has become apparent that human beings differ in their serum-proteins as well as in their red blood-cells. Genetically controlled variations have been described in haptoglobins and transferrins (Smithies and Connell 1959), γ-globulins (Grubb 1957, Ropartz 1960), and in the Gc system definable by immuno-electrophoresis (Hirschfeld and Beckman 1960). Though the chemical differences between the alternative proteins are not yet known, they are generally assumed to be small —analogous to the single aminoacid substitutions that distinguish the different human hæmoglobin half-molecules (Hunt and Ingram 1959) and the small sugar substitutions that distinguish the blood-group substances (Morgan 1959). Nevertheless, the possibility remains that these or other serum-proteins are sufficiently different in structure to induce serum-protein isoimmunisation when blood is transfused.

Such a reaction is already known in rabbits. Oudin (1956), Dray and Young (1959), and others have shown that rabbits can be immunised with serum from other rabbits of different γ-globulin type. In the original experiments the foreign plasma was introduced with adjuvants, but recent observations (S. Dray, personal communication) have shown that adjuvants are unnecessary for immunisation. Three or four intravenous injections of serum from rabbits of another γ-globulin type led to the production in recipients of well-defined precipitating antibodies.

The experiments described here were undertaken to find out whether any similar reaction can be shown in man. Sera were obtained from subjects who had received mul-

2

tiple transfusions and were tested for precipitating antibodies against a panel of sera chosen to represent all the available genetically controlled serum-protein types. A serum has been found which gives strong precipitation with some, but not all, sera; and this reaction defines a polymorphism involving the α_2-globulins. Besides being of genetic interest, this finding raises the possibility that serum-protein isoimmunisation takes place in man and represents a clinical problem in persons receiving blood-transfusions.

Materials and Methods

Post-transfusion sera.—These were obtained from patients at the Clinical Center of the National Institutes of Health, Bethesda, who had received 5 or more transfusions. The indications for transfusion included open-heart surgery, hæmolytic and aplastic anæmias, pelvic tumours, and leukæmias. Wherever possible, blood-samples were drawn between 10 and 30 days after the last transfusion. Sera were stored at −20°C and were tested within 5 days of collection. Many of the subjects had received 30 or more transfusions.

Panel sera.—These were selected from the collection of the Section on Geographic Medicine and Genetics of the National Institute of Arthritis and Metabolic Diseases. The 24 sera included haptoglobin types 1–1, 2–1, 2–2, 2–1M, and O; transferrin types CC, CD, and BC; and γ-globulin types Gm a+, Gm a−, Gm b+, Gm b−, Gm x+, and Gm x−. In addition, 3 sera from patients with definite rheumatoid arthritis having high titres of rheumatoid factor, and 1 serum from a patient with well-developed lupus erythematosus (L.E.) having positive L.E. histological preparations and high titres of anti-nuclear antibodies, were used. The sera had been stored at −20°C for periods varying from 2 days to 2 years. Storage did not alter the reaction to be described. The sera were from different population groups, including American whites (sixteen), American Negroes (three), Micronesians (three), an Alaskan Eskimo, and a Viet-Namese.

Gel-diffusion precipitation technique.—The Ouchterlony procedure was carried out in 5 cm. petri dishes containing 8 ml. of 0·9% 'Oxoid Ionagar' (w/v) in 0·07 *M* sodium phosphate buffer, pH 7·0, with 0·001 *M* sodium ethylenediamine-tetra-acetate and 0·001 *M* sodium azide. With a die, six wells 6 mm. in diameter were cut round a centre well of the same size. The circumferences of the centre well and peripheral wells were 4 mm. apart. After removal of the agar cores from the wells, the bases of the latter were sealed with small volumes of molten agar. The post-transfusion sera were placed in the centre wells and panel sera in the peripheral wells. The plates were stored at room temperature and observed at 18 hours (by which time clear-cut precipitation was usually visible), 48 hours, and 4 days. The precipitation is best seen by oblique illumination from below against a dark background.

3

Immunoelectrophoresis.—This was done by the micromethod of Grabar and Williams (1955) and Scheidegger (1955) using standard lantern-slide plates coated with agar made up in barbitone buffer 0·01 *M*. pH 8·4. Antibody (or antigen) was placed in the side well and examined after reacting for 18 hours, by which time well-defined precipitates were visible. The reaction was continued to 36 hours, after which the plates were washed in saline and stained.

Starch-gel electrophoresis.—Starch gel (Connaught Laboratories, Toronto) was made up in *tris*-(hydroxymethl)-aminomethane buffer (Poulik 1957) and poured into 6 × 80 × 150 mm. trays. Samples were inserted as a starch paste (Smithies 1955) into a slot 3 mm. wide. Electrophoresis was carried out for 5 hours at 6V per cm. The position of the components was identified in a small strip of gel; the various components were eluted by the method of Gordon (1960) and were concentrated by ultrafiltration.

Preparation of γ-globulin components.—Separation of 7S and 19S γ-globulins was achieved by sucrose density-gradient centrifugation (Fudenberg and Kunkel 1957) using 40, 30, 20, and 10% solutions of sucrose.

Bentonite-flocculation test.—The method of Bozicevich et al. (1958) was used.

Latex-fixation test.—This was carried out as described by Singer and Plotz (1956), using the Hyland 'RA-test' kit (Hyland Laboratories, Los Angeles, California).

Observations

A precipitin was found in the serum of a patient who had received many transfusions. He had a long and varied history of illness.

The patient is a 64-year-old white American retired executive of Hungarian birth. At the age of 16 he had albuminuria which lasted for 4 years. In 1934, while still living in Budapest, he had an episode of sudden onset of paræsthesiæ and paralysis of the legs. This resolved spontaneously in 2 months. A second episode, involving complete paralysis up to the waist, weakness of the arms, and loss of sensation and reflexes in the legs, took place in the United States in 1958. Cerebrospinal fluid contained 73 mg. protein and 150 lymphocytes per ml. Polyneuritis of unknown cause was diagnosed, and again remission was spontaneous.

The patient also had symptoms of peptic ulceration dating from 1921, with gastrointestinal hæmorrhage in 1938 and 1940. In 1941 a gastroenterostomy was performed, but symptoms and hæmorrhage have recurred at intervals. The patient was again admitted to hospital in 1958, at which time he was reported to have low red-cell counts, low hæmoglobin levels, and high white-cell and platelet counts. X-ray examination suggested an ulcer crater in the proximal duodenum. The bleeding continued for a month and then ceased, and with the help of blood-transfusions the hæmoglobin was raised to 14 g. per

4

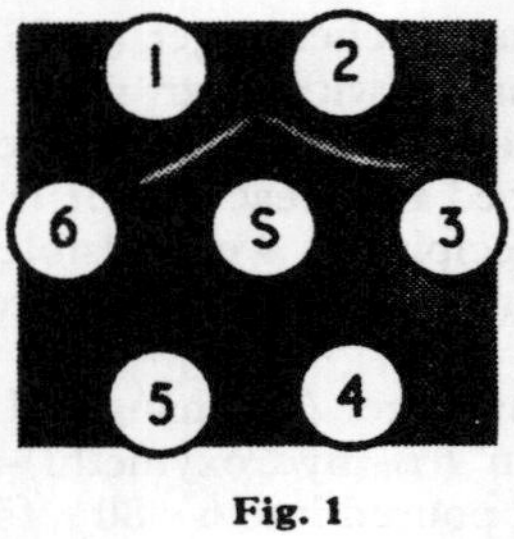

Fig. 1

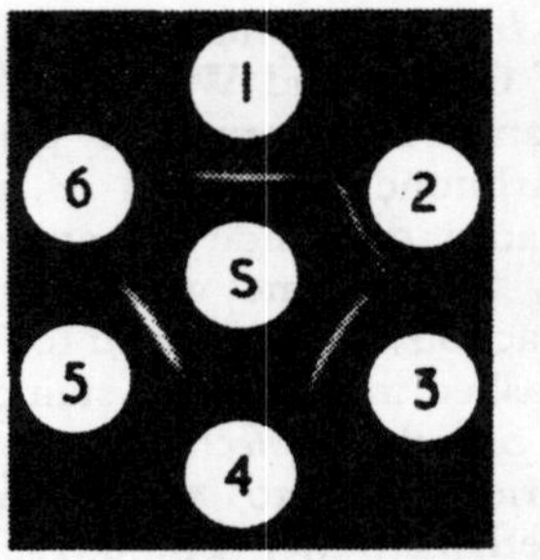

Fig. 2

Fig. 1—The precipitin reaction, seen in an agar-gel Ouchterlony plate, between the serum of the patient in the centre well and some, but not all, members of a panel of sera from normal persons. Unstained plate after 24 hours.

Fig. 2—The precipitin reaction between the serum of the patient in the centre well against some, but not all, members of a panel of sera.

The two degrees of positive reaction are shown: nos. 1, 3, and 5 are 2+ reactors and no. 2 is a weak + reactor.

---

100 ml. Because of the high white-cell count and the presence of numerous megakaryocytes in the bone-marrow and some in the peripheral blood, the possibility that the patient might have leukæmia was raised. No convincing evidence that this is so has been obtained.

Without transfusion the patient's hæmoglobin continued to fall, and in February, 1958, he was admitted to the Clinical Center of the National Institutes of Health. During his time in hospital the patient had persistent anæmia without reticulocytosis. The leucocyte count varied from 10,000 to 15,000 and the platelet count was consistently above 500,000 per c. mm. The differential count showed predominantly neutrophils with a relative increase of immature forms. Bone X-rays showed no abnormalities. Refractory anæmia of unknown cause was diagnosed. The patient was given cells from 4 units of blood, which raised his hæmoglobin to 11 g. per 100 ml., and was discharged.

When the patient first received blood-transfusions in 1958 they were well tolerated. Since that time he has received 47 units of blood. The patient is of blood group AB, Rh+, and there has never been evidence of red-cell incompatibility with donor bloods. In the spring of 1960 the patient began to have transfusion reactions. Within about 1 hour of some, but not all, transfusions there was a fever, rising to 38–39°C (100·4–102·2°F) accompanied by headache and muscle pain, but no other symptoms. The fever and symptoms persisted for about 5 hours, and then slowly subsided. The patient had been discharged from hospital before the precipitin in his blood was discovered, so that it has not yet been possible to ascertain whether only bloods containing the precipitating antigen give transfusion reactions.

5

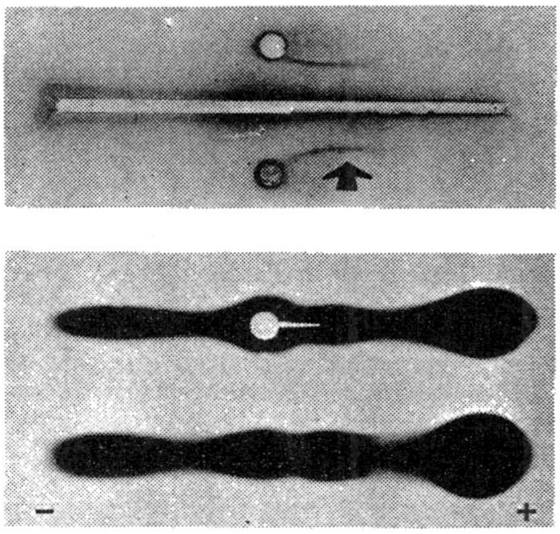

Fig. 3—Agar-gel immunoelectrophoresis to characterise antigen. The serum of a strongly positive reactor was concentrated three times by ultrafiltration and submitted to electrophoresis. The serum of the patient was placed in the rectangular antibody well. The position of the precipitin line (arrow) suggests that the antigen is an α_2-globulin. Above, unstained precipitin lines; below, protein stain.

Precipitation Reactions

The serum of the patient gave well-defined precipitation reactions in agar with some, but not all, panel sera (fig. 1). The reaction was clearly visible at 18 hours and was intense after 2 days. The precipitation line was convex toward the centre (precipitin) well, indicating that the antigen is of relatively high molecular weight (Korngold et al. 1959). Two intensities of reaction were definable in most groups of sera studied (fig. 2). Some sera lacked detectable antigen, the precipitation lines extending into the corresponding wells without deviation (fig. 1). The precipitation lines were clear-cut and quite unlike the diffuse reaction which slowly develops between rheumatoid factor and γ-globulin in some subjects (Franklin 1960a). Rheumatoid-factor reaction was excluded because neither the patient nor most of the positive subjects had rheumatoid factor, as determined by bentonite-flocculation and latex-fixation tests. Some panel subjects with rheumatoid factor gave reactions with the antiserum, but others did not. A second sample of serum from the patient, drawn 3 weeks after the first, at a time when the patient had not received any medication since discharge, gave identical reactions, as did a third sample drawn 1 month later, and a fourth 3 months later.

Support for the interpretation that the precipitin in the

6

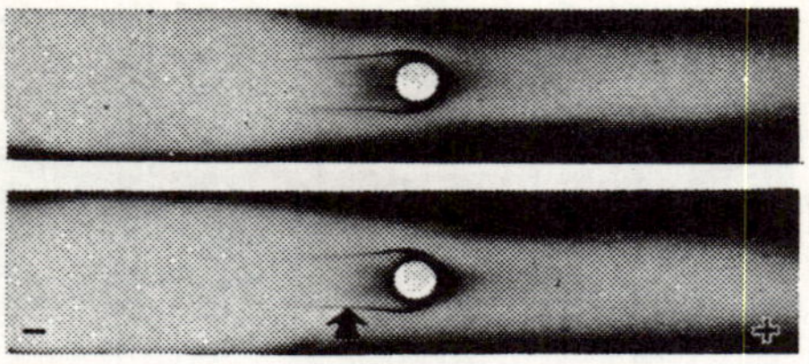

Fig. 4—Agar-gel immunoelectrophoresis to characterise the antibody.

The serum of the patient was placed in the circular wells and submitted to electrophoresis, and a strongly positive serum concentrated three times was placed in the rectangular wells. A broad precipitin line (arrow) is seen in the position of the γ-globulin.

serum of this patient is an antibody came from experiments showing that the active protein is a 7S-γ-globulin. We propose the designation Ag(a+) for sera containing the α-globulin antigen and Ag(a−) for sera that do not contain the antigen. The precipitin in the serum of the patient thus has the specificity anti-Ag(a+).

Characterisation of Antigen

A panel serum giving a strongly positive precipitation line in the gel-diffusion test was concentrated three times by ultrafiltration and used for immunoelectrophoresis. The antibody well was filled with serum from the patient. A well-defined precipitation line developed between the α_2 region and the antibody (fig. 3). This result, along with that in the Ouchterlony plates, suggested that the antigen is a high-molecular-weight α_2-component. The only α_2-macroglobulin so far defined is the 19S-α_2 component described by Brown et al. (1954) and Wallenius et al. (1957), which appears to be equivalent to the slow-α_2 component found on starch-gel electrophoresis (Poulik and Smithies 1958). Purified α_2-macroglobulin (Behringwerke, Marburg) in a wide range of concentrations failed to give precipitation with the patient's serum. Furthermore, when an Ag(a+) serum was submitted to starch-gel electrophoresis, the slow-α_2 fraction recovered by the technique of Gordon (1960) failed to show precipitation with the anti-Ag(a+) serum. The further identification of the antigen present in the α-globulin is currently being undertaken.

Characterisation of Antibody

The patient's serum was submitted to immunoelectrophoresis, and a serum giving a strong reaction in the Ouchterlony tests was placed in the side well. Within 18 hours a well-defined precipitation line appeared in the

7

γ-globulin region (fig. 4). This result, together with the shape of the precipitation curve in the Ouchterlony plate, suggested that the antibody was a relatively low-molecular-weight γ-globulin. This interpretation was confirmed by an experiment in which the 7S- and 19S-γ-globulin components in the patient's serum were separated by density-gradient centrifugation. Precipitation lines in Ouchterlony plates were obtained with low-molecular-weight fractions giving no reaction with immune-rabbit serum against 19S human γ-globulin as described by Franklin (1960b). It can, therefore, be concluded that the antibody is a 7S-γ-globulin. The serum of the patient's daughter showed no reaction when tested against panel sera nor against her father's serum.

Stability of the Antigen

The antigen is unaffected by heating at 57°C for 40 minutes, by hæmolysis with its own or foreign human hæmoglobin, or by storage at room temperature for up to 7 days. The antigen was not lost by dialysis or ultrafiltration. Sera of either type (reacting or non-reacting) stored for as long as 4 years at −20°C were consistently found to give the same reaction as fresh sera from the same subjects. The reaction in the fresh sera, however, was in some cases stronger and better defined than in the stored sera. Repeated freezing and thawing of the sera up to at least fifteen times did not affect the reaction. Blood-samples were withdrawn twice weekly for 4 weeks from two normal volunteers whose sera reacted strongly with that of the patient. The reaction remained unchanged during this period. Sera withdrawn approximately twice yearly over a 2-year period from six subjects, two of whom reacted and four of whom did not, consistently gave the same reaction. Of 2 sera withdrawn in 1955, 1 was a reactor and the other not; the reactions in sera withdrawn from the same subjects in 1959 were unchanged.

Variation in the position of the slow-α_2-globulin after starch-gel electrophoresis has been noted. By means of vertical electrophoresis, using borate buffer at pH 8·6 (Smithies 1959), a slow-moving, a fast-moving, and a double band can be distinguished. It has been suggested that this variation may be due to storage (Smithies 1955); but this may not be the only explanation. Sera from different persons stored for the same length of time under identical conditions may show the variation, and not all stored sera show it (Blumberg 1961). In any event, this does not appear to affect the reaction with the patient's serum. Persons of each of the three slow-α_2-types were both positive and negative reactors.

8

TABLE I—FREQUENCY OF POSITIVE REACTORS

Population	No. studied	No. positive
Parents of white U.S. families	56	32 (*57%*)
U.S. Negroes	21	10 (*48%*)
Micronesians (Rongelap, Marshall Islands) ..	51	50 (*98%*)

It appears, therefore, that the antigen is relatively stable and persistent in any one person.

Distribution of the Antigen in Populations

Because the supply of antibody serum is limited, extensive population surveys have not been possible. However, positive and negative sera were found in all the populations studied. On the basis of preliminary examinations, the frequency seems to vary in different populations. The frequencies of positive reactors in some populations are shown in table I. More extensive studies are planned when micromethods are developed.

Inheritance of the Antigen

126 sera from 29 American families (25 white, 4 Negro) were studied. The results are shown in table II. In addition, 38 sera from 8 Micronesian families were tested. Since the frequency of positive reactors was high in the Micronesians, all of these were Ag(a+) × Ag(a+) matings, and nearly all the offspring were Ag(a+). These are also shown in table II, but are not included in the statistical evaluation, since they are not decisive in testing the genetic hypothesis. In the families from the United States, among the offspring of positive × positive and positive × negative matings, positive and negative offspring were found, whereas among the offspring of negative × negative matings only negative offspring appeared. This result suggests that the inheritance of the antigen follows mendelian segregation, with negative subjects homozygous for a recessive gene *Ag*, and positive subjects homozygous or heterozygous for the allelic gene Ag^A. It is suggested that the Ag^A gene controls the

TABLE II—INHERITANCE OF THE ANTIGEN

Mating type	No. of families	Offspring		Total
		Ag(a+)	Ag(a−)	
United States:				
Ag(a+)×Ag(a+) ..	11	21	5	26
Ag(a+)×Ag(a−) ..	10	9	13	22
Ag(a−)×Ag(a−) ..	8	0	20	20
Micronesian:				
Ag(a+)×Ag(a+) ..	8	21	1	22

9

TABLE III—OBSERVED AND EXPECTED NUMBERS OF Ag(a−) CHILDREN FROM VARIOUS MATINGS

Test	Mating	χ^2
Comparison of observed and expected number of Ag(a−) children in families where there is at least one Ag(a−) child	Ag(a+)×Ag(a−)	0·99
Comparison of observed and expected number of Ag(a−) children in families where there is at least one Ag(a−) child	Ag(a+)×Ag(a+)	1·37
Comparison of observed and expected number of families with at least one Ag(a−) child, assuming an Ag-gene frequency of 0·66	Ag(a+)×Ag(a−)	0·49
Comparison of observed and expected number of families with at least one Ag(a−) child, assuming an Ag-gene frequency of 0·66	Ag(a+)×Ag(a+)	0·03
Comparison of observed and expected number of recessives in Ag(a−)× Ag(a−) matings	Ag(a−)×Ag(a−)	0

$\Sigma\chi^2 = 2{\cdot}88$ with 5 degrees of freedom; $0{\cdot}8 > P > 0{\cdot}7$.

synthesis of the *Ag*(a+) antigen. If a product of the allelic gene is found, the gene could be termed Ag^B and the antigen *Ag*(b+).

The family data were analysed by the methods summarised by Smith (1956). In this, four types of comparison are made, in each of which a correction is included for family size. The comparisons of observed and expected figures involve:

1. The number of recessive Ag(−) children resulting from Ag(+)×Ag(a−) matings, given the number of families in which there is at least one recessive child.
2. The number of recessives resulting from Ag(a+)×Ag(a+) matings in which there is at least one recessive offspring, given the total number of families with at least one recessive offspring.
3. In Ag(a+)×Ag(a−) matings, the total number of families with at least one recessive offspring, given the total number of Ag(a+)×Ag(a−) matings and the gene frequencies in the population from which the families were selected (q = frequency *Ag* = 0·66).
4. In Ag(a+)×Ag(a+) matings, the total number of families with at least one recessive offspring, given the total number of Ag(a+)×Ag(a+) matings and the gene frequency.

The first two computations are independent of the gene frequencies in the population study. For the third and fourth computations, however, an estimate is needed of the gene frequency in the population from which the families were drawn. We were unable to obtain a large random sample from the population from which the families were selected, and the gene frequency was

10

estimated from the frequency of the phenotypes in the parental members of the families (table I). Of 56 parents, 57% were found to be Ag(a+).

The results of these computations are summarised in table III, which includes also the results of Ag(a−) × Ag(a−) matings. Although the numbers are small, this preliminary family analysis is consistent with the genetic hypothesis. On the basis of segregation in this group of families, no evidence of close linkage was found with the following blood-group systems and haptoglobins types: ABO, Rh, MN, Duffy, Kidd, and P. Analyses of 30 twin pairs to be reported elsewhere were also consistent with the genetic hypothesis.

Independence of the Ag and Other Known Inherited Systems of Serum-proteins

The reactions in the panel sera showed that the Ag reactions were independent of rheumatoid factor and of haptoglobin, transferrin, and γ-globulin types. The segregation of the Ag^A and *Ag* genes in families was also independent of these systems. Since the Ag(a +) factor is an α-globulin it is also independent of another recently described inherited γ-globulin system designated In_v by Ropartz (1960). In collaboration with Dr. J. Hirschfeld, of Stockholm, Ag tests were carried out on 113 sera of known Gc types, which are distinguishable by immuno-electrophoresis (Hirschfeld and Beckman 1960). Positive and negative Ag reactions were found in subjects of Gc types 1–1, 2–1, and 2–2, the percentage of reaction being of the same order in all these groups. Hence the Ag system is independent of all other known inherited serum-protein differences. The Ag(a+) antibody is also distinct from the factor in normal human serum giving a precipitate in agar with red-cell hæmolysates (Peetom et al. 1960); the antigen in the latter case is not present in serum and does not react as an α-macroglobulin. The Ag factor is present in too many normal sera to be C-reactive protein; which in any case migrates in the γ-region in immunoelectrophoresis (Zach and Zimmermann 1959).

Discussion

The precipitin in the patient's serum defines the presence of a factor in some sera from all human populations so far tested. The antigen is an α_2-macroglobulin and the precipitin a 7S-γ-globulin. The presence of the antigen appears to be genetically determined, so that a new polymorphic system in human serum-proteins has been revealed.

11

Besides being of genetical and anthropological interest, the finding of a specific precipitin in the patient's serum suggests that isoimmunisation by human serum-proteins may occur. Though there is no direct evidence that the precipitin was induced by the many transfusions which the patient received, this would seem to be the most probable explanation of the findings. In all, 61 sera of subjects who had received multiple transfusions have been tested, and 3 of these have given definite precipitation lines with some of the panel sera. The reaction which is the subject of this paper was the strongest, and the specificity of the others, which are different from the Ag system, will be reported later. Sera of forty persons who had not received transfusions failed to give precipitation lines with the panel sera.

Though our patient initially tolerated transfusions well, during the past 2 years he has had pyrexical reactions after most, but not all, transfusions. Whether these reactions take place only when donor serum is Ag(a+), and represent an antigen/antibody reaction, is now being investigated. Other properties of the precipitin will be described elsewhere. The unusual history of this case may be irrelevant, since the other subjects showing precipitins had typical thalassæmia.

Summary

A serum from an anæmic patient who had received approximately 50 transfusions was found to give strong precipitation with sera of some, but not all, normal subjects.

It appears that an antigen/antibody reaction is involved. The antigen in some normal subjects is an α_2-globulin, and the antibody in the patient is a 7S-γ-globulin. The antigen appears to be inherited according to simple mendelian rules and its frequency varies in different racial groups.

It is possible, though not proven, that antigen/antibody reactions were responsible for severe post-transfusion reactions in the patient, and similar processes may be involved in other unexplained post-transfusion reactions.

We are indebted to Dr. Fred Stohlman and Dr. Archie McKinney for supplying the serum from the patient; to Dr. Sheldon Dray and Dr. Kurt J. Bloch for valuable help and advice; to Dr. J. Hirschfeld, Stockholm, for collaboration with the Gc tests; to Dr. George A. Silver, Montefiore Hospital, New York, for allowing access to families under his care; to Dr. Arthur G. Steinberg for supplying blood from American Negro families and carrying out the γ-globulin typing; and to Dr. Lyman Crittenden for assistance with the linkage studies. Most of this investigation was undertaken while one of us (A.C.A.) was a visiting scientist at the National Institute of Arthritis and Metabolic Diseases.

12

REFERENCES

Blumberg, B. S. (1961) Unpublished.
Bozicevich, J., Bunim, J. J., Freund, J., Ward, S. B. (1958) *Proc. Soc. exp. Biol., N.Y.* 97, 180.
Brown, R. K., Baker, W. H., Peterkofsky, A., Kauffman, D. L. (1954) *J. Amer. chem. Soc.* 76, 4244.
Dray, S., Young, G. O. (1959) *Science*, 129, 1023.
Franklin, E. C. (1960a) *Arth. Rheum.* 3, 16.
— (1960b) *J. Immunol.* 85, 138.
Fudenberg, H. H., Kunkel, H. G. (1957) *J. exp. Med.* 106, 689.
Gordon, A. H. (1960) *Biochim. biophys. Acta*, 42, 23.
Grabar, P., Williams, C. A., Jr. (1955) *ibid.* 17, 67.
Grubb, R. (1957) *Vox Sang.* 2, 305.
Hirschfeld, J., Beckman, L. (1960) *Acta genet.* 10, 48.
Hunt, J. A., Ingram, V. M. (1959) Biochemistry of Human Genetics; p. 347. Boston.
Korngold, L., Van Leeuwen, G., Brener, J. L. (1959) *J. Lab. clin. Med.* 55, 517.
Morgan, W. T. J. (1959) Biochemistry of Human Genetics; p. 237. Boston.
Oudin, J. (1956) *C.R. Acad. Sci., Paris*, 242, 2606.
Paetoom, F., Rose, N., Ruddy, S., Micheli, A., Grabar, P. (1960) *Ann. Inst. Pasteur*, 98, 252.
Poulik, M. D. (1957) *Nature, Lond.* 180, 1477.
— Smithies, O. (1958) *Biochem. J.* 68, 636.
Ropartz, C. (1960) *Rev. franç. Etud. clin. biol.* 5, 933.
Scheidegger, J. J. (1955) *Int. Arch. Allergy*, 7, 103.
Singer, J. M., Plotz, C. M. (1956) *Amer. J. Med.* 21, 888.
Smith, C. A. B. (1956) *Ann. hum. Genet.* 20, 257.
Smithies, O. (1955) *Nature, Lond.* 175, 307.
— (1959) *ibid.* 181, 1203.
— Connell, G. E. (1959) *Biochem. J.* 72, 115.
Wallenius, G., Troutman, R., Kunkel, H. G., Franklin, E. C. (1957) *J. biol. Chem.* 225, 253.
Zach, J., Zimmerman, K. (1959) *Klin. Wschr.* 37, 160.

Printed in Great Britain

The Lancet Office, 7, Adam Street, Adelphi, London, W.C.2

Journal of Clinical Investigation
Vol. 41, No. 10, 1962

A HUMAN LIPOPROTEIN POLYMORPHISM

By BARUCH S. BLUMBERG, DAVID BERNANKE, AND ANTHONY C. ALLISON

(From the Geographic Medicine and Genetics Section, National Institutes of Health, Bethesda, Md., and the National Institute for Medical Research, London, England)

(Submitted for publication March 8, 1962; accepted July 3, 1962)

During the past decade, several groups of inherited human serum proteins constituting polymorphic systems have been discovered. These include the haptoglobins, transferrins, gamma globulin groups, and the group-specific substances (for review see Reference 1). In some of these systems the gene frequencies are such that a patient receiving repeated blood transfusions would have a high probability of receiving serum from an individual of a phenotype different from his own. For example, in a single transfusion, a European of haptoglobin type 1-1 (approximately 15 per cent of a European population) would have a chance of about 1 in 3 of receiving a serum of haptoglobin type 2–2 and a chance of about 1 in 2 of receiving type 2–1. Consideration of some of the other serum systems leads to the same general conclusion: a frequently transfused individual would almost certainly receive transfused plasma containing some proteins slightly different from his own. There is no evidence that any of the known differences among these proteins leads to antibody formation (2, 3). Nonetheless, under some circumstances and in some individuals either the known protein polymorphisms or other still undiscovered polymorphisms may induce isoimmunization. This has been considered possible for some time, and analogous findings are known in animals (4–6).

In a study of human material, Allison and Blumberg (7) collected sera of patients who had required multiple transfusions for a variety of reasons and examined them for precipitating antibodies against normal human sera. With the Ouchterlony technique, a substance which acted as a precipitating antibody was found in the serum of a patient, C. de B., who had received approximately 50 transfusions because of a refractory anemia of unknown cause. The precipitin line was formed between the serum of C. de B. and some, but not all, normal sera (Figure 1). The antibody-like material in the serum of the transfused patient appeared to be 7S gamma globulin (7) and the "antigen" with which it reacted was a low-density beta-lipoprotein (8). Sera of 55 per cent of a U.S. population formed precipitin lines with this serum (see below), suggesting that the reacting "antigen" might be inherited. Preliminary population and family studies were consistent with the hypothesis that individuals homozygous for a recessive autosomal allele, termed *Ag,* were non-reactors, of phenotype Ag(a−). Those homozygous or heterozygous for its dominant allele Ag^A, those of phenotype Ag(a +), did react (7, 9).

Additional studies are presented on the system determined by the C. de B. serum, the Ag(a) system, and further evidence for the genetic hypothesis is advanced. Other sera containing precipitins from transfused individuals have been found, and some of their properties will be described.

MATERIAL AND METHODS

Ouchterlony procedure. The precipitin reaction was carried out in agar gel as previously described (7). In testing the sera of transfused patients with a panel of normal sera, each transfused patient's serum was placed in the center well, and a different panel serum in each of the six peripheral wells.

Post-transfusion sera. The blood of 109 frequently transfused patients was obtained by venipuncture. Each serum was removed from the clotted blood on the day of venipuncture and stored at −20° C until tested. In some cases, plasma separated from acid citrate dextrose-treated blood was used. Specimens were collected at the following institutions: Clinical Center of the National Institutes of Health, Bethesda; New York Hospital—Cornell Medical Center, Frances Delafield Hospital, and Presbyterian Hospital, New York City; Children's Hospital Medical Center, Boston; and Mt. Alto Veterans' Administration Hospital, Washington, D. C. The diagnoses of these 109 patients are listed in Table I. All of these patients had received 5 or more transfusions, and 33 had received approximately 40 or more.

Panel sera. Each transfused serum was tested against a panel of 24 sera obtained from non-hospitalized, ap-

TABLE I
Disease categories of transfused patients

Disease category	No. of patients with 5 or more transfusions	No. of patients with about 40 or more transfusions	Patients with anti-serum protein antibody*
Cancer, all kinds except blood	50	1	1
Leukemia, acute childhood	6	4	0
Thalassemia major	18	17	3
Refractory and aplastic anemia	8	7	1
G.I. bleeding	7	1	0
Coagulation defects	2	2	0
Lupus erythematosus	2	1	0
Other	16	0	0
Total	109	33	5†

* Detectable by the Ouchterlony method.
† All with about 40 or more transfusions.

parently normal individuals, including males, females, children, and adults (Table II). The sera were from U. S. whites (7), Negroes, Indians (10), Eskimos (10), Vietnamese (11), Micronesians (12), and Polynesians (11). The samples for the population studies were selected casually from sera collected for population studies of other inherited traits (7, 10–12). The U. S. white sample was composed of presumably healthy employees of the National Institutes of Health and members of U. S. white families described elsewhere (7). The U. S. Negro sample was obtained during a population study in Evans and Bulloch counties, Georgia.[1] Methods of collection of the remaining sera are given in the references for the other populations; those of family and twin sera will be described in the next section.

RESULTS

Studies on sera of transfused patients. To help rule out the possibility that the antibody was naturally occurring and not related to multiple transfusion, the sera of 163 apparently normal individuals who had received no transfusions were tested for the presence of antibody in the same manner as the 109 sera of transfused patients. Five of the transfused patients, but none of the nontransfused had a precipitin (Table III, $p < .01$). All the antibodies were found in the sera of patients who had received approximately 40 or more transfusions (Table IV).

Some of the family members of three of the antibody-formers were studied. The parents of Ba., the parents and brother of P. C., and the daughter of C. de B. were tested and no antibody was found. The family of P. C. is particularly interesting. Both she and her brother have thalassemia, for which they have received multiple transfusions. P. C. was born November 2, 1942, and the diagnosis of Cooley's anemia made at age 4 years. She has required nearly monthly transfusions and has received about 250 transfusions. Her brother was born September 19, 1945, nearly 3 years after his sister, and the diagnosis of Cooley's anemia was made at 11 months. He also has been treated with monthly transfusions,

TABLE II
Distribution of Ag(a) phenotypes in various populations

Population	Total no.	Ag(a+) No.	Ag(a+) %	Ag(a−) No.	Ag(a−) %
Polynesian, Bora Bora	96	96	100	0	0
Micronesian, Rongelap Atoll	101	95	95	6	5
Eskimo, North Alaska	97	79	81	18	19
Vietnamese	99	70	71	29	29
Athabascan Indian, Alaska	102	69	68	33	32
U. S. Negro	167	92	55	75	45
U. S. white	188	103	54	85	46

TABLE III
Comparison of antibody frequency in transfused and nontransfused subjects

	No. of subjects with no transfusions	5 or more transfusions	Above 40 or more transfusions
Number studied	163	109	33
Number with antibody	0	5	5
Per cent with antibody	0	4.5 $p < .01$	15.1 $p < .001$

[1] In preparation, by A. J. Cooper.

TABLE IV

Transfused patients forming precipitins

Patient	Age	Sex	Diagnosis	No. of transfusions	Year of first transfusion	Date blood collected for antibody study
C.deB.	66	M	Refractory anemia	~50	1957	Aug., 1960
P.C.	18	F	Cooley's anemia	~250	1946	Dec., 1960
Br.	45	M	Zollinger-Ellison syndrome	~40*	1956	Jan., 1961
M.	6	F	Cooley's anemia	~40	1956	Dec., 1960
B.	14	M	Cooley's anemia	~250	1947	Dec., 1960

* Exact number of transfusions received is not known, but is probably 40 or more.

and at the time of testing had received approximately 200. Despite the similarity in disease pattern and the nearly similar number of transfusions given at the same frequency, only P. C. has developed a detectable antibody. It is possible that her brother will develop an antibody later in life.

Repeated transfusions in antibody-formers. With the exception of Br., who died presumably as a result of his primary disease, all the antibody-formers continued to receive frequent transfusions after the discovery of antibody. In no case did they develop sufficiently severe transfusion reactions to require interruption of transfusion therapy. C. de B., whose case has been previously described (7), was observed at the Clinical Center, National Institutes of Health, for the possibility of transfusion reactions. The patient regularly had a moderate rise in temperature every afternoon. After the administration of 2 units of Ag(a−) blood, there was a rise slightly above the usual; a similar rise was seen when 2 units of Ag(a+) blood was administered. Comparable results were obtained with 6 other units of blood. Figure 2 shows a typical study.

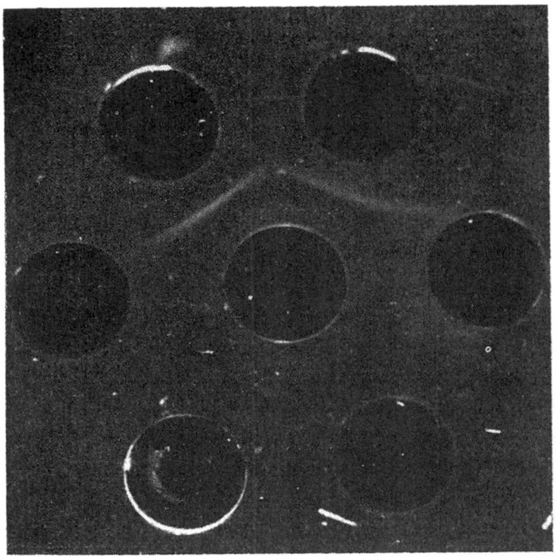

FIG. 1. PRECIPITIN LINE IN OUCHTERLONY PLATE FORMED BETWEEN CENTER WELL, CONTAINING SERUM OF C. DE B., AND 2 OF 6 OUTER WELLS, CONTAINING SERA FROM NORMAL SUBJECTS. Note concavity of line toward the peripheral wells, suggesting an antigen of high molecular weight.

Studies on the Ag(a) system. The first and most clearly reacting antibody was found in the serum of patient C. de B. His serum defined the Ag(a) beta-lipoprotein system (7-9). Further studies on the Ag(a) phenotypes are presented here. With nearly all sera, the precipitin line was concave towards the antigen well and coalesced with the precipitin line of adjoining sera. Occasionally, other kinds of reaction were seen. These included instances where two separate lines were

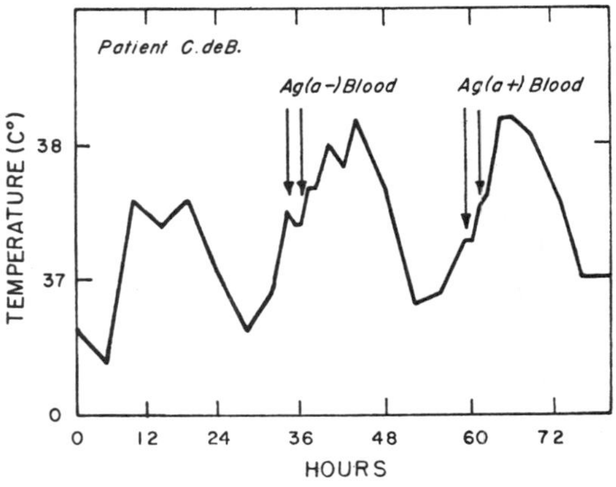

FIG. 2. TEMPERATURE CHART OF C. DE B. IN RELATION TO TRANSFUSIONS OF AG(A+) AND AG(A−) BLOOD.

TABLE V

Frequency of Ag(a+) phenotype by sex in U. S. Negro and white populations

Population	Males			Females			Males and females		
	Total no.	Ag(a+) No.	Ag(a+) %	Total no.	Ag(a+) No.	Ag(a+) %	Total no.	Ag(a+) No.	Ag(a+) %
U. S. Negro	69	38	55	98	54	55	167	92	55
U. S. white	85	41	48	103	62	61	188	103	54
Total U. S.	154	79	51	201	116	58	355	195	55

seen and also where the single precipitin line was qualitatively different from the usual line.

Sex and age. There was no significant sex difference in the frequency of positive reactors, those of Ag(a +) phenotype, in U. S. white and Negro populations (Table V). The distribution of positive reactors by age in 356 Americans (168 U. S. Negroes, 188 U. S. white) is shown in Table VI). Although there appears to be little age nonhomogeneity, the sample is not large enough to rule out small differences. Testing a larger number of people older than 60 might show whether the observed slight drop in the frequency of Ag(a+) in this group is significant.

Twin studies. The twin sera were presented to us by Dr. J. McDonough of the Public Health Service Heart Disease Control Program, and were collected in Evans and Bulloch counties in southern Georgia (13). A census of the area was taken and the total number of twin pairs ascertained. Whenever both twins were still resident locally, they were asked to participate in the study. All those volunteering were examined and blood samples obtained; 56 pairs of twins were located, 23 monozygotic, 27 dizygotic, and 6 undetermined. Blood specimens from a total of 18 identical and 27 nonidentical twin pairs were available for the present study. The diagnosis of zygosity was made with the following criteria: sex; gross physical appearance; A–B–O, Rh_0, rh′, and rh″ red blood cell groups; and haptoglobin serum types. For 4 identical and 4 nonidentical pairs, the following blood group data were also obtained: M–N–S, Fy, Jk, C^W, P, Le, Lu, Diego, and Js.

TABLE VI

Frequency of Ag(a+) phenotype in combined U. S. white and Negro population by age group

Age group	Total	No. Ag(a+) Expected *	No. Ag(a+) Observed
years			
0 to 9	28	15.4	16
10 to 19	79	43.5	40
20 to 29	52	28.6	25
30 to 39	56	30.8	36
40 to 49	87	47.9	50
50 to 59	31	17.1	18
60+	23	12.7	10
Total	356		195
	$0.8 > p > 0.9$		

* Calculated by multiplying total number of group by 0.55, (the fraction of the entire population that is Ag(a+).

Of the 18 supposed identical twin pairs, 17 were concordant and one discordant for the Ag(a) phenotype (Table VII). Of the 27 nonidentical twin pairs, 17 were concordant and 10 discordant. The difference in concordance between the identical and nonidentical twin pairs is highly significant.

Both members of the discordant, presumably identical twin pair, were white males of blood group O Rh_0Rh^1 and haptoglobin type 2–2. Smith and Penrose (14) have provided a method for calculating the probability of zygosity from blood groups and other traits. From the blood-group frequencies in their tables and a frequency of 0.387 for the haptoglobin type one (Hp^1) gene, it was calculated that the chances of dizygosity in this pair are approximately 3 in 10. It is difficult to explain the existence of a discordant pair among the presumably identical twins, but there are several possible explanations: the twins may not have been identical; the phenotype may have been misclassified; or there may be a nongenetic factor affecting the phenotype.

Family studies. Preliminary family studies of the Ag(a) phenotype in 29 U. S. and 8 Micronesian families have been presented previously (7). The results supported the hypothesis that individuals homozygous for the recessive gene Ag did not react with the C. de B. antibody. Those homozygous or heterozygous for its dominant al-

TABLE VII

Twin studies of Ag(a+) phenotype

Twins	Concordant pairs			Discordant pairs	Total
	Ag(a+)	Ag(a−)	Total		
Identical	11	6	17	1	18
Nonidentical	8	9	17	10	27
				Total	45
					.01 <p <.02

lele Ag^A did react. According to this hypothesis, the offspring of Ag(a+) and Ag(a+) matings or of Ag(a +) and Ag(a−) matings could be either Ag(a+) or Ag(a−). The offspring of Ag(a−) and Ag(a−) matings must all be Ag(a−).

Results in an additional 24 U. S. Negro families are presented in this report. These were casually selected from families whose sera were collected in Evans and Bulloch counties, Georgia. Studies on the following red blood cell groups were available on nearly all the family members: A, A_1, B, O, C(Rh′), D(Rh$_0$), E(Rh″), c, d, r, M, N, S, s, Fy^a, Le^a, Le^b, Kp^a, Kp^b, K, C^w, Lu^a, Wr^a, Vw, M^g, and Js. With these, some offspring who were not the children of the supposed parents could be detected, and they were excluded from the computations.

The frequency of the Ag(a+) phenotype in the population studied was 55 per cent. On the assumption that mating is random, that marked selection does not occur, and that the genetic hypothesis is correct, the expected number of each kind of mating and offspring can be calculated. The expected and observed results are shown in Table VIII.

A method summarized by Smith (15) permits more complete use of the data: four types of comparison are made in each of which a correction is included for family size (Table IX). For classes 3 and 4, an estimate of 0.33 was used for the Ag^A gene frequency. The observed results are close to those expected and agree with the previous family studies.

Population studies. These are shown in Table II. The red blood cell groups (listed previously under Family studies) and Ag(a) phenotypes of 96 Georgia Negroes and 79 white inhabitants of Tangier Island, Maryland, were determined. There was no apparent association between Ag(a) and red blood cell phenotypes.

Mother-child studies. Information on the time of life when gene expression is detectable can be obtained by the study of mothers and their newborn children. Eighty-one pairs of mother and newborn sera were obtained from the U. S. Naval Medical Center, Bethesda, and the Johns Hopkins Hospital, Baltimore. The blood was drawn from the mother at term. With the exception of 16 cases where the blood was removed from the infant's heel between the second and fourth days of life, the newborns' blood was obtained from the umbilical cord at the time of delivery. Sera were also obtained from 9 infants approximately 7 weeks old, and from two of these blood was also withdrawn at age 3 months. Sera withdrawn at age 2 years were available from 12 children.

Table X shows the results of the mother-new-

TABLE VIII

Family studies of Ag(a) phenotype in U. S. Negroes

Mating type	No. of families		Ag(a+) children		Ag(a−) children	
	Obs.*	Exp.	Obs.	Exp.	Obs.	Exp.
Ag(a+) × Ag(a+)	9	8.3	29	29.4	6	5.6
Ag(a+) × Ag(a−)	8	11.8	20	19.2	12	12.8
Ag(a−) × Ag(a−)	7	4.9	0	0	18	18
Total	24	24.0	49	48.6	36	36.4

* Obs. = observed, and Exp. = expected.

TABLE IX

Observed and expected number of Ag(a−) children from various matings in U. S. Negro population, by method of C. A. B. Smith (15)

Test comparison	Mating	χ^2	d.f.
Observed no. of Ag(a−) children, given no. of families with Ag(a−) children	Ag(a+) × Ag(a−)	.174	1
Observed no. of Ag (a−) children, given no. of families with Ag(a−) children	Ag(a+) × Ag(a+)	.870	1
Observed no. of families with at least one Ag(a−) child, given total no. of families	Ag(a+) × Ag(a−)	.006	1
Observed no. of families with at least one Ag(a−) child, given total no. of families	Ag(a+) × Ag(a+)	.099	1
		χ^2 = 1.149	d.f. = 4
			0.9>p>0.8

born child studies arranged according to the possible combinations of mother and child phenotypes. Two major hypotheses may be considered to account for the observed results: *1*) The trait is inherited according to the genetic hypothesis stated above, and the genes are expressed at birth; and *2*) the trait is inherited according to the genetic hypothesis, and the genes are expressed not at birth but at varying times thereafter. Two further possibilities follow from the second hypothesis: *a*) The Ag(a) material is not transmitted placentally, and *b*) the Ag(a) material is transmitted placentally in some or all cases.

The data can help in deciding among these possibilities. The frequency of Ag(a+) in the mother was 51 per cent, which was not significantly different from the frequency in the general U. S. population. On the assumptions that the genetic hypothesis is correct and that mating is random for Ag(a) phenotype, the combinations expected according to the first hypothesis can be calculated, and are shown in the fourth column of Table X. There is a large deficiency of the Ag(a−) mother,–Ag(a+) infant combination.

If hypothesis 2a is correct, then the presence of the Ag(a+) material in some of the newborn children could reflect the difference in the times that the gene is expressed. The children of Ag(a+) mothers, however, are much more frequently positive than those of Ag(a−) mothers. In the children of the Ag(a+) mothers, 24 of an expected 29.1 infants are Ag(a+) at birth; in the children of the Ag(a−) mothers, 12 Ag(a+) children would be expected, whereas only one questionable case was seen. Hence, for hypothesis 2a it would be necessary to assume that the children of Ag(a+) mothers express the gene earlier than the children of Ag(a−) mothers.

If hypothesis 2b is correct, then none of the Ag(a −) mothers should have Ag(a +) children. Mothers who are Ag(a+) might or might not have children with reactive serum. In one

TABLE X

Comparison of Ag(a) phenotyes in mother-child pairs

Mother*	Child	Observed no. of pairs	Expected no. of pairs: Hypothesis 1	Expected no. of pairs: Hypothesis 2
Ag(a−)	Ag(a−)	39	28	40
Ag(a−)	Ag(a+)	1	12	0
Ag(a+)	Ag(a−)	17	11.9	41
Ag(a+)	Ag(a+)	24	29.1	41
	Total	81	81	81

* Ag(a+) frequency in mothers = 51 per cent.

TABLE XI
Serial studies of Ag(a) phenotypes in mother-child pairs

No.	Mother	Children			
	Term	Birth	40 to 60 Days	3 Months	2 Years
1	Ag(a+)	Ag(a+)	Ag(a+)		
2	Ag(a−)	Ag(a−)	Ag(a−)		
3	Ag(a+)	Ag(a+)	Ag(a+)		
4	Ag(a−)	Ag(a−)	Ag(a+)		
5	Ag(a+)	Ag(a+)	Ag(a+)	Ag(a+)	
6	Ag(a+)	Ag(a+)	Ag(a+)		
7	Ag(a+)	Ag(a+)	Ag(a+)		
8	Ag(a−)	Ag(a−)	Ag(a−)	Ag(a−)	
9	Ag(a−)	Ag(a−)	Ag(a−)		
10	Ag(a+)	Ag(a−)			Ag(a−)
11	Ag(a+)	Ag(a−)			Ag(a+)
12	Ag(a−)	Ag(a−)			Ag(a+)
13	Ag(a−)	Ag(a−)			Ag(a+)
14	Ag(a−)	Ag(a−)			Ag(a+)
15	Ag(a−)	Ag(a−)			Ag(a−)
16	Ag(a−)	Ag(a−)			Ag(a+)*
17	Ag(a+)	Ag(a+)			Ag(a+)
18	Ag(a−)	Ag(a−)			Ag(a−)
19	Ag(a−)	Ag(a−)			Ag(a+)*

* Double band on Ouchterlony plate.

case, the serum of a child of an Ag(a−) mother gave a broad, diffuse band rather than the sharp, narrow band characteristic of the usual precipitin. The presence of the Ag(a+) material in the children of the Ag(a+) mothers could be explained by placental transmission. This, of course, would not occur in the children of the Ag(a−) mothers.

In order to determine when after birth the Ag^A gene is expressed, blood collected at 40 to 60 days, 3 months, and 2 years after delivery was studied (Table XI). One infant's serum, which had been negative at birth, was positive at 57 days (Table XI, no. 4). Six others negative at birth were positive at 2 years (Table XI, nos. 11–14, 16, 19). If the positive material present in the cord blood was placentally transmitted, then we could expect to see some examples of the conversion of an Ag(a+) to an Ag(a−) phenotype in the offspring of an Ag(a+) mother. This was not observed. There were, however, only 6 cases in which the necessary observations have been made, and in 4 the observation period was less than 60 days. With the gene frequency of Ag^A as 0.33 and with random mating assumed, it can be shown that approximately 30 per cent of Ag(a+) mothers would be expected to have *AgAg* offspring. It is not surprising, therefore, that none was seen. Further follow-up studies are needed to see if such a conversion occurs.

Fetal Blood. Sera from five fetuses whose ages were approximately 72, 73, 84, 109, and 143 days were made available to us through the courtesy of Dr. J. Robbins. One of these gave a positive precipitin reaction with the serum of C. de B. This fetus was estimated to be 72 days old, but was one of the largest (13.0 cm) of all those examined.

Transudates. A. Synovial fluid. Sera and synovial fluid from the knees of four patients with gout was was made available to us through the courtesy of Dr. J. E. Seegmiller. Patient 1 had gout affecting primarily his metatarsal joints. He developed transient inflammation in the left but not the right knee; serum and synovial fluid from both knee joints were Ag(a+), and there was no evidence of gout in either knee. Patients 2, 3, and 4 had gout, but with only moderate involvement of their knees. The sera and synovial fluids of Patients 2 and 3 were Ag(a−); those of Patient 4 were Ag(a−).

B. Cerebrospinal fluid. Both serum and cerebrospinal fluid were available, through the courtesy of Dr. P. Carbone, from ten patients with neoplastic diseases. In nearly all cases, there was no apparent involvement of the meninges. Six of the sera were classified as Ag(a+) and 4 as Ag(a−); all the cerebrospinal fluids were classified at Ag(a−). The cerebrospinal fluids of the six Ag(a+) individuals were pooled and concentrated by ultrafiltration to 25 and 100 times normal. Each concentrated preparation was non-reactive.

DISCUSSION

It is reasonable to conclude tentatively that these antibodies developed in response to multiple transfusions and are comparable to those developed from repeated injections of foreign proteins into experimental animals. In this case, the foreign protein is a normal human serum component, the low-density beta-lipoprotein, of which at least two antigenically different, genetically determined forms are presumed to exist in most populations. The antigenic differences probably derive from small physico-chemical variations of the molecule. Individuals homozygous or heterozygous for the Ag^A gene have a low-density beta-lipoprotein that forms a precipitin with the antibody in the serum of C. de B. *Ag* homozygotes presumably have an antigenic specificity, but to date no antibody has been found that would detect the presence of this hypothesized gene product. C. de B. was presumably of type Ag(a−) and developed an antibody as a result of multiple transfusions with Ag(a+) lipoprotein.

The reaction between a patient's antibody and the antigen in the transfused blood might be the cause of some of the mild transfusion reactions seen in frequently transfused patients. Four of the five patients in whom antibody was detected (see Table IV) continued to receive transfusions. One of the patients, C. de B., had mild subjective symptoms and a moderate rise in temperature after most transfusions, but it appeared to be independent of the Ag(a) phenotype of the blood he received (see Figure 2). There is no information available on the long-term effect of repeated transfusions in patients who are antibody-formers. The reaction between the Ag(a+) protein and C. de B. antibody does not fix complement (7), nor does it give a positive passive cutaneous anaphylaxis test. It is posible, however, that recurrent precipitin reactions may occur *in vivo* and have long-term, chronic pathological consequences, particularly in small blood vessels.

In three of the five antibody-formers, C. de B., P. C., and Br., the antibodies appeared to be qualitatively similar. P. C. and Br. were both Ag(a−), and none of these three sera formed precipitin lines with each other. M. and Ba. were both Ag(a+) and formed precipitin lines with the antigen present in C. de B.'s serum. They appear to determine quite different precipitin systems, and will be described in detail elsewhere.

It is not known why some of the transfused patients developed an antibody and others did not. Those who did received a minimum of approximately 40 transfusions, but a large number of transfusions is not the only requirement for antibody formation. Fifteen of the frequently transfused patients had received more than 100 transfusions, but antibody was detected in only two of these. Of the 15, three had received 300 or more transfusions, but no precipitin was found in any of these. Three of the antibody-formers had thalassemia. Here the antibody production may result from the large number of transfusions these patients receive, or it may reflect some inherent characteristic of the disease.

The Ouchterlony method is a relatively crude means for detecting antibody; it is likely that more sensitive techniques would reveal a larger number of antibody-formers, and studies using such methods are in progress.

The presence of a precipitin band with a different appearance from the usual concave one, and the occasional finding of two separate bands, suggest that the serum of C. de B. can react with more than one antigenic site.

These studies are in general consistent with the original genetic hypothesis. We conclude that the genetic mechanism proposed explains the major variations of this trait, although other genetic and environmental factors may have a bearing on the formation of the specific lipoprotein. This trait is polymorphic in most of the populations studied, and the gene frequency varies widely.

The mother-child studies are consistent with the explanation that the antigen is not generally produced at birth but that synthesis begins at a variable time thereafter. This effect is analogous to that with the serum haptoglobins, where a large percentage of children have no haptoglobins until after the third month of life (16). It is possible that the Ag(a+) material crosses the placenta in sufficient quantity to be detected in some offspring of Ag(a+) mothers.

The Ag(a+) material is present in the synovial fluid of Ag(a+) individuals, implying that the specific lipoprotein either is transmitted across the synovial membrane or is produced in the joint. The Ag(a+) material was not detected in the spinal fluid.

SUMMARY

1. Further studies on the Ag(a) human serum polymorphism, which involves low-density beta-lipoprotein components, are presented. Population, family, and twin studies agree with the hypothesis that the presence of this component, which is identified by a precipitation reaction with the serum of a patient who had received numerous transfusions, is determined by an autosomal, dominantly-transmitted gene designated Ag^A.

2. Examination of the paired sera of mothers and newborn children leads to the conclusion that the Ag(a) phenotype may not be expressed at birth. Mothers may transmit the lipoprotein under discussion across the placenta in some cases, rendering the newborn a phenocopy. From the evidence at hand, the Ag(a+) phenotype can be expressed as early as 57 days after birth.

3. In one of the antibody-forming patients, C. de B., the slight temperature rise in response to transfusion was independent of the Ag(a) phenotype of the transfused blood. After the discovery of antibody, four of the five antibody-formers continued to receive transfusions with no apparent severe reaction.

4. Study of other transfused individuals has detected four subjects who form precipitins with normal sera. All of these patients received approximately 40 or more transfusions.

ACKNOWLEDGMENTS

We are indebted to the following for their kind assistance in collecting blood from transfused patients: Drs. Marion Erlandson, A. J. Cooper, H. C. Goodman, R. H. Aster, D. E. Gentile, R. Osborne, P. Carbone, J. E. Ultman, P. Calabresi, and S. V. Kevy. We are particularly grateful to Drs. C. H. Smith and Marion Erlandson of The New York Hospital for permitting us to obtain several specimens from their patients and providing clinical information on some of these. We appreciate the kind assistance of Dr. H. Pierson of the U. S. Naval Hospital, Bethesda, and of Drs. J. L. Sever and V. E. Anderson of the Perinatal Research Branch, National Institute of Neurological Diseases and Blindness, for providing the mother-child sera. Dr. J. R. McDonough was instrumental in obtaining the sera from Claxton and Bulloch counties, Ga. Sera from the N.I.H. employees were obtained through the courtesy of Dr. J. M. Lynch of the Employees' Health Service.

REFERENCES

1. Blumberg, B. S. Inherited susceptibility to disease and its relation to environment. Arch. environm. Hlth. 1961, **3**, 612.
2. Clausen, J., and Heremans, J. An immunologic and chemical study of the similarities between mouse and human serum proteins. J. Immunol. 1960, **84**, 128.
3. Hirschfeld, J. Shape of the precipitate in immunoelectrophoresis. Nature 1960, **185**, 164.
4. Oudin, J. L'allotypie de certains antigènes protéidiques du serum. C. R. Acad. Sci. (Paris) 1956, **242**, 2606.
5. Dray, S., and Young, G. O. Two antigenically different γ-globulins in domestic rabbits revealed by isoprecipitins. Science 1959, **129**, 1023.
6. Dubiski, S., Dudziak, Z., Skaiba, D. Serum groups in rabbits. Immunology 1959, **2**, 85.
7. Allison, A. C., and Blumberg, B. S. An isoprecipitation reaction distinguishing human-serum protein types. Lancet 1961, **1**, 634.
8. Blumberg, B. S., Dray, S., and Robinson, J. C. Antigen polymorphism of a low-density beta-lipoprotein. An allotypy in human serum. Nature 1962, **194**, 656.
9. Blumberg, B. S., and Allison, A. C. Studies on the isoprecipitin determined human serum polymorphins to "new" phenotypes. Proceedings of the Second International Conference of Human Genetics, Rome, September, 1961.
10. Blumberg, B. S., Allison, A. C., and Garry, Barbara. The haptoglobins and haemoglobins of Alaskan Eskimos and Indians. Ann. hum. Genet. 1959, **23**, 349.
11. Blumberg, B. S., and Gartler, S. M. The urinary excretion of β-aminoisobtuyric acid in Pacific populations. Hum. Biol. 1961, **33**, 355.
12. Blumberg, B. S., and Gentile, Z. Haptoglobins and transferrins of two tropical populations. Nature 1961, **189**, 897.
13. McDonough, J. R., Hames, C. G., Greenberg, B. G., Griffin, L. H., Jr., and Edwards, A. J., Jr. Observations on serum cholesterol levels in the twin population of Evans County, Georgia. Circulation 1962. In press.
14. Smith, M., and Penrose, L. S. Monozygotic and dizygotic twin diagnoses. Ann. hum. Genet. 1955, **19**, 273.
15. Smith, C. A. B. A test for segragation ratios in family data. Ann. hum. Genet. 1956, **20**, 257.
16. Galatius-Jensen, F. On the genetics of the haptoglobins. Acta genet. (Basel) 1958, **8**, 232.

Reprinted from

BULLETIN OF THE NEW YORK ACADEMY OF MEDICINE

Second Series, vol. 40, no. 5, pp. 377-386, May 1964

POLYMORPHISMS OF THE SERUM PROTEINS AND THE DEVELOPMENT OF ISO-PRECIPITINS IN TRANSFUSED PATIENTS*

Baruch S. Blumberg

Chief, Geographic Medicine and Genetics Section
National Institutes of Health, Bethesda, Maryland

Polymorphism, according to Ford,[1] is the occurrence in the same habitat of two or more inherited forms of a species in such proportions that the rarest of them cannot be maintained merely by recurrent mutation. This definition excludes the rare, genetically disadvantageous inherited diseases which are maintained by mutation and selected against by death or infertility. In polymorphic traits two or more of the genotypes determining variation of the trait are common in the population. Although much of the early work on polymorphism was performed on lower animals, it is now clear that humans constitute one of the most favorable species in which to study these systems. In addition, the polymorphisms provide convenient systems for the study of inherited discontinuous biochemical variation in man.

Polymorphisms are thought to arise as a result of selective differences between genotypes. For example, the sickle cell homozygote develops a severe hemolytic anemia which, under natural conditions, is usually fatal. The heterozygote, however, is at a selective advantage compared to the normal homozygote, apparently because of greater resistance to falciparum malaria, and increased fertility. As a consequence of the selection in favor of the heterozygote, the sickle gene is maintained in the population at high levels despite the elimination of genes due to the death of the homozygotes.[2] Other examples of associations of polymorphisms with exogenous disease-producing agents are known. Individuals bearing the sex-linked glucose-6-phosphate dehydrogenase deficiency (G6PD) gene (and probably at least one other trait as well)

* The *Fifth Bernadine Becker Memorial Lecture* of the New York Rheumatism Association, Inc., presented at a combined meeting of the New York Rheumatism Association and the Section on Medicine, held at The New York Academy of Medicine, November 18, 1963.

TABLE I.—SOME POLYMORPHIC TRAITS IN HUMANS

Red Blood Cell Types	*Serum Proteins*
ABO	Haptoglobin
MNS	Transferrin
P	Gamma globulin
Rhesus	Beta-Lipoprotein
Lutheran	Group-specific substance
Kell	Serum cholinesterase
Lewis	Serum esterase
Duffy	
Kidd	
Diego	
Sutter	
X-linked	

Other Cell Types	*Miscellaneous*
Hemoglobins	BAIB urinary excretion
G6PD	
Red cell phosphatase	Secretor of ABH blood group substance
Red cell phosphogluconate dehydrogenase	Taste of PTC
White blood cell antigens	Ear cerumen character
Platelet antigens	

can develop a hemolytic anemia when they eat, or are otherwise exposed to the fava bean. A similar anemia may develop if they ingest any of a large number of drugs. Individuals without this trait will not develop the anemia given the same exposure. Similarly, individuals with the inherited atypical serum cholinesterase trait can develop prolonged apnea after the administration of Suxamethonium or other related drugs (see below). It is probable that disease susceptibility related to a single inherited trait is less common than susceptibility dependent on several inherited traits. The identification of such traits may serve a useful function in identifying individuals who require special protection against specific exogenous agents.

As a consequence of disease and environmental forces, as well as other factors, a large number of polymorphisms may exist in a population. Some may be related to present selective forces, and others to forces which operated in the past, but which are no longer significant. Present gene frequencies may also result from gene mixture between populations.[3]

Table I is a list of some of the biochemical polymorphisms known in man. In the present lecture I would like to discuss the polymor-

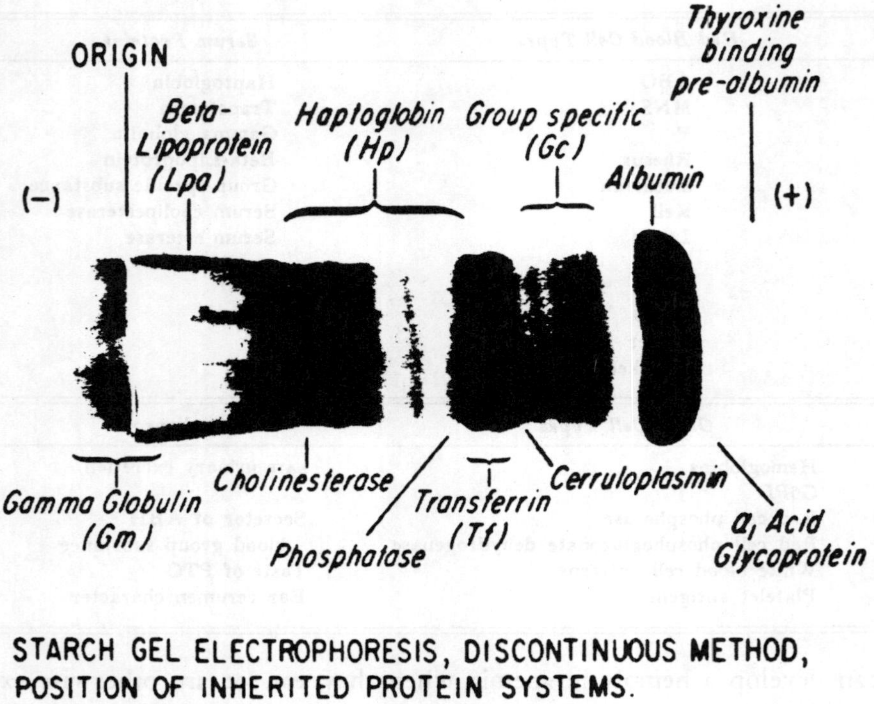

Figure 1. Photograph of starch gel electrophoresis of normal serum to illustrate the position of some of the inherited serum protein systems discussed in the text. (Adapted from reference 19.)

phisms of the serum proteins and how they relate to the problem of blood transfusion.

Serum Protein Polymorphisms (Figure 1)

The haptoglobins. The haptoglobins are a family of serum proteins which bind hemoglobin. The early studies of Jayle and his co-workers demonstrated that the haptoglobins are elevated in a wide variety of inflammatory diseases. In some illnesses associated with hemolysis, the hemoglobin released from the breakdown of the red blood cells combines with the haptoglobin to form a strong complex which appears to be removed mainly by the reticuloendothelial system. If the rate of breakdown of red blood cells is greater than the production of haptoglobin, the serum haptoglobin level will decrease, in some cases, to undetectable levels. If the hemolytic process is arrested, the haptoglobin will be regenerated, presumably in the liver, and the serum levels may return to normal.

When starch gel is used as a supporting medium for electrophoresis, human sera can be separated according to haptoglobin type. Three major types, 1-1, 2-2, and 2-1, can be distinguished. It has been shown that in most populations these are controlled by a pair of allelic autosomal genes designated Hp^1 and Hp^2. Hp^1 homozygotes are type 1-1, Hp^2 homozygotes are type 2-2, and heterozygotes are type 2-1. Recently, Nance and Smithies[4] have shown that when purified serum haptoglobin is subjected to reductive cleavage by mercaptoethanol in the presence of 8 M urea, fast-moving and slow-moving haptoglobin variants can be distinguished. By chemical and genetic studies they demonstrated that there are three genetically determined alpha polypeptide chain variants controlled by five allelic genes Hp^{1F}, Hp^{1S}, Hp^{2FF}, Hp^{2FS}, and Hp^{2SS}. There are other relatively rare genes (i.e., Hp^{2M}) which control uncommon haptoglobin phenotypes.

Allison, Blumberg, and apRees[5] reported that approximately 30 per cent of a West African population had insufficient haptoglobin to permit typing by the starch gel method. Since this original description ahaptoglobulinemia and hypohaptoglobulinemia have been detected in other Africans, in U. S. Negroes, and in some European and Asiatic populations. There is good evidence that some forms of ahaptoglobinemia or hypohaptoglobinemia are inherited,[6] but some cases are due to the hemolysis of red blood cells and subsequent loss of haptoglobin in a variety of hemolytic diseases.

The transferrins. The transferrins (or siderophilins) are a group of serum proteins which bind iron. They can be identified by their mobility on starch gel electrophoresis and by the use of radioactive iron and autoradiography. The phenotypes are controlled by a series of apparently allelic genes, Tf^C, Tf^{BO}, Tf^{B1}, etc. The most common phenotype in U. S. and European population is type C, which corresponds to the genotype Tf^C/Tf^C. The slow-moving D variants have been found in African, American Negro, Australian aboriginal, Chinese, and other populations. Fast-moving types are occasionally found in white, Asiatic, and other populations. In recent years a large number of variants with only minute differences in electrophoretic mobility have been detected. To date no associations with particular illnesses have been found, and there does not appear to be a significant difference in the amount of iron bound by each of the phenotypes. It has been stated that the transferrins have an inhibitory effect on virus multiplication, but it is not clear if this effect has any biological importance.

Gamma globulin groups.[7] Inherited differences in the gamma globulins were detected as a result of a chance observation made while using an immunological test for rheumatoid arthritis. The serum of many patients with rheumatoid arthritis has a gamma globulin of high molecular weight called "rheumatoid factor," which agglutinates blood cells or other particles coated with specially prepared serum gamma globulin. Grubb[8] used a system in which Rh+ human red blood cells were coated with an incomplete anti-Rh antibody. Some sera of patients with rheumatoid arthritis would agglutinate these particles. Grubb found that a gamma globulin present in the serum of some, but not all, normal individuals could inhibit the agglutination reaction. The inhibiting material was inherited in simple Mendelian fashion. Individuals homozygous or heterozygous for a dominant gene Gm^a were inhibitors. Those homozygous for all alternate allele *Gm* did not have inhibiting material in their serum. Using different combinations of coated red blood cells and sera of rheumatoid arthritis patients, different genetically controlled gamma globulins have been detected. Several of them are controlled by a series of allelic genes at the same locus (Gm^a, Gm^b, Gm^x, etc.), and others by genes at a second locus (Inv^a, Inv^b). There is as yet no information on the nature of the physiological or pathological conditions associated with this trait, although it may be connected in some way with the immune process.

Genetic variations of human serum phosphatase. Using starch gel electrophoresis, Arfors *et al.*[9] have shown that there are at least two types of human serum phosphatases. All human sera have one phosphatase band which migrates with a mobility slightly slower than serum transferrin. Some individuals, in addition, have a slower moving phosphatase band. On this basis, human sera were classified in two groups: Group I with a single phosphatase band and group II with two bands. Twin studies were highly consistent with the hypothesis that these types are inherited, but the mode of inheritance is not known. The authors also found a strong association with the Lewis blood group system, and they suggested that one serum phosphatase zone may represent the Le^b substance or a complex with the Le^b substance.

Polymorphism of α_1-acid glycoprotein. Alpha$_1$-acid glycoprotein is a serum protein with a high (40%) carbohydrate content. It corresponds with the slow pre-albumin band on starch gel electrophoresis (Fig. 1). Schmidt and his associates[10] developed a technique for isolating large quantities of this protein from human serum or plasma. The isolated material was concentrated and subjected to starch gel electrophoresis. In comparing material isolated from different individuals, they found different patterns containing varying numbers of bands. Preliminary twin studies indicate that these patterns are inherited.

Group-specific components. Hirschfeld[11] has described a variation in the serum α-globulin which is detected by immunoelectrophoresis and is inherited in simple Mendelian fashion. The α_2-globulin involved is in the postalbumin region on starch gel electrophoresis.[12] Differences in mobility of these bands were noted by Smithies in his early studies using vertical starch gel electrophoresis. No associations with disease have been detected using this system.

Serum cholinesterase.[13] There is a gene (whose frequency in the populations tested is approximately 0.02) which determines the presence of an atypical serum cholinesterase. Individuals who are homozygous for this gene have only the atypical enzyme and are unable to inactivate the muscle relaxant succinylcholine (Suxamethonium) and related drugs with the same rapidity as individuals who are homozygous for the normal gene. When succinylcholine was administred to individuals homozygous for the atypical gene, they developed a prolonged apnea. Heterozygotes have both typical and atypical enzymes. Since the homozygotes are easily identifiable by means of a simple spectrophotometric test requiring less than 0.1 ml. of serum, it is quite practical to pre-test individuals receiving muscle relaxants. This is particularly important in shock therapy where patients often receive a prolonged course of muscle relaxants during their treatment. Other drugs which probably are also differentially metabolized by the typical and atypical enzyme include procaine hydrochloride (Novocain), physostigmine (Eserine), and chlorpromazine. It is, of course, unlikely that the differences in drug susceptibility are responsibile for the maintenance of the polymorphism. However, a number of solanaceous plants contain an inhibitor of cholinesterase. Potato peels, for instance, contain an inhibitor which can distinguish between the three cholinesterase phenotypes. It should be emphasized that under normal conditions the atypical enzyme homozygotes are quite normal; they can only be distinguished from the other

TABLE II.—REACTORS WITH ANTISERA IN DIFFERENT POPULATIONS

	C. de B.		*New York*		*J. B.*		*A. Di B.*		*S. L.*	
	No.	*% Re-actor*	*No.*	*% Re-actor*	*No.*	*% Re-actor*	*No.*	*% Re-actor*	*No.*	*% Re-actor*
Whites (U.S.A.)	120	59	164	100	168	98	144	77	157	97
Micronesians (Rongelap)	194	98	185	45	54	98	45	24	50	24
Naskapi-Montagnais Indians (Labrador)	234	97	103	85	234	93	91	73	93	95

phenotypes when the appropriate drugs are administered or if the cholinesterase types are determined biochemically.

Isoantibodies in transfused patients. The existence of many polymorphisms of serum proteins makes it highly likely that a transfused patient would receive serum proteins slightly different from his own. For example, an individual of haptoglobin type 1-1, which occurs in about 15 per cent of the normal U. S. white population, has 85 of 100 chances of receiving plasma containing haptoglobin -2, if the donor is also U. S. white. Similar possibilities of receiving a different protein occur with many of the other systems. If this is so, then multiply transfused patients would often receive plasma containing "foreign" protein, albeit from the same species. In a study of transfused sera using the relatively insensitive technique of double diffusion in agar gel, a serum from a patient (C. de B.) was found to contain a precipitin which reacted with a protein present in the serum of some but not all normal individuals. The isoprecipitin did not react with any of the known polymorphic proteins, but was directed against a low density beta-lipoprotein. This technique revealed a "new" polymorphic system.[14] The ability of a serum to show a precipitin reaction with the human anti-human antiserum was found to be inherited. Individuals homozygous or heterozygous for a gene Ag^A reacted with antiserum C. de B. and individuals homozygous for its alternate recessive allele did not react. Further studies were initiated using the sera of other transfused patients and more than 20 sera containing isoprecipitins were found. In a study of 11 of these, at least 5 and probably 7 different specificities were identified.[15] Preliminary family studies with two of

TABLE III.

	C. de B.			New York			J. B.		
	Total	% Reactors	?	Total	% Reactors	?	Total	% Reactors	?
Coronary Artery Disease	68	60.3	1	67	88.1	3	57	75.4	13
Control	66	59.1	1	66	77.3	2	56	76.8	11
Diabetes	208	71.6	0	205	96.6	2	205	97.1	2
Control	166	55.4	1	159	98.1	2	154	94.8	5
Rheumatic Fever	165	83.6	0	37	100	0	35	100	0
Control	152	82.2	0	34	100	0	34	100	0

Reactions of three anti-lipoprotein antisera with sera from diseased patients and controls. The coronary artery disease and rheumatic fever patients and their controls were U. S. Whites and the diabetes patients and their controls were U. S. Negroes. The number of sera which could not be typed are shown in the column headed by a question mark. These studies are described in greater detail elsewhere.[17]

these suggest that they are inherited, and it is likely that the other specificities are inherited as well. Most of the antigenic specificities so far studied are common in most populations (see Table II). It appears that most individuals have several inherited antigenic specificities on these lipoproteins.

The isoprecipitins against serum lipoproteins are quite common in multiply transfused individuals. Approximately one third of thalassemia patients and 10 per cent of other patients who receive 35 or more transfusions will develop detectable isoprecipitins. The reasons for the higher frequency in thalassemia patients are not clear.

Patients who develop an isoantibody usually require additional transfusions, and this could presumably lead to antigen-antibody interactions *in vivo* similar to those studied in animals.[16] Approximately 15 patients have been observed for varying lengths of time after the isoprecipitin was discovered. In all cases, the patients have needed frequent transfusions, very often with blood known to contain the proteins which *in vitro* form precipitin bands with the patient's isoprecipitin. In none of these cases has it been possible to associate with the isoprecipitin an acute, severe transfusion reaction. It has not been possible as yet to evaluate the long-term effect of continued transfusions.

The selective or disease factors, if any, associated with this polymorphism are unknown. As a preliminary investigation of possible disease associations, two entities known to be associated with abnormalities of the lipoproteins were studied. In addition, the sera of patients with rheumatic fever, collected for other reasons, were also studied. The frequency of reaction of these patients and controls with three of the antisera are shown in Table III.[17] There is a significantly higher number of reactors with the C. de B. antisera in the diabetic as compared to the control group. No such differences are detected for the other groups. This finding is interesting, but before any importance can be attached to it, this study will have to be repeated in other populations. If it proves to be reproducible, then the higher frequency of reactors might be due to some feature of the disease which converts apparent non-reactors into detectable reactors, or it may possibly be related to an inherited susceptibility factor.

In addition to the isoprecipitins which react with lipoproteins, an isoprecipitin has been found which reacts with an as yet unidentified protein which does not appear to have a large amount (or in some cases, any) lipid. The isoprecipitins are found in multiply transfused hemophilia patients. The protein with which they react is very rare in populations of Western origin, but it is not uncommon in Micronesians, Vietnamese, Formosans and Australian aborigines. The reacting antigen is also quite common in patients with leukemia, and the significance of this is currently under study.

In addition to these precipitating antibodies, Allen and Kunkel[18] have found agglutinating antibodies against gamma globulin (anti-Gm antibodies) in the sera of many multiply transfused thalassemia patients. It is not unlikely that, using the appropriate techniques, antibodies against a very large number of serum proteins will be detected. These may be the cause of some of the transfusion reactions which cannot be ascribed to interaction with the formed elements of the blood.

Summary

The concept of polymorphism is discussed and known biochemical polymorphisms in man are listed. The serum protein polymorphisms and the formation of isoprecipitins in multiply transfused human patients are described.

REFERENCES

1. Ford, E. B. Polymorphism in plants, animals and man, *Nature (London)* *180*:1315-19 (Dec. 14), 1957.
2. Allison, A. C. Aspects of polymorphism

in man, *Cold Spring Harbor Symposia on Quantitative Biology* *20*:239-55, 1955.

3. Blumberg, B. S. Inherited susceptibility to disease. Its relation to environment, *Arch. Environ. Health (Chicago)* *3*:612-36, 1961.
4. Nance, W. E. and Smithies, O. New haptoglobin alleles: A prediction confirmed, *Nature (London)* *198*:869-70 (June 1), 1963.
5. Allison, A. C., Blumberg, B. S. and apRees, W. Haptoglobin types in British, Spanish Basque and Nigerian populations, *Nature (London)* *181*:824-25 (March 22), 1958.
6. Giblett, E. R. and Steinberg, A. G. Inheritance of serum haptoglobin types in American Negroes: Evidence for a third allele, Hp²M, *Amer. J. Hum. Genet.* *12*:160-69, 1960.
7. Fundenberg, H. and Franklin, E. C. Human gamma globulin. Genetic control and its relation to disease, *Ann. Intern. Med.* *58*:171-80, 1963.
8. Grubb, R. Relationship between blood group serology and rheumatoid arthritis serology. Serum protein groups, *Vox Sang.* *2*:305-12, 1957.
9. Arfors, K.-E., Beckman, L. and Lundin, L.-G. Genetic variations of human serum phosphatases, *Acta Genet. (Basel)* *13*:89-94, 1963.
10. Schmid, K. and others. Studies on the structure of α_1-acid glycoprotein and the partial resolution and characterization of its variants, *Biochemistry (Wash.)* *1*:959-66, 1962.
11. Hirschfeld, J. Immune-electrophoretic demonstration of qualitative differences in human sera and their relation to the haptoglobins, *Acta Path. Microbiol. Scand.* *47*:160-68, 1959.
12. Cleve, H., Prunier, J. H. and Bearn, A. G. Isolation and partial characterization of the two principal inherited group-specific components of human serum, *J. Exper. Med.* *118*:711-26, 1963.
13. Kalow, W. and Gunn, D. R. Some statistical data on atypical cholinesterase of human serum, *Ann. Hum. Genet.* *23*:239-50, 1959.
14. Blumberg, B. S., Bernanke, D. and Allison, A. C. A human lipoprotein polymorphism, *J. Clin. Invest.* *41*:1936-44, 1962.
15. Blumberg, B. S. and others. Multiple antigenic specificities of serum lipoproteins detected with sera of transfused patients. *Vox Sang.* *9*:128-45, 1964.
16. Weigle, W. D. Fate and biological action of antigen-antibody complexes, *Advances in Immunology* *1*:283-317, 1961.
17. Blumberg, B. S., Ledbetter, B. B. and Visnich, S. Inherited antigenic differences in serum beta-lipoproteins. Studies in association with disease, *Amer. J. Public Health*, 1964. In press.
18. Allen, J. C. and Kunkel, H. G. Antibodies to genetic types of gamma globulin after multiple transfusions, *Science* *139*:418-19, 1963.
19. Robinson, J. C. and Pierce, J. E. Prevention of background staining of starch gels, *Amer. J. Clin. Path.* *40*:588-90, 1963.

Press of Charles C. Morchand Co., Inc., N. Y. C.

Reprinted from
Annals of the New York Academy of Sciences
Volume 103, Article 2, Pages 1052–1057
May 8, 1963

ISO-ANTIBODIES IN HUMANS AGAINST INHERITED SERUM LOW DENSITY BETA-LIPOPROTEINS, THE AG SYSTEM

Baruch S. Blumberg
National Institutes of Health, Bethesda, Md.

In genetic polymorphic systems two or more inherited forms of a trait may exist in the same population, and all of them can occur in relatively high frequency (Ford, 1957). From theoretical considerations it is reasonable to assume that some of these polymorphisms persist as a result of differences in survival values for each of the genotypes; that is, that natural selection in the Darwinian sense is operating. Such polymorphisms exist in human serum, and several have been described in the past few years. These include the haptoglobins, transferrins, group specific substance, gamma globulin groups (Gm), and others (For review see Blumberg, 1961). Patients transfused with blood cross-matched for red blood cell types, but not for serum types, may receive plasma containing proteins different from their own, and often develop precipitating antibodies against isologous serum proteins (Allison and Blumberg, 1961; Blumberg and Allison, 1961). Of 110 patients who received approximately forty or more transfusions for a variety of reasons, 19 developed a precipitating antibody which reacted with a protein present in the serum of some but not all humans, and some lower animals. Fifty-two of the transfused patients had thalassemia major and 14 reactions were detected among them. Nearly all of these were studied in collaboration with C. Smith and Marion Erlandsen of New York Hospital. The remaining 58 patients had a variety of diseases including leukemia, hemophilia, recurrent GI bleeding and idiopathic anemias. Only 9 percent of these patients developed the antibody. This suggests that there may be some special factor in the population of patients who had thalassemia which leads to the development of antibody, but the methods of patient selection, and other factors, do not rule out other explanations.

In every case the antibody was directed against a serum lipoprotein and in the six antibodies which have been further studied, the antigen was a low density β-lipoprotein (FIGURE 1) (Blumberg *et al.*, 1962a).

It is conceivable that these precipitins represent "natural" isoantibodies similar to those against the ABO blood groups. Evidence is accumulating, however, that they are the result of the multiple transfusions, and are analogous to antibody production in experimental animals injected with isologous proteins (Dray & Young, 1958).

(a) No anti-lipoprotein antibodies were found in the sera of 163 nontransfused subjects (Blumberg *et al.*, 1962b).

(b) Rabbits develop antibodies against beta-lipoproteins when injected with whole serum from other rabbits of the same strain (Dray and Blumberg, unpublished).

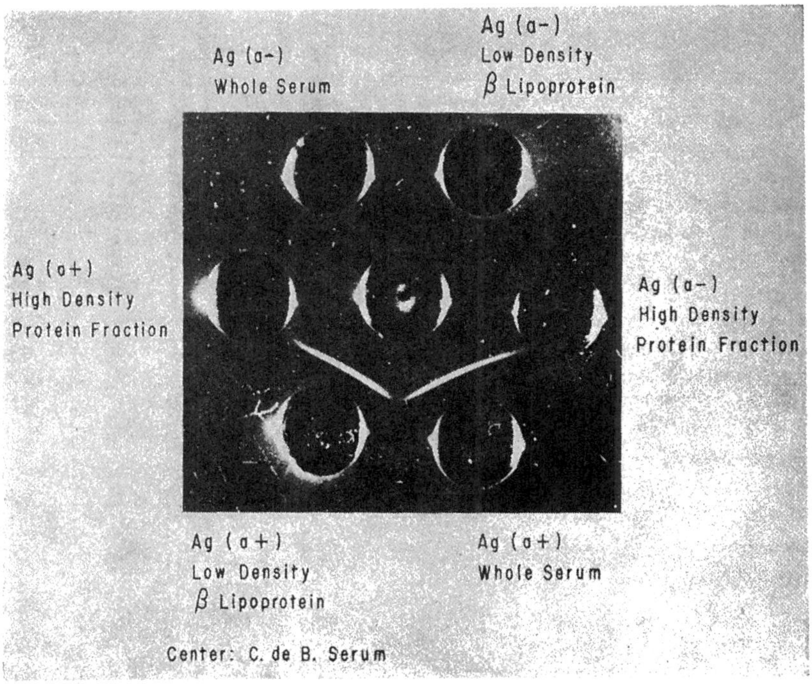

FIGURE 1. Photograph of an Ouchterlony experiment designed to show that the C. de B. antiserum (center well) reacts only with the low density β-lipoprotein fraction of the serum of an Ag(a+) individual (From Blumberg *et al.*, 1962). Reproduced by permission of *Nature* (*London*).

(c) Three patients have been studied who did not have anti-β-lipoprotein antibodies before transfusion, or after only a few transfusions, but did develop antibodies after multiple transfusions (Blumberg & Riddell, unpublished observations).

The serum β-lipoprotein fraction has been shown to contain many enzymatic activities, including esterase, beta glucouronidase, lactic dehydrogenase, isocitric dehydrogenase, and others (Lawrence and Melnick, 1961; Uriel, 1961). Zymogram studies on Ouchterlony plates using the method of Uriel (1961) have demonstrated that the β-lipoprotein isoprecipitin line contains esterase activity. The catalytic activity of the esterase was not eliminated by the antigen-antibody reaction.

The most detailed studies have been done using antisera C. de B. (Allison and Blumberg, 1961; Blumberg & Allison, 1961; Blumberg *et al.*, 1962b), and the New York antisera (Blumberg & Riddell, 1963). The C. de B. serum reacts with 50 per cent to 60 per cent of sera selected from a U.S. population. Twin and family studies have demonstrated that the antigenically different proteins are inherited in a simple Mendellian fashion (TABLE 1). Individuals homozygous or heterozygous for a gene Ag^A have beta-lipo-

TABLE 1

FAMILY STUDIES OF Ag(a+) PHENOTYPE

Mating type	N. of families		Ag(a+) offspr.		Ag(a−) offspr.	
	Obs.	Exp.	Obs.	Exp.	Obs.	Exp.
U.S. White						
Ag(a+) × Ag(a+)	9	7.6	15	14.3	2	2.7
Ag(a+) × Ag(a−)	10	12.4	9	13.2	13	8.8
Ag(a−) × Ag(a−)	6	5.0	0	0	13	13.0
Total	25	25.0	24	27.5	28	24.5
U.S. Negro						
Ag(a+) × Ag(a+)	9	8.3	29	29.4	6	5.6
Ag(a+) × Ag(a−)	8	11.8	20	19.2	12	12.8
Ag(a−) × Ag(a−)	7	4.9	0	0	18	18.0
To'al	24	24.0	49	48.6	36	36.4
Micronesian						
Ag(a+) × Ag(a+)	9	8.6	24	22.5	1	1.5

TABLE 2

Ag BETALIPOPROTEIN POLYMORPHISM.
POPULATION STUDIES.

Population	Location	Total no.	C deB%Pos.	New York % pos.
U.S. white	Maryland	120	59	97
U. S. negro	Georgia	149	68	99
Greeks	Chalkidiki	103	71	90
Greeks	Atalanti	100	73	97
Athabascan Indians	N. Alaska	102	68	—
Sioux Indians	S. Dakota	143	91	78
Quechua Indians	Peru	102	70	86
Eskimos	N. Alaska	97	81	—
Vietnamese	Viet Nam	99	71	—
Micronesians	Rongelap Marshall	189	98	43
Polynesians	Bora Bora	96	100	—

proteins which form precipitins with the C. de B. antiserum. Those who are homozygous for the alternate allele *Ag* do not have these specificities. Ouchterlony and absorption studies indicate that there are at least two antigenically different specificities detected by the C. de B. antiserum (FIGURE 2). The second antiserum (New York antiserum) reacts with the beta-lipoproteins of nearly all U.S. sera. However, only 43 percent of the sera collected from the inhabitants of Rongelap Atoll in the central Pacific reacted. Family studies in this population indicate that the trait is inherited in the same manner as the Ag(a+) lipoprotein, but is controlled by a different gene, tentatively designated Ag^{B}. This gene appears to be allelic to Ag^{A} in some populations, but there may be other alleles. The frequency of the reactors varies considerably in different populations. (TABLE 2). C. de B. and New York antisera react with each other, and the precipitin lines

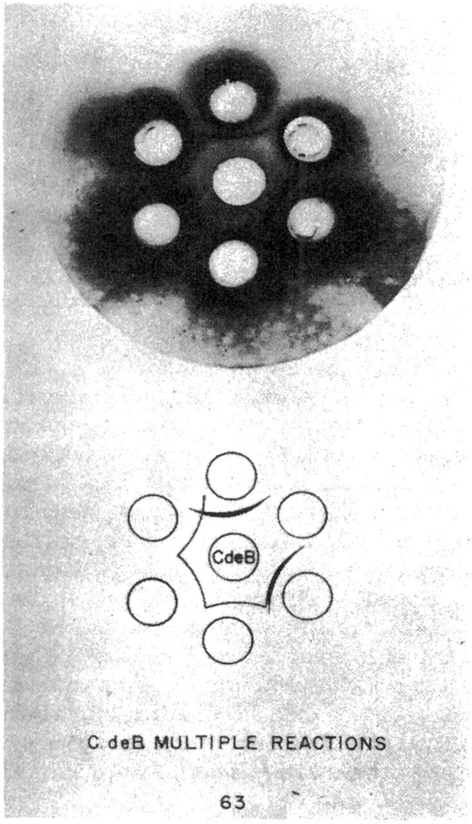

FIGURE 2. Reactions of the C. de B. antiserum with six normal sera. A reaction of partial fusion is seen at the lower right, and of crossing at the upper left. One of the sera is a non-reactor, and the remaining two show fusion of the precipitin lines. Sudan Black stain.

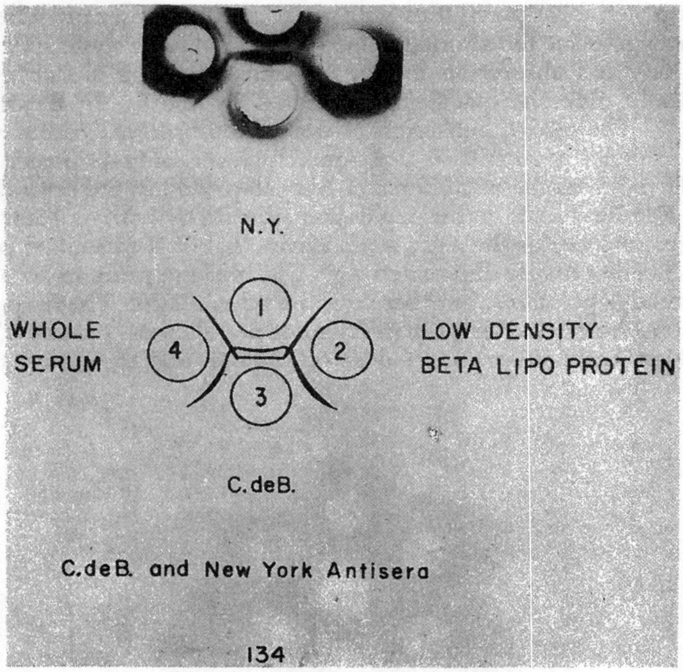

Figure 3. Reaction between New York (hole 1) and C. de B. (hole 3) antisera, and the whole serum (hole 4) and low density β-lipoprotein (hole 2) of a reactor. Sudan Black stain.

cross, indicating that they are not antigenically identical (FIGURE 3) (Blumberg and Riddell 1963). Absorption experiments confirm these findings. The simplest explanation is that C. de B. has an antibody against a beta-lipoprotein antigen present in the New York serum, [designated Ag(a+)] but absent in the serum of C. de B. Conversely, New York antiserum has an antibody against a protein present in C. de B. [Ag(b+)] but absent in the New York antiserum. The specificity of the antibody developed in the multiply transfused patient is determined in part by the specificity of his own inherited antigenic beta-lipoprotein; that is, he does not form an antibody against his own protein but only against the alternate configuration. Preliminary studies using the other antisera indicate that some have the same specificities as the C. de B. antisera, but antibodies with other specificities have also been detected. It appears likely that there are at least six, and possibly more, antigenically distinct inherited specificities in the serum beta lipoprotein.

All the antisera studied have precipitating antibodies against lipoprotein. It would appear, therefore, that under the conditions of clinical transfusion the lipoproteins are the most antigenic serum proteins, as detected by the Ouchterlony agar gel method.

Mother-child studies using the C. de B. and New York antisera suggest that the Ag^A and Ag^B genes do not produce a detectable amount of the specific beta-lipoproteins at birth, but that it is produced in detectable quantities within two to four months after birth. The evidence suggests that in some cases the specific β-lipoprotein produced by the mother may be transmitted placentally to the fetus, although this has not been confirmed by direct observation (Blumberg *et al.*, 1962b; Blumberg & Riddell, 1963). If this is true, then some children would be subjected to this antigenic stimuli *in utero* while others would not; and this may affect their reaction to the same antigenic stimulus later in life.

References

ALLISON, A. C. & B. S. BLUMBERG. 1961. An isoprecipitation reaction distiguishing human serum protein types. Lancet. **1**: 637.

BLUMBERG, B. S. 1961. Inherited susceptibility to disease and its relation to environment. Arch. of Env. Med. **3**: 612–636.

BLUMBERG, B. S. & A. C. ALLISON. 1961. Studies on the isoprecipitin determined human serum polymorphism. Proc. of the Second International Conference of Human Genetics. Rome. Italy.

BLUMBERG, B. S., S. DRAY & J. C. ROBINSON. 1962a. Antigen polymorphism of a low density β-lipoprotein. Allotypy in human serum. Nature. **194**: 656–658.

BLUMBERG, B. S., A. D. BERNANKE & A. C. ALLISON. 1962b. A human lipoprotein polymorphism. J. Clin. Invest. **41**: 1936–1944.

BLUMBERG, B. S. & N. M. RIDDELL. 1963. Inherited antigenic differences in serum human beta-lipoproteins. A second antiserum. J. Clin. Invest. In press.

DRAY, S. & GLENDOWLYN O. YOUNG. 1958. Differences in the antigenic components of sera of individual rabbits as shown by induced isoprecipitins. J. of Immunol. **81**: 2, 142–149.

FORD, E. B. 1957. Polymorphisms in plants, animals and man. Nature. **180**: 1315–1319.

LAWRENCE, S. H. & P. J. MELNICK. 1961. Enzymatic activity related to human serum beta-lipoprotein: histochemical, immuno-electrophoretic and quantitative studies. Proc. Soc. Exp. Biol. and Med. **107**: 998–1001.

URIEL, J. 1961. Caractérisation des cholinestérases et d'autres estérases carboxyliques après électrophorèse et immunoélectrophorèse en gélose. Ann. Inst. Pasteur. **101**: 104–119.

Reprinted From the Journal of the American Medical Association
February 15, 1965, Vol. 191, pp. 541-546
Copyright 1965, by American Medical Association

Reprinted from the *Journal of the American Medical Association*
February 15, 1965, Vol. 191, pp. 541–546
Copyright 1965, by American Medical Association

A "New" Antigen in Leukemia Sera

The "Australia antigen" is found in the sera of some normal individuals from foreign populations. The total absence of the antigen from the sera of normal United States subjects and its relatively high frequency in acute leukemia suggests that the presence of the antigen may be of value in the diagnosis of early acute leukemia. Whether the antigen results from or precedes the leukemia process remains to be seen.

Baruch S. Blumberg, MD, Harvey J. Alter, MD, and Sam Visnich

Patients who receive large numbers of transfusions for anemia and other causes may develop precipitins in their blood. These precipitins may react in agar gel double diffusion experiments with specific human serum lipoprotein found in the blood of other individuals. Since these precipitins were found only in patients who had received transfusions they were thought to be antibodies against serum lipoproteins which developed in the patients as a result of the repeated transfusions. The precipitin is referred to as an isoprecipitin since it develops against a specificity found in an individual from the same species. The antilipoprotein isoprecipitin[1,2] developed in approximately 30% of 47 patients with thalassemia who had received transfusions. Isoprecipitins also developed in smaller number of transfused patients with other diseases. All precipitins stained with sudan black, a dye specific for lipid. Immunoelectrophoretic and ultracentrifugal studies showed that the protein with which the isoprecipitins reacted was a low density lipoprotein. The reactor specificity associated with the beta lipoprotein is inherited as an autosomal-dominant trait and several lipoprotein specificities have been found.[3-5]

In 1963, sera from patients with hemophilia who had received transfusions were tested for the presence of isoprecipitins using a panel of 24 sera from normal individuals, including sera from foreign populations. Two of the hemophilia sera formed a clearly defined precipitin line with one of the panel sera (from an Australian aborigine), but with none of the others. In contrast to the usual findings the precipitin line stained only faintly with sudan black; it did, however, stain with azo carmine, a general protein stain (Fig 1). Subsequent studies have shown that this protein system differs from that detected with the antilipoprotein antisera. The serum protein with which the hemophilia isoprecipitin reacts has not been fully characterized and has been tentatively called the "Australia antigen." This paper will describe the epidemiologic and immunologic aspects of the Australia antigen-isoprecipitin system.

From the Institute for Cancer Research, Philadelphia (Dr. Blumberg), and the National Institutes of Health, Bethesda, Md (Dr. Alter and Mr. Visnich).

Reprint requests to 7701 Burholme Ave, Philadelphia 19111 (Dr. Blumberg).

Materials and Methods

Double diffusion in agar gel was done using a micro-Ouchterlony technique on lantern slides.[5] Each of the two hemophilia sera containing the isoprecipitin was placed in the center well of a seven-hole micro-Ouchterlony pattern. The sera to be tested for presence of Australia antigen were placed in the peripheral wells. When a panel of antigen-containing sera were identified in this manner, they in turn were each placed in the center wells of similar seven-hole Ouchterlony patterns, and the sera of patients who had received transfusions, which were to be tested for the presence of isoprecipitins, were placed in the peripheral wells. In the final testing program two sera-containing Australia antigens which reacted with all the hemophilia antisera first discovered, were selected to screen for the remaining antisera. Two of the strongest hemophilia antisera were used in screening for the sera containing Australia antigen. In screening for antilipoprotein antisera, the sera from patients who had received transfusions were tested using a panel of 24 sera selected from four or more population groups as in previous studies.[2]

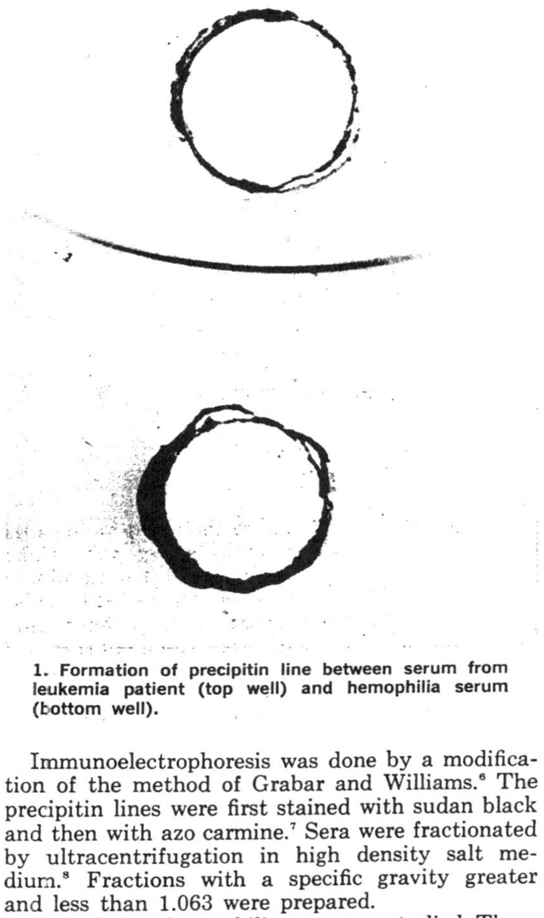

1. Formation of precipitin line between serum from leukemia patient (top well) and hemophilia serum (bottom well).

Immunoelectrophoresis was done by a modification of the method of Grabar and Williams.[6] The precipitin lines were first stained with sudan black and then with azo carmine.[7] Sera were fractionated by ultracentrifugation in high density salt medium.[8] Fractions with a specific gravity greater and less than 1.063 were prepared.

A total of 28 hemophilia sera were studied. These included samples from patients from Mt. Sinai Hospital, NY, New York Hospital, and the Clinical Center of the National Institutes of Health. They were all United States whites and had all received transfusions many times with fresh frozen plasma and whole blood obtained from the hospital blood banks as well as commercial sources. The racial makeup of the donors could not be known with certainty, but they probably included United States whites and Negroes. The hemophilia serum used in the immunoelectrophoresis experiments contained antibodies against both lipoproteins and the Australia antigen. It was obtained from a patient (1) who had received more than 900 transfusions.

The sources of some of the normal sera used in the studies are given in the Tables. The sera from patients were obtained at the Clinical Center, National Institutes of Health, New York Hospital, and elsewhere. The calculation of the probability value was done by using two by two tables and Fisher's χ-square method.

Results

Frequency of the Isoprecipitin.—The sera of 107 patients who had received approximately ten or more transfusions, and of 150 normal individuals who had not received transfusions, were tested for the presence of the Australia isoprecipitin (Table 1). Eleven isoprecipitin containing sera were found in the patient group; none was found in the normal population. Of these 11, eight were patients with hemophilia, one had plasma thromboplastin component deficiency disease (PTC), one thalassemia, and one aplastic anemia. The frequency is highest in the hemophilia and related PTC group, and the difference from the other groups combined is statistically significant ($P<0.001$). It had previously been found that the frequency of antilipoprotein isoprecipitin was very high in thalassemia as opposed to other patient groups.[2] Hemophilia sera containing the Australia isoprecipitin were tested for the presence of antibeta lipoprotein; they were found to have an incidence similar to other patients who did not have thalassemia. Only one of the patients in the latter group had an anti-Australia antigen isoprecipitin.

There were two sera which contained both antilipoprotein and anti-Australia antigen isoprecipitins. This is approximately the number expected by chance.

Frequency of Antigen.—A total of 1,704 sera from nonhospitalized and presumably normal subjects were tested for the presence of Australia antigen using at least two, and generally more antisera (Table 2). Reactors were found only in Oceanic, Oriental, and Mediterranean populations, and in one from the American Indian population. No reactors were found among the approximately 700 United States sera tested including cord serum from 18 newborns. The highest frequencies were

Table 1.—Isoprecipitins in Patients Who Had Received Transfusions Which React With the Australia Antigen and With Lipoproteins

	Antibody to Australia Antigen				Antibody to Lipoprotein			
Disease	Total	Present	Absent	% Present	Total	Present	Absent	% Present
Hemophilia	28	8	20	28.6	31	3	28	9.7
PTC*	2	1	1	....	2	0	2	....
Thalassemia	48	1	47	2.1	47	14	33	29.8
Other Diseases	29	1	28	3.4	33	3	30	9.1
Total	107	11	96	10.3	113	20	93	17.7
Normal controls	150	0	150	0	200	0	200	0

*Plasma thromboplastin component deficiency disease.

found in a small group of native Taiwanese and in Australian aborigines. All of these tests were done on stored sera.

Sera from a total of 659 patients with various illnesses were tested against one or more of the antisera as shown in Table 3. Of the ten reactors found, eight had leukemia (including one patient with a diagnosis of both acute lymphocytic leukemia and multiple myeloma, and one patient with a newly described preleukemia syndrome associated with a missing chromosome in marrow cells,[10]) and two had thalassemia. These patients with thalassemia were siblings whose parents were born in Greece. As noted in Table 2, the frequency of reactors in presumably normal Greeks is 4%. The difference between the frequency in leukemia patients and that in all the disease groups combined is statistically significant ($P<0.001$). All of the patients with the Australia antigen had received transfusions. However, approximately 300 of the patients who did not have the Australia antigen had also received transfusions.

The leukemia sera were studied in greater detail. The antigen was found in all the larger subclassifications with the exception of chronic myelogenous leukemia. The patients with reacting sera were of both sexes, and their ages at the time of the diagnoses varied from 6 to 55 years. Of the ten patients in whom the antigen was found, eight are now dead, including all but one of the eight patients with leukemia, and one of the patients with thalassemia.

A total of 54 patients with acute leukemia were studied. Of these, six had the Australia antigen and were all dead. Of the 49 who did not have the Australia antigen, 19 were dead. The sample size is too small to permit adequate corrections for age, duration and intensity of disease, and other factors, but there does appear to be a gross difference with respect to mortality between these two groups.

Characteristics of Antibody and Antigen.—The original isoprecipitin-containing sera and the reacting sera had been found using specimens which had been collected and stored at −20 C for various lengths of time up to three years. Several specimens collected at different times were subsequently obtained from patients who had received transfusions and whose sera contained the isoprecipitins, and from patients with leukemia and normal subjects whose sera contained the Australia antigen. These studies showed that the isoprecipitin was present in sera or plasmas which had been stored at −20 C for up to 27 months. The Australia antigen was present in fresh sera as well as in sera or plasmas stored for up to six years. These studies also indicate that within the limits of observations, the presence of the Australia antigen is essentially an invariant characteristic of the individual; that is, if it is present at one point in time, it is present when tested subsequently.

The isoprecipitin appears to be a 7S γ-globulin.[9] Australia antigen migrates in the beta and slow α-globulin region on immunoelectrophoresis, but can readily be distinguished from the antigen which reacts with the antilipoprotein isoprecipitin found in the serum of patient 2[1] as shown in Fig 2. (The top basin contains the antilipoprotein antiserum from patient 2,[4] and the second basin serum from hemophilia patient 1. The latter has isoprecipitins directed against both the Australia antigen and lipoprotein. The electrophoresis well contains the serum of a patient with leukemia which reacts with serum from patient 2 and with both of the isoprecipitins in serum from patient 1. Two distinct lines of slightly different mobility were seen between the sera of patients 1 and 3. The lower one, which stains blue with sudan black, corresponds to the patient 2 lipoprotein line. The upper one, which stains only faintly or not at all with sudan black,

Table 2.—Australia Antigen in Normal Populations

Population	No. Tested	Australia Antigen Present	
		No.	%
Aborigines, Australia	208	12	6
Chinese, USA and Taiwan	65	0	0
Eskimo, Alaska[13]	24	0	0
Greeks, Greece[14]	179	8	4
Indians, Canada[15]	78	0	0
Indians, Mexico	100	1	1
Israelis	96	2	2
Japanese, USA	48	0	0
Koreans	1	1	
Micronesians, Rongelap[16]	193	7	4
Negro, USA[17]	241	0	0
Newborn children, white	18	0	0
Polynesians, Bora Bora	24	1	4
Samaritans, Israel[18]	125	2	2
Taiwanese	23	3	13
Tristan da Cunha Islanders	42	0	0
Vietnamese	24	1	4
White, USA (NIH* employees)	215	0	0
Total	1,704	38	

* National Institutes of Health.

Table 3.—Australia Antigen in Patients

Disease	No. Tested	Australia Antigen Present	
		No.	%
Abetalipoproteinemia	4	0	0
Amyotrophic lateral sclerosis	15	0	0
Anemia	18	0	0
Arthritis, various*	15	0	0
Cancer (other than leukemia)	47	0	0
"Connective tissue" disorders†	5	0	0
Diabetes	96	0	0
Hemophilia	24	0	0
Hypercholesterolemia	17	0	0
Leukemia	70	8‡	11.4
Acute myelogenous	17	4	
Acute lymphocytic	38	2	
Chronic myelogenous	10	0	
Chronic lymphocytic	3	1	
45 chromosomes[10]	2	1	
Lupus erythematosis	69	0	0
Multiple myeloma and macroglobulinemia	93	1‡	1.1
Myasthenia gravis	11	0	0
Rheumatic fever	124	0	0
Tangiers Island disease	3	0	0
Thalassemia	48	2	4.2
Total	659	10	

*Includes eight patients with rheumatoid arthritis, three with psoriatic arthritis, and four with Sjogren's disease.
†Other than lupus erythematosis.
‡One patient had both chronic lymphocytic leukemia and multiple myeloma, and is included in both categories.

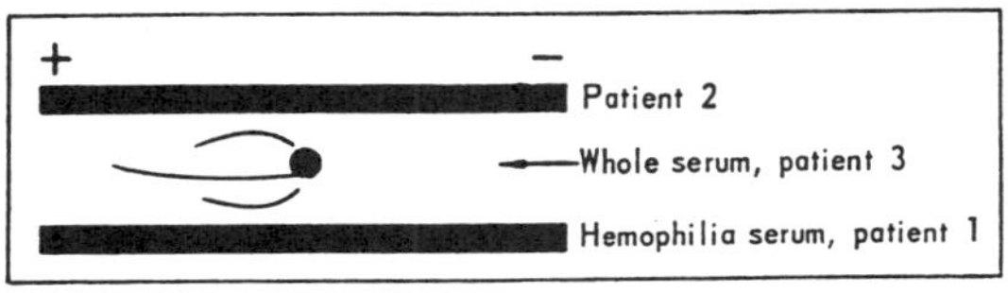

2. Diagram of immunoelectrophoresis experiment: antilipoprotein (patient 2), leukemia (patient 3), and anti-Australia antigen (patient 1).

does stain with azo carmine.) Using the hemophilia serum from patient 1 (see above), two distinct lines with slightly different mobilities can be seen. One of these stains blue with sudan black and corresponds to the lipoprotein antigen which reacts with patient 2. The other, which has a slightly different mobility and a different curvature, does not stain with sudan black, but does stain with azo carmine. In addition, a reaction of nonidentity can be demonstrated by immunodiffusion when the Australia isoprecipitin system is compared with the typical lipoprotein precipitin as shown in Fig 3. (The serum of patient 2 contains an antilipoprotein isoprecipitin. The serum of a patient with thalassemia with an isoprecipitin against Australia antigen is in the top right hand well. The serum of the Australian aborigines has both the specific lipoprotein and the Australia antigen. The patient-2 line stains blue with sudan black and the patient-4 line stains red with azo carmine.) Additional characteristics of the Australia antigen and a method for its partial isolation are described in another paper.[9]

Specificity.—To determine if the antisera had different specificities, they were all tested against the same antigen placed in the center well of a seven-hole-Ouchterlony pattern. No crossing of lines were seen between the antisera. Similar experiments were used to compare the Australia antigen found in the presumably normal subjects with that found in the leukemia patients and no differences could be detected. On the basis of these initial experiments, we have not been able to detect any specific differences between the various antisera, nor between the Australia antigen found in normal persons, and in leukemia patients.

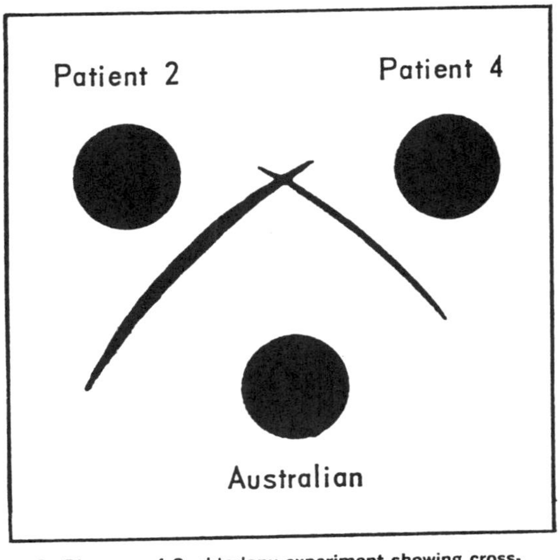

3. Diagram of Ouchterlony experiment showing crossing of antilipoprotein (patient 2) and anti-Australia antigen (patient 4) lines.

Family Studies.—The studies on normal populations were done on sera collected on field trips for other purposes and stored in the Institutes serum bank. In some cases, family sera were available, and the results were analyzed to determine if there was a family aggregation of Australia antigen. Batsheva Bonné examined 125 sera from Samaritans living near Tel Aviv, Israel[18] (Fig 4). This represented nearly all the members of this highly inbred community. Of these, two siblings who were the offsprings of a consanguineous marriage (both of whose parents were double cousins) were the only individuals with detectable antigen. In the Micronesian population, there was one father and son affected, but in the five other cases the individuals were only remotely related or unrelated. The reactors in the Greek study were not apparently closely related. Of the 47 patients with thalassemia studied, only two were positive, and they were siblings. Their parents and one aunt were not reactors. The two brothers of one of the patients with leukemia with the Australia antigen did not react with the antisera.

Comment

A second kind of isoprecipitin system has been revealed by this study. The Australia antigen with which the isoprecipitin reacts differs from the lipoprotein antigens previously described and the distribution of isoprecipitins and reactors in patients and in normal populations is very different from that found for the lipoprotein system.

It has not been firmly established that the isoprecipitin in the hemophilia sera is an antibody nor that the protein with which it reacts is an antigen. However, by analogy with the lipoprotein system, it is probable that this is the case and these terms have been used in describing the system.

In our discussions it has been assumed that the "antibody" is present in the sera of patients with hemophilia and other patients, and the "antigen" in the leukemia and normal sera. This appears likely since the patients with hemophilia and other patients had all received transfusions whereas some of the individuals in whose sera the Australia antigen was found had not. The possibility still remains, however, that the rare normal sera and leukemia sera actually contain an antibody against an antigen present in hemophilia patients. It is hoped that further studies with patients with hemophilia may help resolve this point.

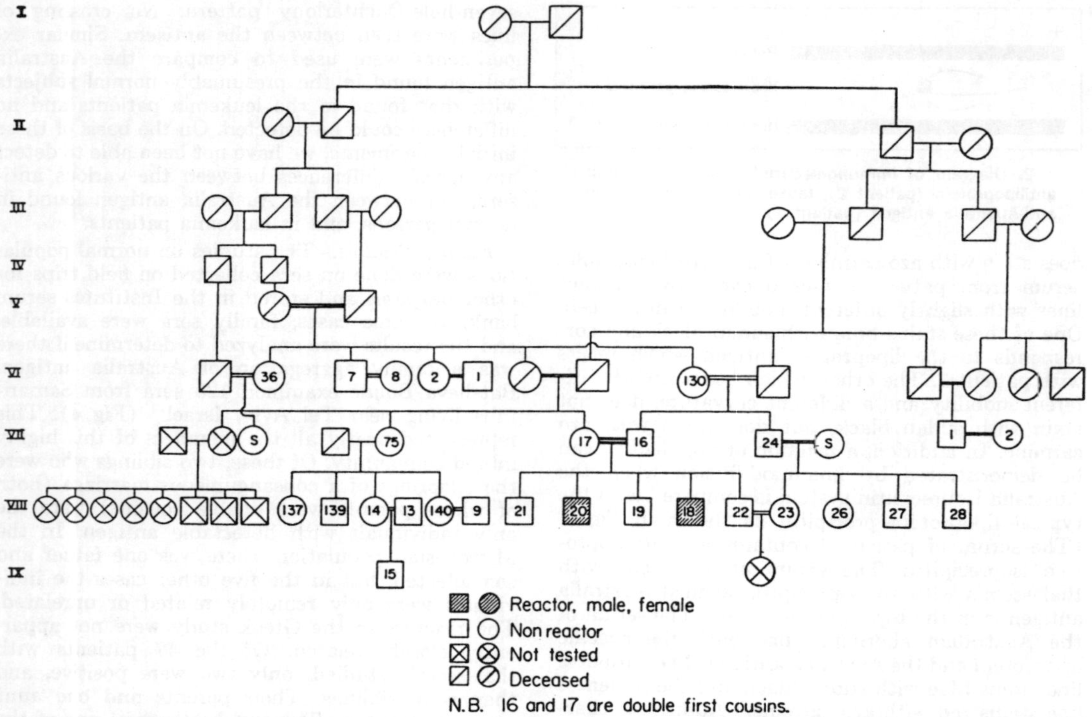

4. Portion of a pedigree from the Samaritan community, representative of nearly all members, showing the presence of the Australia antigen and the close interrelation of the parents.

The high incidence of this isoprecipitin in the hemophilia-patient group suggests that either hemophilia per se predisposes to the formation of the isoprecipitin or that the administration of fresh frozen plasma (which distinguishes the treatment of hemophilia and PTC patients from the other individuals who had received transfusions) particularly predisposes to the formation of the antibody. Preliminary evidence suggests that the Australia antigenic sites may be revealed by a freezing process which leads to the denaturation of lipoprotein. This possibility will be discussed in another paper.[9]

The high incidence of the Australia antigen in the leukemia population as compared with normals or other patient populations suggests either that; (1) persons with the Australia antigen have an increased susceptibility to the development of leukemia, or (2) the antigen itself is a manifestation of the disease process, perhaps secondary to an alteration in some normal serum constituent with a resultant change in antigenic configuration, or (3) the Australia antigen is related to the virus which has been suggested as the cause of leukemia.

The first hypothesis implies that the Australia antigen is present in a patient before he develops signs and symptoms of leukemia. The long-term observations necessary to support this possibility have not yet been made. The Micronesian, Samaritan, and Australian aborigine populations would be suitable for such a study. In the case of the "preleukemic"[10] (Table 4) patient the presence of the Australia antigen predated the development of frank leukemia.

Changes in red blood cell antigens during the course of leukemia lend indirect support to the hypothesis that the Australia antigen results from, rather than precedes the leukemia process. Many authors have demonstrated alterations in red blood cell ABO specificities during the course of leukemia.[11] More recently, it has been shown that there is a loss of I specificity on the red blood cells of certain patients with leukemia.[12] The I antigen is almost universally present in normals and patients with other diseases.

The third hypothesis is being tested in collaboration with workers at the National Institutes of Health.

The total absence of the Australia antigen from normal United States subjects studied and its relatively high frequency in acute leukemia suggests that the presence of the antigen may be of value in the diagnosis of early acute leukemia.

The available data are too few to support a genetic hypothesis, but none of the family studies are inconsistent with simple recessive inheritance of the specificity.

Summary

An isoprecipitin is present in the sera of many patients with hemophilia who have received transfusions. It reacts with a protein (the "Australia antigen") that is found in the sera of some normal individuals from foreign populations but is absent in sera of the United States populations studied. It is found in approximately 10% of patients with leukemia.

R. Rosenfield, MD, and E. Smith, MD, of Mt. Sinai Hospital provided the hemophilia antisera. P. Carbone, MD, E. Cohen, MD, E. J. Freireich, MD, and A. B. Rey, MD, provided the leukemia sera. Other valuable sera used in the experiments were provided by T. D. Dublin, MD, Marion Erlandson, MD, J. Fahey, MD, H. H. Fudenberg, MD, R. Kirk, PhD, L. Laster, MD, L. Rosen, MD, C. Sheba, MD, N. R. Shulman, MD, and A. G. Steinberg, PhD.

Samples of sera on patient 1 were provided by J. M. Hill, MD, of the Wadley Research Institute and Blood Bank, Dallas.

This investigation was supported in part by Public Health Service grants CA-08069-01 and CA-06551-02 from the National Cancer Institute.

References

1. Blumberg, B.S.; Dray, S.; and Robinson, J.C.: Antigen Polymorphism of a Low-Density Beta-Lipoprotein. Allotypy in Human Serum, *Nature* **194**:656-658 (May) 1962.

2. Blumberg, B. S., et al: Multiple Antigenic Specificities of Serum Lipoproteins Detected With Sera of Transfused Patients, *Vox Sanguinis* **9**:128-145 (March-April) 1964.

3. Allison, A.C., and Blumberg, B.S.: Isoprecipitation Reaction Distinguishing Human Serum Protein Types, *Lancet* **1**:634-637 (March 25) 1961.

4. Blumberg, B.S.; Bernanke, D.; and Allison, A.C.: Human Lipoprotein Polymorphism, *J Clin Invest* **41**:1936-1944 (Oct) 1962.

5. Blumberg, B.S., and Riddell, N.M.: Inherited Antigenic Differences in Human Serum Beta Lipoproteins. Second Antiserum, *J Clin Invest* **42**:867-875 (June) 1963.

6. Grabar, P., and Williams, C.A., Jr.: Methode Immuno-electrophoretique d'analyse de Melanges de Substances Antigeniques, *Biochim Biophys Acta* **17**:67-74 (May) 1955.

7. Uriel, J., and Grabar, P.: Emploi de colorants dans l'analyse electrophoretique et immuno-electrophoretique en milieu gélifié, *Ann Inst Pasteur* **90**:427-440 (April) 1956.

8. Havel, R.J.; Eder, H.A.; and Bragdon, J.H.: Distribution and Chemical Composition of Ultracentrifugally Separated Lipoproteins in Human Serum, *J Clin Invest* **34**:1345-1353 (Sept) 1955.

9. Alter, H.J.; Blumberg, B.S.; and Visnich, S.: Further Studies With Australia Antigen: To be published.

10. Freireich, E.J., et al: Refractory Anemia, Granulocytic Hyperplasia of Bone Marrow and Missing Chromosome in Marrow Cells. New Clinical Syndrome? *Clin Res* **12**:284, 1964.

11. Richards, A.G.: Loss of Blood Group B Antigen in Chronic Lymphocytic Leukaemia, *Lancet* **2**:178-179 (July 28) 1962.

12. McGinniss, M.H.; Schmidt, P.J.; and Carbone, P.P.: Close Association of I Blood Group and Disease, *Nature*, to be published.

13. Corcoran, P., et al: Blood Groups of Alaskan Eskimos and Indians, *Amer J Phys Anthrop* **17**:187-193 (Sept) 1959.

14. Barnicot, N.A., et al: Haemoglobin Types in Greek Populations, *Ann Human Genet* **26**:229-236 (Feb) 1963.

15. Blumberg, B.S., et al: Blood Groups of Naskapi and Montagnais Indians of Schefferville, Quebec, *Human Biol* **36**:263 (Sept) 1964.

16. Blumberg, B.S., and Gentile, Z.: Haptoglobins and Transferrins of Two Tropical Populations, *Nature* **189**:897-899 (March 18) 1961.

17. Cooper, A.J., et al: Biochemical Polymorphic Traits in US White and Negro Population, *Amer J Human Genetics* **15**:420-428 (Dec) 1963.

18. Bonné, B.: Samaritans: Demographic Study, *Human Biol* **35**:61-89 (Feb) 1963.

Reprinted from ANNALS OF INTERNAL MEDICINE, Vol. 66, No. 5, May, 1967
Printed in U. S. A.

A Serum Antigen (Australia Antigen) in Down's Syndrome, Leukemia, and Hepatitis

BARUCH S. BLUMBERG, M.D., D.PHIL., F.A.C.P., BETTY JANE S. GERSTLEY, M.D., DAVID A. HUNGERFORD, PH.D., W. THOMAS LONDON, M.D., and ALTON I. SUTNICK, M.D., F.A.C.P.
Philadelphia, Pennsylvania

WE HAVE PREVIOUSLY REPORTED the presence of an isoantigen of human sera, rare or absent in normal U. S. and northern European populations but relatively common in patients with leukemia (1, 2). The "new" isoantigen has been called "Australia antigen" because it was first found in the serum of an Australian aborigine. It appears to be different from the low-density lipoprotein isoantigen system first described in our laboratory (3–5). Hemophilia patients and others who have received multiple transfusions may develop precipitating isoantibodies against Australia antigen. These precipitating isoantibodies react with a protein in the blood of some normal and some sick individuals. This isoantigen (Australia antigen) could not be identified as being any of the known serum proteins.

Australia antigen (Au(1)) has the immunoelectrophoretic mobility of an alpha globulin and a specific gravity of less than 1.21 (6). It stains faintly with the lipid-specific dye Sudan black, indicating the presence of some lipid. It can be distinguished from the serum alpha- and beta-lipoproteins by its staining characteristics, flotation properties, and immunological specificities. Specific antibodies against Australia antigen have been produced in rabbits by immunization with whole serum that contains Australia antigen followed by absorption with the sera of individuals without the protein (7).

Recent studies of blood collected in Cebu, Peru, and Rongelap Atoll show significant family clustering of the trait (8). In addition, the segregation data are consistent with the hypothesis that individuals homozygous for an allele Au^1 (genotype Au^1/Au^1) have detectable Australia antigen (phenotype Au(1)) while those homozygous for an alternative allele (Au/Au) and heterozygotes (Au^1/Au) have no detectable antigen (phenotype Au(0)). Pending the acquisition of additional data, this genetic hypothesis has been proposed to explain the family clustering in the populations tested.

With the use of human and rabbit antisera, the distribution of Au(1) in several normal and patient populations has been determined (2, 8). Au(1) is absent in normal U. S. populations and rare in north European populations (approximately 1/1,000) but is fairly common in some Southeast Asian and Mediterranean populations. It is also relatively common in patients with leukemia, particularly in those with acute granulocytic leukemia (see below).

The original finding of Au(1) in very low frequency in normal individuals and in

Received November 2, 1966; accepted for publication January 24, 1967.

From the Institute for Cancer Research, Fox Chase, Philadelphia, Pa.

This study was supported by research grants CA-06551, CA-08069, and FR-05539, the National Cancer Institute, the National Institutes of Health, Bethesda, Md.

Requests for reprints should be addressed to Baruch S. Blumberg, M.D., The Institute for Cancer Research, 7701 Burholme Ave., Fox Chase, Philadelphia, Pa. 19111.

relatively high frequency in some kinds of leukemia suggested that individuals with this trait (which they may have inherited) may be more susceptible to leukemia than those without it. A corollary of this hypothesis is that individuals with a high risk of developing leukemia would comprise a population having a higher frequency of the trait than subjects drawn from the normal populations. Several such groups are known. These include patients with Down's syndrome (mongolism), polycythemia vera, co-twins of an identical twin having leukemia, individuals exposed to radiation, and others (for a review see Miller (11)). Several of these groups have been tested for the presence of Au(1), and the highest frequency in any U. S. population so far tested has been found in a group of patients with Down's syndrome (9, 10).

In the present paper, we describe our further studies on the prevalence of Au(1) in Down's syndrome patients.

Material and Methods

Blood specimens were obtained from 75 male residents of the State Colony at New Lisbon, N. J., who were admitted with the diagnosis of mongolism. For a control group, blood specimens were obtained from 76 male residents with the diagnosis of "epilepsy." The age distribution was approximately the same as in the Down's syndrome group, and they had been institutionalized fro approximately the same length of time. The admitting diagnoses of residents at New Lisbon State Colony are supplied by the patient's local physician, and these in general are not altered after admission. Chromosome studies were preformed on 17 of the Down's syndrome patients and 4 of the controls (see below).

After the identification of Down's syndrome patients with the without detectable Australia antigen, a second collection was undertaken in whcih blood was withdrawn from 19 residents with Au(1) and 26 residents without it. A series of tests were preformed on these samples. The white blood cell, red blood cell, and platelet counts, the white blood cell differential count, hemoglobin concentration, and hematocrit volume were determined using standard laboratory methods. We are indebted to Dr. R. A. Donato of Jeanes Hospital, Fox Chase, Philadelphia, Pa., for performing these tests. The following red blood cell antigens were identified using the appropriate antisera: B, A_1, A, C, D, D^u, E, c, e, Le^a, Le^b, Kp^a, Kp^b, M, N, S, s, Vel, K, k, Fy^a, Fy^b, Jk^b, Lu^a, Lu^b, Js^a, P, Tj^a, and V^a. We are indebted to Dr. Neva M. Abelson of the Hospital of the University of Pennsylvania for performing these tests. Uric acid levels were determined by the method of Liddle, Seegmiller, and Laster (12) through the courtesy of Dr. L. A. Healey of the University of Washington. The white blood cell glucose-6-phosphate dehydrogenase and galactose-1-phosphate uridyl transferase levels were determined by Drs. W. J. Mellman and F. A. Oski of the University of Pennsylvania using the methods described by them (13). The vitamin B_{12} levels were determined by the method of Hutner, Bach, and Ross (14). We are indebted to Dr. W. J. Williams of the University of Pennsylvania for performing these tests.

Blood specimens from nine Down's syndrome residents (five girls and four boys) of a Maryland institution for mentally retarded children were also tested. Blood from patients with polycythemia vera was obtained at three different clinics. These included 14 patients from Philadelphia, 36 from Portland, Ore., and 32 from San Francisco. Other sera from patients were collected primarily in the Philadelphia area, as noted in the Acknowledgments section. Venous blood was collected into dry Vacutainers (Becton, Dickinson Co.) or test tubes. The blood was allowed to clot overnight and the serum, removed by centrifugation. In some cases, small amounts of sera were separated from the blood immediately after collection and used in the double diffusion tests. Sera were stored at −20 C for later testing.

Double diffusion in agar gels was done by a micro-Ouchterlony technique described previously (15). Double diffusion in agar gel tubes was done by the method of Preer and Preer (16). Chromosome studies were performed on leukocyte cultures from 17 of the Down's syndrome patients using a method recently described (17).

The chi-square determinations were done as described by Snedcor (18). When any of the two-by-two table cells contained numbers lower than five, Fisher's exact method was used. Student's *t* test was used for comparing the quantitative results obtained on the antigen-positive and -negative bloods. Where appropriate, the method for comparing unpaired series of numbers was used.

TABLE 1. Distribution of Australia Antigen in Patients and U. S. Normals

Disease	No. Tested	No. Positive	% Positive
A-beta-lipoproteinemia	6	0	0
Amyotrophic lateral sclerosis	15	0	0
Anemia, various	26	0	0
Arthritis*	70	0	0
Cancer, other than leukemia	95	0	0
Diabetes†	303	0	0
Down's syndrome (mongolism)‡	84	25	29.8
Fanconi's anemia (hypoplasia bone marrow)	2	1	—
Hemophilia	60	3	5.0
Hepatitis, virus§	48	5	10.4
Hodgkin's disease	12	1	8.3
Hypercholesterolemia	17	0	0
Leukemia and related diseases (see Table 2)‖	177	16	9.0
Lupus erythematosus	69	0	0
Multiple myeloma and macroglobulinemia‖	95	1	1.1
Myasthenia gravis	11	0	0
Polycythemia vera	82	0	0
Rheumatic fever	124	0	0
Tangier disease	3	0	0
Thalassemia	84	2	2.4
U. S. normal population**	1,524	0	0
Total	2,907	54	

* Fifty rheumatoid arthritis, 3 psoriatic, 4 Sjogren's, 8 ankylosing spondylitis, 5 "connective tissue" disease.
† Includes 96 from the United States and 207 from Brazil.
‡ Includes 75 male patients from the New Jersey institution, and 5 females and 4 males from the Maryland institution.
§ These sera were from patients suspected of having infectious or serum hepatitis.
‖ One patient with Au(1) had both chronic lymphocytic leukemia and multiple myeloma and is included in both categories.
** This includes 607 Negro and 917 white persons.

Results

DISTRIBUTION OF AUSTRALIA ANTIGEN IN PATIENTS

Nearly 30% of the Down's syndrome patients had detectable Australia antigen in their serum. This includes (Table 1) 19 of the 75 males from the New Jersey institution and 3 of 5 females and 3 of 4 males from the Maryland institution. This represents the highest frequency found in any patient group so far tested. Four of the 76 control patients had Australia antigen. Of these, three had some stigma of Down's syndrome, that is, prominent epicanthal folds, Brushfield's spots, large interpupillary distance, short middle phalanx of the fifth finger, "mongoloid crease" of the palm, high palatal arch, etc. However, none of these had the typical Down's syndrome chromosome pattern on examination of the peripheral blood. As noted below, mosaicism cannot be ruled out by chromosome analysis of a relatively small number of cells from a single source. The difference in frequency of Australia antigen between the Down's syndrome and the control group is highly significant ($\chi^2 = 11.8$, $P < 0.001$). The protein was not found in any of the normal individuals from the United States. Australia antigen has been found in patients with leukemia, hemophilia, and thalassemia as well as in those with Down's syndrome. Forty-eight sera from patients suspected of having either serum or infectious hepatitis were tested. Of these, five were found to have Australia antigen. The findings in hepatitis will be described in greater detail in a subsequent paper. In addition, a single patient with Fanconi's anemia (hypoplasia of the marrow) and one patient (from Finland) with Hodgkin's disease have the antigen (Table 1).

The distribution of Australia antigen in leukemia and related disease groups is given in Table 2. The frequency is relatively high in acue granulocytic leukemia patients and to this date has not been found in any patient with chronic granulocytic leukemia. Australia antigen was found in a single case of the relatively rare chronic reticuloendotheliosis. None of the polycythemia vera patients had detectable Australia antigen.

We have recently concluded a study on Cebu Island, the Philippines, which included blood specimens from 584 patients with lepromatous leprosy, 377 patients with tuberculoid leprosy, and 764 controls without leprosy (19, 20). There was a significantly higher incidence of Au(1) in the lepromatous leprosy cases as compared with either the tuberculoid leprosy cases or the controls (9.4% lepromatous, 3.4% tuberculoid, 4.8% control). This difference is even more striking if only young men (under 20 years) are considered (26.7% lepromatous, 7.9% tuberculoid, 5.5% controls). As noted above, Au(1) is relatively common (about 5%) in nonhospitalized, apparently normal populations in several Asian and Oceanic countries including the Philippines (8).

TABLE 2. Distribution of Australia Antigen in Leukemia and Related Diseases

Disease	Total No. of Cases	Positive for Australia Antigen	
		no.	%
Acute granulocytic leukemia	38	7	18.4
Chronic granulocytic leukemia	41	0	0
Acute lymphocytic leukemia	58	2	3.5
Chronic lymphocytic leukemia	30	4	13.3
Other*	10	2†	—

* Lymphosarcoma, 3, reticulum cell sarcoma, 3, chronic reticuloendotheliosis, 1, plasma cell leukemia, 1, acute stem cell leukemia, 2.

† Chronic reticuloendotheliosis, 1, acute stem cell leukemia, 1.

AGE

The age distribution by 20-year groups of the trait in the Down's syndrome patients is shown in Table 3, in which this distribution is compared with that in several populations of apparently normal people in whom Au(1) occurs and with that in the sera of leprosy patients and controls from Cebu. The normal and patient groups used in the comparison were selected because of the high frequency of Au(1) and the availability of relatively large numbers of samples. The frequency of Au(1) decreases with age in all of these populations. When each of the populations is divided into younger (0 to 39) and older (40+) age groups, the differences in frequency of Au(1)

TABLE 3. The Distribution of Australia Antigen by Age in Patients with Down's Syndrome and in Three Other Populations

Age Group	Down's Syndrome, New Jersey			Marshall Islands, U.S.T.T.P.I.*			Cebu, Philippines†			Manila, Philippines		
	Cases	Positive		Cases	Positive		Cases	Positive		Cases	Positive	
yr	*no.*	*no.*	%	*no.*	*no.*	%	*no.*	*no.*	%	*no.*	*no.*	%
0–19	34	12	35.3	196	16	9.5	348	30	8.6	22	2	9.1
20–39	30	6	20.0	148	12	8.1	948	62	6.5	132	6	4.5
40–59	11	1	9.1	85	4	4.7	378	16	4.2	36	1	2.8
60+	0	0	—	67	1	1.5	128	3	2.3	6	0	0
Total	75	19		496	33	6.2	1,802	111	6.2	196	9	4.6

* U.S. Trust Territory of the Pacific Islands.

† The Cebu population includes persons with and without leprosy.

TABLE 4. Blood Factors Studied in Australia Antigen-positive (Au(1)) and -negative (Au(0)) Sera*

Study†	Positive, Au(1)		Negative, Au(1)		Total Au(1) + Au(1)	
	No. Studied	Mean Value	No. Studied	Mean Value	No. Studied	Mean Value
Hemoglobin, *g/100 ml*	19	15.2	26	15.5	45	15.4
RBC count, *million/mm³*	19	4.48	26	4.65	45	4.58
Hematocrit reading, %	19	43.0	26	43.8	45	43.5
MCV, μ^3	19	96.2	26	94.5	45	95.2
MCH, $\mu\mu g$	19	33.9	26	33.5	45	33.7
MCHC, %	19	35.3	26	35.5	45	35.4
WBC count, */mm³*	19	7,153	26	7,725	45	7,483
Polymorphonuclear cells, %	19	56.3	23	54.0	42	55.0
Stab cells, %	19	3.9	23	4.4	42	4.2
Lymphocytes, %	19	34.1	23	36.9	42	35.7
Monocytes, %	19	2.3	23	2.3	42	2.3
Eosinophils, %	19	2.8	23	2.0	42	2.4
Basophils, %	19	0.4	23	0.5	42	0.5
WBC G-6-PD, *μmoles TPN reduced/10⁶ WBC/min*	9	17.1	9	15.6	18	16.4
WBC G-1-PUT, *μmoles UDPG/hr/10⁹ WBC*	9	29.1	10	26.3	19	27.6
Platelets, */mm³*	19	191,158	26	213,461	45	204,044
Uric acid, *mg/100 ml*	18	5.2	26	5.3	44	5.2
Vitamin B_{12}, *μμg/ml*	17	698.6	23	347.1	40	496.5

* None of the differences were significant with the exception of vitamin B_{12}, where $t = 2.926$ and $P < 0.01$.

† MCV = mean corpuscular volume, MCH = mean corpuscular hemoglobin, MCHC = mean corpuscular hemoglobin concentration, G-6-PD = glucose-6-phosphate dehydrogenase, TPN = triphosphopyridine nucleotide, G-1-PUT = galactose-1-phosphate uridyl transferase, UDPG = uridine diphosphate glucose.

are statistically significant for the Cebu population ($0.02 > P > 0.01$) and nearly so for the Marshall Islands ($0.1 > P > 0.05$) but not for the others.

CHROMOSOME STUDIES

Seventeen patients of the 75 Down's syndrome patients included in this study were selected for cytogenetic examination. Criteria used in the selection were low maternal age at birth or other indication of possible familial transmission of the syndrome, relatively high intelligence, and, in one case (385H), the relatively advanced age of the patient (60 years). One case was chosen at random. The selection was made without prior knowledge of the antigen phenotype. A minimum number of 10 metaphase chromosome counts and 1 karyotype analysis was made on each individual (leukocyte cultures); in those cases in which the modal chromosome number was not 47, additional counts and analyses were made. Fourteen of the 17 patients were found to be trisomic for chromosome 21, the classical finding in Down's syndrome (21). Of the remaining 3, 1 (389H) was a mosaic of normal and trisomy-21 cells, the second (384H) was the carrier of a 13–15/21 translocation and effectively trisomic for 21, and the third (385) had only chromosomes that were entirely normal in number and morphology in the single cell type studied. More extensive cytogenetic analysis might have revealed additional mosaicism in which cells of subordinate clones were present in low frequency, but because it is possible to study the chromosomes of relatively few cell types from a limited number of sites, it is impossible to rule out mosaicism completely in any human. None of these three had detectable Au(1).

Of the 14 typical Down's syndrome patients (that is, with 47 chromosomes), 3 had Australia antigen. Chromosome studies were done on 4 of the control patients. None of these had the typical Down's syndrome trisomy pattern.

The blood factors studies are shown in Table 4. With the exception of the vitamin B_{12} values, there was no significant difference between the patients with and those without Australia antigen. The mean of the B_{12} values for the positive group (Australia antigen present) was 698.6 and for the negative group (Australia antigen absent), 341.3. This gives a t value of 2.926 and $0.01 > P > 0.001$ and indicates that it is highly unlikely that this difference between the means is due to chance. The levels of white blood cell galactose-1-phosphate uridyl transferase and glucose-6-phosphate dehydrogenase are higher on the average in the Au(1) than the Au(0) patients, but this difference is not significant. They are considerably higher than the values given for normal white blood cells in the table published by Mellman and colleagues (13). The hematocrit reading, mean corpuscular volume, mean corpuscular hemoglobin, and mean corpuscular hemoglobin concentration for all the Down's syndrome patients combined are slightly higher than those for normals, confirming the findings of Naiman, Oski, and Mellman (22). The average of the uric acid levels for all patients combined is about the same as that found for male Down's syndrome patients by Mertz, Fuller, and Concon (23) (who used a different technique for measuring uric acid). They noted that young mongoloids had higher levels than did the control patients.

Discussion

There is a significantly higher frequency of Au(1) in the Down's syndrome patients than in the controls. From this, we may conclude that the high frequency is associated with Down's syndrome and that it is not due to institutionalization alone. The frequency appears to decrease with age in the Down's syndrome patients as it does in other populations. There is a higher average vitamin B_{12} level in the Au(1) individuals compared with the Au(0), but this finding should be confirmed before further comment is warranted. All the Down's syndrome patients with Au(1) and whose chromosome complement is known have the usual 21 trisomy.

The present state of knowledge concerning Australia antigen has been summarized in the introduction. The most striking feature of our data is the discovery of Australia antigen in high frequency in patients with Down's syndrome, leukemia (particularly acute granulocytic leukemia and chronic lymphocytic leukemia), and viral hepatitis. These diseases have been linked for other reasons: Patients with Down's syndrome have an increased risk of developing leukemia, particularly acute myelogenous leukemia (11); an association has been found between epidemics of hepatitis and the birth 9 months later of children with Down's syndrome (24); furthermore, it has been reported that sera of patients convalescing from hepatitis cause chromosome aberrations in tissue cultures (25).

Most of the disease associations could be explained by the association of Au(1) with a virus, as suggested in our previous publications (2, 6, 8). The discovery of the frequent occurrence of Au(1) in patients with virus hepatitis raises the possibility that the agent present in some cases of this disease may be Australia antigen or be responsible for its presence. The presence of Australia antigen in the thalassemia and hemophilia patients could be due to virus introduced by transfusions. This, however, could not be the only explanation, since many transfused patients neither have the antigen nor an antibody against it. To our knowledge, none of the Down's patients with Australia antigen have received transfusions, and its presence in these patients may be the result of very early or maternal infection. Associations between Down's syndrome and thyroiditis (26) and rubella (27) have also been suggested, but these have not been studied in respect to Au(1).

As noted in the introduction, a striking family clustering of Au(1) has been found in Filipino and other populations (8). This could be explained on a simple infection hypothesis if the attack rate were higher in children than it is in adults. (See Table 3 of this paper and Table 5 of reference 8.) However, the segregation in the families is consistent with simple autosomal recessive inheritance. If Au(1) is inherited, as these results suggest, this would require an explanation consistent with the infection theory. It is possible that the trait is inherited in some groups such as the Filipinos but not in others. Or, the presence of the appropriate genes may confer an added susceptibility to infection with the virus. In areas where infection is very common, the trait could follow a pattern of mendelian segregation as it appeared to do in the Philippine studies.

Studies on the biochemical and immunologic nature of Au(1) are in progress, as are additional studies on the association with hepatitis. When the results of these are available, a more explicit interpretation may be possible.

Summary

Australia antigen (Au(1)) found in patients with leukemia (particularly those with acute granulocytic leukemia) has now been found in high frequency in patients with Down's syndrome and virus hepatitis. Down's syndrome patients with Australia antigen have, on the average, a higher level of vitamin B_{12} than do the patients without this trait. Several hypotheses are discussed in an attempt to explain the associations.

Acknowledgments

This study was made possible as a result of the enthusiastic support and encouragement of Mr. F. W. Russell, Superintendent, and Dr. J. D. Nelson, Medical Director, of the State Colony at New Lisbon, New Lisbon, N. J. We are also indebted to the residents of the colony who participated in this study and to their parents, who have shown a continuing interest in the progress of our work.

Many of the valuable sera of patients used in this work were presented to us by physicians from the Philadelphia area and elsewhere to whom we are grateful. These include Drs. Erlandson, Erslev, Freireich, Frost, Joseph, Lawrence, Lee, McElfresh, Naiman, Osgood, Ultman, Whereat, Winchell, Woldow, and Wolman. We are particularly indebted to Dr. James E. Prier, Director, Division of Laboratories, Pennsylvania Department of Health, for supplying serum specimens from patients suspected of having viral hepatitis.

Summario in Interlingua

Antigeno Australia, trovate in patientes con leucemia (particularmente in casos de acute leucemia granulocytic) ha essite incontrate con un alte frequentia in patientes con syndrome de Down e hepatitis virusal. Patientes con syndrome de Down e antigeno Australia ha, al media, un plus alte nivello de vitamina B_{12} que patientes sin iste tracto. Es commentate plure hypotheses presentate pro explicar le mentionate associationes.

References

1. Blumberg, B. S.: Polymorphism of serum proteins and the development of iso-precipitins in transfused patients. *Bull. N. Y. Acad. Med.* 40: 377, 1964.
2. Blumberg, B. S., Alter, H. J., Visnich, S.: A "new" antigen in leukemia sera. *JAMA* 191: 541, 1965.
3. Allison, A. C., Blumberg, B. S.: An isoprecipitin reaction distinguishing human serum-protein types. *Lancet* 1: 634, 1961.
4. Blumberg, B. S., Allison, A. C.: Studies on the isoprecipitin-determined human serum polymorphism, in *Proceedings of the International Congress on Human Genetics, 2nd (Rome).* Istituto Gregorio Mendel, Rome, 1961, pp. 733–736.
5. Blumberg, B. S., Dray, S., Robinson, J. C.: Antigen polymorphism of a low-density beta-lipoprotein. Allotypy in human serum. *Nature (London)* 194: 656, 1962.
6. Alter, H. J., Blumberg, B. S.: Studies on a "new" human isoprecipitin system (Australia antigen). *Blood* 27: 297, 1966.

7. Melartin, L., Blumberg, B. S.: Production of antibody against "Australia antigen" in rabbits. *Nature (London)* 210: 1340, 1966.
8. Blumberg, B. S., Melartin, L., Guinto, R. A., Werner, B.: Family studies of a "new" human serum isoantigen system (Australia antigen). *Amer. J. Hum. Genet.* 18: 594, 1966.
9. Blumberg, B. S., Alter, H. J.: Precipitating antibodies against a serum protein ("Australia antigen") in the serum of transfused hemophilia patients (abstract). *J. Clin. Invest.* 44: 1029, 1965.
10. Blumberg, B. S.: An inherited serum isoantigen in leukemia and Down's syndrome (abstract). *J. Clin. Invest.* 45: 988, 1966.
11. Miller, R. W.: Radiation, chromosomes and viruses in the etiology of leukemia. Evidence from epidemiologic research. *New Eng. J. Med.* 271: 30, 1964.
12. Liddle, L., Seegmiller, L. E., Laster, L.: The enzymatic spectrophotomotic method for determination of uric acid. *J. Lab. Clin. Med.* 54: 903, 1959.
13. Mellman, W. J., Oski, F. A., Tedesco, T. A., Maciero-Coelho, A., Harris, H.: Leucocyte enzymes in Down's syndrome. *Lancet* 2: 674, 1964.
14. Hutner, S. H., Bach, M. K., Ross, G. J. M.: A familial form of aplastic anemia supposed to be hereditary; due to a recessive gene. *J. Protozool.* 3: 101, 1956.
15. Blumberg, B. S., Riddell, N. M.: Inherited antigenic difference in human serum beta-lipoproteins. A second antiserum. *J. Clin. Invest.* 42: 867, 1963.
16. Preer, J. R., Preer, L. B.: Gel diffusion on the antigens of isolated components of paramecium. *J. Protozool.* 6: 88, 1959.
17. Hungerford, D. A.: Leucocytes cultured from small inocula of whole blood and the preparation of metaphase chromosomes by treatment with hypotonic KCl. *Stain Techn.* 40: 333, 1965.
18. Snedcor, G. W.: *Statistical Methods.* Iowa State University Press, Ames, Iowa, 1962, p. 534.
19. Blumberg, B. S., Melartin, L.: Conjectures on inherited susceptibility to lepromatous leprosy. *Int. J. Leprosy* 34: 60, 1966.
20. Blumberg, B. S., Melartin, L., Lechat, M., Guinto, R. A.: Studies on leprosy and Australia antigen. *JAMA,* March 1967.
21. LeJeune, J., Turpin, R., Gautier, M.: Le mongolisme, premier exemple d'aberration autosomique humaine. *Ann. Genet. (Paris)* 1: 41, 1959.
22. Naiman, J. L., Oski, F. A., Mellman, W. J.: Phosphokinase activity of erythrocytes in mongolism. *Lancet* 1: 821, 1965.
23. Mertz, E. T., Fuller, R. W., Concon, J. M.: Serum uric acid in young mongoloids. *Science* 141: 535, 1963.
24. Stoller, A., Collman, R. D.: Incidence of infectious hepatitis followed by Down's syndrome nine months later. *Lancet* 2: 1221, 1965.
25. Mella, B., Lang, D. J.: Leucocyte mitosis: suppression in vitro associated with acute infectious hepatitis. *Science* 155: 80, 1967.
26. Failkow, P. J., Uchida, I., Hecht, F., Motulsky, A. G.: Increased frequency of thyroid auto-antibodies in mothers of patients with Down's syndrome. *Lancet* 2: 868, 1965.
27. Robinson, A., Puck, T. T.: Sex chromatin in newborns; presumptive evidence for external factors in human nondisjunctions. *Science* 148: 83, 1965.

Reprinted from ANNALS OF INTERNAL MEDICINE, Vol. 70, No. 1, January 1969
Printed in U. S. A.

Australia Antigen and Acute Viral Hepatitis

W. THOMAS LONDON, M.D., F.A.C.P., ALTON I. SUTNICK, M.D., F.A.C.P., and BARUCH S. BLUMBERG, M.D., F.A.C.P.
Philadelphia, Pennsylvania

SUMMARY **Australia antigen (Au(1)) was found in the sera of 20% of 125 patients with acute viral hepatitis. It was not present in the sera of 138 patients with other diseases affecting the liver (Laennec's cirrhosis, infectious mononucleosis, hepatoma, and so on). These data indicate that the antigen is selectively associated with viral hepatitis and is not a nonspecific manifestation of liver damage.**

Previously we have reported the association of Au(1) with leukemia, lepromatous leprosy, and Down's syndrome, and in 1 to 20% of some tropical populations. In these groups the antigen occurs more frequently in young males; once present, it tends to persist for long periods (years). In the acute hepatitis patients there was no significant difference in distribution between the sexes; the antigen was not found in any patient less than 19 years old, and with two exceptions it was present in the sera for short intervals (days and weeks). Host factors related to immunologic competence may be responsible for these differences.

WE HAVE PREVIOUSLY described the characteristics of an unusual antigen, Australia antigen (Au(1)) (1–3). When present in serum the antigen forms a precipitin with specific human and rabbit antisera (4). With the use of the micro-Ouchterlony technique the sera from large numbers of persons with and without disease have been tested. The antigen occurs frequently in tropical populations (1 to 20%) but rarely in normal North Americans and Europeans (2). In the United States we have found increased frequencies of the antigen in association with certain diseases, that is, acute and chronic lymphocytic leukemias, acute granulocytic leukemia, Down's syndrome, and viral hepatitis (3).

In the same U. S. study those patients with Down's syndrome and Australia antigen had significantly higher levels (mean and median) of serum glutamic-pyruvic transaminase (SGPT) than did age- and sex-matched patients with Down's syndrome without the antigen. Liver biopsies from eight such patients with Au(1) showed chronic hepatitis (5).

The antigen has been isolated from the sera of five patients by ultracentrifugation on a sucrose gradient. Electron microscopy showed that these isolates contained particles 20 mμ in diameter, and these agglutinated upon the addition of specific antisera against Au(1) (6).

We have considered two alternative hypotheses to explain the disease associations and physical characteristics of Australia antigen: [1] Australia antigen is released into the circulation as a result of liver injury and its presence in serum is a nonspecific reflection of hepatic damage; and [2] Australia antigen is intimately associated with a virus causing hepatitis, and its presence in serum indicates the presence of that virus. Results compatible with the second hypothesis are reported here.

MATERIALS AND METHODS

Serum was collected from 125 patients with viral hepatitis and 2 or more samples were ob-

From the Institute for Cancer Research, Fox Chase, Philadelphia, Pa.

This study was supported in part by research grants CA-06551 and CA-08069, National Cancer Institute, National Institutes of Health, Bethesda, Md.; and by an appropriation from the Commonwealth of Pennsylvania.

Requests for reprints should be addressed to W. Thomas London, M.D., The Institute for Cancer Research, 7701 Burholme Ave., Fox Chase, Philadelphia, Pa. 19111 tained from 56 of these patients. Seventy-five sera were provided by Drs. Prier and Strachen of the Pennsylvania State Department of Health Laboratories. The remainder were obtained from patients at affiliated hospitals of the University of Pennsylvania; the Jeanes Hospital, Fox Chase, Philadelphia; and Bellevue Hospital, N. Y. In the hospital group the diagnosis of hepatitis was established by serum glutamic-pyruvic transaminase (SGPT) elevation, or liver biopsy, or both. With respect to the sera obtained by the State Department of Health, we attempted to substantiate the diagnosis by contacting the patients' physicians 6 months to a year after the acute illness. The patients included here are those in whom the diagnosis was confirmed. Of the 125 patients, 41 had received a prior transfusion or were considered by their physicians to have had serum hepatitis; we have classified these as posttransfusion hepatitis and all the remaining cases as infectious hepatitis.

Sera were obtained from 57 patients with Laennec's cirrhosis at the affiliated hospitals of the University of Pennsylvania and at the Veterans Administration Hospital, Brooklyn, N. Y. Sera from 55 patients with heterophil-positive infectious mononucleosis (IM) were kindly provided by Dr. Esko Kaarsalo of Turku, Finland. An additional 12 sera from patients with IM were collected at the Student Health Clinic of the University of Pennsylvania. Sera from 14 patients with other diseases involving the liver were tested. These included three patients with granulomatous hepatitis, two with jaundice secondary to choledocholithiasis, one with carcinoma of the common bile duct, five with hepatoma, two with acute cholecystitis, and one with Wilson's disease. All of these were patients at hospitals in Philadelphia; the diagnoses were made on aspiration liver biopsy or at laparotomy with biopsy.

In order to obtain an estimate of the frequency of the antigen in the normal population of the United States, 2,120 sera from presumed normal persons were tested. Of these, 563 (323 Negroes and 240 whites, ages 15 to 74 years) were from a sample of the population of Evans County, Ga. (7); 215 were white employees of the National Institutes of Health, Bethesda, Md.; 180 were negro employees of the same institution; 80 were white residents of Tangier Island, Md., and 187 were negro residents of Sapelo Island, Ga. Others included were 518 blood donors at Jeanes Hospital, 114 members of the staff of Pennhurst State School and Hospital, Spring City, Pa., 42 sets of twins from Evans County, Ga., 43 members of the staff of the Institute for Cancer Research, Philadelphia, Pa. and their families, and 105 specimens from various other normal persons.

A broad sample of the patient population in Philadelphia was obtained by testing 974 consecutive sera (486 individuals) collected at the clinical laboratory of the Jeanes Hospital and 1,064 consecutive sera (556 individuals) collected at the Pepper Laboratory of the Hospital of the University of Pennsylvania.

The sources of patients with Down's syndrome (Tables 2, 3, and 4) have been reported in detail previously (5).

All sera were tested for Australia antigen with specific rabbit antiserum (4) by the micro-Ouchterlony technique of Blumberg and Riddell (8). Semiquantitative estimates of the titer of Au(1) were performed by double diffusion in capillary tubes using the method of Preer and Preer (9). Contingency tables and chi-square determinations were done as described by Snedecor (10).

TABLE 1. Frequency of Australia Antigen (Au(1)) in the United States in Patients with Viral Hepatitis, and Other Liver Diseases and in Controls

	Normal	Au(1)	Percent Au(1)
		no.	
U. S. normal population	2,120	2	0.09
U. S. hospital inpatients	1,024	1	0.19
Infectious mononucleosis	67	0	0
Laennec's cirrhosis	57	0	0
Other liver diseases	14	0	0
Viral hepatitis	125	25	20.0
Infectious	84	11	13.1
Posttransfusion	41	14	34.1

RESULTS

Twenty percent of 125 patients with hepatitis had Au(1) on one or more occasions (Table 1), whereas Australia antigen occurs in less than 0.1% of the normal population. (Of the two presumed normal persons with Au(1), one had had a previous transfusion and the other had been exposed to hepatitis.) The only hospitalized patient with Au(1) had acute posttransfusion hepatitis at the time her blood was drawn. The antigen was not found in 138 patients with liver diseases other than viral hepatitis. Of particular significance is its absence in the

TABLE 2. Frequency of Australia Antigen (Au(1)) by Sex in Patients with Viral Hepatitis and Institutionalized Patients with Down's Syndrome

	Hepatitis			Down's Syndrome*		
	Normal	Au(1)	Percent Au(1)	Normal	Au(1)	Percent Au(1)
	no.			*no.*		
Male	63	14	22.2	118	41	34.7
Female	62	11	18.0	92	16	18.4
Total	125	25	20.0	210	57	27.1

* From two Institutions housing both males and females. (χ^2 = 10.97, P = <0.025).

infectious mononucleosis group, since these patients have a type of hepatitis thought to be caused by a virus (11).

The hepatitis patients are divided into two categories: [1] posttransfusion hepatitis—those who had a prior transfusion or were considered by their physicians to have serum hepatitis; and [2] infectious hepatitis—the remaining hepatitis cases. The difference in the frequencies of Au(1) in posttransfusion hepatitis (34.1%) and infectious hepatitis (13.1%) is statistically significant ($\chi^2 = 6.46$, $P < 0.025$).

Other populations with an increased prevalence of Au(1) (Down's syndrome, lepromatous leprosy, and "normal" persons from the Philippines and Micronesia) have higher frequencies of the antigen in males than in females and higher frequencies in younger than in older persons (2, 3). Patients with Down's syndrome who are institutionalized are the most thoroughly studied of such populations in the United States (5) and have been used here for comparison with the hepatitis patients, but patients with lepromatous leprosy and "normal" populations from the Philippines and Micronesia show approximately the same distributions of Au(1) (2, 3). In the hepatitis patients the frequency of Au(1) is essentially the same in both sexes: 22% in males and 18% in females. This distribution of Au(1) by sex is significantly different from that found in institutionalized patients with Down's syndrome where the prevalence in males is almost twice that in females ($\chi^2 = 10.97$, $P = < 0.025$, Table 2). Table 3 shows the distribution of Au(1) by age in the hepatitis patients and institutionalized patients with Down's syndrome. Australia antigen has not been found in any hepatitis patient younger than 19 years old, whereas among the patients with Down's syndrome the highest frequency of the antigen (43.3%) is in the 0 to 19 year age group. These distributions differ significantly ($\chi^2 = 20.77$, $P < 0.005$).

TABLE 3. Frequency of Australia Antigen (Au(1)) by Age in Hepatitis and Institutionalized Patients with Down's Syndrome

Age	Hepatitis			Down's Syndrome*		
	Normal	Au(1)	Percent Au(1)	Normal	Au(1)	Percent Au(1)
yr.	*no.*			*no.*		
0–19	22	1	4.4	67	29	43.3
20–30	51	14	27.5	109	32	29.4
40–59	28	6	21.4	34	10	29.4
60+	9	2	22.2	12	0	0

* (χ^2 =20.77, P = <0.005)

Another characteristic of Australia antigen in other diseases and population groups is its persistence (2, 5). In hepatitis, however, the antigen is transient. In Table 4 the duration of Au(1) in patients with hepatitis and patients with Down's syndrome is compared. Of 15 patients with hepatitis and Au(1) on first testing, the antigen disappeared from 13 after 2 to 3 weeks. In one patient Au(1) was present for only a single day, but another hepatitis patient had Au(1) for 7 months (his SGPT's remained over 100 during that time). In

TABLE 4. Comparison of the Chronicity of Australia Antigen (Au(1)) in Hepatitis and Down's Syndrome

		Hepatitis, Time 2*		Down's Syndrome, Time 2†	
		Au(1)	Au(0)	Au(1)	Au(0)
Time 1‡	Au(1)	2	13	10	2
	Au(0)	1	40	1	29

* Time 2 for the hepatitis patients is 2 to 3 weeks after the initial testing (Time 1).
† Time 2 for the Down's syndrome patients is 1 year after the initial testing (Time 1); (χ^2 =31.50, P <.001).
‡ Initial testing.

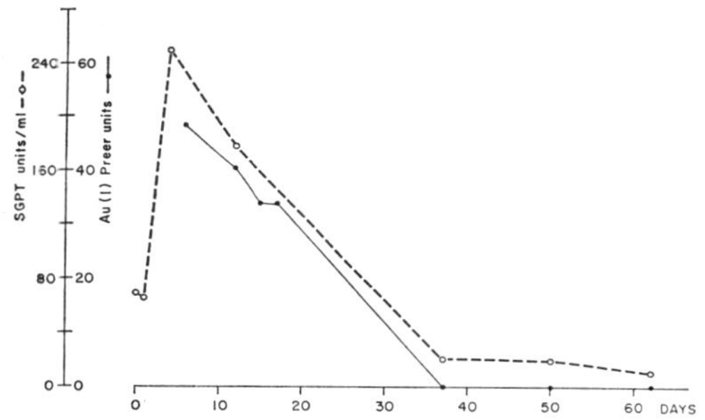

FIGURE 1. Australia antigen (*Au(1)*) and serum glutamic-pyruvic transaminase (*SGPT*) levels of a patient with posttransfusion hepatitis. Au(1) disappeared as SGPT fell to normal.

contrast, of 12 patients with Down's syndrome and Au(1) on first testing, 10 still had the antigen when tested a year later. These differences in persistence are significant ($\chi^2 = 31.59$, $P = < 0.001$). Figure 1 gives the variation of Au(1) titer, estimated by the Preer technique, and the SGPT levels in a patient with acute posttransfusion hepatitis. In this patient the Australia antigen titer parallels the fall in SGPT. Such a direct parallel between SGPT and Au(1) titer has not been found in patients with Down's syndrome.

Discussion

The data indicate that Australia antigen is associated with viral hepatitis, confirming our earlier studies (3, 12). Furthermore, Au(1) is not associated with other diseases affecting the liver; that is, the presence of the antigen in serum is not a nonspecific manifestation of liver damage.

Okochi and Murakami (13) have confirmed the association of Australia antigen with hepatitis in Japan. They found 44 of 358 patients with hepatitis to have Australia antigen. They also screened all of the blood and blood donors passing through the blood transfusion service of the Tokyo University Hospital and found 141 of 11,820 blood donors to have Australia antigen. SGPT levels were slightly but significantly higher in sera containing Au(1). They further demonstrated that persons receiving blood containing Australia antigen could either develop antibody to Au(1) or the antigen itself. Those in whom the antigen appeared had posttransfusion hepatitis.

We have pointed to a number of characteristics of Australia antigen in patients with hepatitis different from those in other groups where Au(1) is present in high frequency, such as Down's syndrome, lepromatous leprosy, and "normal" persons living in tropical areas. Persistence, higher frequency in males than females, and higher frequencies in the first two decades of life have been consistent findings in these groups. In contrast, transience, lack of sex predilection, and absence of Au(1) in patients under the age of 19 are features of the hepatitis group. The biologic meaning of these differences is not clear, but they may reflect the immunologic status of the host. That is, the patients who develop viral hepatitis do not have an underlying abnormality and are able, in a short time, to clear their blood of antigen and to effect complete recovery. The patients who retain the antigen chronically have an immunologic defect and therefore do not rid themselves of antigen nor do they completely eradicate the inflammatory process (14, 15). The patients with Down's syndrome illustrate this hypothesis.

In large state institutions for the mentally retarded, viral hepatitis is a very common disease; probably every patient is exposed to the virus (16). Yet Australia antigen occurs much more frequently in the patients with Down's syndrome than in the other patients in the same institutions. Furthermore, patients with Down's syndrome and Au(1) have elevated SGPT's and have evidence of hepatitis on liver biopsy (5).

In this study we found a significantly higher frequency of Au(1) in posttransfusion hepatitis than in infectious hepatitis. Okochi and Murakami (13) did not find this difference. Further studies are necessary before we can determine whether the prevalence of Au(1) is greater in posttransfusion hepatitis.

Isolated Australia antigen is a particle 20 mμ in diameter with an appearance compatible with that of a virus (6). Currently, we are inoculating animals and tissue cultures in an effort to determine whether it has other virus characteristics.

Addendum

Since the submission of this manuscript, an additional confirmation of the association of hepatitis with Australia antigen has appeared. The SH antigen described by Prince (17, 18) is Australia antigen.

Acknowledgments

We wish to thank Dr. J. Prier and Dr. S. Strachen (Pennsylvania State Department of Health Laboratories) for sera from hepatitis patients, Dr. R. Donato (Jeanes Hospital), Dr. D. Arvan (Hospital of the University of Pennsylvania), Dr. E. Kaarsalo, and Dr. C. Hames for sera and diagnostic information.

References

1. Blumberg, B. S., Alter, H. J., Visnich, S.: A "new" antigen in leukemia sera. *JAMA* 191: 541, 1965.
2. Blumberg, B. S., Melartin, L., Guint, R. A., Werner, B.: Family studies of a human serum isoantigen system (Australia antigen). *Amer. J. Hum. Genet.* 18: 594, 1966.
3. Blumberg, B. S., Gerstley, B. J. S., Hungerford, D. A., London, W. T., Sutnick, A. I.: A serum antigen (Australia antigen) in Down's syndrome, leukemia, and hepatitis. *Ann. Intern. Med.* 66: 924, 1967.
4. Melartin, L., Blumberg, B. S.: Production of antibody against "Australia antigen" in rabbits. *Nature (London)* 210: 1340, 1966.
5. Sutnick, A. I., London, W. T., Gerstley, B. J. S., Cronlund, M. M., Blumberg, B. S.: Anicteric hepatitis associated with Australia antigen. Occurrence in patients with Down's syndrome. *JAMA* 205: 108, 1968.
6. Bayer, M. E., Blumberg, B. S., Werner, B.: Particles associated with Australia antigen in the sera of patients with leukemia, Down's syndrome and hepatitis. *Nature (London)* 215: 1057, 1968.
7. Cooper, A. J., Blumberg, B. S., Workman, P. L., McDonough, J. R.: Biochemical polymorphic traits in the U. S. white and negro population. *Amer. J. Hum. Genet.* 15: 420, 1963.
8. Blumberg, B. S., Riddell, N. N.: Inherited antigenic differences in human serum beta-lipoproteins. A second antiserum. *J. Clin. Invest.* 42: 867, 1963.
9. Preer, J. R., Preer, L. B.: Gel diffusion studies on the antigens of isolated components of paramecium. *J. Protozool.* 6: 88, 1959.
10. Snedecor, G. W.: *Statistical Methods,* 5th ed., Iowa State University Press, Ames, Iowa, 1956, p. 219.
11. Hoagland, R. J., McCluskey, R. T.: Hepatitis in mononucleosis. *Ann. Intern. Med.* 43: 1019, 1956.
12. Sutnick, A. I., London, W. T., Blumberg, B. S.: Australia antigen, Down's syndrome and hepatitis. *J. Clin. Invest.* 46: 1122, 1967.
13. Okochi, K., Murakami, S.: Observations on Australian antigen in Japanese. *Vox Sanguinis.* In press, 1968.
14. Miller, D. G.: Hodgkin's disease, lymphosarcoma, and chronic lymphocytic leukemia, in *Immunological Diseases,* edited by Samter, M., Little, Brown and Co., Boston, 1965, p. 372.
15. Waldorf, D. S., Sheagren, J. N., Trautman, J. R., Block, J. B.: Impaired delayed hypersensitivity in patients with lepromatous leprosy. *Lancet* 2: 773, 1966.
16. Krugman, S., Giles, J. P., Hammond, J.: Infectious hepatitis. Evidence for two distinctive clinical, epidemiological, and immunological types of infection. *JAMA* 200: 365, 1967.
17. Prince, A. M.: An antigen detected in the blood during the incubation period of serum hepatitis. *Proc. Nat. Acad. Sci.* 60: 814, 1968.
18. Prince, A. M.: Relation of Australia and SH antigens. *Lancet* 2: 462, 1968.

Reprinted From The Journal of The American Medical Association
September 2, 1968, Vol. 205, pp. 670-674

Anicteric Hepatitis Associated With Australia Antigen

Occurrence in Patients With Down's Syndrome

Alton I. Sutnick, MD; W. Thomas London, MD; Betty Jane S. Gerstley, MD; Malcolm M. Cronlund, MS; and Baruch S. Blumberg, MD, PhD

The frequency of Australia antigen is high in patients with Down's syndrome in large institutions (27.7%) and is low in patients without this syndrome in the same institutions (3.2%). It is rare in patients with Down's syndrome in small institutions (1.5%) and absent in newborn patients and those who are not institutionalized. Serum glutamic pyruvic transaminase levels were studied in age- and sex-matched groups of 16 patients with Down's syndrome and Australia antigen [phenotype Au (1)], 16 with Down's syndrome but without Australia antigen [Au(0)] and 16 Au(0) mental defectives without Down's syndrome. The findings indicated active liver cell breakdown in Au(1) patients, and were confirmed in an extensive study including 581 individuals. The findings are best explained by the operation of an environmental factor, probably infectious, present in the large institutions, and a host susceptibility factor present in association with Down's syndrome.

In recent reports[1-5] we have described the characteristics of an unusual antigen present in the blood of certain people. The antigen was originally detected by micro-Ouchterlony technique with a precipitating antibody found in the blood of a transfused hemophiliac (Fig 1). It was termed "Australia antigen," since it was first described as appearing in the serum of an Australian aborigine. It is extremely rare in normal Americans (two in 2,412) but is relatively common in patients with acute myelogenous leukemia (18.4%), chronic lymphocytic leukemia (13.3%), Down's syndrome (9.4 to 29.8%),[4,6] and nodular (lepromatous) leprosy (9.4%).[5] We have been searching for features in common among this apparently heterogeneous group of diseases. The discovery of a strong statistical association between Australia antigen and viral hepatitis[4,7] and the identification of Australia antigen as a particle[8] suggest that hepatitis might be the common factor leading to the high frequency of Australia antigen in persons with these diseases.

In this communication we describe our studies of serum glutamic pyruvic transaminase (SGPT) levels in patients with Down's syndrome, with and without Australia antigen, in institutions of various sizes and in patients living at home.

From the Institute for Cancer Research, Fox Chase, Philadelphia.
Reprint requests to 7701 Burholme Ave, Fox Chase, Philadelphia 19111 (Dr. Sutnick).

Methods

This study included 377 patients with Down's syndrome; 34 were not institutionalized, and the rest were from nine separate institutions. Four were large state institutions for mental defectives, and five were small, private institutions. Institution 1 (State Colony at New Lisbon, NJ) has about 1,200 male residents, all of whom are 6 years of age or older. All of the 144 patients with Down's syndrome in the institution were tested. The diagnosis of Down's syndrome had been made clinically prior to admission. The patients with this syndrome are not housed separately, but live with other residents in several cottages. All patients eat in a common dining hall. Institution 2 (Pennhurst State School and Hospital, Spring City, Pa) has a total population of about 3,000 mentally retarded patients. About 250 have Down's syndrome, of whom 113 have been tested. The clinical diagnosis is made at the admission examination. The patients with this syndrome are not segregated from other mentally retarded residents, but male patients and female patients live in different residences on the 700-acre tract of land. There are separate kitchens and dining halls, and male and female patients do not generally intermingle. The other two state institutions were of similar size. Each of the five private institutions had total patient populations of less than 50. The diagnosis of Down's syndrome was confirmed with chromosome preparations in 82 of the 375 patients. In every preparation, the prior clinical diagnosis was proven correct.

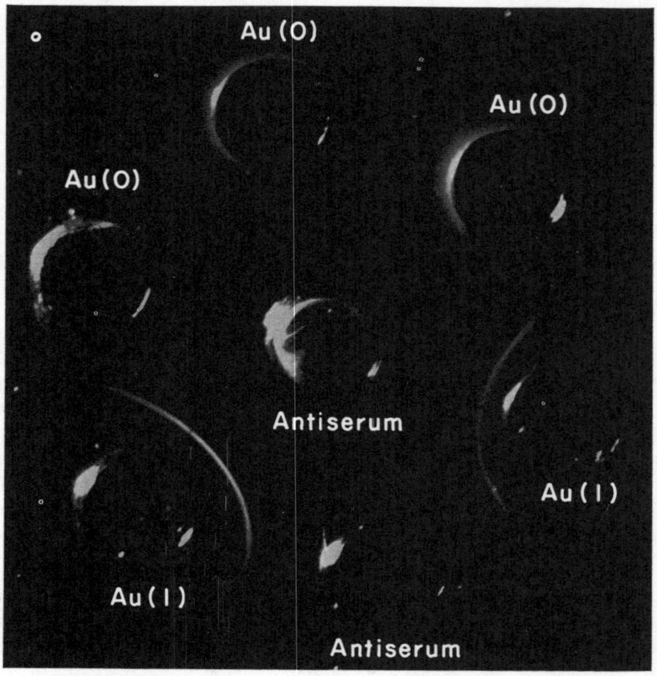

1. Precipitin line indicating Australia antigen in two of lower wells [Au(1)] detected by antiserum in central and bottom well of double diffusion Ouchterlony plate. Upper three wells contain control sera [Au(0)].

From institutions 1 and 2, 188 mental defectives without Down's syndrome were tested. This included 76 patients from institution 1 who were classified as "epileptics" and 112 from institution 2 with various classifications of mental retardation. The testing also included 114 staff personnel from institution 2. Sera from 14 newborn patients with Down's syndrome, confirmed by cytogenetic studies, were tested. In addition, 20 patients with Down's syndrome who were not institutionalized were tested. Fourteen of these attended the day school of the Philadelphia Association for Retarded Children (Table 1).

Forty-eight male patients from institution 1 were selected for a controlled study of SGPT levels in relation to Australia antigen. This included 16 patients with Down's syndrome and Australia antigen [phenotype Au(1)], 16 patients with Down's syndrome without the antigen [phenotype AU(0)], and 16 patients without Down's syndrome. The three groups were matched for age and sex. Following this controlled study, a more extensive investigation was undertaken to include 101 patients with Down's syndrome in institution 1 [38 were Au(1) and 63 Au(0)], 113 patients with Down's syndrome from institution 2 [34 Au(1) and 79 Au(0)], and 112 controls from institution 2 [all Au(0)]. Within each of the institutions, the Au(1), the Au(0), and, in the case of Institution 2, the patients without Down's syndrome, were comparable in respect to sex and age.

Sera were tested for Australia antigen with both human and rabbit antisera[2] by the micro-Ouchterlony technique of Blumberg and Riddell.[9] Quantitative studies were done with the methods of double diffusion in agar gel tubes described by Preer and Preer.[10] The SGPT level was determined by the method of Wroblewski and Lą Due[11] at 89.6 F (32 C) with a recording spectrophotometer. Sera were stored at −13 F (−25 C) until the SGPT measurements were completed. All SGPT determinations were completed within seven days of drawing the blood. The enzyme is known to be stable at this temperature for this period of time.[12]

Liver biopsy was performed for diagnostic purposes on four patients with Down's syndrome, Australia antigen, and an elevated SGPT level. These were open biopsies, done with the patients under general anesthesia, because adequate cooperation could not be obtained for a needle biopsy. Because of the element of risk involved, only those patients in whom the procedure was of clinical diagnostic importance and for whom proper consent from informed parents or guardians could be obtained, were subjected to this procedure. The specimens were fixed in formaldehyde solution, stained with hematoxylin and eosin, and examined by light microscopy.

Table 1.—Distribution of Australia Antigen in Mentally Retarded Patients and Institution Staff Personnel

Group Tested	No. Tested	No. Positive	% Positive
Down's syndrome (large institutions)			
Institution 1 (all males over 6 years old)	144	41	28.5
Institution 2			
Male	68	26	38.2
Female	45	8	17.8
Two other large institutions	53	11	20.8
Total	**310**	**86**	**27.7**
Down's syndrome (other)			
Five small institutions	33	1	3
Patients who were not institutionalized	20	0	0
Newborn	14	0	0
Total	**67**	**1**	**1.5**
Mentally retarded (unclassified, from large institutions)			
Institution 1	76	4	5.3
Institution 2			
Male	69	1	1.4
Female	43	1	2.3
Total	**188**	**6**	**3.2**
Staff personnel (large institution)	114	0	0

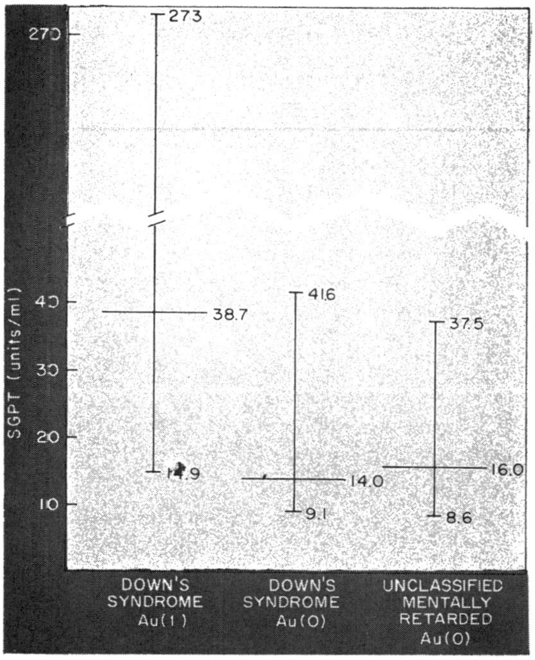

2. Serum glutamic pyruvic transaminase (SGPT) levels in age- and sex-matched groups of patients with Down's syndrome [Au(1) and Au(0)], and other mental defectives [Au(0)]. Medians and range indicated.

Statistical comparisons were done with Student's *t* test and an MN Chi computer program, providing an age- and sex-corrected comparison of two groups regarding frequency of the Au(1) phenotype.

Results

Australia antigen is common in patients with Down's syndrome in large institutions, and is extremely rare in patients without Down's syndrome in the same institutions (Table 1). The difference is highly significant when an age- and sex-corrected comparison is made with the MN Chi program ($P < 0.001$). This confirms our previously reported findings.[4] The frequency is much lower in smaller institutions. It is entirely absent in newborn patients with Down's syndrome, in those who live at home, and in the staff personnel of institution 2 (Table 1). The frequency of Australia antigen is age-dependent in the Down's syndrome group (highest frequency in the youngest patients), as it is in other populations which have been tested.[3-5] It is also more common in males than in females, but this difference is not significant when an age- and sex-comparison is made with the MN Chi program ($P = 0.1057$). A similar sex-related difference has been reported in other populations.[3-5]

There is a clear association between the presence of Australia antigen and an elevated SGPT level

Table 2.—SGPT Values in Persons With [Au(1)] and Without [Au(0)] Australia Antigen

	Unclassified Mentally Retarded [Au(0)]	Patients With Down's Syndrome Au(0)	Patients With Down's Syndrome Au(1)
Age- and sex-controlled study			
No. tested	16	16	16
Median value	16	14	38.7 ($P<0.001$)
Institution 1*			
No. tested	17	63	38
Median value	16.3	14.2	38.2 ($P<0.001$)
Institution 2			
No. tested	112	79	34
Median value	14.6	15.4	39.4 ($P<0.001$)
Total (Institutions 1 and 2)			
No. tested	**129**	**142**	**72**
Median value	**14.7**	**15.3**	**37.8 ($P<0.001$)**
Outpatients			
No. tested		14	
Median value		11.5	
Staff Personnel Institution 2		**Normal Au(0)**	
No. tested		114	
Median value		13.0	

*Including age- and sex-controlled study.

(Table 2). This is seen in the age- and sex-matched comparison in Fig 2 [Down's Au(1), 38.7 units/ml; Down's Au(0), 14 units/ml; and non-Down's Au(0), 16 units/ml], and in the subsequent larger study [Down's Au(1), 39.4 units/ml; Down's Au(0), 15.4 units/ml; and non-Down's Au(0), 14.6 units/ml]. The findings are the same in both institutions (Fig 3). The association is so close that, in general, the presence of Australia antigen can be used as an indication of an elevated SGPT level. The median SGPT values in the patients with Down's syndrome who were not institutionalized (11.5 units/ml) and the staff (13 units/ml) are also significantly lower than the levels in the Au(1) patients with Down's syndrome.

Inflammatory changes of various degrees were found in the liver of each of the four patients who had Down's syndrome, Australia antigen, an elevated SGPT level, and who had a biopsy performed. Patient 1 had periportal inflammation with lymphocytes and plasma cells, minimal hydropic change and binucleation of liver cells, and an increase in the number of Kupffer's cells with the presence of ceroid pigment. Patient 2 showed marked hydropic degeneration with balloon-like distension of the cells, liver cell breakdown with loss of margin, cellular debris, and periportal infiltration of lymphocytes and plasma cells (Fig 4, *top*). In this patient, there was also nodule formation indicative of postnecrotic cirrhosis (Fig 4, *bottom*). Patient 3 had increased collagenous connective and fibrous tissue in the hepatic triads, and infiltration by lymphocytes and plasma cells. There was bile duct proliferation and focal liver cell necrosis with cytoplasmic degeneration, acidophilic bodies, and cellular reactive infiltrate. The Kupffer's cells contained ceroid pigment and blue-staining granular matter which has remained unidentified. Patient 4 showed periportal lymphocyte, plasma cell, and histiocyte infiltration, advanced albuminous degeneration, large nuclei, irregular liver cell size, cell destruction, and cellular debris. Histologically, this appeared to be a more active case of hepatitis.

Comment

Australia antigen is associated with an elevated SGPT value in patients with Down's syndrome. The four Au(1) patients with elevated SGPT levels who underwent biopsy had histological evidence of hepatitis. There was a small number of liver biopsies performed, and no control biopsies were done in patients without Australia antigen. Consequently, this cannot be considered proof that Australia antigen is associated with liver damage, but is consistent with this hypothesis. The antigen, with a concomitant raised SGPT value, is rare or absent in other mental retardates and in staff personnel from the same institutions. High SGPT levels are due to release of the enzyme as a result of liver cell damage or death.[11] Australia antigen thus appears closely related to some agent responsible for liver cell injury.

The increased prevalence of Australia antigen in patients with Down's syndrome in large institutions is striking. The confirmation of the clinical diagnosis in all of the 82 patients whose chromosome preparations were studied makes it seem unlikely that major discrepancies occur in the remaining patients. Even if several erroneous diagnoses were included, the statistical correlation is so great that the conclusions would remain unchanged. Although the frequency of the antigen in patients with Down's syndrome is high in large institutions, it has not been found in patients who are not institutionalized and occurs in very low frequency in small institutions. (In addition to the 20 sera we examined of patients with Down's syndrome who were not placed in institutions, L. J. Old, MD, of New York told us [in June 1967] that Australia antigen was not found in sera from 25 such patients. These sera were collected by J. Storm, MD, of Manhasset, NY.) This suggests that a factor present in large institutions may be responsible for the presence of Australia antigen. It is reasonable to suppose that this factor is an infectious agent, such as a virus, which involves the liver and is disseminated by frequent and intimate contacts between individuals in large and crowded populations. Many chemicals and drugs may cause liver damage, but these seem less likely to be implicated in this situation. There is additional evidence consistent with the hypothesis of infection. Australia antigen clusters in families.[3] It occurs in high frequency in tropical areas where the people have poor sanitary practices, and its age-related distribution (higher frequency in younger than in older people) is similar to the attack rate of viral hepatitis.[13]

The following are two remarkable features of these findings: (1) limited distribution of the antigen in institutions, occurring nearly exclusively in patients with Down's syndrome, and (2) absence of the antigen in patients with Down's syndrome who live at home. This strongly suggests that the following two factors are in operation: (1) an environmental factor, for example, an infectious agent, present in the large institutions, and (2) a host factor which makes the patients with Down's syndrome more susceptible than other inhabitants of the same environment. Patients with Down's syndrome are either more susceptible to infection with the postulated agent or retain the agent chronically. In either event, this implies impairment of the immune mechanism in patients with Down's syn-

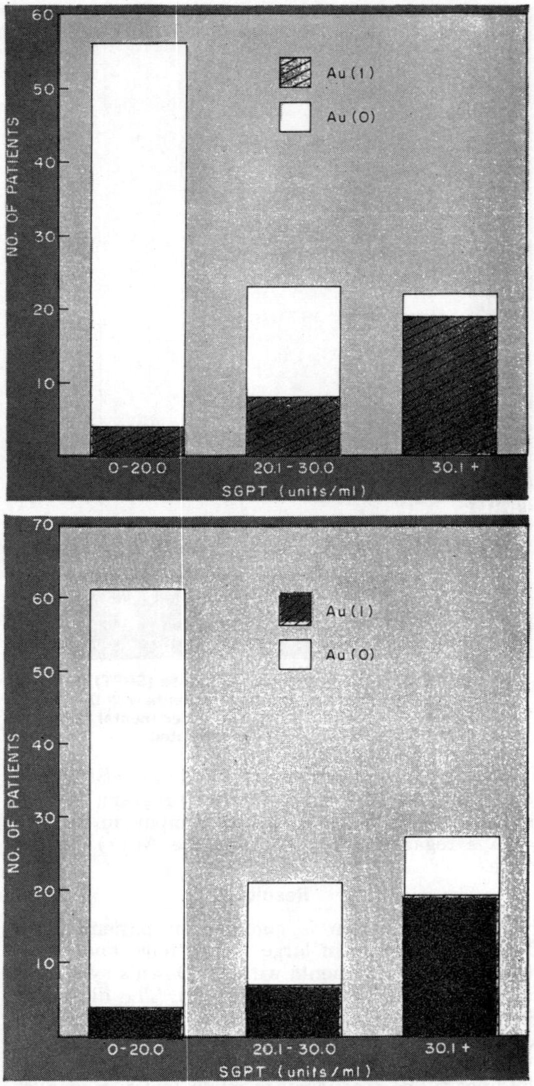

3. Australia antigen and serum glutamic pyruvic transaminase (SGPT) level associated with Down's syndrome at institution 1 (*top*) and institution 2 (*bottom*).

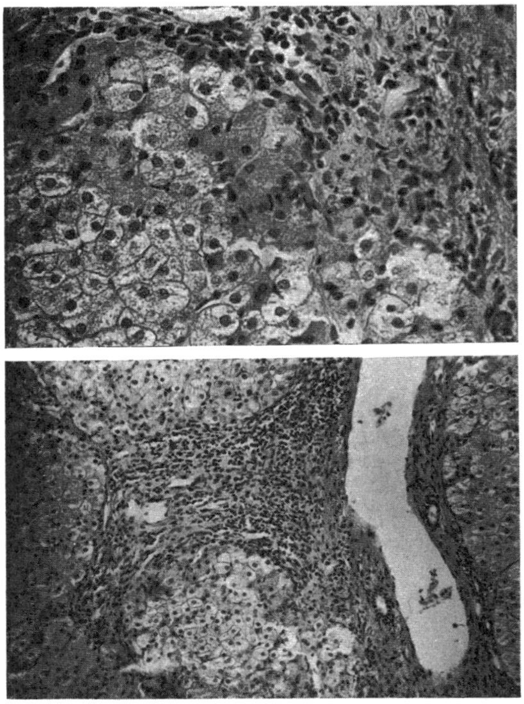

4. Liver specimens of patient 2. *Top*, Hydropic degeneration (hematoxylin and eosin, ×250). *Bottom*, Indications of early postnecrotic cirrhosis (×100).

drome.[14] Inherited susceptibility to this postulated agent may explain previously reported family data which are consistent with simple autosomal recessive inheritance of Australia antigen.[3]

Australia antigen is also commonly associated with other chromosome abnormalities. Four of 16 patients with a variety of autosomal and sex chromosome abnormalities have the antigen (unpublished observations). Leukemia is also relatively common in patients with Down's syndrome[15] and in patients with certain other chromosome abnormalities.[16-27] We have previously reported the high frequency of Australia antigen associated with some types of leukemia.[1-4] The definition of the mechanism linking these findings is the subject of our current research.

This study was supported by Public Health Service research grants CA-06551, CA-06927, and CA-08069 from the National Cancer Institute, and by an appropriation from the Commonwealth of Pennsylvania.

Arrangements were made and facilities provided for testing the mentally retarded patients at State Colony at New Lisbon, NJ, by F. W. Russell and J. D. Nelson, MD; at Pennhurst State School and Hospital, Spring City, Pa, by L. A. Potkonski, MD, and B. H. Marshall, MD. Outpatients were provided by Edith Taylor and John W. Halpin, Philadelphia Association for Retarded Children. The following individuals and institutions provided serum specimens and permitted observation of some patients in detail: Rainier State School, Maryland Child Center, Melmark School, Beaver Run School, Donegal Spring School, Elwyn School, Clifford Nursing Home; L. Gerstley III, MD; W. J. Mellman, MD, who provided 12 serum samples from newborn patients with Down's syndrome; and F. A. Oski, MD. P. J. Grotzinger, MD; S. N. Herschberg, MD; and V. E. Zavatone, MD, were instrumental in the management of the inpatients. Liver biopsy slides were reviewed by R. A. Donato, MD. J. Ipsem, PhD, devised the computer program.

References

1. Blumberg, B.S.; Alter, H.J.; and Visnich, S.: A "New" Antigen in Leukemia Sera, *JAMA* **191**:541-546 (Feb 15) 1965.
2. Melartin, L., and Blumberg, B.S.: Production of Antibody Against "Australia Antigen" in Rabbits, *Nature* **210**:1340-1341 (June 25) 1966.
3. Blumberg, B.S., et al: Family Studies of a Human Serum Iso-antigen System (Australia Antigen), *Amer J Hum Genet* **18**:594-608 (Nov) 1966.
4. Blumberg, B.S., et al: A Serum Antigen (Australia Antigen) in Down's Syndrome, Leukemia and Hepatitis, *Ann Intern Med* **66**:924-931 (May) 1967.
5. Blumberg, B.S., et al: Association of Lepromatous Leprosy With a Serum Antigen (Australia Antigen): Studies in Cebu, *Lancet* **2**:173-176 (July 22) 1967.
6. Melartin, L., and Panelius, M.: Occurrence of Australia Antigen in Finnish Mongolism Patients, *Ann Med Exp Fenn* **45**:157-158, 1967.
7. Sutnick, A.I.; London, W.T.; and Blumberg, B.S.: Australia Antigen, Down's Syndrome and Hepatitis, *J Clin Invest* **46**:1122 (June) 1967.
8. Bayer, M.E.; Blumberg, B.S.; and Werner, B.G.: Particles Associated With Australia Antigen in the Sera of Patients With Leukemia, Down's Syndrome, and Hepatitis, *Nature*, to be published.
9. Blumberg, B.S., and Riddell, N.M.: Inherited Antigenic Differences in Human Serum β-lipoproteins: A Second Antiserum, *J Clin Invest* **42**:867-875 (Feb) 1963.
10. Preer, J.R., and Preer, L.B.: Gel Diffusion Studies on the Antigens of Isolated Components of Paramecium, *J Protozool* **6**:88-100 (Feb) 1959.
11. Wroblewski, F., and La Due, J.S.: Serum Glutamic Pyruvic Transaminase in Cardiac and Hepatic Disease, *Proc Soc Exp Biol Med* **91**:569-571 (Jan) 1956.
12. Juul, P.: Stability of Plasma Enzymes During Storage, *Clin Chem* **13**:416-422 (May) 1967.
13. Von Bormann, F.: Hepatitis Epidemica, *Ergebn Inn Med Kinderheilk* **58**:201-284, 1940.
14. Blumberg, B.S., et al: Down's Syndrome: Two Types Distinguished by a Serum Antigen (Australia Antigen), *Ann Intern Med* **66**:1040 (May) 1967.
15. Miller, R.W.: Radiation, Chromosome and Viruses in the Etiology of Leukemia: Evidence From Epidemiologic Research, *New Eng J Med* **271**:30-36 (July 2) 1964.
16. Schroeder, T.M.; Anschütz, F.; and Knopp, A.: Spontane Chromosomenaberrationen bei familiärer Panmyelopathie, *Humangenetik* **1**:194-196, 1964.
17. Schmid, W., et al: Chromosomenbruchigkeit bei der familiären Panmyelopathie (Typus Fanconi), *Schweiz Med Wschr* **95**:1461-1464 (Oct) 1965.
18. Bloom, G.E., et al: Chromosome Abnormalities in Constitutional Aplastic Anemia, *New Eng J Med* **274**:8-14 (Jan 6) 1966.
19. Swift, M.R., and Hirschhorn, K.: Fanconi's Anemia: Inherited Susceptibility to Chromosome Breakage in Various Tissues, *Ann Intern Med* **65**:496-503 (Sept) 1966.
20. German, J.; Archibald, R.; and Bloom, D.: Chromosomal Breakage in Rare and Probably Genetically Determined Syndrome of Man, *Science* **148**:506-507 (April 23) 1965.
21. Sawitsky, A.; Bloom, D.; and German, J.: Chromosomal Breakage and Acute Leukemia in Congenital Telangiectatic Erythema and Stunted Growth, *Ann Intern Med* **65**:487-495 (June) 1966.
22. German, J., and Crippa, L.P.: Chromosomal Breakage in Diploid Cell Lines From Bloom's Syndrome and Fanconi's Anemia, *Ann Genet* **9**:143-154 (Dec) 1966.
23. Hecht, F., et al: Leukaemia and Lymphocytes in Ataxia-Telangiectasia, *Lancet* **2**:1193 (Nov 26) 1966.
24. Mamunes P., et al: Acute Leukaemia in Klinefelter's Syndrome, *Lancet* **2**:26-27 (July 1) 1961.
25. Tough, I.M., et al: Chronic Myeloid Leukaemia: Cytogenetic Studies Before and After Splenic Irradiation, *Lancet* **2**:115-120 (July 21) 1962.
26. Lewis, F.J.W.; Poulding, R.H.; and Eastham, R.D.: Acute Leukaemia in an XO/XXX Mosaic, *Lancet* **2**:306 (Aug 10) 1963.
27. Bousser, J., and Tanzer, J.: Syndrome de Klinefelter et leucemie aiguë, *Nouv Rev Franc Hemat* **3**:194-197 (March-April) 1963.

[Reprinted from THE JOURNAL OF EXPERIMENTAL MEDICINE, February 1, 1970, Vol. 131, No. 2, pp. 307–319]
Printed in U.S.A.

THE LOCALIZATION OF AUSTRALIA ANTIGEN BY IMMUNOFLUORESCENCE*

BY VERONICA E. COYNE(ZAVATONE), M.D., IRVING MILLMAN, PH.D., JAMES CERDA, M.D., B. J. S. GERSTLEY, M.D., THOMAS LONDON, M.D., ALTON SUTNICK, M.D., AND BARUCH S. BLUMBERG, M.D.

(From The Institute for Cancer Research, Fox Chase, Philadelphia, Pennsylvania 19111)

(Received for publication 8 September 1969)

Since the discovery of Australia antigen, Au(1), by Blumberg et al. in 1965 (1) the association of this antigen with viral hepatitis has been well established (2–4). Au(1) is present in 58% of patients with posttransfusion hepatitis and in 38% of patients with infectious hepatitis (5). When patients are tested serially after blood transfusion, the incidence of Au(1) was found to be 74% (6). In most acute cases, the antigen appears in the blood early in the disease and disappears within 2–3 wk as clinical improvement occurs.

In addition to the epidemiologic association, there are several findings which provide strong support for the hypothesis that Au(1) is an antigenic determinant of a virus capable of causing hepatitis. (*a*) Purified Au(1), isolated from serum of patients with hepatitis, consists of virus-like particles, 200 A in diameter, which are agglutinated by specific anti-Au(1) antiserum (7). (*b*) Patients who have received blood containing Au(1) have developed hepatitis with Au(1) in their blood or remained healthy and developed antibody to Au(1) (4). (*c*) With the use of a specific fluorescein-conjugated antibody to Au(1) we were able to demonstrate fluorescent granules in or on the surface of nuclei of cells from liver tissue in four patients with hepatitis and Au(1) in their serum. Five control patients who had neither hepatitis nor serum Au(1) did not have fluorescent granules (8).

The purpose of the present study is to determine whether the fluorescent particles in liver tissue are associated with Au(1) in the blood and whether the particles are present in tissues other than liver. This study involves 61 liver biopsy specimens, 25 specimens of other human tissues, cells from 8 samples of bile or duodenal drainage and peripheral lymphocyte cultures from 8 patients. This study confirms our previous results and establishes: (*a*) that the fluorescent particles in the cells are related to Au(1) in the serum and to the diagnosis of hepatitis. (*b*) The fluorescent particles are present in tissues other than liver in some patients.

* This work was supported in part by United States Public Health Service Grants CA-08069, CA-06927, FR-05539, FR-0040, and 5 MO1 RR40 from the National Institutes of Health, the Brian William Donovan Memorial Fund, and by an appropriation from the Commonwealth of Pennsylvania. V.C. was in receipt of a Damon Runyon Cancer Research Fellowship.

Materials and Methods

Collection of Specimens.—61 liver biopsy specimens were obtained from 58 patients in hospitals in the Philadelphia area and one from elsewhere.[1] In 57 cases needle biopsies were performed using a Menghini needle. Two liver specimens were obtained at surgery and two at autopsy. Two patients had a second biopsy performed 4–6 wk after the initial biopsy. The second specimen in each case was treated and recorded as a separate specimen for statistical analysis. Eight bone marrow specimens were obtained by needle aspiration of the sternum. The other tissues were obtained from organs removed at surgery for therapeutic reasons. Buffy coat smears and peripheral lymphocytes were obtained from patients in our Clinical Research Unit. Seminal fluid was obtained from one patient with acute hepatitis who also had Au(1) in his serum. A blood sample was obtained from each patient at the time of liver biopsy or collection of other specimens.

One patient in our Clinical Research Unit was studied in detail and will be reported elsewhere (9). He was a 21 year old male (D.M.) with Down's syndrome and chronic hepatitis and Au(1) in his blood who later developed acute myelogenous leukemia. A bone marrow specimen was obtained 3 wk prior to death, and immediately after death specimens of liver, spleen, mesentery, and testis were taken.

Preparation of Tissues for Examination.—Each tissue specimen was placed in a small volume of McCoy's 5a tissue culture medium (Grand Island Biological Co., Grand Island, N.Y.) containing 30% newborn calf serum (Hyland Laboratories, Los Angeles, Calif.) and was minced to produce a cell suspension.[2] A drop of the cell suspension was placed on a clean slide, air-dried and fixed in acetone at room temperature for 10 min.[3] At the same time, a piece of tissue was placed in formalin and processed for routine histologic examination by the clinical pathology departments of the respective hospitals. The bone marrow specimens were prepared by placing a drop of the marrow directly on a glass slide, air-drying and fixing as above. Cells were obtained from bile, duodenal drainage, and seminal fluid by centrifugation for 10 min in a clinical centrifuge, washing the cell pellet in Earle's balanced salt solution (Grand Island Biological Co.), resuspending in McCoy's medium, and preparing slides as above. Buffy coat smears were made from peripheral blood and fixed in acetone for 10 min. Peripheral lymphocyte cultures were prepared by the method of Hungerford (10) with the omission of colchicine.

Preparation of Reagent and Staining of Slides.—Fluorescein-conjugated rabbit anti-Au(1) antiserum was prepared and direct fluorescent staining of slides was performed by the methods previously published (8). Replicate slides were stained with fluorescein-conjugated goat vs. human antiserum and goat-vs.-rabbit antiserum to test the specificity of the reaction. Indirect staining was performed by flooding the slides with unconjugated rabbit anti-Au(1) antiserum, washing in buffered saline, and then applying fluorescein-labeled goat-vs.-rabbit anti-globulin antiserum. Control slides for the indirect test were treated first with normal rabbit serum in place of the anti-Au(1) antiserum before the application of conjugated goat-vs.-rabbit anti-globulin antiserum.

---

[1] These include: The American Oncologic Hospital, Children's Hospital of Philadelphia' the Hospital of the University of Pennsylvania, Jeanes Hospital, Jefferson Medical College Hospital, Philadelphia General Hospital, Pennsylvania Hospital, Temple University Hospital, the Philadelphia Veterans Administration Hospital, and Woman's Medical College Hospital. One case was obtained from the Strong Memorial Hospital, Rochester, New York.

[2] McCoy's medium alone or buffered saline, pH 7.4, were used in some cases and produced similar results.

[3] We have tried direct tissue impressions on many liver specimens in the past and found that they were unsatisfactory.

Testing of Serum.—Serum from each patient was tested for the presence of Au(1) by double diffusion in agar gel (Ouchterlony method), using a panel of anti-Au(1) antisera consisting of rabbit, mouse, and human antisera (11).

Examination of Slides.—Slides were examined for fluorescence in dark field using a Zeiss microscope with the HBO 200 watt illuminator, BG 12 exciter filter, and 53 barrier filter. All slides were read by one of us (V.C.) without knowledge of the histologic diagnosis or the serum results for Au(1). The reproducibility of results was established by examining duplicate slides from the same patients and by reading the same slides on different days. To test the accuracy of reading, 10 slides were chosen at random and coded. They were read by V.C. and I.M. independently. The correlation between the two observers was 100%.

RESULTS

I. Liver Biopsies

Appearance of Fluorescence.—When tested by the direct method, the liver biopsies could be divided into two groups on the basis of fluorescence. In the first group, which we called negative, the cytoplasm of the cells usually appeared dull orange-brown almost identical to the autofluorescence seen in unstained preparations. The nuclei of cells in this group always appeared completely dark (Fig. 1). In the second group, which we called positive, we could detect apple-green fluorescence in, or on the surface of, the nucleus and occasionally in the cytoplasm. The nuclear fluorescence appeared in three different forms: (*a*) discrete granules varying in size and number, but usually averaging one to three per cell (Fig. 2); (*b*) diffuse finely granular fluorescence of the entire nucleus (Fig. 3a); (*c*) fluorescence of the nuclear rim (Fig. 3b). The first type was the most common, but all three types could be seen in the same specimen in some cases. In the positive cases, most of the cells on the slide showed some form of fluorescence. In some cases discrete fluorescent granules were seen in the cytoplasm as well as in the nucleus. Control slides stained with fluorescein-conjugated goat-vs.-human or goat-vs.-rabbit antisera, not expected to react with Au(1), showed neither intranuclear nor discrete cytoplasmic fluorescence. Occasionally we observed a pale green diffuse cytoplasmic fluorescence in both positive and negative slides as well as in some control slides. This was considered nonspecific.

The indirect method of staining demonstrated fluorescent granules only in those preparations that were positive by the direct stain. However, the granules were located to a large extent in the cytoplasm and intranuclear particles were few. In some cases the nucleus was disrupted and fluorescent granules were dispersed in the cytoplasm (Fig. 4a, b). We interpreted this to indicate that the particles had been dislodged from the nucleus by the prolonged washing procedure. Therefore, we confined our observations to the direct method.

Classification of Patients.—Table I lists the patients on whom liver biopsies were performed according to their clinical diagnosis, the histologic diagnosis made on the liver specimen, the presence of Au(1) in the serum, and the presence of fluorescence. 33 patients had a clinical and histologic diagnosis of viral hepatitis including acute infectious, posttransfusion, chronic recurrent, and

TABLE I

Classification of Patients Whose Liver Biopsies Were Studied

Clinical diagnosis	Histologic diagnosis	Number of patients	Number with Au(1)	Number with positive fluorescence
Infectious hepatitis	Viral hepatitis	9	5	8
Posttransfusion hepatitis	Viral hepatitis	3	3	3
Hepatitis in drug users (exposure to needles)	Viral hepatitis	6	4	4
Chronic alcoholism and drug use	Laënnec's cirrhosis + acute and chronic inflammation	2	0	2
Chronic recurrent hepatitis	Chronic active hepatitis	2	1	2
Chronic anicteric hepatitis and Down's syndrome	Mild inflammatory changes to early postnecrotic cirrhosis	12	10	12
Chronic anicteric hepatitis, Down's syndrome, and leukemia	Chronic hepatitis	1	1	1
Postnecrotic cirrhosis	Postnecrotic cirrhosis	1	0	1
Mental retardation (institutionalized)	Normal liver	3	1	3
Biliary cirrhosis	Biliary cirrhosis	1	0	1
Biliary atresia	Biliary atresia	1	0	1
Chronic renal disease, (received many blood transfusions)	Hemosiderosis and chronic inflammation	1	1	1
Chronic renal disease	Normal liver	1	0	0
Drug toxicity	Toxic hepatitis	3	0	0
Hepatoma	Hepatoma	1	0	0
Metastatic malignancy	Carcinoma of colon metastasized to liver	1	0	0
Hepatomegaly and jaundice	Adenocarcinoma, ? of bile duct origin	1	0	1
Renal malignancy	Carcinoma of kidney metastasized to liver	1	0	1
Fever of unknown origin	Nonspecific necrosis	2	0	0
Acute alcoholic	Acute alcoholic hepatitis	4	0	0
Chronic alcoholism	Laënnec's cirrhosis	4	0	0
Sickle cell disease	Hemosiderosis and intravascular sickling	1	0	0
Totals		61	26	41

chronic anicteric hepatitis. Two patients were chronic alcoholics and drug users who had been exposed to needles. One patient had postnecrotic cirrhosis which was presumed to have resulted from previous viral hepatitis. One patient with chronic renal disease had received many blood transfusions. The 24 remaining patients had a variety of other diagnoses.

Relationship between Fluorescence and Serum Au(1).—Table II A gives the comparison between the occurrence of fluorescence in the liver cells and the presence of Au(1) in the serum. There were 26 patients who had Au(1) in their blood. 24 had a diagnosis of viral hepatitis, one was a patient with chronic renal disease who had received many blood transfusions, and the other was an institutionalized mentally retarded patient with both Au(1) and antibody to Au(1) in his serum, but a histologically normal liver. All 26 patients with

TABLE II

Comparison of Immunofluorescence with Presence of Au(1) and Pathologic Diagnosis

A. Fluorescence vs. serum Au(1)

Fluorescence	Number of patients with serum Au(1) +	Number of patients with serum Au(1) 0	Total
+	26	15	41
0	0	20	20
Totals	26	35	61

Chi square = 19.58; *P* value <0.001

B. Fluorescence vs. hepatitis (histologic diagnosis)

Fluorescence	Number of patients with hepatitis +	Number of patients with hepatitis 0	Total
+	30	11	41
0	3	17	20
Totals	33	28	61

Chi square = 16.05; *P* value <0.001

Au(1) in their blood had specific fluorescence in the nuclei of cells obtained from the liver biopsy. In addition, 15 of the 35 patients who did not have Au(1) in their blood had fluorescent granules in their liver cells. Six of these patients had a histologic diagnosis of viral hepatitis. Two were alcoholics and drug users; one was a patient with postnecrotic cirrhosis; two were residents of an institution for the mentally retarded and had been exposed repeatedly to patients with known hepatitis and Au(1). Two were patients with malignancies involving the liver and had undergone surgery. One patient had biliary atresia and the other primary biliary cirrhosis.

Relationship between Fluorescence and Hepatitis.—Table II B gives the comparison between the occurrence of fluorescence and the histologic diagnosis of viral hepatitis. 30 out of 33 patients with evidence of viral hepatitis (91%)

showed nuclear fluorescence. The three patients with hepatitis who did not have fluorescence were in the recovery phase of their illness and did not have Au(1) in their serum.

Results of Repeat Biopsies.—Both patients on whom a repeat biopsy was performed had Au(1) in their serum and acute hepatitis at the time of the first biopsy. In both cases fluorescent granules were present in the first specimen. In one patient fluorescence was not seen on the second specimen, taken 4 wk after the first biopsy, coincident with clinical improvement and disappearance

TABLE III

Tissues and Fluids Tested for Fluorescence

Type of specimen	Number of patients	Number with Au(1) in serum	Number with positive fluorescence
Kidney	2	1	0
Jejunal loop	4	4	0
Lymph node	2	2	0
Spleen	2	1	1
Bone marrow	8	6	2
Testis	1	1	1
Mesentery	1	1	1
Adrenal	1	0	0
Ovary	1	0	0
Gall bladder	1	0	0
Lung	1	0	0
Cancer of lung	1	0	0
Bile and duodenal drainage	8	4	0
Seminal fluid	1	1	0
Total	34	21	5

of Au(1) from the serum. The other patient had a more prolonged illness with persistence of Au(1) in his blood for at least 3 months. Fluorescent granules were still present in his second biopsy specimen, obtained 6 wk after the initial biopsy.

II. Tissues Other Than Liver

Table III lists the tissues and cells studied for fluorescence. All the tissues studied from D.M. (the patient with Down's syndrome and acute myelogenous leukemia), taken immediately after death, showed fluorescence. The fluorescence in the liver, testis, and mesenteric connective tissue is similar to that found in the liver preparation from patients with hepatitis in that the three patterns of fluorescence were found in all these tissues (Fig. 5a, b, c). In the spleen, however, only a small percentage of the cells showed fluorescence which was mostly in the cytoplasm. The bone marrow specimen, obtained 3 wk prior to death and consisting almost entirely of myeloblasts, did not have fluorescence (Fig. 5d).

Two of eight bone marrow specimens from patients with acute or chronic hepatitis showed fluorescence confined to the cytoplasm of the cells. These were from a patient with acute infectious hepatitis with Au(1) in the blood, and from a mentally retarded institutionalized patient who had both Au(1) and antibody to Au(1). In both of these cases only a small percentage of the cells on the slide were involved.

None of the other tissues studied including buffy coat smears, cells obtained from bile, duodenal drainage, and seminal fluid showed fluorescence. Peripheral lymphocyte cultures in six patients with chronic hepatitis and Au(1) in their serum also were negative.

DISCUSSION

By means of the fluorescent antibody technique, using a specific anti-Au(1) antiserum, we have demonstrated fluorescent particles mainly in the nuclei and occasionally in the cytoplasm of cells from liver biopsies. Fluorescent particles were present in 100% of patients who had Au(1) in their serum. Fluorescent granules occurred in 15 patients who did not have Au(1) in their blood. Six of these had definite clinical and histologic evidence of acute or chronic hepatitis, and 5 others probably had been exposed to viral hepatitis (Table I). We interpret these results to indicate that the fluorescence test is more sensitive in detecting Au(1) than is the precipitin test. An alternative explanation is that in some patients there is little or no cellular necrosis and consequently little or no antigen is released into the peripheral blood; however, it can be detected in the cells.

The finding of fluorescent granules in one patient with biliary atresia and another patient with biliary cirrhosis (Table I) raises the question of possible involvement by viral hepatitis in the pathogenesis of these diseases of the biliary system. Biliary atresia as an end result of neonatal hepatitis has been discussed by Alpert et al. in their study of patients with trisomy 17–18 (12).

The presence of three forms of fluorescence indicates that Au(1) is not restricted to one location within the nucleus. In most cases the fluorescence appeared as one to three small masses within the nucleus corresponding morphologically to the site of the nucleoli. The finding of fluorescence in the region of or in the nucleolus is interesting in view of the studies of Bearcroft and Peachey (13, 14). Using cytological and cytochemical methods and electron microscopy they found that there was swelling of the nucleolus and an increase in nucleolar ribonucleoprotein early in viral hepatitis. They postulated that these changes may be concerned with viral synthesis. The three forms of fluorescence which we observed could indicate stages of viral replication within the nucleolus, proceeding to the dispersal of granules throughout the nucleus, attachment to the nuclear membrane, and eventual appearance in the cytoplasm.

The appearance of fluorescence in a few cells in two bone marrow specimens could be explained on the basis of phagocytosis of circulating antigen. The widespread presence of fluorescent particles in the tissues of D.M. may represent either massive invasion by virus in a patient who was critically ill and debilitated or a viral proliferation which was directly related to the leukemic process.

We believe that the fluorescent antibody used in this study specifically identifies Au(1) for the following reasons. (*a*) The immunologic specificity of the antisera was confirmed by immunodiffusion and immunoelectrophoretic analysis (8). (*b*) Fluorescein conjugates of antisera which did not contain antibody to Au(1) failed to produce fluorescence in the positive liver preparations. (*c*) The association between fluorescent particles in the liver and the presence of Au(1) in the serum is highly significant ($P < 0.001$). (*d*) Most of the liver specimens negative for fluorescence were from abnormal livers, suggesting that the fluorescent antibody does not detect simply a nonspecific product of cell injury.

SUMMARY

We have studied the localization of Australia antigen, a particulate substance associated with hepatitis, by means of the fluorescent antibody technique. Preparations were made from 61 liver biopsy specimens taken from patients with infectious hepatitis, serum hepatitis, and a variety of other diseases. When tested with fluorescein-conjugated rabbit anti-Au(1) antisera all 26 patients who had Au(1) in their serum had specific fluorescence in their liver cells. The fluorescence appeared in three forms: as discrete particles within the nucleus, diffuse fluorescence of the entire nucleus, and fluorescence of the nuclear rim. Occasionally there were also fluorescent particles in the cytoplasm. Other specimens were tested with the fluorescent antibody including a variety of human tissues, buffy coat smears, peripheral lymphocyte cultures, and cells obtained from bile and duodenal drainage. Among these specimens, fluorescence was found in the cytoplasm of a few cells in the bone marrow of two patients with hepatitis and Au(1) in their serum, and in the liver, spleen, mesentery, and testis of one patient with leukemia, chronic hepatitis, and Au(1) in his serum. We have shown that the presence of fluorescent particles in the liver cells is strongly associated with the presence of Au(1) in the serum and the diagnosis of viral hepatitis. We believe that this study adds support to the hypothesis that Australia antigen is an antigenic determinant of a virus capable of causing hepatitis.

BIBLIOGRAPHY

1. Blumberg, B. S., H. J. Alter, and S. Visnich. 1965. A "new" antigen in leukemia sera. *J. Amer. Med. Ass.* **191**:541.

2. Blumberg, B. S., B. J. S. Gerstley, D. A. Hungerford, W. T. London, and A. I. Sutnick. 1967. A serum antigen (Australia antigen) in Down's syndrome, leukemia and hepatitis. *Ann. Intern. Med.* **66**:924.
3. Blumberg, B. S., A. I. Sutnick, and W. T. London. 1968. Hepatitis and leukemia: their relation to Australia antigen. *Bull. N. Y. Acad. Med.* **44**:1566.
4. Okochi, K., and S. Murakami. 1968. Observations on Australia antigen in Japanese. *Vox Sang.* **15**:374.
5. Blumberg, B. S., A. I. Sutnick, W. T. London, B. J. S. Gerstley, I. Millman, V. E. Zavatone, J. R. Senior, and J. J. Cerda. 1969. Current status of the Australia antigen hepatitis studies. *Gastroenterology.* **56**:1212.
6. Hirschmann, R. J., N. R. Shulman, L. F. Barker, and K. O. Smith. 1969. Virus-like particles in sera of patients with infectious and serum hepatitis. *J. Amer. Med. Ass.* **208**:1667.
7. Bayer, M. E., B. S. Blumberg, and B. Werner. 1968. Particles associated with Australia antigen in the sera of patients with leukemia, Down's syndrome and hepatitis. *Nature (London).* **218**:1057.
8. Millman, I., V. E. Zavatone, B. J. S. Gerstley, and B. S. Blumberg. 1969. Australia antigen detected in the nuclei of liver cells of patients with viral hepatitis by the fluorescent antibody technique. *Nature (London).* **222**:181.
9. Millman, I., L. A. Loeb, M. E. Bayer, and B. S. Blumberg. 1970. Australia antigen (hepatitis associated antigen): purification and physical properties. *Fed. Proc.* In press.
10. Hungerford, D. A. 1965. Leukocytes cultured from small inocula of whole blood and the preparation of metaphase chromosomes by treatment with hypotonic KCl. *Stain Technol.* **40**:333.
11. Melartin, L., and B. S. Blumberg. 1966. Production of antibody against Australia antigen in rabbits. *Nature (London).* **210**:1340.
12. Alpert, L. I., L. Strauss, and K. Hirschhorn. 1969. Neonatal hepatitis and biliary atresia associated with trisomy 17–18 syndrome. *N. Engl. J. Med.* **280**:16.
13. Bearcroft, W. G. C., and R. D. C. Peachey. 1962. Cytological and cytochemical studies on the liver in infective hepatitis. *J. Pathol. Bacteriol.* **83**:373.
14. Bearcroft, W. G. C., and R. D. C. Peachey. 1962. Electron microscopic studies on the liver in infective hepatitis. *J. Pathol. Bacteriol.* **83**:383.

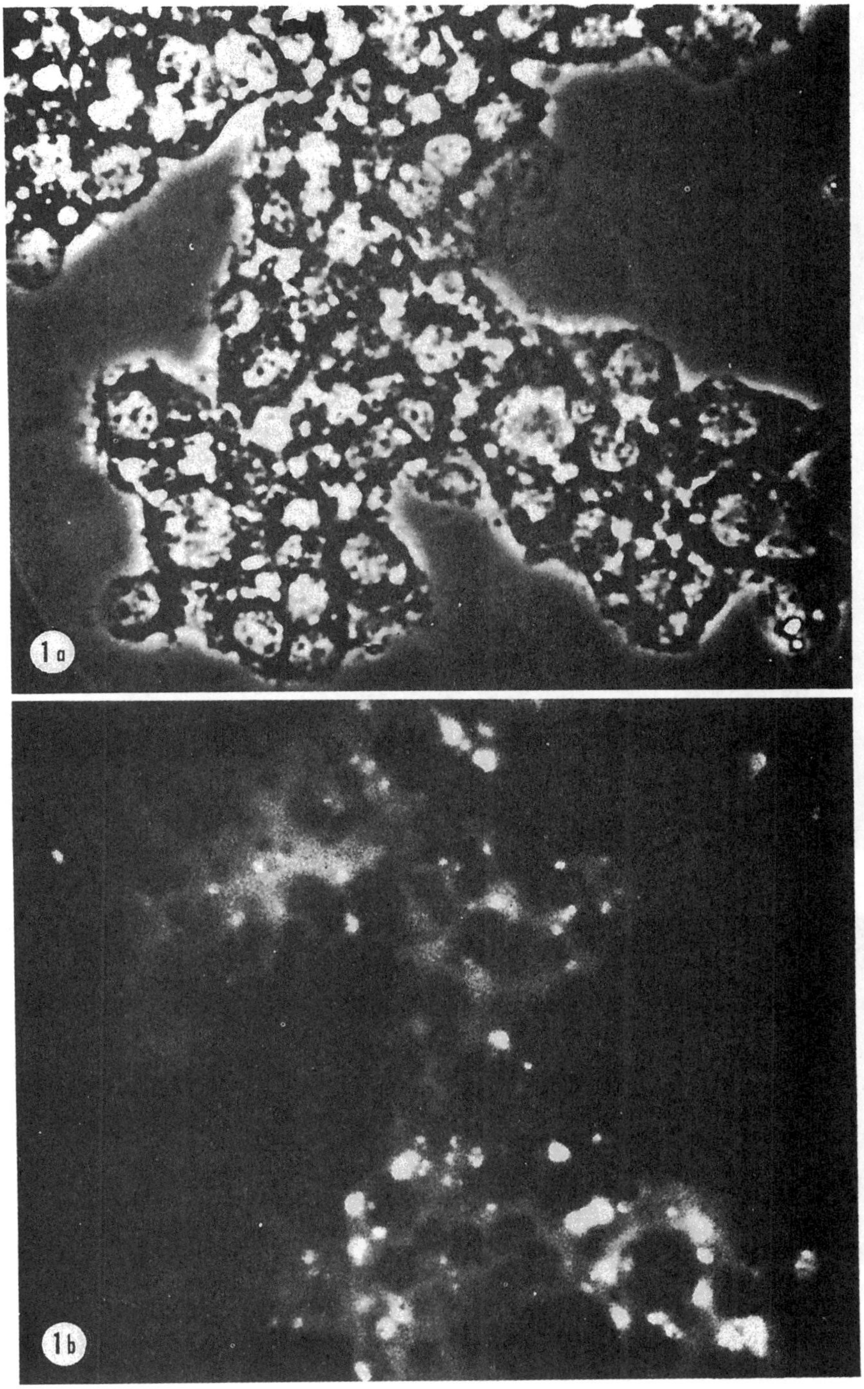

Fig. 1a and b. Cells from liver biopsy of a patient with Laënnec's cirrhosis stained with conjugated rabbit anti-Au(1) antiserum. This patient did not have Au(1) in the peripheral blood and had no clinical or histologic evidence of hepatitis. Nuclei appear dark. Cytoplasmic particles due to orange autofluorescence of lipids. This is scored as negative for fluorescence. (a) phase; (b) ultraviolet. Magnification, × 1700.

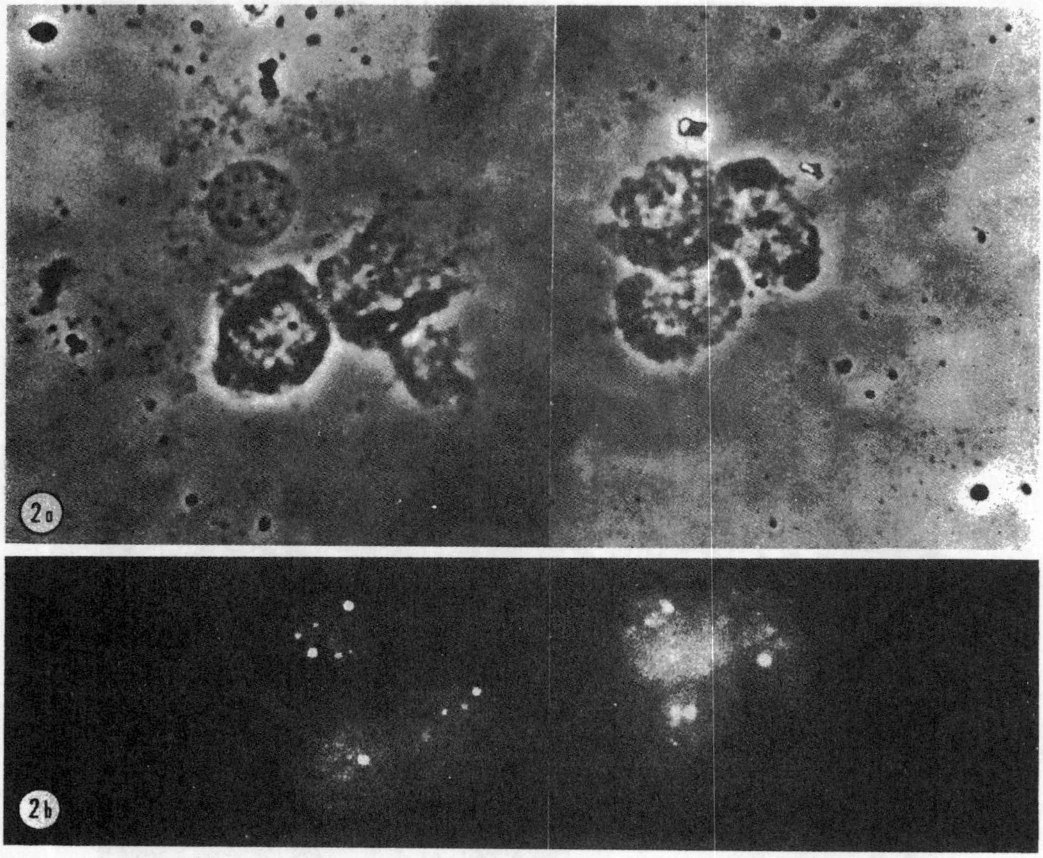

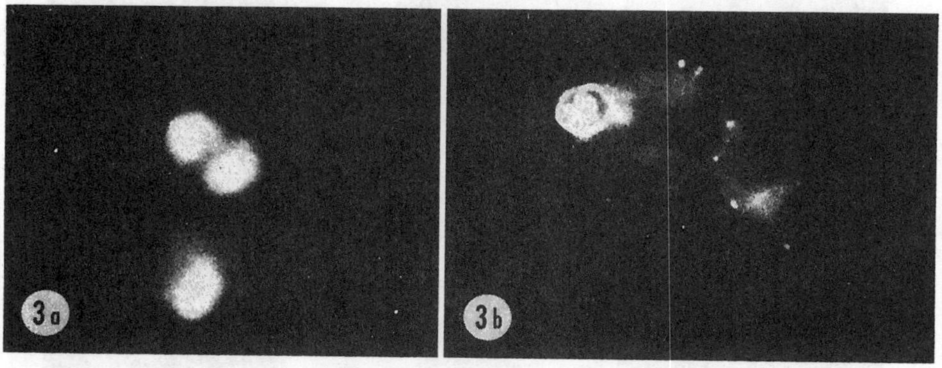

FIG. 2a and b. Cells from liver biopsy of a patient with chronic anicteric hepatitis who had Au(1) in his peripheral blood. The biopsy is stained with conjugated rabbit anti-Au(1) antiserum, and shows intranuclear fluorescent particles. This is scored as positive fluorescence. (a) phase; (b) ultraviolet. × 1700.

FIG. 3. Liver preparations from two patients with viral hepatitis who had Au(1) in their peripheral blood. The biopsy is stained with conjugated rabbit anti-Au(1) antiserum. (a) Diffuse fluorescence of entire nucleus. (b) Fluorescence of nuclear rim, as well as intranuclear particles. × 1350.

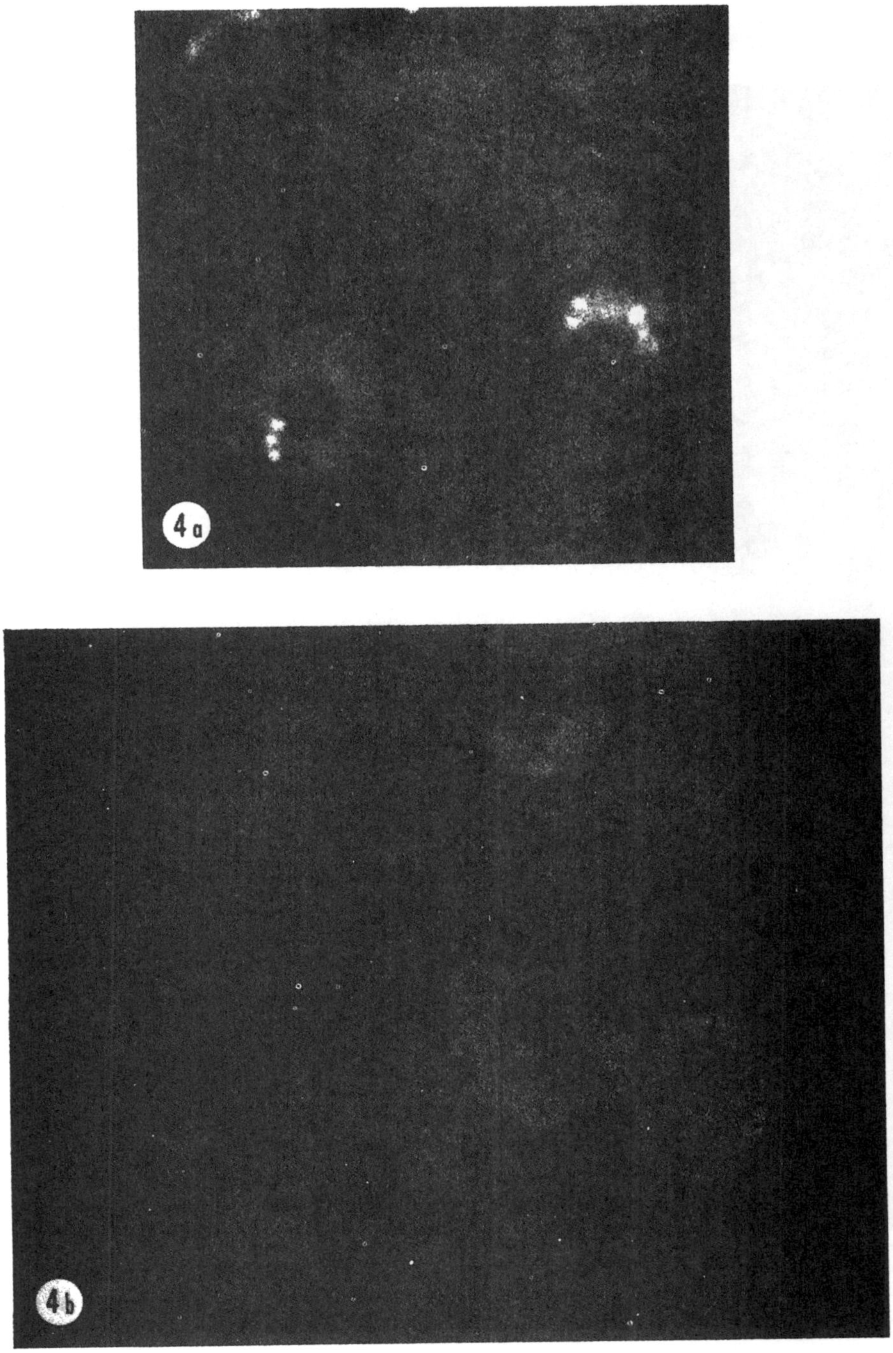

FIG. 4. Liver preparation from a patient with viral hepatitis stained by the indirect method. (a) Stained with unconjugated rabbit anti-Au(1) antiserum. (b) Stained with unconjugated normal rabbit serum. × 1700.

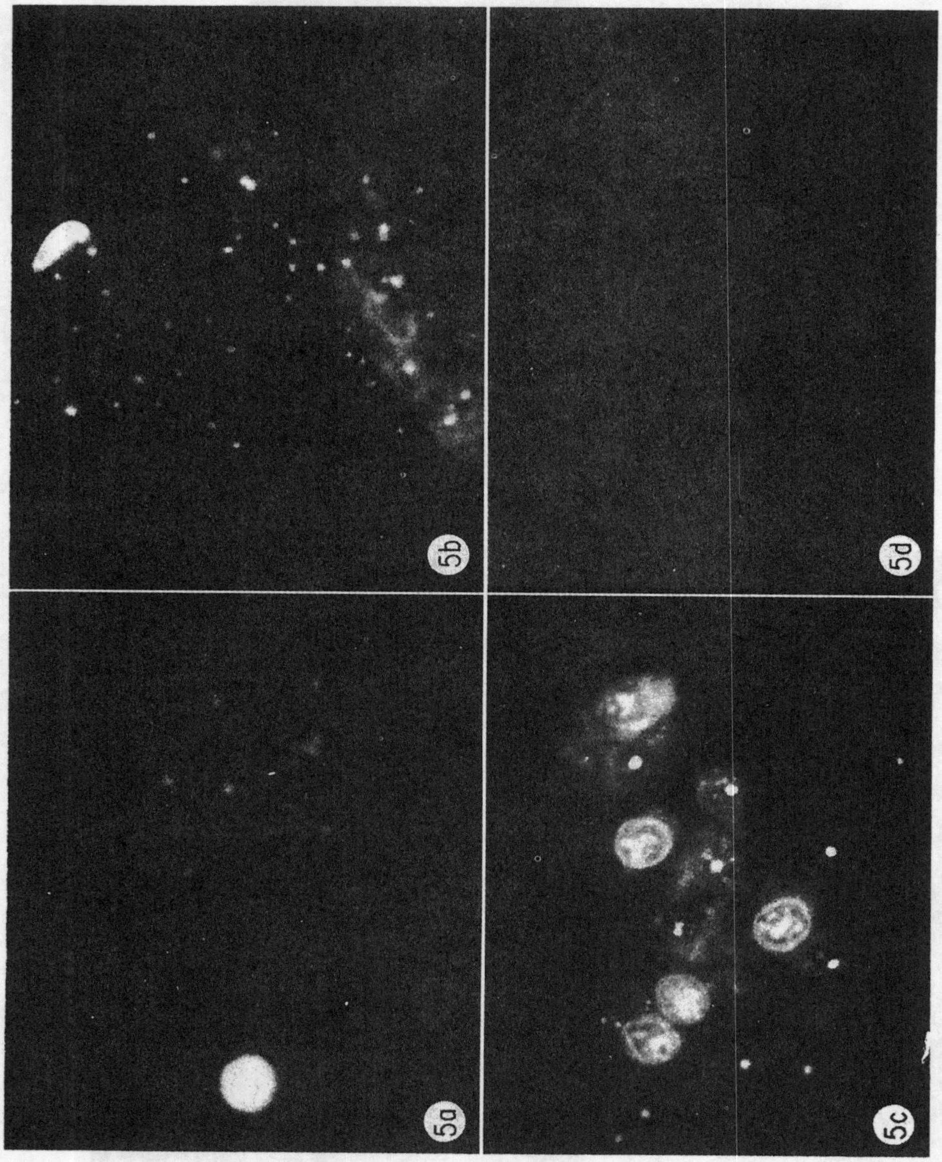

FIG. 5a, b, c, and d. Tissues from D.M. (a patient with Down's syndrome, chronic anicteric hepatitis and acute myelogenous leukemia) stained with conjugated rabbit anti-Au(1) antiserum. (a) liver; (b) mesentery; (c) testis; (d) bone marrow. × 1350.

Reduction of post-transfusion hepatitis by exclusion of Australia antigen from donor blood in an urban public hospital

JOHN R. SENIOR
ALTON I. SUTNICK
EUGENE GOESER
W. THOMAS LONDON
MIRIAM B. DAHLKE
BARUCH S. BLUMBERG

Abstract: Post-transfusion hepatitis occurred in ten of 56 (17.9 per cent) patients in 1963-1964, and in 14 of 78 (17.9 per cent) patients in 1968-1969 at Philadelphia General Hospital (PGH). Since November 1969 all donor blood has been tested for Australia antigen (Au) by immunodiffusion and later by counterelectrophoresis, and no positive units have been transfused. All patients were screened by stringent criteria to exclude any pre-existing liver disease and had repeated follow-up examinations after transfusion, including clinical examination, as well as serial serum glutamic pyruvic transaminase (SGPT) and Au determinations. Among 204 patients receiving Au-negative donor blood who could be followed adequately for six months, 12 (5.9 per cent) hepatitis patients (four icteric) were detected. These results indicate a two thirds reduction from the 18 per cent incidence in post-transfusion hepatitis at PGH found on two previous studies. Au testing to exclude positive donors and administering only Au-negative blood, and the changes in composition of the donor population which resulted, were effective in reducing the incidence of post-transfusion hepatitis.

KEY INDEXING TERMS

Australia antigen — *Hepatitis* — *Transfusion*
Liver disease — *Post-transfusion hepatitis*

It has been recognized that hepatitis following blood transfusion occurs more frequently[1,2] than earlier reports had indicated.[3,4] The source of blood probably is the most important determinant of the hepatitis risk after transfusion.[5-7] Since the discovery of Australia antigen (Au, HAA, HBHg),[8,9] it has been possible to identify high or low risk sources.[10] Blood from a group with a high percentage of Au-positive donors carries a high risk of transmitting hepatitis.[11]

In a study performed in 1963-1964 at the Philadelphia General Hospital (PGH),[1] we observed prospectively patients on the obstetrical-gynecological service for occurrence of hepatitis after blood transfusion. A hepatitis frequency

From the Institute for Cancer Research, The Fox Chase Center for Cancer and Medical Sciences, Philadelphia, Pennsylvania; and University of Pennsylvania Medical Division, Philadelphia General Hospital, Philadelphia, Pennsylvania. Dr. Senior is presently at the Clinical Research Center, Graduate Hospital of the University of Pennsylvania, Philadelphia, Pennsylvania.

This work supported by USPHS grants CA-06551, CA-06927, and RR-05539 from the National Institutes of Health; Contract No. N00014-67-A-046-005 from the Office of Naval Research; and an appropriation from the Commonwealth of Pennsylvania.

Requests for reprints should be addressed to Alton I. Sutnick, M.D., The Institute for Cancer Research, 7701 Burholme Avenue, Fox Chase, Philadelphia, Pennsylvania 19111.

Reprinted from THE AMERICAN JOURNAL OF THE MEDICAL SCIENCES Vol. 267(3)171-177.

REDUCTION OF POST-TRANSFUSION HEPATITIS

of 17.9 per cent was found. In an effort to confirm and extend these earlier observations, recipients of whole blood, plasma, packed red blood cells, and nontransfused control patients were followed at the Hospital of the University of Pennsylvania (HUP) and PGH. These two hospitals serve distinctly different population segments. The PGH patients come almost exclusively from the poor, inner city population, while the patients at HUP are referred from a wide area around Philadelphia. Post-transfusion hepatitis in patients on medical, surgical, and gynecologic services at PGH during the repeat study in 1968-1969 confirmed the almost 18 per cent incidence of hepatitis (Table I).

We report our experience with a program of testing donor blood units at these hospitals, particularly the effect of withholding all positive units in producing a significant reduction in the incidence of post-transfusion hepatitis at PGH.

Materials and Methods

Beginning in September 1968 and continuing until April 1971, patients were accepted for study from all services in both hospitals, male and female, of any age, and with a variety of disease entities (Table I). We followed prospectively after transfusion 551 patients who were willing to be followed, who had no history of prior transfusion, hepatitis contact, heavy alcohol intake, drug addiction, exposure to hepatotoxic agents or drugs, and no history or clinical indication by physical examination of past or present liver disease, malignancy or sickle cell disease. Control patients from neighboring beds on the same services, of similar age, sex, and disease, but not transfused, were followed concurrently in the same manner, as was done previously.[1] Because of the suggestion that packed red blood cells might be safer than whole blood,[12] other groups of recipients of only packed cells were also followed.

Determinations of serum glutamic pyruvic transaminase (SGPT), serum glutamic oxalacetic transaminase (SGOT), bilirubin, alkaline phosphatase, protein electrophoresis, and testing for Au and antibodies to Au (AuAb) were performed prior to or at the time of transfusion, and only patients with normal values were followed. All tests on patients from both hospitals were performed at the Pepper Clinical Chemistry Laboratory of HUP. Follow-up of patients was done at intervals of 1, 2, 3, 4, 6, 8, 10, 13, 17, 21, and 26 weeks after transfusion; at these times history, physical examination, and laboratory determination of serum glutamic pyruvic transaminase (SGPT) Au and AuAb were performed.

The criterion for the diagnosis of anicteric hepatitis was an elevation of SGPT above 100 International Units (I.U.), otherwise unex-

TABLE I

HEPATITIS AFTER TRANSFUSION OF BLOOD PRODUCTS

Study	Product	Recipients	Hepatitis (Icteric)	%
At PGH:				
1963-1964	Whole blood, plasma	56	10 (0)	17.9
1968-1969	Whole blood, plasma	78	14 (4)	17.9
1968-1969	Packed red cells only	29	3 (0)	10.3
	After Au Screening-Exclusion Program			
1970-1971	Whole blood, plasma	204	12 (4)	5.9
	Nontransfused Control Patients Followed Similarly			
1963-1964		25	0	0
1968-1970		62	0	0
At HUP:				
1968-1969	Whole blood, plasma	64	3 (1)	4.7
1968-1970	Frozen, washed red cells	33	0	0

plained, occurring on at least subsequent occasions in a patient with previously normal values and persisting for at least three weeks. Hospitalization for further studies and liver biopsy was recommended for these patients. All patients with this degree of serum enzyme activity also showed findings of hepatitis when liver biopsy could be performed, and transient elevations in bromsulfonphthalein (BSP) retention, as previously reported.[1]

Donor blood from HUP and PGH was tested for Australia antigen at The Institute for Cancer Research, Fox Chase, by the immunodiffusion method.[13] Results were communicated to the blood banks by telephone, and positive units, if not yet used, were immediately removed from the blood bank inventory.

Based on early observations that about half the recipients of donor blood containing Au developed hepatitis,[14,15] we decided in July 1969 that no donor blood found to contain Au should be administered, and that mechanisms would be set up to examine all donor units. By the fall of 1969 complete testing of all donor blood at PGH was being performed and units were released for transfusion only if negative for Au. Thus PGH probably became the first hospital in the United States to implement a full Au screening and exclusion policy. As more test sera became available in 1970, the blood bank at PGH was able to set up its own testing, with both the immunodiffusion and the counterelectrophoresis[16] methods; duplicate specimens were sent to Fox Chase for processing as controls.

Results

In 1969, before the policy of testing and exclusion was established, a total of 57 patients had received at least one unit of blood containing Australia antigen. Of these patients (Table II), 13 at HUP and seven at PGH died of underlying disease within a few weeks after transfusion; four at HUP and six at PGH did not wish to be re-examined; 14 at HUP and 13 at PGH could be followed satisfactorily for a six-month period. Twelve of the 27 patients (44 per cent) followed for six months showed evidence of hepatitis (three icteric). Seven patients in whom hepatitis developed showed transient detectable Au; one developed antibodies to Au, and four patients had hepatitis without demonstrating either Au or antibodies. Of the 15 patients who did not have evidence of hepatitis, seven developed antibodies to Au, while eight patients had no abnormalities.

TABLE II

HEPATITIS IN RECIPIENTS OF AU-CONTAINING BLOOD, 1969

Subject					No.
Died of other disease					20
Refused follow-up					10
Followed (observed for full six months)					27
	No.	Au	AuAb*	Neither	
Hepatitis	12 (44%)	7	1	4	
No hepatitis	15 (56%)	0	7	8	

**AuAb, antibodies to Au, as detected by immunodiffusion.*

Eight of the Au-positive donors implicated in possibly transmitting hepatitis were available for further evaluation. All eight showed transaminase abnormalities in the range of SGPT 70-510 I.U. and SGOT 75-240 I.U., as well as elevation in the immunoglobulins (mainly in immunoglobulin G) varying from 2000 to 3800 mg per cent. All had been clinically asymptomatic and denied exposure to hepatitis or drug use when evaluated prior to blood donation.

Early in the study we had noticed a marked difference in the overall frequency of post-transfusion hepatitis between the two hospitals. In 1968-1969, there were 14 patients with hepatitis, of whom four were icteric, at PGH among 78 patients (17.9 per cent) followed for a full six-month period. Hepatitis did not occur in 55 nontransfused control patients. During the same time period, only three cases of hepatitis after transfusion, one icteric, were seen among 64 patients (4.7 per cent) followed at HUP. The correlation of post-transfusion hepatitis with donor antigen frequency became evident when we obtained more results of Au testing of the various donor groups. During 1969-1970, more than 26,000 donor units were tested at both hospitals. Testing was incomplete

REDUCTION OF POST-TRANSFUSION HEPATITIS

TABLE III

AU FREQUENCY IN DONOR BLOOD UNITS, AT HUP AND PGH

	1969			1970		
	Tested	Au	%	Tested	Au	%
HUP						
Donor room, club	1735	6	0.35	5008	11	0.21
American Red Cross	2322	4	0.17	4907	16	0.32
AABB*	1502	2	0.13	4200	3	0.07
Total	5559	12	0.22	14115	30	0.21
PGH						
Donor room	634	6	0.95	699	8	1.14
Prisoners	1307	41	3.14	1555	23	1.48
American Red Cross	768	0	0	1009	5	0.50
Commercial, professional	133	2	1.50	930	0	0
Total	2842	49	1.72	4193	36	0.86

**Units obtained from other hospitals in the region through the registry of the American Association of Blood Banks.*

TABLE IV

AU FREQUENCY IN DONOR BLOOD AT PGH

	1969	1970	1971
Volunteer donors	0.95%	1.14%	1.31%
	634/1084*	699†	153†
Prison donors	3.14%	1.48%	1.48%
	1307/2340	1555	1279
American Red Cross	0	0.50%	0.06%
	768/987	1009	1798
Commercial donors	1.5%	0	0
(including professional)	133/420	930	875
Total	1.72%	0.86%	0.54%
	2842/4831	4193	4105
Units used (as whole blood, packed cells, fresh frozen plasma, etc.)	4334	4074	3860

**Tested/Collected.*
†Collected and tested.

in 1969; only 38 per cent of all donor units collected for use at HUP and 58 per cent of all units for PGH were screened, while in 1970 all units were tested before transfusion (Table III). The overall frequency of Au was markedly higher in the PGH donor populations than in HUP populations. By far the highest frequency was observed in the PGH paid prisoner donors (1969, 3.14 per cent; 1970, 1.48 per cent). Volunteer donations by friends and relatives of PGH patients showed a much higher frequency (1969, 0.95 per cent; 1970, 1.14 per cent) than the comparable group at HUP (1969, 0.35 per cent; 1970, 0.21 per cent).

During 1970 and 1971, after donor screening and exclusion of Au-positive units had been fully instituted, only antigen-negative blood was transfused at PGH. It was already evident that screening and excluding Au-positive donors resulted in a marked change in the prevalence of Au in the PGH donor population, prisoner donors (Tables III and IV). We have been able to follow a total of 204 patients who received only tested Au-negative blood with serial SGPT and Austrialia antigen determinations for six months after transfusion. Twelve cases of hepatitis (5.9 per cent) occurred in this group, four of which were icteric (Table I), a significant decrease from the previous level ($p = 0.000035$). The use of packed erythrocytes resulted in only a slightly lower prevalence of hepatitis, and the difference was not significant (10.3 per cent, $p = 0.145$).

When liver biopsy was performed, all patients with three SGPT elevations to 100 I.U. or more showed findings consistent with mild viral hepatitis and BSP retention greater than 10 per cent at 45 minutes, as had been observed in the previous study.[1] Most anicteric patients were either completely asymptomatic or had only vague complaints of anorexia and fatigue. Of the four icteric patients only one required hospitalization. This patient had received 15 units of whole blood; his transaminase rose to 1528 I.U. of SGPT and 1908 I.U. of SGOT, and his serum total bilirubin reached a high of 22.6 mg/100 ml. None of the other three icteric patients showed a total bilirubin of more than 4.0 mg/100 ml; their clinical courses were mild. Maximum transaminase (SGPT) elevation ranged from 235 to 1530 I.U., the average being 607 I.U. for the entire group. Liver histopathology was available from six anicteric patients, and the findings were interpreted as viral hepatitis in each patient by the PGH pathologists, who were without knowledge of the study. Inflammatory cell infiltrates of mild to moderate degree in the periportal and intralobular areas, mild to moderate degenerative liver cell changes manifested as granularity, ballooning and occasional focal necrosis as well as presence of hyaline (Councilman) bodies were the consistent findings in the tissue specimens. The clinical picture, the laboratory findings, and the biopsy results showed no significantly different features as compared to the findings of previously studied patients who had presented with icteric and anicteric hepatitis after transfusion of blood containing Australia antigen. None showed serum Australia antigen during the follow-up period. In a nontransfused control group of 62 patients from neighboring beds who were selected by the same criteria and followed concurrently the same way, no case of hepatitis was found.

Discussion

In two independently conducted similar studies, separated by a five-year period, we found the frequency of icteric and anicteric hepatitis after blood transfusion to be about 18 per cent among patients at Philadelphia General Hospital.[1] A comparable number of patients selected in the same manner and followed during the same time period at the adjacent Hospital of the University of Pennsylvania showed a frequency of less than 5 per cent. The major effort of 1970-1971 was directed toward determining to what extent post-transfusion hepatitis could be reduced at PGH by an Au screening and exclusion program. The 5.9 per cent overall incidence of hepatitis after blood transfusion observed at PGH during 1970-1971 was a marked reduction from the 18 per cent observed in two previous groups of patients. It was comparable to the less than 5 per cent incidence previously observed at HUP in a group of patients who had received untested blood, mainly from donors with a much lower frequency of Au.

As had been shown empirically before Au testing was available, there are certain population groups in which the hepatitis-inducing agent or agents are present more frequently.[2,6,17] In early 1969 we began to collect data about the sources of blood used for transfusion at both hospitals. Blood from paid donors from three Philadelphia prisons, a source upon which PGH is dependent to obtain sufficient blood for emergency needs, had the highest frequency of positive units. Although known addicts were excluded as donors, it is well known that drug use and addiction are high among prisoners and, since they donate for payment, it is likely that many deny drug use, previous hepatitis, or exposure to hepatitis. The cramped, close living conditions in the penal institutions would also facilitate the nonparenteral spread of hepatitis

REDUCTION OF POST-TRANSFUSION HEPATITIS

from asymptomatic carriers. Thus, prisons appear to constitute a significant hepatitis virus pool in this community.

A marked difference was present in volunteer-donated units. Friends and family members donating for PGH patients had a fivefold higher frequency of Au units than the same kind of donors at HUP. The volunteer group at PGH had a different socioeconomic background from the HUP donor population; PGH patients were almost exclusively from the poor, predominantly black, inner city population. The substandard housing and hygiene, cramped living conditions, and poorer overall health care in this group are other factors that could facilitate the spread of infection. Blood obtained from the American Red Cross showed approximately equal frequencies of positive units at both hospitals.

Our studies and those of others have shown that administration of blood containing Au produces observable hepatitis in almost 50 per cent of recipients. Another 20 to 30 per cent produce antibodies to Au, an indication of some active immunologic response to the agent transfused. The evidence indicates that these antibodies may protect these patients against the disease. A significantly ($p < 0.05$) lower incidence of hepatitis was observed in those developing antibodies after blood containing Au was transfused (Table II). The patients at PGH who developed hepatitis in the combined 1968-1969 and 1970-1971 studies each received an average of 4.3 units of blood, while, in the entire group followed, the average patient received just above 2.5 units. It appears that most of the Au-negative donor blood causing hepatitis originated from the same donor population that had been shown to cause hepatitis in nearly 50 per cent of recipients of Au-containing blood. Sixty-five per cent of the blood transfused to the hepatitis group had originated from prison donors and volunteer PGH donations.

In a recent study of 126 patients at the National Institutes of Health,[18] by excluding both commercial donors and all units from Au-positive donors, post-transfusion hepatitis has been reduced to only 3.7 cases per 1000 units transfused. Efforts to reduce further the incidence of transfusion-related hepatitis should proceed in the direction of excluding Au-positive blood and interdicting transfusion of blood from demonstrated high-risk donor groups.

References

1. Hampers CL, Prager D, Senior JR: Post-transfusion anicteric hepatitis. New Eng J Med **271**:747-754, 1964.
2. Walsh JH, Purcell RH, Morrow AG et al: Post-transfusion hepatitis after open heart operations—incidence after the administration of blood from commercial and volunteer donor populations. JAMA **211**:261-265, 1970.
3. Allen JG, Sayman WA: Serum hepatitis from transfusions of blood: epidemiologic study. JAMA **180**:1079-1085, 1962.
4. Grady GF, Chalmers TC: Risk of post-transfusion viral hepatitis. New Eng J Med **271**:337-342, 1964.
5. Senior JR: Reflections upon the incidence of post-transfusion hepatitis in various parts of the world. Amer J Gastroenterol **49**:298-303, 1968.
6. Creutzfeldt W: Die Transfusionshepatitis und ihre Verhutung. Internist **7**:1-8, 1966.
7. Okochi K, Murakami S: Observations of Australia antigen in Japanese. Vox Sang **15**:374-385, 1968.
8. Blumberg BS, Alter HJ: Precipitating antibodies against a serum protein ("Australia antigen") in the serum of transfused hemophilia patients. J Clin Invest **44**:1029, 1965.
9. Blumberg BS, Gerstley BJS, Hungerford DA et al: A serum antigen (Australia antigen) in Down's syndrome, leukemia, and hepatitis. Ann Intern Med **66**:924-931, 1967.
10. Taswell HF, Shorter R, Poncelet TV et al: Hepatitis-associated antigen in blood donor populations. JAMA **214**:142-144, 1970.
11. Gocke DJ, Greenberg HB, Kavey NB: Correlation of Australia antigen with post-transfusion hepatitis. JAMA **212**:877-879, 1970.
12. Kliman A: No hepatitis after packed red cells? New Eng J Med **279**:1290, 1968.
13. Blumberg BS, Byrne RJ, Chanock RM et al: Viral hepatitis and tests for the Australia (hepatitis-associated) antigen and antibody. Bull Wld Hlth Org **42**:957-992, 1970.
14. Blumberg BS, Sutnick AI, London WT et al: Current status of the Australia antigen viral hepatitis studies. Gastroenterology **56**:1212, 1969.
15. Goeser E, London WT, Sutnick AI et al: Short incubation post-transfusion hepatitis in recipients of blood containing Australia antigen. Gastroenterology **58**:1026, 1970.

16. Pesendorfer F, Krassnitzky O, Wewalka F: Immunoelektrophoretischer Nachweis von Hepatitis-Associated Antigen (Au/SH-Antigen). Klin Wschr **48**:58-59, 1970.
17. Cherubin CE, Prince AM: Serum hepatitis specific antigen (SH) in commercial and volunteer sources of blood. Transfusion **11**:25-27, 1971.
18. Alter HJ, Holland PV, Purcell RH, et al: Post-transfusion hepatitis after exclusion of commercial and hepatitis-B antigen-positive donors. Ann Intern Med 77:691-699, 1972.

British Blood Transfusion Society

NEWSLETTER No. 19 JANUARY 1990

Feature Article

James Blundell Award Lecture: 1989

BLOOD TRANSFUSION AND THE DISCOVERY OF VIRUSES

Baruch S. Blumberg

INTRODUCTION

I am very honoured to have been granted the James Blundell Award of the British Blood Transfusion Society. It is gratifying to know that the results of our research have been widely applied to the safeguarding of the blood supply and that it has proven useful to those, such as the members of your Society, who are responsible for this important aspect of health care.

There were two features of blood banking practices that were used in the research which led to our discovery of the hepatitis B virus. One was the concept of polymorphism, that is, inherited biochemical variation, which was studied extensively in humans by blood transfusionists. The second is that blood-borne viruses can infect some transfused patients and lead to the development of antibodies against them in others.

Immunology began as the study of the reaction of a putative host to an infectious agent. This became more generalised to the study of reactions to any foreign agent and, as such, provided the biological models for allergic diseases. This was later extended to the understanding of reactions to host antigens (possibly altered from their native state) as an explanation for autoimmune diseases. Early in the history of immunology an important application derived from a study of reaction to inherited antigens from the same species as the host, and this became the biological model for blood transfusion. Most of the incompatibility reactions in transfusion could be controlled by the identification of a small number of systems, i.e. ABO and Rh. However, the continuous discovery of new antisera usually in the blood of transfused persons greatly increased the number of polymorphisms which could be studied in humans and the number of categories in which individual humans could be classified. The complexity of classification was enhanced by the study of other polymorphisms, usually of blood, including the rich systems in the gammaglobulins, the low density lipoproteins and, particularly the antigens of white blood cells, the HLA system.

Possibly the ultimate in individual characterisation has now arrived with the use of the RFLP system which allows classification at the genetic level and which can distinguish a single person as unique. We can now envision a classification system in which each person is in his or her own category.

DISCOVERY OF HEPATITIS B VIRUS

Hepatitis B virus was discovered as a consequence of a population study designed to identify inherited immunological traits in humans. (A detailed account of these events, with references, is given elsewhere [1,2]). Beginning in 1955, with the introduction of the starch gel electrophoresis method by Smithies, it became clear that there was a large number of inherited biochemical polymorphisms of human serum proteins. Allison and Blumberg (3) had postulated that if any of these polymorphisms involved antigenic proteins, then patients who received even a small number of transfusions would be likely to be exposed to proteins or other biochemical entities which they themselves had not inherited or acquired, but which were present in the donors' blood. If this were to happen, then the transfused patient might develop a precipitating antibody against this "foreign" protein, and the antisera so formed could be used to detect the polymorphic protein system.

This hypothesis was tested by using the serum from transfused patients as a reagent in agar gel immunodiffusion experiments (3). We soon identified a previously unknown polymorphism of the serum low density lipoproteins (termed the Ag system) which has proven to be of interest in genetics, disease studies and forensic medicine (4). After this initial discovery, we continued to test the hypothesis that we could identify serum protein polymorphisms by the transfused-patient technique and discovered a "new" precipitin system that was distinctly different from the first (5,6). The initial reaction which we studied was between the serum of a transfused haemophilia patient from New York City and an Australian aborigine. We termed the antigenic material present in the aborigine "Australia antigen" (Au) and designed a series of studies to determine the biological significance of this unusual constituent.

We found that Australia antigen was very rare in the United States and northern European populations (about 0.1%) but common in tropical and Asian groups (5-15%). It was also found to be common in patients with leukemia (6), an observation which arose as a consequence of systematic studies of the distribution of Au in a variety of diseases. Based on the Au/leukemia disease association, we made the hypothesis that patients who were likely to develop leukemia were also likely to have Au. A series of populations with a high likelihood of developing leukemia were identified and tested. These included persons with Down's syndrome who were known to have a much greater risk than the general population of developing leukemia in childhood. Our prediction was fulfilled; more than 30% of institutionalised patients with Down's syndrome were found to have the Australia antigen in their blood.

At about the time of this discovery (1966), a series of observations had been made which directed our attention to the possibility that Australia antigen might be associated with a hepatitis virus. The most important of these, in retrospect, was an observation made on J.B., one of our patients with Down's syndrome. In our original series of observations on people with Au, we had seen that its presence or absence appeared to be a persistent characteristic. J.B. did not have Au in his serum in the initial analysis; contrary to expectation, Au was found in his blood on subsequent examinations. Concomitant with the appearance of Au, he developed enzyme elevations and findings on liver biopsy of hepatitis. We then made the hypothesis that Au was associated with a hepatitis virus and tested this association hypothesis with another series of epidemiological studies. We found that Au was more common in patients with acute and chronic hepatitis than in healthy people or patients with other diseases (7). These observations were soon confirmed by Okochi and Murakami (8). Prince (9) and Vierucci et al (10). Prince (9) focussed attention on the specific relation of Australia antigen to serum hepatitis.

Based on these population studies, we formulated a biological model; namely that Au was, or was on, a virus which could cause hepatitis. This was tested by a series of experimental laboratory studies. Eventually the virus particle was isolated by column separation, ultracentrifugation and enzymatic digestion of remaining serum proteins (11). It was visualised under the electron microscope (12,13), and its transmission by transfusion (14,15) and animal inoculation was shown (16).

One of the most important applications of the diagnostic powers of the original test for Au (also called hepatitis-associated antigen, HAA and currently hepatitis B surface antigen, HBsAg) and the sensitive derivatives from it was in the prevention of post-transfusion hepatitis by testing blood donor units and removing those which contained HBsAg. This application was primarily based on the data supporting the hypothesis that Australia antigen was on the hepatitis virus, but there were a series of interesting organisational and administrative procedures which led to the final acceptance of these procedures in the United States (2). I understand that there were a similar series of events in Great Britain and elsewhere, although, it is likely that there were unique features in each country. It provides, I believe, an instructive example of how basic research finds its way into practical application.

The series of experiments which led to the hypothesis that Australia antigen (Au) was, or was on, a hepatitis virus has been described. These results generated the hypothesis that "Au" in the blood of a donor signified the presence of the virus in a carrier; but this had to be tested directly, and we designed a study to do so at the Philadelphia General Hospital. Before this project was complete, our Japanese colleague, Dr. Kazuo Okochi, visited our laboratory and showed us the results of his own investigation. He had tested the same hypothesis that we had proposed, that patients who received donor blood containing Au were much more likely to develop evidence of hepatitis than those who had been given blood which did not contain Au. His study had progressed more rapidly than ours since Au positive donors are ten times more common in Japan than in the U.S.

We realized that his data showed a significant difference, clearly supporting the hypothesis. We decided to stop our experiment at Philadelphia General Hospital but to continue testing the blood for Au. It was at this moment that research turned into application, at least in Philadepha.

What happened between the time that we decided that the evidence was sufficient to warrant medical application and the time that this notion was generally accepted in the medical and public health community? There was very little new scientific evidence published. Gocke (14) reported eight cases that developed evidence of hepatitis infection following transfusion of blood containing Australian antigen, but, to my knowledge, there were no additional controlled trials. Despite that, the machinery of medical organisation began to react rapidly and effectively.

In June of 1969 (17), we published a paper suggesting that testing of donor blood should be done. There was a reaction to this by members of the blood banking community who requested that withdraw the paper since it might cause major and expensive changes in blood transfusion practice.

Before I could react to this, events began to sweep away the objections. In October/November of 1969, a committee of the U.S. National Research Council did not recommend testing, but it did recommend the further development of tests. However, the position that testing should be deferred was not held by all. Paul Schmidt, who was in charge of the Blood Bank at the Clinical Center of the prestigious government research laboratory, the National Institutes of Health, recommended that all blood banks capable of doing so should test donor blood. Their paper was publshed in **Lancet** on 18 July 1970 (18) and included a disclaimer from the U.S. Government stating that Schmidt and the others spoke for themselves and did not represent Goverment opinion. This was a very unusual step since the N.I.H. usually stood behind the publications of its employees.

A few days later, the **New York Times** reported on Schmidt recommendation and his views spread beyond the limited community of physicians interested in hepatitis.

Late in 1969, we began to hear that legal action for malpractice was being taken by patients or their families who had received blood transfusions and developed hepatitis. They claimed that hospitals, blood banks, and physicians were negligent because they had not tested donor blood for Australia antigen. We were also contacted by hospitals and others who wanted to establish the test and we had the sense that the threat of expensive law suits was shifting the decision making groups towards accepting testing.

Events now proceeded more quickly. The National Research Council committee reconvened on October 15, 1970 and now recommended testing. The report was provided to the American Association of Blood Banks within a few days and, by October 29, 1970 they too recommended testing. It had taken little more than a year from our decision to test in Philadelphia to arrive at the agreement that testing should be general.

Within the next few years, different jurisdictions passed laws, enacted regulations or used other devices to ensure compliance with high standards for testing.

RETROVIRUSES AND TRANSFUSION

Hepatitis B virus has many characteristics in common with retroviruses. One phase of its replication process involves the production of DNA from an RNA template through the action of a reverse transcriptase. There are also many similarities in the mechanism of transmission of HBV and other retroviruses, in particular, HIV. Soon after the onset of the AIDS epidemic it was, tragically, learned that HIV could be transmitted by transfusion of blood and blood products. Immediately after the discovery of HIV it was applied to the screening of blood donors which quickly controlled its spread by this mechanism. HTLV-1, whose disease consequences may not be as severe, is also spread by transfusions, and donor bloods are now also tested in the U.S. and elsewhere. This testing will probably lead to the discovery of other retroviruses in the blood.

The sera of transfused people provide a mechanism by which the effects of HTLV-1 are amplified and can be used for studying the epidemiology of the virus. As an example, we recently determined the frequency of HTLV-1 in the sera of patients who were transfused (in Philadelphia) in 1969 and found it to be about the same as in a study conducted on blood collected in 1976 (19).

CONCLUSION

Blood transfusionists have access to biological material of great interest for the study of blood borne infections of humans. They also have the capability of studying the details of human immunological and biochemical individually. The specimens collected and stored in blood bank facilities are a rich resource for future research on infectious disease and on variation.

REFERENCES

1. Blumber, B.S. Australia antigen and the biology of hepatitis B. Science **197**: 17-25 (1977).
2. Blumberg, B.S. Keynote address: The Australia antigen story; in Millman, Eisenstein, Blumberg, Hepatitis B: the virus, the disease and the vaccine. (Plenum Publishing Corp., New York 1984).
3. Allison, A.C. Blumberg, B.S. An isoprecipitation reaction distinguishing human serum protein types. Lancet **1**: 634-637 (1961).
4. Blumberg, B.S., Dray, S., Robinson, J.C. Antigen Polymorphism of a low-density beta-lipoprotein. Allotype in human serum. Nature, Lond. **194**: 656-658 (1962).
5. Blumberg, B.S. Polymorphisms of serum proteins and the development of isoprecipitins in transfused patients. Bull. N.Y. Acad. Med. **40**: 377-386 (1964).
6. Blumberg, B.S., Alter, H.J., Visnich, S. A "new" antigen in leukemia sera. J. Am. Med. Assoc. **191**: 541-546 (1965).
7. Blumberg, B.S., Gerstley, B.J.S., Hungerford, D.A., London, W.T., Sutnick, A.I. A serum antigen (Australia antigen) in Down's syndrome, leukemia and hepatitis. Ann. Intern. Med. **66**: 924-931 (1967).
8. Okochi, K., Murakami, S. Observations on Australia antigen in Japanese. Vox Sang **15**: 374-385 (1968).
9. Prince, A.M. An antigen detected in the blood during the incubation period of serum hepatitis. Proc. Natl. Acad. Sci. USA **60**: 814-821 (1968).
10. Vierucci, A., Bianchini, A.M., Morgese, G., Bagnoli, F., Messina, G. L'antigene Australia. I. Rapporte con l'epatite infettiva e da siero. Una ricerca in pazienta pediatrici. Pediatria int. **18**: 4, 3-11 (1968).
11. Millman, I., Leob, L.A., Bayer, M.E., Blumberg, B.S. Australia antigen (a hepatitis-associated antigen). Purification and physical properties. J. Exp. Med. **131**: 1190-1199 (1970).
12. Bayer, M.E., Blumberg, B.S., Werner, B. Particles associated with Australia antigen in the sera of patients with leukemia, Down's syndrome and hepatitis. Nature, Lond. **218**: 1057-1059 (1968).
13. Dane, D.S., Cameron, C.H., Briggs, M. Virus-like particles in serum of patients with Australia antigen-associated hepatitis. Lancet **1**: 695-698 (1970).
14. Gocke, D.J., Greenberg, H.B., Kavey, N.B. Hepatitis antigen. Detection of infectious blood donors. Lancet **2**: 248-249 (1969).
15. Okochi, K., Murakami, S., Nomoniya, K., Kaneko, M. Australia antigen, transfusion and hepatitis. Vox Sang **18**: 289-300 (1970).
16. London, W.T. Transmission of Australia antigen to man and nonhuman primates. Proc. Natl. Acad. Sci. USA **66**: 235-236 (1970).
17. Blumberg, B.S., London, W.T., Sutnick, A.I. Relation of Australia antigen to virus of hepatitis. The Bulletin of Pathology **10**: 164 (1969).
18. Alter, H.J., Holland. P.V. and Schmidt, P.J. Hepatitis-associated antigen: to test or not to test? Lancet **2**: 142-143 (1970).
19. Blumberg, B.S., Blanton, S.H., Shrager, D.I., Blattner, W.A. HTLV-1 (Philadelphia, 1969). Lancet **2**: 327-328, (1989).

Baruch S. Blumberg
Fox Chase Cancer Center,
77011 Burholme Avenue,
Philadelphia, U.S.A.

Present address: Balliol College, Oxford, England, OX1 3BJ.

Reprinted from PROC. SOC. EXP. BIOL. AND MED. Volume 138, Number 1 October 1971

PROCEEDINGS OF THE SOCIETY FOR EXPERIMENTAL BIOLOGY AND MEDICINE 138, 249–257 (1971)

Radioimmunoprecipitation Assay for Australia Antigen, Antibody, and Antigen–Antibody Complexes[1] (35872)

JOHN A. COLLER,[2] IRVING MILLMAN, THERESA C. HALBHERR, AND BARUCH S. BLUMBERG

The Institute for Cancer Research, Fox Chase, Philadelphia, Pennsylvania 19111

The association of Australia antigen, Au(1), with viral hepatitis is well established (1–5). The immunodiffusion (ID) (Ouchterlony) method is widely used (6) and permits specificity control by means of lines of identity with known specimens; however, relatively large amounts of antigen and antibody are required to form a visible precipitin band. "Cross over" or "reaction" electrophoresis described by Lang (7) and used by Krassnitsky *et al.* (8) and Pesendorfer *et al.* (9) for the detection of Au(1) is also contingent upon a visible precipitin reaction. Although more rapid, it is only slightly more sensitive than immunodiffusion, and the advantage gained by the rapidity is compromised by the inability to maintain a specificity comparison. Complement fixation (CF) for the detection of Au(1) first described by Purcell *et al.* (10) is a sensitive test but it also lacks the built in specificity comparison of immunodiffusion since any antigen–antibody combination will permit the fixation of complement.

Although 60–90% of patients developing posttransfusion hepatitis have detectable Au(1) at some time during the course of their illness, only 20–30% of them have received blood from Au(1) positive donors as determined by immunodiffusion or complement fixation. The majority of the patients have therefore contracted hepatitis either from a source other than blood transfusion or from Au(1) negative donors. If transmission has in fact been from Au(1) negative donors, then the inability to detect Australia antigen in these blood samples may be a result of the insufficient sensitivity of the current methods.

Our current investigation, an extension of an earlier report (11)[3] was directed toward the development of a highly sensitive detection method for Au(1) and human antibody to Au(1) without loss of immunologic specificity. Specificity is preserved through the use of purified Au(1) as a constant reactant. Whereas many radioimmunossay systems precipitate immune complexes either as a double antibody complex or by gamma-globulin precipitation with 50% saturated ammonium sulfate the procedure described below combines both methods to accomplish specific immunoprecipitation of double antibody–antigen complexes with a less concentrated ammonium sulfate solution.

Materials and Methods. Purification of Au(1). Susceptible normal protein components of an Au(1)-containing plasma specimen were degraded by enzyme treatment (13). Au(1) was then purified by sequential Sephadex G-200 column gel filtration, sucrose and cesium chloride gradient centrifugation. That fraction which displayed immunoreactivity with human and heterologous anti-Au(1) antisera, but not with antinormal human sera antisera, was used as the antigen

[1] This work was supported by U.S. Public Health Service Grant 5TO1-GM01540-05; by U.S. Public Health Service Grants CA-06551, CA-08069, CA-06927, and RR-05539 from the National Institutes of Health; by an appropriation from the Commonwealth of Pennsylvania and by contract No. NIH 70-2334.

[2] Department of Surgery, Hospital University of Pennsylvania.

[3] At the time of our first disclosure (11), Walsh *et al.* (12) described a radioimmunoassay for the detection of Australia antigen. Their method differed from the one reported here in the manner of determining bound and unbound radioactive precipitate.

source in these experiments. The details of the purification procedure have been described elsewhere (13).

Purified Au(1) conjugation with ^{125}I. Conjugation of purified Au(1) with ^{125}I-iodide was accomplished with chloramine-T using a modification of the method of Greenwood *et al.* (14). The reactants, in small volumes, consisted of 100 μg of purified Au(1), 1 mCi of ^{125}I, 100 μg of Chloramine-T, 244 μg of sodium metabisulfite. Following removal of unconjugated ^{125}I-iodide by both gel filtration and dialysis, 90–95% of the remaining radioactivity was precipitated in 5% trichloroacetic acid. Seventy to 92% of the protein-associated radioactivity was precipitable by human anti-Au(1) in the method described below.

Immunoprecipitation of ^{125}I-labeled Au(1). Human antibody to Au(1) was obtained from a blook-bank employee who had never had recognized clinical or chemical hepatitis but had persistent anti-Au(1) antibody activity as determined by use of immunodiffusion. The fraction of this serum that was precipitated in 50% saturated ammonium sulfate was equilibrated against a 0.1 *M* borate–0.85% NaCl, pH 8.4 buffer without concentration and designated the standard antibody preparation. This preparation has been used extensively for the routine clinical immunodiffusion detection of Australia antigen in our laboratory and is immunoreactive in titers up to 1:8. Immunoprecipitation of ^{125}I-labeled Au(1) with this preparation was performed in the following manner:

Twofold serial dilutions of the standard antibody from 1:40 to 1:1,280,000 were made using 0.1 *M* borate–0.85% NaCl buffer with 0.04% bovine serum albumin (BSA), pH 8.4. Five hundred microliters of each dilution was incubated with 500 μl of ^{125}I-labeled Au(1) (approx 15,000 cpm on the day of labeling), at 37° for 2 hr and 4° for 22 hr. Thereafter, 50 μl of rabbit antihuman IgG (H-chain specific) was added, followed by incubation at 37° for 2 hr and 4° for 18 hr. Centrifugation of immune complexes was performed at either 8000 rpm for 1 hr or at 5000 rpm for 30 min after the addition of sufficient 60.9% saturated ammonium sulfate to provide a final saturation of 25%.

The qualitative detection of anti-Au(1) in an unknown human serum was accomplished by substituting for the standard antibody preparation a 10-μl volume of undiluted serum. Titers of positive specimens were determined using 500-μl dilutions in the same manner as described for the standard antibody preparation.

The detection of Au(1) in human sera by coprecipitation-inhibition is based on the capacity of added unlabeled Au(1) to compete with labeled Au(1) for a predetermined and constant amount of the standard anti-Au(1) preparation and thereby decrease the amount of precipitable radioactivity. In this procedure, 10 μl of a 1:10 or greater dilution of the test specimen is incubated with 500 μl of the standard antibody preparation (generally a 1:20,000 dilution) at 37° for 2 hr and 4° for 22 hr prior to incubation with the ^{125}I-labeled Au(1) and rabbit antihuman IgG and separation with 25% saturated ammonium sulfate. All dilutions are made in 0.1 *M* borate–0.85% NaCl buffer with 0.04% BSA, pH 8.4.

Comparisons were made of sera which were Au(1) positive by immunodiffusion; Au(1) negative but containing antibody to Au(1) by immunodiffusion; and sera found by use of immunodiffusion to contain neither antigen nor antibody. The latter specimens were from first year medical students with no known hepatic dysfunction as determined by historical, clinical, and chemical investigations. In each case, these specimens did not react with a panel of 20–30 Au(1) and anti-Au(1) sera.

Results. Standard anti-Au(1) antibody preparation. The precipitin curve of increasing dilutions of the standard antibody preparation with constant ^{125}I-labeled Au(1) and rabbit antihuman IgG is shown in Fig. 1. There is an area of equivalence between 1:10,000 and 1:20,000 dilution of standard antibody. A nearly identical curve is obtained when whole serum dilutions of standard antibody are used in place of the 50% saturated ammonium sulfate precipitable fraction of standard antibody.

In preliminary investigations, the human

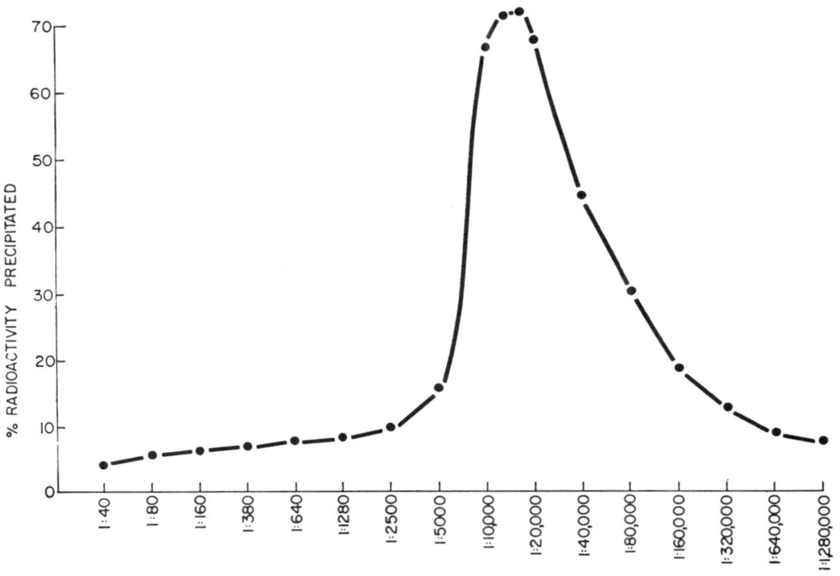

Fig. 1. Standard anti-Au(1) antibody precipitin curve: Dilutions of the standard antibody were incubated with ^{125}I-labeled Au(1), rabbit antihuman IgG and centrifuged at 8000 rpm for 1 hr. The percentage of precipitable radioactivity is plotted.

anti-Au(1) antibody by itself did not form an immune complex with labeled Au(1) that could be separated from free unbound labeled Au(1) by centrifugation. Concentrations of ammonium sulfate that efficiently precipitate gamma globulin and thus any bound labeled antigens were also unsatisfactory in delineating immune complexes. We found that the bulk of free Au(1) is also precipitated at such concentrations. Therefore, a method was developed to detect the immune complexes in the area of antibody excess by salt precipitation of double antibody complexes. A concentration of ammonium sulfate was used that was inadequate to precipitate the individual components but that could quantitatively precipitate the intact double antibody complex. A final concentration of 25% saturated ammonium sulfate followed by centrifugation resulted in the selective precipitation of the double antibody-Au(1) complex.

Following the addition of ammonium sulfate to provide a final saturation of 25%, maximal precipitable radioactivity equal to, but not exceeding, that observed at equivalence is obtained at the dilutions corresponding to anti-Au(1) excess (Fig. 2).

Similar dilutions of normal human sera used in place of the standard antibody preparation failed to precipitate more than 3.5% of the total radioactivity at any dilution. At a standard antibody dilution of 1:640,000 the precipitable radioactivity of the standard antibody dilution exceeds that of the normal sera by 3 SD. There was no significant precipitable radioactivity in the absence of rabbit antihuman IgG.

Detection of anti-Au(1) antibody activity in human sera. Serial dilutions of whole serum with known anti-Au(1) and Au(1) activity as determined by immunodiffusion were made. After sequential incubation with labeled Au(1) and rabbit antihuman IgG, centrifugation was performed at 5000 rpm following the addition of ammonium sulfate. A comparison of titers as determined with immunodiffusion, complement fixation, and radioimmunoprecipitation (RIP) is given in Table I. Antibody was detected by radioimmunoprecipitation in three of five specimens which contained only Au(1) by immunodiffusion or complement fixation; in one there was a moderately high titer (1:30,000).

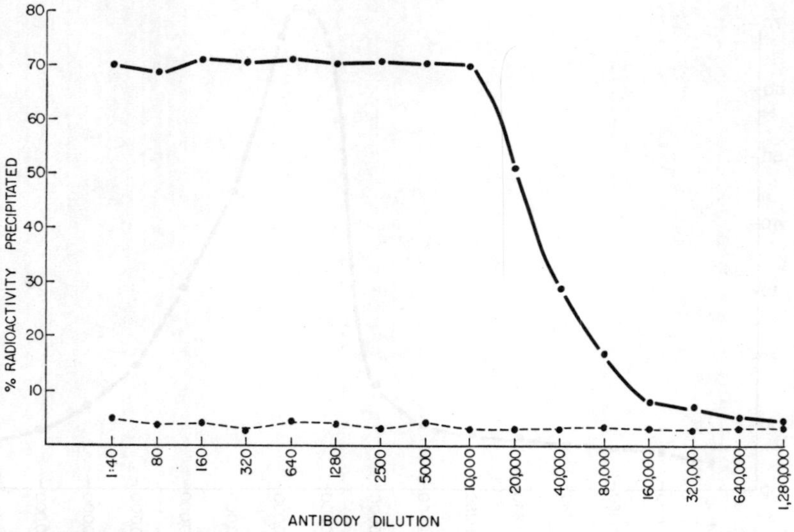

FIG. 2. Ammonium sulfate precipitation of double antibody complexes: Dilutions of the standard antibody were incubated with ^{125}I-labeled Au(1), and rabbit antihuman IgG. Ammonium sulfate to a final saturation of 25% was added before centrifugation at 5000 rpm for 30 min. The percentage of precipitable radioactivity for the standard antibody (●—); and for normal human sera (● - -) is plotted.

Detection and quantitation of purified Au(1) by coprecipitation inhibition. In Fig. 3 is shown the inhibition of precipitable radioactivity as a function of added unlabeled purified Au(1). Ten-microliter dilutions of purified Au(1) are incubated with 1:20,000

TABLE I. Anti-Au(1) Antibody in Human Sera: Comparison of Immunodiffusion, Complement Fixation, and Radioimmunoprecipitation as Assay Methods.[a]

Specimen	Titer ID	Titer CF	Titer RIP
Au(1) negative[b]			
66993	**1:8**	**1:8**	**1:640,000**
41177	**1:16**	**1:32**	**1:160,000**
66821	**1:4**	**1:8**	**1:160,000**
66820	**1:8**	**1:16**	**1:320,000**
67003	**1:8**	**—**	**1:40,000**
Au(1) positive[c]			
66707	**—**	**—**	**—**
45417	**—**	**—**	**1:50**
66751	**—**	**—**	**1:200**
42467	**—**	**—**	**—**
66835	**—**	**AC**	**1:30,000**

[a] ID = immunodiffusion; CF = complement fixation; RIP = radioimmunoprecipitation; AC = anticomplementary.

[b] 66993, blood bank nurse, Hospital University of Pennsylvania; 41177, aplastic anemia, female; 66821, hemophiliac, male; 66820, hemophiliac, male; 67003, Cooley's anemia, female.

[c] 66707, hemodialysis for renal failure, male; 45417, Down's syndrome, male; 66751, infectious viral hepatitis, female; 42467, serum hepatitis, male; 66835, hepatitis of unknown origin, female.

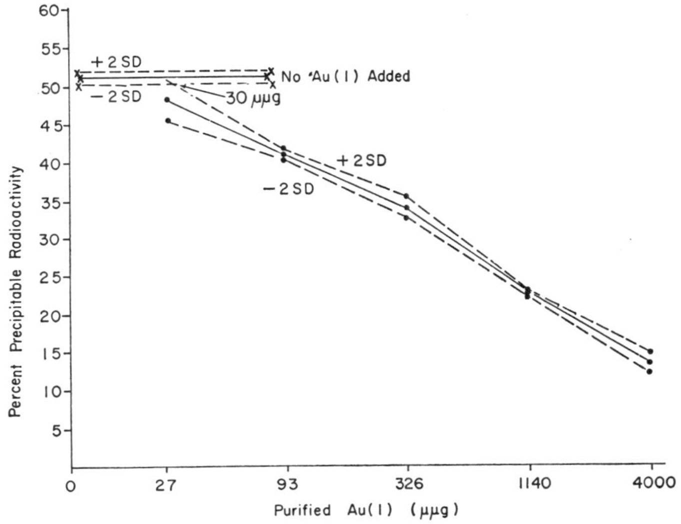

FIG. 3. Inhibition of precipitable radioactivity as a function of added unlabeled purified Au(1): A 1:20,000 dilution of standard antibody was sequentially incubated with unlabeled purified Au(1), ^{125}I-labeled Au(1), and rabbit antihuman IgG, followed by ammonium sulfate separation.

standard antibody prior to incubation with labeled Au(1), rabbit antihuman IgG and separation with ammonium sulfate. In the absence of added unlabeled Au(1), the 1:20,000 antibody dilution precipitates approximately 50% of the immunologically reactive radioactivity. Thirty picograms (micromicrograms) of purified Au(1) caused a significant inhibition of precipitable radioactivity.

Detection and quantitation of Au(1) in human sera by coprecipitation inhibition. The detection of Au(1) in human sera, unlike the inhibition obtained with purified unlabeled Au(1) may be complicated by the presence of other serum components. Principally, this includes the possible coexistent presence of anti-Au(1) antibody and the competition for rabbit antihuman IgG by varying amounts of nonspecific gammaglobulin.

For the most part, nonspecific interference due to components of the test specimen other than Au(1) or anti-Au(1) can be eliminated by the selection of an appropriate serum dilution. In the absence of added normal human serum a 1:20,000 dilution of standard antibody preparation precipitates approximately 50% of the ^{125}I-labeled Au(1). If this standard antibody dilution is preincubated with 10 μl of a 1:10 dilution of normal human sera prior to the addition of ^{125}I-labeled Au(1), the mean precipitable activity is 44.1% ± 7.8 (1 SD). Greater dilutions result in a much smaller deviation among normal sera, *e.g.*, 1:20 dilution permits recovery of 54.1% ± 1.9 (1 SD) precipitable activity.

In Fig. 4, the inhibiting capacity of 5 sera which contain Au(1) according to immunodiffusion is compared with that of 5 normal sera. The mean precipitable activity ± 2 standard deviations for each 10-μl dilution of the normal specimens is indicated. The titer of Au(1) in each specimen is the greatest serum dilution that results in precipitable activity of more than 2 SD below the corresponding dilution of normal serum.

The quantity of serum Au(1) (μg of protein/ml of serum) is determined by comparing, under identical conditions, the degree of inhibition of an appropriate dilution of the test specimen with that of purified Au(1).

Immunodiffusion and complement fixation testing is compared in Table II with the antigen titer and Au(1) protein concentration as determined by radioimmunoprecipitation for the same Au(1) positive specimens that were tested for antibody (Table I).

Detection of anti-Au(1) antibody by co-

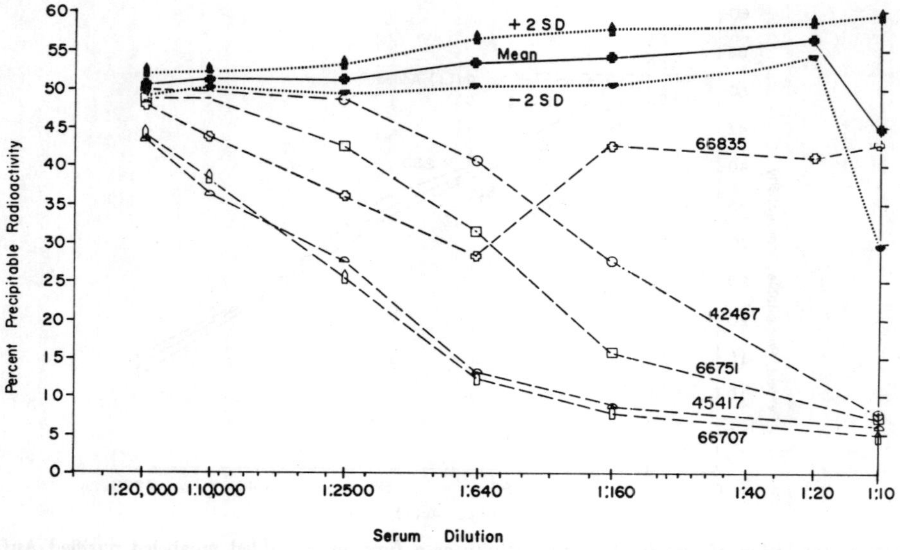

FIG. 4. Detection of Au(1) in human sera by coprecipitation-inhibition: Dilutions of 10-μl volumes of either normal human sera (—); or immunodiffusion Au(1) positive specimens (- - -); were incubated sequentially with the standard antibody (1:20,000), ^{125}I-labeled Au(1), rabbit antihuman IgG and separated with ammonium sulfate.

precipitation enhancement. The method of antigen detection employs a concentration of standard antibody that is sufficient to precipitate about one half of the ^{125}I-labeled Au(1). If the test specimen contains antibody rather than antigen, the total amount of antibody in the system will be increased and the final percentage of precipitable ^{125}I-labeled Au(1) will exceed the 50% control level. With increasing dilutions of the test specimen, less antibody is added and the precipitable activity will decrease to the 50% level.

We refer to this method of detecting predominantly antibody activity in a specimen being tested for antigen as "coprecipitation enhancement."

Delineation of antigen–antibody complex in human serum. 1. Antigen excess. Antigen testing of specimen 66835 at the initial 10-μl dilution of 1:10 (Fig. 4) fails to demonstrate Au(1) by inhibition. As greater dilutions are examined, however, inhibition is clearly seen, increasing first to a maximum before decreasing to a nondetectable level. That this nonlinear inhibition curve is related to the previously detected antibody activity is shown by comparison of the amount of test

TABLE II. Comparison of Immunodiffusion, Complement Fixation, and Radioimmunoprecipitation in the Detection of Au(1) in Human Sera.

Specimen[a]	Titer: ID	Titer: CF	RIP: Titer	RIP: Protein (μg/ml)
66707	1:16	1:128	1:30,000	174
45417	1:32	1:256	1:26,000	(149)[b]
66751	1:4	1:32	1:5000	(26)[b]
42467	1:8	1:16	1:2000	10
66835	1:8	1:256	1:15,000	(60)[b]

[a] See Table I.
[b] Antigen–antibody complex, see Table I.

specimen added relative to the constant amount of standard antibody used. Ten-microliter dilutions greater than 1:600 of No. 66835 do not contribute appreciable antibody in terms of the 1:20,000 standard antibody and the degree of inhibition is predominantly a function of antigen activity. On the other hand, even though the test specimen contains a large quantity of antigen, dilutions less than 1:600 result in the addition of significant amounts of antibody of test specimen origin thereby increasing the amount of antibody available for subsequent complexing of labeled antigen. Although the actual influence of the additional antibody on inhibition in this region of greater concentration is determined by several factors, *e.g.*, antigen–antibody ratio in the test specimens, antibody affinity, dissociability, etc., the net effect is usually less inhibition than expected for the amount of antigen. Significant inhibition may not be discernible unless the specimen is tested in dilution.

2. *Antibody excess.* In most circumstances, 10 μl (1:10 dilution) of human serum with ID detectable antibody will, when added to the standard antibody (1:20,000), precipitate 95–100% of the immunologically reactive ^{125}I-labeled Au(1) that is subsequently added. A major exception to obtaining full precipitation occurs when coexistent antigen is present in the test specimen. In this situation, a relatively constant level of precipitation, short of full precipitation, may occur for several dilutions before the progressive decrease of the 50% level begins.

While it is evident that an unknown specimen may be evaluated in a single test by observing coprecipitation inhibition when antigen is present or coprecipitation enhancement when antibody is present, this technical simplification should not be relied upon to provide complete characterization of the specimen because both antigen and antibody may frequently be present in the same specimen. If complexes of antigen and antibody are present, only the predominant activity will be manifested at any single dilution; and, if the components of these complexes are at equivalence (equivalent amounts of antigen and antibody), detection of either antigen or antibody may be missed unless tested further.

We have therefore found it necessary to test each specimen for antibody directly (method in Sect. 2 of Results) as well as for antigen by inhibition. If antibody is present in the direct test and coprecipitation enhancement is not observed when testing for antigen then it is probable that there is coexistent antigen that will become evident only when antigen is tested for in dilution.

Discussion. In view of the established association of Australia antigen with viral hepatitis it is apparent that a highly sensitive and specific method for its detection would have both pragmatic and experimental utility. In addition to the obvious advantage of lowering the threshold level of detection, such an assay might provide the means for characterizing the nature of the Au(1) association with different "types" of viral hepatitis as well as with other disease states in which it has been implicated. (i) Whereas Australia antigen is frequent in parenterally derived viral hepatitis it is usually absent in most point-source epidemics of hepatitis. Recently Del Prete *et al.* (15) reported the occurrence of an unstable antigen in the blood of patients involved in hepatitis epidemics. The antigen is detected with an antiserum which has antibody to Au(1) and the precipitin reaction shows a line of partial identity with Au(1). In view of conflicting data and impressions concerning the similarity of the agents responsible for serum and infectious hepatitis, it is important to determine whether epidemics are associated with Australia antigen. (ii) A qualitative association of Au(1) with Down's syndrome, leukemia, lepromatous leprosy, chronic renal disease, and certain population groups has been established (16) but the direct quantitative relationships of Au(1) with these entities as well as with viral hepatitis have not been investigated. (iii) The current methods are particularly insensitive with respect to the detection of antibody. Previous investgations have failed to demonstrate an antibody incidence or titer in convalescent sera or among large populations that approaches that which might be anticipated in response to a viral infectious agent.

In general, immunodiffusion and complement fixation procedures do not detect antigen–antibody complexes. Such complexes, at equivalence, may remain precipitated and undetected in an immunodiffusion plate well or cause anticomplementary activity by complement fixation test. Shulman and Barker (17) observed a very high incidence of anticomplementary activity among sera from patients with hepatitis during the incubation stage and early course of the disease. Millman *et al.* (18) have shown that the supernatant fraction following ultracentrifugal pelleting of Au(1) derived from plasma, may possess antibody activity capable of reacting with the pellet. Presumably, centrifugation permitted sedimentation of the heavier Au(1) from free antibody after promoting the separation of an antigen–antibody complex.

The degree of sensitivity of the radioimmunoprecipitation assay method for the detection of human antibody to Au(1) and antigen–antibody complexes compared with immunodiffusion and complement fixation has been illustrated (Table I). Although undiluted specimens 66993, 41177, 66820 produce strong precipitin bands for antibody, reactivity is lost above dilutions of 1:8 to 1:32. In contrast radioimmunoprecipitation assay evidence of antibody activity persists up to dilutions of 1:160,000–1:640,000. Antibody titers do not always correlate. The antibody titer of 66993 is four times higher than 41177 when tested by RIP but only ¼ when tested by CF. The reason for this discrepancy may be due to differences in the specificities of antibodies used in each test. Specimen 66993 from a blood bank nurse contained an unusually high RIP antibody titer. Work is currently in progress to determine whether this is a common situation among these nurses. Three of the five Au(1) positive specimens (45417, 66751, 66835) demonstrate concurrent antibody activity only when tested by radioimmunoprecipitation assay. Specimen 66835, though not having detectable antibody titer by immunodiffusion, has a 1:30,000 radioimmunoprecipitation assay titer.

The degree of sensitivity of radioimmunoprecipitation assay for the detection of Australia antigen compared with immunodiffusion and complement fixation has been illustrated (Table II). Among the specimens used as little as a 1:30,000 dilution of 10 μl of serum contains sufficient Australia antigen to cause coprecipitation inhibition. In comparison of equal volumes of Au(1) containing sera, detection by radioimmunoprecipitation assay is possible at dilutions that are 1000 times greater than that which is detectable by immunodiffusion. The concentration of Au(1) (μg/protein/ml) may be determined by relating the degree of inhibition of purified Au(1) to that of an appropriate dilution of serum.

The radioimmunoprecipitation assay has been found to be not only highly sensitive, but also reliable and reproducible. When Au(1) is present *in vivo* as an immune complex the radioimmunoprecipitation assay can be relied upon to detect both components of such a complex. This ability to detect Australia antigen and antibody in extremely small quantities whether or not they exist as complexes may prevent the transmission of hepatitis by the transfusion of sera that are negative by other methods of assay.

Summary. Purified Australia antigen [Au(1)], conjugated with ^{125}I, was reacted with and quantitatively precipitated by human anti-Au(1) in the presence of heterologous antihuman IgG and "dilute" ammonium sulfate. The sensitivity of the devised radioimmunoprecipitation assay procedure proved to be up to 1000 times more sensitive for the detection of antigen and up to 20,000 times more sensitive for the detection of antibody when compared with either immunodiffusion or complement fixation assays. This method is extremely valuable not only in the quantitative detection of antigen and antibody but also in the delineation of naturally occurring antigen–antibody complexes.

The authors thank Dr. Scott Mazzur who performed the complement fixation assays; and acknowledge the generous assistance of Drs. Alton Sutnick and W. Thomas London.

1. Blumberg, B. S., Gerstley, B. J. S., Hungerford, D. A., London, W. T., and Sutnick, A. I., Ann.

Intern. Med. **66**, 924 (1967).

2. Sutnick, A. I., London, W. T., Gerstley, B. J. S., Cronlund, M. M., and Blumberg, B. S., J. Amer. Med. Ass. **205**, 670 (1968).

3. Sutnick, A. I., London, W. T., and Blumberg, B. S., Amer. J. Dig. Dis. **14**, 189 (1969).

4. Blumberg, B. S., Sutnick, A. I., and London, W. T., J. Amer. Med. Ass. **207**, 1895 (1969).

5. London, W. T., Sutnick, A. I., and Blumberg, B. S., Ann. Intern. Med. **70**, 55 (1969).

6. Blumberg, B. S., and Riddell, N. M., J. Clin. Invest. **42**, 867 (1965).

7. Lang, N., Protides Biol. Fluids, Proc. Colloq. **14**, 517 (1966).

8. Krassnitsky, O., Pesendorfer, F., and Wewalka, F., Deut. Med. Wochenschr. **95**, 1 (1970).

9. Pesendorfer, F., Krassnitsky, O., and Wewalka, F., Klin. Wochenschr. **98**, 58 (1970).

10. Purcell, R. H., Holland, P. V., Walsh, J. H., Wong, D. C., Morrow, A. G., and Chanock, R. M., J. Infec. Dis. **120**, 383 (1969).

11. Coller, J. A., and Millman, I., J. Clin. Invest. **49**, 20a (1970).

12. Walsh, J., Yallow, R., and Berson, S., J. Infec. Dis. **121**, 550 (1970).

13. Millman, I., Loeb, L. A., Bayer, M. E., and Blumberg, B. S., J. Exp. Med. **131**, 1190 (1970).

14. Greenwood, F. C., Hunter, W. M., and Glover, J. S., Biochem. J. **89**, 114 (1963).

15. Del Prete, S., Constantino, D., Doglia, M., Graziina, A., Ajdukiewicz, A., Dudley, F. J., Fox, R. A., and Sherlock, A., Lancet **2**, 579 (1970).

16. Blumberg, B. S., Friedlaender, J. S., Woodside, A., Suntick, A. I., and London, W. T., Proc. Nat. Acad. Sci. U.S.A. **62**, 1108 (1969).

17. Shulman, N. R., and Barker, L., Science **165**, 304 (1969).

18. Millman, I., London, W. T., Sutnick, A. I., and Blumberg, B. S., Nature (London) **226**, 83 (1970).

Received April 26, 1971. P.S.E.B.M., 1971, Vol. 138.

[Reprinted from THE JOURNAL OF EXPERIMENTAL MEDICINE, June 1, 1970,
Vol. 131, No. 6, pp. 1190–1199]
Printed in U.S.A.

AUSTRALIA ANTIGEN (A HEPATITIS-ASSOCIATED ANTIGEN)

PURIFICATION AND PHYSICAL PROPERTIES*

BY IRVING MILLMAN, PH.D., LAWRENCE A. LOEB, M.D.,
MANFRED E. BAYER, M.D., AND BARUCH S. BLUMBERG, M.D.

(from the Institute for Cancer Research, Fox Chase, Philadelphia, Pennsylvania 19111)

(Received for publication 16 December 1969)

Australia antigen, Au(1), was first detected by Blumberg in 1964 in the serum of an Australian aborigine (1). Work concerning the geographic distribution, disease association, genetics, and physical and chemical characteristics of this unusual antigen has been rcviewed (2). In one of the early studies, specific anti-Au(1) antisera agglutinated particles 180–210 A in diameter which were isolated from the sera of patients with viral hepatitis (3). Furthermore, antisera to the Au(1) antigen coupled with fluorescein isothiocyanate stained intranuclear granules in the liver cells of patients with hepatitis (4). This indicates that cells of the liver of patients with hepatitis contain an antigen in common with the serum particles. All this strongly suggests that Au(1) may be, or may be part of, a hepatitis virus. The association of Au(1) with acute and chronic hepatitis has already been well documented (5–15).

One of the approaches which could shed light on the question of whether Au(1) is a virus or not is to study the physical properties of purified Au(1) particles to determine whether they are consistent with those known for viruses. It is also important to ascertain whether the particles contain nucleic acids. It is generally accepted that the hepatitis virus is small, 260 A in diameter or less, and that it survives at 56°C for 30 min and freezing at -10°C to -20°C for 1 yr or longer. It is resistant to the action of enzymes, ether, and other chemical agents (16), and is present in the blood of infected individuals since hepatitis can be transferred by needle puncture. The Au(1) antigen is also found in high concentrations in the blood of individuals suffering from chronic hepatitis, and has many of the properties of the postulated hepatitis virus, as this report will show.

Materials and Methods

Source of Au(1).—D. M. was a 21 year old male with Down's syndrome who had chemical and histological evidence of chronic hepatitis and Au(1) in his serum when first seen by us in

* This work was supported by U.S. Public Health Service grants CA-06551, CA-08069, CA-06927 and FR-05539 from the National Institutes of Health, and by an appropriation from the Commonwealth of Pennsylvania.

1967. These findings persisted. In early 1969 he developed acute granulocytic leukemia and was admitted to our Clinical Research Unit. During the course of his treatment he was given 1.0 mCi of $NaH_2{}^{32}PO_4$ by intravenous injection and 24 hr later 250 ml of plasma was obtained by plasmapheresis. The patient died of leukemia 6 days later.

Assay of Au(1).—Immunodiffusion assays were performed by the Ouchterlony method using the modifications of Blumberg and Riddell (17). Immunoelectrophoresis was performed on 3¼ x 4¼ inch photographic plates, using 1.1% agarose dissolved in sodium veronal buffer, pH 8.2, ionic strength 0.038.

Electron Microscopy.—Samples containing Au(1) were dialyzed for 4–10 hr against distilled water at 2–5°C. For negative staining a loopful of material was placed on a carbon-formvar–coated copper grid to which an equal amount of staining solution was added. The excess fluid was immediately blotted with filter paper and the grid dried in the vacuum of the microscope. For negative staining 2% uranyl acetate (w/v) in water or 4% silicotungstate (w/v) in solution adjusted to pH 7 with 1 N NaOH was employed. Each technique gave equivalent results. The micrographs were taken with a modified Siemens Elmiskop I (Siemens America, New York) electron microscope, using double condenser, 80 kv acceleration voltage, and 50 μ objective apertures at magnifications of 40,000–80,000.

Counting.—Radioactivity was determined by recording the Čerenkov radiation emitted by ^{32}P in a conventional liquid scintillation counter as described by Haberer (18). In preliminary experiments it was shown that the presence of 1 M sucrose, 4 M cesium chloride, and a variety of other inorganic salts did not diminish the amount of radioactivity detected. Since the radioactivity determination was carried out at 0°–2°C, without the addition of any scintillation solution, samples up to 20 μl were completely recoverable for subsequent purification.

Resistance of Au(1).—(*a*) *Enzymes:* A serum fraction containing Au(1) was partially purified by the method of Alter and Blumberg (19) and standardized to a concentration which produced a sharp precipitin line when assayed by the immunodiffusion technique. 25-μl portions of this fraction were allowed to react with a series of enzymes under optimal conditions as described by the supplier (Worthington Biochemical Corporation, Freehold, N. J.) of all but one of the enzymes. These enzymes included trypsin, wheat germ lipase, neuraminidase, alpha amylase, phospholipase C, ribonuclease, and deoxyribonuclease. In addition, pronase (Calbiochem Corporation, Los Angeles, Calif.), was dissolved in distilled water in a concentration of 2 mg/ml and 5 μl of this enzyme solution was allowed to react with 25 μl of Au(1) fraction. All enzyme reactions were carried out at 37°C for 1 hr.

(*b*) *Other agents:* The effect of elevated temperatures was determined by incubating Au(1) in 0.05 M tris(hydroxymethyl)aminomethane (Tris) maleate buffer, pH 7.5, at 56°, 85°, and 100°C for 1 hr. A similar solution containing Au(1) was exposed to ether and chloroform for 1 hr at room temperature. These solvents were removed by evaporation before immunoreactivity was determined.

Chemical Methods.—Deoxyribonucleic acid was assayed by the method of Dische (20) and the fluorometric method of Kissane and Robins (21). Ribonucleic acid was determined by the method of Albaum and Umbreit (22) and protein by the method of Lowry et al. (23).

RESULTS

Preliminary results indicated that treatment with the enzymes did not affect the immunoreactivity or structure of Au(1). An isolation procedure was then devised in which enzymatic digestion was used to degrade other constituents without apparently altering Au(1).

Purification of Au(1).—Step 1. Centrifugation: 100 ml of plasma was clarified by centrifugation at 0°–2°C for 1 hr at 2000 *g*. The supernatant was re-centrifuged at 300,000 *g* for 18 hr. Au(1) was detected only in the pellet; this was dispersed in 10 ml of 0.15 M NaCl.

Step 2. Enzyme treatment: To the above suspension were added 4 mg of amylase, 0.4 mg of lipase, 0.2 mg of neuraminidase and 0.2 mmoles of Tris-HCl, pH 7.5. The total volume (11.4 ml) was incubated for 15 min at 37°C. Thereafter 0.6 ml of a solution containing 0.1 mmoles of $CaCl_2$, 4.0 mg of trypsin, and 0.1 mg of phospholipase C was added. The mixture then gelled. 2 ml of pronase (10 mg/ml) was added and the mixture incubated for 1 hr. The gel disappeared and an additional 2 ml of pronase was added. The incubation was continued for an additional hour and then the mixture was cooled in an ice bath.

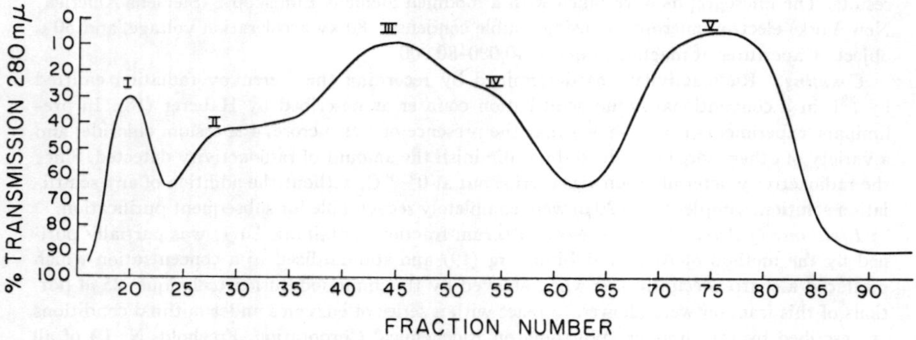

FIG. 1. Gel filtration. 5 ml of the enzyme-treated suspension was placed on a 650 x 32 mm column of Sephadex G-200 gel which had previously been equilibrated with 0.85% NaCl. Elution was with the same solution and 5 ml fractions were collected. All recovered Au(1) material eluted with the void volume (peak *I*).

Step 3. Gel filtration: Portions (5 ml) of the enzyme-treated suspension were filtered through Sephadex G-200. The absorption profile is shown in Fig. 1. All fractions comprising the first peak (*I*), which contained Au(1), were combined. Fig. 2 shows an immunoelectrophoresis pattern of each of the separation stages. These patterns were developed by adding horse anti-human serum to the troughs. There is a decrease in the number and intensity of precipitin bands following enzyme treatments and gel filtration.

Step 4. Sucrose gradient centrifugation: The combined fractions of peak *I* from step 3 were dialyzed against 0.01 M KCl and sedimented through sucrose gradients. Fig. 3 shows an analysis of the gradient for ^{32}P incorporation and immunoreactivity. Fractions 16–22 with major Čerenkov radioactivity and reactivity were then combined.

Step 5. Centrifugation in cesium chloride: Fractions 16–22 from step 4 were sedimented in cesium chloride gradients. Fig. 4 shows the resultant profile

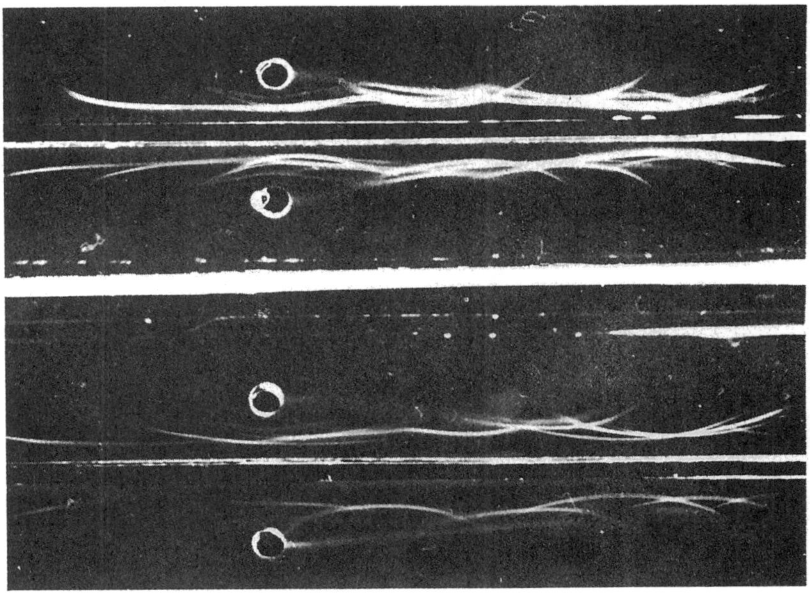

FIG. 2. Immunoelectrophoretic patterns of fractions at different stages of purification. Pattern *1* is starting plasma; pattern *2* is pellet after centrifugation (step 1); pattern *3* is enzyme-treated pellet suspension (step 2); pattern *4* is enzyme-treated pellet suspension passed through Sephadex G-200 (step 3). Troughs contained horse anti-human serum which does not contain anti-Au(1).

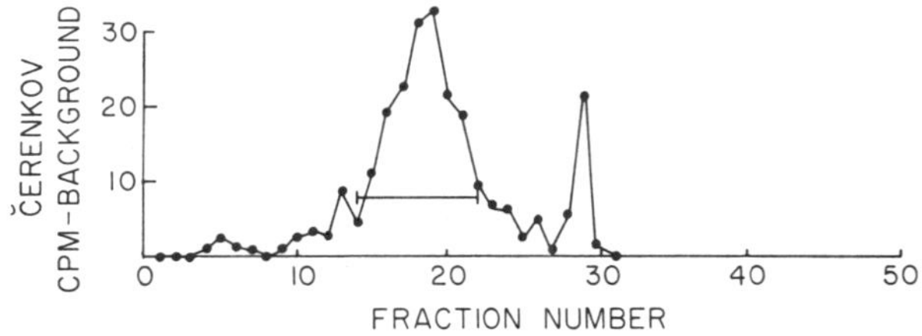

FIG. 3. Sucrose gradient profile. Sucrose gradients were prepared and sampled by the method of Martin and Ames (26). Portions (1.0 ml) of peak *I* in step 3 were layered over 32.5 ml linear sucrose gradients ranging from 10 to 30% by volume, in 0.01 M Tris maleate buffer pH 7.4. The gradients were centrifuged in a Spinco SW 27 rotor (Beckman Instruments, Inc., Fullerton, Calif.), at 57,000 g for 20 hr at 2–5°C. Forty-drop fractions were removed by puncturing the bottom of the tubes. Radioactivity was determined by counting each entire fraction for 100 min; background was 20.4 cpm. Immunological reactivity of each fraction was determined as detailed in Materials and Methods. The line within the radioactivity peak represents the Au(1) positive range according to immunodiffusion.

analyzed for ^{32}P content, absorption at 280 mμ, density, and reactivity with Au(1) antiserum. Immunological reactivity was found in fractions 15–21 with greatest activity in fraction 16. The relative antigenicity of each fraction was determined by twofold dilution titration. That fraction which could produce a precipitin line at the highest dilution was considered the peak fraction. The

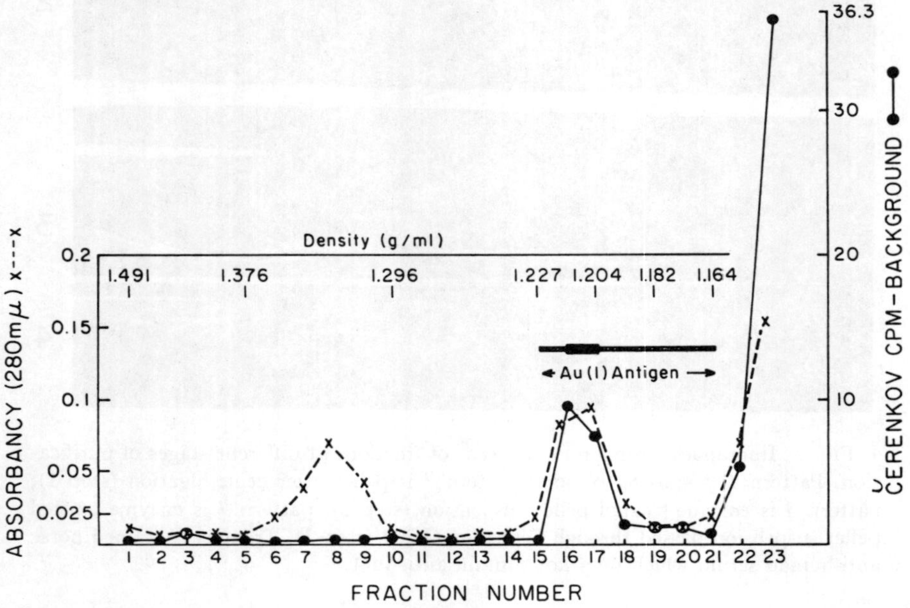

FIG. 4. The pooled immunoreactive fractions from the sucrose density gradients (Fig. 3) were dialyzed against 0.01 M KCl at 5°C for 18 hr and lyophilized. The dried Au(1) fraction was then dissolved in a small volume of distilled water and mixed with saturated cesium chloride solution to produce a final density of 1.3. The resultant solution (5.0 ml) was centrifuged in a Spinco SW 50 rotor at 222,000 g to equilibrium (40 hr). 10 drop fractions were collected through the bottom of the tube. Bar represents Au(1) activity measured by immunodiffusion assay. Solid area within bar indicates highest antigenic activity.

density of the peak fraction (fraction 16) is 1.21 and it falls within both the 280 mμ and Čerenkov peaks. The 280 mμ peak seen in fraction 8 (Fig. 4) is not coincident with ^{32}P or Au(1) activity. Its absorption profile is not characteristic of nucleic acid. Fractions 15–18 were combined and after dialysis against 0.01 M KCl were used for subsequent studies.

Analysis for Immunoreactivity.—Fig. 5 shows an immunoelectrophoretic analysis of the most purified fractions (15–18) derived from step 5. There is no reactivity with horse anti-human serum (pattern 2), indicating the absence of serum proteins (pattern 3). In contrast, Au(1) is seen in pattern 1 (arrow), in which the purified material was reacted with human anti-Au(1). A precipitin band appears to be in the same position and has the same configuration as the band formed by unpurified Au(1) (not shown). The presence of Au(1) in the most purified fractions (step 5) indicates resistance of the antigen to the enzyme treatments and to gradient manipulations. The absence of detectable serum components in purified fractions 15–18 was confirmed by immunodiffusion test.

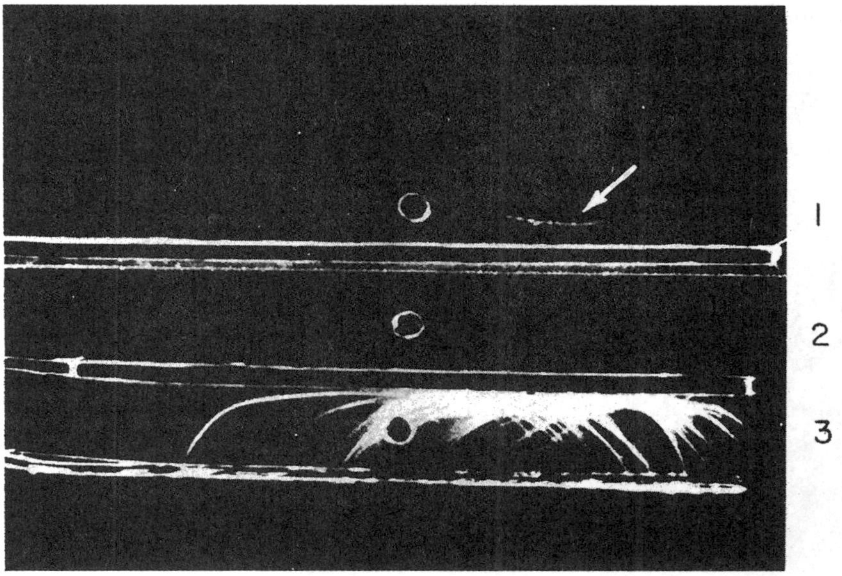

FIG. 5. Immunoelectrophoretic patterns of purified Au(1) fractions 15–18 (see Fig. 4). Pattern 1—Au(1) band (arrow) developed with anti-Au(1) in trough below. Pattern 2—Au(1) reacted with horse anti-human serum placed in trough below. Pattern 3— Whole plasma, containing Au(1), developed with horse anti-human serum in trough above.

Extraction with Phenol or Ether.—The small amount of ^{32}P in the most purified fraction prior to phenol extraction does not appear to be associated with nucleic acids of high molecular weight. 2 mg of calf thymus deoxyribonucleic acid as a carrier was added to the combined fractions 15–18 of step 5. After extraction with phenol by the procedure of Thomas and Abelson (24) and centrifugation of the aqueous phase in a cesium chloride gradient, no nucleic acid as measured by ^{32}P radioactivity was detected. It is possible that phenol did not dissociate nucleic acid from the protein. In a parallel isolation procedure in which enzyme treatment was omitted, 70% of the ^{32}P radioactivity in the most purified fraction was extracted with ether. There was no nucleic acid detected chemically. The sensitivities of these methods indicated that if nucleic acids were present, the concentration of DNA was less than 1% and the concentration of RNA less than 10% of the total weight of protein.

Effect of Other Agents.—Heat treatment at 85° and 100°C destroyed detectable immunological reactivity as revealed by immunodiffusion test. There was no apparent decrease in reactivity after heating Au(1) solution to 56°C for 1 hr and the reactivity was retained in the unheated control. After ether and chloroform treatments, two precipitin bands appeared where only one had been

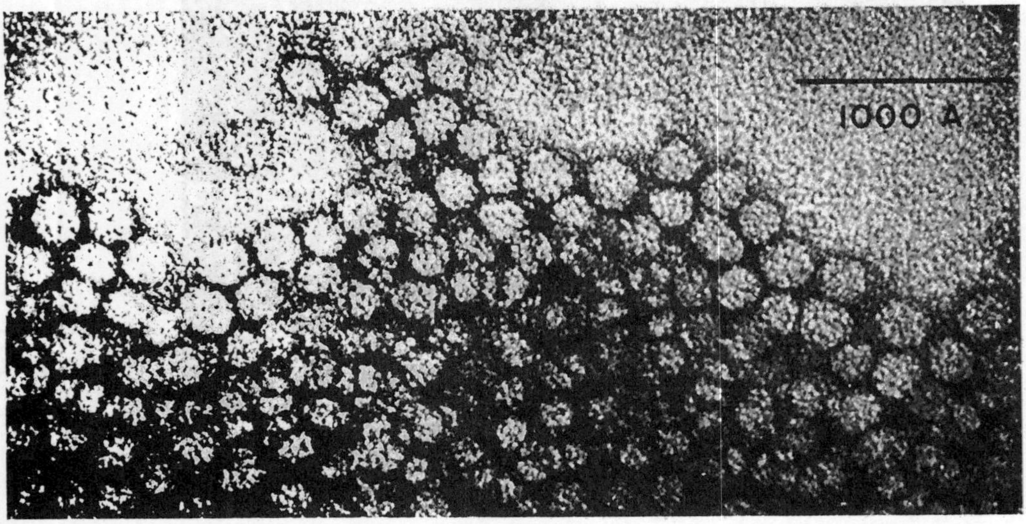

FIG. 6. Negatively stained particles from purified Au(1) fractions 15–18 of the cesium chloride gradient. Magnification 300,000. Bar indicates 1000 A.

seen previously. This may indicate that these treatments uncover other antigenic groups. One of the precipitin bands of the ether-treated Au(1) fraction appeared to be different from one of the two bands of chloroform-treated Au(1) fraction, since they crossed, i.e., did not form a line of identity.

Electron Microscopy of Purified Au(1).—Fig. 6 is an electron micrograph of Au(1) present in fractions 15–18 of the cesium chloride gradient. In some of the particles a central "core" was observed. Particles were found in all those cesium chloride fractions which produced a precipitin band by immunodiffusion (Fig. 4, fractions 15–21). The greatest concentration of particles appeared to be in fraction 16. Heat (100°C), ether, or chloroform did not affect the appearance of Au(1) as seen under the electron microscope. However, the particles disappeared completely after treatment with Carnoy's solution (3 parts ethanol: 1 part glacial acetic acid) followed by pronase digestion.

DISCUSSION

Australia antigen was apparently not affected by enzyme digestion as determined by its immunological characteristics and its appearance under the electron microscope. The remarkable stability despite such treatment made it possible to remove most other human serum proteins as a preliminary step in the purification of Au(1). The absence of host-related material (i.e. serum proteins) was demonstrated by a sensitive criterion, namely, the lack of immunological reactivity with horse anti-human serum in immunoelectrophoretic and immunodiffusion analysis.

The buoyant density of purified Au(1) was 1.21. This value confirms the earlier result of Alter and Blumberg (19). This density is lower than that reported for most picorna- and parvoviruses (which are comparable in size to Au(1) particles); these range between 1.32 and 1.4 (25). The lower density of purified Au(1) could result from the absence of nucleic acid and/or the presence of lipids. An external lipid coat has not been reported in viral agents of this size range. The presence of lipids in Australia antigen has previously been postulated on the basis of the staining reactions (19). Exposure to ether and chloroform appears to uncover previously masked antigenic groups; when the treated Au(1) was tested against both human and rabbit anti-Au(1) an additional precipitin band was detected.

The morphology of the particles isolated from the cesium chloride gradient, as revealed by the electron microscope, does not differ from that already reported by Bayer *et al.* (3). The particles are 180–210 A in diameter and in most cases spherical in shape. They have central cores and appear to be composed of subunits. A few elongated particles with the same diameter (3) have been observed in the present material. Considering the size of these particles they could be compared to the adeno- associated (AAV), parvo- , and picornavirus groups (25). In the most purified fractions there is no evidence for the presence of significant amounts of nucleic acid. Thus, it is possible that the ^{32}P seen in the gradients containing Au(1) may have been associated with lipids or lipoproteins which were subsequently removed by the phenol extraction.

The tentative hypothesis is that the bulk of material identified as Australia antigen may be an incomplete virus or virus capsid associated with viral hepatitis. Studies are currently directed at isolating an infectious particle.

SUMMARY

Australia antigen [Au(1)], a particle associated with viral hepatitis, was isolated from the plasma of a patient with chronic anicteric hepatitis and leukemia who had received radioactive phosphorus.

We have found that the immunoreactivity and appearance of Au(1) in the electron microscope were not altered by treatment with enzymes including trypsin, pronase, lipase, phospholipase C, ribonuclease, deoxyribonuclease, amylase, and neuraminidase. In contrast, other serum constituents were degraded by these enzymes. Therefore, treatment of the patient's plasma with many enzymes was exploited as an initial step for the isolation of Au(1). Subsequently, Au(1) was purified from the enzyme-treated ^{32}P-labeled plasma by gel filtration through Sephadex G-200 and centrifugation through sucrose and in cesium chloride gradients. There were no detectable human serum components in the purest fractions, as tested by immunoelectrophoresis and immunodiffusion. The density of the purified Au(1) was 1.21 in CsCl. The particle measured about 200 A in diameter, was predominantly spherical in shape and appeared to be composed of subunits. Nucleic acids were not detected by spectrophotometric, radiochemical, and chemical analyses. Immunoreactivity of purified Au(1) was destroyed by heating for 1 hr at 85°C but was stable at 56°C. Treatment with Carnoy's solution (3 parts ethanol: 1 part glacial acetic acid) followed by pronase disrupted the particles as seen with the electron microscope. These findings, combined with other published information on Australia antigen and viral hepatitis, suggest that the bulk of Australia antigen in the blood of this patient is an incomplete virus or virus capsid.

BIBLIOGRAPHY

1. Blumberg, B. S. 1964. Polymorphisms of serum proteins and the development of isoprecipitins in transfused patients. *Bull. N.Y. Acad. Med.* **40**:377.
2. Blumberg, B. S., A. I. Sutnick, and W. T. London. 1968. Hepatitis and leukemia. Their relation to Australia antigen. *Bull. N.Y. Acad. Med.* **44**:1566.
3. Bayer, M. E., B. S. Blumberg, and B. Werner. 1968. Particles associated with Australia antigen in the sera of patients with leukemia, Down's syndrome, and hepatitis. *Nature (London)*. **218**:1057.
4. Millman, I., V. Zavatone, B. J. S. Gerstley, and B. S. Blumberg. 1969. Australia antigen in the nuclei of liver cells of patients with viral hepatitis by the fluorescent antibody technique. *Nature (London)*. **222**:181.
5. Blumberg, B. S., B. J. S. Gerstley, D. A. Hungerford, W. T. London, and A. I. Sutnick. 1967. A serum antigen (Australia antigen) in Down's syndrome, leukemia, and hepatitis. *Ann. Intern. Med.*, **66**:924.
6. Sutnick, A. I., W. T. London, B. J. S. Gerstley, M. M. Cronlund, and B. S. Blumberg. 1968. Anicteric hepatitis associated wtith Australia antigen. Occurrence in patients with Down's syndrome. *J. Amer. Med. Ass.* **205**:670.
7. Okochi, K. and S. Murakami. 1968. Observations on Australia antigen in Japanese. *Vox Sang.* **15**:374.
8. Prince, A. M. 1968. An antigen detected in the blood during the incubation period of serum hepatitis. *Proc. Nat. Acad. Sci. U.S.A.* **60**:814.
9. Prince, A. M. 1969. Relation of Australia and SH antigens. *Lancet.* **2**:462.
10. Hirschman, R. J., N. R. Shulman, L. F. Barker, and K. O. Smith. 1969. Virus-like

particles in sera of patients with infectious and serum hepatitis. *J. Amer. Med. Ass.* **208**:1667.

11. Gocke, D. J. and N. B. Karey. 1969. Hepatitis antigen. *Lancet.* **1**:1055.
12. Nordenfelt, E. and L. Kjellen. 1969. Presence and persistence of Australia antigen in a Swedish hepatitis series. *Acta Pathol. Microbiol. Scand.* In press.
13. Editorial. Australia antigen and hepatitis. 19 July 1969. *Lancet.* **2**.
14. Almeida, J. D., A. J. Zuckerman, P. E. Taylor, and A. P. Waterson. 1969. Immune electron microscopy of the Australia-SH (serum hepatitis) antigen. *Microbios.* **2**:117.
15. London, W. T., A. I. Sutnick, and B. S. Blumberg. 1969. Australia antigen and acute viral hepatitis. *Ann. Intern. Med.* **70**:55.
16. Kissling, R. E. 1965. Laboratory status of infectious hepatitis agent. *In* Transmission of Viruses by the Water Route. Gerald Berg, editor, Interscience Publishers Inc. New York. 337.
17. Blumberg, B. S. and N. M. Riddell. 1963. Inherited antigenic differences in human serum beta-lipoproteins. A second antiserum. *J. Clin. Invest.* **42**:867.
18. Haberer, E. K. 1965. Messung von beta-aktivitäten an wässrigen proben auf grund der Čerenkov-Strahlung. *Atomwirtschaft.* **10**:36.
19. Alter, H. J. and B. S. Blumberg. 1966. Studies on a "new" human isoprecipitin system (Australia antigen). *Blood J. Hematol.* **27**:297.
20. Dische, Z. 1955. Color reactions of nucleic acid components. *In* The Nucleic Acids. E. Chargaff and J. N. Davidson, editors. Academic Press, New York. **1**:287.
21. Kissane, J. M. and E. Robbins. 1958. The fluorometric measurement of deoxyribonucleic acid in animal tissues with special reference to the central nervous system. *J. Biol. Chem.* **233**:184.
22. Albaum, H. G. and W. W. Umbreit. 1947. Differentiation between ribose-3-phosphate and ribose-5-phosphate by means of the orcinol-pentose reaction. *J. Biol. Chem.* **167**:369.
23. Lowry, O. H., N. J. Rosenbrough, A. L. Farr, and R. J. Randall. 1951. Protein measurement with the Folin phenol reagent. *J. Biol. Chem.* **193**:265.
24. Thomas, C. A. and J. Abelson. 1966. The isolation and characterization of DNA from bacteriophage. *In* Procedures in Nucleic Acid Research. G. L. Cantoni and D. R. Davies, editors. Harper & Row, New York. 557.
25. Fenner, F. The Biology of Animal Viruses. 1968. Academic Press, New York. **1**:12.
26. Martin, R. G. and B. N. Ames. 1961. A method for determining the sedimentation behavior of enzymes: Application to protein mixtures. *J. Biol. Chem.* **236**:1372.

Res. Virol.
1997, 148, 91-94

The current state of the prevention of HBV infection, the carrier state and hepatocellular carcinoma

B.S. Blumberg

Fox Chase Cancer Center, 7701 Burholme Avenue, Philadelphia, Pennsylvania 19111 (USA)

Introduction

Hepatitis B vaccine was first conceived in our laboratory in 1969 (Blumberg, 1977), and after the completion of the research and the development phase, became available for use in the early 1980's. Regional and national vaccination programs were started at about this time, and according to Dr. Mark Kane of the World Health Organization, there are now about seventy national programs. In many places hepatitis B vaccine is included as part of the routine paediatric vaccinations which, in some cases, are compulsory and financed by public funds. An example of this is the program of the Ministry of Health of the Republic of Italy. Law number 165, dated 27/5/91, promulgated during the period when Professor Giuliano Da Villa, the Secretary of this Congress, was the advisor to the Ministry, mandates vaccination and provides public funds to implement the program. A universal childhood vaccination program was instituted in Taiwan soon after the availability of the vaccine and has now been in place for more than ten years. Extensive programs are in place in Peoples Republic of China, Korea, Singapore, Hong Kong, The Gambia, and many other small and large jurisdictions.

By the mid 1970's, evidence was accumulating that hepatitis B virus (HBV) was aetiologically related to hepatocellular carcinoma (HCC), and several research groups discussed its relevance to the prevention of HBV. In the introduction to an article published in 1975 entitled "The relation of infection with the hepatitis B agent to primary hepatic carcinoma" (Blumberg *et al.*, 1975) we stated:

> *During recent years, there have been parallel developments in understanding, on the one hand, the pathogenesis of primary hepatic carcinoma (PHC, hepatoma) and, on the other, the biology of Australia antigen (hepatitis B surface antigen) and the infectious agent, hepatitis B virus (HBV), to which it is intimately related. Recently, the paths of these developments have begun to converge, and from this it is possible to design a preliminary strategy which would, if the interpretations of these data are correct, result in the prevention of many, and perhaps most, cases of one of the most widespread and deadly cancers of humans.*

The evidence supporting the role of HBV is now more substantial, and it has been estimated that about 80% of HCC is caused by HBV. Hepatitis C virus, by itself and as a coinfection with HBV is also related to many of the cases of HCC. Dr. Kane has stated that HBV vaccination for the prevention of hepatocellular carcinoma is now considered to be the second most important cancer control program, exceeded only by the campaign against cigarette smoking. Even though vaccination has been available for only ten years, there are measurable effects on the incidence of HCC in some regions.

Prevention of infection and carrier state

Several reports were presented at the IX Triennal International Symposium on Viral Hepatitis and Liver Disease held in Rome in 1996, and the publication of abstracts of these papers provides a body of data that can be used to summarize the current status (IX Triennal International Symposium on Viral Hepatitis and Liver Disease, abstract volume, April 21-25, 1996). In addition there are several published reports with more complete data. Some of these will be cited here.

There are estimated to be about 100 million carriers of HBV in China. Vaccination programs have

Received November 5, 1996.

been in progress since 1986 (Xu *et al.*, 1996). In a study of a population of about 2 million people from five provinces, 25,703 blood specimens were collected. The seropositivity has been reduced from 16.3% to about 1 to 2%. A very interesting finding was that the prevalence of carriers was also lower in the population that had not been vaccinated. The authors predicted that the prevalence of carriers might drop to about 1% in only a single generation! This represents a remarkable example of the value of public health vaccination programs.

A vaccination program was initiated in 1983 in Afragola, a community in southern Italy, near Naples, selected because of its very high rates of HBV carriers, and of morbidity and mortality from liver diseases (Da Villa *et al.*, 1992, 1995, and Da Villa, personal communication). There were major decreases in HBV prevalence in the vaccinated population; for example, the prevalence of HBsAg (hepatitis B surface antigen) in males up to 12 years dropped from 10.51% in 1978 to 0.8% in 1993 after the vaccination program was in place for ten years. The prevalence of anti-HBc in the same group dropped from 52.6% to 1.18%. There were similar decreases in other age groups. Of particular interest was the drop in the prevalence of HBsAg in the mostly unvaccinated population i.e., males age 13-60. HBsAg dropped from 17.97% in 1978 to 5.5% in 1989. This implies that the reduction of the carrier prevalence in the vaccinated group has an indirect effect on unvaccinated carriers and susceptibles, and suggests that there may be an amplification characteristic to the vaccination program.

A study in Cagliari, Sardinia (Mastio *et al.*, 1996) compared the prevalence of HBV markers in 1989 and 1994. The compulsory vaccination program had been initiated in 1991. There was a striking increase in the number of children with anti-HBs (from 9% to 44.9%), the protective antibody, and a 50% decrease in markers of HBV infection (from 21.6% to 11.2%) during this period.

In the Liguria region of Italy (Lo Monaco *et al.*, 1996) the vaccination program for children (12-23 months) and adolescents began in 1988. There was a striking decrease between 1988 and 1995 in reported cases of acute HBV infection in the 15-24-year age group, which was primarily impacted by the vaccination program. The incidence decreased from 22.9 to 8.0 per 100,000 inhabitants. There were no cases reported in the vaccinated persons. There were also decreases in the older, non-vaccinated group, but the differences were not significant.

In Taiwan (Chang *et al.*, 1996, Hsu *et al.*, 1988 and Chen *et al.*, 1996), where universal childhood vaccination has been in use for more than ten years, the prevalence in the 1-2-year age group decreased from 10.7% in 1984 to 1.5% in 1989, and the decrease since that time has been even greater. There is substantial funding for this program, and it encompasses a very large percentage of the children born since the inception of vaccination.

A universal childhood vaccination program was undertaken between 1981 and 1983 in Native Americans in Alaska (McMahon *et al.*, 1996). Acute hepatitis B infection dropped from 215 cases/100,000 before the vaccination program, to 7 to 14 cases in 1993 after the program was in place. In 1995 no cases were reported.

In the province of Alberta, Canada (Larke *et al.*, 1996), a program of screening for HBsAg of all pregnant women in the province was instituted in 1985. The effectiveness of the program was evaluated ten years later. The prevalence of HBV carriers was 0.4% of 520,485 sera that were tested. The offspring of carrier mothers were given immunoglobulin (HBIG) and HBV vaccine postnatally. Of these 654 (95.5%) children of the carrier mothers received HBIG and vaccine and 576 were tested for HBsAg after birth. Only 13 (2.3%) became carriers. Although control figures were not given in the abstract, this is far less than would be expected based on historical data.

An extensive vaccination program and research project in The Gambia, West Africa (Viviani *et al.*, 1996) was launched in July 1986. The vaccine efficacy (that is the efficiency of protection) was found to be 84% for infection and 94% for the development of the carrier state, by the age of four years.

Prevention of hepatocellular carcinoma

There are already indications that the incidence of hepatocellular carcinoma, much of which is due to chronic infection with HBV, is also decreasing. At the 1996 Rome meeting, it was reported that the incidence of HCC has significantly decreased in the vaccinated population from Taiwan (Chen, 1996 and Chang *et al.*, 1996). During the period 1981-1991 the annual incidence of HCC in children from ages 6-14 years ranged from 4.48-7.13 cases per 100,000 population. In 1992, after vaccination had been in place for nearly ten years, it was 1.74 and in 1993 it was 2.39. In a study from Korea (Lee, 1995)

HBIg = anti-HB immunoglobulin.
HBsAg = hepatitis B surface antigen.
HBV = hepatitis B virus.
HCC = hepatocellular carcinoma.

not only the impacted vaccinated group but also several unvaccinated cohorts have a lower incidence of HCC.

Prospects for the eradication of HBV

Vaccination programs have been in effect for ten years or less, and the eventual control of HBV appears possible. There have already been discussions of eradication; see, for example, the reports of the International Conference on Prospects for the Eradication of Hepatitis B, held in Geneva in 1989 (Blumberg, 1990). In view of the excellent results that have been achieved, the scientific community and public health authorities will now want to open this issue again.

Several of the reports cited above note a decrease in the prevalence of carriers in *unvaccinated* groups in populations where a sizable proportion of individuals have been vaccinated. If this observation is validated then it appears that there is an accelerating effect of vaccination which could hasten the process of control. This could be a consequence of decreased transmission, but it could also mean that the carrier state is less likely to be sustained in an individual if there are fewer carriers in his or her environment. It would be interesting to consider the molecular biology of this phenomenon.

Despite the success of the vaccination programs, there are still about 300 million HBV carriers worldwide. In addition to its value to the patient or carrier, effective treatment could also hasten the control of infection. Even if treatment did not completely eliminate the virus from a host, it could decrease viral levels to the extent that the carrier would be far less likely to transmit the virus to others.

There are hopeful reports on HBV anti-virals, and they are discussed elsewhere in this conference. The value of interferon in the treatment of carriers will be defined in the next few years. Lamivudine is currently under evaluation, and initial results are very encouraging. A variety of enzyme inhibitors, — reverse transcriptase, DNA polymerase, protease — are under development and test. A novel therapeutic possibility based on the inhibition of the terminal glycosylation of the M protein of the HBV surface antigen has been reported (Block *et al.*, 1994). It appears to inhibit the assembly and exports of the virus from liver cells, but there is no information on its possible role as a therapy.

A large body of data is accumulating on environmental and other factors which affect the risk of developing the HBV carrier state, and of carriers developing HCC (Blumberg, 1995). These include the genetics of the host, host gender, iron storage levels in the host, aflatoxin in the diet, genetic polymorphism in the host which controls detoxification of aflatoxin, the RsaI polymorphism at the cytochrome P4502E1 locus (Ladero *et al.*, 1996), arsenic, co-infection with other hepatitis viruses and other organisms and other factors. Intervention is possible for several of these and could lead to further decreases in HBV infection and its pathological consequences.

Conclusion

There has been significant progress in the control of HBV infection, and the prospects for the future are promising. It may be possible to effectively control or even eradicate this virus, which has caused so much disease and sorrow over the centuries. It is particularly gratifying to those of us who have been involved in research on HBV and its application since the 1960's to witness these remarkable results. One day, perhaps not too far in the future, we may hold another International Congress on Prevention of Cirrhosis and Hepatocellular Carcinoma here in Capri, or another beautiful location in Italy, in which success is announced and a beginning is made on the control of another disease.

Key-words: HBV, Hepatitis B, Vaccination; Carrier state, HCC, Prevention, Review.

Acknowledgements

I dedicate this paper to Dottore Antonino Bondi of Catania and Naples, my daughter-in-law's father, who attended this meeting, and, sadly, died shortly afterwards.

This work was supported by USPHS grant CA-06927 from the National Institutes of Health and by an appropriation from the Commonwealth of Pennsylvania.

References

Block, T.M., Lu, X., Platt, F.M., Foster, G.R., Gerlich, W.H., Blumberg, B.S. & Dwek, R.A. (1994), Secretion of human hepatitis B virus is inhibited by the imino sugar N-butyldeoxynojirimycin. *Proc. Natl. Acad. Sci.* USA, 91, 2235-2239.

Blumberg, B.S., Larouze, B., London, W.T., Werner, B., Hesser, J.E., Millman, I., Saimot, G. & Payet, M. (1975), The relation of infection with the hepatitis B agent to primary hepatic carcinoma. *Am. J. Pathol.*, 81, 669-682.

Blumberg, B.S. (1977), Australia antigen and the biology of hepatitis B. *Science,* 197, 17-25.

Blumberg, B.S. (1990), Proc. International Conference on Prospects for Eradication of Hepatitis B Virus, Vaccine 8, Introduction S5, Conclusion S139.

Blumberg, B.S. (1995), Complex interactions of hepatitis B virus with its host and environment. *J. R. Coll. Physicians Lond.*, 29, 31-40.

Chang, M.H., Chen, C.J., Lai, M.S., Kong, M.S., Wu, T.C., Liang, D.C., Hsu, H.M. & Chen, D.S. (1996), Hepatitis B mass vaccination program decreases childhood hepatocellular carcinoma in Taiwan. IX Triennal International Symposium on Viral Hepatitis and Liver Disease, Rome, Italy, April 1996, abstract C243.

Chang, M.H., Chen, C.J., Lai, M.S., Kong, M.S., Wu, T.C., Liang, D.C., Hsu, H.M. & Chen, D.S. (1996), Hepatitis B mass vaccination program decreases childhood hepatocellular carcinoma in Taiwan. Proc. IX Triennal International Symposium on Viral Hepatitis and Liver Disease, April 1996, Rome, Italy (abstract C243).

Chen, D.S., Hsu, H.M., Bennett, C.L., Pajeau, T.S., Blumberg, B., Chang, P.Y., Nishoka, K., Huang, A. & Sung, J.L. (1996), A program for eradication of hepatitis B from Taiwan by a 10 year, four-dose vaccination program. *Cancer Causes Control,* 7, 305-311.

Chen, D.S. (1996), Hepatitis B vaccines: status report on long-term efficacy. Proc. IX Triennal International Symposium on Viral Hepatitis and Liver Disease, April 1996, Rome, Italy (abstract 147).

Da Villa, G., Piazza, M., Iorio, R., Picciotto, L., Peluso, P., De Luca, G. & Basile, B. (1992), A pilot model of vaccination against hepatitis B virus suitable for mass vaccination campaigns in hyperendemic areas. *J. Med. Virol.,* 36, 274-278.

Da Villa, G., Picciotto, L., Elia, S., Peluso, F., Montanaro, F. & Maisto, T. (1995), Hepatitis B vaccination: universal vaccination of newborn babies and children at 12 years of age versus high risk groups. A comparison in the field. *Vaccine,* 13, 1240-1243.

Hsu, H.M., Chen, D.S., Chuang, C.H., Lu, J.C.F., Jwo, D.M., Lee, C.C., Lu, H.C., Cheng, S.H., Wang, C.C., Lo, K.J., Shih, C.J. & Sung, J.L. (1988), Efficacy of a mass vaccination program in Taiwan. *JAMA,* 260, 2231-2235.

Ladero, J.M., Agundez, J.A.G., Rodriguez-Lescure, A., Diaz-Rubio, M. & Benitez, J. (1996), RsaI polymorphism at the cytochrome P4502E1 locus and risk of hepatocellular carcinoma. *Gut,* 39, 330-333.

Larke, R.P.B., Waters, J.R. & Pagnucco, J.S. (1996), A decade of success for Canada's earliest province-wide program to prevent mother-to-infant transmission of HBV infection. Proc. IX Triennal International Symposium on Viral Hepatitis and Liver Disease, April 1996, Rome, Italy (abstract C261).

Lee, M.S. (1995), A cohort study on the causal association between hepatitis B virus infection and hepatocellular carcinoma with reference to the protective effect of vaccination against hepatitis B (Thesis). Seoul, Korea, Postgraduate School, Seoul National University.

Lo Monaco, R., Mentore, B., Rebora, M., Pezzano, D. & Crovari, P. (1996), Hepatitis B vaccination coverage and epidemiology of HBV related disease in Liguria, Italy. Proc. IX Triennal International Symposium on Viral Hepatitis and Liver Disease, 1996, Rome, Italy (abstract C282).

McMahon, B., Mandsager, R., Wainwright, K., Wainwright, R., Williams, J. & Singleton, R. (1996), The Alaska native hepatitis B control program. Proc. IX Triennal International Symposium on Viral Hepatitis and Liver Disease, April 1996, Rome, Italy (abstract 74).

Mastio, M.P., Palomba, A., Garau, M.G., Loi, I., Liguori, M., Crociani, C. & Piu, G. (1996), Hepatitis B virus infection: effect of vaccinal prophylaxis in children. Proc. IX Triennal International Symposium on Viral Hepatitis and Liver Diseases, April 1996, Rome, Italy (abstract C254).

Viviani, S., Jack, A.D., Maine, N., Montesano, R., Hall, A.J., Mendy, M. & Whittle, H. (1996), The Gambia hepatitis intervention study. Proc. IX Triennal International Symposium on Viral Hepatitis and Liver Disease, April 1996, Rome, Italy (abstract C264).

Xu, Z.Y., Cao, H.L., Liu, C.B., Wang, S.S., Yan, T.Q., Zhao, S.J. & Wu, W.S. (1996), Control of hepatitis B in China. Proc. IX Triennal International Symposium on Viral Hepatitis and Liver Disease, April 1996, Rome, Italy (abstract 68).

Reprinted from THE AMERICAN JOURNAL OF PATHOLOGY

Published by Medical Department
HARPER & ROW, PUBLISHERS

Printed in the U.S.A.

The Relation of Infection With the Hepatitis B Agent to Primary Hepatic Carcinoma

Baruch S. Blumberg, MD, Bernard Larouzé, MD, IHP, W. Thomas London, MD, Barbara Werner, PhD, Jana E. Hesser, PhD, Irving Millman, PhD, Gerard Saimot, MD, and Maurice Payet, MD

In Asia, Africa, and other tropical areas, primary hepatic carcinoma (PHC) is associated with liver cirrhosis of the postnecrotic (macronodular) type. Chronic viral hepatitis is likely to be the cause of this cirrhosis in many patients from regions where chronic infection with the hepatitis B virus (HBV) is common. More than 95% of patients with hepatoma (in Mali and Senegal) have evidence of infection with HBV, a much higher frequency than in controls. Thirty-nine of 62 patients with PHC had hepatitis B surface antigen (HB_sAg) (controls, 8 of 98) and 56 of 63 (controls, 26 of 100) had antibody against hepatitis B core antigen (anti-HB_c). In earlier studies, we demonstrated a maternal effect of HB_sAg. If the mother has the antigen and the father does not, the children are much more likely to also have HB_sAg than if the father has the antigen and the mother does not (93/161 = 57.8% when mother is positive *vs.* 28/135 = 20.7% when father is positive: $P = 0.6 \times 10^{10}$). Studies in Greece and in the Solomon Islands show that presence of HB_sAg in parents affects the sex ratio of the offspring of the mating. This implies that the presence of the agent in a parent can affect the fetus early in life. Parental studies in the west African hepatoma patients showed that there is a very high frequency of HB_sAg in mothers (71.6%), while the frequency in fathers (18.5%) is significantly less. This suggests that the development of hepatoma in offspring is related to infection in parents. Several years ago, we described a vaccine which may be useful in preventing infection with hepatitis B. Strategies are discussed which might be effective in preventing the development of carriers with, it is hoped, a consequent decrease in the frequency of HBV carriers, chronic hepatitis, and primary hepatic carcinoma. The strategy would employ methods for decreasing the frequency of the agent in the environment by the application of public health methods including the vaccination of appropriate newborns and other members of the population. (Am J Pathol 81:669–682, 1975)

DURING RECENT YEARS, there have been parallel developments in understanding, on the one hand, the pathogenesis of primary hepatic carcinoma (PHC, hepatoma) and, on the other, the biology of Australia antigen (hepatitis B surface antigen) and the infectious agent, hepatitis B virus (HBV), to which it is intimately related. Recently, the

From The Institute for Cancer Research, Fox Chase Cancer Center, Philadelphia, and Institut de Medécine et d'Épidémiologie Africaines, Hospital Clande Bernard, Paris, France.

Presented at the Fifty-ninth Annual Meeting of the Federation of American Societies for Experimental Biology, Atlantic City, New Jersey, April 14–18, 1975.

Supported by Grants CA-06551, RR-05539, and CA-06927 from the National Cancer Institute, and by an appropriation from the Commonwealth of Pennsylvania.

Address reprint requests to Dr. Baruch S. Blumberg, The Institute for Cancer Research, Fox Chase Cancer Center, 7701 Burholme Avenue, Philadelphia, PA 19111.

paths of these developments have begun to converge, and from this it is possible to design a preliminary strategy which could, if the interpretations of these data are correct, result in the prevention of many, and perhaps most, cases of one of the most widespread and deadly cancers of humans. These developments will be discussed under five categories, and in a sixth section suggestions for the use of this information will be given. The subject headings are: a) The relation of hepatoma to hepatitis in Africa and Asia. b) The relation of HBV to hepatoma. [Hepatitis B surface antigen (HB_sAg); antibody to the surface antigen (anti-HB_s); antibody to hepatitis B core antigen (anti-HB_c)]. c) Family clustering of HB_sAg. Maternal (and possibly paternal) transmission of HBV. d) Presence of HB_sAg in the mothers of patients with hepatoma. e) The development of a vaccine against HBV. f) Considerations of strategies to prevent hepatoma in areas where it occurs in high frequency.

These studies may also be useful in investigating other agents which may be related to the pathogenesis of other cancers.

The Relation of Hepatoma to Hepatitis B in Asia and Africa

Primary hepatic carcinoma is the most common form of cancer occurring in males in certain areas of the world with large populations, e.g., China, Africa, the Philippines. Thus, although PHC is a relatively uncommon disease in the United States, it is an international health problem of major significance. For example, it has been stated that in Taiwan it is the second most common cause of death.[1]

Hepatoma is associated with liver cirrhosis of the postnecrotic (macronocular) type. Hepatoma often arises in a liver affected by long standing cirrhosis (70 to 80%), but hepatoma and cirrhosis may develop simultaneously or, less commonly (10%), hepatoma may occur in an otherwise normal liver.[2] Since postnecrotic cirrhosis is the response of the liver to an agent (or agents) which causes necrosis of hepatic cells, much of the research on hepatoma has centered on studies of hepatotoxic agents which are common in areas where hepatoma is common. For example, considerable effort has been expended on the studies of cycads, mycotoxins, and nitrosamines. The best-documented cause of postnecrotic cirrhosis, however, is viral hepatitis. In the late 1950s and early 1960s, Payet [3] and Steiner and Davies [4] suggested, on epidemiologic and pathologic grounds, that viral hepatitis was the major cause of hepatic cirrhosis and hepatoma in east Africa (Uganda) and west Africa (Senegal). Since, at that time, no tests were available for identifying the viral agent of hepatitis, this hypothesis could not be tested adequately. With the discovery of Australia antigen and its specific association with hepatitis B, this became possible.[5]

This hypothesis states that there is a relation between infection with the hepatitis B agent, the development of chronic hepatitis and chronic liver disease, and the development of hepatoma. These relations need not be linear, and it is also obvious that other factors must be involved. Not everyone infected with HBV develops hepatitis or becomes a carrier (some develop anti-HB_s, and in some persons exposed to HBV the effect is not measurable). Nor does everyone who develops chronic hepatitis develop postnecrotic cirrhosis, and of those who do, even in high-incidence areas, only a fraction develop hepatoma. From this, it is obvious that other factors may also be needed to develop hepatoma. If there are many factors essential to cause an illness, the prevention of only one of them could prevent the development of the disease. Complete knowledge of all the etiologic factors is useful, but not essential, in designing prevention strategies.

In general, carriers of HB_sAg are common in areas where hepatoma is common. Do some of these carriers have chronic hepatitis? There is evidence that this is the case. In Table 1 (taken from Economidou *et al.*[6]) a summary of studies on liver chemistry abnormalities in carriers is shown. In most of these locations some of the carriers have evidence of liver malfunction. In a matched pair study in the village of Plati, Greece, a high prevalence area, it was shown that there is a significantly higher level of SGPT in the carriers than in matched individuals who are not carriers.[6,7]

Table 1—SGPT Levels in Blood Donors with HB_sAg Reported in Published Studies

			SGPT determination	
Source	No. of donors	Prevalence of HB_sAg	No. HB_sAg carriers tested	Abnormal* (%)
Okochi and Murakami	3,131	1.78	56	9
Soulier *et al.*	18,084	0.43	35	11
Otto-Servais *et al.*	6,656	0.15	9	(6)†
Wewalka			113	38
Sanwald and Sauerbruch	2,053	0.83	17	0
Klinman	52,429	0.1	23	39
Cazal and Robinet-Levy	52,966	0.28	40	35
Szmuness *et al.*	208,000	0.15	197	11
				12
Economidou and Binopoulou	6,606	2.9	143	16
				8

* The definition of abnormal is generally an elevation of SGPT greater than 50 IU.
† Number with abnormal tests.

Reproduced from Economidou *et al.*[6] (Res Commun Chem Pathol Pharmacol 10:703, 1975) by permission of PJD Publications.

From this we may conclude that a proportion of carriers in these regions have chronic liver malfunction and it is reasonable to conclude that, over the course of many years of the carrier state, their livers could be significantly affected.

The Relation of Hepatitis B Virus to Hepatoma

In the early studies of association of HB_sAg with hepatoma, insensitive detection techniques were used. These showed no or weak associations of HB_sAg with hepatoma. In recent years, however, sera from patients with hepatoma have been tested by very sensitive methods, immune adherence hemagglutination (IAHAA) and radioimmunoassay (RIA), and high frequencies of HB_sAg among the patients with hepatoma from different parts of the world have now been found (Table 2). Although an association of HB_sAg with hepatoma has not been documented in American and northern European populations, the increased frequency of HB_sAg in hepatoma detected by RIA and IAHAA is not paralleled by a similar increase in detection frequency among the general population of the same area. This suggests that HB_sAg in hepatoma patients is either present in small amounts or that the antigen present in patients with hepatoma is antigenically similar but not the same as HB_sAg. There are other interrelationships of HB_sAg, cirrhosis, and hepatoma which also suggest that these three entities may be causally interconnected. For example, all three are more common in males than females and all may cluster in families.

Recently it has become possible to test not only for HB_sAg but also for anti-HB_s, antibody against the surface antigen, and anti-HB_c, antibody against the core antigen (HB_cAg) of the hepatitis B agent. These presumably all indicate that the individual has been infected with HBV. Preliminary results of our study in Mali and Senegal are shown in Table 3 (adapted from Maupas *et al.*[13] and Larouzé *et al.*[14]). From this it can be said that nearly all of the hepatoma patients have some evidence of infection. The frequencies of HB_sAg and anti-HB_c are higher in the hepatoma patients than in controls. The frequency of anti-HB_s is higher in the controls in Senegal (also true in Japan, see Table 2), suggesting that the hepatoma patients have a different biologic response to HBV than the controls.

Family Clustering of HB_sAg

Among the earliest investigations of Australia antigen were studies on family distribution. The initial studies supported the hypothesis that HB_sAg in carriers segregated as if it were controlled by an autosomal

Table 2—HB_sAg and Anti-HB_s in Primary Hepatic Carcinoma

Location	PHC: Technique	PHC: No. tested	PHC: Percent HB_sAg positive	Controls: Type	Controls: No. tested	Controls: Percent HB_sAg positive	P*
Singapore[8]	CEP	156	1.9	Male	1516	4.1	NS
				Blood donors	1516	7.5	1×10
	IAHAA	156	35.3	Nasopharyngeal cancers	207	5.8	9×10
Japan[8]	IAHAA	260	37.3	Blood donors	4387	2.6	1×10
				Other cancers	249	2.4	1×10
Taiwan[8]	IAHAA	86		Blood donors	278	12.1	7×10^{14}
				Other cancers	347	13.5	1×10
France[9]	CEP	21	4.7		314	0.01	NS
	RIA	21	4.7	Blood donors	314	0.01	NS
Spain[10]	ID	21	34.6	Other patients	900	0.3	9×10
Senegal[9]	CEP	117	48	Blood	99	10	1×10
	RIA	117	64	donors	99	13	1×10
Uganda[11]	CF	90	40	Other patients	224	3.1	2×10
	Technique	**No. Tested**	**Percent anti-HB_s**		**No. Tested**	**Percent anti-HB_s**	
Japan[12]	HA	80	7	General population	376	18.4	2×10
Uganda[11]	HA	90	30	Other patients	244	31	NS

* P = Fisher's exact 2×2 test.

PHC = primary hepatic carcinoma, CEP = counter electrophoresis, ID = immunodiffusion, CF = complement fixation, RIA = radioimmunoassay, HA = hemagglutination.

Table 3—Hepatitis B Infection in Primary Hepatic Carcinoma in Mali and Senegal

		HB_sAg (RIA)		Anti-HB_c (CEP, ID)		Anti-HB_s (HA)		HB_sAg anti-HB_c or anti-HB_s	
	N	No. positive	Percent positive	No. positive	Percent positive	No. positive	Percent positive	No. positive	Percent Positive
Mali									
PHC patients	21	10	47.6	15/20	75.0	8	38.1	19	90.5
Controls	40	2	5.0	10/39	25.6	17	42.5	25	62.5
Senegal									
PHC patients	43	29/41	70.7	41	95.3	9	20.9	42	97.7
Controls	61	6/58	10.3	16	26.2	27	44.3	39	63.9

Adapted from Maupas *et al.*[13] and Laouzé *et al.*[14]

recessive gene (for review see Blumberg [15]). This simple hypothesis has been formally rejected by the finding that some offspring of matings in which both parents are positive are not carriers. There are, in addition, some other exceptions to simple genetic control. There is a sex effect (in general, males are more often carriers than females) and an age effect (the frequency is low in the very young, increases to a peak in early decades of life, and then decreases in later years). In addtion, there is a striking maternal effect and this will be reviewed here.

In completed (or nearly completed) families, if the mother is positive and the father negative, many more of the offspring have HB_sAg than in families where the father is a carrier and the mother is not, or when neither parent is a carrier [15] (Table 4). This distribution varies from population to population and in some (for example, in the village of Plati, Greece) this trend is not seen.[16]

In another study on maternal effect, Mazzur *et al.*[17] used data on the subtypes of HB_sAg to investigate its family distribution. They examined families from Bougainville, Papua, New Guinea, in which HB_sAg was segregating and where several different subtypes of HB_sAg were present. In families in which the mother was positive, essentially all the children had the same subtypes as the parent, but they had variable subtypes when the father was the positive parent or when neither parent was positive. This suggests, as does the previous data, that transmission can occur directly from the mother to the offspring either *in utero*, early in life, or later during the contact between mother and children.

There is also more direct information on mother–child transmission. Schweitzer [12] and others have shown that when mothers have hepatitis and HB_sAg in their blood at the time of delivery, their offspring are very likely to become carriers within weeks or months of birth. With the availability of more sensitive methods for detection of HB_sAg, the antigen has now been found in low levels in some cord bloods and newborn children.[12] Nishioka and his colleagues in Japan [8] have shown that mothers who are carriers but do not have apparent hepatitis at the time the child is delivered are also likely to transmit the antigen to their children. This is also true in Taiwan, as Stevens and his colleagues have shown,[18] but it is not necessarily the case in western countries such as United States. In Taiwan, the probability of a child's developing antigenemia increased in direct proportion to the levels of HB_sAg in the mother's blood.[18]

Taken together, it appears that there is a maternal effect. Are there any detectable effects of parental carriage of HB_sAg, in addition to the development of carriers among the offspring? An effect, found by Hesser and

Table 4—Maternal Effect

Location	Father positive				Mother positive				P
	No. of families	Children			No. of families	Children			
		No.	$HB_ssAg(+)$	$HB_sAg(-)$		No.	$HB_sAg(+)$	$HB_sAg(-)$	
Cebu, etc.	7	20	5	15	2	7	6	1	0.87×10^{2}
Bougainville	28	68	8	61	36	86	52	34	0.17×10^{9}
Turin	7	26	9	17	2	7	5	2	0.09
Lau, Malaita	6	17	5	12	9	29	16	13	0.08
Baegu, Malaita	2	3	1	2	11	32	14	18	0.61
Total	50	135	28	107	59	161	93	68	0.55×10^{10}

Number of children with HB_sAg and without HB_sAg in matings in which at least one parent has HB_sAg. The matings in which father is $HB_sAg(+)$ and mother $HB_sAg(-)$ are shown under "father positive" and the alternative matings under "mother positive." In every population, the $HB_sAg(+)$ mother has a larger number of $HB_sAg(+)$ children than the $HB_sAg(-)$ mother (from Blumberg, Bull Acad Med Toronto 45:45-51, 1972).

her colleagues,[19,20] may, if substantiated by subsequent studies, indicate that the action of the HBV agent does have an effect before the birth of the child. They found that the presence of HB_sAg in either parent results in an alteration of sex ratio. In Plati, Greece, the sex ratio (100 × number of males:number of females) of matings in which either parent has HB_sAg (185) is significantly higher than in matings in which neither parent has HB_sAg (112) (Table 5). In subsequent studies in four communities in Bougainville and the Solomon Islands, the effect of HB_sAg in the parent on the sex ratio of offspring was confirmed.[20] In this case, however, the direction of the change was different than in Plati. When HB_sAg was present in the mother, then there was a decrease in the number of males born; this finding was consistent in each of the communities (Table 6).

An additional effect has been reported by Kukowski *et al.*[21] A small but significant increase in gestation time was found in mothers who were HB_sAg carriers. If this finding is confirmed, then it would indicate a further profound effect of HBV on the biology of parturition.

Presence of Hepatitis B Virus in the Mothers of Patients With Hepatoma

Ohbayashi and his colleagues [22] reported studies on three Japanese families of patients with hepatoma. In these families, the mothers were HB_sAg carriers, and the implication has been drawn that transmission through the mother—as discussed in the previous section—may have an effect on the development of hepatoma in some of the children. We have recently completed controlled studies in Mali and Senegal, west Africa, on patients with hepatoma, their mothers, and other members of their families. These studies will be reported in detail shortly;[14] preliminary results are given here (Table 7). A large percentage of the mothers of these patients are carriers of HB_sAg, supporting the notion that a maternal effect may be significant in the development of hepatoma.

Table 5—Sex Ratio of Offspring from Matings of Parents With and Without HB_s Ag in a Greek Population

	No. of Mothers	No. of offspring		Sex ratio
		Male	Female	
Mother $HB_sAg(+)$	27	49	28	175
Father $HB_sAg(+)$	20	36	18	200
Total, either parent $HB_sAg(+)$	47	85	46	185
Parents $HB_sAg(-)$	172	287	255	112

Reprinted from: Hepatitis-associated (Australia) antigen, fertility and sex ratio: Implications for health planning. Hum Biol 47, 1975 (In press) by Hesser, Blumberg, and Drew by permission of Wayne State University Press.

Table 6—Sex Ratio of Offspring From Matings of Parents With and Without HB_sAg in Four Melanesian Populations

	Aita			Lau			Baegu			Bougainville		
	No. mothers	Offspring M/F	Offspring SR	No. mothers	Offspring M/F	Offspring SR	No. mothers	Offspring M/F	Offspring SR	No. mothers	Offspring M/F	Offspring SR
Mother $HB_sAg(+)$	12	24/35	69	21	23/26	89	12	17/26	65	21	54/70	77
Father $HB_sAg(+)$	14	28/26	108	5	14/6	233	2	3/0		20	51/47	109
Parents $HB_sAg(-)$	37	70/84	83	33	65/72	90	45	75/69	109	226	593/552	107

SR = Sex ratio

Reprinted from: Hepatitis-associated (Australia) antigen, fertility and sex ratio: Implications for health planning. Hum Biol 47, 1975 (in press) by Hesser, Blumberg, and Drew by permission of Wayne State University Press.

Vol. 81, No. 3
December 1975

Table 7—HB_sAg in Parents of Patients With Primary Heptic Carcinoma and Parents of Matched Controls*

	No.	No. $HB_sAg(+)$ (RIA)	Percent $HB_sAg(+)$		No.	No. $HB_sAg(+)$ (RIA)	Percent $HB_sAg(+)$
PHC patients	28	22	78.6	Matched controls	28	16	57.1
Fathers	27	5	18.5	Fathers	27	5	18.5
Mothers	28	20	71.6	Mothers	28	4	16.6

2 × 2 tests†	
PHC patients *vs.* matched controls	0.15
Fathers of PHC *vs.* fathers of controls	NS
Mothers of PHC *vs.* mothers of controls	3.1×10^{5}
Mothers of PHC *vs.* fathers of PHC	1.8×10^{3}
Mothers of controls *vs.* fathers of controls	NS

* Adapted from Larouzé *et al.*[14]
† Fisher's exact test.

The Development of a Vaccine Against Hepatitis B Virus

Shortly after our discovery of Australia antigen and its association with hepatitis, it became apparent that the antigen found in carriers could be used to prepare a vaccine which might be protective against HBV. We devised a method using ultracentrifugation to separate the antigenic material from possibly infectious particles, and techniques for removing excess serum proteins and insuring that any remaining infectious material would be inactivated.[23,24] This was a unique method for producing a vaccine. The antigen is removed from the blood of carriers who have large amounts of the material and given in small quantities to other individuals who require protection.

A series of successful animal trials using chimpanzees and some preliminary human studies have now been completed in several laboratories and they provide encouraging support for the use of the vaccine as a preventative for hepatitis and, probably, for the prevention of the carrier state. Trials of the vaccine in humans are now being considered.[25]

Consideration of Strategies to Prevent Hepatoma in High-Incidence Areas

If the vaccine is indeed protective, then several strategies for control of hepatoma can be suggested.

There has been considerable discussion on the possibility of developing a vaccine against cancer. A frequent objection to this procedure is that the number of cases of a given cancer (say leukemia) is very small compared to the number of individuals who are theoretically at risk and would require vaccination. In the case of the hepatitis B vaccine we have described, its use would be justified on the basis that it prevents acute and chronic hepatitis, chronic liver disease, and the development of carriers. If it also prevents hepatoma, then this would be an added important advantage of the vaccination program. In areas with a high endemicity of HBV, hepatitis and chronic liver disease are common in the population, and the use of the vaccine could be medically and economically justified on this basis.

The simplest strategy may be to vaccinate large segments, or conceivably all, of the population in high endemicity areas in Africa, Asia, or elsewhere. If this were the case, then it would first be neccessary to know what effect this vaccine would have on carriers of HB_sAg, since it is conceivable that it might be disadvantageous. If so, then screening of the population for carriers would be required.

Another strategy might be to specifically treat the newborn offspring of HB_sAg-positive mothers. In some parts of the world it has been estimated that most carriers originate from carrier mothers, and the data from Japan and Africa indicate that these children may have a higher risk of developing hepatoma in later life. It has been found that high titer anti-HB_s immunoglobulin given to newborn offspring of positive mothers can prevent the development of the carrier state in these children.[26] Hence, a anti-HB_s procedure might be developed to provide immunoglobulin shortly after birth and vaccine later in life. Since the endemic areas infection with HBV is likely to be common, recurrent stimulation of the immune system would produce anti-HB_s.

Any vaccine program should be coordinated with general sanitary and public health measures to eliminate HBV from the general environment, i.e., improvement in waste elimination, control of arthropod vectors, protection of food from contamination, etc.

Discussion

The advances in recent years in our understanding of the biology of Australia antigen and its recognition as the surface antigen of the hepatitis B agent have led to consideration of the feasibility of the immunologic prevention of a common cancer. In considering such an extensive public health program, attention must be given to the possible adverse consequences to the community and population. The removal of an agent such as HBV in highly endemic areas may have consequences which cannot be foreseen. It is for this reason that we are attempting to learn as much as possible about the ecologic interactions of this agent. For example, our findings of an association of HBV in parents with the sex ratio of their offspring raise the question whether removal of the agent may affect the population dynamics and sociologic structure of a community. It may be possible to anticipate consequences of this nature in the overall planning of a prevention campaign.

References

1. Beasley RP: Personal communication
2. Davies JNP: Hepatic neoplasm. The Liver. Edited by EA Gall, FK Mostofi. Baltimore, Williams and Wilkins, 1973, pp 361–369
3. Payet M, Camain R, Pene P: Le cancer primitif du foie, etude critique a propos de 240 cas. Rev Intern Hepatol 4:1–20, 1956
4. Steiner PE, Davies JNP: Cirrhosis and primary liver carcinoma in Uganda Africans. Br J Cancer 11:523–534, 1957
5. Blumberg BS, Gerstley BJS, Hungerford DA, London WT, Sutnick AI: A serum antigen (Australia antigen) in Down's syndrome, leukemia and hepatitis. Ann Intern Med 66:924–931, 1967
6. Economidou I, Hadziyannis S, Paraskevas E, Binopoulou A, Hesser JE, Lustbader E, Blumberg BS: Australia antigen (HB_sAg) carriers in a Greek community: Studies of transaminase (SGPT) levels. Res Commun Chem Pathol Pharmacol 10:703–713, 1975
7. Blumberg BS, Hesser JE, Economidou I, Hadziyannis S, Gioustozi A, Heliakis P, Livadas D: The variety of responses within a community to infection with Australia (hepatitis B) antigen. Dev Biol Stand Vol. 30 (In press)
8. Nishioka K, Mayumi M, Okochi K, Okada K, Hiraymam T: Natural history of Australia antigen and hepatocellular carcinoma. Analytic and Experimental Epidemiology of Cancer. Edited by W Nakahara, T Hirayama, K Nishioka, H Sugano. Baltimore, University Park Press, 1973, pp 137–146
9. Saimot G, Coulaud JP, Payet M, Brochard C, Schlegel N, Menache D: Antigene australie son apport dans l'etude etiologique du cancer primitif du foie. Ann Med Interne (Paris) 1975 (In press)
10. Teres J, Guardia J, Bruguera M, Rodes J: Hepatitis-associated antigen and hepatocellular carcinoma [Letter]. Lancet 2:215, 1971
11. Vogel CL, Anthony PP, Sadikali F, Barker LF, Peterson MR: Hepatitis-associated antigen and antibody in hepatocellular carcinoma: Results of a continuing study. J Natl Cancer Inst 48:1583–1588, 1972
12. Schweitzer IL, Wing A, McPeak C, Spears RL: Hepatitis and hepatitis-associated antigen in 56 mother–infant pairs. JAMA 220:1092–1095, 1972
13. Maupas P, Larouzé B, London WT, Werner B, O'Connell A, Blumberg BS, Saimot G, Payet M: Antibody to hepatitis B core antigen in patients with primary hepatic carcinoma. Lancet 2:9–11, 1975
14. Larouzé B, Blumberg BS, London WT, Saimot G, Payet M: Maternal transmission of hepatitis B infection and primary hepatic carcinoma. (Unpublished data)
15. Blumberg BS: Australia antigen: The history of its discovery with comments on genetic and family aspects. Viral Hepatitis and Blood Transfusion. Edited by GN Vyas, HA Perkins, R Schmid. New York, Grune & Stratton, 1972, pp 63–83
16. Hesser JE, Hadziyannis S, Economidou I, Blumberg BS: Family studies of HB_sAg in Plati, Greece. (Unpublished data)
17. Mazzur S, Blumberg BS, Friedlaender JS: Silent maternal transmission of Australia antigen. Nature 247:41–43, 1974

18. Stevens CE, Beasley RP, Tsui J, Lee WC: Vertical transmission of hepatitis B antigen in Taiwan. N Engl J Med 292:771–774, 1975
19. Hesser JE, Economidou J, Blumberg BS. HB_sAg in parents and sex ratio of offspring in a Greek population. Hum Biol 1975 (In press)
20. Hesser JE, Blumberg BS, Drew JS: Hepatitis-associated (Australia) antigen, fertility and sex ratio: Implications for health planning. Hum Biol 1975 (In press)
21. Kukowski K, London WT, Sutnick AI, Kahn M, Blumberg BS: Comparison of progency of mothers with and without Australia antigen. Hum Biol 44:489–499, 1972
22. Ohbayashi A, Okochi K, Mayumi M: Familial clustering of asymptomatic carriers of Australia antigen and patients with chronic liver disease or primary liver cancer. Gastroenterology 62:618–625, 1972
23. Blumberg BS: Viral hepatitis, Au antigen, and hope for a vaccine. Gastroenterology (Medical World News), pp 14–18, 1972
24. Blumberg BS, Millman I: Vaccine against viral hepatitis and process. US Patent Office, Patent No. 3,636,191. Jan 18, 1972 (and other patents)
25. Maugh TH: Research news: Hepatitis B—A new vaccine ready for human testing. Science 188:137–139, 1975
26. Kohler PF, Dubois RS, Merrill DA, Bowes WA: Prevention of chronic neonatal hepatitis B virus infection with antibody to the hepatitis B surface antigen. N Engl J Med 291:1378–1380, 1974

0270-9139/82/0202-010S$02.00/0
HEPATOLOGY

Vol. 2, No. 2, p. 10S, 1982
Printed in U.S.A.

A Cellular Model of the Role of Hepatitis B Virus in the Pathogenesis of Primary Hepatocellular Carcinoma

W. THOMAS LONDON AND BARUCH S. BLUMBERG

Institute for Cancer Research, Fox Chase Cancer Center, Philadelphia, Pennsylvania 19111

We have devised a model to explain the role of hepatitis B virus (HBV) in the pathogenesis of acute and chronic liver disease and primary hepatocellular carcinoma. The model postulates that the liver contains two populations of cells—R and S cells. R cells are resistant to productive infection with HBV, whereas S cells (susceptible cells) actively replicate virus when infected. As a result, infected S cells have a shortened survival; their death stimulates the growth (regeneration) of R cells. The model also proposes that R cells are less mature than S cells. They can divide and produce R cells or divide and differentiate into S cells, but S cells can only give rise to other S cells. Integration of viral DNA may occur in either R or S cells, but only integration of HBV DNA in R cells is of potential tumorigenic significance. R cells are already being stimulated to proliferate by the death of S cells; the site or arrangement of integrated DNA may increase the expression of one or more cellular genes providing an additional growth advantage for a particular R cell over other proliferating R cells. As this clone expands, it may develop the phenotypic characteristics of a neoplasm. A clinical tumor will not result, however, unless the stimulus to cell division is maintained by the death of S cells. Integration of HBV DNA sequences into S cells is not of tumorigenic significance. S cells have a limited capacity for cell division, and integration of viral DNA does not increase that capacity.

In this report, the model is used to explain various phenomena associated with HBV infection, and new data presented at the Kroc conference are used to test the model.

Infection with hepatitis B virus (HBV) can have many outcomes. Transient infections can result in subclinical disease, acute anicteric hepatitis, acute icteric hepatitis, or fulminant hepatitis. Persistent infections may give rise to no detectable liver disease, chronic persistent hepatitis, chronic active hepatitis, postnecrotic cirrhosis, or primary hepatocellular carcinoma (PHC). The mechanisms by which one virus can produce so many different effects in the same organ, the liver, are poorly understood. We have devised a model that provides at least a partial explanation of these and other phenomena (listed below) associated with HBV infection. The purpose of the model is to generate testable hypotheses of the role of HBV in the pathogenesis of acute and chronic liver diseases and PHC.

A pathogenetic model of HBV infection should attempt to explain the following phenomena:

(i) PHC is associated with persistent but not with transient HBV infection (1).

(ii) Persistent infection is also associated with chronic hepatitis and cirrhosis (2, 3).

(iii) Acute hepatitis is usually associated with transient rather than persistent infection (2, 3).

(iv) There is a long "incubation" period (20 to 40 years) between the onset of persistent HBV infection and the development of PHC (2, 3).

(v) Hepatitis B surface antigen (HBsAg) is commonly found in nonneoplastic cells, but only rarely in the neoplastic cells of a liver containing PHC (4).

(vi) Sequences of HBV DNA are found integrated into the genomes of both tumor and nontumor cells of livers containing PHC (5, 6).

(vii) The concentration of HBsAg in the serum of patients with PHC is low (2, 3).

(viii) In areas where PHC is endemic, the incidence of liver cancer in males is 4 to 9 times that in females (7).

(ix) Infants born to carrier mothers do not become HBsAg(+) until 2 to 3 months of age. Fetuses do not produce HBsAg (8).

(x) Infants frequently develop persistent infections

This work was supported by USPHS Grants CA-06551, RR-05539, CA-22780, and CA-06927 from NIH and by an appropriation from the Commonwealth of Pennsylvania.

Address reprint requests to: W. Thomas London, M.D., Institute for Cancer Research, Fox Chase Cancer Center, 7701 Burholme Avenue, Philadelphia, Pennsylvania 19111.

with HBV, whereas adults usually develop transient infections (8, 9).

THE MODEL

Our model (2, 3) proposes that the liver contains two populations of hepatocytes: (a) cells that are susceptible to productive infection with HBV, hereafter called S cells; and (b) cells that are resistant (or relatively resistant) to productive infection, hereafter called R cells. Integration of HBV genomes can occur in either R or S cells. We assume that infected S cells have a shortened survival. This could occur by several mechanisms. Virus production probably compromises host-cell metabolism sufficiently to eventually kill the cell. Secondly, there may be an immune response to viral-cellular antigen complexes on the surface of hepatocytes replicating virus (10). Thirdly, infected S cells may be more vulnerable to the effects of toxic oxygen radicals than uninfected cells. [Vierucci et al. have shown that HBV proteins cause the release of superoxide anions from polymorphonuclear leukocytes (11).]

In the liver, the stimulus for cell division is the death or removal of hepatocytes (12). Hence, persistent infection with HBV leads to continuous cell death and to continuous regeneration. Infected S cells have the greatest risk of dying; R cells, on the other hand, are not productively infected and are, therefore, likely to increase in number.

Integration of viral DNA into an R cell may have tumorigenic significance. Since R cells are already being stimulated to proliferate by the death of S cells, the site or arrangement of integrated DNA and the associated rearrangements of cellular DNA may increase the expression of one or more cellular genes providing an additional growth advantage for a particular R cell over other proliferating R cells. As this clone expands, it may, by the process of clonal evolution, develop the phenotypic characteristics of a neoplasm. The point here is that integration, expression of a cellular oncogene, and phenotypic transformation are insufficient to produce a clinical tumor unless the stimulus to cell division is maintained. There must, of course, be some stage at which the R cell clone (or subclone) acquires sufficient autonomy to not require external stimuli, but we assume that this is quite late in the process.

This model requires two populations of hepatocytes, one susceptible and one relatively resistant to HBV infection. Such cell populations have not yet been identified in the human liver. However, it is known that the liver will respond to repeated cytotoxic insults with repeated rounds of regeneration eventually leading to liver failure (12). Therefore, it is likely that there are less mature cells which can divide and differentiate further, and fully differentiated cells which either cannot divide or can only divide 1 or 2 times. We have proposed that R cells are the less mature, and S cells are the fully differentiated hepatocytes. According to this scheme, an R cell can divide and yield two R cells, or one R cell and one S cell, or two S cells. An S cell, if it can divide at all, can only produce S cells (Figure 1). Since the selective pressure of HBV infection is applied mainly to S cells, R cells

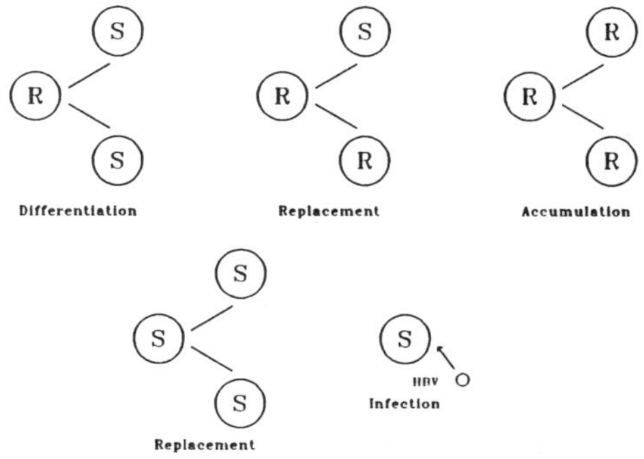

FIG. 1. The liver contains two kinds of cells, less mature cells that are resistant to productive infection with HBV, called R cells and terminally differentiated cells that are susceptible to HBV infection, called S cells. R cells can divide into R or S cells, but S cells can only give rise to other S cells.

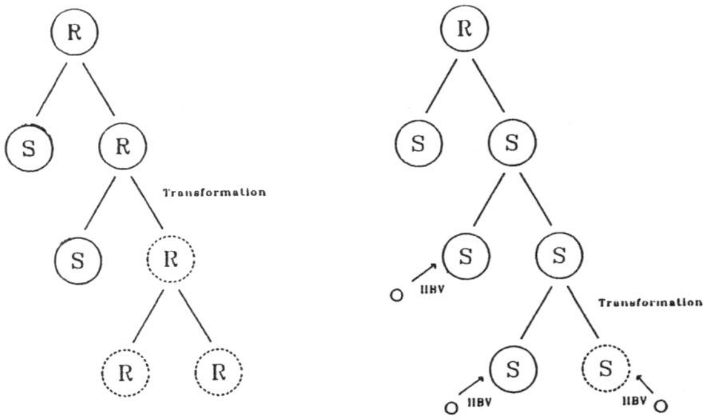

FIG. 2. Integration of HBV DNA and neoplastic transformation can occur in either R or S cells, but S cells have a limited capacity to divide and are subject to productive infection with HBV. Therefore, transformation of an S cell will not lead to the development of a clinically significant tumor.

will gradually accumulate (Figure 2). The model hypothesizes that integration of an HBV genome into an R cell decreases the probability of differentiation into S cells. Phenotypic transformation of an R cell ensures that it will only give rise to other R cells. The integration event may, in addition, result in increased (or decreased) expression of a cellular gene (or genes) which may provide a further growth advantage (R cell surrounded by *dashed line* in Figure 2). If this occurs, the clone arising from that R cell will soon outgrow all other R cells. Even at this stage, the model postulates that this clone will not become a clinically significant tumor unless growth stimulation is maintained by selective pressure against S cells. Further steps, collectively known as clonal evolution, are required before a clone or subclone can become a clinical tumor.

Integration of viral DNA and malignant transformation of an S cell would not result in a clinically significant tumor both because of the limited capacity of an S cell to divide and its susceptibility to productive infection with HBV.

EVIDENCE FOR R AND S CELLS

Is there evidence that hepatocytes differ in their susceptibility to HBV infection? Several investigators have attempted, using serial liver biopsies and immunofluorescence or immunoperoxidase techniques, to identify the patterns of HBV infection within the liver. Studies in adult humans have, in general, shown that at the time of diagnosis of acute hepatitis B [when alanine aminotransferase (ALT) levels are elevated], very few cells demonstrate either HBsAg or hepatitis B core antigen (HBcAg) (13, 14). HBsAg is present in the plasma prior to the rise of liver enzymes. Observations of the pattern of HBV infection in the liver during the incubation period of hepatitis have been made on experimentally infected chimpanzees (15–17). In these animals, liver enzymes rise 6 to 10 weeks after the appearance of HBsAg in plasma. Initially, the cell membranes of a few hepatocytes show a low-intensity staining for HBsAg in a "honey-comb"-like pattern. Cells showing this pattern of staining increase in number and are accompanied by the appearance of cells whose nuclei stain for HBcAg and less commonly other cells which demonstrate HBsAg in their cytoplasm. Very few cells have both cytoplasmic HBsAg and nuclear HBcAg. Eventually, 90% or more of hepatocytes demonstrate membrane staining of HBsAg and up to 50% nuclear staining of HBcAg; diffuse cytoplasmic staining of HBsAg is rarely seen. Cells demonstrating staining for HBV proteins begin to disappear from the liver prior to the rise in ALT and continue to disappear until very few are demonstrable at the time ALT level reaches its peak.

When a virus infects a cell, it must adsorb to the cell surface, penetrate the cell membrane and reach some specific intracellular site before viral replication can begin. Very little is known about these events with respect to HBV. One interpretation of the immunofluorescence studies, following acute infection, is that most hepatocytes have receptors for HBV, as demonstrated by HBsAg staining of cell membranes. A smaller population of cells replicate virus as shown by nuclear staining for HBcAg or cytoplasmic staining for HBsAg. Whether the rapid disappearance of nuclear- and membrane-stained cells coincident with the ALT rise indicates elimination of these cells or a combination of suppression of viral replication and cell death is unclear.

In persistent HBV infections in humans, the immunohistology is more explicit and better documented than in acute infections (18, 19). In patients with chronic persistent hepatitis or chronic carriers without histologic evidence of liver disease, hepatocytes showing diffuse cytoplasmic staining for HBsAg are scattered throughout the liver. In some areas, aggregates of HBsAg(+) cells are seen. With hematoxylin and eosin stains, the cytoplasm of these cells frequently has a "ground-glass" appearance. Precise counts of stained and unstained cells in the liver of chronic carriers have not been reported, but the impression is that 30 to 80% of cells are HBsAg(+). HBcAg is found in only a few per cent of hepatocytes in "normal" carriers, whereas in carriers receiving immunosuppressive drugs (renal transplant patients), 50 to 90% of cells may contain core antigen.

In patients with chronic active hepatitis, HBcAg may appear in up to 60% of liver cells together with membranous or cytoplasmic HBsAg. The positive cells tend to appear in clusters. In patients with postnecrotic cirrhosis, clusters of cells are again seen showing cytoplasmic staining for HBsAg. By orcein and immunoperoxidase stains of formalin-fixed, paraffin-embedded tissue, HBsAg is seen in the form of discrete cytoplasmic bodies adjacent to, or surrounding, the nucleus. HBcAg has been detected in as many as 40% of hepatocytes, but more commonly it is identified in 5% or fewer cells (19).

In patients with PHC, HBsAg can frequently be found in the cytoplasm of some clusters of nontumor cells usually in the form of large discrete bodies. Cells showing membrane staining or multiple small granules of HBsAg in the cytoplasm are sometimes seen. The tumor cells generally do not show HBsAg. In our experience, when tumor cells contain HBsAg, it is in the small granular form. HBcAg is either not detectable or is present in less than 5% of cells (4, 19).

Collectively, these observations are consistent with the view that the majority but not all cells present in the liver of normal adults are mature hepatocytes, S cells, and that they are susceptible to infection with HBV. It is uncertain whether the membranous staining for HBsAg, without nuclear HBcAg or cytoplasmic HBsAg, seen in the early stages of acute HBV, indicates adsorption of virus particles on the cell surface or expression of viral protein in the cell membrane. In patients with persistent infection, as the liver disease progresses over time, fewer liver cells express viral protein. This is compatible with the hypothesis that S cells are selected against, and R cells accumulate.

APPLICATIONS OF THE MODEL

According to the model, the liver in the fetus and newborn would be composed mostly or entirely of R cells and would not be able to support replication of HBV (Figure 3). After birth, further division and differentiation occur giving rise to S cells which can permit viral replication. Infection with HBV during the first 6 months of life generally results in a persistent infection (Figures 3–5). Again, according to the model, infection is taking place at a time when the R cell population is decreasing and the S cell population is increasing as a result of normal processes of growth and development. Thus, as S cells become infected and gradually die, new S cells are continuously available to maintain the infection. Later in life, the reverse is true; most cells in the liver are S and few are R (Figures 3, 4). Infection may result in the death of large numbers of S cells (acute hepatitis), and too few S cells are available to maintain the infection.

The male-female difference in tumor risk is probably related to sex differences in the risk of persistent infection with HBV. Females are less likely than males to develop a persistent infection following initial exposure to HBV and more likely to develop anti-HBs. Among individuals infected for more than 1 year, females are more likely than males to eventually clear the infection. In terms of the model, selective pressure against S cells is maintained for a longer time in males than in females. One hypothesis

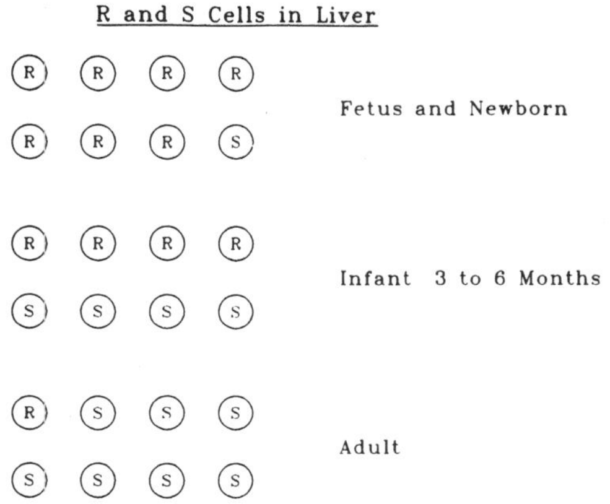

FIG. 3 The proportion of R and S cells varies at different stages of life.

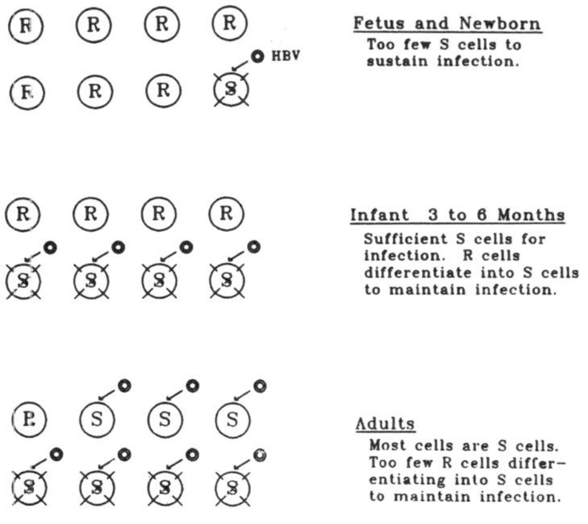

FIG. 4 The availability of S cells at different stages of life influences the kind of infection and liver disease which develops.

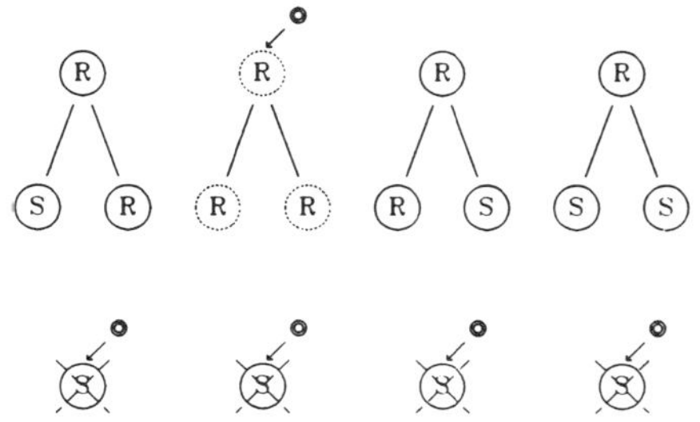

FIG. 5. Chronic infection with HBV results in elimination of S cells, generation of new S cells by division and differentiation of R cells, and the gradual accumulation of R cells. R cells containing integrated HBV DNA sequences are unable to differentiate into S cells and, therefore, accumulate more rapidly than other R cells.

derived from this model is that males maintain a higher proportion of R cells in their livers throughout their lives than females. That is, the distribution of R and S cells in adult male livers is closer to the 3- to 6-month-old infant than the female adult pattern (Figures 3, 4).

TESTS OF THE MODEL

New information was presented at this conference. These data, since they were acquired after the model was formulated, can be used as informal tests of the model. The studies of Bréchot et al. (20), Shafritz (21), and Summers and Mason (22) confirmed earlier reports from their laboratories (5, 6) that integrated sequences of HBV DNA can be found in tumor and nontumor cells in the livers of HBV-infected humans or woodchuck hepatitis virus-infected woodchucks. Since their preliminary reports were used in modifying earlier versions of the model, the new results cannot be used to test it. Nevertheless, it is important to reiterate that according to the model, integration occurs in both R and S cells, but only R cells are at risk of giving rise to tumors.

The studies of Burrell et al. (23) concerning *in situ* hybridization of HBV DNA in human liver biopsy specimens are pertinent to the model. His method was capable of detecting cells containing 10 or more HBV genomes, suggesting that only cells actively replicating virus were identified. This was supported by the findings that, in 90% of the cells positive for hybridization, the HBV DNA sequences (grain counts) were localized to the cytoplasm and that most of the grain counts were contributed by sequences of single-stranded HBV DNA. As predicted by the model, the cells actively replicating virus (positive for HBV DNA) were focally distributed and, histologically, were undergoing degeneration. According to the model, the cells which produce virus are S cells and their survival is shortened as a result of the infection. It is the death of S cells that stimulates the growth of R cells.

A further test of the model is provided by the work of Eddleston (10). Using the patients' own lymphocytes and hepatocytes, he found that the target antigen for HBV-infected liver cells was the core, not the surface antigen. Other studies, including those of Burrell et al. (23) presented here, suggest that the hepatocytes which express HBcAg are the ones that are actively replicating virus. That is, the host immune response is directed only against the cells actively replicating virus, not against cells with integrated HBV DNA that are not producing HBV. Again, according to the model, cells actively replicating virus and subject to immune killing are S cells and the cells not producing core antigen (not replicating virus) would be R cells.

Thus, our model can explain most of the phenomena previously associated with acute and chronic HBV infections and many of the new observations. Our research is now focused on isolating and characterizing R and S cells in woodchuck livers and testing directly the hypothesis that tumors will arise from R cells in woodchucks persistently infected with woodchuck hepatitis virus.

Acknowledgment: We wish to thank Dr. Edward Lustbader for his many contributions to our thinking about this model.

REFERENCES

1. Beasley RP, Lin C-C, Hwang L-Y, et al. Hepatocellular carcinoma and hepatitis B virus. A prospective study of 22,707 men in Taiwan. Lancet 1981; 2:1129–1132.
2. London WT, Blumberg BS. Hepatitis B virus and primary hepatocellular carcinoma. In: Burchenal JH, Oettgen HF, eds. Cancer: Achievements, challenges and prospects for the 1980s, Vol I. New York: Grune and Stratton, 1981: 161–183.
3. London WT, Blumberg BS. Hepatitis B and related viruses in chronic hepatitis, cirrhosis and hepatocellular carcinoma in man and animals. In: Soloway RD, ed. Chronic active liver disease, New York: Churchill Livingstone, 1982 (in press).
4. Thung SN, Gerber MA, Sarno E, et al. Distribution of five antigens in hepatocellular carcinoma. Lab Invest 1979; 41:101–105.
5. Brechot C, Scotto J, Charnay P, et al. Detection of hepatitis B virus DNA in liver and serum: a direct appraisal of the chronic carrier state. Lancet 1981; 2:765–767.
6. Shafritz D, Shouval D, Sherman HI, et al. Integration of hepatitis B virus DNA into the genome of liver cells in chronic liver disease and hepatocellular carcinoma. N Engl J Med 1981; 305:1067–1073.
7. Szmuness W. Hepatocellular carcinoma and hepatitis B virus: evidence for a causal association. Prog Med Virol 1978; 24:40–69.
8. Stevens CE, Beasley RP, Tsui J, et al. Vertical transmission of hepatitis B antigen in Taiwan. N Engl J Med 1975; 292:771–774.
9. Stevens CE, Beasley RP, Szmuness W, et al. Efficacy of hepatitis B immune globulin in prevention of perinatally transmitted hepatitis B: results of a second clinical trial in Taiwan. In: 1981 International Symposium on Viral Hepatitis. Philadelphia: Franklin Institute Press (in press).
10. Eddleston ALWF. Hepatitis B core antigen: a target antigen for T cell cytotoxicity in HBV infection. Hepatology 1982; 2(Suppl):
11. Vierucci A, DeMartino M, Graziani E, et al. A mechanism for liver cell injury in viral hepatitis: effects of hepatitis B virus on neutrophil function in children with chronic active hepatitis. Ped Res 1982 (in press).
12. Becker FF. Regeneration. In: Becker FF, ed. The liver: Normal and abnormal functions, New York: Marcel Dekker, 1974: 69–81.
13. Gudat F, Bianchi L, Sonnabend W, et al. Pattern of core and surface expression in liver tissue reflects state of specific immune response in hepatitis B. Lab Invest 1975; 32:1–9.
14. Krawczynski K, Nazarewicz T, Brzosko WJ, et al. Cellular localization of hepatitis associated antigen in livers of patients with different forms of hepatitis. J Infect Dis 1972; 126:372–377.
15. Barker LF, Chisari FV, McGrath PP, et al. Transmission of type B viral hepatitis to chimpanzees. J Infect Dis 1973; 127:648–662.
16. Berquist KR, Peterson JM, Murphy BL, et al. Hepatitis B antigens in serum and liver of chimpanzees acutely infected with hepatitis B virus. Infect Immun 1975; 12:602–605.
17. Hoofnagle JH, Michalak T, Nowoslawski A, et al. Immunofluorescence in experimentally induced, type B hepatitis in the chimpanzee. Gastroenterology 1978; 74:182–187.
18. Omata M, Ashcavai M, Liew C-T, et al. Hepatocellular carcinoma in the USA, etiologic considerations. Localization of hepatitis B antigens. Gastroenterology 1979; 76:279–287.
19. Bianchi L, Gudat F. Viral antigens in liver tissue and type of inflammation in hepatitis B. In: Bianchi L, Gerok W, Sickinger K, et al. eds. Virus and the liver, Lancaster: MTP Press Ltd., 1980: 197–204.
20. Bréchot C, Pourcel C, Hadchouel M, et al. State of hepatitis B virus DNA in liver disease. Hepatology 1982; 2(Suppl):27S–34S.
21. Shafritz D. Status of hepatitis B virus DNA molecules in the liver of HBsAg carriers and in liver and tumor from patients with hepatocellular carcinoma. Hepatology 1982; 2(Suppl):35S–41S.
22. Summers J, Mason WS. Properties of the hepatitis B-like viruses related to their taxonomic classification. Hepatology 1982; 2(Suppl):61S–66S.
23. Burrell CJ, Gowans EJ, Jilbert AR, et al. Hepatitis B virus DNA detection by *in situ* cytohybridization: implications for viral replication strategy and pathogenesis of chronic hepatitis. Hepatology 1982; 2(Suppl):85S–91S.

J. theor. Biol. (1992) 159, 215–240

Analysis of a Cellular Model to Account for the Natural History of Infection by the Hepatitis B Virus and its Role in the Development of Primary Hepatocellular Carcinoma

ROBERT J. H. PAYNE†, MARTIN A. NOWAK‡ AND BARUCH S. BLUMBERG§

† *Centre for Mathematical Biology, Mathematical Institute, University of Oxford, Oxford OX1 3LB,* ‡ *Department of Zoology, University of Oxford, OX1 3PS, and* § *Balliol College, Oxford OX1 3BJ, U.K.*

(*Received on* 5 *December* 1991, *Accepted on* 9 *May* 1992)

Infection with the hepatitis B virus (HBV) can have many different outcomes. Transient infection may result in acute hepatitis or may remain subclinical. Persistent infection may also be subclinical, or may involve chronic active hepatitis, and can finally lead to the development of primary hepatocellular carcinoma. A mathematical model is given to account for the many different outcomes of HBV pathogenesis. The model is based on the assumption that the liver contains two cell populations with differing abilities to support active HBV replication and/or viral integration into the genome. The model helps account for the relationship of the different clinical courses of HBV infection to the age when the disease is acquired, together with the state of the immune system of the patient.

1. Introduction

The hepatitis B virus (HBV) is estimated to be carried by more than 300 million people worldwide, of whom 25–50% may develop primary hepatocellular carcinoma (PHC) (Beasley *et al.*, 1981). HBV is therefore one of the most common endemic viral pathogens and one of the most common causes of death (Maynard, 1990), its importance as a known human carcinogen coming second only to tobacco. The discovery of the virus in 1967 and the subsequent development of a protective vaccine by the early 1980s gave rise to the hope of controlling, and possibly eradicating, the virus by mass vaccination (Blumberg, 1989). Currently available vaccines are protective against new infection, but no appropriate treatment is as yet available for people who have already acquired the virus (Hollinger, 1989). Alpha-interferons offer a 50% chance of long-term inhibition of HBV replication in patients who acquire the infection as an adult, but appear to be ineffective when the infection is acquired very young (Alexander *et al.*, 1987; Jacyna & Thomas, 1990). Although the main target cells of HBV are hepatocytes the integrated virus DNA can be found not only in tumour and non-tumour cells of the liver, but has also been observed in

‡ Author to whom correspondence should be addressed.

0022–5193/92/220215+26 $08.00/0

the kidney, spleen, pancreas, skin, bone marrow and circulating blood cells (Tiollais & Buendia, 1991). In addition there are several hepatitis-like viruses, known as hepadnaviruses (Pugh & Bassendine, 1990), that are found in a range of animals, including the woodchuck, the ground squirrel and the Peking duck, which have been used to provide useful "models" for the infection of humans by HBV (Summers *et al.*, 1978; Marion *et al.*, 1980).

The consequences of HBV infection may vary considerably between different individuals, ranging from a subclinical and anicteric course, through acute clinical hepatitis, to chronic hepatitis, cirrhosis and PHC. Acute hepatitis develops after a rather variable incubation period (McIntyre, 1990), with an average of 70–80 days, after which the patient will normally be expected to recover completely. This is the most usual observed outcome of infection in adults, although the actual most common outcome is probably transient infection with no perceptible disease, with subsequent development of detectable antibodies against the surface antigen of HBV. In rare cases fulminant hepatitis can occur, a situation that is often fatal (Fagan & Williams, 1990). In some cases—frequently in infants, occasionally in children, but less commonly in adults—viral antigen will not have cleared from the blood within 6 months and the patient will become an infectious carrier (Dudley *et al.*, 1972). Such cases of chronic infection are often asymptomatic, and the patient may appear healthy and often will even be unaware of having been infected. Chronic carriers may develop chronic active hepatitis, leading to cirrhosis and to PHC (Beasley *et al.*, 1981; Viola *et al.*, 1981; Dusheiko, 1990), although the cancer will not usually manifest until after a 30–50 year latency period. The mechanism for the pathogenesis of PHC is as yet poorly understood, the molecular biology of the process being the subject of much recent research (de The *et al.*, 1987; Hsu *et al.*, 1988; Chisari *et al.*, 1989; Fourel *et al.*, 1990).

The virus itself, which has a genome of only about 3200 base pairs, is the smallest known DNA virus that is a human pathogen. The virus is a 42 nm spherical particle with a host-cell derived membrane made up of "small", "medium" and "large" proteins, which together constitute the viral surface antigen, HBsAg (Ganem & Varmus, 1987; Gerlich *et al.*, 1990). The viral core antigens, HBcAg and HBeAg, form a nucleocapsid including the double-stranded viral DNA and the viral DNA polymerase, an enzyme that also has reverse transcriptase and RNAse H activity. The virus genome contains four open reading frames (Tiollais *et al.*, 1981) encoding for the envelope and nucleocapsid proteins, the DNA polymerase, and a fourth protein, called the HBx protein, that acts as a transcriptional control element to upregulate the expression of other viral genes (Seto *et al.*, 1990). The presence of the *HBx* gene has been associated with the induction of liver cancer in transgenic mice (Kim *et al.*, 1991).

In some ways the structure and the life cycle of HBV are remarkably similar to those of retroviruses, such as HIV for example (Tiollais *et al.*, 1985; Tiollais & Buendia, 1991). After fusion with the cell membrane the envelope and core proteins are removed and the circular double-stranded DNA appears in the cell nucleus. This DNA can be integrated into the host cell genome (Dejean *et al.*, 1984) where it serves as a template for the production of viral messenger RNAs (which lead to viral protein synthesis), and for the full RNA pre-genome, which is reversely transcribed into new HBV DNA by the virus encoded polymerase (Ganem & Varmus, 1987). The new virus particles are then assembled in the cytoplasm and released from the host cell. This process appears to be of only weak cytotoxicity and so it is assumed that the increased cell death of infected hepatocytes is due mainly to cytotoxic T cell responses, and possibly other immune effects, directed against infected cells (Alexander, 1990). A detailed mathematical model of the initial immune response to HBV infection has recently been published by Marchuk *et al.* (1991*a*, *b*). It has been suggested that the rate of virus replication may depend on the differentiation state of the host cell, and it is a model based on this assumption that we shall be considering in this paper.

The first viral antigen to be detected in the blood is usually HBsAg (1–10 weeks after infection) which is found to be produced by virus-infected cells even in the absence of detectable virus replication. HBxAg, the product of the "X" reading frame, may sometimes be detected before HBsAg. Testing for HBxAg is not, however, routine. The appearance of HBsAg may be followed a few days later by that of HBeAg, whose presence of the serum is indicative of active virus replication. The HBc antigen can be detected in the liver but is not found free in sera.

The first measurable antibody response (anti-HBc) is directed against the HBc antigen (Krugman *et al.*, 1974), although occasionally anti-*pol*, the antibody against the *POL* gene product, may be found prior to anti-HBc (anti-*pol* is not a routine test). It is followed next by anti-HBe and then by anti-HBs. The antibodies directed against the viral surface antigen are neutralizing and protective. Approximately 1–10% of infected adults, 20–30% of young children, and 90% of neonates vertically infected by a carrier mother will not develop anti-HBs. The failure to mount an immune response against the viral surface protein appears to increase the probability of developing chronic hepatitis infection. Men have a greater likelihood of becoming carriers than women and, in addition, immunosuppressed and immunodeficient patients are more likely to develop chronicity. The chronic carrier state can be asymptomatic (without active viral replication, no HBeAg, low infectivity) or symptomatic [virus replication, HBeAg present, anti-HBe undetectable, high infectivity (Realdi *et al.*, 1980; Hoofnagle *et al.*, 1981)].

One of the most important questions is why some people do not clear the virus. It seems that in some cases HBV infection may cause the host to become tolerant to one or more of the viral antigens: infection can lead to the production of large amounts of HBsAg, which is found in non-infectious particles that have no internal structure and occur as small spheres or long filaments of 22 nm radius, and are over a thousand times more abundant in the serum of carriers than the virus itself (Ganem & Varmus, 1987). It may be that this prevents an effective immune response by anti-HBs against the whole virus particles. However, it still seems difficult to understand why carriers are negative for anti-HBs despite the large amounts of surface antigen present. The release of HBeAg by virus replicating cells has been discussed as a strategy for tolerizing circulating T cells that would normally recognize epitopes in the viral core proteins and thereby mount an immune response against infected cells, the suggestion being that when T cells become exposed to high concentrations of antigens, rather than the small amounts offered to them by antigen-presenting cells, they may lose their ability to respond to their specific antigen. It is possible, therefore, that tolerization could be the strategy employed by HBV to establish a persistent infection.

Nonetheless this still would not account for the considerable variability in the outcome of infection in different individuals, and it is with this variability in mind that we present our analysis of a cellular model, first put forward by London & Blumberg in 1982. In our analysis we show that it is only necessary to have two generalized populations of cells with differing behaviour, assumed to be undifferentiated and differentiated hepatocytes, in order to account for many of the different outcomes of infection by HBV. In addition we relate our model to the role of HBV in the pathogenesis of PHC. Recent experiments with transgenic mice (Kim *et al.*, 1991) have shown that the viral *HBx* gene can induce liver cancer. These transgenic animals contain the *HBx* gene and its regulatory elements in the germline, with the HBx protein being observed in liver, kidney and testis. Interestingly, and consistent with the London–Blumberg model, the *HBx* gene is expressed only in a subset of hepatocytes that are suggested by Kim *et al.* to be related to a specific differentiation state, the histological changes in those hepatocytes being shown to be associated with the *HBx* gene expression.

2. The Model

According to the London & Blumberg (1982) model the liver is considered to contain two populations of cells, which were designated R cells (for resistant to replication) and S cells (for susceptible to replication). The R cells are assumed to be less mature, undifferentiated cells, whereas the S cells are more mature and differentiated hepatocytes. Thus it is taken that the R cells can divide and differentiate further, into either R cells or S cells, but the S cells, being more mature, can only divide one or two times, and only into S cells. The important point to note in this model is the different response of the two cell types upon infection by the hepatitis B virus: it was proposed that R cells are resistant to productive HBV infection, while S cells can allow active virus replication to take place. Thus it is assumed that the S cells will have a greatly reduced survival probability, not only because they may incur a stronger cytotoxic T cell response but also because virus production may compromise the host-cell metabolism. This destruction of S cells provides a stimulus for further cell division, mostly by R cells. Integration of HBV DNA can occur in both R cells and S cells, but because of the reduced life-expectancy of S cells it is of greater importance in R cells, where it may be of tumourigenic significance. The implication is that the increased proliferation of the R cells, stimulated by the death of infected S cells, may allow the integrated DNA to increase the expression of a gene that creates a long-lived infected cell with a large capacity to divide. Secondarily, rapid division of the cells with integrated HBV genomes may increase the probability of other mutational events. The integrated viral DNA may at some later stage then lead to a transformation of some R cells into tumour cells. In addition we assume that phenotypic transformation of an infected R cell will prevent it from thereafter differentiating into an S cell.

Because of the different nature of the two cell types, we expect that the liver in the fetus and the newborn will consist mostly of R cells, and thus would not support active replication of HBV. Cell division and differentiation in children leads to more and more S cells, until in the adult the liver will be composed mostly of S cells that would allow active HBV replication to take place. In this paper we show that this assumption may explain the different outcome of HBV infection in children and in adults.

3. Mathematical Analysis

3.1. NOTATION

t, time.
R, concentration of uninfected R cells.
S, concentration of uninfected S cells.
R', concentration of infected R cells.
S', concentration of infected S cells.
$N=R+S+R'+S'$, total concentration of cells.
V, concentration of free virus.
$F(N)$, natural growth rate of R cells.
$G(N)$, natural growth of S cells.
E, as subscript denotes equilibrium state.
$F_E=F(N_E)$.
$G_E=G(N_E)$.
$F'_E=\mathrm{d}F/\mathrm{d}N|_{N=N_E}$.
$G'_E=\mathrm{d}G/\mathrm{d}N|_{N=N_E}$.
μ_1, natural mortality rate of R cells.
μ_2, natural mortality rate of S cells.
b, transmission coefficient (probability of cell infection by the free virus).
m_1, additional death rate of infected R cells caused by immune responses.
m_2, additional death rate of infected S cells caused by immune responses.
h_1, replication rate of virus in infected R cells.
h_2, replication rate of virus in infected S cells.
n, natural degeneration rate of the free virus.
S_0, initial concentration of uninfected S cells.
S_C, critical value of S_0, determining initial immunity.
S_A, critical value of S_0, determining occurrence of an acute attack.

3.2. THE EQUATIONS AND RELEVANT CONDITIONS

Our model consists of five ordinary differential equations describing the rates of change at time t of the five variables $R=R(t)$, $S=S(t)$, $R'=R'(t)$, $S'=S'(t)$ and $V=V(t)$, representing the concentration of uninfected R cells, uninfected S cells, infected

R cells, infected S cells and free virus, respectively:

$$\frac{dR}{dt} = (F(N) - \mu_1)R - bVR, \tag{1}$$

$$\frac{dS}{dt} = (G(N) - \mu_2)S - bVS, \tag{2}$$

$$\frac{dR'}{dt} = bVR + (F(N) - \mu_1)R' - m_1R', \tag{3}$$

$$\frac{dS'}{dt} = bVS + (G(N) - \mu_2)S' - m_2S', \tag{4}$$

$$\frac{dV}{dt} = h_1R' + h_2S' - bV(R+S) - nV. \tag{5}$$

The natural growth rates (cell division) of the R and S cells are represented by $F(N)$ and $G(N)$, respectively, and their natural mortality rates by the coefficients μ_1 and μ_2. The cells cannot reproduce unboundedly, so in order to make their growth rates density limited we choose F and G to be monotonically decreasing functions of $N = N(t)$, where $N = R + S + R' + S'$ is the total cell density at time t. R cells are infected by the virus at a rate bVR, and S cells at a rate bVS, where the transmission coefficient b is taken to be the same for both cell types. The shortened survival probability of infected cells is represented by the terms with coefficients m_1 and m_2, for the infected R and infected S cells, respectively, and the replicative ability of the virus in each cell type by the terms with coefficients h_1 and h_2, respectively. The natural degeneration of the virus when outside the cells is represented by the term with coefficient n. The parameters μ_1, μ_2, m_1, m_2, h_1, h_2, b and n are all positive.

Without loss of generality we take the cell and virus concentrations to be scaled such that in the uninfected system the total cell concentration is $N = 1$. This can be achieved by setting

$$F(1) = \mu_1, \tag{6}$$

$$G(1) = \mu_2. \tag{7}$$

This also gives us our initial conditions:

$$R(0) + S(0) = 1, \tag{8}$$

$$R'(0) = 0 = S'(0), \tag{9}$$

$$V(0) = V_0. \tag{10}$$

We will denote the initial concentration of S cells by $S(0) = S_0$.

In our model we are assuming that the survival expectancy of the infected S cells is greatly reduced compared to that of infected R cells, and also that S cells allow a far higher rate of active replication of the virus to occur than in R cells. Thus we have

$$m_1 \ll m_2, \tag{11}$$

$$h_1 \ll h_2. \tag{12}$$

In addition we are assuming that S cells are more mature than R cells, being more fully differentiated hepatocytes, and thus that the capacity for S cells to divide further is limited compared to that of R cells. This means that in our model we take

$$G(N) \ll F(N). \tag{13}$$

3.3. EQUILIBRIUM STATES

Let $N=(R, S, R', S')$ be the vector of cell concentrations, thus enabling the five dynamic variables to be represented by $(N; V)$. The possible steady states are as follows:

(i) $(N; V)=(R_E, S_E, 0, 0; 0)$
(ii) $(N; V)=(0, 0, R'_E, 0; V_E)$
(iii) $(N; V)=(R_E, 0, R'_E, 0; V_E)$
(iv) $(N; V)=(0, 0, 0, S'_E; V_E)$
(v) $(N; V)=(0, S_E, 0, S'_E; V_E)$
(vi) $(N; V)=(R_E, 0, R'_E, S'_E; V_E)$
(vii) $(N; V)=(0, S_E, R'_E, S'_E; V_E)$
(viii) $(N; V)=(R_E, S_E, R'_E, S'_E; V_E)$,

where the subscript E denotes the equilibrium state, and a zero indicates that particular cell type is not present. We will now show that the majority of these states can be discounted on grounds of either instability or dynamical inaccessibility.

First note that since $F(N)$ and $G(N)$ are monotonically decreasing functions and because of the conditions (6) and (7) we have

$$\left.\begin{aligned} F(N)-\mu_1>0 \\ G(N)-\mu_2>0 \end{aligned}\right\} N<1, \qquad \left.\begin{aligned} F(N)-\mu_1<0 \\ G(N)-\mu_2<0 \end{aligned}\right\} N>1. \tag{14}$$

Consider the total number of cells

$$\frac{dN}{dt}=\frac{dR}{dt}+\frac{dS}{dt}+\frac{dR'}{dt}+\frac{dS'}{dt}$$
$$=(F(N)-\mu_1)(R+R')+(G(N)-\mu_2)(S+S')-m_1R'-m_2S'. \tag{15}$$

Using eqn (14), and remembering that initially $N=1$, we can see that for all the time while the system is infected we must have $N<1$. We assume that the net reproductive rate for R cells is faster than for S cells, that is

$$G(N)-\mu_2<F(N)-\mu_1.$$

But we have just shown that as long as the system is infected $N<1$, and thus for all time while the system is infected we know that

$$0<G(N)-\mu_2<F(N)-\mu_1. \tag{16}$$

It is now clear that as long as the virus is present

$$\frac{1}{S}\frac{dS}{dt}<\frac{1}{R}\frac{dR}{dt}. \tag{17}$$

This means that given our initial conditions it is not possible for us to have an equilibrium in which $S>0$ and $V>0$. This removes (v), (vii) and (viii) from our list of possible steady states.

Now compare the dynamics of S' and R'. Taking $S\to 0$ as just discussed we have

$$\begin{aligned}\frac{1}{S'}\frac{dS'}{dt}&=(G(N)-\mu_2)-m_2\\ &<(F(N)-\mu_1)-m_2 \quad \text{by (16)}\\ &<(F(N)-\mu_1)-m_1 \quad \text{by (11)}\\ &<(F(N)-\mu_1)-m_1+bV\frac{R}{R'}\\ &=\frac{1}{R'}\frac{dR'}{dt}.\end{aligned}$$

Thus, once S has become small, we have

$$\frac{1}{S'}\frac{dS'}{dt}<\frac{1}{R'}\frac{dR'}{dt}, \tag{18}$$

and therefore it is not possible for us to have an equilibrium in which $S'>0$. This excludes (iv) and (vi) from being allowable steady states, leaving us only with the following cases:

(i) $(N;V)=(R_E, S_E, 0, 0; 0)$
(ii) $(N;V)=(0, 0, R'_E, 0; V_E)$
(iii) $(N;V)=(R_E, 0, R'_E, 0; V_E)$.

In the next section we shall investigate the existence and stability of these remaining states.

3.4. LINEAR STABILITY ANALYSIS

(i) $(N;V)=(R_E, S_E, 0, 0; 0)$,

where

$$R_E+S_E=1. \tag{19}$$

We investigate the linear stability by considering small perturbations to the system in the vicinity of the steady state. Let

$$\begin{aligned} R &= R_E + r, \\ S &= S_E + s, \\ R' &= R'_E + r', \\ S' &= S'_E + s', \\ V &= V_E + v. \end{aligned} \tag{20}$$

where r, s, r', s' and v are small. Substituting into eqns (1–5) and retaining only linear terms gives us the linearized system

$$\frac{dr}{dt} = (F_E - \mu_1)r + F'_E R_E(r + s + r' + s') - bV_E r - bR_E v, \tag{21}$$

$$\frac{ds}{dt} = (G_E - \mu_2)s + G'_E S_E(r + s + r' + s') - bV_E s - bS_E v, \tag{22}$$

$$\frac{dr'}{dt} = (F_E - \mu_1)r' + F'_E R'_E(r + s + r' + s') + bV_E r + bR_E v - m_1 r', \tag{23}$$

$$\frac{ds'}{dt} = (G_E - \mu_2)s' + G'_E S'_E(r + s + r' + s') + bV_E s + bS_E v - m_2 s', \tag{24}$$

$$\frac{dv}{dt} = h_1 r' + h_2 s' - bV_E(r + s) - b(R_E + S_E)v - nv, \tag{25}$$

where

$$F_E = F(N_E); \qquad G_E = G(N_E),$$

$$F'_E = \left.\frac{dF}{dN}\right|_{N = N_E}; \qquad G'_E = \left.\frac{dG}{dN}\right|_{N = N_E}.$$

In our current case we can use the fact that $R' = S' = V = 0$, and the conditions (6), (7) and (19), to reduce the system to

$$\frac{dr}{dt} = F'_E R_E(r + s + r' + s') - bR_E v, \tag{26}$$

$$\frac{ds}{dt} = G'_E S_E(r + s + r' + s') - bS_E v, \tag{27}$$

$$\frac{dr'}{dt} = bR_E v - m_1 r', \tag{28}$$

$$\frac{ds'}{dt} = bS_E v - m_2 s', \tag{29}$$

$$\frac{dv}{dt} = h_1 r' + h_2 s' - (b+n)v. \tag{30}$$

Note that the last three equations decouple from the other two: for the moment we will look at these three equations on their own. We look for solutions of the form

$$(r', s', v) \propto e^{\lambda t}. \tag{31}$$

Hence

$$\begin{pmatrix} \lambda+m_1 & 0 & -bR_E \\ 0 & \lambda+m_2 & -bS_E \\ -h_1 & -h_2 & \lambda+b+n \end{pmatrix} \begin{pmatrix} r' \\ s' \\ v \end{pmatrix} = \begin{pmatrix} 0 \\ 0 \\ 0 \end{pmatrix}. \tag{32}$$

The solvability condition for (32) is

$$\begin{vmatrix} \lambda+m_1 & 0 & -bR_E \\ 0 & \lambda+m_2 & -bS_E \\ -h_1 & -h_2 & \lambda+b+n \end{vmatrix} = 0, \tag{33}$$

which can be written as the cubic

$$\lambda^3 + a\lambda^3 + c\lambda + d = 0, \tag{34}$$

where

$$a = m_1 + m_2 + b + n,$$
$$c = (m_1+m_2)(b+n) - b(h_1 + S_E(h_2 - h_1)) + m_1 m_2,$$
$$d = m_1 m_2 (b+n) - b(h_1 m_2 + S_E(h_2 m_1 - h_1 m_2)).$$

For the case $d<0$ there will always be instability with respect to perturbations in R', S' and V. For the case $d>0$ it is easy to show that we will also have $c>0$, and that $ac>d$, and hence that there are no roots with positive real part. Thus we have that if $d>0$ there will be stability with respect to perturbations in R', S' and V.

Now we look at eqns (26) and (27). The only case of interest is when the perturbations in R', S' and V are stable, and thus we have the system

$$\frac{dr}{dt} = F'_E R_E (r+s),$$

$$\frac{ds}{dt} = G'_E S_E (r+s).$$

Note that

$$\frac{d(r+s)}{dt} = (F'_E R_E + G'_E S_E)(r+s),$$

and thus, since $F'_E<0$ and $G'_E<0$, that $r+s\to 0$. We therefore have neutral stability in the $R+S=1$ plane. Hence overall we have that if

$$1+\frac{n}{b}>\frac{h_1}{m_1}+S_E\left(\frac{h_2}{m_2}-\frac{h_1}{m_1}\right) \tag{35}$$

then the system is stable with respect to the $R+S=1$ plane (i.e. the virus will be wiped out), and if

$$1+\frac{n}{b}<\frac{h_1}{m_1}+S_E\left(\frac{h_2}{m_2}-\frac{h_1}{m_1}\right) \tag{36}$$

then the system is unstable with respect to the $R+S=1$ plane (i.e. the virus is able to invade).

(ii) $(N;\ V)=(0, 0\ R'_E, 0;\ V_E)$,

where

$$V_E=\frac{h_1}{n}R'_E,$$

$$F_E=\mu_1+m_1. \tag{37}$$

We have previously established that for any equilibrium when the system is infected the variables S and S' will both converge to zero, thus in this case the linearized system (21–25) reduces to

$$\frac{dr}{dt}=m_1r-bV_Er, \tag{38}$$

$$\frac{dr'}{dt}=bV_Er+F'_ER'_E(r+r'), \tag{39}$$

$$\frac{dv}{dt}=h_1r'-bV_Er-nv. \tag{40}$$

Note that if $r=0$ then, since $F'_E<0$, $r'\to 0$, and thus that $v\to 0$ also. Hence the stability depends only on the dr/dt equation: the state will be stable if $m_1<bV_E$, and unstable if $m_1>bV_E$. We expect that $R'_E=N_E\approx 1$, and so we know that the steady state in which only infected R cells and virus are present will be

$$\text{stable if}\quad \frac{n}{b}<\frac{h_1}{m_1},$$

$$\text{unstable if}\quad \frac{n}{b}>\frac{h_1}{m_1}. \tag{41}$$

(iii) $(N;\ V)=(R_E, 0, R'_E, 0;\ V_E)$,

where we need to satisfy

$$0 = F_E - \mu_1 - bV_E, \tag{42}$$

$$0 = bV_E R_E + (F_E - \mu_1) R'_E - m_1 R'_E, \tag{43}$$

$$0 = h_1 R'_E - (bR_E + n) V_E. \tag{44}$$

Substituting (42) into (43) gives

$$bV_E(R_E + R'_E) = m_1 R'_E.$$

We expect that $R_E + R'_E = N_E \approx 1$, and so we have

$$V_E \approx \frac{m_1}{b} R'_E. \tag{45}$$

Upon substituting this into (44) we obtain

$$R_E \approx \frac{h_1}{m_1} - \frac{n}{b}, \tag{46}$$

and hence that

$$R'_E \approx 1 + \frac{n}{b} - \frac{h_1}{m_1}, \tag{47}$$

and

$$V_E \approx \frac{m_1}{b}\left(1 + \frac{n}{b} - \frac{h_1}{m_1}\right). \tag{48}$$

Since we require that $0 < R_E < 1$ this equilibrium state will only exist if

$$\frac{n}{b} < \frac{h_1}{m_1} < 1 + \frac{n}{b}. \tag{49}$$

We will now demonstrate that this particular equilibrium is always unstable.

As before we know that the S and S' variables will always converge to zero, and so in this case the linearized system (21–25) reduces to

$$\frac{dr}{dt} = F'_E R_E (r + r') - bR_E v, \tag{50}$$

$$\frac{dr'}{dt} = bR_E v + bV_E r + (bV_E - m_1) r' + F'_E R'_E (r + r'), \tag{51}$$

$$\frac{dv}{dt} = h_1 r' - bV_E r - (bR_E + n) v. \tag{52}$$

Again we look for solutions of the form

$$(r', s', v) \propto e^{\lambda t}. \tag{53}$$

Hence

$$\begin{pmatrix} \lambda - R_E F'_E & -R_E F'_E & bR_E \\ -bV_E - R'_E F'_E & \lambda + m_1 - bV_E - R'_E F'_E & -bR_E \\ bV_E & -h_1 & \lambda + bR_E + n \end{pmatrix} \begin{pmatrix} r \\ r' \\ v \end{pmatrix} = \begin{pmatrix} 0 \\ 0 \\ 0 \end{pmatrix}. \tag{54}$$

The solvability for (54) is

$$\begin{vmatrix} \lambda - R_E F'_E & -R_E F'_E & bR_E \\ -bV_E - R'_E F'_E & \lambda + m_1 - bV_E - R'_E F'_E & -bR_E \\ bV_E & -h_1 & \lambda + bR_E + nn \end{vmatrix} = 0. \tag{55}$$

This expression can be written in the form of the cubic

$$\lambda^3 + a\lambda^2 + c\lambda + d = 0, \tag{56}$$

where, after a little algebra [using eqns (45) and (47), and the approximation $N_E \approx 1$],

$$a = n + (b + m_1)R_E - N_E F'_E,$$

$$c = (n + bR_E)(m_1 R_E - N_E F'_E) - m_1 R_E F'_E - bR_E(h_1 + bV_E),$$

$$d = m_1 b R_E (nV_E + R'_E F'_E).$$

Note that, since $F'_E < 0$, $a > 0$. Also we know that the concentration of the virus is usually far smaller than that of the cells, and so we may take that $d < 0$. Thus this equilibrium state will always be unstable.

3.5. PARAMETER REGIMES

Equation (36), the condition for the initial growth of the infection, can be used to define an initial concentration of S cells,

$$S_C = \left(1 + \frac{n}{b} - \frac{h_1}{m_1}\right) \bigg/ \left(\frac{h_2}{m_2} - \frac{h_1}{m_1}\right), \tag{57}$$

which is critical to the outcome of the disease. Using this together with the results so far obtained we can predict the system behaviour for various parameter values.

(i) $h_1/m_1 > h_2/m_2$.

- (a) $S_C < 0$:
  the virus cannot invade.
- (b) $0 < S_C < 1$:
  if $S_0 < S_C$ the virus will invade, and because of the condition (17) the system will always end up with only infected R cells present; if $S_0 > S_C$ the virus will not invade.

(c) $1 < S_C$:
the virus will always invade, and the system will always end up with only infected R cells present.

(ii) $h_1/m_1 < h_2/m_2$.

(a) $S_C < 0$:
the virus can always invade and the system will always end up with only infected R cells present.

(b) $0 < S_C < 1$ *and* $n/b < h_1/m_1$:
if $S_0 < S_C$ the virus will not invade;
if $S_0 > S_C$ then the equilibrium when the system is infected and the equilibrium when the system is uninfected are both stable. Which of the two the system will end up in cannot be predicted with certainty, but we expect that as $h_1/m_1 \to n/b$ it is more likely to rid itself of the virus, and as $h_1/m_1 \to 1 + n/b$ it is more likely to become persistently infected. Note that this is the regime where the unstable steady state (iii) exists. This state is a form of saddle point and so it is possible that the system may under certain circumstances spend some time in the vicinity of this intermediary state.

(c) $0 < S_C < 1$ *and* $n/b > h_1/m_1$:
if $S_0 < S_C$ the virus will not invade;
if $S_0 > S_C$ the virus can invade, but because of the condition (41) the

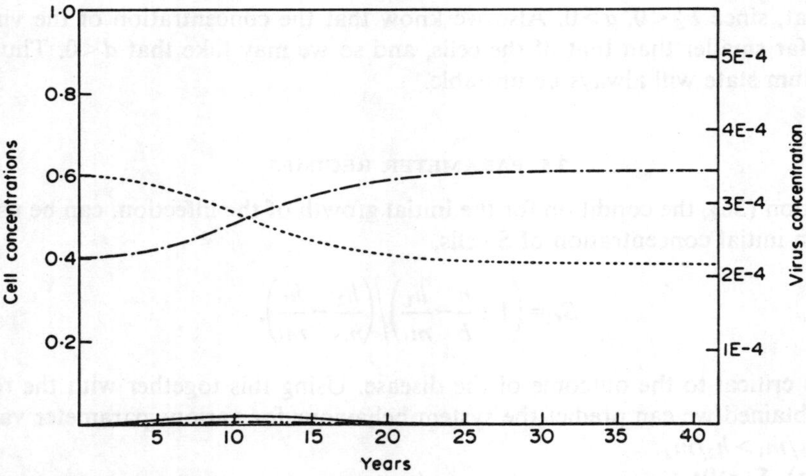

FIGS 1–6. Simulation results showing the main types of behaviour exhibited by the model. Parameter values used: $n = 0{\cdot}03$, $b = 0{\cdot}02$, $h_1 = 3 \times 10^{-5}$, $m_1 = M \times 10^{-5}$, $h_2 = 5 \times 10^{-3}$, $m_2 = M \times 10^{-3}$, $F(N) = 0{\cdot}1/(1{\cdot}0 + N^2)$, $G(N) = 10^{-3}/(1{\cdot}0 + N^2)$, $\mu_1 = 5 \times 10^{-2}$, $\mu_2 = 5 \times 10^{-4}$. FIG. 1. Subclinical recovery. $M = 1{\cdot}6$, $S_0 = 0{\cdot}6$. The virus initially invades but is then successfully eliminated by the immune system. The maximum virus concentration during the course of infection is not enough to cause clinical symptoms. After having recovered the patient is then immune to further infection. (——), free virus; (· · · ·), S cells; (— · —), R cells; (- - -), infected S; (— · · · —) infected R.

system will always eventually rid itself of the infection. Note that when this happens it leaves the system in an immune state—the virus cannot reinvade.

(d) $1 < S_C$:
the virus cannot invade.

3.6. THE ACUTE CASE

Since our model was designed in part to explain the increased incidence of acute hepatitis in older subjects it is worth considering whether we, by using our model, can predict the occurrence of an acute attack.

Acute cases always occur soon after infection, compared to the incubation period for symptomatic chronic disease. This suggests that we may make the following approximations: $R \approx 1 - S_0$; $S \approx S_0$; $N \approx 1$. In addition we assume that R' and S' are of the same magnitude and thus, because of the condition (12), that $h_1 R' \ll h_2 S'$. Hence we consider the system

$$\frac{dR}{dt} = -b(1 - S_0)V, \tag{58}$$

$$\frac{dS}{dt} = -bS_0 V, \tag{59}$$

$$\frac{dR'}{dt} = b(1 - S_0)V - m_1 R', \tag{60}$$

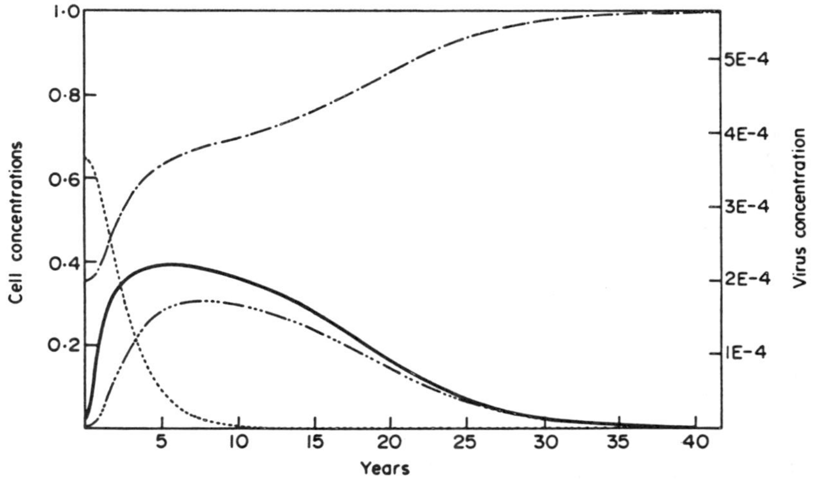

FIG. 2. Recovery preceded by relapse. $M = 1{\cdot}4$, $S_0 = 0{\cdot}65$. The virus initially invades but is then successfully eliminated by the immune system. The maximum virus concentration is higher than in Fig. 1, possibly being large enough to cause clinical symptoms. See Fig. 1 legend for key.

$$\frac{dS'}{dt} = bS_0V - m_2S', \tag{61}$$

$$\frac{dV}{dt} = h_2S' - (b+n)V. \tag{62}$$

Note that the last two equations decouple, and will determine the dynamics of the system.

In order to have the positive feedback that will give the explosion in virus concentration that is characteristic of an acute attack of HBV we require that $dS'/dt > 0$ and $dV/dt > 0$. This will be the case if

$$\frac{S'}{V} < \frac{b}{m_1} S_0$$

and

$$\frac{n+b}{h_2} < \frac{S'}{V}.$$

Combining these we obtain the criterion that:

if $S_0 < S_A$ the disease will not have an acute phase;

if $S_0 > S_A$ the disease will become acute soon after infection,

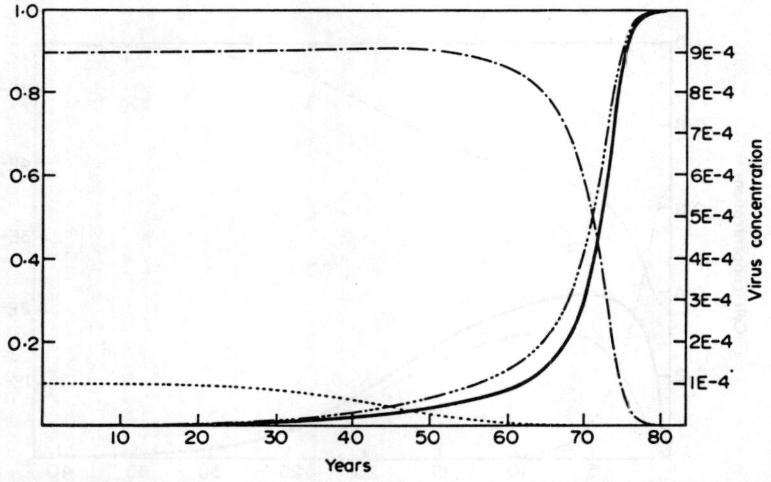

FIG. 3. Persistent preceded by a latent period. $M = 1{\cdot}2$, $S_0 = 0{\cdot}1$. In this example the system eventually consists of nothing but infected R cells. This is the situation with which we associate the occurrence of PHC. Before this occurs there is a long period of time during which there is negligible noticeable activity, so that the patient may well be unaware of having the virus prior to the onset of PHC. See Fig. 1 legend for key.

where

$$S_A = \frac{m_2}{h_2}\left(1 + \frac{n}{b}\right). \tag{63}$$

4. Numerical Results

The model eqns (1–5) were solved numerically using a simple Euler scheme with initially no infected cells present. For the examples given here the forms used for $F(N)$ and $G(N)$ were $F(N) = f_1/(f_2 + N^2)$ and $G(N) = g_1/(g_2 + N^2)$. Other qualitatively similar forms were also tried, the same results being obtained. The simulation was started by adding a small concentration of virus into the system, so that

$$V(0) = 10^{-3},$$
$$R(0) = 1 - S_0,$$
$$S(0) = S_0,$$
$$R'(0) = 0,$$
$$S'(0) = 0,$$

where, since we are assuming that the initial proportion of infected S cells is not fixed, S_0 was allowed to vary between 0 and 1. All the example plots shown here

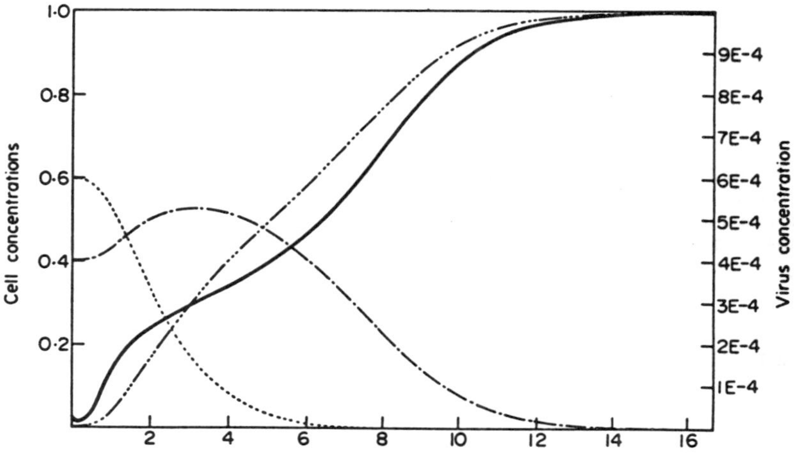

FIG. 4. Persistent without a latent period. $M = 1{\cdot}3$, $S_0 = 0{\cdot}6$. The system eventually becomes overwhelmed by infected R cells. However, unlike in Fig. 3, the saturation with infected R cells occurs relatively quickly, without being preceded by a long latent period. See Fig. 1 legend for key.

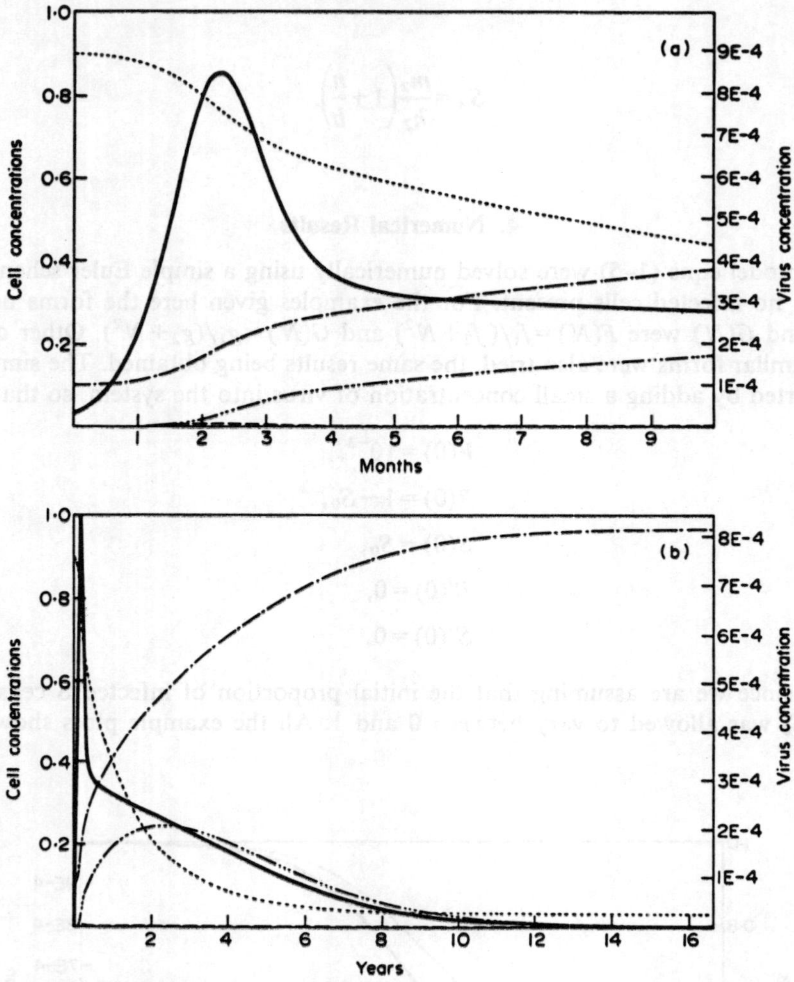

FIG. 5. Acute followed by recovery. (a) Initial behaviour. (b) Overall behaviour. $M=1{\cdot}6$, $S_0=0{\cdot}9$. Initially there is a sudden sharp increase in the virus concentration, which peaks and then drops again. This occurs on a time scale of a few months, as compared to the tens of years necessary for the onset of cancer shown in Figs 3 and 4. After this clinically detectable attack the virus is then eliminated as in Fig. 2, leaving the patient immune to further infection. See Fig. 1 legend for key.

were produced using the following parameter values:

$$n=0{\cdot}03; \qquad h_1=3\times10^{-5}; \qquad h_2=5\times10^{-3};$$

$$b=0{\cdot}02; \qquad m_1=M\times10^{-5}; \qquad m_2=M\times10^{-3};$$

$$f_1=0{\cdot}1; \qquad f_2=1{\cdot}0; \qquad \mu_1=5\times10^{-2};$$

$$g_1=10^{-3}; \qquad g_2=1{\cdot}0; \qquad \mu_2=5\times10^{-4}.$$

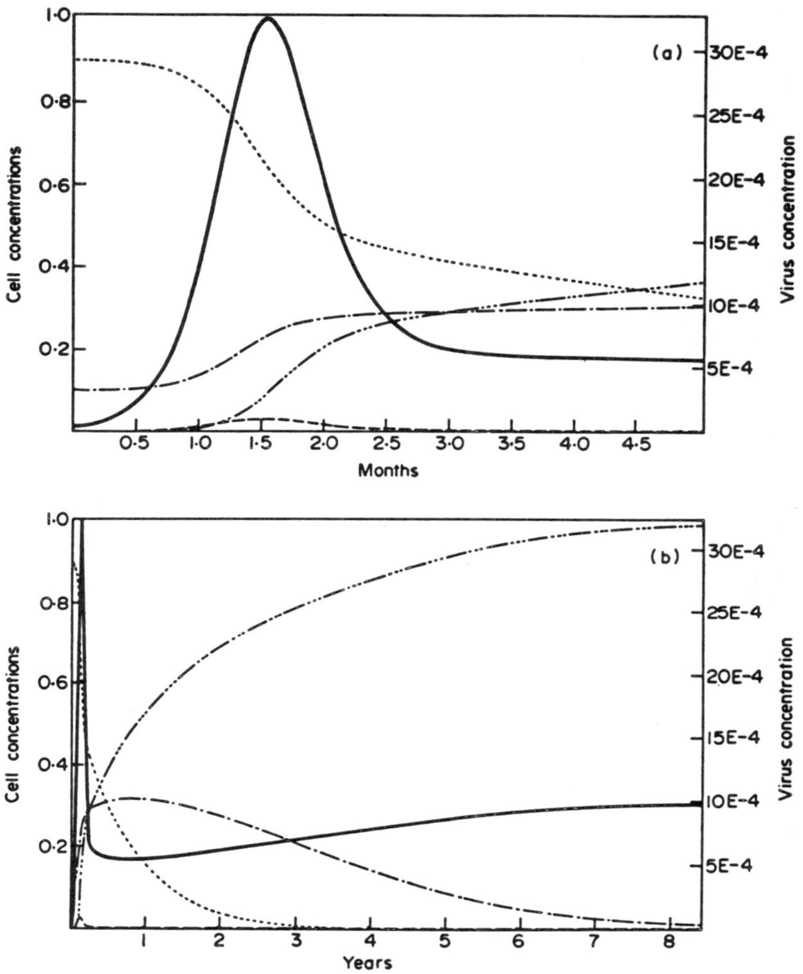

FIG. 6. Acute followed by persistent. (a) Initial behaviour. (b) Overall behaviour. $M=1{\cdot}4$, $S_0=0{\cdot}9$. Initially there is a sudden sharp increase in the virus concentration, which peaks and then drops again. However, unlike in Fig. 5, after the initial attack the system then becomes saturated by infected R cells as in Fig. 4. A severe attack is never followed by a long latent period of the sort shown in Fig. 3. See Fig. 1 legend for key.

The parameter M is used as a means of varying the immune response, the assumption being that in younger subjects the immune system is less fully developed so that the magnitude of both m_1 and m_2 will be less than in fully grown adults.

The different behaviour exhibited by the simulation under different parameter regimes [Figs 1–6] was as predicted by our analysis, and may be summarized by two phase diagrams, one for $h_1/m_1 < h_2/m_2$ and the other for $h_1/m_1 > h_2/m_2$. These are

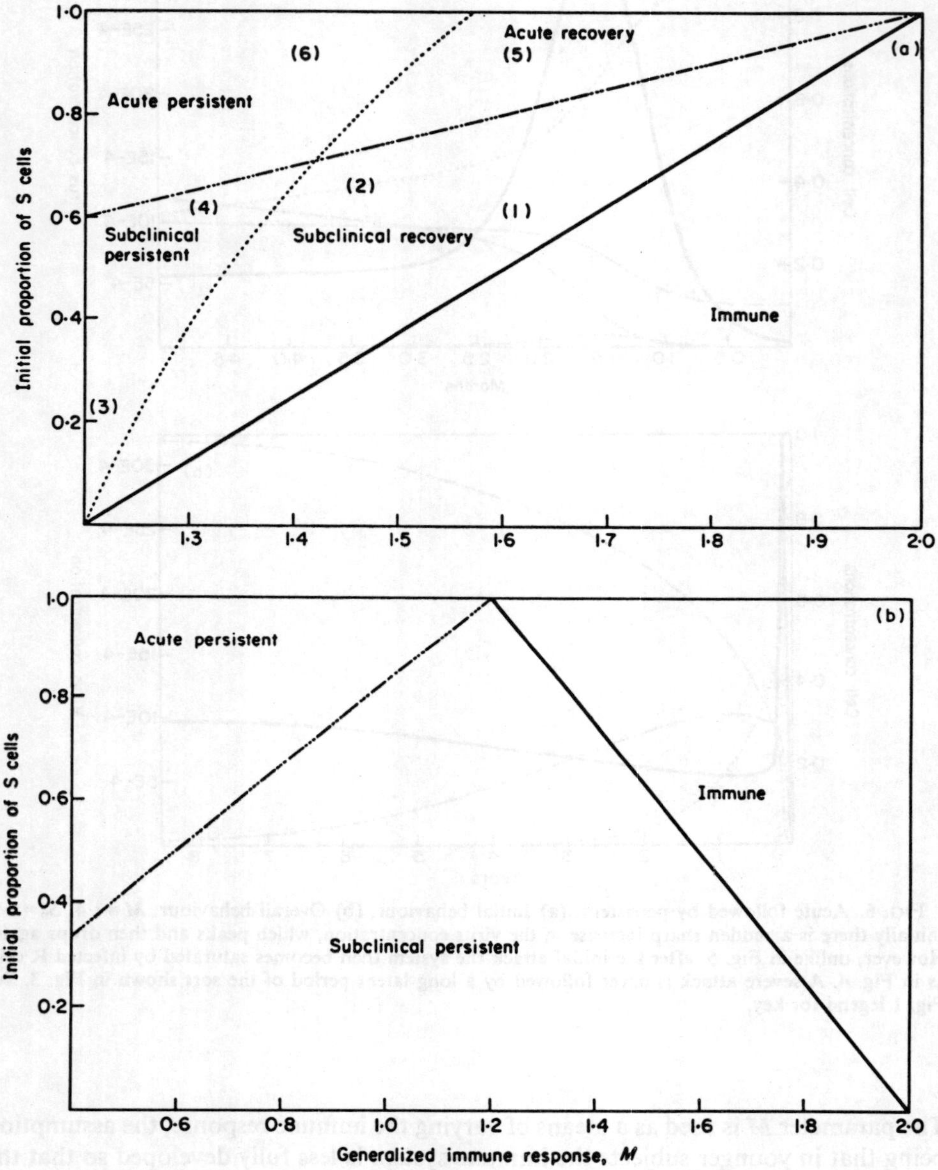
(a)
Acute recovery
(5)
(6)
Acute persistent
(2)
(4)
(1)
Subclinical persistent
Subclinical recovery
Immune
(3)
Initial proportion of S cells
1·0
0·8
0·6
0·4
0·2
1·3
1·4
1·5
1·6
1·7
1·8
1·9
2·0
(b)
Acute persistent
Immune
Subclinical persistent
0·6
0·8
1·0
1·2
1·4
1·6
1·8
2·0
Generalized immune response, *M*

FIG. 7

shown in Figs 7(a) and (b). Note however that these phase diagrams are two-dimensional representations of a three-dimensional phase space, the immune responses m_1 and m_2 having been tied together in the variable M. In the full three-dimensional diagram the persistent/transient dividing line will depend only on h_1/m_1, and the acute/non-acute dividing line will depend only on h_2/m_2. The independence of these two effects is demonstrated in the next phase diagrams, given in Figs 8(a) and (b), where the behaviour of the system as a function of the immune response against the R and S cells is shown for different fixed values of S_0. Note that the word persistent is being used in this context to describe those cases when the final state of the system is the infected equilibrium, that is, those cases when the system will end up with only infected R cells present. The different regions marked on the phase diagrams are described in the captions to Figs 1–6. In Figs 1–6 note that the magnitude of the cell concentrations are given on the left-hand axis, and that the concentration of the free virus is given on a different scale on the right-hand axis.

5. Discussion

The mathematical analysis presented in this paper helps to explain how the London–Blumberg model can be used to explain the many different outcomes of infection with HBV. Central to the model is the assumption that the liver contains two cell types with differing abilities to support HBV replication, the undifferentiated R cells being less permissive for active virus replication than the more fully differentiated S cells. The virus itself is said to be only weakly cytopathic, and thus most of the damage to the liver is mediated by cytotoxic T lymphocytes directed at infected S cells. The extensive death of S cells then allows for the proliferation of the immature R cells. Virus integration can occur in both R and S cells, but may be relevant only in those R cells that have a longer life span and a higher potential to divide, and it therefore seems likely that transformation of the integrated genome occurs in R cells.

FIG. 7. Diagrams showing the expected outcome of infection for differing initial concentrations of S cells, and differing strengths of the overall immune response. In this case the immune responses against the infected R and S cells have been taken to be proportional: explicitly, $m_1 = M \times 10^{-5}$ and $m_2 = M \times 10^{-3}$. (a) Phase diagram for $h_1/m_1 < h_2/m_2$. The numbered areas on the diagrams refer to Figs 1–6. The assumption of our model is that the more fully differentiated S cells will be scarce in the newborn and in infants. Together with a weak immune system (low M) this places them on the lower left-hand area of the diagram, where the expected outcome of infection is a subclinical chronic carrier state that will eventually develop into PHC. As the age of first infection increases the proportion of S cells increases, and the immune response becomes more effective. This leads us across the diagram moving to the right and upwards, until for adults we are in the area on the top right-hand side (almost all S cells and a strong immune response) for which the most likely outcome is an acute attack followed by complete recovery. During infection the number of infected S cells drops, so that if recovery occurs the patient is then left in the area of the diagram characterized by immunity to further infection. (Parameter values used are as for Figs 1–6.) (b) Phase diagram for $h_1/m_1 > h_2/m_2$. In this parameter region recovery from infection is impossible. The probability that the virus establishes an infection decreases as the initial proportion of S cells increases. This happens because $h_1/m_1 > h_2/m_2$ implies that more free virus particles are released from R cells than from S cells. This seems biologically unlikely, since we expected h_1 to be very small. The diagram is only included for the sake of completeness. (Parameter values used are as for Figs 1–6, except that $h_1 = 5 \times 10^{-5}$ and $h_2 = 3 \times 10^{-3}$.)

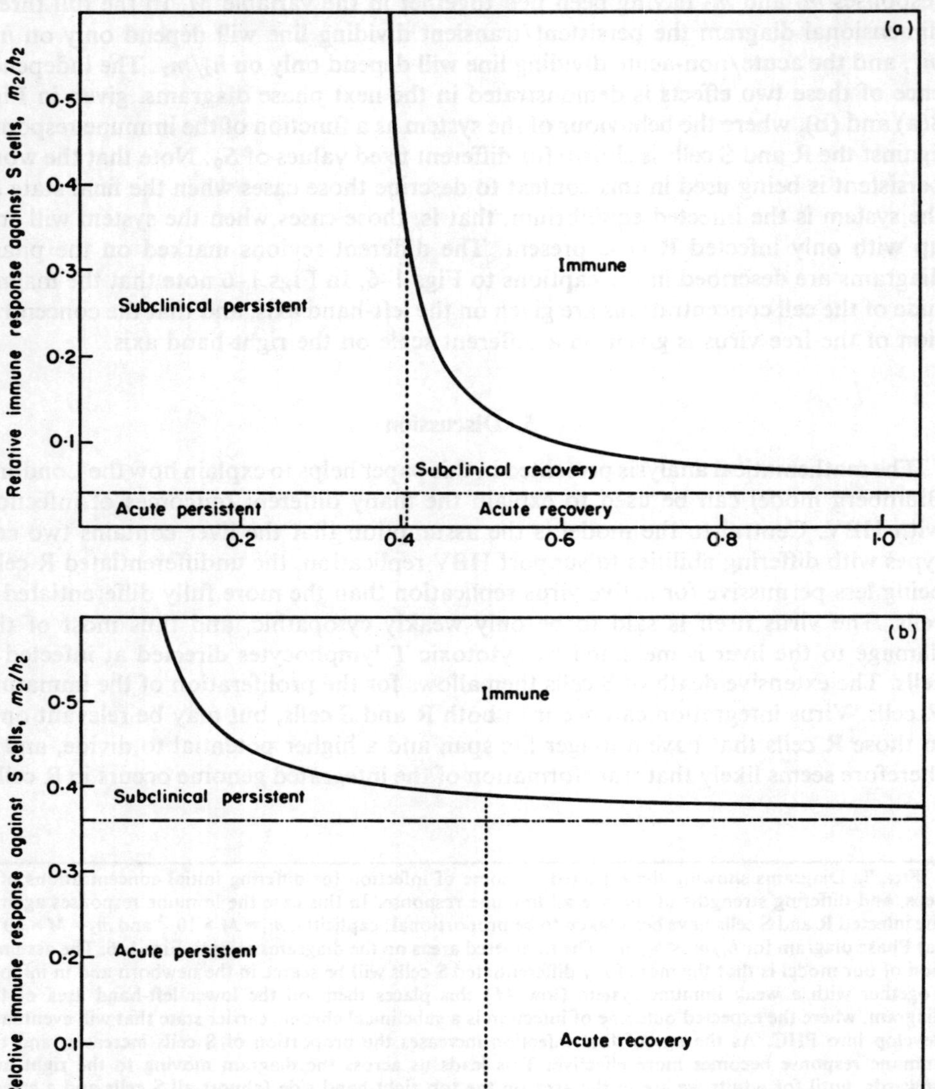

(a)
Immune
Subclinical persistent
Subclinical recovery
Acute persistent
Acute recovery
Relative immune response against S cells, m_2/h_2
0·1
0·2
0·3
0·4
0·5
0·2
0·4
0·6
0·8
1·0
(b)
Immune
Subclinical persistent
Acute persistent
Acute recovery
Relative immune response against S cells, m_2/h_2
Relative immune response against R cells, m_1/h_1

FIG. 8

The phase diagram shown in Fig. 7(a) illustrates the many different outcomes of HBV infection depending on the initial ratio of R and S cells, and the general strength of the immune response to HBV. We expect both of these factors to be dependent on the age of the patient when the infection is acquired: according to the London–Blumberg model the proportion of S cells increases with increasing age, levelling out sometime late in childhood, and in general the ability to mount immune responses will be a one-humped pattern as a function of age, being lower for the young and the elderly.

The fetus and newborn may have a liver that contains only undifferentiated R cells, and this liver cannot support any HBV replication. With increasing age the proportion of S cells increases and we enter the parameter space that leads to a subclinical infection followed by a chronic carrier state. In Asian populations this appears to be at about 3 months, but it may be different in other populations. This outcome is frequent in infected children. Increasing age further leads again to an increased proportion of S cells, and also to an increasingly effective immune response. This places us either in the parameter regime characterized by a subclinical course followed by recovery and immunity, or, for higher proportions of S cells and weaker immune responses (possibly due to some immunodeficiency), in the region of acute hepatitis followed by chronic disease. Most adults however will be able to mount effective immune responses and will have a very high proportion of S cells, which puts them in the region characterized by an acute attack with subsequent recovery. For this case we find that during the course of the disease the reserves of uninfected S cells become depleted, so that after the virus has been eliminated the patient is then left in the parameter space for which there is immunity to further infection. Such immunity may only be temporary, although this is not included explicitly in our current mathematical formulation. This is because after recovery we expect the proportion of S cells to start increasing again, which may possibly lead to the patient moving back into the non-immune parameter regime. In addition the worse the initial attack the greater the immunity conferred upon the patient afterwards. Thus it can be seen that our model has the potential to account for the age dependence of the main different clinical pictures of HBV replication. In addition, both transient and persistent HBV infection may be subclinical, or may show clinical symptoms.

FIG. 8. Diagrams showing the expected outcome of infection for specific initial concentrations of S cells and for differing immune responses against the infected R and the infected S cells. The parameter values used are as for Fig. 7(a), except that m_1 and m_2 are allowed to be independent variables. (a) Phase diagram for $S_0=0{\cdot}1$. This is the situation for infants, or for a patient who has recovered from a previous attack. For infants the immune responses against both R cells and S cells are not well developed. This places us in the lower left-hand corner of the diagram, for which the most likely outcome of the disease is to be subclinically but persistently infected. For older patients who have recovered from a previous infection their proportion of S cells becomes reduced to the level normally found in children, while still maintaining an effective immune system, and are thus placed in the region for which immunity is conferred. (b) Phase diagram for $S_0=0{\cdot}9$. This is the expected situation for previously uninfected adults. In this case the immune responses are more effective so that both m_1/h_1 and m_2/h_2 will be larger than for infants, and thus the most likely outcomes are either immunity, or a subclinical infection followed by recovery, or an initially acute course followed by recovery.

This is reflected in our model by the different levels of virus concentration and of infected cells during infection.

The phase diagrams plotted for constant S_0 [Figs 8(a) and (b)] show the dependency of the model behaviour on the relative immune responses against the infected R and S cells for subjects of a specific age. Figure 8(a) shows the situation for a patient with a low initial concentration of uninfected S cells. This will be the case both for infants, in which case the immune responses will not be well developed and the most likely outcome will be a subclinical course followed either by recovery or by persistent infection, and for older patients who have recovered from a previous attack, in which case the immune system will remain effective and the subject will be unable to support any further invasion by the virus. As previously mentioned, such immunity may only be temporary. It is interesting to note that this immunity is not a result of any specific change in the immunological status of the patient, but is a "dynamical immunity": the situation is no different to that prior to infection except that the reduction in the proportion of uninfected S cells places the subject in a different area of the phase diagram.

Figure 8(b) shows the situation for a patient with a high initial concentration of uninfected S cells. This will be the case for previously uninfected adults. Since the immune system will be more fully developed the most likely outcomes are either immunity, or a subclinical infection followed by recovery, or an initially acute course followed by recovery. An immunodeficient adult will most likely have an acute attack followed by the onset of persistent infection. Figures 8(a) and (b) also demonstrate clearly the independence of the two main effects: whether the infection is transient or persistent depends only on the behaviour of the R cells; whether the initial attack is acute or not depends only on the behaviour of the S cells. This may have implications in the targeting of treatment. In particular it is worth noting that the increase in the proportion of S cells with age does not increase the probability of recovery (in fact the opposite)—it is only the increase in the immune response against the R cells that makes the difference.

According to our model, primary hepatocellular carcinoma occurs by transformation of R cells, and potentially by the viral *HBx* gene. Therefore we assume that the probability of a patient developing liver cancer will be proportional to the total amount of infected R cells. In our model the concentration of infected R cells increases with time in those patients with chronic disease (Figs 3, 4 and 6). In some cases (Fig. 3) the increase in infected R cells may only be very slow, which may also help to contribute to the long "incubation" period between the initial infection and the development of PHC. An additional prediction of the model is that patients who have acquired infection in adulthood will display a shorter incubation period for PHC.

In addition our model allows for complete elimination of the virus after a period of chronic HBV infection (Fig. 2). It may be that this could explain those cases where PHC develops in people who have been infected, but are found to have little or no replicating virus at the time when liver cancer is diagnosed. It could also explain how evidence of the replicating virus and of the HBs antigen can disappear many years after first infection, as occurs in about 10–30% of chronic carrier cases.

REFERENCES

ALEXANDER, G. J. M. (1990). Immunology of hepatitis B virus infection. *Br. Med. Bull.* **46(2),** 354–367.

ALEXANDER, G. J. M., BRAHM, J., FAGAN, E. A., SMITH, H. M., DANIELS, H. M., EDDLESTON, A. L. W. F. & WILLIAMS, R. (1987). Loss of HBsAg with interferon therapy in chronic hepatitis B virus infection. *Lancet* **ii,** 66–68.

BEASLEY, R. P., HUANG, L., LIN, C. & CHIEN, C. (1981). Hepatocellular carcinoma and hepatitis B virus: a prospective study of 22707 men in Taiwan. *Lancet* **ii,** 1129–1133.

BLUMBERG, B. S. (1989). Feasibility of controlling or eradicating the hepatitis B virus. *Am. J. Med.* **87 (Suppl. 3A),** 2S–4S.

CHISARI, F. V., KLOPCHIN, K., MORIYAMA, T., PASQUENELLI, C., DUNSFORD, H. A., SELL, S., PINKERT, C. A., BRINSTER, R. L. & PALMITER, R. D. (1989). Molecular pathogenesis of hepatocellular carcinoma in hepatitis B virus transgenic mice. *Cell* **59,** 1145–1156.

DE THE, H., MARCHIO, A., TIOLLAIS, P. & DEJEAN, A. (1987). A novel steroid thyroid hormone receptor-related gene inappropriately expressed in human hepatocellular carcinoma. *Nature, Lond.* **330,** 667–670.

DEJEAN, A., SONIGO, P., WAIN-HOBSON, S. & TIOLLAIS, P. (1984). Specific hepatitis B integration in hepatocellular carcinoma DNA through a viral 11-base-pair direct repeat. *Proc. natn. Acad. Sci. U.S.A.* **81,** 5350–5354.

DUDLEY, F. I., SCHEUR, P. J. & SHERLOCK, S. (1972). Natural history of hepatitis-associated antigen-positive chronic liver disease. *Lancet* **ii,** 1388–1393.

DUSHEIKO, G. M. (1990). Hepatocellular carcinoma associated with chronic viral hepatitis. *Br. Med. Bull.* **46(2),** 492–511.

FAGAN, E. A. & WILLIAMS, R. (1990). Fulminant viral hepatitis. *Br. Med. Bull.* **46(2),** 462–480.

FOUREL, G., TREPO, C., BOUGUELERET, L., HENGLEIN, B., PONZETTO, A., TIOLLAIS, P. & BUENDIA M. (1990). Frequent activation of *N-myc* genes by hepadnavirus insertion in woodchuck liver tumours. *Nature, Lond.* **347,** 294–298.

GANEM, D. & VARMUS, H. E. (1987). The molecular biology of the hepatitis B virus. *Ann. Rev. Biochem.* **56,** 651–693.

GERLICH, W. H., DEEPEN, R., HEERMAN, K., KRONE, B., LU, X., SEIFER, M. & THOMSSEN, R. (1990). Protective potential of hepatitis B virus antigens other than the *S* gene protein. *Vaccine* **8,** S63–S68.

HOLLINGER, F. B. (1989). Factors influencing the immune response to hepatitis B vaccine, booster dose guidelines, and vaccine protocol recommendations. *Am. J. Med.* **87 (Suppl. 3A),** 36S–40S.

HOOFNAGLE, J. H., DUSHEIKO, G. M., SEEFF, L. B., JONES, E. A., WAGGONER, J. G. & BALES, Z. B. (1981). Seroconversion from hepatitis B e antigen to antibody in chronic type B hepatitis. *Ann. Intern. Med.* **94,** 744–748.

HSU, T. Y., MÖRÖY, T., ETIEMBLE, J., LOUISE, A., TREPO, C., TIOLLAIS, P. & BUENDIA, M. (1988). Activation of *c-myc* by woodchuck hepatitis insertion in hepatocellular carcinoma. *Cell* **55,** 627–635.

JACYNA, M. R. & THOMAS, H.C. (1990). Antiviral therapy: hepatitis B. *Br. Med. Bull.* **46(2),** 368–382.

KIM, C., KOIKE, K., SAITO, I., MIYAMURA, T. & JAY, G. (1991). *HBx* gene of hepatitis B virus induces liver cancer in transgenic mice. *Nature, Lond.* **351,** 317–320.

KRUGMAN, S., HOOFNAGLE, J. H., GERETY, R. J., KAPLAN, P. M. & GERIN, J. L. (1974). Viral hepatitis, type B: DNA polymerase activity and antibody to hepatitis B core antigen. *N. Engl. J. Med.* **290,** 1331–1335.

LONDON, W. T. & BLUMBERG, B. S. (1992). A cellular model of the role of hepatitis B virus in the pathogenesis of primary hepatocellular carcinoma. *Hepatology* **2(2),** 10S–14S.

MARCHUK, G. I., PETROV, R. V., ROMANYUKHA, A. A. & BOCHAROV, G. A. (1991*a*) Mathematical model of antiviral immune response, I. *J. theor. Biol.* **151,** 1–40.

MARCHUK, G. I., PETROV, R. V., ROMANYUKHA, A. A. & BOCHAROV, G. A. (1991*b*). Mathematical model of antiviral immune response, II. *J. theor. Biol.* **151,** 41–69.

MARION, P. L., LYNDON, S. O., REGNERY, D. C., SCULLARD, G. H. & ROBINSON, W. S. (1980). A virus in Beechey ground squirrels that is related to hepatitis B virus of humans. *Proc. natn. Acad. Sci. U.S.A.* **77,** 2941–2945.

MAYNARD, J. E. (1990). Hepatitis B: global importance and need for control. *Vaccine* **8,** S18–S20.

MCINTYRE, N. (1990). Clinical presentation of acute viral hepatitis. *Br. Med. Bull.* **46(2),** 533–547.

PUGH, J. C. & BASSENDINE, M. F. (1990). Molecular biology of hepadnavirus replication. *Br. Med. Bull.* **46(2),** 329–353.

REALDI, G., ALBERTI, A., RUGGE, M., BORTOLOTTI, F., RIGOLI, A. M., TREMOLADA, F. & RUOL, A. (1980). Seroconversion from hepatitis B e antigen to anti-HBe in chronic hepatitis B virus infection. *Gastroenterology* **79,** 195–199.

SETO, E., MITCHELL, P. J. & YEN, T. S. B. (1990). Transactivation by the hepatitis B virus X protein depends on AP-2 and other transcription factors. *Nature, Lond.* **344**, 72–74.
SUMMERS, J., SMOLEC, J. M. & SNYDER, R. (1978). A virus similar to human hepatitis B virus associated with hepatitis and hepatoma in woodchucks. *Proc natn. Acad. Sci. U.S.A.* **75**, 4533–4537.
TIOLLAIS, P. & BUENDIA, M. (1991). Hepatitis B virus. *Sci. Am.* **264**(4), 48–54.
TIOLLAIS, P., CHARNEY, P. & VYAS, G. N. (1981). Biology of hepatitis B virus. *Science* **213**, 406–411.
TIOLLAIS, P., POURCEL, C. & DEJEAN, A. (1985). The hepatitis B virus. *Nature, Lond.* **317**, 489–495.
VIOLA, L. A., BARRISON, I. G., COLEMAN, J. C., PARADINAS, F. J., FLUKER, J. L., EVANS, B. A. & MURRAY-LYON, I. M. (1981). Natural history of liver disease in chronic hepatitis B surface antigen carriers. *Lancet* **ii**, 1156–1159.

Proc. Natl. Acad. Sci. USA
Vol. 93, pp. 6542–6546, June 1996
Medical Sciences

The dynamics of hepatitis B virus infection

ROBERT J. H. PAYNE*, MARTIN A. NOWAK*, AND BARUCH S. BLUMBERG†

*Department of Zoology, University of Oxford, Oxford OX1 3PS, United Kingdom; and †Fox Chase Cancer Center, Philadelphia, PA 19111

Contributed by Baruch S. Blumberg, February 12, 1996

ABSTRACT **We consider a cellular model of infection by the hepatitis B virus and describe how it may be used to account for two important features of the disease, namely (*i*) the wide variety of manifestations of infection and the age dependence thereof, and (*ii*) the typically long delay before the development of virus-induced liver cancer (primary hepatocellular carcinoma). The model is based on the assumption that the liver is comprised of both immature and mature hepatocytes, with these two subpopulations of cells responding contrastingly upon infection by the virus.**

Although replication of the hepatitis B virus (HBV) is only weakly cytopathic toward the host cell, HBV infection is nonetheless the major cause of liver disease in man (1). The extent of the suffering caused by the virus is a consequence not just of its high prevalence, there being an estimated 300 million carriers worldwide, but also of the variety of different clinical manifestations that can result from infection. Disease in adults usually leads to recovery and immunity, most commonly subsequent to acute hepatitis, but also sometimes following subclinical infection (2); whereas if the hepatitis B virus surface antigen (HBsAg), which can be detected even in the absence of active replication, has not been cleared within 6 months, the disease is termed chronic and infection is likely to persist. The progression to a chronic carrier state is age-dependent, being common in infants, occasional in children, and rare in adults (3, 4). Persistent infection may be either asymptomatic or symptomatic—the mean age of asymptomatic carriers is lower than that of symptomatic carriers (5)—and can lead to various outcomes, including chronic persistent hepatitis, chronic active hepatitis, and postnecrotic cirrhosis (6). Some 25–50% of carriers go on to develop primary hepatocellular carcinoma (PHC), typically around 30–50 years after initial infection (7, 8). The destruction of infected hepatocytes is believed to be a consequence of the host's immune responses, and primarily results from the action of cytotoxic T cells (9, 10).

In this report we describe how a simple cellular model for the pathogenesis of HBV may be used to account for the variety of manifestations of infection, the outcome depending only on the age and immunological status of the patient. In addition, the model provides an explanation for various features of the relationship between HBV and PHC, including a possible dynamical explanation for the characteristic long delay before the onset of virally-induced neoplasms.

The London–Blumberg Model

London and Blumberg (11) assumed that the liver is comprised of two subsets of hepatocytes that respond contrastingly to infection by the virus and that are in different stages of maturation, the less differentiated cells being referred to as R cells ("resistent" to viral replication) and the more differentiated as S cells ("susceptible" to viral replication). The R cells have the capacity to divide, being able to differentiate into either R cells or S cells, but the S cells do not undergo significant further division; the proportion of S cells in an uninfected liver therefore increases with increasing age. It is supposed that the integration of HBV DNA into the genome of both cell types can alter gene expression that, in combination with increased cell proliferation in chronic hepatitis, thereby increasing the probability of mutational events, may increase the probability of carcinogenesis. In addition, the phenotypic transformation of infected R cells is hypothesised to prevent any further differentiation into S cells, thus providing them with a selective advantage. A number of conditions specify the behavior of the two cell types in response to infection by the virus. London and Blumberg suggested that the rate of viral replication must be greater in the S cells than in the R cells, and that the rate of cell death of infected S cells should likewise be much greater than that of the infected R cells. We show that the S cells must also be required to have a higher infectivity than the R cells. A mathematical formulation of the London–Blumberg model is given in Payne *et al.* (12).

There is observational and experimental evidence for the basic scenario of the model. Kim *et al.* (13) showed that the presence of the *HBx* gene [the HBx antigen uses a tumor promoter pathway (14)] is associated with the development of liver cancer in transgenic mice, and found the gene to be expressed in only a subset of hepatocytes, thought to relate to a specific state of differentiation. Thung *et al.* (15), who investigated the presence of antigens in hepatocellular carcinoma, concluded that although tumor cells are histologically monomorphic they are functionally heterogeneous. Brechot *et al.* (16) suggested that viral replication may involve only the more differentiated tumor cells. The existence of stem cells in the liver is controversial (17, 18), but recent work suggests that "oval" cells, which are thought to be progeny of hepatic stem cells, can act as a precursor to liver tumors. These oval cells can proliferate and are capable of differentiating *in vivo* into both mature hepatocytes and bile epithelium cells (19), but when transplanted subcutaneously can produce aggressively growing tumors. Interestingly, the tumorigenic potential of transformed cells transferred intrahepatically is dependent on the microenvironment, being suppressed in situations where hepatocytic differentiation is induced (20). Oval cells therefore seem to correspond to the R cells of our model, with S cells being equivalent to fully mature hepatocytes.

Explaining the Variety in Outcomes of HBV Infection

Computer simulations of the model exhibit dynamics reflecting the main observed outcomes of infection. In this section we use five different examples to demonstrate the variety of possible behaviors. The parameter values used in the simulations (R.J.H.P., unpublished data) were estimated using a combination of analytical predictions and clinical data. Note that all of the figures given here represent possible outcomes in a single patient, varying only in the age at first infection (only

The publication costs of this article were defrayed in part by page charge payment. This article must therefore be hereby marked "*advertisement*" in accordance with 18 U.S.C. §1734 solely to indicate this fact.

Abbreviations: HBV, hepatitis B virus; HBsAg, hepatitis B surface antigen; PHC, primary hepatocellular carcinoma; HBeAg, hepatitis B e core antigen.

the strength of the immune responses and the initial proportion of S cells are varied). Fig. 1*a–e* are displayed in order of decreasing age at infection. Let X represent the proportion of uninfected S cells in the liver. It can be shown that there is a critical level in the proportion of uninfected S cells, X_A say, for which if $X_A < X$ then the overall concentration of free virus will increase, and if $X < X_A$ then it will decrease. The critical concentration X_A is defined by the strength of the immune response against the infected S cells, but is effectively independent of the immune response against the infected R cells, the rate of viral replication in the R cells being too slow to be significant in this context (for "immune response" we should more properly say "elevated cell death rate resulting from infection," this including not just the immune response itself but also any additional cell death, for example from interference with the host cell metabolism, lysis of infected cells by mechanisms other than the immune system, etc.).

Initial Phase: Acute or Subclinical? If $X_A < X$ at the time of infection, then the virus will prosper and X decreases as susceptible cells become infected. At the same time the infected cells will be dying, mostly as a result of attack by cytotoxic T cells. Importantly, however, the dying S cells are replaced not by more S cells, but by the less differentiated, faster-proliferating R cells. Consequently the proportion of uninfected S cells continues to decrease until such time as $X < X_A$, whereupon the level of active viral replication starts to fall. Such behavior always occurs at the start of infection, but its magnitude may vary widely. The severity of the damage sustained by the liver during this initial phase is controlled by two opposing factors: the number of uninfected S cells that

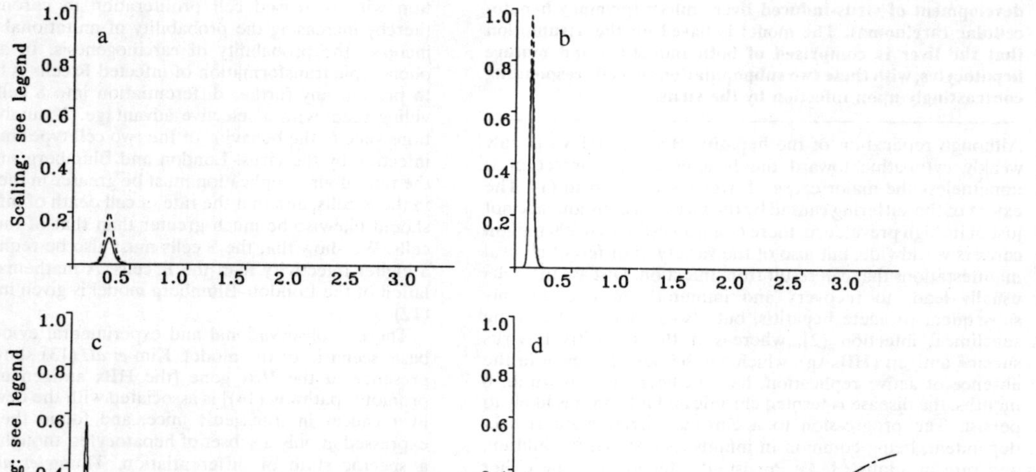

FIG. 1. Computer simulation of the model, based on mathematical formulation and parameter values in Payne *et al.* (12) and Payne (unpublished data). The variables followed are (*i*) concentration of free virus (mean viral particles per hepatocyte ÷ 0.25); (*ii*) the combined excess hepatocyte mortality resulting from infection (mean total cell deaths per hepatocyte per month ÷ 1.5); (*iii*) replication rate of infected R cells (mean infected R cell divisions per hepatocyte per month ÷ 50.0). If the probability of a phenotypic transformation leading to the initiation of PHC is small for each cell division then the overall instantaneous probability of such a carcinogenic event is approximately proportional to the rate of replication of the infected R cells. *a–e* are given in order of decreasing immune effectiveness, reflecting decreasing age at first infection. (*a*) Transient subclinical infection. (*b*) Transient acute hepatitis. (*c*) Recurring acute hepatitis. (*d*) Asymptomatic persistent infection likely to lead to PHC. (*e*) Symptomatic persistent infection likely to lead to PHC.

have to be lost before X becomes smaller than the critical value X_A, and the rate at which infected cells are destroyed. If the immune response against the infected S cells is strong (but still small enough for the virus to take hold), and X_A therefore large, then few cells have to die before X drops beneath X_A, and thus the disease will be subclinical, as in Fig. 1*a*. At the other end of the scale, if the response is very weak then the rate of cell death is slow and this also will be subclinical, as in Fig. 1 *d* and *e*. Between these two extremes, however, there is the possibility of having the immune response against the S cells both small enough that many cells have to become infected before X falls below X_A, and large enough that the rate of cell death will be high; in this case significant damage will be inflicted on the liver and acute hepatitis will occur, as in Fig. 1 *b* and *c*.

Transient or Persistent? Once X becomes smaller than X_A, the viral concentration falls and the liver begins to return to a more normal state, except with a higher proportion of R cells than prior to infection. The differentiation of these R cells into S cells then causes the proportion of S cells to start to increase, until eventually X is again larger than the critical value X_A. At this point one of two things can happen: if there is no virus remaining then recovery will be complete, as in Fig. 1 *a* and *b*; if, however, there is still some free virus present then viral replication starts again and chronic infection sets in. Which of these two courses is followed depends on the ratio of uninfected to infected R cells, which in turn depends on the strength of the immune response against the infected R cells. Thus those patients with less effective immune systems, most probably the very young, are more likely to develop persistent infection. The manner in which the resurgence of viral infection manifests itself depends on the degree of "overshoot" as X rises past X_A. If there is significant overshoot then separate waves of active replication will occur, and recur (compare Fig. 1 *c*, *d*, and *e*). Less extreme oscillations may account for those cases of unresolved acute hepatitis that subsequently have fluctuating levels of serum transaminases with intermittent jaundice. Note, therefore, that the appearance of these recurrent bouts of disease does not necessarily require reinfection, viral mutation, or changes in the patient's immunological status.

Chronic Disease: Asymptomatic or Symptomatic? We consider two features of interest in persistent disease: whether the chronic carrier state is asymptomatic or symptomatic, and whether or not PHC is seen to develop. The latter is discussed in the next section. Asymptomatic chronic disease is characterized by absence of viral replication, low infectivity, and the presence of antibodies against the hepatitis B e core antigen (HBeAg); symptomatic chronic disease is characterized by active viral replication, high infectivity, and failure to seroconvert to anti-HBeAg (21). Although transition to persistent infection is determined by the properties of the R cells, the level of active replication apparent once persistence is established is determined by the properties of the S cells. If the immune response against the infected S cells is large then the chronic carrier state will be asymptomatic, as in Fig. 1*d*, whereas if it is small then the chronic state will be symptomatic, as in Fig. 1*e*. This implies that the mean age of HBeAg-positive patients should be lower than that of those who are HBeAg-negative, a prediction in agreement with the clinical observations of Realdi *et al.* (5).

A Dynamic Origin for the Latency of PHC

The London–Blumberg model was designed to help account for various features of HBV-induced PHC, principally: (*i*) PHC is an outcome of persistent but not transient infection (7), (*ii*) tumor cells do not usually contain HBsAg (15), (*iii*) the concentration of HBsAg in patients with PHC is low (22), (*iv*) HBV DNA can be observed integrated into both tumor and nontumor cells (16), (*v*) HBV DNA can insert into many different sites in the host genome (23), (*vi*) the presence of a carcinogenic cofactor does not appear to be necessary for the occurrence of PHC (24), and (*vii*) there is usually a long (30–50 years) latency period between infection and the development of PHC (22). Examination of the figures shows that most of these observations can, in dynamical terms, be accounted for if the probability of initiating PHC is somehow dependent on the concentration of infected R cells. Such a correlation could occur in two ways: either the random integration of the HBV DNA into the host cell genome may have a small probability of causing a relevant phenotypic transformation, or the transformation could occur by a mutation of the infected cell DNA during cell division. Both the absence of an essential cofactor and the observation that viral DNA integration, random in nontumor cells, is nonrandom in malignant cells (25) would suggest the former scenario to be more likely. The infected S cells could also be potentially tumorigenic, but their limited capacity for cell division means that their role in the promotion of PHC would not be significant. This means that we do not need to look for any differences between the function (or dysfunction) of R and S cells upon integration of the HBV DNA—the differing responses of the two cell types to active replication is sufficient for the working of the model.

Although it should be clear how items *i*–*vi* are satisfied by the assumptions of our model, it is less obvious why there should be a delay before the rise in the concentration of infected R cells. In the numerical simulations this delay was achieved by choosing the infectivity of the R cells to be much smaller than that of the S cells. To see why this is necessary, bear in mind that the concentration of infected R cells may increase by two different mechanisms: infection and regeneration. If infection were the dominant process then high concentrations of infected R cells would correlate with high levels of free virus. Since, however, active viral replication is usually highest at the start of the disease, this is in contradiction with our required dynamics; in addition, it is in contradiction with item *iii* above. Thus the gradual build-up in numbers of infected R cells must instead be primarily a result of cell division, with the large increase some years after infection coming about when positive feedback takes control of the system, as the rate of cell death becomes overwhelmed by the rate of regeneration. This notion is in accord with the observed focal nature of pre-cancerous lesions in HBV infected transgenic mice (13), and in woodchuck hepatitis virus infected woodchucks (26). Such behavior can only occur if the infectivity of the R cells is small enough for the overall rate of infected R cell division, although low, to be larger than the rate of cell–cell infection. This reasoning indicates how the increase in the concentration of infected R cells can be delayed in one individual, but its extension to cover the overall epidemiological picture is more subtle.

Given that the duration of the delay is controlled by the balance between the addition of infected R cells by cell division and the removal of these cells by immune attack, the question is why the domination of the liver by infected R cells does not occur much more rapidly in those patients with very weak immune responses. Although the level of active viral replication has already been discarded as a determining factor for the initiation of PHC, the S cells do, nonetheless, play a crucial role. The important point to note is that the R cells, whether infected or not, can only increase by multiplying to fill the space left vacant by dying S cells. In the (hypothetical) situation of there being no immune response of any sort against the infected S cells, no extra S cells would die as a result of the disease, the infected R cells would not have an opportunity to prosper, and the likelihood of PHC would be negligible. By considering this scenario we can see that the most rapid accumulation of infected R cells will occur when there is a positive immune response against the infected S cells, for which the long-term mortality of infected S cells is maximized (in contrast to the immune response against the infected R cells, which would need to be zero). This means that there is a limit to the possible rate of increase of infected R cells, and, when considered over a population of individuals, there is thus a minimum duration to the latency period that precedes the dramatic rise in the incidence of PHC. Given a small enough infectivity of the R cells this minimum delay can be of the order of decades, as demonstrated in Fig. 2. (It should not be forgotten, however, that we are dealing with probabilities, so the possibility of rare cases with an early onset of PHC is not excluded.) This provides us with a purely dynamical explanation for the long latency period characteristic of HBV-associated PHC.

Summary and Implications of the Model

In this report we have described how the London–Blumberg model may be used to provide an understanding of the dynamics of HBV infection—the presence of two subpopulations of cells in differing stages of maturation being crucial in accounting for the variety of manifestations. The severity of the initial phase of disease is determined by the immune response against the infected S cells, which also controls the asymptomatic/symptomatic character of persistent hepatitis, whereas possible progression to persistent disease is determined by the immune response against the infected R cells. This separation of mechanisms is in accord with the observation that protective antibodies against the HBsAg cannot usually be detected until some months after the passing of acute disease (27), the process for the initial subjugation of disease apparently being distinct from that for the elimination of the virus. Although the risk of oncogenesis is proportional to the number of infected R cells, we find that both cell types play a role in determining the expected duration of latency before the onset of PHC. Crudely put, this is because the total number of R cells present depends on the death of the infected S cells, whereas of those R cells present, the proportion that are infected depends on the balance between the infected and the uninfected R cells. The interaction of the two cell types can also lead to oscillatory behavior, which may account for the intermittent symptoms of hepatitis seen in some patients. Oval liver cells, which we have speculated to correspond to the R cells of the London–Blumberg model, have been shown to coexpress transforming growth factor α (TGF-α) and HBsAg (28). Since the gene for TGF-α can lead to hepatic oncogenesis in transgenic mice this supports our inference that R cells act as the principal site for the development of HBV-induced PHC.

Our analysis explains the observed age dependence of the various manifestations of infection (recall that Fig. 1 *a–e* are displayed in order of decreasing strength of immune effectiveness, reflecting decreasing age at first infection). Transient infection is most likely in adults, subclinical chronic infection is most likely in infants, asymptomatic carriers have a lower mean age than symptomatic carriers, and the probability of progression to PHC is higher the lower the age at infection. Cirrhosis of the liver is also a known risk factor for the development of PHC (29), regardless of whether the cirrhosis is caused by alcohol, HBV infection, or other active hepatitis. In our model this can be attributed to the cirrhotic cell necrosis providing further room for the proliferation of infected R cells.

The differing roles of the two cell types has implications for the treatment of HBV infection. For example, to alleviate acute hepatitis the S cells should be targetted, whereas for the prevention of progression to a carrier state it is the R cells that are pertinent. Reducing the likelihood of developing PHC is a more complex matter, with both cell types being of importance, the most appropriate target for treatment depending to a large extent on context. Note, however, that although it would in all cases be beneficial to boost the host's immune response against the infected R cells, tampering with the immune response against the infected S cells could prove to be a double-edged sword: increasing the death rate of infected S cells with the intention of reducing the amount of viral replication during chronic hepatitis is, in the absence of other factors, liable to increase the probability of initiating PHC.

Our model is able to mimic the wide variety of disease behavior without having to include the detailed complexities of immune responses. For example, following transient infection the patient will be immune regardless of any change in immunological status (the subsequent differentiation of R cells into S cells means that such immunity may only be temporary), and the reappearance of disease following apparently transient infection, such as that in Fig. 1*c*, may occur for dynamical reasons rather than from reinfection. Similarly, it is possible to have a resurgence of disease even in the absence of viral mutation or changes in immunological status. That is not to say that these other possibilities will not happen, just that they are not necessary in accounting for the disease dynamics. More significantly, our explanation for the delay in the onset of PHC is also based on purely dynamical considerations and does not depend on either the existence of a carcinogenic cofactor or the occurrence of an early event with a delayed manifestation [but does not preclude the effect of additional carcinogens, such as aflatoxin (30)]. Provided that the infectivity of the R cells is very small compared with that of the S cells, no

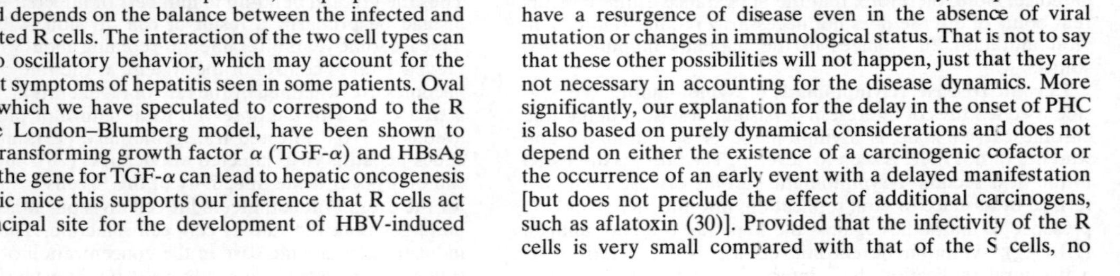

FIG. 2. Duration of PHC latency period in a population. Change with time of (*a*) proportion of liver made up of infected R cells and (*b*) cumulative probability of developing PHC (approximately proportional to the time integral of the number of infected R cells), taken over a range of *M* (a measure of the immune response against the infected S cells). Considered over a population—that is, over the range of all possible *M*—there is a minimum duration of the delay before the proportion of infected R cells in the liver begins to rise significantly. If, as here, this minimum delay is of the order of decades, then the typical latency period before PHC in a population of carriers will, as required, also be of the order of decades. To achieve this the infectivity of the R cells must be very small.

additional mechanisms, over and above the basic assumptions of the London–Blumberg model, are necessary to account for the typically long delay before the development of HBV-induced PHC. It would be interesting to know whether the type of dynamic behavior described here could also be the reason for the delayed development of other virus-associated tumors.

We thank W. Thomas London for his critical reading of the manuscript. This work was partially supported by the Science and Engineering Research Council and the Natural Environment Research Council of the United Kingdom (R.J.H.P.), by a Wellcome Trust Senior Research Fellowship and the E. P. Abraham Junior Research Fellowhsip at Keble College (M.A.N.), and by Public Health Service Grant CA-06927 from the National Institutes of Health and by an appropriation from the Commonwealth of Pennsylvania (B.S.B.).

1. Maynard, J. E. (1990) *Vaccine* **8** (Suppl.), S18–S20.
2. Sherlock, S. (1990) *Vaccine* **8** (Suppl.), S6–S9.
3. Dudley, F. J., Scheur, P. J. & Sherlock, S. (1972) *Lancet* **ii,** 1388–1393.
4. Edmunds, W. J., Medley, G. F., Nokes, D. J., Hall, A. J. & Whittle, H. C. (1993) *Proc. R. Soc. London B* **253,** 197–201.
5. Realdi, G., Alberti, A., Rugge, M., Bortolotti, F., Rigoloi, A. M., Tremolada, F. & Ruol, A. (1980) *Gastroenterology* **79,** 195–199.
6. Viola, L. A., Barrison, I. G., Coleman, J. C., Paradinas, F. J., Fluker, J. L., Evans, B. A. & Murray-Lyon, I. M. (1981) *Lancet* **ii,** 1156–1159.
7. Beasley, R. P., Huang, L., Lin, C. & Chien, C. (1981) *Lancet* **ii,** 1129–1133.
8. Dusheiko, G. M. (1990) *Br. Med. Bull.* **46,** 492–511.
9. Mondelli, M., Vergani, G. M., Alberti, A., Vergani, D., Portmann, B., Eddelston, A. L. W. F. & Williams, R. (1982) *J. Immunol.* **192,** 2773–2778.
10. Alexander, G. J. M. (1990) *Br. Med. Bull.* **46,** 354–367.
11. London, W. T. & Blumberg, B. S. (1982) *Hepatology* **2** (Suppl.), 10S–14S.
12. Payne, R. J. H., Nowak, M. A. & Blumberg, B. S. (1994) *Math. Biosci.* **123,** 25–50.
13. Kim, C., Koike, K., Saito, I., Miyamura, T. & Jay, G. (1991) *Nature (London)* **351,** 317–320.
14. Kekulé, A. S., Lauer, U., Weiss, L., Luber, B. & Hofschneider, P. H. (1993) *Nature (London)* **361,** 742–745.
15. Thung, S. N., Gerber, M. A., Sarno, E. & Popper, H. (1979) *Lab. Invest.* **41,** 101–105.
16. Brechot, C., Scotto, J., Charney, P., Hadchouel, M., Degos, F. & Trepo, C. (1981) *Lancet* **ii,** 765–767.
17. Sell, S. (1993) *Int. J. Dev. Biol.* **37,** 189–201.
18. Thorgeirsson, S. S. (1993) *Am. J. Pathol.* **142,** 1331–1333.
19. Factor, V. M., Radaeva S. A. & Thorgeirsson, S. S. (1993) *Am. J. Pathol.* **145,** 409–422.
20. Coleman, W. B., Wennerberg A. E., Smith G. J. & Grisham, J. W. (1993) *Am. J. Pathol.* **142,** 1373–1382.
21. Hoofnagle, J. H., Dusheiko, G. M., Seeff, L. B., Jones, E. A., Waggoner, J. G. & Bales, Z. B. (1981) *Ann. Intern. Med.* **94,** 744–748.
22. London, W. T. & Blumberg, B. S. (1981) in *Cancer: Achievments, Challenges and Prospects for the 1980s,* eds. Burchenal, J. H. & Oettgen, H. F. (Grune & Stratton, New York), Vol 1, pp. 161–183.
23. Shafritz, D., Shouval, D., Sherman, H. I., Hadziyannis, S. J. & Kew, M. C. (1981) *N. Engl. J. Med.* **305,** 1067–1073.
24. Tiollais, P. & Buendia, M. (1991) *Sci. Am.* **264,** 48–54.
25. Slagle, B. L., Zhou, Y.-Z., Butel & J. S. (1992) *Cancer Res.* **51,** 49–54.
26. Toshkov, I., Hacker, H. J., Roggendorf, M. & Bannosch, P. (1990) *J. Cancer Res. Clin. Oncol.* **116,** 581–590.
27. Deinhart, F. (1982) in *Progress in Liver Diseases,* eds. Popper, H. & Schaffner, F. (Grune & Stratton, New York), Vol. 7, pp. 451–467.
28. Hsia, C. C., Thorgeirsson, S. S. & Tabor, E. (1994) *J. Med. Virol.* **43,** 216–221.
29. Zaman, S. N., Melia, W. M., Johnson, R. D., Portmann, B. C., Johnson, P. J. & Williams, R. (1985) *Lancet* **i,** 1357–1360.
30. Fujimoto, Y., Hampton, L. L., Wirth, P. J., Wang, N. J., Xie, J. P. & Thorgeirsson S. S. (1994) *Cancer Res.* **54,** 281–285.

THE LANCET, JULY 5, 1975

Preliminary Communication

ANTIBODY TO HEPATITIS-B CORE ANTIGEN IN PATIENTS WITH PRIMARY HEPATIC CARCINOMA

PHILIPPE MAUPAS, BARBARA WERNER, BERNARD LAROUZÉ, IRVING MILLMAN, W. THOMAS LONDON, ANNA O'CONNELL, BARUCH S. BLUMBERG

Institute for Cancer Research, Fox Chase Cancer Center, Philadelphia, Pennsylvania, U.S.A., and Department of Microbiology, University of Tours, Tours, France

GERARD SAIMOT, MAURICE PAYET

Institut de Medécine et d'Épidémiologie Africaine, Hôpital Claude Bernard, Paris, France

Summary Antibody to hepatitis-B core antigen (anti-HB_c) was assayed in the serum of patients with primary hepatic carcinoma (P.H.C.) and controls from Hong Kong, West Africa, and the United States. In each region the prevalence of anti-HB_c was higher in P.H.C. patients than in controls, ranging from 70 to 95% in the patients and from 20 to 68% in the controls from Asia and Africa; 24% of P.H.C. patients and 4% of controls from the U.S. had anti-HB_c. These data support the hypothesis that chronic infection with hepatitis-B virus is ætiologically related to P.H.C., especially in Asia and Africa, although other factors must also be involved.

INTRODUCTION

PRIMARY hepatic carcinoma (P.H.C.) is the most common malignant neoplasm in several of the world's major populations (Chinese, Africans, Filipinos). Therefore, it may be the most common cancer in man.

On the basis of necropsy studies showing an association between chronic hepatitis and/or macronodular post-necrotic cirrhosis with P.H.C., Payet[1] and Steiner and Davies[2] suggested that viral hepatitis may be the major cause of P.H.C. in Africa. When tests for infection with hepatitis-B virus (H.B.V.) became available it was possible to evaluate the association of H.B.V. with P.H.C. Several studies in Africa and Asia showed an increased prevalence (35–80%) of Australia antigen (hepatitis-B surface antigen, HB_sAg) among patients with P.H.C. These high prevalences were generally detected only when sensitive methods (radioimmunoassay, immune adherence hæmagglutination) for assaying HB_sAg were used.[3,4] This observation suggested that either HB_sAg was present in decreased quantity in P.H.C. patients or there was some antigenic difference between the HB_sAg in P.H.C. and that found in HB_sAg carriers.[4] No increase in the prevalence of antibody to hepatitis-B surface antigen (anti-HB_s) has been reported in P.H.C. patients.[5]

If patients with P.H.C. are infected with H.B.V. but are producing small quantities of HB_sAg, there may be defects in viral assembly or viral replication in the malignant cells. Hoofnagle et al.[6] have suggested that antibody to hepatitis-B core antigen (anti-HB_c) is a better indicator of intracellular replication of H.B.V. than anti-HB_s or HB_sAg. In this paper we report

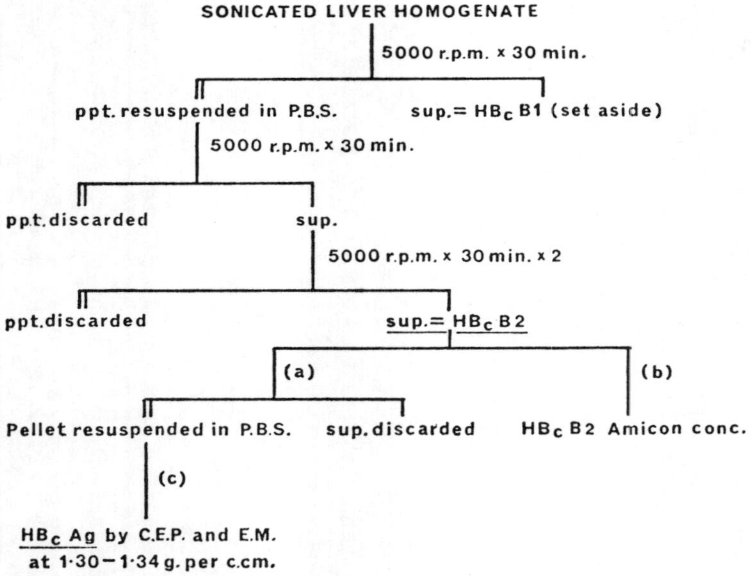

Scheme for isolation and purification of HB_cAg from liver of a human HB_sAg carrier.

(a) 1/2 HB_cB2 centrifuged 106,000×g×2 hr. in Beckman L2-65B.

(b) 1/2 HB_cB2 dialysed against 0·1*M* P.B.S. pH 7·2×24 hr. concentrated ×10 in Amicon apparatus (SM-300 90 mm. filter).

(c) Isopycnic centrifugation in $CsCl_2$ (1·316 g./c.cm.) 108,000×g×72 hr.

ppt. = precipitate. sup. = supernatant. C.E.P. = counter-electrophoresis. E.M. = electron microscopy.

the detection of anti-HB_c in a very high proportion of patients with P.H.C. These data support the hypothesis that chronic infection with H.B.V. is at least one ætiological factor in P.H.C. and suggest a method of preventing this form of cancer.

MATERIALS AND METHODS

HB_cAg was extracted by the method of Hoofnagle et al.[7] from the liver of a patient who had chronic hepatitis, was positive for HB_sAg, and had received a renal transplant. Electron microscopy of a biopsy specimen of this patient's liver showed numerous 27–29 nm. particles in the nuclei of parenchyma cells. (Attempts by the methods described below to extract HB_cAg from the livers of four other HB_sAg carriers and one liver from an HB_sAg-negative person were unsuccessful.) The liver was removed at necropsy, perfused with 0·25*M* sucrose, and frozen in dry ice. For preparation of core antigen the liver was thawed and homogenised in a Waring blender as a 25% suspension in cold phosphate-buffered saline. The homogenate was sonicated twice for 60 seconds each at maximum setting in a Bronson W 185E sonicator (Heat Systems Inc.) with special microtip. The procedures used to concentrate and isolate HB_cAg are shown in the figure.

HB_cAg in the liver homogenate extract (HB_cB2) reacted by C.E.P. and immunodiffusion (I.D.) with the rhesus anti-HB_c and the precipitin in I.D. produced a line of identity with anti-HB_c in the serum of HB_sAg carriers. All sera from the P.H.C. patients and controls were tested first by C.E.P. and I.D. against HB_cB2. C.E.P. was carried out in 1·1% agarose in barbitone buffer, pH 8·2. I.D. was performed in 0·9% Noble agar in barbitone buffer, pH 8·2. HB_cB2 was placed in the centre well and known anti-HB_c was placed in the first and fourth wells of a seven-well I.D. pattern. Sera were considered to contain anti-HB_c if they showed a line of identity with known anti-HB_c by I.D. and a precipitin by C.E.P. Sera which gave equivocal results were tested by C.E.P. against HB_cB2 Amicon concentrate and the gradient fractions containing HB_cAg. If a precipitin was observed, such sera were scored anti-HB_c positive. In general, C.E.P. and I.D. using the HB_cB2 fraction gave definitive results.

ANTI-HB_c IN PATIENTS WITH P.H.C.

Location*	No.	Anti-HB_c+ (%)	P†
(1) *Dakar, Senegal-A*			
P.H.C.	43	41 (*95%*)	$2{\cdot}3\times10^{-13}$
Controls	62	16 (*26%*)	
(2) *Dakar, Senegal-B*			
P.H.C.	28	25 (*89%*)	0·055
Controls	28	18 (*64%*)	
(3) *Bamako, Mali*			
P.H.C.	24	17 (*71%*)	$7{\cdot}8\times10^{-4}$
Controls	40	10 (*25%*)	
(4) *Hong Kong*			
P.H.C.	37	26 (*70%*)	$2{\cdot}3\times10^{-3}$
Controls	58	21 (*36%*)	
(5) *Philadelphia, U.S.A.*			
P.H.C.	21	5 (*24%*)	0·041
Controls	50	2 (*4%*)	
Total			
P.H.C.	153	114 (*75%*)	2×10^{-19}
Controls	238	67 (*28%*)	

* (1) Patients with P.H.C. in Hôpital Le Dantec, Dakar. Controls are 1 or 2 persons without P.H.C. from Senegal matched for each case by age, sex, and ethnic group.
(2) Patients with P.H.C. at Hôpital Le Dantec, Dakar. Controls are individuals without P.H.C. matched for each case by age, sex, and ethnic group.
(3) Patients with P.H.C. at Hôpital Point G, Bamako. Controls are 1 or 2 persons from Mali without P.H.C. matched for each case by age, sex, and ethnic group.
(4) Patients with P.H.C. collected by Dr David Todd at the Queen Mary Hospital, Hong Kong. Controls are 58 patients without P.H.C. from the same hospital with a variety of other diseases. 7 of the 58 controls had cirrhosis; 5 of these had anti-HB_c.
(5) Sera from patients with P.H.C. were submitted by physicians at several hospitals in Philadelphia. Controls were 25 healthy volunteer blood-donors and 25 prisoner blood-donors.
† Calculated from Fisher's exact 2 × 2 test.

PATIENTS AND CONTROLS

The P.H.C. patients and controls tested for anti-HB_c are listed in the table. The diagnosis of P.H.C. was accepted only if patients had histological evidence of P.H.C. on biopsy or at necropsy or if their clinical picture was compatible with P.H.C. and they had α-fetoprotein detectable in their serum by C.E.P. or I.D.

RESULTS

The table shows that, of 153 patients with P.H.C. from three continents, 114 had anti-HB_c detected by the relatively insensitive C.E.P. and I.D. tests. In Senegal over 90% of the patients had anti-HB_c, whereas in Mali and Hong Kong about 70% were positive. Of the 5 anti-HB_c-positive American P.H.C. patients, 1 was Chinese and another was from Puerto Rico.

DISCUSSION

The prevalence of anti-HB_c among these P.H.C. patients is higher than the reported prevalence of HB_sAg or anti-HB_s among P.H.C. patients.[3–5] The frequency of these factors in P.H.C. patients will be reported elsewhere in studies describing the variation of response of West Africans to H.B.V. infection. Here we wish to emphasise the likelihood that virtually all P.H.C. patients in West Africa and South-East Asia are (or have been) infected with H.B.V. In the U.S.A., H.B.V. is associated with a considerable proportion of, but not all, cases of P.H.C.

Furthermore, anti-HB_c must be present in high titre to be detectable by C.E.P. and I.D., whereas HB_sAg is present only at low concentrations in P.H.C. patients.[4,5] This suggests that in P.H.C. there is a continuous production of core antigen and limited synthesis of HB_sAg. Hadziyannis [8] has reported that, by immunofluorescent techniques, HB_sAg is found only in the parts of the liver unaffected by P.H.C. On the basis of these observations and the data reported here, we suggest that both the unaffected and malignant liver cells are producing HB_cAg.

The data do not in themselves indicate a causal relationship between hepatitis B and P.H.C., but, taken together with previous findings, they support the hypothesis that H.B.V. infection is ætiologically related to P.H.C. Since infection with H.B.V., as detected by the presence of HB_sAg, anti-HB_s, or anti-HB_c, is common in the general populations of West Africa and Hong Kong, additional factors must be involved in the causation of P.H.C. Some of these are under investigation and will undoubtedly be identified in due course. No matter what other factors are involved, however, if H.B.V. infection is one necessary factor then prevention of infection with H.B.V. could lead to prevention of P.H.C. With the development of a vaccine against H.B.V., prevention of many cases of P.H.C. may be a realistic expectation.

We thank Dr S. N. Huang for supplying the human liver for the source of HB_cAg, Dr David Todd for the sera from the Chinese patients and controls, Dr Jay Hoofnagle for the

THE LANCET, JULY 5, 1975 11

rhesus anti-HB$_e$ antibody, and Barbara Peticolas, Ellen Halpern, Theresa Halbherr, and Francis Kern for technical assistance. This work was supported by U.S. Public Health Service grants CA-06551, RR-05539, and CA-06927 from the National Institutes of Health, and by an appropriation from the Commonwealth of Pennsylvania.

Requests for reprints should be addressed to W. T. L., Institute for Cancer Research, Fox Chase Cancer Center, Philadelphia, Pennsylvania 19111, U.S.A.

REFERENCES

1. Payet, M., Camain, R., Pene, P. *Revue int. Hépat.* 1956, **4**, 1.
2. Steiner, P. E., Davies, J. N. P. *Br. J. Cancer*, 1957, **11**, 523.
3. Vogel, C. C., Anthony, P. P., Mody, N. J., Barker, L. F. *Lancet*, 1970, ii, 621.
4. Nishioka, K., Hirayama, T., Sekine, T., Okochi, K., Mayumi, M., Juei-Low, S., Chen-Hui, L., Tong-Min, L. *in* Alpha-Fetoprotein and Hepatoma (edited by H. Hidematsu and T. Miyaji); p. 167. Baltimore, 1973.
5. Nishioka, K., Mayumi, M., Okochi, K., Okada, K., Hirayama, T. *in* Analytic and Experimental Epidemiology of Cancer (edited by W. Nakahara, T. Hirayama, K. Nishioka, and H. Sugano); p. 137. Baltimore, 1974.
6. Hoofnagle, J., Gerety, R., Hi, L., Barker, L. *New Engl. J. Med.* 1974, **290**, 1336.
7. Hoofnagle, J., Gerety, R., Barker, L. *Lancet*, 1973, ii, 869.
8. Hadziyannis, S. *Proc. natn. Acad. Sci. U.S.A.* (in the press).

Hypothesis

SCHIZOPHRENIA, EPILEPSY, CANCER, METHIONINE, AND FOLATE METABOLISM PATHOGENESIS OF SCHIZOPHRENIA

REFAEL N. LEVI SAMUEL WAXMAN

Departments of Psychiatry and Medicine, Mount Sinai School of Medicine of City University of New York, New York, N.Y. 10029, U.S.A.

WE base our hypothesis upon the seemingly unrelated observations that folate-responsive schizophrenic-like syndromes develop in some epileptics treated with anticonvulsant drugs, the low incidence of carcinoma in schizophrenics, and the absolute need for methionine for the growth of malignant tissues in vitro.

FOLATE-RESPONSIVE ACQUIRED SCHIZOPHRENIC-LIKE SYNDROMES

The relation between schizophrenia and epilepsy was first observed by Meduna [1] who said that schizophrenia and epilepsy were biologically antagonistic diseases. Reynolds [2] formulated Meduna's observation in biochemical terms when he postulated that the biochemical disturbances in untreated epileptics render them less prone to schizophrenia, and, similarly, that the schizophrenic is less liable to develop epilepsy. However, schizophrenic-like psychoses in treated epilepsy are seen,[3] and Reynolds [4,5] attributed these to the effect of anticonvulsants on folate metabolism.

Anticonvulsants, especially diphenylhydantoin, lower serum-folate in significant numbers of patients, and cause frank megaloblastic anæmia in a few.[6] Reynolds et al.[6] suggested that this antifolate effect is important in controlling the epilepsy but may provoke mental symptoms which resemble schizophrenia. When these patients were treated with folic acid and vitamin B_{12}, the anticonvulsant drug did not control the epilepsy, but it did ameliorate the schizophrenic-like syndrome.[7]

SCHIZOPHRENIA AND CANCER

Mentally ill populations, especially patients with schizophrenia, seem to be at relatively low risk of cancer. Rassidakis et al.[8] found that the percentage of mental patients who died from cancer was considerably lower than that of the general population; 15% of deaths were caused by malignant neoplasms in the general population compared with 4·9% among the mentally ill. Other randomly selected causes of death (cardiovascular, diabetes) showed no appreciable differences. In England and Wales about 20% of deaths are caused by neoplasms compared with 6·9% of deaths in mentally ill populations; for Scotland the figures are 17% and 5%, respectively. Other investigations suggest that schizophrenic patients may be even more resistant to neoplasms than patients with other forms of mental illness. Of 2145 State hospital deaths over fifteen years, 16 were neoplastic and only 6 of these were among schizophrenics. In another large mental hospital, of a total of 520 deaths during a two-year period, only 4 schizophrenics died of cancer. The Kashenko Hospital, in Moscow, reports the incidence of deaths from malignant neoplasms in schizophrenics to be only 0·1–0·2% from a yearly population of 2500. Katz and colleagues [9] also reported lower cancer mortality in mental patients than that in matched segments of the general population. The low incidence of cancer is particularly evident among patients who have been in mental hospitals for over ten years. The reason is unknown. Lindelius and Kay,[10] studying 355 male schizophrenics admitted to the hospital from 1947 to 1958, reported 32 deaths; only 1 was due to cancer.

Csatary [11] suggests that specific environmental factors and, above all, a peculiarity in the as yet unravelled biochemistry of mental diseases, especially schizophrenia, may hold an important clue to the understanding of host/tumour relationships.

FORMULATION OF BIOCHEMICAL HYPOTHESIS

Methionine is an essential aminoacid which plays an important role in transmethylation processes which provide a major catabolic pathway for biogenic amines.[12,13] L-methionine cannot provide labile methyl

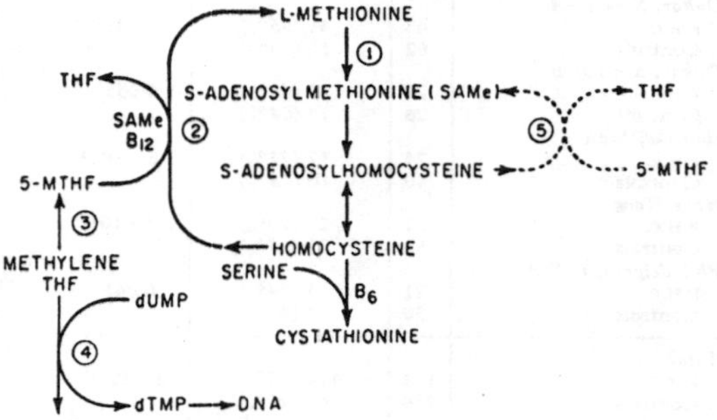

Metabolic pathways for production of labile methyl groups.

Reaction 1: methionine adenosyltransferase pathway for formation of SAMe.

Reaction 2: homocysteine transmethylation pathway through demethylation of 5-M.T.H.F. and resynthesis of L-methionine.

Reaction 3: methylene T.H.F. reductase (inhibited by SAMe).

Reaction 4: thymidylate synthetase pathway for de-novo D.N.A. synthesis.

Reaction 5: proposed methylation of S-adenosylhomocysteine through demethylation of 5-M.T.H.F. for endogenous resynthesis of SAMe. This reaction should not require the presence of methionine adenosyltransferase.

Solid lines represent proven pathways. Broken lines represent proposed pathways for endogenous SAMe formation.

Reprinted from THE LANCET, *September 11, 1976, p.534-538*

HOST RESPONSES TO HEPATITIS-B INFECTION IN PATIENTS WITH PRIMARY HEPATIC CARCINOMA AND THEIR FAMILIES
A Case/Control Study in Senegal, West Africa

B. LAROUZE W. T. LONDON
G. SAIMOT B. G. WERNER
E. D. LUSTBADER M. PAYET
B. S. BLUMBERG

Institute for Cancer Research, Fox Chase Cancer Center, Philadelphia, Pennsylvania 19111, U.S.A., and Institut de Médecine et d'Épidémiologie Africaine, Hôpital Claude Bernard, Paris, France

Summary A case/control study of patients with primary hepatic carcinoma (P.H.C.) and their families was carried out in Dakar, Senegal. 28 P.H.C. cases were matched by age, sex, and ethnic group with 28 controls. Serum was collected from cases, controls, parents (28 mothers, 27 fathers) of cases, parents of controls, 71 siblings of cases, and 58 siblings of controls. Assays of their sera for hepatitis-B surface antigen (HBsAg), antibody to HBsAg (anti-HBs) and antibody to hepatitis-B core antigen (anti-HBc) produced the following results. (1) Nearly all P.H.C. cases (97%) and controls (93%) had some evidence of infection with hepatitis-B virus (H.B.V.), but the cases were more likely to be anti-HBc(+) and less likely to be anti-HBs(+) than the controls. (2) Most of the mothers of the cases were HBsAg(+) (71%), whereas only 14% of the mothers of controls were HBsAg(+). Lower titres of anti-HBs were more common in the mothers of the cases. (3) None of 27 fathers of cases had detectable anti-HBs, but 13 (48%) of the fathers of controls were anti-HBs(+). (4) Siblings of the P.H.C. cases were more likely to have anti-HBs than either their sibs with P.H.C. or the sibs of the controls. However, sibs of P.H.C. cases had lower titres of anti-HBs than the sibs of the controls. These data are consistent with the hypothesis that the P.H.C. cases were infected with H.B.V. by their mothers and that there was an environmental factor which affected the immunological response of all family members to H.B.V. Infection with H.B.V. and the mode of response to that infection among members of families appear to be major factors in the ætiology of P.H.C. in West Africa.

Corrections:

Summary: Line 17 - Only 3 mothers of the cases had anti-HB_s and these were at low titer.

2

Introduction

INFECTION with hepatitis-B virus (H.B.V.) is associated with post-necrotic cirrhosis and primary hepatic carcinoma (P.H.C.). Hepatitis-B surface antigen (HBsAg) is more prevalent in areas where P.H.C. is common, and HBsAg titres are higher in P.H.C. patients than in controls from the same area when assayed by the most sensitive techniques.[1-3] In Senegal and Mali, we found that 90% and 75%, respectively, of the P.H.C. patients had antibody to hepatitis-B core antigen (anti-HBc) in their blood,[4] and that nearly all the patients had some evidence of hepatitis-B infection (i.e., HBsAg, anti-HBc, and/or anti-HBs). These frequencies are much higher than in matched controls from the same countries (anti-HBc 27·6% in Senegal and 16% in Mali; any evidence of hepatitis-B infection, 65% in Senegal and 61·7% in Mali).[5 6]

Carriers of HBsAg cluster in families[7-12] and several workers have demonstrated transmission of H.B.V. from mother to offspring.[13-18] The risk of individuals infected in utero or shortly after birth of becoming chronic carriers is apparently high.

Several investigators have described a clustering of HBsAg carriers and chronic hepatic disease in families of patients with P.H.C.[19-21] Ohbayashi et al. reported three such families, each with an HBsAg(+) mother, and suggested that maternal transmission of H.B.V. may be an important factor in hepato-carcinogenesis.[3 21] Beasley tested the mothers of 6 Taiwanese patients with P.H.C.; all were HBsAg(+).[22]

Taken together, these observations suggest that: (1) different modes of transmission of, and different host responses to, hepatitis-B infection cluster in families, and (2) particular kinds of transmission and host response are associated with P.H.C. If this is so, such modes of transmission and types of host response would constitute risk factors for P.H.C. In this study we have tested these hypotheses, identified risk factors, and calculated the relative and absolute risk for P.H.C. contributed by these factors.

Material and Methods

The patients were from the department of medicine of the Le Dantec Hospital, Dakar, Senegal.

The diagnosis of P.H.C. was based on the following criteria: (1) palpable mass in the liver, or an enlarged, hard liver; (2) histological examination of a transparietal liver biopsy speci-

Corrections:

Introduction: Line 5 - HB_sAg is more common in PHC patients than in controls from the same area.

3

men (24 of 28 cases); and (3) detection of alpha-fetoprotein (A.F.P.) by immunodiffusion (I.D.). The four cases in which biopsies were not taken were positive for A.F.P. by I.D.

Each case of P.H.C. was individually matched for age (mean age 32), sex (22 males, 6 females), and ethnic group with an apparently healthy individual living in the same urban neighbourhood or the same rural village as the patient. Blood was collected from all the mothers of the cases and controls, from 27/28 fathers of cases and controls, and from as many siblings as possible—71 in the P.H.C. families (40 males, 31 females) and 58 from 22 of the 28 control families (27 males, 31 females).

Blood was collected by venepuncture into dry glass tubes. The sera were separated from the red cells, frozen immediately at −4°C, and were eventually shipped to Paris and then to Philadelphia where laboratory tests were performed. The sera were kept frozen in dry ice during transportation until the time of testing. All tests were done using coded specimens and the results were recorded without knowledge of the classification of the sera. HBsAg was determined by radioimmunoassay ('Ausria' II, Abbott Laboratories)[23] and I.D., anti-HBs by hæmagglutination,[24] and anti-HBc by counter-immunoelectrophoretic and I.D. techniques.[4] The data were coded and stored in computer disc files from which the tables were computed and analysed by Fisher's exact test,[25] χ^2,[25] Z score,[26] and the Mann-Whitney test.[25]

Results

The frequency and distribution of factors related to H.B.V. infection in the P.H.C. cases and controls and their parents is shown in table I. Essentially all the P.H.C. patients and controls and most of their mothers showed some evidence of infection with H.B.V. However, the nature of the responses of the patients and their parents to H.B.V. infection differed radically from the responses of the controls and their parents. In the P.H.C. patients anti-HBc was more common and anti-HBs less common than in the controls. 20 of 28 mothers of cases were HBsAg(+) and anti-HBc(+), whereas only 4 of the mothers of controls had HBsAg and 9 (including the 4 HBsAg [+]) had anti-HBc. The remaining 5 anti-HBc(+) control mothers also had anti-HBs. The frequency and titres of anti-HBs were much higher in the control mothers than in the mothers of the P.H.C. patients (the 3 anti-HBs[+] mothers of cases had titres of 1/4, 1/8, and 1/8).

Although frequencies of HBsAg and anti-HBc were similar in the fathers of the cases and the fathers of the

4

TABLE I—EVIDENCE OF HEPATITIS-B INFECTION IN P.H.C. PATIENTS AND THEIR PARENTS, AND CONTROLS AND THEIR PARENTS

Evidence	P.H.C. cases No.	P.H.C. cases +	P.H.C. cases %+	Controls No.	Controls +	Controls %+	P
HBsAg(+) I.D.	28	9	32·1%	28	5	17·9%	0·35
HBsAg(+) R.I.A.	28	22	78·6%	28	16	57·1%	0·15
anti-HBc(+)	28	25	89·2%	28	18	64·3%	0·05
anti-HBs(+)	28	7	25·0%	28	18	64·3%	6×10^{-3}
HBsAg(+), anti-HBc(+), and/or anti-HBs(+)*	28	27	96·4%	28	26	92·9%	0·99

Evidence	Mothers of P.H.C. cases No.	Mothers of P.H.C. cases +	Mothers of P.H.C. cases %+	Mothers of controls No.	Mothers of controls +	Mothers of controls %+	P
HBsAg(+) I.D.	28	15	53·6%	28	3	10·7%	1×10^{-3}
HBsAg(+) R.I.A.	28	20	71·4%	28	4	14·3%	3×10^{-5}
anti-HBc(+)	28	20	71·4%	28	9	32·1%	$6{\cdot}9\times10^{-3}$
anti-HBs(+)	28	3	10·7%	28	15	53·6%	1×10^{-3}
HBsAg(+), anti-HBc(+), and/or anti-HBs(+)*	28	21	75·0%	28	19	67·9%	0·76

Evidence	Fathers of P.H.C. cases No.	Fathers of P.H.C. cases +	Fathers of P.H.C. cases %+	Fathers of controls No.	Fathers of controls +	Fathers of controls %+	P
HBsAg(+) I.D.	27	2	7·4%	27	3	11·1%	0·99
HBsAg(+) R.I.A.	27	5	18·5%	27	5	18·5%	1·00
anti-HBc(+)	27	5	18·5%	27	8	29·6%	0·52
anti-HBs(+)	27	0	0	27	13	48·1%	3×10^{-5}
HBsAg(+), anti-HBc(+), and/or anti-HBs(+)*	27	5	18·5%	27	18	66·6%	7×10^{-6}

* Any evidence of infection with H.B.V.

5

controls, none of the 27 fathers of the P.H.C. patients had anti-HBs compared with 13 anti-HBs(+) among 27 fathers of controls.

These data suggest that the mothers of P.H.C. patients are carriers of H.B.V. and the fathers have an inadequate (or different) humoral immunological response to this infectious agent; and further that the mothers transmitted the hepatitis-B infection to their affected offspring. To test the last hypothesis we compared the frequencies of HBsAg in the mothers of anti-HBc(+) cases and anti-HBc(+) controls (table II). Anti-HBc(+) P.H.C. patients

TABLE II—ASSOCIATION OF HBSAG IN MOTHERS OF P.H.C. PATIENTS WITH ANTI-HBC* IN P.H.C. PATIENTS AND NO ASSOCIATION OF HBSAG IN MOTHERS OF CONTROLS WITH ANTI-HBC IN CONTROLS

—	HBsAg(+)	HBsAg(−)
Mothers of P.H.C. cases:		
P.H.C. anti-HBc(+)	17	8
P.H.C. anti-HBc(−)	3	0
Mothers of controls:		
Control anti-HBc(+)	2	16
Control anti-HBc(−)	2	8

* Anti-HBc is used here because it is the H.B.V.-related factor with the highest frequency in cases and controls.
2×4 table, $\chi^2=19\cdot99$; $P=1\cdot7\times10^{-4}$

were significantly more likely to have an HBsAg(+) mother ($P=4\times10^{-4}$) than were anti-HBc(+) controls. In every instance either the P.H.C. patient, his mother, or both showed evidence (HBsAg[+] and/or anti-HBc[+]) of current and presumably chronic infection with H.B.V.

A related question is, what is the relative risk of P.H.C. which can be attributed to a mother being a chronic carrier of HBsAg? Since each control was carefully matched with each case, a matched-pair analysis was done. A pair consists of a mother of a case and the mother of the matched control. There were 4 pairs in which both the mother of the case and the mother of the control were HBsAg(+), 16 pairs in which the mother of the case was HBsAg(+) and the mother of the control HBsAg(−), and 8 pairs in which both mothers were HBsAg(−). In no instance was the mother of the control HBsAg(+) and the mother of the case HBsAg(−). Relative risk (R) for matched pairs was calculated by the number of mother pairs in which the mothers of cases were HBsAg(+)/the number of mother pairs in which the mother of the control was HBsAg(+). This yields a ratio of 20/4 or a relative risk of 5/1. An alternative

6

method of calculating, R,[27] which does not utilise the matched-pair design but holds constant the other factors contributing to risk of P.H.C. (absence of anti-HBs in fathers and absence of anti-HBs in cases or controls), yields a relative risk of 12·3 if the mother is HBsAg(+).

The other factors which show the most significant differences in table I give lower relative risks (R) by matched-pair analysis. For fathers not having anti-HBs, R=1·95, and for the case being anti-HBs(−), R=2·1.

The power of each of these risk factors to discriminate between the P.H.C. and control matched pairs is shown in a branching diagram (see accompanying figure). At each branch point there was a different distribution of case control pairs according to the particular risk factor. In any pair, if the case or the parent of the case was distinguished from the control or the parent of the control by that factor, the branching stopped. For example, in 16 pairs the mother of the case was HBsAg(+) and the mother of the control was HBsAg(−). For these 16 pairs, HBsAg(+) was a distinguishing factor. The remaining 11 pairs were then examined to determine whether the fathers were anti-HBs(+) or (−). In 6 pairs the father of the case was anti-HBs(−) and the father of the control was anti-HBs(+), &c. Ultimately, two pairs remained that were not distinguished by this procedure. These 2 controls had all three risk factors and in addition were HBsAg(+) and anti-HBc(+). From this analysis we conclude that these individuals are at high risk of developing P.H.C.

A matched-pair analysis of the P.H.C. cases and their parents, and the controls and their parents, for the three risk factors is shown in table III. If all cases were correctly classified by the three factors, they would be entered above the diagonal. Misclassifications, control families having more risk factors than cases, would appear below the diagonal. Cases which cannot be distinguished from controls would fall on the diagonal. In this study there were no misclassifications but there were 5 families of cases which have the same number of risk factors as their matched control families. We devised a scoring system (see legend to table III) for evaluating the significance of this distribution of risk factors and by this analysis the risk factors were found to be distributed differently among the case than among the control families.

The analysis of the siblings of the cases and the controls was complicated by the variable size of the families

7

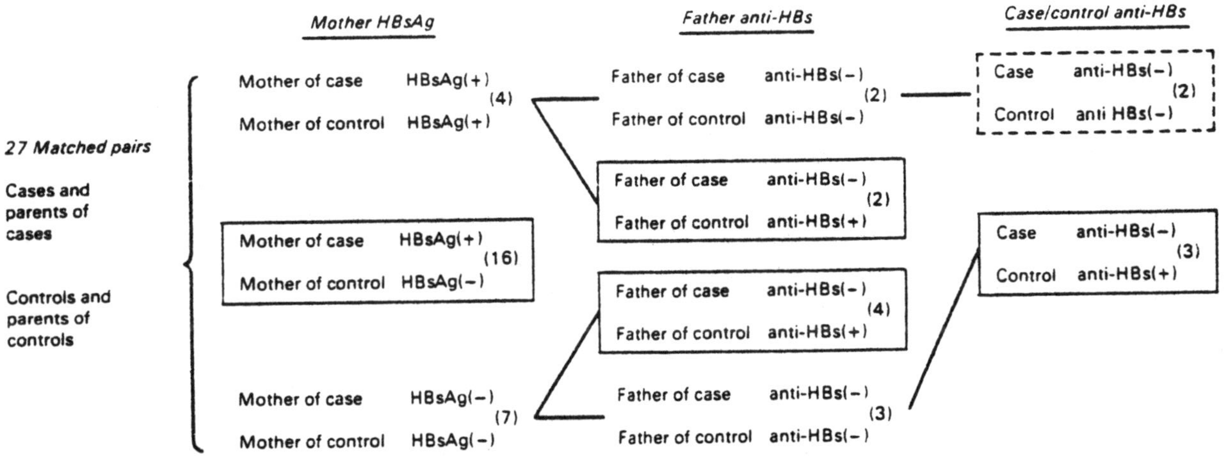

Branch diagram of matched-pairs analysis.

Number of pairs is shown in parentheses.

8

TABLE III—MATCHED-PAIR ANALYSIS OF RISK FACTORS

No. of risk factors* in controls and their parents:	No. of risk factors* in cases and their parents: 0	1	2	3	*Totals*
0	0 (*3/6*)	1 (*4/6*)	3 (*5/6*)	4 (*6/6*)	8
1	0 (*2/6*)	1 (*3/6*)	5 (*4/6*)	7 (*5/6*)	13
2	0 (*1/6*)	0 (*2/6*)	2 (*3/6*)	2 (*4/6*)	4
3	0 (*0/6*)	0 (*1/6*)	0 (*2/6*)	2 (*3/6*)	2
Totals	0	2	10	15	27

* Risk factors are: mother HBsAg(+); father anti-HBs(−); case or control anti-HBs(−).

Scores.—A risk factor which is present in a P.H.C. family adds one point and each risk factor that the matched control family lacks also adds one point to the matched pair. If a case family has 3 risk factors and a control family 0, then the score is 6/6. If a control family has 3 risk factors and a case family 0, then the score is 0/6. For each of the possible combinations in between an appropriate score between 1/6 and 5/6 is given in italics the upper righthand corner of each box. The best total score if case families all had 3 risk factors and control families 0 is 27. If the scores were randomly distributed the expected score would be 13·5. The observed score was significantly different from 13·5. Total score=20·16; Z=2·56; $P<0·01$.

and the failure to collect blood from all siblings. (Sera were not obtained from siblings in 6 of the control families.) With these reservations in mind, certain features are apparent (table IV). Most of the siblings of the P.H.C. patients had been exposed to hepatitis-B infection; 67 of 73 had either HBsAg, anti-HBc, or anti-HBs in their blood. They differed from their affected siblings (the P.H.C. cases) in that their response to hepatitis-B infection was manifested mainly by the production of anti-HBs (82%), and less commonly by the presence of HBsAg or anti-HBc (211%). That is, given parents who confer on their offspring a high risk of developing P.H.C., the ability to produce anti-HBs substantially reduces that risk.

The siblings of the controls had frequencies of HBsAg, anti-HBc, and anti-HBs that were lower than their own brothers or sisters selected as the controls for the P.H.C. cases but were similar to those observed in the general population of Senegal.[5 6] Comparison of the frequencies of H.B.V.-related antigens and antibodies in the sibs of the cases with those in the sibs of the controls revealed a significantly higher ($P=1·6\times10^{-5}$) frequency of anti-HBs in the sibs of the cases. Since these individ-

9

TABLE IV—COMPARISONS OF EVIDENCE OF H.B.V. INFECTION IN P.H.C. CASES AND CONTROLS WITH THEIR SIBLINGS

Evidence	P.H.C. cases			Siblings of P.H.C.* cases			
	No.	+	%+	No.	+	%+	P
HBsAg I.D.	28	9	32%	71	7	9·8%	0·02
HBsAg R.I.A.	28	22	78·6%	73	7	9·6%	$5{\cdot}7\times10^{-11}$
anti-HBc	28	25	89·2%	72	8	11·1%	$2{\cdot}6\times10^{-13}$
anti-HBs	28	7	25%	73	60	82·1%*	$2{\cdot}5\times10^{-7}$
HBsAg(+), anti-HBc(+), and/or anti-HBs(+)	28	27	96·4%	73	67	91·7%	0·48
	Controls			**Siblings of controls***			
Evidence	No.	+	%+	No.	+	%+	P
HBsAg I.D.	28	5	17·8%	58	7	12%	0·67
HBsAg R.I.A.	28	16	57·1%	58	8	14%	$1{\cdot}3\times10^{-4}$
anti-HBc	28	18	64·2%	58	13	22·4%	4×10^{-4}
anti-HBs	28	18	64·2%	58	26	41·8%	0·14
HBsAg(+), anti-HBc(+), and/or anti-HBs(+)	28	26	92·8%	58	43	76·1%	0·70

* None of the comparisons of siblings of P.H.C. cases with siblings of controls yield significant differences except for anti-HBs ($P=1{\cdot}6\times10^{-3}$) and HBsAg(+), anti-HBc(+) and/or anti-HBs(+) ($P=0{\cdot}01$). This last difference is accounted for entirely by the increased frequency of anti-HBs in the sibs of cases.

uals were living, or had lived, with a mother and/or a sibling (in whom P.H.C. developed) who were carriers of HBsAg, this is not surprising. However, the titres of anti-HBs/ay and anti-HBs/ad in the sibs of cases were lower than the titres in the sibs of the controls (based on only those individuals who have detectable antibody). The difference was most significant ($P=0 \cdot 001$) for antibody against HBsAg/ay, the major antigenic subtype in West Africa.

Discussion

The results of this study are consistent with the hypothesis that a particular mode of transmission—e.g., infection of offspring early in life by mothers who are carriers of H.B.V., is an important (and perhaps necessary) factor in the development of P.H.C. 20 of 28 mothers of P.H.C. patients were HBsAg(+). Since their affected children were born as long as 40 years before this study was done, two interpretations must be considered: (1) the mothers had been infected by the offspring with P.H.C. and HBsAg; and (2) the mothers of P.H.C. patients were chronic carriers of HBsAg (and presumably H.B.V.) and transmitted the H.B.V. infection to the P.H.C. case at birth or early in life.

P.H.C. mothers were probably chronic carriers of HBsAg (and H.B.V.). In point prevalence studies, chronic carriers are much more likely to be detected than are individuals who are transiently HBsAg(+). Furthermore, it is unlikely that these women reached old age in Senegal without having been exposed to H.B.V. infection earlier in life. There was evidence of H.B.V. infection (HBsAg, anti-HBc, or anti-HBs) in about 65% of the population in West Africa, in 60% of women in the reproductive age-group, and in ~ 50% of the 10–15 year age-group (unpublished data). Therefore, we think it is unlikely that the P.H.C. mothers (women now in their 40s, 50s, and 60s) were infected by the P.H.C. cases.

If the mothers we detected as being HBsAg(+) were positive at the time their child was born, then they have been HBsAg(+) for 20–40 years. Since chronic carriers are often not carriers forever, it is quite possible that the 8 HBsAg(−) mothers were HBsAg(+) when their affected children were born. (The 8 offspring with P.H.C., of the 8 HBsAg(−) mothers, are currently anti-HBc(+) and 6 of the 8 are HBsAg(+).)

Although maternal transmission of H.B.V. may be a

necessary factor in the development of P.H.C., it is not sufficient. Otherwise, the incidence of P.H.C. would be much higher than has been recorded. In this study we observed a paternal effect, the failure of fathers of P.H.C. patients to produce anti-HBs. In Senegal we would expect that at least 50% of males in this age-group would have anti-HBs. Although the fathers of P.H.C. patients had been heavily exposed to H.B.V. in their homes and in the community, their immunological response to the agent was inadequate (or, at least, different from other men in Senegal). The nature of the factor in the fathers which affects their immunological response is unknown and will be investigated further. Some insight into the nature of this factor is provided by the observations that the frequencies of anti-HBs are also lower in the cases and the mothers of the cases than in their respective controls, and the titres of anti-HBs/ay are lower in the sibs of the cases than in the sibs of the controls. These findings suggest that an environmental factor(s), present in the household of the cases, to which the fathers are maximally exposed and to which other family members are exposed to a lesser degree, may be responsible for suppressing the antibody response to HBsAg. Such a factor(s) could be one of the hepatic carcinogens proposed by others (aflatoxins, nitrosamines, cycads).[28]

Irrespective of why the mothers have HBsAg and the fathers lack anti-HBs, the use of these markers as risk factors has identified individuals at high risk of developing a particular cancer. The relative risks we have calculated are conservative estimates. The odds ratio, (AD/BC in a 2×2 table), is the usual method for calculating relative risk.[26] Using this approach the relative risk (R) of P.H.C. given by a mother being HBsAg(+) was 15 and for a father lacking anti-HBs, R=59. (Because there are no fathers of cases with anti-HBs (bin B=0), 0·5 was added to each bin to permit estimation of R.) R for absence of anti-HBs in the P.H.C. cases was 5·4. In addition, control subjects in this study had much higher frequencies of HBsAg (57%) and anti-HBc (64%) than we have found in studies of general populations in West Africa or in age, sex, ethnic group matched controls selected for other patients with P.H.C.[5 6] The controls selected for this study live in the same neighbourhoods as the P.H.C. patients, and they may be at higher risk of developing P.H.C. than controls outside this neighbourhood.

12

Nevertheless, very few factors associated with common cancers have given anything like the 5 to 12 fold relative risk or the 71·6% absolute risk of P.H.C. associated with HBsAg(+) mothers reported here. Anderson[29] identified a small group of women, with a sister and mother with premenopausal bilateral breast cancer, who have a 47-fold (by odds ratio) relative risk of developing breast cancer, but they account for only 1 or 2% of the breast-cancer population. Smoking cigarettes is associated with a high absolute risk of lung cancer. Yet the relative risk (using similar methods of calculation) for smokers of more than one pack of cigarettes a day is about 10-fold that of non-smokers.[26]

The results reported here are based on a small sample. The conclusions we have drawn require confirmation, and the hypothesis we have stated with respect to risk factors must be tested. Retrospective and prospective studies have been designed, and the field work is in progress. Should these studies confirm these results it would suggest that prevention or control of infection with H.B.V. by public-health measures, hepatitis-B hyperimmune globulin, or hepatitis-B vaccine would be followed by a decline in the incidence of P.H.C.

We thank Prof. M. Sankalé and Prof. B. Diop of the University of Dakar Medical School and Le Dantec Hospital for their assistance and cooperation with this project.

This work was supported by U.S.P.H.S. grants CA-06551, RR-05539, and CA-06927 from the National Institutes of Health and by an appropriation from the Commonwealth of Pennsylvania. B.L. holds a fellowship from the Ministere Français des affaires étrangèrs.

Requests for reprints should be addressed to W.T.L., Institute of Cancer Research, 7701 Burholme Avenue, Fox Chase, Philadelphia, Pennsylvania 19111, U.S.A.

REFERENCES

1. Vogel, C. L., Anthony, B. P., Sadikali, F., Barker, L. F., Peterson, M. R. *J. natn. Cancer Inst.* 1972, **48**, 1583.
2. Schlegel, N., Menache, D., Saimot, G., Payet, M. *Digestion*, 1972, **6**, 302.
3. Nishioka, K., Hirayama, T., Sekine, T., Okochi, K., Mayumi, M., Low, S. J., Hui, L. C., Ming, L. T. *in* Japanese Cancer Association Symposium on Alphafetoprotein and Hepatoma (edited by H. Hirai and T. Miyaji); p. 167. Maryland, Baltimore, 1973.
4. Maupas, P., Werner, B., Larouzé, B., Millman, I., London, W. T., O'Connell, A., Blumberg, B. S., Saimot, G., Payet, M. *Lancet*, 1975, ii, 9.
5. Blumberg, B. S., Larouzé, B., London, W. T., Werner, B., Hesser, J. E., Millman, I., Saimot, G., Payet, M. *Am. J. Path.* 1975, **81**, 669.
6. Larouzé, B., Blumberg, B. S., London, W. T., Lustbader, E. D., Payet, M. Unpublished.
7. Blumberg, B. S., Alter, H. J., Visnich, S. *J. Am. med. Ass.* 1965, **191**, 541.
8. Blumberg, B. S., Melartin, L., Guinto, R. A., Werner, B. *Am. J. hum. Genet.* 1966, **18**, 594.

13

9. Blumberg, B. S., Friedlaender, J. S., Woodside, A., Sutnick, A. I., London, W. T. *Proc. natn. Acad. Sci. U.S.A.* 1969, **62**, 1108.
10. Cepellini, R., Bedairdo, G., Carbonara, A. O., Trinchieri, G., Filippi, G. Antigene Australia et Epatite Virale (Atti Convegni Farmitalia). p. 53., Turin, Italy, 1970.
11. Helske, T. *Scand. J. Hæmat.* 1974, **22**, 65.
12. Cazal, P., Robinet-Levy, M., Lemaine, J. M. *Revue fr. Transf.* 1972, **25**, 477.
13. Merrill, D. A., Du Bois, R. S. Kohler, P. F. *New Engl. J. Med.* 1972, **287**, 1280.
14. Blumberg, B. S. *Bull. Acad. Med. Toronto,* 1972, **45**, 45.
15. Schweitzer, I. L., Dunn, A. E. G., Peters, R. L., Spears, R. L. *Am. J. Med.* 1973, **55**, 762.
16. Mazzur, S., Blumberg, B. S., Friedlaender, J. S. *Nature,* 1974, **247**, 41.
17. Papaevangelou, G., Hoofnagle, J., Kremastinou, J. *Lancet,* 1974, ii, 746.
18. Stevens, C. F., Beasley, R. B., Tsui, J., Lee, W. C. *New Engl. J. Med.* 1975, **292**, 771.
19. Denison, E. K., Peters, R. L., Reynolds, T. B. *Ann. intern. Med.* 1971, **74**, 391.
20. Velasco, M., Sorensen, D., Daiber, R., Carmona, A., Katz, R. *Lancet,* 1971, i, 1183.
21. Ohbayashi, A., Okochi, K., Mayumi, M. *Gastroenterology,* 1972, **62**, 618.
22. Beasley, R. P. *Am. J. med. Sci.* 1975, **270**, 57.
23. Ling, C. M., Overby, L. R. *J. Immun.* 1972, **109**, 834.
24. Vyas, G. N., Shulman, N. R. *Science,* 1970, **170**, 332.
25. Siegel, S. Non-parametric Statistics for the Behavioral Sciences; pp. 104 and 116. New York, 1956.
26. Hildebrand, D. K., Laing, J. P., Rosenthal, H. L. Prediction Analysis of Cross Tabulations. New York, 1976.
27. Mantel, N., Haenszel, W. *J. natn. Cancer Inst.* 1959, **22**, 719.
28. Davies, J. N. P. Hepatic Neoplasm In The Liver (edited by E. A. Gall and F. K. Mostofi); p. 361. Baltimore, Maryland, 1973.
29. Anderson, D. E. *Cancer,* 1974, **34**, 1090.

Reprinted from *J Natl Cancer Inst*
Vol. 58, No. 6, June 1977

Forecasting the Development of Primary Hepatocellular Carcinoma by the Use of Risk Factors: Studies in West Aftica[1,2,3]

Bernard Larouzé,[4,5,6] Baruch S. Blumber,[5,7] W. Thomas London,[5] Edward D. Lustbader,[5] Marc Sankalé,[8] *and* Maurice Payet[6,9]

ABSTRACT—An association between hepatitis B virus (HBV) and primary hepatocellular carcinoma (PHC) has been found in several studies in Africa, Asia, and elsewhere. In this paper we considered the interrelations between several events related to HBV infection, which include the presence of: 1) hepatitis B surface antigen (HBsAg), 2) antibody to hepatitis B core antigen (anti-HBc), 3) antibody to the surface antigen (anti-HBs), 4) chronic liver disease, 5) elevated alpha-fetoprotein, and 6) PHC. With the use of preliminary epidemiologic data, risk factors related to these events were calculated. We suggested that the interactions between these events and HBV infection in parents be used to estimate the risk of PHC for an individual in this environment.—J Natl Cancer Inst 58: 1557–1561, 1977.

PHC is the most common form of cancer in mails in many parts of Africa and the Far East (*1*). For many years, clinical and epidemiologic studies have suggested that viral hepatitis may be related to the pathogenesis of the disease (*2–4*). With the availability of the test for the hepatitis B surface antigen (HBsAg, Australia antigen), it became possible to test this hypothesis; a strong association between PHC and the presence of HBsAg has been found (*5–10*). Recent studies (*11*), including data on antibody to the core antigen (anti-HBc) of HBV, have given additional support to the hypothesis.

The biologic interpretation of this relationship has led to the theory of a sequential pathogenesis for PHC that begins with HBV infection. The progression of events may start at an early age or even prenatally and proceed through chronic hepatitis, postnecrotic cirrhosis, and then PHC (text-fig. 1).

In this paper we describe a retrospective study that was intended to estimate the frequencies of the events comprising the sequential pathogenesis pathway. These events could then be classified as risk factors for PHC, and relative risk could be estimated. Retrospective studies suffer from the difficulty of ascertaining whether the risk factors were antecedents of the disease. In PHC, for instance, a high level of AFP in the serum is a critical diagnostic tool in identifying PHC cases. Whether the elevated AFP is a precursor to PHC, and hence a risk factor, cannot be conclusively determined from a retrospective study. However, we chose a retrospective study design because our primary objective was to identify risk factors; for this purpose, a retrospective study is the most sensitive means of detection.

In previous papers we reported studies on the frequency of HBsAg, anti=HBc, and anti-HBs in patients and controls from Mali and Senegal, West Africa, and in families and control families from Senegal (*9–11*). In this paper we present data on the quantitative relations of the risk factors (HBV infection, elevated AFP, and CLD) to each other and to PHC. The data do not reject the hypothesis that certain carriers of HBV develop CLD, elevated levels of AFP, and ultimately PHC. A prospective study to quantitatively test this hypothesis is proposed.

Patient population.—Two groups of patients with PHC and one group with CLD were studied. Control groups selected for each patient group were individuals from the same region matched as closely as possible by age (±5 yr), sex, and ethnic origin. One PHC group, designated Senegal, was composed of 39 inpatients and outpatients from the Department of Internal Medicine of Le Dantec Hospital in Dakar, Senegal. These patients included all such patients seen by this department between May and September 1974. Patients with both PHC and tuberculosis were not included, but no other PHC patients were excluded. (Some patients seen only once in the clinic may have had PHC but did not have blood drawn; the clinical impression could not be confirmed.) The diagnosis of PHC was established on clinical grounds, including palpation of a tumor mass and/or an enlarged liver and by the presence of AFP as detected by immunodiffusion. In a few instances, the diagnosis was confirmed by histologic examination. One or 2 controls (if suitable matches could be obtained) were selected for each PHC patient from healthy people living in villages near Dakar (No. = 11) and from inpatients and outpatients from Le Dantec Hospital (No. = 47). None of the controls had diagnoses of hepatic or renal disease, cancer, leprosy, or tuberculosis.

The Mali PHC group included all patients identified between January and July 1974 (excluding May) by the same criteria used for the Senegal patients (No. = 21). They were from the Point G and Gabriel Toure Hospitals in Bantako, Mali, and from the hospital of Kati, a village 5 km from Bamako. Forty apparently healthy

ABBREVIATONS USED: PHC = primary hepatocellular carcinoma; HBV = hepatitis B virus; AFP = alpha-fetoprotein; CLD = chronic liver disease; RIA = radioimmunoassay.

[1]Received August 2, 1976; accepted December 21, 1976.
[2]Supported by Public Health Service (PHS) grants CA06551 and CA06927 from the National Cancer Institute, by PHS grant RR05539 from the Division of Research Resources, and by an appropriation from the Commonwealth of Pennsylvania.
[3]Presented in part at the Third International Symposium on the Detection and Prevention of Cancer, April 30, 1976, in New York, N.Y., and at the meeting of the American Association for Cancer Research, May 7, 1976, in Toronto, Canada.
[4]Supported in part by a fellowship from Le Ministere Français des Affaires Etrangères.
[5]Institute for Cancer Research, The Fox Chase Cancer Center, 7701 Burholme Ave., Philadelphia, Pa. 19111.
[6]Institut de Médecine et d'Epidemiologue Africaines, Hôptial Calude Bernard, Paris, France.
[7]*Address reprint requests to* Dr. Blumberg.
[8]Clinique Médicale, Université Dakar, Senegal, West Africa.
[9]We thank the Ministers of Health of Mali and Senegal for their permission to conduct this study and for their assistance, and we are grateful to Professors B. Diop and M. Sow and the physicians of the Bamako and Kati hospitals for their help.

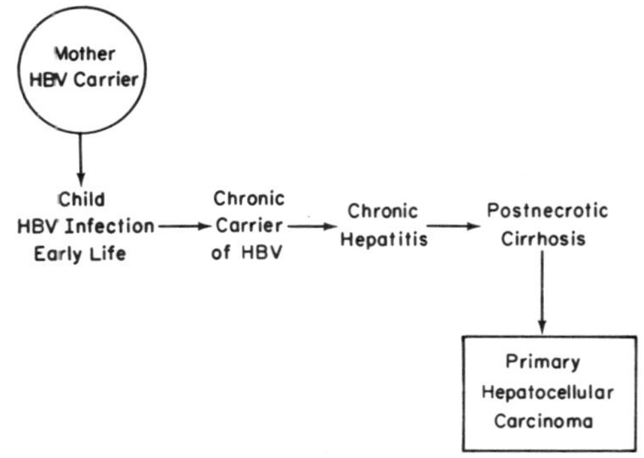

TEXT-FIGURE 1.—Pathogenesis of PHC. Data supporting this proposed sequence of events were reviewed by Blumberg et al. (*9*) and were based on the earlier work of Payet et al. (*2*) and Steiner and Davies (*3*) and recent observations from Japan (*6*) and West Africa (*10*). Dietary carcinogens have also been proposed as etiologic factors in PHC (*12–14*). They could interact with this pathway by influencing the progression from any one stage to the next.

individuals (2 controls for each case matched as in Senegal) were selected from villages near Bamako.

Patients with CLD were identified in the same hospitals in Mali, between January and July 1974. The criteria for selection were an enlarged hard liver with or without clinical manifestations of hepatocellular insufficiency and portal hypertension. All patients meeting these criteria were included (No.=40). AFP was not present by immunodiffusion. The patients were matched with controls in the same manner as were the PHC patients and controls; 81 controls were selected from apparently normal people from villages in the Bamako area.

The age, sex, and ethnic distributions of the patients and controls are shown in tables 1 and 2. The ethnic distribution was representative of the general distribution of these groups in the Dakar and Bamako areas, with the possible exception of a slightly higher frequency of Sarakole people in the Mali sample.

Serologic tests.—HBsAg was determined by RIA with the use of the "Ausria II" test (Abbott Laboratories, Chicago, Ill.). Anti-HBs was detected by passive hemagglutination (*15*). The tests were performed against ay- and ad-coated cells separately and scored as positive when the titers were equal to or greater than 1:4 and could be inhibited by an HBsAg(+) serum. Anti-HBc was determined by the counter electrophoresis and immunodiffusion methods (*16*). All tests were performed at The Institute for Cancer Research, The Fox Chase Center, except the AFP level by RIA (*17*), which was done at Princeton Laboratories (Princeton, N. J.) by the courtesy of Dr. Nicholas Starkovsky.

RISK FACTOR FREQUENCIES

HBV infection.—The results of the serologic tests for HBV infection are given in table 3. The high frequency of HBsAg and anti-HBc in the PHC patients as compared with controls confirmed again the frequently reported association of HBV with PHC. Further, the low frequency of anti-HBs in these patients compared with the controls suggests that the immunologic status of the PHC patients was such that they responded differently to infection with HBV than did individuals without PHC. Moreover, the increased frequency of anti-HBc compared with controls may have reflected the continued activity of the infectious agent within liver cells (*16*). Individuals positive for at least one of the tests for HBsAg, anti-HBc, or anti-HBs were defined as being

TABLE 1.—*Age and sex distribution of disease and control populations*[a]

Study group	No. in each group	Mean age	SD of age	No. in age groups, yr							Sex ratio, male/female
				5–9	10–19	20–29	30–39	40–49	50–59	60+	
Senegal PHC	39	41.48	11.68			6	11	12	5	5	2.9
Controls	59	38.79	12.75			15	13	15	7	7	2.2
Mali PHC	21	49.76	13.16			1	5	4	4	7	9.5
Controls	40	47.67	12.58			2	11	9	8	10	12.3
Mali CLD	42	37.85	14.65		4	11	3	16	5	3	2.8
Controls	81	37.90	14.08		7	8	15	26	9	6	3.2

[a] In a few instances, the ages of the patients and controls were unknown or difficult to estimate; thus they were omitted.

TABLE 2.—*Ethnic origins of the populations studied*

Study group	No. in ethnic groups									Total
	Bambara	Sarakole	Peuhl	Malinke	Woloff	Serere	Toucouleur	Sosso	Miscellaneous	
Senegal PHC			4		25	3	6		1	39
Controls			6		39	6	7		1	59
Mali PHC	12	4	1	3					1	21
Controls	21	7	1	9					2	40
Mali CLD	15	7	5	8					7	42
Controls	45	10	3	18					5	81

TABLE 3.—*Frequency of HBsAg, anti-HBc, and anti-HBs in cases and controls*

Study group	Test	Patients[a]				Controls[a]				P[b]
		No. tested	+	−	%+	No. tested	+	−	%+	
Senegal PHC	HBsAg RIA	39	31	8	79.4	53	6	47	11.3	4×10^{-11}
	Anti-HBc	39	35	4	89.7	58	16	42	27.6	1×10^{-9}
	Anti-HBs	39	8	31	20.5	58	26	32	44.8	0.02
	Total exposed	39	37	2	94.8	58	38	20	65.1	8×10^{-4}
Mali PHC	HBsAg RIA	21	10	11	47.6	38	2	36	5.2	4×10^{-4}
	Anti-HBc	20	15	5	75.0	40	10	30	25.0	5×10^{-4}
	Anti-HBs	21	8	13	38.0	40	17	23	42.5	0.95
	Total exposed	21	19	2	90.4	40	25	15	62.0	0.03
Mali CLD	HBsAg RIA	41	19	22	46.0	80	4	76	5.0	2×10^{-7}
	Anti-HBc	42	25	17	59.0	71	12	59	16.0	8×10^{-6}
	Anti-HBs	42	11	31	26.1	81	28	53	34.5	0.46
	Total exposed	42	35	7	83.0	81	34	47	41.7	1×10^{-5}

[a] + = No. positive; − = No. negative; %+ = percent positive.
[b] The two-tailed probability obtained from Fisher exact test.

Senegal

		HBsAg +	HBsAg −	Anti-HBc +	Anti-HBc −	Absence of Anti-HBs +	Absence of Anti-HBs −
PHC	+	31 A	8 B	35	4	31	8
	−	6 C	47 D	16	42	32	26

Relative risk = $\frac{(A)\ (D)}{(B)\ (C)}$ = 30 | 23 | 3

Approx. 95% Conf. Limits: (11, 114) | (9, 101) | (1.1, 9)

Mali

		HBsAg +	HBsAg −	Anti-HBc +	Anti-HBc −	Absence of Anti-HBs +	Absence of Anti-HBs −
PHC	+	10	11	15	5	13	8
	−	2	36	10	30	23	17

Relative risk = 16 | 9 | 1.2

Approx. 95% Conf. Limits: (3, 165) | (2, 38) | ——

TEXT-FIGURE 2.—Calculation of relative risk for PHC by the factors: 1) presence of HBsAg, 2) presence of anti-HBc, and 3) absence of anti-HBs in Senegal and Mali studies.

exposed to HBV. In every population, the frequency of HBV exposure was higher in the patient population than in the controls.

The relative risks of PHC for individuals with HBsAg, anti-HBs, and anti-HBc are given in text-fig. 2. The relative risk was estimated by the odds ratio formula. The risk of PHC for individuals with HBsAg was in the range of the risk of lung cancer for cigarette smokers.

TABLE 4.—*Frequency and levels of AFP tested by RIA in CLD patients and CLD and PHC controls*

Patients or controls	No. tested	+[a]	%+[b]	Range[c] ng/ml
CLD patients				
Total	40	7[d]	17.5	37–205
Males	31	7[d]	22.5	37–205
Controls				
Of Senegal males	40	1	2.5	16
Of Mali PHC males	39	1	2.5	18
Of Mali CLD males	58	2	3.4	27 and 29,500
Total	137	4	2.9	

[a] + = No. positive.
[b] %+ = percent positive.
[c] A level of >10 ng/ml is positive.
[d] All males.

The risk factors accounted for a high proportion of the PHC cases. For example, anti-HBc occurred in 75–90% of the patients.

AFP.—The PHC patients were selected on the basis of the presence of AFP by immunodiffusion; hence all had high levels of AFP by RIA. The low frequencies of AFP elevation in the PHC controls (table 4) reaffirmed the diagnostic value of AFP in identifying PHC cases. AFP elevations, detectable only by RIA, were found in 17.5% of the CLD patients and in approximately 3% of the male controls. One of the Mali controls had an extremely high AFP level, providing indirect evidence that increase in AFP may be a precursor to PHC.

CLD.—In Mali, the frequencies of HBsAg and anti-HBc were nearly the same in the patients with CLD as in the PHC patients. Also, the ratio of HBsAg to anti-HBs was similar for both PHC and CLD patients. From these findings, it appeared that the patients who developed CLD had a pattern of immune reaction to HBV and a frequency of infection with HBV similar to those of patients with PHC.

INTERACTION OF THE RISK FACTORS WITH PHC

In the CLD group, 17.5% had elevated AFP levels by RIA (table 4) and all were males. In the control population, 2.9% of the males had elevated AFP. (Females were excluded, since they might have had elevated AFP due to pregnancy.) Elevations of AFP are particularly frequent in patients with PHC from areas where the disease is frequent (*5, 6, 8, 18, 19*) and where it is associated with HBsAg. This finding is also true for postnecrotic cirrhosis of the liver (*8, 19*). These observations are compatible with a hypothesis that patients with cirrhosis, who have elevated AFP, have a very high risk of developing PHC (*20*). The simultaneous existence of all the risk factors suggested that a cross-sectional static analysis could yield insight into the risk factor interactions. Text-figure 3 shows the risk factor frequencies illustrated as circles of appropriately scaled areas. We adopted the following symbolic notation: presence of HBsAg (H); CLD (L); elevated AFP (A); PHC (C). The intersections of the circles with each other indicated the interactions among the risk factors.

The magnitudes of the interactions were appraised by standard set theory notation (*see* "Appendix"). From this study and previous reports, the following frequencies have been estimated and used in the construction of the diagram.

H: Frequency of HBsAg in the general population: 5–11% (derived from the controls in the Senegal and Mali PHC and Mali CLD studies; 10% was used in the calculations).

A: Frequency of elevated AFP in the general population: 3% [derived from the male controls in Senegal and Mali PHC and Mali CLD (table 4)].

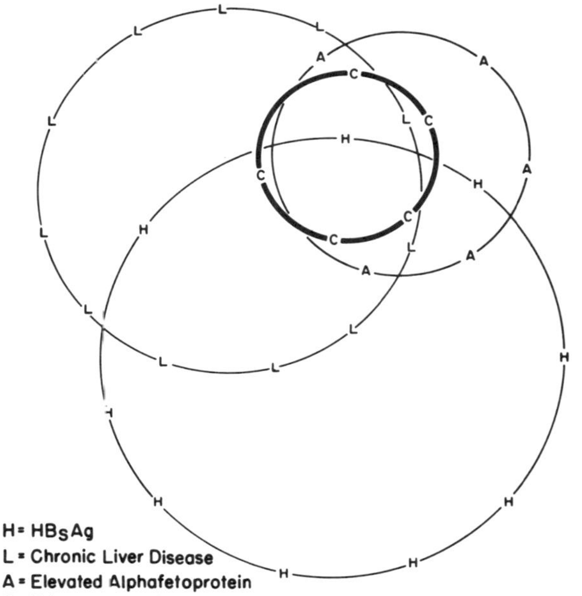

TEXT-FIGURE 3.—Illustration of the relations of risk factors to PHC. The relative sizes of the circles are to scale, except that the size of the circle for the PHC (circle C) is 280 times scale (increased from 0.005% to 1.4%) to maintain the quantitative relations of the intersections. The relative size of the intersections are also to scale but, as noted in the text, are based in some cases on tentative frequency estimates.

C: Frequency of PHC in the general population: a minimum of 5/100,000, but probably much higher [reliable epidemiologic data not available (*1*)].

H/L: Frequency of HBsAg in CLD: 46% (from table 3).

H/A: Frequency of HBsAg in individuals with AFP in the general population: approximately 40% (based on very small numbers).

H/C: Frequency of HBsAg in PHC: 48–80% (from table 3; 60% was used in calculations).

A/C: frequency of elevated AFP by RIA in PHC: approximately 90% (estimates for West Africa).

A/L: Frequency of elevated AFP in CLD: 23% in males only (from table 4).

L/H: Frequency of CLD in individuals with HBsAg: approximately 30% [no accurate data available, but summary of some estimates given in (*21*)].

L/C: Frequency of CLD in patients with PHC: approximately 85%.

Using these data and set theory probability estimates, we calculated the values for the remaining intersections (*see* "Appendix"):

L: Frequency of CLD in the general population: 6.5%.

A/H: Frequency of elevated AFP in patients with HBsAg: 12%.

C/H: Frequency of PHC in individuals with HBsAg: 0.03%.

L/A: Frequency of CLD in individuals with elevated AFP: 50%.

C/A: Frequency of PHC in individuals with elevated AFP: 0.15%.

C/L: Frequency of PHC in individuals with CLD: 0.065%.

The intersecting circles neither require nor imply the validity of the sequence of events hypothesis. However, the magnitude of the interactions are useful in estimating approximate sample sizes for the prospective study to follow. The diagram can also be used for comparison with subsequent epidemiologic studies; by an iterative process, more exact frequencies in this and other populations can be estimated. It should be emphasized that some of these estimates are based on small numbers or inadequate surveys and may be changed considerably in the future.

Quantitative aspects of the sequential hypothesis could be tested if an individual or a group of individuals could be followed for a sufficient time to observe the occurrence of all of the events. But this would take a long time, e.g., 30 years or more. Logically, the hypothesis could also be supported if individuals in each of the categories of the sequence could be identified and a prediction made as to the probability of an individual's proceeding to the next stage. It appears reasonable that if a prediction could be made as to which individuals are more likely to proceed from one category to the next, then a prediction can be made that an individual in the the first category would proceed to the development of PHC. Since individuals in all categories exist simultaneously and the time to go from one category to another is less than the time to go through the entire process, the hypothesis could be substantiated in a shorter time.

We have proposed the study of a program for the prevention of CLD and PHC by the control of hepatitis B infection using public health measures, including a vaccine developed in our laboratory (*9*). If our analysis of these data is appropriate and the sequential hypothesis is confirmed in prospective studies, then it should be possible to estimate the effectiveness of public health measures on the prevention of the various diseases associated with HBV infection. Hence, the possible effectiveness of the hepatitis B vaccine to prevent PHC could be determined without having to wait for a major effect on the decrease in cancer incidence.

APPENDIX

If we have two events that can identify an individual, then any individual who possesses both identifiers is labeled as $E_1 E_2$. The probability that an individual will be labeled $E_1 E_2$ depends on the probability that the event E_1 occurs and also the probability that the event E_2 takes place, given that E_1 has already occurred. In symbols:

$$Pr\,(E_1\,E_2)=Pr\,(E_1)\cdot Pr\,(E_2/E_1), \quad [1]$$

where the label E_2/E_1 represents the occurrence of the event E_2, given that E_1 has already occurred. Individuals can also be labeled as E_1E_2 in a second fashion. Namely, E_2 occurs first followed by E_1; then

$$Pr\,(E_1\,E_2)=Pr\,(E_2)\cdot Pr\,(E_1/E_2). \quad [2]$$

Combining these two relations yields the fundamental identity that

$$Pr\,(E_1)\cdot Pr\,(E_2/E_1)=Pr\,(E_2)\cdot Pr\,(E_1/E_2). \quad [3]$$

Knowledge of any three of the probabilities in [3] immediately yields an equation for solving for the unknown fourth probability (*22*).

REFERENCES

(*1*) HUTT MS: Epidemiology of human primary liver cancer. *In* Liver Cancer. Proceedings of International Agency for Research on Cancer, Lyon: IARC, 1969, pp 21–29

(*2*) PAYET M, CAMAIN R, PENE P: Le cancer primitif du foie. Etude critique à propos de 240 cas. Rev Intern Hepatol 4:1–20, 1956

(*3*) STEINER P, DAVIES JN: Cirrhosis and primary liver carcinoma in Ugandan Africans. Br J Cancer 11:523–534, 1957

(*4*) MIKAYE M: Primary hepatic cancer and liver cirrhosis in Japan. Acta Un Int Contra Cancr 17:892–897, 1961

(*5*) VOGEL C, ANTHONY P, SADIKALI F, et al: Hepatitis-associated antigen and antibody in hepatocellular carcinoma: Results of a continuing study. J Natl Cancer Inst 48:1583–1588, 1972

(*6*) NISHIOKA K, HIAYAMA T, SEKINE T, et al: Australia antigen and hepatocellular carcinoma. Gann Monogr 14:167–183, 1973

(*7*) PAYET, M, SAIMOT G, BROCHARD C, et al: Cancer primitif du foie et HB antigen (Australia antigen): Epidémiologie et étude génétique. Ann Med Interne (Paris) 126:269–271, 1975

(*8*) SAKURAI M, MIYAJI T: α-Fetoprotein and hepatitis B antigen in hepatocarcinogenesis. Carcinofetal proteins: Biology and chemistry. Ann NY Acad Sci 259:156–167, 1975

(*9*) BLUMBERG BS, LAROUZÉ B, LONDON WT, et al: The relation of infection with the hepatitis B agent to primary hepatic carcinoma. Am J Pathol 81:669–682, 1975

(*10*) LAROUZÉ B, LONDON WT, SAIMOT G, et al: Host response to hepatitis B infection in patients with primary hepatic carcinoma and their families. A case-control study in Senegal, West Africa. Lancet 2:534–538, 1976

(*11*) MAUPAS P, LAROUZÉ B, LONDON WT, et al: Antibody to hepatitis B core antigen in patients with hepatocellular carcinoma. Lancet 2:9–11, 1975

(*12*) WOGAN GN: Mycotoxins and liver injury. *In* The Liver. International Academy of Pathology Monograph (Gall EA, Mostofi FK, eds.). Baltimore: Williams & Wilkins, 1973, pp 161–181

(*13*) TOMATIS L, MAGEE P, SHUBIK P: Induction of liver tumors in the Syrian golden hamster by feeding dimethylnitrosamine. J Natl Cancer Inst 33:341–345, 1964

(*14*) MUGERA GJ, NDERITO PC: Cycads as hepatic carcinogens in Kenya. *In* Cancer in Africa. Nairobi, Kenya (Clifford B, Linsell A, Timms GL, eds.). Nairobi: East African Publ House, 1968, pp. 323–326

(*15*) VYAS GN, SHULMAN NR: Hemagglutination assay for antigen and antibody associated with viral hepatitis. Science 170:332–333, 1970

(*16*) HOOFNAGLE J, OJERCTY R, HI L, et al: Antibody to hepatitis B core antigen. A sensitive indicator of hepatitis B virus replication. N Engl J Med 290:1336–1340, 1974

(*17*) RUOSLAHTI E, SEPPALA M: Studies of carcino-fetal proteins. III. Development of a radioimmunoassay for α-fetoprotein. Demonstration of α-fetoprotein in serum of healthy human adults. Int J Cancer 8:374–383, 1971

(*18*) MASSEYEFF R, SANKALE M, ONDE M, et al: Valeur de la recherche de l'α-fetoproteine sérique pour le diagnostic du cancer primitif du foie. Bull Soc Med Afr Noire Lang Fr 13:537–548, 1968

(*19*) NISHI S, HIRAI H: Radioimmunoassay of α-fetoprotein in hepatoma, other liver diseases and pregnancy. Gann Monogr 14:79–87, 1973

(*20*) PURVES LR, MANSO C, TORRES FO: Serum α-fetoprotein levels in people susceptible to primary liver cancer in Southern Africa. Gann Monogr 14:51–66, 1973

(*21*) ECONOMIDOU I, HADZIYANNIS S, PARASKEVAS E, et al: Australia antigen (HBsAg) carriers in a Greek community. Studies of transaminase (SGPT) levels. Res Commun Chem Pathol Pharmacol 10:703–713, 1975

(*22*) WALPOLE RE: Introduction to Statistics. New York: Macmillan, 1974

Reprinted from THE JOURNAL OF PEDIATRICS, St. Louis
Vol. 106, No. 5, pp. 843-849, May, 1985 (Printed in the U.S.A.)

Lack of perinatal transmission of hepatitis B virus infection in Senegal, West Africa

Between 1977 and 1980, 1442 pregnant women in Thies, Senegal, were tested for serologic markers of hepatitis B virus (HBV) infection. Of these, 9.8% were $HB_sAg(+)$, 59.9% were anti-$HB_s(+)$, and 15.6% had anti-HB_c alone. Of 116 $HB_sAg(+)$ pregnant women, only 19.8% were $HB_eAg(+)$, a much lower proportion of infectious carriers than seen in Asian populations. Cord blood from 1353 babies was $HB_sAg(-)$, implying that the babies were not infected prior to birth. Four hundred sixty-two babies, including 88 born to $HB_sAg(+)$ mothers, were observed for 2 weeks to 38 months after birth. In contrast to observations in Asia, none of the babies became $HB_sAg(+)$ before 5 months of age, and only three of the 16 born to $HB_eAg(+)$ mothers became $HB_sAg(+)$ within the first year of life; all three developed chronic infections (i.e., $HB_sAg(+)$ for $\geq$6 months. In the second year of life, six of 34 babies born to $HB_sAg(+)$, $HB_eAg(-)$/anti-$HB_e(-)$ mothers became infected with HBV, and four of the six developed chronic infections. During the first 3 years of life, infections occurred at a higher rate in infants born to $HB_sAg(+)$ (17%) than to $HB_sAg(-)$ (4%) women. The latter group of infants included 4.0% of those born to anti-$HB_s(+)$ mothers, 4.6% born to anti-$HB_cAg(+)$, and 3.2% born to uninfected women. These observations indicate that HBV infections in Senegal usually do not occur perinatally, but do occur at high incidence later in infancy and childhood. Such infections can be prevented by the use of hepatitis B vaccine alone; administration of hepatitis B immune globulin should not be needed. (J PEDIATR 106:843, 1985)

Evelyne Marinier, M.D., M.S., Veronique Barrois, M.D., Bernard Larouze, M.D., W. Thomas London, M.D., Ardean Cofer, B.A., Lamine Diakhate, M.D., and Baruch S. Blumberg, M.D.
Paris, France, Philadelphia, Pennsylvania, and Dakar, Senegal, West Africa

HEPATITIS B VIRUS INFECTION and chronic carriers of the virus are common in many tropical countries, particularly Asia and sub-Saharan Africa.[1,2] A high incidence of infection may also occur in Arctic communities.[3,4] There is now substantial evidence[5] (recently reviewed) that chronic infection with HBV is etiologically associated with the development of primary hepatocellular carcinoma. In areas endemic for HBV, PHC is the most common or one of the most common malignant tumors. In a study in Taiwan of chronic HBV carriers and controls, the relative risk of PHC in carriers was >200 times that of noncarriers.[6] More than 95% of cases of PHC have developed in chronic carriers.[6,7]

From Institut de Medecine et d'Epidemiologie Tropicales, Hopital Claude Bernard, Paris; Institute for Cancer Research, Fox Chase Cancer Center, Philadelphia; and Centre National de Transfusion Sanguine, Dakar.

Supported by Grants CA-06551, RR-05539, and CA-06927 from the National Institutes of Health, by an appropriation from the Commonwealth of Pennsylvania, and by Inserm 119006 to the Institut de Medecine et d'Epidemiologie Tropicales.

Submitted for publication June 7, 1984; accepted Sept. 28, 1984.

Reprint requests: Baruch S. Blumberg, M.D., Institute for Cancer Research, 7701 Burholme Ave., Philadelphia, PA 19111.

See related article, p. 777.

HBV	Hepatitis B virus
PHC	Primary hepatocellular carcinoma

In a study in Senegal, we found that mothers of patients with PHC were more likely to be carriers (70%) than mothers of matched individuals who did not have PHC (10%).[8] This finding was subsequently confirmed in Korea[9] and in Taiwan.[7] From this observation[8] we inferred that many persons who eventually develop PHC have been infected by their mothers early in life.

Table I. Number of children's sera collected at different ages according to mother's HBV response

Status of mothers	*Age of babies* 1 wk to 3 mo	3 to 5 mo	5 to 12 mo	1 to 2 yr	2 to 3 yr	Total
$HB_sAg(+)$	26	16	40	42	18	142
Anti-$HB_s(+)$	153	55	94	52	26	380
Anti-$HB_c(+)$ only	42	13	28	11	6	100
Negative	36	12	27	9	5	89
Total	257	96	189	114	55	711

Some babies are represented more than once if specimens were obtained from same child at different ages; 462 babies are included.

Table II. Effect of mothers' HBV status on babies' HBV status at birth

HBV status (cord blood) of babies	*HBV status of mothers* $HB_sAg(+)$	Anti-$HB_s(+)$	Negative	Total
$HB_sAg(+)$	0*	0*	0	0*
Anti-$HB_s(+)$	8	707	30	745
Negative	124	108	376	608
Total	132	815	406	1353

Association is highly significant: $P < 0.0001$.
*At beginning of study, cord blood sera of seven babies were found to be $HB_sAg(+)$. This apparently was caused by contamination of baby's cord blood sample with maternal blood. Procedures for collection of cord blood were revised and carefully followed. No positive cord blood sera were detected thereafter, and these seven mother-baby pairs were excluded.

Studies in Asia,[10, 11] Africa,[12, 13] and elsewhere[14-17] have shown that maternal transmission plays a large role in transmission of HBV and in the development of the carrier state. Maternal transmission, however, does not necessarily mean that infection takes place in the prenatal period. Carrier mothers who are also $HB_eAg(+)$ are more likely to transmit HBV to their newborn infants than those who are anti-$HB_eAg(+)$ or do not have either HB_eAg marker,[18, 19] but little is known about this relation in African families.[20]

Infection at an early age, particularly maternally transmitted infection, may increase the risk of developing chronic liver disease and PHC, and decrease the age of onset. Hence, the mother-child relationship in respect to HBV infection is very important in understanding the pathogenesis of chronic liver disease and PHC, and in developing methods for the control of infection, particularly use of the newly available hepatitis B vaccine. We report a prospective study of mother-child-HBV relationships conducted in Senegal, West Africa.

METHODS

Blood specimens were obtained from mothers visiting a mother-child clinic and from their offspring at about the time of birth and at periods up to 3 years after delivery. The clinic is in Thies, the second largest urban center in Senegal, located 80 km east of Dakar, the capital. About 7000 deliveries per year are performed in the clinic in-patient facility. Between April 1977 and April 1980, 1442 women admitted for delivery were interviewed, a questionnaire completed, and specimens of maternal and cord blood obtained. Only time and personnel available determined which women were selected for study. The information obtained included name, address, ethnicity, age, and previous history of jaundice, and birth date, birth weight, and sex of the offspring.

Maternal sera were collected and immediately tested for HB_sAg at the Centre National de la Transfusion Sanguine in Dakar by reverse passive hemagglutination. All babies born to $HB_sAg(+)$ mothers were included in the follow-up. Follow-up was also conducted in $HB_sAg(-)$ mothers and their offspring. The intention was to choose as controls the next two $HB_sAg(-)$ mothers who lived in the same district as the $HB_sAg(+)$ women. As the study proceeded, many babies could not be followed, and we ultimately accepted as controls all the mothers who agreed to keep their babies in the study after the first (15 day) follow-up visit, regardless of the district. Four times as many $HB_sAg(-)$ as $HB_sAg(+)$ women were studied. Controls came from all districts where $HB_sAg(+)$ women lived. Blood collections from the children were planned for intervals of approximately 15 days; 3, 6, and 12 months; and then yearly for up to 3 years; and from the mothers at 6 months after delivery (Table I). Most babies submitted two or more follow-up samples; in some, all the planned samples were collected.

All sera, including those tested in Dakar, were tested at the Institute for Cancer Research in Philadelphia for HB_sAg by radioimmunoassay (Ausria II; Abbott Laboratories, North Chicago, Ill.). Anti-HB_s was determined by passive hemagglutination[21] against red blood cells coated with HB_sAg subtypes ad and ay, separately, and were scored as positive if titers were ≥1:4 and could be inhibited by $HB_sAg(+)$ serum (Electro-Nucleonics, Bethesda, Md.). Titers of 1:4 and 1:8 were categorized as low; ≥1:16 but <1:128 as medium; and ≥1:128 as high. Anti-HB_c and HB_eAg were determined by RIA (Corab; Abbott Laboratories) (Abbott Laboratories).[22] Anti-HB_e was only tested by a micro-micro immunodiffusion method developed in our laboratory (personal communication). All sera were assayed for HB_sAg and anti-HB_s. Those negative for both were then tested for anti-HB_cAg. One hundred sixteen of the mothers' sera positive for HB_sAg were tested for HB_eAg and anti-HB_eAg.

The "survival" analyses were done by the method of Kaplan and Meier.[23]

RESULTS

Mothers and children studied. A total of 462 babies (242 boys), including 88 born to $HB_sAg(+)$ mothers, were observed for 2 weeks to 38 months. One to six blood specimens were obtained for each baby (Table I). Babies born to 88 $HB_sAg(+)$ and 374 $HB_sAg(-)$ women were observed. Ten infant deaths were reported during the perinatal period (six babies of carriers and four born to $HB_sAg(-)$ women); the cause of death was not recorded.

A total of 1442 pregnant women were tested. Of these, 141 (9.8%) were $HB_sAg(+)$, 864 (59.9%) had anti-HB_s, and 225 (15.6%) had anti-HB_c alone. Only 212 (14.7%) did not have any evidence of infection. The responses did not differ significantly by ethnic group. Total rate of HBV infection increased with age (P <0.0016), but the prevalence of HB_sAg remained about the same (Fig. 1). Only one case of clinical liver disease was found; a woman who was $HB_sAg(+)$ on first testing had PHC and died 2 months after the delivery of her child.

The presence of HB_eAg and anti-HB_e were determined in 116 of the mothers who were $HB_sAg(+)$. Of these, 23 (19.8%) were $HB_eAg(+)$, 24 (20.7%) were anti-$HB_e(+)$, and the remaining 69 (59.5%) had neither. Because HB_eAg was tested by RIA, but anti-HB_e was tested by the micro-micro immunodiffusion method, the high proportion of $HB_eAg(-)$, anti-$HB_e(-)$ women probably reflects the relative insensitivity of the micro-micro immunodiffusion method.

Findings at birth. There were 1353 pairs of mothers and their newborn babies, but the questionnaire information was not complete for all pairs. Cord blood samples of all babies were $HB_sAg(-)$, implying absence of in utero infection (Table II). There were 815 mothers who had anti-HB_s, and 707 (86%) of their offspring also had

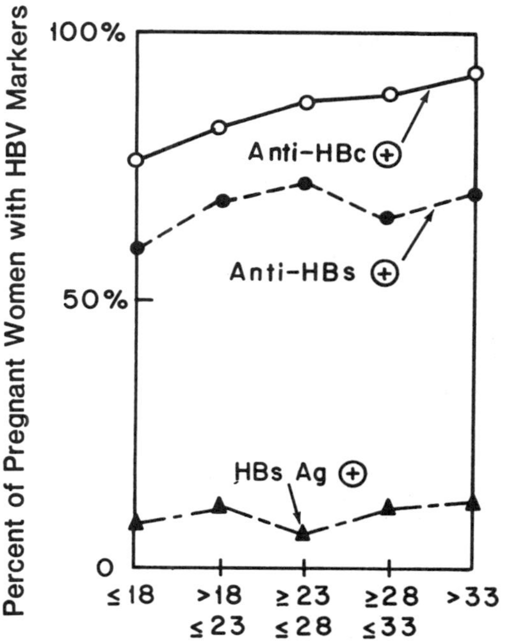

Fig. 1. Cumulative HBV infection rate in 1009 pregnant women related to age in years. Percentage of women who remained uninfected in five age categories are 24%, 20%, 14%, 13%, 8%.

Table III. Relation of anti-HB_s titers in mothers to anti-HB_s titers in babies at birth

Anti-HB_s titers in babies	*Anti-HB_s titers in mothers*			
	Low	*Medium*	*High*	*Total*
Low	187	75	5	267
Medium	31	213	35	279
High	1	29	131	161
Total	219	317	171	707

Low titer, 1:4 to 1:8 by passive hemagglutination; medium titer, 1:16 to 1:64; high titer, ≥128. $P < 0.0001$.

anti-HB_s (Table II). Further, the titers of anti-HB_s in mothers and their children were similar (Table III); 531 of the children and mothers had about the same titer. For the remaining 176, the children's antibody titers were higher or lower than the mothers', but in most instances the differences were equal to or less than twofold, which could be accounted for by variations in the assay system. Thirty children with $HB_sAg(-)$, anti-$HB_s(-)$ mothers had anti-HB_s (Table II). In all 30, the anti-HB_s titers were low (≤1:16), suggesting to us that anti-HB_s below the level of detectability was probably present in the maternal sera.

Events after birth Babies who were anti-$HB_s(+)$ at

Table IV. Transient* and chronic† infections in children at different times after birth according to HBV status of mother

		HBV status of mothers										
		Noncarrier				*Carrier*						
		Uninfected‡		*anti-HB$_c$ (+)§*		*anti-HB$_s$(+)*		*HB$_s$Ag(+)*		*Total*		*Cumulative*
Age of babies	*Infection events*	*n*	*%*	*n*	*%*	*n*	*%*	*n*	*%*	*n*	*%*	*total*
1 wk to 5 mo	Transient	0		0		0		0		0		0
	Chronic	0		0		0		0		0		0
Total sera collected		62		65		247		88		462		462
5 to 12 mo	Transient	0		0		1		0		1		1
	Chronic	0		1		2		3		6		6
Total sera collected		33		30		123		78		264		462
12 to 24 mo	Transient	1		0		2		2		5		6
	Chronic	0		1		2		4		7		13
Total sera collected		14		15		68		55		152		462
24 to 38 mo	Transient	0		0		0		2		2		8
	Chronic	1		1		3		4		9		22
Total sera collected		5		5		25		17		52		462
Total												
	Transient	1	1.6	0		3	1.2	4	4.5	8	1.7	
	Chronic	1	1.6	3	4.6	7	2.8	11	12.5	22	4.7	
	Transient + Chronic	2	3.2	3	4.6	10	4.0	15	17.0	30	6.4	
Total sera collected		62		65		247		88		462		

In each time period, only new events are recorded, and babies infected are not counted in next period. Total collected, number of babies at risk of HBV infection collected during indicated period.

*Child anti-HB$_s$(−) at birth, and subsequently anti-HB$_s$(+) at titer ≥1:4 or anti-HB$_s$(+) after HB$_s$Ag(+) test. Babies anti-HB$_s$(+) at birth who had twofold increases in titer in two subsequent samples or who became anti-HB$_s$(+) after two anti-HB$_s$(−) tests.

†Child became HB$_s$Ag(+), and remained positive on all subsequent tests or on last sample tested.

‡HB$_s$Ag(−), anti-HB$_s$(−), anti-HB$_c$(−).

§Anti-HB$_s$(−), HB$_s$Ag(−).

birth lost this antibody within 6 to 9 months of age (Table IV). As might be expected, the children with high titers were slowest to lose antibody.

There were no infections prior to 5 months of age, and this was independent of the mothers' response category.

During the follow-up periods (maximum 38 months), 30 infections (eight transient and 22 chronic) occurred. The carrier mothers had children with the highest frequency of both chronic and transient infection events. Seventeen percent of the offspring of carrier mothers became infected, whereas only 4% were infected in the other categories ($P < 0.0001$).

The children of mothers with any marker became infected earlier and in higher frequency than the offspring of mothers who were uninfected (Fig. 2). Most of the infections occurred in the children of carrier mothers. In the mothers who were anti-HBs(+), children did not become infected until 9 months of age. (The only exception was probably a laboratory error, in a child who had a high titer of anti-HB$_s$ in cord blood was anti-HB$_s$(−) at 15 days, and developed a low titer of anti-HB$_s$ at 5 months of age.)

HB$_e$Ag and anti-HB$_e$ were determined in 87 HB$_s$Ag(+) mothers whose babies returned for follow-up. The offspring of the HB$_e$Ag(+) mothers became infected before the children of carrier mothers with neither HB$_e$Ag nor anti-HB$_e$, and the offspring of carriers with anti-HB$_e$ were the last to become infected. None of 16 babies born to HB$_e$Ag(+) women became HB$_s$Ag(+) before 5 months of age. By the end of the 38 months, 25% of the babies born to HB$_e$Ag(+) mothers had become infected, and all of these were chronic infections (Table V). The anti-HB$_e$(+) and HB$_e$Ag(−)/anti-HB$_e$(−) mothers had fewer chronically infected babies than the HB$_e$Ag(+) mothers, but more than HB$_s$Ag(−) mothers (8% compared with 4%).

DISCUSSION

The frequency of HBV markers in our study population of mothers is similar to that found in other African females.[12, 13, 24, 25] The frequency is lower than in Taiwan[10] but higher than in Japan[11] and in western countries.[15-17, 26, 27]

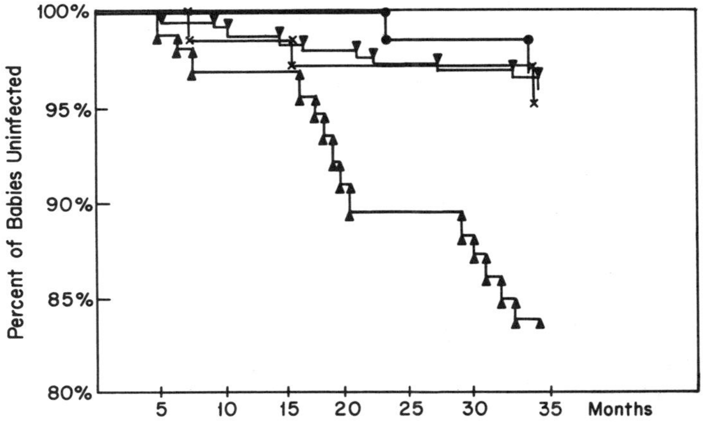

Fig. 2. Survival curves showing percentage of children remaining uninfected at successive time intervals after birth according to HBV status of mothers. Infected mothers had more infected offspring than did uninfected mothers, and most infections occurred in offspring of carrier mothers. ▲, Babies born to $HB_sAg(+)$ mothers (83.0% uninfected at 35 months); X, babies born to anti-$HB_c(+)$ mothers (95.4% uninfected at 35 months); △, babies born to anti-$HB_s(+)$ mothers (95.6% uninfected at 35 months); ●, babies born to uninfected mothers (96.8% uninfected at 35 months).

Table V. Infection of offspring of carrier ($HB_sAg(+)$) mothers according to time after birth and HB_eAg, anti-HB_e response of mothers

Age of babies	Infection events	e status of $HB_sAg(+)$ mothers: $HB_eAg(+)$ n	%	anti-$HB_e(+)$ n	%	$HB_eAg(+)$, anti-$HB_e(-)$ n	%	Total n	%	Cumulative total
1 wk to 5 mo										
	Transient	0		0		0		0		0
	Chronic	0		0		0		0		0
	Risk(n)	16		20		51		87		87
5 to 12 mo										
	Transient	0		0		0		0		0
	Chronic	3		0		0		3		3
	Risk(n)	15		15		47		77		87
12 to 24 mo										
	Transient	0		0		2		2		2
	Chronic	0		0		4		4		7
	Risk(n)	8		12		34		54		87
>24 mo										
	Transient	0		2		0		2		4
	Chronic	1		1		1		3		10
	Risk(n)	3		5		8		16		87
Total										
	Transient	0	0	2	10.0	2	3.9	4	4.6	
	Chronic	4	25.0	1	5.0	5	9.8	10	11.5	
	Transient + chronic	4	25.0	3	15.0	7	13.7	14	11.1	
	Risk(n)	16		20		51		87		

The presence of HB_eAg in carriers (an indicator of infectivity) is significantly lower in our African study population than in populations from the Far East.[18, 19, 28] There are several possible explanations for these differences.

First, there may be differences in the sensitivity of the technique we used for measuring HB_eAg and that of other investigators. That is, we may not have detected some significant number of $HB_eAg(+)$ sera. However, this is unlikely, because the commercial RIA (Abbott Laboratories) is a highly sensitive, reliable test.[22] Moreover, the low rate of transmission of HBV from $HB_sAg(+)$, $HB_eAg(-)$ mothers to their babies is consistent with the assay for HB_eAg being accurate.

Second, HB_eAg is associated with higher HB_sAg titers,[29] viral DNA polymerase,[29] HBV DNA,[30] and probably the amount of whole virus produced. Therefore, infected Africans may produce less virus than Asians and others. If this is true, then there may be geographic or genetic differences in virus cell interactions.

The offspring of carrier mothers were more likely to become infected than the children of the other mothers; this confirms previous studies in Senegal and elsewhere.[13-17] Many of the offspring of non-HB_sAg carrier mothers also became infected, usually later, presumably from contact with other sibs or persons outside the home whom they encountered more frequently with increasing age.

Among the $HB_sAg(+)$ mothers, there were apparent differences in infectivity for their babies in the following order: (1) $HB_eAg(+)$, (2) neither $HB_eAg(+)$ nor anti-$HB_e(+)$, and (3) anti-$HB_e(+)$. This is consistent with studies in other parts of the world. The role of HB_eAg in predicting infectivity appears to be independent of its lower frequency in this population. However, the frequency of transmission of HBV from $HB_eAg(+)$ mothers to their offspring is less in Africa than in Asia. In Asia, about 90% of babies born to $HB_eAg(+)$ women become $HB_sAg(+)$,[18, 19, 31] whereas in the Senegalese population in our study, only 25% were infected during the entire period of observation (maximum 36 months, average about 2 years).

None of the infants had serologic markers of HBV infection before 5 months of age. This is somewhat different from babies in western populations[15, 16, 26, 32] and particularly from babies in Asian populations.[10, 11] In Taiwan, for example, more than 90% of babies born to $HB_sAg(+)$ mothers became $HB_sAg(+)$ by 3 months of age.

These results can now be viewed from the standpoint of recommending strategies for mounting a prevention program including the use of the hepatitis B vaccine. The offspring of carrier mothers become infected earlier than other children, but the offspring of other mothers also have a high likelihood of infection later in life. Mothers with anti-HB_s will transmit the antibody to their children, particularly if it is in high titer in the mothers. However, anti-HB_s appears to offer only a minimal amount of protection, and for only a few months. An observation of great importance is that none of the children we observed became infected before about 5 months of age. Therefore, vaccination before that age is advisable; it is known that children can muster an effective response to the vaccine during the first few months of life.[33] Studies initiated by the late Professor Maupas and his colleagues indicated that such vaccinations prevented almost all HBV infections in the first 2 years of life.[33] It appears that in Africa it will not be necessary to screen mothers to detect antigen or anti-HB_s, because all children in high-risk areas should be vaccinated regardless of the response of their mothers, and this vaccination should be done between 1 and 3 months of age. Furthermore, because infection does not occur until $\geq$5 months after birth, the use of hepatitis B immune globulin in babies born to $HB_eAg(+)$ mothers is not warranted.

Before any decision on the use of this vaccine is considered, it is necessary to evaluate the possible benefits against risks and costs of vaccination. The risk appears to be small. There is now experience with more than 500,000 vaccinations, and there do not appear to be any side effects of consequence. In West Africa, although hepatitis infection and its sequelae are important medical problems, malaria, schistosomiasis, infectious diarrheas, onchocerciasis, malnutrition, and other diseases may have higher priorities for public health resources.[34] However, although hepatitis may not be the most important disease problem in West Africa, it may be one of the most important solvable disease problems. Population-based information of the kind provided by our study should be useful for planning the detailed tactics and strategy of prevention programs.

We thank Dr. Soumah, Mr. Souleymane Sy, Mr. Mamadou Sene, Dr. Bathilly, and Dr. Mustapha Sow for help and support.

REFERENCES

1. Blumberg BS, Melartin L, Guinto RA, Werner B: Family studies of a human serum isoantigen system (Australia antigen). Am J Hum Genet **18**:594, 1966.
2. Szmuness W, Prince AM, Diebolt G, LeBlanc L, Baylet R, Masseyeff R, Linhard J: The epidemiology of hepatitis B infection in Africa: Results of a pilot survey in the Republic of Senegal. Am J Epidemiol **98**:104, 1973.
3. Skinhøj P, McNair A, Andersen ST: Hepatitis and hepatitis B antigen in Greenland. Am J Epidemiol **99**:50, 1974.
4. Barrett DH, Burks JM, McMahon B, Elliott S, Berquist KR, Bender TR, Maynard JE: Epidemiology of hepatitis B in two Alaskan communities. Am J Epidemiol **105**:118, 1977.
5. Blumberg BS, London WT: Hepatitis B virus: Pathogenesis and prevention of primary cancer of the liver. Cancer **50**:2657, 1982.
6. Beasley RP, Lin C-C, Hwang L-Y, Chien C-S: Hepatocellular carcinoma and hepatitis B virus: A prospective study of 22,707 men in Taiwan. Lancet **2**:1129, 1981.
7. Beasley RP: Hepatitis B virus as the etiologic agent in hepatocellular carcinoma: Epidemiologic considerations. Hepatology **2**:21S, 1982.
8. Larouze B, London WT, Saimot G, Werner BG, Lustbader ED, Payet M, Blumberg BS: Host responses to hepatitis B infection in patients with primary hepatic carcinoma and their

families: A case/control study in Senegal, West Africa. Lancet **2**:534, 1976.
9. Hann HL, Kim CY, London WT, Whitford P, Blumberg BS: Hepatitis B virus and primary hepatocellular carcinoma: Family studies in Korea. Int J Cancer **30**:47, 1982.
10. Stevens CE, Beasley RP, Tsui J, Lee W: Vertical transmission of hepatitis B antigen in Taiwan. N Engl J Med **292**:771, 1975
11. Okada K, Yamada T, Miyakawa Y, Mayumi M: Hepatitis B surface antigen in the serum of infants after delivery from asymptomatic carrier mothers. J PEDIATR **87**:360, 1975.
12. Barrois V, Larouze B, Drew JS, London WT, Blumberg BS, Payet M: Hepatitis B antigen and its relation to alpha fetoprotein. *In* Fishman WH, Sell S, editors: Onco-development gene expression. New York, 1976, Academic Press.
13. Barin F, Perrin J, Chotard J, Denis F, N'Doye R, DiopMar I, Chiron JP, Coursaget P, Goudeau A, Maupas P: Cross-sectional and longitudinal epidemiology of hepatitis B in Senegal. Prog Med Virol **27**:148, 1981.
14. Derso A, Boxall EH, Tarlow MJ, Flewett TH: Transmission of HB_sAg from mother to infant in four ethnic groups. Br Med J **1**:949, 1978.
15. Papaevangelou G, Hoofnagle JH: Transmission of hepatitis B virus infection by asymptomatic chronic HB_sAg carrier mothers. Pediatrics **63**:602, 1979.
16. Skinhøj P, Cohn J, Bradburne AF: Transmission of hepatitis type B from healthy HB_sAg positive mothers. Br Med J **1**:10, 1975
17. Schweitzer IL, Mosley JW, Ashcavai M, Edwards V, Overby LB: Factors influencing neonatal infection by hepatitis B virus Gastroenterology **65**:277, 1973.
18. Stevens CE, Neurath RA, Beasley RP, Szmuness W: HB_eAg and anti-HB_e detection by radioimmunoassay: Correlation with vertical transmission of hepatitis B virus in Taiwan. J Med Virol **3**:237, 1979.
19. Beasley RP, Trepo C, Stevens CE, Szmuness W: The e antigen and vertical transmission of hepatitis B surface antigen. Am J Epidemiol **105**:94, 1977.
20. Prince MA, White T, Pollock N, Riddle J, Brotman B, Richardson L: Epidemiology of hepatitis B infection in Liberian infants. Infect Immun **32**:675, 1981.
21. Vyas GN, Shulman NR: Hemagglutination assay for antigen and antibody associated with viral hepatitis. Science **170**:332, 1970
22. Mushawar IK, Overby LR, Frosner G, Deinhardt F, Ling CM: Prevalence of hepatitis B e antigen and its antibody as detected by radioimmunoassays. J Med Virol **2**:77, 1978.
23. Kaplan EL, Meier P: Non-parametric estimation from incomplete observations. J Am Stat Assn **53**:457, 1958.
24. Larouze B, Feret E, London W, Marinier E, Lustbader E, Diop B, Blumberg B: Epidemiology of hepatitis B in Gawan, a rural community in Senegal, in preparation.
25. Drucker J, Barin F, Chiron JP, Coursaget P, Goudeau A: Hepatitis B virus infection in infants of anti-HB_s positive mothers. Lancet **2**:259, 1981.
26. Papaevangelou G, Hoofnagle J, Kremastinou J: Transplacental transmission of hepatitis B virus by symptom free chronic carrier mothers. Lancet **2**:746, 1974.
27. Dupuy JM, Giraud P, Dupuy C, Drouet J, Hoofnagle J: Hepatitis B in children. J PEDIATR **92**:200, 1978.
28. Shiraki K, Yoshihara N, Sarukai M, Eto T, Kawana T: Acute hepatitis B in infants born to carrier mothers with the antibody to HB_eAg. J PEDIATR **97**:768, 1980.
29. Nordenfelt E: Dane particles, DNA polymerase and e antigen in two different categories of hepatitis B antigen carriers. Intervirology **5**:225, 1975.
30. Werner B, O'Connell AP, Summers J: Association of e antigen with Dane particle DNA in sera from asymptomatic carriers of hepatitis B surface antigen. Proc Natl Acad Sci USA **74**:2149, 1977.
31. Okada K, Kamiyama I, Inomata M, Imai M, Miyakawa Y, Mayumi M: e Antigen and anti-e in the serum of asymptomatic carrier mothers as indicators of positive and negative transmission of hepatitis B virus to their infants. N Engl J Med **294**:746, 1976.
32. Gerety RJ, Schweitzer IL: Viral hepatitis type B during pregnancy, the neonatal period, and infancy. J PEDIATR **90**:368, 1977.
33. Barin F, Goudeau A, Denis F, Yvonnet B, Chiron JP, Coursaget P, Diop Mar I: Immune response in neonates to hepatitis B vaccine. Lancet **1**:251, 1982.
34. Larouze B, Froment A, London W, Feret E, Marinier E, Lustbader E, Diop B, Blumberg B: Hepatocellular carcinoma: A public health problem in tropical countries. *In* Maupas P, Guesry P, editors: Hepatitis B vaccine. INSERM Symposium 18. Amsterdam, 1980, Elsevier/North Holland Biomedical Press, p 195.

AMERICAN JOURNAL OF EPIDEMIOLOGY
Vol. 125, No. 1
Printed in U.S.A.

EPIDEMIOLOGY OF HEPATITIS B VIRUS INFECTION IN THE RURAL COMMUNITY OF TIP, SENEGAL

ELISABETH FERET,[1,2] BERNARD LAROUZE,[2] BIRON DIOP,[3] MOUSTAPHA SOW,[4] W. THOMAS LONDON,[1] AND BARUCH S. BLUMBERG[1]

Feret, E., B. Larouze, B. Diop, M. Sow, W. T. London, and B. S. Blumberg (Fox Chase Cancer Center, Philadelphia, PA 19111). Epidemiology of hepatitis B virus infection in the rural community of Tip, Senegal. *Am J Epidemiol* 1987;125:140–9.
The distribution of responses to hepatitis B virus infection in the rural community of Tip, Senegal, has been determined. Children are rarely persistently infected before two years of age, at which time about 25% of the population are infected. Uninfected members continue to be at risk. Vaccination programs instituted at an early age could prevent infection in most of the population and considerably decrease the risk of chronic liver disease and primary cancer of the liver in adults.

hepatitis B virus; hepatoma; vaccination

Postnecrotic cirrhosis and primary hepatocellular carcinoma are major causes of morbidity and mortality in the adult, particularly the male, populations of sub-Sahara Africa and east Asia. Persistent infection with hepatitis B virus plays a central role in the etiology and pathogenesis of both of these diseases (1). A large, ongoing prospective study in Taiwan by Beasley et al. (2) and Beasley and Hwang (3) indicates that chronic carriers of hepatitis B have a relative risk of developing primary hepatocellular carcinoma that is about 200 times that of noncarriers. More than 98 per cent of the cases of primary hepatocellular carcinoma (attributable risk) have occurred in carriers in the study by Beasley and Hwang. These investigations have estimated that the lifetime risk for a carrier to develop and die of primary hepatocellular carcinoma is about 40 per cent and that the risk of dying of cirrhosis is about 15 per cent (3). Thus, in Taiwan, and probably in other endemic areas, if persistent infection with hepatitis B could be prevented, almost all cases of primary hepatocellular carcinoma would also be prevented.

A safe and effective vaccine against hepatitis B virus introduced from our laboratory (4) is now available and is being used in the United States and other countries to protect persons at high risk of infection. Although clinical trials in Senegal have demonstrated the efficacy of the vaccine in an endemic area (5), optimal strategies for

Received for publication October 28, 1985, and in final form April 24, 1986.

Abbreviations: anti-HBc, antibody to hepatitis B core antigen; anti-HBs, antibody to hepatitis B surface antigen; HBsAg, hepatitis B surface antigen.

[1] Division of Clinical Research, Institute for Cancer Research, Fox Chase Cancer Center, Philadelphia, PA 19111. (Reprint requests to Dr. Baruch S. Blumberg.)
[2] Department d'Epidemiologie, Institut de Medecine et d'Epidemiologie Africaines, Hopital Claude Bernard, Paris, France.
[3] Clinique Medicale de l'Hôpital de Dantec, Faculte de Medecine et de Pharmacie de Dakar, Dakar, Senegal.
[4] Service de Medecine Interne, Hôpital Abbas N'Dao, Faculte de Medecine et de Pharmacie de Dakar, Dakar, Senegal.

This work was supported by USPHS grants CA-96551, RR-05539, CA-06927, and CA-22780 from the National Institutes of Health, by an appropriation from the Commonwealth of Pennsylvania, and by INSERM Programme de Recherches Coordonnees, Domaine du Cancer, Recherches Cliniques Contrat 119006.

its use in those areas of the world where persistent hepatitis B infection is a significant public health problem remain to be devised. To develop such strategies, more needs to be known about the epidemiology of hepatitis B in such areas. To that end, we have carried out extensive studies of risk factors related to hepatitis B infection and the types of responses to infection in a geographically defined population in a rural area of Senegal, West Africa, where both primary hepatocellular carcinoma and hepatitis B infection are endemic.

Materials and methods

The study was conducted in the rural community of Tip, which is 15 km from the town of Touba and 200 km east of Dakar. The Tip community included the village of Tip and five other villages within a 5-km radius (Boustane, Nassrou, Tewel, Tao-Fekh, and Ourcogne). From preliminary surveys in the nearby area of Gawan, Senegal, we established that 1) both hepatitis B infection and primary hepatocellular carcinoma were endemic to this region of Senegal; 2) the population was stable and ethnically representative of the majority of villages in the region; and 3) there were few people who drank alcoholic beverages, thereby eliminating a common confounding factor in other studies (Larouze et al., in preparation).

Population and environment

The vegetation of the Tip area, which consisted of virgin bush forest at the beginning of the century, is now a sparse, thorny woodland on sandy soil. There is no surface water. All the rain occurs in sporadic storms from July–September.

The modern settlement of the study area was by members of the Woloff ethnic group who came from about 200 km away and belonged to the Islamic sect of Mourides. The settlers arrived around 1920, in the area where the Peulh nomadic shepherds were raising cattle. During the same period, when Christianity was adopted by some of the Serer, those who remained animists also moved to this area and now live in hamlets around the Mouride villages. The social cohesion of the Mouride sect is strengthened by their religious organization and is focused on the motto "religion and work." In Mouride villages, the role of religious authorities is very important (6).

Most of the Mouride villages are built according to the same basic plan, a system of square compounds arranged around a central square in which the mosque, the meeting place, and the well or fountain are located. Each compound is enclosed by a high fence made of millet straw. Each family, which is usually polygamous, has its own compound. Within the compound, each woman has a hut built of millet straw in which she and her children live. The father, the head of the family, has his own house, often built of mud-brick with an iron roof. There is a supply of running water from an artesian well, and all the villages studied, except Boustane, are connected to this water supply.

The only staple crops grown in this dry country are peanuts and millet. Two-thirds of the land is used for growing peanuts, which provide the economic base for the region. Most of the peanuts are sold for the production of peanut oil, and the others are eaten by the inhabitants. Given the local methods of storage, infection of peanuts with *Aspergillus flavus* probably is common in the area, and it is likely that the entire population is exposed to aflatoxins (7). One-third of the land is used for raising millet, which is a staple of the daily diet. In addition, the people eat rice, some vegetables, dried fish, and, after the rainy season, wild leaves. Mangoes grown in the west of the county are the only fruit available and then only for two months a year. A few sheep, goats, and chickens are raised, but their meat is reserved for feasts.

As in similar areas of Senegal, malnutrition, sickle cell anemia, and infectious diseases (including tuberculosis, leprosy, and falciparum malaria), are common. In childhood, diarrhea and measles frequently lead to malnutrition and increased mortality.

The three ethnic groups in the study area—Woloff, Peulh and Serer—live separately; have different languages, origins, and cultures; and do not intermarry. The Peulh live in small communities outside of the Woloff villages. Their huts are in the middle of the fields and are not enclosed by fences. The men herd cattle, and only a few Peulh cultivate land. Their protein intake may be higher because of drinking milk. Milk is a part of their regular diet, except at the end of the dry season. They are Moslem, but do not belong to the Mouride sect.

Most of the Serer of the area are animists who live in hamlets around the village of Tao-Fekh. They are poorer than the Woloff or the Peulh. Their compounds, which are enclosed by a high fence made of millet straw, contain more people in less space than is found in the living areas of the other ethnic groups.

Survey methods

The population in the Tip area was investigated between 1977 and 1979 and again in February 1981. All living members of the population were identified in 1977. For each person, we recorded the name, date of birth, ethnic group, identity of the parents, and relationship with other families. A complete obstetric history was obtained from each woman who had reached reproductive age. In February 1981, one of us (E. F.) returned to Tip and conducted a new census of the population. All persons who had died after the time of the first census was recorded.

Blood samples were collected yearly between 1977 and 1979, at the end of the dry season, using 10-cc dry vacutainers for adults and 5-cc vacutainers for children. Serum was separated from clotted blood on the evening of the day of collection. The sera were stored initially at 4 C but were placed at −20 C within two days, and the cold chain was not interrupted thereafter. Serum samples were shipped in dry ice to Philadelphia. All serologic tests and the statistical analyses were performed at the Institute for Cancer Research in Philadelphia.

Blood samples were collected at least once from a total of 3,034 persons. This was approximately 70 per cent of the population living in the six villages studied. For the two principal villages, Tip and Boustane, blood was obtained from 1,130 persons belonging to 85 families. This included 92 per cent of those aged five years or more and 50 per cent of the children aged less than five years. The age and sex distribution of the living populations for these two villages (figure 1) was similar to that of the total Senegalese population recorded in the 1974 national census. The distribution of the tested people by ethnic group is similar to the proportion of these groups in the population of the study area. Included were 2,870 Woloffs, 100 Peulhs, 63 Serers, and one person of unknown ethnicity.

Hepatitis B surface antigen (HBsAg) and antibody to hepatitis B core antigen (anti-HBc) were tested by radioimmunoassay (Abbott Laboratories, North Chicago, Il), and antibody to hepatitis B surface antigen (anti-HBs) was assayed by passive hemagglutination with HBsAg-coated red blood cells (obtained from Electronucleonics, Bethesda, MD). Titers of anti-HBs of 1:4 or 1:8 were considered positive only if they were completely inhibited by an HBsAg(+) serum. Such sera were categorized as "low titer" anti-HBs(+) for the subsequent analyses. Sera with titers ≥1:16 were classified as high-titer anti-HBs(+).

Pearson's chi-square and Fisher's exact test were performed to estimate the significance of differences observed.

RESULTS

The prevalence of serologic responses to hepatitis B infection by ethnic group is shown in table 1. The Serer show the highest rate of HBsAg positivity (17.5 per cent), and the Peuhl show the lowest (7 per cent); Woloffs are intermediate. Although the overall distribution of hepatitis B markers differs significantly by ethnic group, the

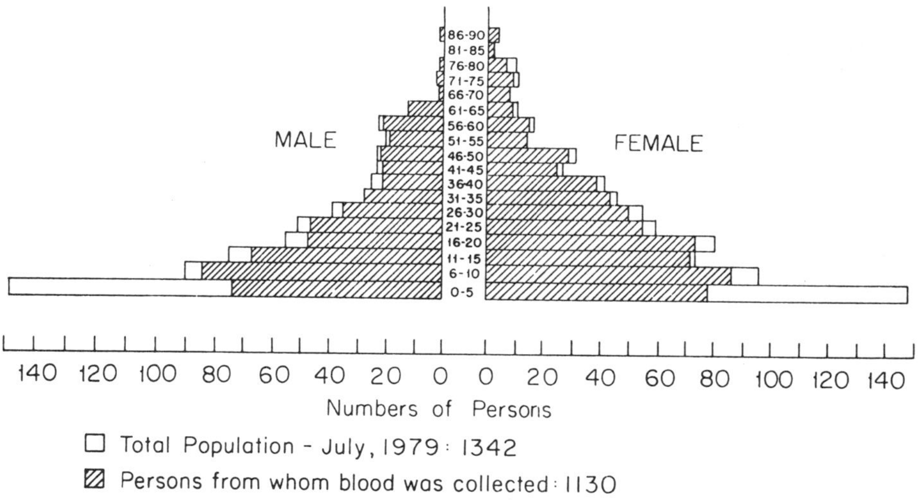

FIGURE 1. Distribution by age and sex of the total population and of the persons from whom blood was collected in the principal villages of Tip and Boustane, Senegal, 1977–1981.

TABLE 1

Prevalence of serologic markers of hepatitis B infection by ethnic group, Tip community, Senegal, 1977–1981

Ethnic group	HBsAg(+)		Anti-HBs(+)		Anti-HBc(+) alone		Uninfected		Total
	n	%	*n*	%	*n*	%	*n*	%	
Peulh	7	7	47	47	20	20	26	26	100
Woloff	344	12	1,327	46.2	599	20.9	600	20.9	2,870
Serer	11	17.5	21	33.3	11	17.5	20	31.7	63

differences for HBsAg positivity alone are not significant.

Table 2 shows the distribution of hepatitis B markers in six villages. There are significant variations not accounted for by age, sex, or ethnic differences in the population. The prevalence for HBsAg varies from 9.7 per cent in Tip to 14.3 per cent in Nassrou (p = not significant), and the prevalence for anti-HBs varies from 33.7 per cent in Tao-Fekh to 50.2 per cent in Tip ($p < 0.001$). The percentages of non-infected (HBsAg(−), anti-HBs(−), and anti-HBc(−)) persons vary from 15.7–26.1 per cent ($p < 0.05$).

The frequency of hepatitis B markers by age and sex is shown in figure 2. The presence of at least one serologic marker of hepatitis B infection is found with about the same frequency in males and females, but the distribution of markers is different. Males, as found in most other populations (8), are more likely to be HBsAg(+) (14.3 vs. 9.7 per cent, $p < 0.001$). The prevalence of hepatitis B markers by age is shown in figures 3 and 4. (Figure 4 shows the distribution in the younger people in greater detail.) The proportion of infected persons increases rapidly with age. Fifty per cent of the population is infected by age seven years, and 80 per cent is infected by age 15 years. Thereafter, the prevalence of infection rises gradually to age 60 years. The annual incidence of infection among the previously uninfected members of the population is 6 per cent per year.

The decline in HBsAg prevalence with age is shown more clearly when the proportion of HBsAg(+) persons is examined among those showing one or more markers

TABLE 2

Prevalence of serum markers of hepatitis B infection by village, Tip community, Senegal, 1977–1981

Village	HBsAg(+)		Anti-HBs(+)		Anti-HBc(+) alone		Noninfected		Total
	n	%	*n*	%	*n*	%	*n*	%	
Tip	82	9.7	425	50.2	125	14.8	215	25.4	847
Boustane	30	10.6	128	45.2	51	18	74	26.1	283
Nassrou	109	14.3	334	43.9	165	21.7	153	20.1	761
Tewel	17	10.6	76	47.2	39	24.2	29	18	161
Tao-Fekh	50	13.8	122	33.7	111	30.7	79	21.8	362
Ourcogne	49	10.8	226	49.9	107	23.6	71	15.7	453

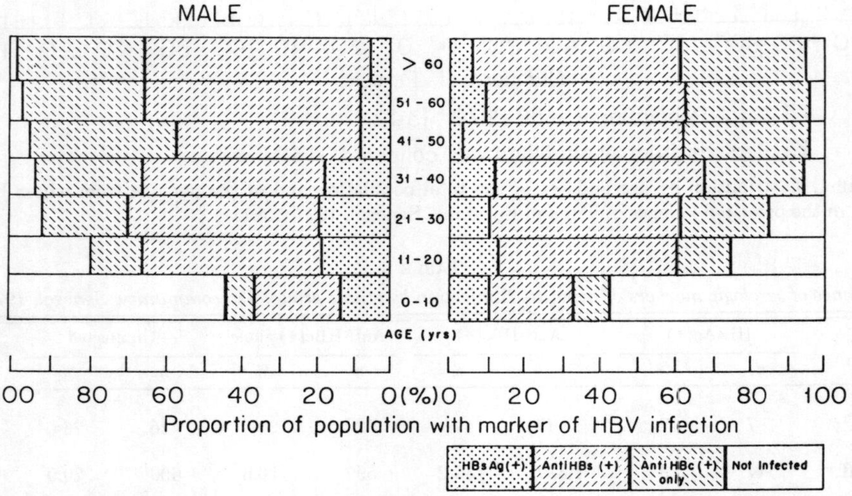

FIGURE 2. Frequency of markers of hepatitis B infection by age and sex, Tip community, Senegal, 1977–1981.

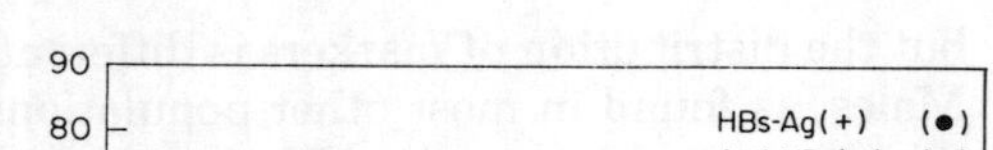

FIGURE 3. Prevalence of hepatitis B serologic responses by age. $n = 3{,}034$ persons. Tip community, Senegal, 1977–1981.

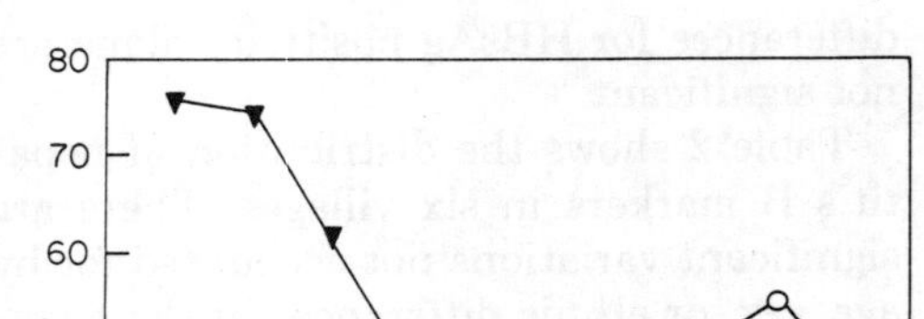

FIGURE 4. Prevalence of hepatitis B serologic responses by age for the population 20 years old or less. $n = 1{,}511$ persons. Tip community, Senegal, 1977–1981. ●, HBsAg(+); ○, anti-HBs(+); Δ, anti-HBc(+) alone; ▼, uninfected.

of hepatitis B infection (figure 5). The rate of anti-HBs positivity is 12 per cent at one year and rises sharply after age six years to 45 per cent by age 10 years (figure 4). The

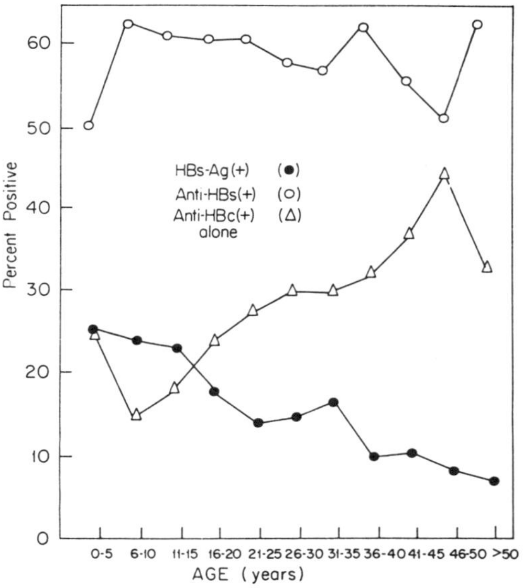

FIGURE 5. Responses to hepatitis B infection by age in persons with at least one hepatitis B marker. Tip community, Senegal, 1977–1981.

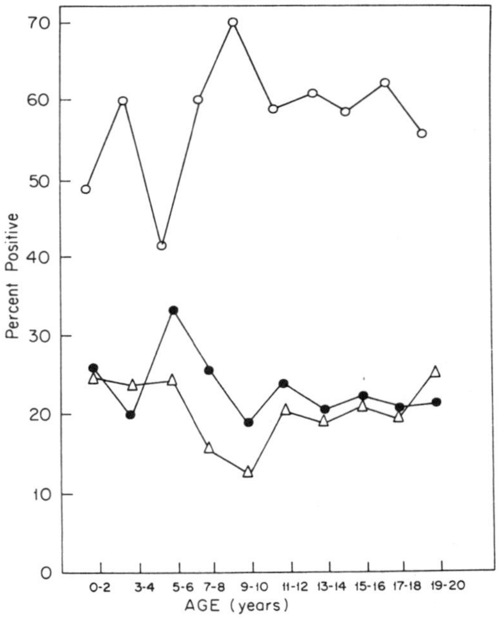

FIGURE 6. Responses to hepatitis B infection by age for the population less than 20 years old and having at least one marker of hepatitis B infection. Tip community, Senegal, 1977–1981. ●, HBsAg(+); ○, anti-HBs(+); Δ, anti-HBc(+) alone.

frequency of anti-HBc alone increases gradually with age from 1–5 years (figures 3 and 4).

If we look at serologic responses only among those persons demonstrating one or more markers of hepatitis B infection (figures 5 and 6), anti-HBc alone has a prevalence of 15–25 per cent from ages 2–20 years. Thereafter, it rises progressively to 35–45 per cent in the older age groups. For anti-HBs, even in the youngest age groups, 50–60 per cent of children mount a good immune response to hepatitis B.

In figure 7, we separated the anti-HBs(+) persons into those with low (1:4–1:8) or high (≥1:16) titers. The proportion of persons with high titers of anti-HBs declines with age, and the proportion of those with low titers rises with age. That is, in this population, it appears that in many persons with low titers of anti-HBs, the level of this antibody falls below the level of detectability, but anti-HBc persists.

To examine this possibility, we collected a second blood sample from 1,145 persons one or two years after the first sample. Among 313 persons who had low-titer anti-HBs in their first sample, 52 (16.6 per cent) were anti-HBs(−) in their second sample and had anti-HBc alone. By comparison, of 235 persons who were anti-HBc(+) alone on first testing, 57 (24.3 per cent) had low titer anti-HBs on their second test. Putting the longitudinal and cross-sectional information together, there is fluctuation of anti-HBs titers above and below the level

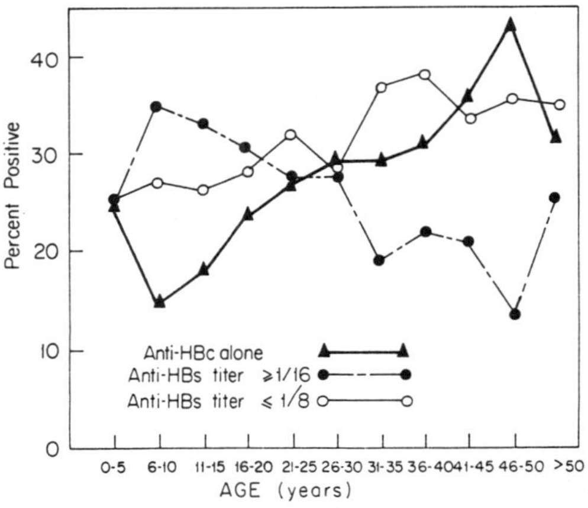

FIGURE 7. Prevalence of high and low titers of anti-HBs, and of anti-HBc alone, Tip community, Senegal, 1977–1981.

of detectability. Over longer time intervals, anti-HBs titers tend to fall. When only persons who show some marker of hepatitis B infection are considered, those with anti-HBc alone appear to be the antithesis of those with high titers of anti-HBs (figure 7).

Mortality during the four-year interval from 1977–1981 was examined in relation to hepatitis B serologic status in 1977. Serologic and follow-up information was available on 3,022 persons. Sixty-eight of these persons died during the interval, of whom 14 (20.6 per cent) were HBsAg(+) when first tested (table 3). Mortality was higher in HBsAg(+) persons than in all other members of the population (p = 0.04). The increased deaths among HBsAg(+) persons were concentrated in males below the age of 40 years. The mortality rate in this subgroup of HBsAg(+) males was 3,846/100,000 per year compared with 297/100,000 per year for HBsAg(−) males in the same age category.

Discussion

The most important observation in this study is that in rural Senegal infection occurs early in life, but not so early as to preclude a vaccination program against hepatitis B infection. About 25 per cent of the population has been infected by two years of age, and most of the people who are going to become chronic carriers in the population have become carriers by six years of age. That is, a vaccination program would have to be instituted before age two years to be maximally effective.

From the studies of Maupas et al. (5), we know that infants in Senegal respond well to the vaccine. From prospective studies of mother-to-child transmission in Senegal by Marinier et al. (9) and cross-sectional studies from Senegal (10), Liberia (11), Gambia (12), Namibia (13), and Zambia (14), we know that serologic evidence of persistent infection does not appear until five months of age at the earliest, and that few babies (<10 per cent) are infected in the first year of life. This is in contrast with Asia where transmission of hepatitis B from carrier mothers to their babies probably occurs at or near the time of birth, and most babies exposed to maternal virus show serologic evidence of infection by three months of age (15, 16).

TABLE 3

Mortality between 1977 and 1981 by hepatitis B response in 1977, Tip community, Senegal

Sex and age (years)	Status in 1981	HBsAg(+)	Anti-HBs(+)	Anti-HBc(+) (alone)	Uninfected	Total
Male						
≤39	Dead	7	5	0	6	18
	Alive	175	448	164	298	1,085
≥40	Dead	2	16	5	0	23
	Alive	32	195	125	11	363
Female						
≤39	Dead	2	7	4	3	16
	Alive	126	511	205	315	1,157
≥40	Dead	3	4	3	1	11
	Alive	21	200	114	14	349
Total	Dead	14	32	12	10	68
	Alive	354	1,354	608	638	2,954
	Mortality per 100,000	3,084	2,363	1,974	1,567	2,302
	Male ≤39	3,846	1,104		2,013	1,632
	Female ≤39	1,562	1,351	1,914	943	1,364

Uninfected members of the Tip population continue to be at risk of infection throughout their lives. The calculated risk of infection at any age is about 6 per cent per year. From this observation and those made of other African populations, recommendations can be made for the design of vaccination programs. The first priority would be to vaccinate all infants at 2–3 months of age when other pediatric vaccines are given. In areas of lower endemicity for hepatitis B, vaccination programs are usually based on the identification of carrier mothers and the immunoprophylaxis of their babies. In hyperendemic areas, such as Senegal, screening of mothers is not indicated because maternal transmission early in life is less important and children often are infected from sibs and others in the community. An additional program, if funds and facilities are available, would be to vaccinate children under seven years of age; after that age, more than half would have been infected, and vaccination would be less cost-effective. A further program, again if resources could be found, would be to extend the vaccination program to age 15 years after which 80 per cent of the population has been infected. For older children, it may be economic to pretest the populations for anti-HBs (and possibly HBsAg) to identify those persons who are still susceptible to infection.

There is a suggestion from the distribution of hepatitis B markers by ethnic group that crowding may influence the response to infection. The Serer, who tend to have more people living in compounds which are more closely crowded together, have a carrier rate of 17.5 per cent. On the other hand, the Peulh live in scattered, less crowded compounds outside of the Woloff villages, and they have a lower carrier rate (7 per cent). In the nearby community of Gawan, the Peulh also have a lower frequency of carriers than do the Woloff or the Serer (27).

Another difference between the Peulh and the other ethnic groups is that they raise cattle and drink milk all year, except towards the end of the dry season. This supplemental source of protein and calories may protect Peulh children from protein-calorie malnutrition, and this might have an effect on infection and carrier rates (17).

We also observed differences in the distribution of hepatitis B serologic markers by village that could not be explained by variations from village to village in age, sex, or ethnic group. No environmental differences between villages were obvious to us. Nevertheless, Nassrou and Tao-Fekh had higher frequencies of carriers (HBsAg(+)); and Tao-Fekh, but not Nassrou, had a lower prevalence of anti-HBs(+) and a higher prevalence of anti-HBc persons than did the other villages. Some of the intervillage differences may be explained by clustering within families of similar responses to hepatitis B.

In the published literature, there is confusion about the meaning of the serologic status when anti-HBc(+) is present without other markers of hepatitis B infection. Considerable attention has been given to the possibility that such persons are recovering from a recent infection with hepatitis B; that is, HBsAg has disappeared and anti-HBs has not yet appeared (the "window" hypothesis) (18). Some investigators have suggested that persons with anti-HBc alone are carriers of infectious virus (19). From our observations, we suggest that in Tip most persons with anti-HBc alone were anti-HBs(+) in the past and that this antibody has declined in titer to undetectable levels. This view is supported by the decline in anti-HBs titer with age, the increase in frequency of anti-HBc alone with age, and the direct observation of the conversion of persons with low titer anti-HBs to anti-HBc(+) alone during a one- to two-year follow-up.

The prevalence of HBsAg in males in the age group 11–40 years is about 20 per cent. Thereafter, the frequency falls to about 10 per cent. There are at least three tenable explanations for this observation. One possibility is that this is a cohort effect; that is, hepatitis B infection was less common

40–60 years ago than it has been in recent decades. Secondly, this decline may be simply a loss of HBsAg with age (12). Thirdly, it may reflect a selective loss (death) of carriers from the population. These explanations are not mutually exclusive and each may be partly correct, but biologically the last possibility is the most important.

Cirrhosis and primary hepatocellular carcinoma are common causes of death in Senegal after age 40 years (20, 21). We do not have direct information on causes of death in the Tip population, but we observed an overall increase in mortality of HBsAg(+) persons which was highest among males below the age of 40 years.

In a study performed in the Solomon Islands, London et al. (22) and Stevens et al. (23) recorded the serologic responses of persons to hepatitis B between 1966 and 1970 and compared these responses with their mortality as of 1978. In the oldest age group, the highest death rate, 27.5 per cent, was observed in persons who were HBsAg(+) at the first time point, and the lowest death rate, ~10 per cent, was observed in persons who had had high titers of anti-HBs when first studied. The causes of death in the Solomon Islands population were not known. Beasley et al. (2) and Beasley and Hwang (3) have been following a cohort of 40- to 60-year-old Chinese men in Taiwan who were classified as HBsAg(+) or HBsAg(−) on entry to the study. They have found a significant increase in mortality among the hepatitis B carriers that was entirely explained by an excess of deaths caused by cirrhosis or liver cancer. From these two studies and our observations of increased deaths among HBsAg(+) persons in this population, it seems reasonable to conclude that the decreased prevalence of carriers over age 40 years is, in part, due to an increase in mortality from chronic liver disease.

Field trials with the hepatitis B vaccine (24) indicate that it is safe, elicits production of anti-HBs antibodies in more than 95 per cent of the subjects, and protects such persons from infection with hepatitis B (25, 26). Hence, vaccination would probably prevent chronic liver disease and most of the consequent excess deaths. In underdeveloped areas of the world such as Tip, most of the shortened life expectancy is caused by excess infant mortality. Hepatitis B vaccine, although administered in childhood, has its major preventive effect on the mortality of adults. The persons who live to adulthood are the sources of wisdom and continuity in these and other societies. Preventing the deaths of many adults and thus preserving their societal contributions would be a major contribution to the stability of these communities.

REFERENCES

1. Blumberg BS, London WT. Hepatitis B virus: pathogenesis and prevention of primary cancer of the liver. Cancer 1982;50:2657–65.
2. Beasley RP, Lin C-C, Hwang L-Y, et al. Hepatocellular carcinoma and hepatitis B virus. A prospective study of 22,707 men in Taiwan. Lancet 1981;2:1129–32.
3. Beasley RP, Hwang L-Y. Epidemiology of hepatocellular carcinoma. In: Vyas GN, Dienstag JL, Hoofnagle JH, eds. Viral hepatitis and liver disease. New York: Grune & Stratton, 1984:209–24.
4. Blumberg BS. Australia antigen and the biology of hepatitis B. Science 1977;197:17–25.
5. Maupas P, Chiron J-P, Barin F, et al. Efficacy of hepatitis B vaccine in prevention of early HBsAg carrier state in children: controlled clinical trials in an endemic area (Senegal). Lancet 1981; 1:289–92.
6. O'Brien DBC. The Mourides of Senegal. Oxford, England: Oxford University Press, 1971.
7. Peers FG, Linsell CA. Dietary aflatoxin and human primary liver cancer. In: Jemmali M, ed. Mycotoxins in foodstuffs. Paris: CNRS, 1976:1005–18.
8. Blumberg BS, Sutnick AI, London WT, et al. Sex distribution of Australia antigen. Arch Intern Med 1972;130:227–31.
9. Marinier E, Barrois V, Larouze B, et al. Lack of perinatal transmission of hepatitis B virus infection in Senegal, West Africa. J Pediatr 1985;106:843–9.
10. Barin F, Perrin J, Chotard J, et al. Cross-sectional and longitudinal epidemiology of hepatitis B in Senegal. Prog Med Virol 1981;27:148–62.
11. Prince AM, White T, Pollock N, et al. Epidemiology of hepatitis B infection in Liberian infants. Infec Immun 1981;32:675–80.
12. Whittle HC, McLaughlan K, Bradley AK, et al. Hepatitis B infection in two Gambian villages. Lancet 1983;1:1203–6.
13. Botha JF, Ritchie MJJ, Dusheiko GM, et al. Hepatitis B virus carrier state in black children in Ovamboland: role of perinatal and horizontal infection. Lancet 1984;1:1210–12.

14. Tabor E, Bayeley AC, Cairns J, et al. Horizontal transmission of hepatitis B virus among children and adults in five rural villages in Zambia. J Med Virol 1985;15:113–20.
15. Stevens CE, Beasley RP, Tsui J, et al. Vertical transmission of hepatitis B antigen in Taiwan. N Engl J Med 1975;292:771–4.
16. Okada K, Kamiyama I, Inomata M, et al. e Antigen and anti-e in the serum of asymptomatic carrier mothers as indicators of positive and negative transmission of hepatitis B virus to their infants. N Engl J Med 1976;294:746–9.
17. Suskind RA, Olson LC, Olson RF. Protein calorie malnutrition and infection with hepatitis. Pediatrics 1973;51:525–30.
18. Hoofnagle JH, Seef LBN, Bales ZB, et al. Serologic responses in hepatitis B. In: Vyas GN, Cohen, SN, Schmid R, eds. Viral hepatitis. Philadelphia, PA: Franklin Institute Press, 1978:219–42.
19. Hoofnagle JH, Gerety RJ, Ni LY, et al. Antibody to hepatitis B core antigen: a sensitive indicator of hepatitis B virus replication. N Engl J Med 1974;290:1336–40.
20. Szmuness W, Prince AM, Diebolt G, et al. The epidemiology of hepatitis B infections in Africa: results of a pilot survey in the Republic of Senegal. Am J Epidemiol 1973;98:104–10.
21. Larouze B, Blumberg BS, London WT, et al. Forecasting the development of primary hepatocellular carcinoma by the use of risk factors: studies in West Africa. JNCI 1977;58:1557–61.
22. London WT, Stevens RG, Shofer FS, et al. Effects of hepatitis B virus on the mortality, fertility and sex ratio of human populations. In: Szmuness W, Alter HJ, Maynard JE, eds. Viral hepatitis. Philadelphia, PA: Franklin Institute Press, 1982:195–202.
23. Stevens RG, Kuvibidila S, Kapps M, et al. Iron-binding proteins, hepatitis B virus, and mortality in the Solomon Islands. Am J Epidemiol 1983;118:550–61.
24. Blumberg BS, Millman I. Vaccine against viral hepatitis and process. Serial no. 864,788, filed October 8, 1969. Patent 3,636,191 issued January 18, 1972.
25. Szmuness W, Stevens CE, Harley EJ, et al. Hepatitis B vaccine. Demonstration of efficacy in a controlled clinical trial in a high-risk population in the United States. N Engl J Med 1980;303:833–41.
26. Stevens CE, Taylor PE, Tong MJ, et al. Hepatitis B vaccine: an overview. In: Vyas GN, Dienstag JL, Hoofnagle JH, eds. Viral hepatitis and liver disease. New York: Grune & Stratton, 1984:275–91.
27. Larouze B, Feret E, Marinier E. Epidemiology of hepatitis B virus in Gawan, a rural community in Senegal, West Africa. (Unpublished manuscript.)

Guest Editorial

Baruch S. Blumberg[1] *and* W. Thomas London[1]

Hepatitis B Virus and the Prevention of Primary Cancer of the Liver[2,3]

A convincing body of evidence now exists to support the hypothesis that chronic infection with the HBV is required for the development of most cases of primary cancer of the liver in humans. The plasma-derived hepatitis B vaccine, which is now available, has in many field trials been found to be highly effective and very safe. Thus, in time, there should be a marked reduction or possibly a total elimination of primary cancer of the liver (PHC).

In this editorial we will review the evidence in favor of the hypothesis that persistent HBV infection is a necessary "cause" for very many cases of PHC. This does not rule out the possibility that other environmental factors (e.g., aflatoxins) and host factors may also play a significant role in its pathogenesis. We will then discuss several forms of prevention, the first two currently in use and a third possibly applicable in the future. [These features are detailed in (*1, 2*).] The forms of prevention include the following: 1) primary prevention—prevention of PHC by preventing infection with HBV, 2) secondary prevention—"prevention" of PHC in people who are already HBV-infected by detection of very early tumors and removal of them surgically, and 3) prevention by delay. We have introduced this term to describe possible ways in which asymptomatic carriers of HBV could be managed so that the onset of symptomatic chronic liver disease and PHC might be delayed for a sufficiently long period to enable the carrier to live out his or her normal life-span without experiencing many of the detrimental effects of hepatitis B infection, including cancer.

In the 1950's, Payet et al. (*3*), Steiner and Davies (*4*), and other investigators in Africa suggested that viral hepatitis might be the cause of primary cancer of the liver, on the basis of clinical, pathologic, and observational evidence. At that time it was not possible for this hypothesis to be tested directly; but after 1967, with the introduction of better methods for the detection of HBV (*5*), tests of this notion were undertaken.

In 1971, at a meeting on cancer in Africa held in Uganda, several studies showing a striking association between HBV and PHC were presented. This event appeared to be a turning point in the development of interest in this relationship, and the intensity of the work in the field increased. In 1975 we said (*6*):

> During recent years, there have been parallel developments in understanding, on the one hand, the pathogenesis of primary hepatic carcinoma . . . and, on the other, the biology of Australia antigen . . . and the infectious agent, hepatitis B virus (HBV), to which it is intimately related. Recently, the paths of these developments have begun to converge, and from this it is possible to design a preliminary strategy which could, if the interpretations of these data are correct, result in the prevention of many, and perhaps most, cases of one of the most widespread and deadly cancers in humans.

The evidence that was then available and that supported the statements was presented. We concluded by discussing the vaccine we had introduced (*7*) and possible prevention strategies. In the relatively short time since that article appeared, prevention programs have begun in Alaska (among Native Americans), Korea, Senegal, northern Italy, Taiwan, Japan, and China and among east Asians living in Los Angeles and Philadelphia. There is considerable hope that the objectives may be achieved.

PERSISTENT INFECTION WITH HBV REQUIRED FOR THE DEVELOPMENT OF MOST CASES OF PHC

Carriers of HBV are common in the parts of the world where PHC is most common, including such surprising locations as Alaska where HBV and PHC are common in the Eskimo population (*8*). A large series of case-control studies, in which the frequency of the virus in patients with PHC was compared to the frequency in controls, has been completed [reviewed in (*1, 2*)]. In some of these studies, which utilized the most sensitive methods for the

---

ABBREVIATIONS USED: AFP=α-fetoprotein(s); AIDS=acquired immune deficiency syndrome; CAH=chronic active hepatitis; HBV=hepatitis B virus; DHBV=duck HBV; HBeAg=hepatitis B e-antigen; HBeAg(+)=positive HBeAg; HBsAg=hepatitis B surface antigen; PHC=primary hepatocellular carcinoma; WHV=woodchuck hepatitis virus.

---

[1] Division of Clinical Research, Fox Chase Cancer Center, 7701 Burholme Ave., Philadelphia, PA 19111.

[2] Received July 17, 1984.

[3] Supported by Public Health Service (PHS) grants CA-06551 and CA-06927 from the Division of Extramural Activities, National Cancer Institute; by PHS grant RR-05539 from the Division of Research Resources, National Institutes of Health; and by an appropriation from the Commonwealth of Pennsylvania.

Editor's note: Periodically, the Journal publishes solicited guest editorials as a means of transmitting to investigators in cancer research the essence of current work in a special field of study. The Board of Editors welcomes suggestions for future editorials that succinctly summarize current work toward a clearly defined hypothesis regarding the causes, cure, or prevention of cancer.

detection of the virus, the frequency of infection in the cases approached 100%.

The virus can be identified by immunohistologic methods in the liver tissue of people with PHC (*9*). An interesting feature of this observation, shown by Popper and others (*10*) who have pioneered this work, is that the virus is most abundant in the cells that do not appear to have undergone malignant transformation. The neoplastic cells themselves often have no or sparse evidence of viral protein (*10*).

Currently, one of the most exciting areas of research is the molecular biology of HBV. The entire base sequence of the virus has been determined and confirmed by several groups (*11, 12*). HBV DNA is integrated into the DNA of the host liver cells in a very large percentage of PHC cases—in some studies, all the cases. However, there is also integration of HBV DNA in the liver cells of patients without PHC, including those with chronic liver disease and asymptomatic carriers of HBV. Hence, integration itself cannot explain the pathogenesis of the cancer. The site of integration of HBV DNA appears to be the same in every cell in an individual tumor (i.e., clonal), but the insertion locations of viral DNA are not the same in different tumors (*13-17*).

The most impressive evidence supporting the HBV-induced PHC hypothesis is epidemiologic. In one of the most convincing studies on cancer etiology, Beasley and his co-workers (*18, 19*) in Taiwan are studying prospectively 22,707 male government workers between the ages of 40 and 59 years. Recruitment into the study population began in November of 1975 and was concluded in June of 1978. Of the total, 3,435 men were asymptomatic carriers of HBV and the remainder were not. As of December 31, 1983, 116 cases of PHC had developed in the study population and 113 of these were in HBV carriers. Even the three tumors in noncarriers occurred in persons who showed evidence of prior HBV infection. The annual incidence for the whole population was 82.4/100,000, but for the carrier group it was 527.7/100,000 (*18, 19*). Beasley and Hwang (*19*) estimated the relative risk for the carrier men to be 217 times that of the noncarriers. The relative risk showed little variation from year to year and no secular trend. This is probably the highest odds ratio for any known environmental cause for a common cancer. On the basis of life table projections, they estimated the lifetime risk of death from PHC or cirrhosis for a Chinese male carrier to be between 40 and 50%, a remarkably high value.

Beasley's study established that the association between HBV and PHC was closer than that for any other virus and a site-specific cancer, probably for any environmental factor and a specific cancer. It has been shown also that chronic infection with the virus, rather than cirrhosis, carries the risk for cancer. Patients with cirrhosis due to causes other than HBV have a low risk for PHC, while those with cirrhosis due to HBV have a much higher risk (*19, 20*).

Prospective studies similar to Beasley's are now in progress with Eskimos in Alaska, with Chinese in Hong Kong and Singapore, and with Japanese in Tokyo. Early reports confirm Beasley's findings, i.e., a greatly increased risk of PHC in the carriers. In 1971 (*21*) we had proposed, on the basis of unusual population distribution and clinical and physical characteristics of HBV, that it represented the first identified member of a new group of infectious agents. We termed these "Icrons." Using his knowledge of the molecular characteristics of HBV, Summers (*22*), working at the Institute for Cancer Research, Philadelphia, PA, was searching for viruses similar to HBV. He subsequently discovered two of these (in woodchucks and ducks), and another has been identified in California ground squirrels (*22*).

The finding of the WHV provided another convincing piece of evidence for the hypothesis that HBV and similar viruses cause cancer of the liver. Woodchucks (*Marmota monax*), which develop persistent infection with WHV, predictably (≥80%) develop PHC, whereas uninfected animals or transiently infected animals never or rarely develop tumors (*22*). Another example is seen in ducks found in China. The DHBV is similar to HBV and WHV and it is also associated with primary cancer of the liver, at least in China (*23, 24*).

It has been shown both indirectly and directly that many of the individuals who go on to develop PHC are infected at a very young age (*25*). In some parts of the world, particularly Asia, the transmission may be maternal. In other locations, transmission is more likely to be horizontal. In either situation, since infection at an early age increases the probability of persistent infection, protection by vaccination in infancy appears to be indicated.

VACCINE

Infectious agents may produce many antibodies in the infected host. Some of these antibodies protect against subsequent infection and others do not; production of an appropriate vaccine requires the identification of the protective antibody. At the time of the very first observations on what was then called Australia antigen, it was obvious that the virus and the antibody against its surface antigen (anti-HBs) rarely occurred in the same serum. These observations were further supported by the early studies of Okochi and his colleagues (*26*), who were able to demonstrate in their first transfusion studies that the patients who had anti-HBs had a lower risk of infection with HBV than those who did not have the antibody.

These observations, scanty as they were, were sufficient to stimulate the development of a vaccine. In October of 1969, Blumberg and Millman (*7*) and Millman (*27*) described the methods for extracting surface antigen from the blood of carriers and its use as the source of the vaccine. This vaccine was based on the then unique concept that the viral protein could be extracted directly from human plasma by methods that would make it safe for inoculation in humans and would preserve its antigenicity (*27*).

Experiments in humans were not undertaken at the time of its initial proposal, and the value of this kind of vaccine was questioned by some authorities. Later (in

1970), Krugman and his colleagues (*28*) inoculated children with a preparation of HBV-positive serum that had been boiled for 1 minute. In challenge experiments the heated serum provided protection against HBV infection (*28*). This experiment had the effect of convincing manufacturers and researchers that the Blumberg-Millman vaccine would probably work, and large batches of the vaccine were prepared for use.

In an elegantly designed and executed field trial, the late Wolf Szmuness and his co-workers (*29*) demonstrated the safety and effectiveness of the vaccine. He used as his study population a group of male homosexuals in New York City at high risk for HBV infection (*29*). There have subsequently been several other field trials that confirm the effectiveness of the vaccine. It has gained the approval of the Food and Drug Administration in the United States, as well as that of licensing authorities in other jurisdictions. It is probable that about 2 million doses of plasma-derived HBV vaccines have now been administered and there have not, to our knowledge, been any significant adverse effects associated with their use.

PRIMARY PREVENTION OF PHC

Prevention programs will be based primarily, but not exclusively, on the use of the vaccine. The vaccine that is currently used is derived from plasma. This vaccine may, in due course, be replaced by vaccine produced by recombinant DNA techniques or by synthetic antigens. The price for the vaccine is currently very high, even for countries with large public health budgets. It is likely that the price of the vaccine will decrease as more is required, as more companies and countries begin to manufacture the product, and as new methods of manufacture are developed. This has already happened in South Korea where a plasma-derived vaccine is being produced for about one-third the cost of the American or French vaccines (Kim CY: Personal communication). The duration of the efficacy of the vaccine is also not known. In the United States the current recommendations are for revaccination after 5 years. As more experience is gained, it will be possible to design more accurately the timing of revaccinations if they are required. Protection at an early age is most important since it is during this period that infection is most likely to result in the development of the carrier state. Infection in later life is more likely to lead to development of antibody against the virus and, therefore, to less risk of chronic liver disease and PHC.

Primary prevention programs can be classified in three categories.

1) *Populations with low prevalence of HBV infection.*—Northern European countries, most of North America, and other nontropical or subtropical areas fit into this category. Vaccination is recommended in these regions for people in high-risk groups. These include health care workers (particularly those with frequent exposure to human blood), travelers to endemic areas, military personnel, male homosexuals, individuals working on renal dialysis units, possibly dialysis patients, drug abusers, and other persons frequently exposed to HBV carriers.

Despite the demonstrated safety of the vaccine, there has been a development that has affected its acceptance. Since 1980, AIDS has been identified in certain susceptible populations, primarily in the United States. Epidemiologic evidence indicates that AIDS can be transmitted by blood transfusion, plasma products (factor VIII), or contaminated needles. Although the vaccine manufacturing process is lethal to all known viruses, some individuals have avoided immunization with the vaccine because of fear of contracting AIDS. There is no evidence that this can occur. Recently, the probable etiologic agent of AIDS has been identified as a retrovirus. The process of hepatitis B vaccine manufacture has been shown to inactivate this virus.

2) *Populations of moderate prevalence in which maternal transmission is important.*—In Japan, the frequency of carriers is about 3%. Evidence indicates that 30–50% of the carriers occurs as a consequence of maternal transmission. Dr. Nishioka and his co-workers in Toyko (*20*), among others, have been screening expectant mothers for HBsAg for several years. Those women identified as positive for HBsAg are then screened for the presence of HBeAg. For those found to be HBeAg(+), immunization of the newborn is recommended. The infant is given high-titer anti-HBsAg at birth, a second dose of anti-HBsAg is given at about 2 months of age, and vaccination is started at about 3 months of age. Several studies show that the vaccine is very effective in newborns. The vaccination programs can be started at birth or within 3 months of age, when other pediatric vaccines are administered.

In the future, more extensive vaccination programs may be used in these countries to eliminate the remaining horizontally transmitted infections. These programs could include all newborns as well as adults who are still susceptible. The extension of the program to these other groups will probably depend on the availability and cost of the vaccine.

3) *Populations with a high frequency of HBV carriers.*—The frequency of HBV carriers in these populations ranges from 5 to 15%. These populations include much of China, other regions in Asia, and sub-Saharan Africa. Although maternal transmission occurs commonly, particularly in Asia, it may be relatively less important than transmission from virus carriers among siblings, other family members, and members of the general community. In these populations, universal childhood vaccination beginning at birth or within the first 3 months of age has been proposed and is being used in some regions (e.g., South Korea). It is believed that public health authorities in the People's Republic of China have also decided on a program to vaccinate all newborns. If vaccine becomes more available and less expensive, then susceptible adults may also be vaccinated.

An interesting problem arises in those low-frequency areas for HBV carriers that contain large populations from high-frequency areas of HBV carriers. For example, in the United States (particularly in California) there are many immigrants from regions in Asia where HBV is endemic. In Philadelphia, we are testing pregnant women in this population for the presence of HBsAg. An immunization program for the newborns of the mothers positive for this antigen is now in progress. In due course, vaccination may also be made available to all newborns and to susceptible adults in these populations.

Similarly, the Eskimo and Indian populations of Alaska have a high frequency of HBV carriers and an increased risk of developing PHC (*30*). A program has been proposed to immunize all susceptible individuals in this relatively small population (≈60,000) as well as all newborns (Rhoades E: Personal communication).

Extensive immunization programs, therefore, are in progress already, and more are planned. In addition to the vaccine, however, there may be other measures that can be used to control hepatitis B transmission. General improvement in sanitation may have a beneficial effect. There is substantial evidence that insects, particularly bedbugs, are heavily infected with HBV (*31-34*). Possibly, insect control could contribute to decreasing HBV infection. Also, some human behaviors, particularly those between mothers and newborn children, could influence transmission (*35*).

We can now consider what can be done for the approximately 210 million people in the world who are already carriers of HBV.

SECONDARY PREVENTION OF PHC

If carriers of HBV are at a high risk of developing PHC, is there some way of detecting cancer at an early stage? Small resectable tumors can be found by following HBV carriers prospectively with periodic testing of AFP and possibly other fetal proteins (α-glutamyl transferase, acidic ferritin). Sun, as well as the Coordinating Group for Research on Liver Cancer in China, has had extensive experience using mass surveys for elevations of AFP (*36*). Persons with elevated AFP are investigated with imaging procedures (particularly ultrasound) which can localize small masses. If a tumor is visualized, surgery is recommended. Using these methods, Tang's group (*37*), at the Shanghai First Medical College, reported a significant improvement in survival. In some cases, more than 1 tumor was detected by continued surveillance of AFP levels in patients who had already undergone hepatic resection. Similar studies are being conducted in Eskimo populations in Alaska and persons of Asian origin residing in Los Angeles and Philadelphia. In Alaska at least 4 early cancers have been detected in HBV carriers monitored for AFP elevations [(*38*); Heyward W: Personal communication].

During the course of the next few years the value of this approach can be appraised as the results from different centers become available. The question of screening of carriers for AFP in low-frequency HBV areas is still at issue. Studies now in progress may allow the identification of a subgroup of carriers who are at particularly high risk of developing tumors.

In collaboration with Beasley in Taiwan, we are currently evaluating the role of the iron-binding proteins ferritin and transferrin in predicting the development of liver cancer in HBV carriers. A series of predictors may eventually be identified that could focus on those subgroups of HBV carriers at highest risk of PHC (*39*).

MODEL FOR PATHOGENESIS OF PHC

There are certain features of the HBV-PHC relationship that require an explanation not provided by conventional theory. We have introduced a model to explain these findings (*40*). The observations used to generate the model are reviewed and the model itself is described. New studies are now in progress to test the model, but these will not be considered here.

PHC is associated with chronic infection with HBV, not with an acute, transient infection. PHC usually occurs in a liver affected with chronic hepatitis and/or cirrhosis. Hepatitis A virus infections, which do not have a chronic form, are not associated with liver cancer. Most people who develop the carrier state of HBV and, subsequently, PHC do not experience an antecedent, acute disease. The "incubation period" for the development of PHC appears to be very long; people who develop PHC may be infected as early as the first few months of life, but the disease usually does not appear until very much later, i.e., the fifth or sixth decade. Hence, any model has to take into account the very long period of exposure of liver cells to HBV. A simple mutational model does not seem to explain all the observed phenomena.

Histologic studies of the liver tissues of patients with PHC reveal that the neoplastic cells do not contain large amounts of HBsAg, hepatitis B core antigen, or whole virus, while the surrounding cells often contain large quantities of viral components (*41, 42*). The amount of HBsAg in the serum of patients with PHC is low compared to that in healthy HBV carriers and individuals with other forms of chronic HBV infection. Further, the amount of HBsAg in the serum decreases as the cancer progresses.

From the molecular studies HBV DNA is known to integrate into the genome of tumor cells (*14-17*). HBV may also be integrated into the DNA of cells that are not transformed or in the hepatocytes of individuals with chronic infections who do not have cancer (*14, 17*). The integrations in nontumor tissue often appear to be clonal. Finally, chronic infection with HBV causes not only PHC but also various forms of chronic liver disease; therefore, the model must explain how the same agent can cause more than one disease.

Our model postulates that the livers of mature humans [and animals that can be infected with a similar virus (*22*)] contain at least two populations of cells. The first and larger component readily becomes infected with HBV, synthesizes viral proteins, and can produce infectious virus. We have referred to this population of cells as "susceptible" or S-cells.

The second and, in adults, smaller population of cells is relatively resistant to replicative infection with HBV.

We have called this group of cells "resistant" or R-cells. If these cells become infected, integration of viral DNA may occur, but they produce little or no intact virus.

In individuals chronically infected with HBV, S-cells have a shortened survival. This fact may be due to an immune response to viral and cellular antigens on the surface of the virus-infected hepatocytes, or it may be that cells that are actively replicating virus are metabolically compromised. The death of S-cells, which in a chronically infected liver would be continuous, stimulates the division of R-cells. Hence there is a continuous process of death, regeneration, and, possibly, scarring of the liver. It is this continuous process that leads to the picture of chronic liver disease and posthepatic cirrhosis, which is often associated with chronic HBV infection.

The R-cells are relatively immune to death from HBV infection, since HBV replication does not occur, and host immune responses are not actively evoked; hence, R-cells are at a selective advantage compared to the S-cells. As a result, the S-cell population gradually decreases and the R-cell population increases.

Eventually, integration of HBV DNA into the DNA of the host cells may occur. If integration occurs in an S-cell, this probably has little consequence since the S-cell will eventually die as a result of the HBV replication and the host response. If the integration occurs in an R-cell, it may lead to a tumor. The R-cell population, because of the continuing death of S-cells, is continually expanding. Integration of viral DNA may give a further advantage to a clone of R-cells originating from a cell into which HBV DNA has been integrated. The integration of the HBV DNA into the R-cell may arrest its maturation and, thus, when it divides it will produce only other R-cells. Integration of the DNA may also stimulate growth factors or oncogenes in the particular clone of cells, which will further increase growth potential. As the clone expands, it may, by the process of clonal evolution, develop the characteristics of a malignant neoplasm.

What are the origins of the R- and S-cells? We have proposed that R-cells are less mature cells and S-cells are fully differentiated hepatocytes. The R-cells are the predominant cell form in early fetal life, decreasing in number but never totally disappearing in the maturing liver. S-cells are the predominant form in the normal mature liver.

According to this scheme, under normal circumstances, a dividing R-cell will give rise to 1 R- and 1 S-cell. Under abnormal circumstances an R-cell can divide and yield 2 R-cells or it can divide and differentiate into 2 mature S-cells. The more mature S-cell, if it can divide at all, will produce only other S-cells. Since the selective pressure of HBV infection is applied mainly to S-cells, R-cells will gradually accumulate. As already noted, integration of HBV DNA into the R-cell genome could hinder further differentiation and ensure that only R-cells are produced. This will further increase the rate of expansion of the R-cell component. If the integration of DNA results in the turning on of a cellular gene, which alters the cell phenotype in some way that provides a growth advantage, the resulting clone will have an even greater growth and selective advantage. In due course this effect becomes a clinical cancer perceptible to the host.

This model can also be used to explain certain features of liver disease other than PHC. Fulminant hepatitis may occur when livers with no or few R-cells become infected. If all the S-cells are killed by replicating HBV, there would be no resistant R-cells from which new liver cells could be regenerated. This consequence would suggest some unusual therapeutic measures.

Recent studies have shown that steroid therapy may be effective for CAH not associated with HBV but may be detrimental for CAH associated with HBV infection. In terms of the model, the steroids would depress the immune response by affecting lymphocytes. This event would lead to a temporary reduction in the destruction of the hepatocytes. However, the steroids would prolong the life of the infected S-cells, encourage the chronicity of infection, and thereby impede the rate of recovery.

The model could also explain the observation that the offspring of carrier mothers who are also HBeAg(+) rarely, if ever, have evidence of infection until about 3 months of age, after which they are highly likely to become persistently infected. Newborns would have relatively few S-cells and therefore would not produce sufficient quantities of virus to be detected by current means. By 6 weeks to 3 months of age a sufficient number of differentiated (S) cells would be available to synthesize enough HBV for detection by conventional assays.

Young people rather than adults are more likely to become chronically infected. In terms of the model this likelihood could be explained by the relative abundance of R-cells in the child. When S-cells are killed by the reaction to the HBV infection, sufficient R-cells are present to provide a new source of S-cells. In the adult, however, there is a relative scarcity of R-cells. When S-cells are killed by the infection, insufficient numbers of R-cells exist to provide S-cells at a high enough rate to maintain the infection.

PREVENTION BY DELAY

There are some interesting features of this model that can be used to discuss what we have termed "prevention by delay." It should be clear that this form of prevention is at present speculative.

The model posits that the driving force in the growth of the cancer is the death of S-cells as the result of replication of HBV. Hence, if we could devise methods to stop or even slow down this process, it might be possible to prevent the development of serious liver disease during the normal life-span of the individual. If this intervention could be started before a large component of R-cells had accumulated, then very mild procedures could be used over a long period rather than severe treatments over a relatively short time, such as are generally used in cancer therapy.

Most forms of cancer therapy are directed toward the killing of as many cancer cells as possible. In the process, because cancer cells and normal (particularly developing) cells have much in common, many normal cells are also affected. This places a limit on the aggressiveness of cancer therapy. The procedures recommended by this model are directed toward preserving the health of the S-cells rather than promoting the destruction of the infected R-cells.

The suggested "therapies" have a limited goal. It is not necessary to eliminate all the R-cells but only to keep their numbers at a sufficiently low level as to be imperceptible to the HBV carrier. Further, it is not even necessary to eliminate completely the HBV infection, which may be very difficult if integration has occurred. What is required is to slow down the process to such an extent that the presence of the virus is not perceived by the patient, that is, to prolong the asymptomatic phase of the carrier state while at the same time to attempt to decrease the risk of the HBV carrier developing chronic hepatitis and PHC.

Another aspect of this form of prevention is that it can be applied very early, although at present it is hard to say how early. It would be helpful to identify those HBV carriers who will remain persistently infected from infancy to adulthood. Methods for doing this should be uncovered with further research. Again, what is proposed is prevention rather than therapy since we hope to be able to develop a benign form of management that could be used by HBV carriers who are not sick.

What can be done to slow down the process of cancer and liver disease pathogenesis? The answer to this is not known now, but there are investigations in our laboratory and others that could have a bearing on these procedures. The goals would be to decrease the replication of HBV and to prevent the entry of newly synthesized virus into uninfected liver cells. These goals should be achieved at an early stage before R-cells accumulate significantly and before integration leading to rapid division of the R-cells commences.

Evidence exists that liver cells with high concentrations of the iron-binding protein ferritin, as well as possibly high levels of iron, are more likely to become infected and to replicate HBV. We are trying to understand this process sufficiently well to see if some management of iron or iron-binding proteins can favorably alter this process (*39, 43*).

A variety of chemicals, many derived from plants, exist that alter the HBsAg characteristics in a manner that affects their immune reactions (*44*). Such materials may have a role in preventing hepatocyte infection. There are also materials that may affect HBV replication by altering its DNA polymerase, and such chemicals may help to ameliorate infection.

HBV, like other viruses, must enter the cell through specific attachment sites. If these sites can be blocked, then invasion of uninfected hepatocytes may be prevented. A series of investigations on polymerized serum albumin have indicated a possible role for this material in the process (*45*). In addition, superoxide anion, peroxides, and hydroxyl radicals may play a role in liver cell destruction. The formation of superoxide anions is related to HBV infection and may be controllable with antioxidants (*46*).

We realize that these approaches are still speculative, but we believe that such lines of investigation may eventually lead to the prevention of chronic active hepatitis, cirrhosis, and liver cancer by delaying their occurrences beyond the life-spans of HBV carriers.

CONCLUSIONS

As indicated in this paper, several national and international prevention programs for the control of HBV infection are in progress and more are contemplated. These programs should lead to a considerable decrease in the frequency of primary cancer of the liver.

More experience with these programs will be needed before deciding on the best manner to proceed, but a campaign to eliminate completely the virus eventually might be considered. This campaign could have considerable consequences for human biology and ecology [*see* (*47*) for discussions of the role of HBV in human mortality, sex determination, and fertility], which must be considered as the prevention programs are evolving.

The price of HBV vaccine is high, but predictably the cost will decrease. Even when the cost decreases, the programs will be expensive, and they are most needed in areas of the world where public health budgets are low. The program for the control of HBV therefore may be conducted as an internationally funded, worldwide effort. A committee of the World Health Organization has recommended such a program.

Infection with HBV is costly in terms of both morbidity and mortality from PHC and chronic liver disease. HBV infection is not, however, the most important medical problem in those regions of the world where it is common; bacterial and viral diseases, malaria, schistosomiasis, filariasis, malnutrition, and other diseases also exert a terrible toll. It would be unfortunate if an anti-HBV campaign detracted from useful work on the control of these diseases. However, although HBV infection may not be the most important disease in Africa, it may well be the most important disease for which control appears feasible. This circumstance argues for the rapid implementation of a vaccination program. Hepatitis vaccination could possibly be combined with other pediatric vaccinations, thereby decreasing the effort of implementation.

For heuristic reasons, we have considered the R-S (cells) developmental model only in terms of HBV and PHC. In time, however, its application to other forms of cancer may be appropriate. The model recommends, we believe, a new and encouraging approach to cancer control in which individuals at high liver cancer risk can be detected. Prevention programs directed toward supporting the health of infected cells, rather than the destruction of malignantly transformed cells, could then be instituted.

REFERENCES

(*1*) LONDON WT, BLUMBERG BS. Hepatitis B and related viruses in chronic hepatitis, cirrhosis and hepatocellular carcinoma in man and animals. In: Cohen S, Soloway RD, eds. Chronic active liver disease. New York: Churchill-Livingstone, 1982:147-170.

(2) Blumberg BS, London WT. Primary hepatocellular carcinoma and hepatitis B virus. Curr Probl Cancer 1982; 6:3–23.
(3) Payet M, Camain R, Pene P. Le cancer primitif du foie, étude critique à propos de 240 cas. Rev Int Hepatol 1956; 4:1–20.
(4) Steiner PE, Davies JN. Cirrhosis and primary liver carcinoma in Uganda Africans. Br J Cancer 1957; 11:523–534.
(5) Blumberg BS, Gerstley BJ, Hungerford DA, London WT, Sutnick AI. A serum antigen (Australia antigen) in Down's syndrome, leukemia and hepatitis. Ann Int Med 1967; 66: 924–931.
(6) Blumberg BS, Larouze B, London WT, et al. The relation of infection with the hepatitis B agent to primary hepatic carcinoma. Am J Pathol 1975; 81:669–682.
(7) Blumberg BS, Millman I. Vaccine against viral hepatitis and process. Serial No. 864,788 filed 10/8/69; Patent 36 36 191 issued 1/18/72. Washington, DC: U.S. Patent Office, 1972.
(8) Heyward WL, Lanier AP, Bender TR, et al. Primary hepatocellular carcinoma in Alaskan natives, 1969-1979. Int J Cancer 1981; 28:47–50.
(9) Nayak NC, Dhark A, Sachdeva R, et al. Association of human hepatocellular carcinoma and cirrhosis with hepatitis B virus surface and core antigens in the liver. Int J Cancer 1977; 20:643–654.
(10) Thung SN, Gerbert MA, Sarno E, Popper H. Distribution of five antigens in hepatocellular carcinoma. Lab Invest 1979; 41:101–105.
(11) Galibert F, Mandart E, Fitoussi F, Tiollais P, Charnay P. Nucleotide sequence of hepatitis B genome (subtype ayw) cloned in *E. coli.* Nature 1979; 281:646–650.
(12) Pasek M, Gotot GW, Zink B, et al. Hepatitis B virus genes and their expression in *E. coli.* Nature 1979; 282:575–579.
(13) Brechot C, Pourcel C, Louise A, Rain B, Tiollais P. Presence of integrated hepatitis B virus DNA sequences in cellular DNA of human hepatocellular carcinoma. Nature 1980; 286:533–535.
(14) Brechot C, Pourcel C, Hadchoel M, et al. State of hepatitis B virus DNA in liver diseases. Hepatology 1982; 2:27S–34S.
(15) Gerin JL, Shih JW, Hoyer BH. Biology and characterization of hepatitis B virus. In: Szmuness W, Alter HJ, Maynard JE, eds. Viral hepatitis, 1981. Philadelphia: Franklin Inst Press, 1982: 49–56.
(16) Shafritz DA, Kew MC. Identification of integrated hepatitis B virus DNA sequences in human hepatocellular carcinomas. Hepatology 1981; 1:1–8.
(17) Shafritz D, Shouval D, Sherman HI, Hadziyannis SJ, Kew MC. Integration of hepatitis B virus DNA into the genome of liver cells in chronic liver disease and hepatocellular carcinoma. N Engl J Med 1981; 305:1067–1073.
(18) Beasley RP, Lin C-C, Hwang L-Y, Chien C-S. Hepatocellular carcinoma and hepatitis B virus. A prospective study of 22,707 men in Taiwan. Lancet 1981; 2:1129–1132.
(19) Beasley RP, Hwang L-Y. Hepatocellular carcinoma and hepatitis B virus. Semin Liver Dis 1984; 4:113–121.
(20) Obata H, Hayashi N, Motoike Y, et al. A prospective study on the development of hepatocellular carcinoma from liver cirrhosis with persistent hepatitis B virus infection. Int J Cancer 1980; 25:741–747.
(21) Blumberg BS, Millman I, Sutnick AI, London WT. The nature of Australia antigen and its relation to antigen-antibody complex formation. J Exp Med 1971; 134:320–329.
(22) Summers J, Smolec JM, Werner BG, Kelley TJ, Tyler GV, Snyder RL. Hepatitis B virus and woodchuck hepatitis virus are members of a novel class of DNA viruses. In: Essex M, Todaro G, zur Hausen H, eds. Viruses in naturally occurring cancers. Vol 7. Cold Spring Harbor Conferences on Cell Proliferation. Cold Spring Harbor, NY: Cold Spring Harbor Laboratory, 1980:459–470.
(23) Zhou YZ. A virus possibly associated with hepatitis and hepatoma in ducks. Shanghai Med J 1980; 3:641–644.
(24) Omata M, Uchiumi K, Ito Y, et al. Duck hepatitis B virus and liver diseases. Gastroenterology 1983; 85:260–267.
(25) Larouze B, London WT, Saimot G, et al. Host responses to hepatitis B infection in patients with primary hepatic carcinoma and their families. A case/control study in Senegal, West Africa. Lancet 1976; 2:534–538.
(26) Okochi K, Murakami S, Nonomiya K, Kaneko M. Australia antigen, transfusion and hepatitis. Vox Sang 1970; 18:289–300.
(27) Millman I. The development of the hepatitis B vaccine. In: Millman I, Eisenstein TK, Blumberg, BS, eds. Hepatitis B: The virus, the disease and the vaccine. New York: Plenum Press. In press.
(28) Krugman S, Giles JP, Hammond J. Viral hepatitis, type B (MS-2 strain): Studies on active immunization. JAMA 1971; 217:41–45.
(29) Szmuness W, Stevens CE, Harley EJ, et al. Hepatitis B vaccine. Demonstration of efficacy in a controlled clinical trial in a high risk population in the United States. N Engl J Med 1980; 303:833–841.
(30) Schreeder MT, Bender TR, McMahon BJ, et al. Prevalence of hepatitis B in selected Alaskan Eskimo villages. Am J Epidemiol 1983; 118:543–549.
(31) Brotman B, Prince AM, Godfrey HR. Role of arthropods in transmission of hepatitis-B virus in the tropics. Lancet 1973; 1:1305–1308.
(32) Wills W, Larouze B, London WT, et al. Hepatitis-B virus in bedbugs (*Cimex hemipterus*) from Senegal. Lancet 1977; 2: 217–219.
(33) Ogston CW, Wittenstein FS, London WT, Millman I. Persistence of hepatitis B surface antigen in the bedbug *Cimex hemipterus* (Fabr.). J Infect Dis 1979; 140:411–414.
(34) Wills W, Saimot G, Brochard C, et al. Hepatitis B surface antigen (Australia antigen) in mosquitoes collected in Senegal, West Africa. Am J Trop Med Hyg 1976; 25:186–190.
(35) Dickie ER, Merten C. Ethnographic observations on child care and the distribution of hepatitis B virus in the nuclear family. Med Anthropol 1982; 6:21–36.
(36) The Coordinating Group for Research on Liver Cancer, China. Alpha fetoprotein assay in primary hepatocellular carcinoma, mass survey and followup studies. 11th international congress on cancer, Florence, Italy, 1974.
(37) Tang Z-Y, Yu Y-Q, Zhou X-D, Zhou N-Q. Surgical treatment of hepatocellular carcinoma (HCC) and its ultimate outcome. J Exp Clin Cancer Res 1983; 3:261–268.
(38) Heyward WL, Lanier AP, Bender TR, et al. A case report: Early detection of primary hepatocellular carcinoma by screening for alpha-fetoprotein in high-risk families. Lancet 1983; 2:1161–1162.
(39) Lustbader ED, Hann HL, Blumberg BS. Serum ferritin as a predictor of host response to hepatitis B virus infection. Science 1983; 220:423–425.
(40) London WT, Blumberg BS. A cellular model of the role of hepatitis B virus in the pathogenesis of primary hepatocellular carcinoma. Hepatology 1982; 2:10S–14S.
(41) Gudat F, Bianchi L, Sonnabend W, Thiel G, Aenishaenslin W, Stalder GA. Pattern of core and surface expression in liver tissue reflects state of specific immune response in hepatitis B. Lab Invest 1975; 32:1–9.
(42) Krawczynski K, Nazarewicz T, Brzosko WJ, Nowoslawski A. Cellular localization of hepatitis associated antigen in livers of patients with different forms of hepatitis. J Infect Dis 1972; 126:372–377.
(43) Blumberg BS, Lustbader ED, Whitford PL. Changes in serum iron levels due to infection with hepatitis B virus. Proc Natl Acad Sci USA 1981; 78:3222–3224.
(44) Millman I, McMichael JC. Glycoproteins of natural origin with an affinity for hepatitis B surface antigen. Infect Immun 1978; 21:879–885.
(45) Franklin SG, Millman I, Blumberg BS. Hepatitis B surface antigen and polymerized albumin binding activity in sheep serum. Proc Natl Acad Sci USA 1984; 564–567.
(46) Vierucci A, de Martino M, Graziani E, Rossi ME, London WT, Blumberg BS. A mechanism for liver cell injury in viral hepatitis: Effects of hepatitis B virus on neutrophil function in vitro and in children with chronic active hepatitis. Pediatr Res 1983; 17:814–820.
(47) London WT, Stevens RG, Shofer FS, Drew JS, Brunhofer JE, Blumberg BS. Effects of hepatitis B virus on the mortality, fertility, and sex ratio of human populations. In: Szmuness W, Alter HJ, Maynard JE, eds. Viral hepatitis, 1981. Philadelphia: Franklin Inst Press, 1982:195–202.

International Conference on Prospects for Eradication of Hepatitis B Virus

Geneva 23–24 February 1989

Introduction

This is a very exciting time in the history of preventive medicine for we now face the prospect of controlling, and possibly eradicating, one of the most common and deadly of the infectious agents that afflict humans, the hepatitis B virus.

This volume which contains the proceedings of an international conference held in Geneva, includes discussion of the epidemiology of hepatitis B virus and its impact on morbidity and mortality, the modes of transmission of the virus, the efficacy of vaccines given at different schedules and to different populations, and a discussion of recombinant vaccine. One special feature is the report on control strategies for hepatitis B virus presented by members of global committees selected from scientists working in six regions of the world: Africa, Asia, Europe, Latin America, the Middle East and North America. Each of these reports represents the efforts of separate committees made up of scientists representing countries from each of these regions. Each of the committee members and chairpersons were asked to compile for their country and the region an estimate of: the prevalence and incidence of hepatitis B virus carriers, HBV-related disease and primary cancer of the liver; cost-benefit calculations for HBV vaccination programmes; current status and future plans for national vaccination programmes; the involvement of national, regional and international institutions in the vaccination programmes. They were also asked to discuss posisble mechanisms of funding for the anticipated vaccinaiton programmes. They have, to a remarkable extent, satisfied this request and as a result we have the beginnings of a comprehensive, quantitative estimate of the infection and disease load caused by this very common virus. Several calculations of cost-benefit have been made.

It is hoped that the information generated for and communicated at this meeting will be of help to those responsible in public health for making judgements on when, how and where the national and regional programmes shall be instituted. They hope that this meeting and its published proceedings will stimulate national health officials to establish vaccination programmes. Finally, I would like to encourage those participating in the vaccination programmes to consider them both as applied and research projects. As physicians, we know that each time we attempt to solve a medical or public health problem, we generate other problems and these are usually problems of which we were unaware before the project began. Research is the best way of detecting problems early, and of generating the data which can be used for their solution. In addition, research projects related to the vaccination programmes can help us to understand the biology and ecology of hepatitis B virus, one of the most interesting of the infectious agents which have interacted with humans.

Baruch S. Blumberg

Conclusion

From the proceedings of this conference, we have accumulated a comprehensive and substantial body of data on the prevalence of carriers of HBV infection throughout the world, in addition to information on morbidity, mortality, vaccination programmes, etc. This can be considered as the beginning of a more detailed compilation. It can be extended in subsequent meetings of the same nature and in meetings of the WHO Global Committee. With this information, a cost-benefit analysis for mass vaccination programmes in many countries and regions can now be computed. In locations where estimates of morbidity, mortality and prevalence have not been made, rough estimates may be possible using data from adjacent countries and from regions where such data exist. It seems likely, therefore, that in the near future cost-benefit estimates will be available from a large number of countries.

The possible strategies for vaccination have been considered and public health authorities in different countries can design programmes that best suit their specific requirements (i.e. universal neonatal vaccination, neonatal vaccinations plus identification of HBeAg + mothers, surveys of all HBsAg + mothers and vaccination of their children, identification of high-risk groups, vaccination of their children, identification of high-risk groups, vaccination of older children and adolescents, etc). There is a special problem relating to immunization programmes in high prevalence communities which exists within developed countries, for example, the native Americans in Alaska, USA. They are one of the first communities in which universal vaccination may soon be complete.

It has been demonstrated that universal vaccination programmes in large populations are practical and possible. In the People's Republic of China, all new born children will soon be vaccinated. In Taiwan, universal childhood vaccination programmes have been in place for several years. Large progammes exist in other countries in Asia and Africa, e.g. in The Gambia. The advisability of incorprating these programmes in the EPI programmes has been presented and steps are being taken to implement this in several regions. The economics of vaccine production indicate that as more is used, the cheaper the price becomes; essentially any needed amount of the vaccine can be produced. It has been stressed by several authors that each vaccination programme, when possible, should include research projects whose results could help subsequent vaccination programmes. These research projects could also teach us more about the biology of this fascinating virus and indirectly help in practical applications. Another advantage of research associated with the vaccination programmes is that they will help to detect problems early in their evolution. They can also add to our store of knowledge about the virus and this, in due course, may help in the introduction of therapies which may eventually be essential if eradication is to be feasible.

What are the next steps? It will be necessary to make the information from this conference and its proceedings known to the general public, to physicians who will have to administer the vaccine, to public health authorities, to philanthropic agencies. It is possible to help in the development of methods of funding. In many cases, such programmes are supported from tax funds. There are other interesting possible solutions including bilateral arrangements between developed and developing countries. In The Gambia the government of Italy is funding the vaccination programme and the vaccine has been donated by a commercial company. The advantage of this approach is that a price tag for a vaccination programme for a particular country can be established and the appropriate funding groups can make a clear decision between possible benefits and their cost.

I would like to thank the organizers of this conference who have planned and executed a complex and remarkably effective meeting. I would also like to thank the speakers, and especially the members of the study groups, for providing this extraordinary quantity of information which required a large investment of time and energy.

There has also been valuable contributions from the audience by way of question and comment. In summary, this meeting has made an important contribution to the future eradication of hepatitis B virus.

Baruch S. Blumberg

Reprinted from *Viruses, Immunity, and Immunodeficiency*
Eds. Andor Szentivanyi and Herman Friedman
Plenum Publishing Corporation, 1986

HEPATITIS B VIRUS, IRON AND IRON-BINDING PROTEINS

Baruch S. Blumberg

Institute for Cancer Research
Fox Chase Cancer Center
Philadelphia, Pennsylvania

INTRODUCTION

Contemporary discussions of immunity in man usually center on various forms of cellular and humoral response. However, bacteria, fungi, protozoa and tumor cells need iron for their growth. A response of the mammalian host to infection and possibly also to the growth of tumor cells is to deny to the invading agent and replicating cells the iron present in the host. Weinberg (1) has prepared a comprehensive analysis of the relation of iron and iron-binding proteins to what has been referred to as "nutritional immunity." Additional references will be found in his paper.

The purpose of the present review is not to marshal the evidence to support or reject the argument that serum and storage iron levels are important in the control of infection and tumor growth; Weinberg's paper (1) is, in part, intended to do that. It is presented, rather, as a compilation of observational and experimental data that may be useful in understanding a series of observations related to hepatitis B virus (HBV), iron and iron-binding proteins that have been made over the past ten years.

Several mechanisms affect the relation between host iron and invading agent. The host can mobilize iron-binding materials at or near the site of infection to withhold humoral and tissue iron and prevent its utilization by the bacteria. The first of these materials to be discovered was conalbumin, a powerful binder of iron and a major constituent of egg white. Lactoferrin, an iron-binding protein originally found in milk, is also present in tears, nasal exudate, saliva, mucus, bile and seminal fluid; it accounts for much of the antibacterial action of these secretions. The specific granules of polymorphonuclear neutrophilic leucocytes contain large amounts of lactoferrin which has a profound effect on their ability to counteract infection by depriving microorganisms of iron.

Transferrin is the major iron-binding protein of serum. Its current name focuses attention on its role in transporting iron, but its original name, siderophilin (2), emphasized its ability to bind iron and its subsequent effect on infection. The probability of infection may increase if transferrin levels are lowered, as is the case in protein malnutrition (i.e. kwashiorkor) or when iron saturation is increased, allowing invading bacteria to remove host iron more readily.

Ferritin is the major iron binding protein of cells and is found in large amounts in the liver, spleen and bone marrow (3). Its function is to store iron which can then be made available for the synthesis of hemoglobin and other iron-containing proteins. Men have up to about two grams of storage iron. Women, during the childbearing period, may have little storage iron; the levels usually increase after menopause. Ferritin is also present in serum, and, in an otherwise normal individual, the serum levels reflect the total iron stores of the body. This relation, however, may not be seen with inflammation, infection and malignancy where increased serum ferritin levels can be independent of iron stores.

Crystalline ferritin has been isolated from liver and spleen. It consists of a protein shell (apoferritin) which covers a core made up of a maximum of 4000 atoms of iron. Apoferritin has a molecular weight of about 450,000 and is composed of 24 subunits; two classes of subunits, 19,000 (L) and 21,000 (H) can be distinguished (4). Ferritin from different sources may be made up of different ratios of the subunits. The ratio of subunits is related to the isoelectric points (pI) of the ferritin molecule and the rate of iron uptake. Investigators have found that ferritin from certain tumors has acidic pI, with a higher H/L ratio of subunits and lowered iron content.

A second general mechanism which the body uses to respond to infection and to the presence of cancer cells is to withhold iron. The intestinal absorption of iron is decreased and iron is shifted to the tissue compartment (e.g. liver, bone marrow) where it binds to ferritin and is stored. Extremely low levels of iron may require treatment in order to allow the immune system to function, but excess therapy with iron during infection can be detrimental.

A major mechanism for denying iron is to block the release of iron contained in macrophages, which have acquired the metal from decaying erythrocytes. The increased amount of iron remaining in the macrophages then stimulates the production of ferritin whose levels may then rise in tissue and in the blood. In malignancies, a ferritin which does not bind iron well may also be produced in large quantities. A key regulator of the iron metabolism in the cell is the hormone leukocytic endogenous mediator (LEM, also termed interleukin-1) which is released by monocytes activated by infection or immune responses.

At the site of infections, large quantities of iron could become available to the invading bacteria due to the breakdown of red blood cells. However, the hemoglobin-binding protein haptoglobin is increased under these circumstances, and the resultant hemoglobin-haptoglobin complex is taken up by the reticuloendothelial system and cleared by it.

An additional, important mechanism for the control of iron during infection is the fever response. The temperature elevation decreases the ability of the iron-binding proteins (siderophores) of the gram-negative bacteria to take up host iron. There is a competition between the binding proteins of the invading organism and those of the host for the available iron. Many of the protective effects described favor the retention of the iron by the host and its denial to the invading organism.

Weinberg (1) has described a variety of mechanisms by which alterations in iron levels and binding can improve the response to infection, that is the enhancement of "nutritional immunity," and counter-measures which diminish the effect. Moderate restrictions of iron may enhance immunity, but severe restrictions can be detrimental; there appears to be an optimal level of available iron which is most advantageous to the host. In an actual clinical situation it may be difficult to know exactly what this is. A variety of observations in re-feeding camps for victims of famine in Africa showed that sudden improvements in nutrition with augmented iron intake could actually increase the frequency and severity of infections including tuberculosis, brucellosis and malaria.

Various bacterial endotoxins and cell wall extracts can decrease available iron and enhance host protection. There are a large series of observations and experiments that indicate that acute and chronic infections as well as therapeutic injections of cell wall components can have an antineoplastic effect. Pearl (5) found that, in a survey of autopsies, bodies that had evidence of active tuberculosis had much lower incidence of concurrent cancer than did the bodies which did not have such infection. A popular form of cancer therapy for many years has been the use of non-viable extracts of bacterial products (Streptococcus pyogenes and Serratia marcescens) in an attempt to cause tumor regressions (see review by Nauts (6)). This form of therapy has not gained wide acceptance, but there is experimental evidence to support the clinical impressions. These bacterial extracts can result in a hypoferremia which might decrease tumor growth.

There is experimental evidence that some iron chelators can serve to combat infection or decrease carcinogenesis while others actually make the iron available to microorganisms and tumor cells and enhance their growth. The addition of excess amounts of iron can enhance infection or tumor growth. Patients with inherited or acquired hemochromatosis have an increased susceptibility to infection as well as to tumor generation and growth. Ingestion of alcohol increases iron levels. African males who consume large quantities of home brewed alcoholic beverages develop siderosis, and this may be related to an increased frequency of cancer of the liver in these individuals.

Pathological organisms and tumor cells may have siderophores which act to bind the iron made available to them by the host. The amount and affinities of these siderophores vary from organism to organism and, presumably, in various tumor cells.

This short review of the effects of iron and iron-binding proteins on microorganisms and tumor cells will, it is hoped, provide a background to consider a series of studies on the relation of iron and iron-binding proteins to HBV. We will review our work in this field and then suggest how phenomena related to iron may explain some curious observations that have been made on the behavior of HBV. Possible uses of this information in therapy and prevention will also be discussed.

Serum Iron Levels in Carrier and Non-carrier Patients with Down's Syndrome

Our introduction to the iron problem came as part of the original investigation of the nature of the "Australia antigen" (Au, HBsAg). Au was contained in a precipitin band found in an immunodiffusion reaction between the serum of an Australian aborigine and a much transfused patient with hemophilia. We were testing the hypothesis that Au was located on a hepatitis virus (7).

One test of the hypothesis was based on an extensive study of patients with Down's syndrome, many of whom had Au in their blood. We postulated that the patients with Au (who had no obvious symptoms of clinical hepatitis) would have chemical evidence of mild liver dysfunction that was more abnormal than the equivalent measurements in the Down's syndrome patients who did not have Au (8). We measured "liver chemistries" including SGPT (now ALT), SGPT, cephalin flocculation, thymol turbidity, BSP, total protein, and serum albumin, all of which showed that the patients with Au were more abnormal than those without it. There was no significant differences in bilirubin, prothrombin time and alkaline phosphatase. Since serum iron levels were known to be elevated in patients with acute hepatitis, its measurement was included in the studies.

Some time in 1973, our colleague Dr. Alton I. Sutnick brought to our attention an article published by a group of Dutch investigators which reported a curious observation on patients with end stage kidney disease undergoing renal dialysis treatment (9). A group of 11 patients had developed anicteric hepatitis in all but one case associated with HBsAg. They experienced a spontaneous rise in hemoglobin levels and required fewer transfusions than a "control" group of patients who did not have hepatitis B.

At that time we could not study patients on renal dialysis, but we had extensive information on hepatitis in Down's syndrome patients. Sutnick et al. (10) then devised and executed a comparison study of 20 patients with Down's syndrome who had HBsAg in their serum and an additional 20 who did not have the antigen. Measurements were made of serum iron and related factors as well as SGPT as an indication of liver dysfunction.

We found, as predicted from the Dutch observation, that the serum iron, hemoglobin, and hematocrit were significantly higher in the HBsAg(+) patients than in those who were HBsAg(-). The total iron binding capacity (essentially a measurement of the binding capacity of the serum transferrin that is not already occupied by iron) was also higher in the hepatitis group. In a series of correlation studies we found that the elevation of iron was not correlated with elevations of the SGPT and, therefore, the elevation of the serum iron in the HBsAg(+) patients could not be accounted for by the breakdown of liver cells alone but was in some way a function of the presence of the virus itself. It also suggested the possibility that the carrier state was advantageous to the patients in that it allowed them to maintain higher levels of serum iron, hemoglobin and hematocrit. We also predicted that there would be similar findings in renal dialysis patients which would then explain the Dutch observations.

Iron Levels and Host Response to HBV Infection

The iron project remained dormant but not forgotten until mid-1978 when our attention was focused on nutritional problems. We, and other workers in the field, had by this time concluded a large body of studies testing the hypothesis that persistent infection with the HBV is necessary for the development of primary cancer of the liver (11,12). It became apparent that HBV was required for the development of PHC, but that other factors were also involved. We wanted to understand as much as possible about pathogenesis in order to affect the course of disease in people who had already become carriers of HBV.

In humans, nearly all iron is derived from eating, therefore it can be considered nutritional. Iron had the further advantage of being manipulatable, that is, if we learned that it was involved in the pathogenesis of chronic liver disease and primary cancer of the liver, it was conceivable that iron levels could be altered to favorably affect outcome.

Our second study included a group of 201 patients receiving renal dialysis, including 39 carriers of HBV, 76 with anti-HBs, and 86 without evidence of HBV infection (13). There were 113 males and 88 females. We first tested the hypothesis derived from the Down's syndrome study and confirmed that serum iron was higher in carriers of HBV than in non-carriers.

We confirmed other observations made in the first study (Table 1). The percent iron saturation and SGPT were also higher in the carriers than in those with anti-HBs, but this difference was only significant in the males. Ironically, we did not find any difference between the carriers and the others in respect to hemoglobin, hematocrit or red blood cell count which had been expected from the original Dutch studies and from the Down's syndrome study.

Table 1. Mean Values for Groups of Renal Dialysis Patients

	SI, mg/dl			TIBC, mg/dl			% Iron saturation			SGPT, international units/liter		
	Male	Female	Total	Male	Female	Total	Male	Female	Total	Male	Female	Total
HBsAg(+)	114	214	145	278	305	285	42.4	46.5	43.6	28.1	20.8	25.8
	**		**				**		**	**	**	**
Anti-HB(+)	95	107	102	304	284	294	29.6	36.3	33.0	11.5	6.2	10.8
											***	***
No evidence of infection	103	95	99	283	281	282	34.6	33.4	34.1	14.8	21.4	17.2
Total	103	117		288	286		34.9	36.4		16.9	15.9	

	RBC, 10^6 cells/mm^3			HB, g/dl			Hematocrit, %		
	Male	Female	Total	Male	Female	Total	Male	Female	Total
HBsAg(+)	2.6	2.2	2.5	6.8	6.7	6.8	22.4	21.8	22.2
Anti-HBs(+)	2.6 *	2.3	2.5	7.1 **	6.3	6.7	23.5 *	20.8	22.1
No evidence of infection	2.4	2.3	2.3	6.7	6.4	6.6	21.9	20.8	21.5
Total	2.5 **	2.3		6.9 *	6.4		22.6 *	20.0	

Differences were assessed by the Mann-Whitney test. % iron saturation = serum iron (SI)/total iron-binding capacity (TIBC). RBC = red blood cells. * is $p = 0.05$; ** is $p = 0.01$; *** is $p = 0.001$.

After completing the analysis of single factors we did a logistic regression analysis to determine the interactions between the variable serum iron, sex, SGPT levels and HBV status (HBsAg, anti-HBs or no response) considered together (Table 2). This was an inductive analysis, that is, the data were analyzed without any preset hypothesis to re-test and complement the original deductive analysis.

We found that interactions of the variables taken two at a time could account for the distribution of the data, and it was not necessary to consider three-way or high order interactions. The inductive study confirmed the original observations: carriers had higher serum iron levels than non-carriers, carriers had higher SGPT levels than non-carriers, and the HBV response was sex related, that is, males were more likely to become carriers and females to develop anti-HBs (14,15). We also found that there were independent interactions between serum iron and SGPT and between HBV response and SGPT, that is, the elevation of iron could be associated with liver cell breakdown but that it also was elevated as a consequence of the carrier state independent of any significant liver cell damage (as measured by SGPT elevation).

We had confirmed, in large part, our original observation that serum iron was raised as a consequence of the HBV carrier state independent of liver cell breakdown. This was found to be true despite the possible masking effects in dialysis patients, or iron therapy, the anemia of chronic renal disease and the decrease in iron absorption due to the use of aluminum hydroxide. A series of questions was raised by these findings.

(1) We suggested that these studies should be repeated in other populations, including non-hospitalized populations in regions of the world with high frequencies of carriers. Our first studies had been done in patient groups known to have immunological and other abnormalities, and we wanted to know if these iron differences had an impact at the population level.

(2) Were the raised iron levels a consequence of the infection with HBV or did they precede them and make the putative carriers more prone to chronic infection with HBV?

Table 2. Distribution of Data According to HBV Response, SGPT Level, Serum Iron Level, and Sex. Renal Dialysis Patients.

	Low iron		Middle iron		High iron	
	Male	Female	Male	Female	Male	Female
	Low SGPT ($\leq$15 international units/liter)					
HBsAg(+)	3	2	6	2	5	3
Anti-HBs(+)	12	11	11	14	6	11
No evidence of infection	19	9	10	9	6	3
	High SGPT (>15 international units/liter)					
HBsAg(+)	1	0	1	1	11	4
Anti-HBs(+)	4	0	1	0	2	3
No evidence of infection	1	3	4	3	8	9

(3) Were HBV carriers more likely to become infected with bacteria as a consequence of their raised iron levels?

(4) We had recently found that incubations of HBV with certain species and strains of Pseudomonas resulted in a loss of HBsAg antigenicity (16). We questioned if this were in some way related to the HBV-iron interaction.

(5) Millman et al. (17) found that transferrin was firmly bound to HBsAG. We suggested that the binding of iron via transferrin to the virus and the liver cell transferrin receptor might facilitate the entry of HBV into cells and thereby promote its infectivity.

(6) We made the hypothesis that increased iron stores might increase the possibility of the development of primary hepatocellular carcinoma in carriers, that is, that iron is involved in the pathogenesis of the cancer and that alterations in iron or iron storage could be used to favorably alter the outcome.

Most of these hypotheses were tested during the next several years, and the consequences of these studies will now be discussed.

Serum Iron Levels in Non-hospitalized Asymptomatic Carriers

In 1977 we had collected sera from members of the community living in the village complex of Tip in an area of central Senegal about 200 kilometers east of Dakar as part of an extensive study of primary cancer of the liver and HBV (18). From this collection we selected 193 sera for the iron study, including 40 carriers, 59 with anti-HBs and 94 who had no evidence of infection. Measurements similar to those completed on the Down's syndrome and renal dialysis patients were undertaken, with the exception that the SGPT determinations were performed on sera that had been stored frozen for about three months. The enzyme loses its activity under these conditions and, although the values could be used for comparison within the Tip population, they could not be compared directly to the earlier studies.

The results were, in general, similar to those of the first two studies (Table 3). The iron levels overall were lower in the Tip study, probably a reflection of the low iron intake of the population. Serum iron levels were higher in the carriers than in the non-carriers. We did an inductive analysis of the data including the variables serum iron, sex, SGPT levels and HBV status similar to that done in the renal dialysis study. Again, only two factor interactions were required to model the distribution. Two of the interactions found in the dialysis study were confirmed. (1) The serum iron was increased in carriers independent of SGPT elevation and (2) the SGPT was increased in the carriers compared to the other HBV responses. Two additional interactions were also observed: between SGPT and sex and between serum iron and sex. The latter was particularly interesting; the serum iron differences between those who were HBsAg(+) and those HBsAg(-) was greatest for the females.

The elevation in serum iron in carriers seen in the two patient groups had now been confirmed in a non-hospitalized population group. It raised again the possibility of advantage to the carriers who could retain more iron, particularly under conditions such as those in Africa with a low dietary intake of the element. This was seen particularly in the females who, in any case, would most benefit from increased iron retention during the period they were losing blood regularly through menstruation. This concept of advantage to the carrier state was consistent with the notion advanced in some of our earliest studies that the carrier trait behaves as a genetic polymorphism.

Table 3. Mean Serum Iron Levels (mg/dl) in Down's Syndrome Patients, Renal Dialysis Patients and Non-hospitalized People from Tip, Senegal.

	Down's Syndrome			Renal Dialysis			Tip, Senegal		
	Male	Female	Total	Male	Female	Total	Male	Female	Total
HBsAg(+)	164	-	164	114	214	145	77	81	79
HBsAg(-)	84	_	84	98	100	100	75	58	65
N	40		40	103	117	220	81	112	193

In each case, the serum iron is higher in carriers (HBsAg(+)) than in non-carriers (HBsAg(-)). N = number of people tested. Females were not tested in the Down's syndrome study. The HBsAg(-) values for the renal dialysis patients are approximate.

Serum Iron Levels Before and After Infection

In an effort to control the spread of HBV, we had, since 1972, been monitoring a group of renal dialysis patients in the Philadelphia area. Specimens of blood were collected two times a month and tested for HBV. At the time our surveillance of this group began, HBV was extremely common in such units and there was a very high probability that patients would become infected, some to remain carriers and others to rid themselves of the infection and develop anti-HBs. (An important consequence of this surveillance program was the complete control of HBV infection. Hepatitis due to HBV is now rare in the dialysis unit we served as well as in most others where the epidemiologic measures were used.)

Two groups of patients were identified; 1) those who, after infection became carriers and, 2) those who became transiently infected. Serum iron levels and other measurements were made on both of these groups one month before, at the time of the and 6-12 months after infection (19). In the patients destined to become carriers, serum iron levels were not elevated before the infection but arose immediately after infection and remained high in the late follow-up period. In the transients, the serum iron levels were not elevated prior to the infection, rose at the time of infection, but decreased in the follow-up period (Table 4). Hence, our question had been answered: the iron in the carrier state was not elevated prior to exposure to HBV but remained elevated after infection.

A distinction, however, could be made between the putative carriers and the transients. Among the carriers, if iron levels had decreased between the first and second time periods, then all of these individuals had increases during the post-conversion period, and if there was a decrease in the post-conversion period there would have been an increase in the pre-conversion period. The carriers never had decreasing iron levels in both time periods, whereas the transients often did. There appeared to be differences in binding and/or storage proteins between the carriers and the transients, and these were investigated in subsequent studies.

In the discussion of this chapter we reaffirmed the hypothesis that bacterial infection and death might be more common in the carriers compared to non-carriers at certain times in their lives. We also suggested that

Table 4. Mean Iron Levels in Carriers and Transients Before (sample 1), Immediately After (sample 2), and 6-12 Months After (sample 3) Seroconversion.

HBV status	Mean iron levels, µg/dl Sample 1	Sample 2	Sample 3
Carrier	97.7	115.3	119.6
Transient	100.7	115.8	101.4

For carriers, the mean iron levels before seroconversion were significantly less than the other two values ($p < 0.05$).

the relation between HBV and the sex ratio of offspring and overall fertility that we had observed during the course of several large studies (20,21) might also be linked to the iron levels, possibly through the intermediary of the bacterial infection.

We elaborated further on the iron-PHC hypothesis. Edmondson and Steiner had, in 1954 (22), found that in cases of PHC there is little or no iron present in the malignantly transformed liver cells but that large quantities of iron may be present in the cells surrounding the cancer which do not appear to be transformed. Similarly, whole HBV virus, surface and core antigen can be found in the surrounding cells but are much less frequent in the cancer cells. Fernandez-Pol (23) had shown experimentally that removal of iron (using a chelating agent prepared from a plant substance) decreases the growth of live tumors in tissue culture. We postulated that the HBV-infected liver cells surrounding the malignant cells would supply iron to the cancer cells which did not have a sufficient amount and thereby encourage or stimulate their growth. Carriers with elevated storage iron and iron-binding proteins (e.g. ferritin) would be at greater risk of developing PHC than other carriers. We had, in effect, made a model to explain a portion of the pathogenetic mechanism, and various direct and indirect studies have been designed to test it. We will return to this matter later.

In all these comparisons of carriers and non-carriers, the differences in serum iron levels, ferritin and transferrin measurements are small; most measurements fall into what is usually considered to be normal limits. However, these differences exist for many years, and in some cases decades, and could, over a long period of time, have a profound effect on the pathogenesis of a chronic disease.

McGlynn et al. (24) have reported a curious finding which could be explained by the interactions of microorganisms and host iron. They found that, among a group of southeast Asians living in Philadelphia, HBV carriers who were tuberculin (PPD) positive were much less likely to also be HBeAg(+) than those negative for PPD and had lower levels of ALT elevation than carriers who were HBeAg(-). This suggests that either the presence of the Mycobacterium or the ability to immunologically react to it (or both) inhibits the replication of HBV and its ability to kill infected liver cells. This is reminiscent of the findings of Pearl and those reviewed by Nauts (see above). It could be associated, in part, with the ability of Mycobacterium to effect the binding and metabolism of iron, and this notion can be tested directly.

Serum Ferritin as a Predictor of Host Response to HBV Infection

Lustbader et al. (25) studied the role of serum ferritin as a predictor of host response to HBV infection. Our attention was, in part, focused on ferritin as a consequence of the extensive studies reported on ferritin and prognosis in neuroblastoma in children. Hann and her colleagues (26,27) found that elevated levels of ferritin predicted a poor outcome in patients with advanced neuroblastoma. In addition, the studies reviewed above (19) suggested that iron binding proteins were likely to distinguish between those who, after infection, were likely to become carriers from those who would develop anti-HBs. We predicted that individuals fated to be carriers would have higher ferritin levels before infection, than those who were to become transiently infected.

Three factors were already known that could be used to predict the development of the carrier state in an infected individual.

(1) Sex. Males were more likely to become carriers than females, and females were more likely to be only transiently infected and then develop anti-HBs.

(2) Age. Younger people (but over the age of about three months) were more likely to become carriers than adults.

(3) Geography. Certain populations were more likely to become carriers than others.

The ferritin study design was similar to that already described for the iron study. Dialysis patients who had at least one serum sample positive for HBsAg were selected. Two other criteria had to be satisfied for inclusion in the study: (1) the serum sample collected two months prior to the first detection of HBsAg had to be negative for both HBsAg and anti-HBs, and (2) a serum sample had to be obtained 12 months after the positive serum. Two possible outcomes, persistent infection or transient infection and five variables: sex, serum iron, serum ferritin, transferrin, and ALT (formerly SGPT) were considered. Data obtained on the serum sample collected prior to the positive sample were to be used to predict the outcomes (persistent or transient infection).

A difference in the direction was predicted in the sera collected before infection occurred in the mean log serum ferritin values of transiently and persistently infected individuals (Table 5). Persistently infected individuals had higher ferritin values than the transients. The difference was significant in females ($p < .001$) but not in males ($p = .11$). All the variables were then considered together in a stepwise linear discriminant analysis to determine the best combinations to predict outcome. As expected, sex was the most significant variable. Log serum ferritin was also significant (p .05). The best model to describe the predictive capacities of the variables was

$$y = 0.412 - 1.931\ S + 1.257\ F$$

where $S = 1$ for males, $S = 2$ for females, and F is the log (base 10) of serum ferritin concentration. The value of y predicts if the infected person will become transiently or chronically infected. Values of y greater than 0 predicted that the subject would develop a persistent infection, and values of y less than 0 that the subject would develop a transient infection.

Hence, the development of this quantitative formula allowed one to predict if a person would become a carrier or only transiently infected based on their sex and serum ferritin value. It also unexpectedly allowed

Table 5. Serum Ferritin as a Predictor of Host Response to HBV Infection

Classification	N	Serum iron (μg/dl)	Transferrin (mg/dl)	F	ALT	S
			Males			
Persistent	23	85.0±32.7	309± 86.9	1.98±0.62	8.83± 3.97	1
Transient	8	87.3±31.4	313±110	1.60±0.37	10.9 ± 6.45	1
			Females			
Persistent	6	132 ±63.6	310± 95.6	2.42±0.57	10.7 ±10.3	2
Transient	17	98.6±46.4	313± 98.1	1.64±0.51	7.24± 6.48	2
			All Patients			
Persistent	29	94.7±44.0	303± 87.0	2.07±0.63	9.21± 5.66	1.21
Transient	25	95.0±41.8	313± 99.7	1.62±0.46	8.40± 6.54	1.68

Data collected two months before HBsAg was found in the serum of patients undergoing hemodialysis treatment. F, log serum ferritin (serum ferritin was measured in nanograms per milliliter); ALT, alanylaminotransferase, measured in Karmen units (14); S, arbitrary value of 1 for males and 2 for females. Results are given as means ± standard deviations. Only the difference in F values for females was significant at the .05 level. The classification of infection as persistent or transient was made one year after HBsAg was first detected.

predictions of the ability of an infected individual to muster an antibody response to the surface antigen of the virus (anti-HBs) after becoming rid of HBsAg. This is described in the original paper; briefly, the analysis indicated that low levels of ferritin predisposed to the development of anti-HBs.

Does ferritin also have a role in determining if individuals vaccinated against HBV will or will not develop anti-HBs? In the field trials about 95% of vaccinated individuals develop antibody. The frequency is very much lower in dialysis patients. There is little known about why some inoculees do not develop antibody, and the possible role of ferritin in this process can be tested directly. It seems unlikely that the process would be the same as natural infection since killed vaccine is used and viability in liver cells is unlikely to be an issue in determining antibody production.

These findings made an important contribution to the pathogenetic model for primary hepatocellular carcinoma we were developing. Birgegard (28) had reported that the increased levels of ferritin seen during acute viral infections were the consequence of increased synthesis of ferritin. The model recommended by Lustbader et al. (25) postulates that the cells with an increased propensity to ferritin production are also more likely to become infected with HBV and to provide a suitable environment for replication. Infection with HBV, in turn, stimulates the production of even more ferritin in a positive feedback mechanism. This accounts for the increased levels of ferritin seen prior to infection and the even higher levels after infection has become persistent.

Iron, HBV, Death, and Disease

In the mid-1960's, the late Albert Damon conceived and executed a study of populations living in the Solomon Islands, some of which had only recently come into contact with western civilization and its medical baggage. Although the studies concentrated on hypertension and cardiovascular disease, unusually comprehensive medical and serological studies were also done. Near-total village populations of several Solomon Island communities were surveyed and followed up in subsequent years. The existence of this data base allowed a study of the long term mortality effects of HBV infection and also the interaction between HBV, iron binding protein levels and mortality. Accurate data or cause of death was not available in the Solomon Islands study, and only the end point of death was used in the analysis.

Early in the study of HBsAg we completed a cross sectional analysis of 1802 people resident on the island of Cebu in the central Philippines (29). The frequency of carriers was found to decrease with age. We offered three possible explanations: (1) There is a higher mortality in the carriers of HBsAg; (2) there is a decrease in the titer of HBsAg with increasing age; (3) both of these mechanisms are operating. There is considerable evidence that the titer of HBsAg does decrease with age, and the Solomon Islands population allowed us to examine the first explanation.

Details of this analysis are given in the original publication (30), and only general conclusions will be given here. Information on the HBV status was obtained from sera collected between 1966-1970, at the initiation of the surveys, and the mortality of individuals was obtained on revisits to the islands up to 1980. We compared HBV status at the initiation of the survey with the subsequent mortality experience. In both males and females, children (15 and under) with high titer anti-HBs are at decreased risk of death compared to children with no evidence of infection (chi-squared = 10.4, 1 d.f., $p < 0.05$). In the breakdown of the data by the different village communities the strongest association of anti-HBs with decreased mortality is seen in the communities with the highest infection rates for HBV. This suggests that the protective effect is primarily against HBV itself, although it may also protect against death from other causes.

Beasley and his colleagues in Taiwan have been following prospectively a group of male government workers with a high incidence of HBV infection. They found that HBsAg carriers had a much higher mortality risk (about three times greater than non-carriers) and that the risk was almost all attributable to death from PHC. Taken overall, the Solomon Islands study did not show any excess risk of death among the carriers. However, the Solomon Islands population is very much younger than that studied by Beasley, who selected, initially, workers 40-59 years of age.

To compare this Solomon Islands study with the Taiwan study, an analysis was done only of males who were aged 40-59 at the beginning of the study. There were 11 deaths among 31 HBsAg carriers compared to 20 deaths among 90 men who were not carriers, a relative risk of 1.8. These numbers are small, but they suggest that the excess mortality associated with HBsAg carriers does not occur until later in life and that the protective effect of the anti-HBs acts primarily earlier in life. This is compatible with the known pathogenesis of HBV in which the effects of chronic liver disease, cirrhosis and PHC are associated with long term exposure to HBV. Mortality in the young may be associated with acute events, including HBV infection, but also probably other childhood diseases.

<u>Iron Binding Proteins, HBV and Mortality in the Solomon Islands</u>

It is widely recognized that insufficient iron intake may result in iron deficiency anemia which is associated with the impairment of the immune response and increased susceptibility to infection. Since the end of the second World War, nutritional programs for the fortification of foods to increase iron intake have been instituted in some countries. This has been done in the expectation that they could, at a population level, decrease the frequency of anemia and enchance resistance to infections, among other beneficial effects. Crosby (31) has objected to these programs warning against the dangers of excess iron in members of the population who have hemochromatosis and other possible forms of toxicity.

As already discussed, excess amounts of iron may also increase the susceptibility to bacterial (and probably other) infections, and there are several clinical and epidemiologic studies which support this notion (see Weinberg (1)). The possible danger of excess iron was already perceived in 1868 by Trousseau, a professor of clinical medicine in Paris, who warned against excessive iron therapy because of the fear that it would induce clinical episodes of tuberculosis (32). Iron overload and a consequent increased susceptibility can occur in humans, as Weinberg point out, by (1) increased intake of iron from exogenous sources; (2) destruction of liver cells in which iron is stored bound to ferritin; (3) hemolysis which occurs in malaria, leukemias, lymphomas and hemoglobinopathies; and (4) decreased synthesis of iron binding protein (transferrin) as a result of poor nutrition, as in kwashiorkor.

In the Solomon Islands, infection has been a major cause of death. We tested the hypothesis that individuals who died during the observation period would be more likely to have increased iron stores at the time of entry than those who had survived (33). In normal people, serum ferritin levels are directly and transferrin levels indirectly related to body iron stores. Specifically, two hypotheses were tested: (1) the 1966-70 serum ferritin levels of people who had died by 1978 (group D) were higher than those of a control group who had survived to 1978 (group A); and (2) the serum transferrin levels in group D were lower than those in group A. A total of 156 deaths had occurred in the population of 2500. Each death was compared with a live control matched for age, sex, subpopulation group and HBV serology (HBsAg(+), anti-HBs, no response). There was sufficient serum available to constitute 105 pairs to test the transferrin hypothesis, and 94 pairs the ferritin hypothesis.

The differences were in the direction predicted for both transferrin and ferritin (Table 6), and the results were significant at the 0.05 level by a parametric test, non-parametric test, or both. An analysis of possible confounding factors (i.e. nutritional status, existing illness at the time the sera were collected, or the years at which the death occurred) indicated that these could not account for the differences observed. There was, however, an interaction with HBV status; the differences in ferritin were most striking in the HBsAg(+) carriers.

In order to study the joint effects of ferritin, transferrin and HBV on mortality, a conditional logistic regression model was applied to the data. The conclusions supported the findings from the single variable analysis, that is, that lower transferrin and higher ferritin (hence, probably higher iron stores) were associated with increased mortality and, in addition, showed that the effect was most evident in HBsAg carriers rather than in the other HBV response groups. The HBV interaction was seen only in females in this analysis. There was also an interaction with hemoglobin; individuals with lower hemoglobin levels at the beginning of the study had a greater probability of death. This combination of low hemoglobin and high iron stores is seen in chronic infections and chronic illness of various kinds.

These findings were taken to support the hypothesis that increased iron, particularly in HBsAg carriers predisposed to death, which by inference would have included many deaths from infection. We concluded the paper by stating that we did not think that the results warranted any change in nutritional practice, but we hoped that it would stimulate additional studies to learn if iron supplementation could, in some circumstances, be harmful to certain members of the population.

Neutrophil Function, HBV and Liver Cell Injury

Recent studies by Vierucci et al. (34) suggest an additional model which could reinforce this process. They studied neutrophil function in a group of nine children with chronic active hepatitis due to HBV and compared them to an equal number of controls. The patient observations were then reinforced by in vitro studies in which isolated neutrophil function was studied with and without the addition of (1) serum containing HBV; (2) isolated HBsAg; and (3) isolated HBcAg. They found profound alterations in the neutrophils of the patients compared to controls and in the neutrophils treated with the HBV preparations. There was an increase in oxidative metabolism in the resting cells of HBsAg carriers demonstrated by an increase in the production of superoxide anion generating activity and an increase in NBT reduction in unstimulated cells. The superoxide formation was reduced after the cells of the carriers were stimulated with zymosan. There was also a reduced capacity of the cells of the HBsAg carriers to kill ingested bacteria after 60 minutes of incubation. The decreased bactericidal activity was shown by studies on phagocytosis and killing of E. coli and Candida albicans. There was one effect seen in vitro but not in vivo, a decrease in chemotaxis. They reasoned that, in the intact host, some compensatory mechanism existed which overcame the direct effects of HBV on the cells.

The most significant effect of the HBV infection was the induction of the respiratory burst in the neutrophils, apparently as a consequence of

Table 6. Comparison of 1966-1970 Serum Ferritin and Transferrin Levels of Solomon Islanders Dead by 1978 (Group D) and Alive in 1978 (Group A).

	Ferritin	Transferrin
No. of pairs	94	105
Group D		
Mean	62.6 ng/ml	242.4 mg/dl
Range	2.2-290 ng/ml	109-394 mg/dl
SEM*	5.4	5.5
Group A		
Mean	51.6 ng/ml	269.4 mg/dl
Range	1.9-408 ng/ml	120-460 mg/dl
SEM*	5.5	5.6
Mean difference	11.0	-27.0
t statistic	1.5	3.2
One-tailed p value	0.07	0.001
Wilcoxon rank test	1728.5	1563.0
One-tailed p value	0.03	0.001

*SEM = standard error of the mean

the presence of the virus or of an antigen-antibody complex with HBV. Similar effects have been seen with virus-antibody complexes with influenza virus and syncytial virus. Presumably, the production of the oxygen radicals could aid host defense by destruction of the virus. It could be due to either increased production or decreased detoxification because of some inadequacy of enzymes such as superoxide dismutase or catalase. In due course, it might also adversely affect the phagocytic activity of the neutrophil.

In addition to the effects on invading organisms and on the neutrophils themselves, the oxygen radicals may also cause inflammation and tissue damage. They have been found to induce peroxidation in lysosomal membranes, depolymerize hyaluronate, affect myoblasts and activate platelet function. It has been suggested that high iron levels in synovial fluid in patients with rheumatoid arthritis can catalyze the formation of free radicals and cause damage to joint tissue (35). The combination of high iron levels and HBV in liver tissue may, therefore, cause severe cellular damage through the intermediary of the oxygen radicals.

A Model for Iron, Iron Binding Proteins and HBV Infection

It is now possible to construct a model based on the observations on iron and HBV in an attempt to explain in part the pathogenesis of HBV infection and disease and to recommend procedures that could affect their course.

Some individuals infected with HBV will have liver cells that contain relatively large amounts of ferritin or an augmented ability to generate the protein, and this could be under genetic control. Within a single liver, cells will be heterogeneous in respect to their ferritin content and potential to generate it. Cells with higher amounts of ferritin or the capacity to synthesize it would be more likely to become infected with HBV. This in turn would produce further increased amounts of ferritin. Infection with HBV would increase the iron content of the liver cells, and when this exceeds the binding capabilities of the ferritin, free iron levels would increase in the cell (as already suggested, the excess iron could come from the released storage iron in the liver cells which are killed by the infecting HBV, from macrophages and other inflammatory cells, from the iron accumulation associated with alcohol intake, and probably by other mechanisms). This would stimulate the production of oxygen radicals with consequent damage to the liver cells. Cell death would be followed by regeneration and scarring, leading to the pathological picture of advanced liver disease and cirrhosis.

The increased probability of HBV infection of the cells containing the high ferritin and iron levels may be a consequence of the large number of transferrin binding sites (transferrin receptors) on the surface of these cells. This notion is based on the testable hypothesis that HBV-cell attachment and entry points are the same as or related to the transferrin binding sites through which iron enters the cell.

Additional tissue destruction could result from the effects of HBV and iron on the phagocytic cells as proposed by Vierucci et al. (34). It would also diminish the host's ability to resist bacterial and other infections, as discussed above, and add to their mortality risks.

Iron, Iron Binding Proteins, HBV and the Development of PHC

The model can be extended to explain the relation of iron and iron binding proteins to HBV and PHC.

An important feature of the pathology of PHC is the absence of stainable iron in the neoplastic cells of the tumor; high levels of stainable iron may be seen in the non-malignant cells which surround them (22). Tumor cells contain a form of ferritin, acidic ferritin, which, according to some studies (3), does not bind iron well, while the non-tumor cells contain ferritin with strong iron binding characteristics (basic ferritin). In addition, the malignantly transformed cells, those with little or no iron, do not contain large quantities of or, in many cases, any whole HBV or its components. The surrounding non-malignant cells, those with large amounts of storage iron, may contain whole virus, HBsAg, and HBcAg. All cells, including tumor cells, require iron to grow and divide. The model suggests that the iron-containing cells, infected with replicating HBV, can supply the iron to be malignant cells and thereby stimulate their growth and division. This suggests a possible mechanisms for inhibiting the growth of the tumor by, in some way, inhibiting the transport of iron from the infected iron-containing cells to the malignant cells or by decreasing the overall supply of iron.

London and Blumberg (12,36) have proposed a model for the development of PHC in individuals who are carriers of HBV. They postulate the existence of two kinds of liver cells. The R cells are less differentiated cells common in the fetus but rare in the mature liver. When infected with HBV, the R cells may integrate the HBV-DNA into their own, but the virus does not replicate in the cells. The S cells are more fully differentiated cells which, when infected with HBV, may or may not undergo integration, but the virus can replicate in the cell.

The less differentiated R cells can divide to form either (1) two S cells during the process of differentiation; (2) an S cell and an R cell when destroyed liver cells are being replaced; (3) two R cells in the early fetus or at other times when the liver is growing at a rapid rate. The fully differentiated S cell, on the other hand, does not divide very readily and, if it does, can produce only other S cells. When an S cell is infected, replication of the virus will ensue and, in due course, the cell will die as a consequence of the immune responses of the host, the oxygen radical mechanisms already described, and other processes. Hence, the S cells are at a selective disadvantage compared to the R cells. The death of the S cells will stimulate the division of the R cells until a point is reached when uncontrolled division occurs with the production of two R cells each time an R cell divides. This may occur after the HBV-DNA becomes integrated into the R cell DNA, thus controlling its division. When the R cell mass becomes sufficiently large, it is perceived by the patient and recognized as clinical cancer.

It is possible to relate the London-Blumberg R/S developmental model for PHC to the iron-HBV model already described. The less differentiated R cells would be the equivalent of the malignant cells which contain little or no stainable iron. Whole virus and parts of the virus would be absent from these cells, although integration may have occurred. The S cells would correspond to the cells with the basic ferritin and large amounts of stored iron which are readily invaded by HBV and in which viral replication occurs. They are killed by the oxygen radicals and other processes as a consequence of the long HBV infection. The death of the S cells stimulates the division of the R cells. The continuous process of death, scarring, cell division and the increase of the R cell component over time results in the pathological and clinical picture of chronic liver disease, cirrhosis and primary cancer of the liver.

<u>Altering Pathogenesis</u>

This model recommends an approach to "therapy" which is different from that usually used in cancer treatment. Most cancer treatments are directed to the killing of as many cancer cells as possible. Since normal cells, particularly developing and dividing cells, are similar to cancer cells, they will also be killed in the process, thus limiting the aggressiveness and extent of therapy.

It is now known that carriers of HBV (at least in some geographical regions) are at a very high risk of developing PHC, and it may, in due course, be possible to identify those at greatest risk. Hence, measures should be developed to slow down the pathological process to allow the carrier to remain asymptomatic until he or she can live out a normal life span. We have referred to this (37) as "prevention by delay." Several possibilities recommend themselves that could be applied to the carriers before symptoms develop and before uncontrolled division (of the R cells) commences.

Can the replication of the virus be decreased by inhibiting the amount of iron (and/or ferritin) available to the host? Does HBV enter the cells in the same place or in the same manner as iron? Is there some measure that can be used to competitively inhibit this entry? Is the polymerized albumin binding site (38) such a location? Can replication of HBV be inhibited by inactivating its polymerase in a manner which is not also detrimental to the host?

It is hoped that as these models are tested and developed, practical intervention measures to prevent the development of disease in carriers can be discovered and implemented.

ACKNOWLEDGMENTS

This work was supported by USPHS grants CA-06551, RR-05539 and CA-06927 from the National Cancer Institute and by an appropriation from the Commonwealth of Pennsylvania.

REFERENCES

1. E. D. Weinberg, Iron withholding: a defense against infection and neoplasia, Physiological Reviews 64:65 (1984).
2. A. L. Schade and L. Caroline, An iron-binding component in human blood plasma, Science 104:340 (1946).
3. M. Worwood, Ferritin in human tissues and serum, Clinics in Hematology 11:275 (1982).
4. J. W. Drysdale, T. G. Adelman, P. Arosio, D. Casarcale, P. Fitzpatrick, J. T. Hazard, and M. Yokota, Human isoferritins in normal and disease states, Seminars in Hematology 14:71 (1977).
5. R. Pearl, Cancer and tuberculosis, Am. J. Hyg. 9:97 (1929).
6. H. C. Nauts, The apparently beneficial effects of bacterial infection on host resistance to cancer: end results in 435 cases, Cancer Res. Inst. Monograph 8:1 (1958).
7. B. S. Blumberg, B. J. S. Gerstley, D. A. Hungerford, W. T. London, and A. I. Sutnick, A serum antigen (Australia antigen) in Down's syndrome, leukemia and hepatitis, Ann. Int. Med. 66:924 (1967).
8. A. I. Sutnick, W. T. London, B. S. Blumberg, and B. J. S. Gerstley, Persistent anicteric hepatitis with Australia antigen in patients with Down's syndrome, Am. J. Clin. Path. 57:2 (1972).
9. A. J. Kolk-Vegter, E. Bosch, and A. M. Leeuwen, Influence of serum hepatitis on haemoglobin level in patients on regular haemodialysis, Lancet 1:526 (1971).
10. A. I. Sutnick, B. S. Blumberg, and E. Lustbader, Elevated serum iron levels and persistent Australia antigen (HBsAg), Ann. Int. Med. 81:855 (1974).

11. B. S. Blumberg, B. Larouze, W. T. London, B. Werner, J. E. Hesser, I. Millman, G. Saimot, and M. Payet, The relation of infection with the hepatitis B agent to primary hepatic carcinoma, Am. J. Pathol. 81:669 (1975).
12. W. T. London and B. S. Blumberg, Hepatitis B and related viruses in chronic hepatitis, cirrhosis and hepatocellular carcinoma in man and animals in: "Chronic Active Liver Disease," S. Cohen and R. D. Soloway, eds., Churchill Livingstone, New York (1982).
13. C. Felton, E. D. Lustbader, C. Merten, and B. S. Blumberg, Serum iron levels and response to hepatitis B virus, Proc. Natl. Acad. Sci. USA 76:2438 (1979).
14. A. Vierucci, W. T. London, B. S. Blumberg, A. I. Sutnick and F. Ragazzini, Australia antigen and antibody in transfused children with thalassemia, Arch. Dis. Childhood 47:760 (1972).
15. W. T. London and J. S. Drew, Sex differences in response to hepatitis B infection among patients receiving chronic dialysis treatment, Proc. Natl. Acad. Sci. USA 74:2561 (1977).
16. S. Mazzur, J. Corbett, and B. S. Blumberg, Loss of immunologic reactivities of Australia antigen after incubation with bacteria, Proc. Soc. Exp. Biol. Med. 142:327 (1973).
17. I. Millman, H. Hutanen, F. Merino, M. E. Bayer, and B. S. Blumberg, Australia antigen: physical and chemical properties, Res. Commun. Chem. Path. Pharm. 2:667 (1971).
18. E. Feret, B. S. Blumberg, P. Whitford, and E. Lustbader, Serum iron levels and the response to infection with hepatitis B virus in the population of Tip, Senegal, West Africa, in preparation (1984).
19. B. S. Blumberg, E. D. Lustbader, and P. L. Whitford, Changes in serum iron levels due to infection with hepatitis B virus, Proc. Natl. Acad. Sci. USA 78:3222 (1981).
20. J. S. Drew, W. T. London, E. D. Lustbader, J. E. Hesser, and B. S. Blumberg, Hepatitis B virus and sex ratio of offspring, Science 201:687 (1978).
21. J. Drew, W. T. London, B. S. Blumberg, and S. Serjeantson, Hepatitis B virus and sex ratio on Kar Kar Island, Human Biology 54:123 (1982).
22. H. A. Edmondson and P. E. Steiner, Primary carcinoma of the liver, Cancer 7:462 (1954).
23. J. A. Fernandez-Pol, Iron: possible cause of the G_1 arrest induced in NRK cells by picolinic acid, Biochem. Biophys. Res. Commun. 78:136 (1977).
24. K. A. McGlynn, E. D. Lustbader, and W. T. London, Tuberculin skin test reactivity and hepatitis B e antigen status, in preparation (1984).
25. E. D. Lustbader, H. L. Hann, and B. S. Blumberg, Serum ferritin as a predictor of host response to hepatitis B virus infection, Science 220:423 (1983).
26. H. L. Hann, A. E. Evans, I. J. Cohen, and J. E. Leitmeyer, Biologic differences between neuroblastoma stages IV-S and IV. Measurement of serum ferritin and E-rosette inhibition in 30 children, N. Engl. J. Med. 305:425 (1981).
27. H. L. Hann, A. E. Evans, S. E. Siegel, H. Sather, D. Hammond, and R. Seeger, Serum ferritin levels as a guide to prognosis in patients with stage IV neuroblastoma, Proc. Am. Soc. Clin. Oncol. 2:72 (Abstract #C-282) (1983).
28. G. Birgegard, The source of serumferritin during infection. Studies with concanavalin A-Sepharose absorption, Clin. Sci. 59:385 (1980).
29. B. S. Blumberg, L. Melartin, R. A. Guinto, and B. Werner, Family studies of human serum isoantigen system (Australia antigen), Am. J. Human Genet. 18:594 (1966).
30. R. G. Stevens and B. S. Blumberg, Hepatitis B in the Solomon Islands, in: "Health Effects of Westernization: Biomedical Survey in the Solomon Islands," J. Rhoads, W. Howells, and J. Friedlaender, eds., Oxford University Press, in press (1984).

31. W. H. Crosby, Editorial: Fortification of food with carbonyl iron, Am. J. Clin. Nutr. 31:572 (1978).
32. A. Trousseau, True and false chlorosis, in: "Lectures on Clinical Medicine," Vol. 5, Lindsay and Blakiston, Philadelphia (1872).
33. R. G. Stevens, S. Kuvibidila, M. Kapps, J. Friedlaender, and B. S. Blumberg, Iron-binding proteins, hepatitis B virus, and mortality in the Solomon Islands, Am. J. Epidemiol. 118:550 (1983).
34. A. Vierucci, M. de Martino, E. Graziani, M. E. Rossi, W. T. London, and B. S. Blumberg, A mechanism for liver cell injury in viral hepatitis: effects of hepatitis B virus on neutrophil function in vitro and in children with chronic active hepatitis, Pediatric Res. 17:814 (1983).
35. D. R. Blake, N. D. Hall, P. A. Bacon, P. A. Dieppe, B. Halliwell, and J. M. C. Gutteridge, The importance of iron in rheumatoid disease, Lancet 2:1142 (1981).
36. B. S. Blumberg and W. T. London, Hepatitis B virus: pathogenesis and prevention of primary cancer of the liver, Cancer 50:2657 (1982).
37. B. S. Blumberg and W. T. London, Hepatitis B virus pathogenesis and prevention o. primary cancer of the liver, in: "Accomplishments in Cancer Research 1981, General Motors Cancer Research Foundation," J. G. Fortner and J. E. Rhoads, eds., J. B. Lippincott Company, Philadelphia (1983).
38. S. G. Franklin, I. Millman, and B. S. Blumberg, Hepatitis B surface antigen and polymerized albumin binding activity in sheep serum, Proc. Natl. Acad. Sci. USA 81:564 (1984).

Reprinted from
Proc. Natl. Acad. Sci. USA
Vol. 76, No. 5, pp. 2438–2441, May 1979
Medical Sciences

Serum iron levels and response to hepatitis B virus

(renal dialysis/aspartate aminotransferase/cancer/hepatitis carrier/antibody against hepatitis B)

CRAIG FELTON, EDWARD D. LUSTBADER, CYNTHIA MERTEN, AND BARUCH S. BLUMBERG

The Institute for Cancer Research, The Fox Chase Cancer Center, Philadelphia, Pennsylvania 19111

Contributed by Baruch S. Blumberg, February 12, 1979

ABSTRACT Response to hepatitis B virus (HBV) infection [HBV surface antigen (HBsAg) and antibody to HBsAg (anti-HBs)], serum iron, total iron-binding capacity, hematological status (erythrocytes, Hb, and hematocrit), and evidence of liver damage (serum glutamic pyruvic transaminase; aspartate aminotransferase, L-aspartate:2-oxoglutarate aminotransferase, EC 2.6.1.1) were determined for 201 patients on chronic renal dialysis. Four factors—serum iron level, transaminase level, sex, and HBV response [i.e., infected-HBsAg(+) (HBsAg positive), anti-HBs(+) (anti-HBs positive), or no response]—were analyzed simultaneously to test the hypothesis that serum iron is higher in those with HBsAg in their serum than in those without HBsAg, independent of the transaminase level. Four independent, statistically significant two-factor interactions were identified. (*i*) Serum iron is higher in those HBsAg(+). (*ii*) Serum iron is higher in those with increased transaminase. (*iii*) Transaminase is higher in those HBsAg(+). (*iv*) Males are more likely to be HBsAg(+) and females are more likely to be anti-HBs(+). Also, those who are HBsAg(+) have significantly higher percent iron saturation (serum iron/total iron-binding capacity). That is, the hypothesis was supported by the findings. Several additional biological hypotheses are suggested, including a possible role of increased iron levels in susceptibility and response to HBV infection and the possible relationship between higher iron levels and the likelihood of HBV infection progressing to primary hepatocellular carcinoma. In addition, further tests of the initial hypothesis in nonhospitalized populations with endemic HBV infection are proposed.

The role of iron in susceptibility and response to infectious disease (primarily bacterial but possibly viral) has been discussed for some years and has recently been summarized (1). Serum iron becomes elevated in hepatitis, presumably released from hepatocytes as a result of cellular damage. We found that males with Down syndrome who had hepatitis B surface antigen (HBsAg) in their blood had significantly higher levels of serum iron than those who did not have HBsAg (2). However, the serum iron level was not correlated with the level of serum glutamic pyruvic transaminase (SGPT) (aspartate aminotransferase; L-aspartate:2-oxoglutarate aminotransferase, EC 2.6.1.1), which is considered to be an indication of liver cell damage. Therefore, it appeared that the higher serum iron in the HBsAg(+) (HBsAg positive) individuals was not due only to the release of cellular iron which accompanies liver damage. Based on these observations, we decided to study a second group of individuals at high risk of hepatitis B virus (HBV) infection. Specifically, we tested the hypothesis that the serum iron level is higher in patients on chronic renal dialysis who have HBsAg in their blood than those who have antibody to HBsAg (anti-HBs) or those who lack both HBsAg and anti-HBs (no evidence of infection).

There are additional reasons for learning more about the relationship of iron to hepatitis infection. Tumor cells may require iron for effective growth and spread (3). Infection with HBV is significantly associated with chronic liver disease and primary cancer of the liver in Asia, Africa, and other regions where these are common diseases. The relation may be etiologic, and we are currently studying strategies for the control of HBV infection. In connection with this, it is important to identify interactions that might contribute to the development of cancer. In the tropics, iron metabolism in humans is affected by a large number of environmental and pathologic conditions, including nutrition, methods of food preparation, hemolytic anemias, and bacterial and viral infection (3, 4). Further study of iron in relation to hepatitis B infection may contribute to an understanding of this complex problem.

METHODS

Sera were collected in iron-free tubes over a 4-day period from patients in two commercial renal dialysis clinics operated by Biomedical Applications of Philadelphia. Serum iron, total iron-binding capacity, SGPT, HBsAg, and anti-HBs were determined at The Institute for Cancer Research. Serum iron and total iron-binding capacity were determined colorimetrically with Hyland Ferro-chek II kits (Travenol, Costa Mesa, CA), HBsAg by the Ausria II method (Abbott), and anti-HBs by direct hemagglutination with HBsAg-coated erythrocytes (Electronucleonics Laboratories, Bethesda, MD) (5). Sera were considered positive for anti-HBs if they agglutinated cells at a titer of 1/4 or greater. SGPT was determined by a modification of the method of Henry *et al.* (6). All other laboratory tests were performed by the commercial laboratory (Metpath, Philadelphia, PA) that routinely provides services for the dialysis clinic.

PATIENT POPULATION

A total of 201 sera were collected. The distribution of patients by HBV response and sex is shown in Table 1. The patients included in this study were receiving dialysis three times each week. All of the patients routinely receive iron therapy, either Fe_2SO_4 (325 mg three times a day) or intravenous iron (Imferon). The response of the patient to hepatitis B infection is not considered in determining the iron therapy. Further, renal dialysis patients have a chronic anemia apparently unrelated to iron deficiency.

RESULTS AND DISCUSSION

Examination of the present data (Table 2) began with an analysis similar to that of the Down syndrome study (2). Since only males were available in the Down syndrome study, just the male dialysis patients were used for this analysis. There was a significantly higher level of serum iron in male patients with HBsAg than in those without HBsAg. SGPT levels in HBsAg(+)

The publication costs of this article were defrayed in part by page charge payment. This article must therefore be hereby marked "*advertisement*" in accordance with 18 U. S. C. §1734 solely to indicate this fact.

Abbreviations: HBV, hepatitis B virus; HBsAg, HBV surface antigen; anti-HBs, antibody to HBVsAg; HBsAg(+), HBsAg positive; anti-HBs(+), anti-HBs positive; SGPT, serum glutamic pyruvic transaminase.

Table 1. Distribution of renal dialysis patients by sex and response to HBV infection

	Male	Female	Total
HBsAg(+)	27	12	39
Anti-HBs(+)	36	40	76
No evidence of infection	50	36	86
Total	113	88	201

patients were, as expected, higher than the levels in HBsAg(−) patients. The measurements of total iron-binding capacity, Hb, and hematocrit in patients with HBsAg were also compared with those of the patients without HBsAg, and no significant differences were found. These findings are consistent with those of the Down syndrome study.

In this investigation more data was available than in the Down syndrome study because females were included and anti-HBs status was determined. The differences in serum iron found between those who were HBsAg(+) and the other HBV response groups occurred primarily in the males. Percent iron saturation was higher in HBsAg(+) patients than in all others, but the only statistically significant difference occurred in the males. The SGPT levels were significantly lower in patients with anti-HBs than in the HBsAg(+) and apparently uninfected patients. There were also significant differences between males and females in erythrocyte, Hb, and hematocrit levels (males had higher values). Among the HBV response groups the difference was significant only in anti-HBs(+) patients. Some interesting biological models relating to differences between males and females in response to hepatitis B infection could be made if these findings are supported in subsequent investigations.

We then proceeded to search for interactions between the four variables: serum iron, sex, HBV response, and SGPT level. Iron levels were scored as low, middle, or high based on a division of the data into three groups containing approximately equal numbers of individuals. The distribution of the data according to these four variables is shown in Table 3.

The testing of a series of hypotheses indicated that the data could be explained by considering only two-factor interactions. (Details of this analysis are given in the *Appendix*.) Pairwise

Table 3. Distribution of data according to HBV response, SGPT level, serum iron level, and sex

	Low iron		Middle iron		High iron	
	Male	Female	Male	Female	Male	Female
Low SGPT (≦15 international units/liter)						
HBsAg(+)	3	2	6	2	5	3
Anti-HBs(+)	12	11	11	14	6	11
No evidence of infection	19	9	10	9	6	3
High SGPT (>15 international units/liter)						
HBsAg(+)	1	0	1	1	11	4
Anti-HBs(+)	4	0	1	0	2	3
No evidence of infection	1	3	4	3	8	9

Three of the individuals from Table 1 had incomplete results and are not included.

interactions of these four variables were then evaluated to see how much each contributed to the explanation of the data. The best model consisted of the iron and HBV, iron and SGPT, HBV and SGPT, and sex and HBV interactions.

To determine whether any of these interactions were due to the effect of transfusions and intravenous iron therapy received by some patients, we removed all data from patients receiving either treatment within 1 month prior to iron and HBV testing and reanalyzed the remaining data. HBV response, SGPT level, and serum iron were considered, but not sex due to insufficient numbers. The interactions between iron and HBV and between iron and SGPT remained significant, while the interaction between HBV and SGPT was only marginally significant. The best model included the interactions between iron and SGPT and between iron and HBV and not the interaction between HBV and SGPT.

The interactions included in these models are consistent with previous observations. London and Drew (7) have shown that, in renal dialysis patients (which include the subjects of this study), males are significantly more likely to become carriers of HBsAg and females to develop anti-HBs. The virus and SGPT interaction has been known for many years (see for example, ref. 8) and was one of the criteria for the establishment of the association between Australia antigen and hepatitis. An

Table 2. Mean values for groups of renal dialysis patients

	SI, mg/dl			TIBC, mg/dl			% iron saturation			SGPT, international units/liter		
	Male	Female	Total	Male	Female	Total	Male	Female	Total	Male	Female	Total
HBsAg(+)	114	214	145	278	305	285	42.4	46.5	43.6	28.1	20.8	25.8
	**		**				**		**	**	**	***
Anti-HBs(+)	95*	107	102**	304	284	294	29.6*	36.3	33.0**	11.5*	6.2	10.8
											***	***
No evidence of infection	103	95	99	283	281	282	34.6	33.4	34.1	14.8	21.4	17.2
Total	103	117		288	286		34.9	36.4		16.9	15.9	

	RBC, 10^6 cells/mm^3			Hb, g/dl			Hematocrit, %		
	Male	Female	Total	Male	Female	Total	Male	Female	Total
HBsAg(+)	2.6	2.2	2.5	6.8	6.7	6.8	22.4	21.8	22.2
Anti-HBs(+)	2.6	* 2.3	2.5	7.1	** 6.3	6.7	23.5	* 20.8	22.1
No evidence of infection	2.4	2.3	2.3	6.7	6.4	6.6	21.9	20.8	21.5
Total	2.5	** 2.3		6.9	* 6.4		22.6	* 20.0	

Differences were assessed by the Mann–Whitney test. % iron saturation = serum iron (SI)/total iron-binding capacity (TIBC). RBC, red blood cells. *, $P = 0.05$; **, $P = 0.01$; ***, $P = 0.001$.

increase of serum iron in association with liver damage (i.e., elevation of SGPT) has been clinically recognized in patients with hepatitis and liver disease. The interaction between iron and HBV was shown in the Down syndrome study and confirmed in this study. However, it was not recognized until this analysis that all four interactions act independently and simultaneously and, in particular, that the presence of HBsAg by itself, independent of SGPT elevation, is associated with elevations of serum iron.

Implications

This study indicates that the association of elevated serum iron with the presence of HBsAg in the serum is due not only to the breakdown of hepatic cells, but also to the presence of HBsAg itself. It is remarkable that these differences can be detected in this patient population given the possible masking effects of iron therapy, the anemia of chronic dialysis, and the decrease of iron absorption due to the use of aluminum hydroxide (9).

We now wish to determine if these same associations exist in other populations with a high frequency of HBV infection—for example, nonhospitalized individuals in Africa, Asia, or Oceania. If our findings are supported in these populations, then there are some interesting biological and pathological questions that can be raised. All living cells require iron. In humans, a method of defense after microbial infection is to decrease the available iron by decreasing intestinal absorption and increasing liver storage. Do HBsAg carriers have a different response when faced with this stress? Are the increased iron levels a consequence of the HBV infection, or are individuals with high iron levels (or a propensity to develop them) more likely to become HBsAg carriers? Recently some relationships between HBV and bacteria have been observed. Incubation of HBsAg with *Pseudomonas* results in the inability to detect HBsAg by immunological methods, although the appearance of the surface antigen is not markedly changed (ref. 10; L. K. Weng and W. T. London, unpublished data). Millman and McMichael (11) have identified a glycoprotein present in various animal sera that can bind to HBsAg and prevent reaction with anti-HBs. Could the bacteria or glycoprotein contribute to increased levels of iron by preventing the recognition of HBsAg by the immune system? Millman *et al.* (12) have reported that transferrin is one of the host proteins associated with HBsAg. Transferrin saturated with iron (as opposed to unsaturated transferrin) is selectively bound to liver cells (13). The transferrin is incorporated (by pinocytosis), and iron is released into the cell. Iron-laden transferrin on the surface antigen of HBV may facilitate the entry of the virus into the hepatocyte. It would be interesting to know if iron is detectable in appreciable amounts in purified HBsAg. Perhaps HBV interacts with normal erythropoietic mechanisms to alter utilization of iron. If there are additional stores of iron in HBsAg carriers, does that enhance their propensity to develop or foster the growth of primary hepatocellular carcinoma, and, if so, could this information be useful in prevention or treatment of this disease?

APPENDIX ON STATISTICAL METHODS

The second part of the data analysis presented here was inductive in that there was no predetermined order imposed on the variables. Hence, the tests of significance ought to be considered as conjectures or new hypotheses rather than conclusions. An independent set of data is necessary to verify the conjectures. Nonetheless, a systematic approach to identifying the variables or combinations of variables that seem to present the most satisfactory explanation of the data is useful in providing clues to further research.

The data consist of the number of individuals having complete results on the four variables—iron (I), sex (S), HBV response (V), and SGPT (T). I and V have three possible categories, while S and T have two. Hence, a 3 × 2 × 3 × 2 contingency table results (Table 3). Logarithmic linear models are used to estimate expected values for each bin of the table. Then these are compared by maximum likelihood procedures to the observed values (14). The models are hierarchal; that is, if the IV interaction is present in the model, then I and V individually are included. The inclusion of a variable or combination of variables in a model implies that the estimated expected values are constrained to have the same marginal totals as the observed values.

The analysis consisted of a sequence of hypotheses. First, a series of general hypotheses were tested in order to determine the highest order of interaction among the variables (d.f. is the degrees of freedom).

	χ^2	d.f.	P
H1: All bins equally likely	60.83	6	<0.0001
H2: No two-factor interactions	62.28	13	<0.0001
H3: No three-factor interactions	11.83	12	>0.74
H4: No four-factor interactions	6.12	4	>0.15

The nonsignificance of H3 and H4 implies that attention should be directed to the two-factor interactions. Marginal and partial associations were computed for all the two-factor effects in order to decide which of these seem to be necessary (15).

		Marginal association		Partial association	
	d.f.	χ^2	P	χ^2	P
H5: No IS effect	2	2.33	>0.25	1.19	>0.5
H6: No IV effect	4	9.16	>0.05	12.5	<0.02
H7: No IT effect	2	26.52	<0.0001	30.5	<0.0001
H8: No SV effect	2	5.34	>0.05	4.84	>0.05
H9: No ST effect	1	0.12	>0.5	0.26	>0.5
H10: No VT effect	2	11.57	<0.005	16.19	<0.001

Combining the results of testing H6, H7, and H10, a model consisting of two-factor interactions—IV, IT, and VT—was tested. The maximum likelihood χ-squared was 28.05 with 22 d.f. for a $P > 0.1$. Thus, the expected values did not deviate significantly from the observed, and an apparently significant model had been obtained.

For each bin of the table, the contribution to the Pearson goodness-of-fit χ-square was computed. Surprisingly, the bin identified by anti-HBs(+) males with low serum iron and high SGPT contributed nearly 50% to the total χ-squared statistic. This value was excluded and hypotheses H1–H10 were reanalyzed as H11–H20. The results were as follows:

	χ^2	d.f.	P
H11: All bins equally likely	60.37	6	<0.0001
H12: No two-factor interactions	72.1	13	<0.0001
H13: No three-factor interactions	4.47	12	>0.5
H14: No four-factor interactions	3.67	4	>0.25

		Marginal association		Partial association	
	d.f.	χ^2	P	χ^2	P
H15: No IS effect	2	1.75	>0.25	0.83	>0.5
H16: No IV effect	4	13.18	<0.02	9.7	<0.05
H17: No IT effect	2	34.0	<0.0001	33.6	<0.0001
H18: No SV effect	2	4.03	>0.1	6.83	<0.05
H19: No ST effect	1	0.22	>0.5	0.12	>0.5
H20: No VT effect	2	19.19	<0.0001	19.46	<0.0001

Again, the model containing IV, IT, and VT was fit, yielding a maximum likelihood χ-square of 17.79 with 21 d.f. and $P > 0.7$. The contribution of each bin to the goodness-of-fit χ-square revealed no individually significant results. However, the pattern of deviations suggested that a further improvement might be obtained by inclusion of an interaction containing the sex variable. Note that H18 is ambiguous as to the importance of the SV effect. Hence, the model containing IV, IT, VT, and SV was fit. The maximum likelihood χ-square was 9.51 based on 18 d.f. with $P > 0.95$. These four two-factor interactions thus represent a quite satisfactory explanation of the data with the stipulation that one bin is aberrant.

This work was supported by U.S. Public Health Service Grants CA-06551, RR-05539, CA-06927, and CA-22780 from the National Institutes of Health and by an appropriation from the Commonwealth of Pennsylvania.

1. Weinberg, E. D. (1974) *Science* **184,** 952–956.
2. Sutnick, A. I., Blumberg, B. S. & Lustbader, E. D. (1974) *Ann. Int. Med.* **81,** 855–856.
3. Jacobs, A. (1977) *Semin. Hematol.* **14,** 89–113.
4. Mazzur, S., Corbett, J. & Blumberg, B. S. (1973) *Proc. Soc. Exp. Biol. Med.* **142,** 327–332.
5. Vyas, G. N. & Schulman, N. R. (1970) *Science* **170,** 332–333.
6. Henry, R. J., Chiamori, N., Golub, O. J. & Berkman, S. (1960) *Am. J. Clin. Pathol.* **34,** 381–398.
7. London, W. T. & Drew, J. S. (1977) *Proc. Natl. Acad. Sci. USA* **74,** 2561–2563.
8. Sutnick, A. I., London, W. T., Gerstley, B. J., Cronlund, M. M. & Blumberg, B. S. (1968) *J. Am. Med. Assoc.* **205,** 670–674.
9. Rastogi, S. P., Padilla, F. & Boyd, C. M. (1975) *Am. Soc. Nephrol.* **8,** 21.
10. Weng, L. K., Bayer, M. E. & London, W. T. (1975) *Infect. Immun.* **12,** 180–186.
11. Millman, I. & McMichael, J. C. (1978) *Infect. Immun.* **21,** 879–885.
12. Millman, I., Hutanen, H., Merino, F., Bayer, M. E. & Blumberg, B. S. (1971) *Res. Commun. Pathol. Pharmacol.* **2,** 667–686.
13. Aisen, P. & Brown, E. B. (1977) *Semin. Hematol.* **14,** 31–53.
14. Goodman, L. A. (1971) *J. Am. Stat. Assoc.* **66,** 339–344.
15. Brown, M. B. (1976) *Appl. Stat.* **25,** 37–46.

Reprinted from *Science*
Vol. 220, pp. 423–425

Serum Ferritin as a Predictor of Host Response to Hepatitis B Virus Infection

Abstract. *With hemodialysis patients, a high serum ferritin before there was serological evidence of hepatitis B virus infection increased the likelihood that the infection would be persistent. This finding suggested that hepatitis B virus is likely to infect and actively replicate in liver cells with the propensity for increased ferritin synthesis. The virus itself could stimulate the synthesis of ferritin in a cyclic positive feedback mechanism that increases intracellular ferritin concentration and, eventually, intracellular iron. Transformed liver cells have low iron content, do not replicate hepatitis B virus, and require iron for growth. Infected, nonmalignant liver cells could supply iron to the transformed cells and nourish their expansion.*

Variation in the response of humans to infection by hepatitis B virus (HBV) is one of the most perplexing aspects of the biology of this virus. Females are more likely than males to make antibody to the viral surface antigen (HBsAg), whereas males are more likely to become carriers and to have HBsAg persisting in their blood. The prevalence of antibody to HBsAg increases with age, whereas surface antigen frequency increases initially and then declines after the teenage years in most populations.

To identify factors that might contribute to the carrier state, we examined males with Down's syndrome because they are highly susceptible to HBV infection (*1*). We measured serum iron, in addition to other variables, because iron is often increased in patients with hepatitis. We found that serum iron and transferrin saturation values were higher in the patients who were carriers than in those who were not carriers. These findings were confirmed in a study of male and female patients undergoing hemodialysis treatment (*2*).

Those studies established that the association of increased serum iron with the HBV carrier state was independent of the increased serum iron resulting from liver cell breakdown as measured by alanylaminotransferase (ALT; formerly referred to as SGPT). However, it was not possible to ascertain whether the increased serum iron was a consequence of HBV infection or, alternatively, whether the carrier state occurred more frequently in individuals with increased serum iron.

In a retrospective cohort study of patients treated with hemodialysis, changes in iron level were measured during a period in which HBV infection was known to have occurred (*3*). Three conditions had to be met for a patient to be included in the study: (i) the patient had to have at least one serum sample that was positive for HBsAg; (ii) the serum sample obtained 2 months before HBsAg was first detected had to be negative for both HBsAg and its antibody; and (iii) a serum sample had to be obtained 12 months after HBsAg was first detected (*4*). Fifty-four patients who met these criteria during the period 1974 to 1981 were included in the analysis. The key finding was that the amount of iron in the serum before the initial finding of HBsAg was no greater in individuals destined to be carriers than in those who proved to be transiently infected.

In blood, iron is transported by transferrin and is stored primarily in the liver and spleen in the form of ferritin. Although most of the ferritin in the body is intracellular, it is found in low concentrations in the serum. In normal individnals, serum ferritin is highly correlated with total body iron stores (*5*). For this reason, the analysis of the hemodialysis patients was expanded to include transferrin and serum ferritin levels.

Data collected before HBsAg was first detected in the serum were used to predict the outcome 12 months after this finding. Two outcomes were predicted: persistent infection or transient infection. Five variables were considered: sex, serum iron, serum ferritin, transferrin, and ALT (*6*).

Table 1 shows a significant difference for mean log serum ferritin values of transiently and persistently infected females ($P < .001$); the corresponding values for males gave $P = .11$. Thus serum ferritin appeared to be the best candidate for improving the simple prediction that males will be persistently infected and females will be transiently infected.

A stepwise linear discriminant analysis was undertaken to search for the best combination of variables that predict outcome. All of the data were combined and, as expected, sex was the most significant variable. In addition, log serum ferritin was also significant ($P < .05$). The best model was $y = 0.412 - 1.931\ S + 1.257\ F$, where $S = 1$ for males, $S = 2$ for females, and F is the log (base 10) of the serum ferritin concentration. Values of $y > 0$ predict that the subject has a persistent infection and $y < 0$ that the subject has a transient infection.

If the mean values of S and F from the last two rows of Table 1 are used in the model, y is 0.68 for carriers and −0.79 for those with transient infections. This suggests a criterion for dividing the population into four groups: (i) those with y values less than −0.79, (ii) those with y values between −0.79 and 0.00, (iii) those with y values between 0.00 and 0.68, and (iv) those with y values greater than 0.68.

The model makes no formal prediction about the presence or absence of antibody to HBsAg. Nevertheless, the patients known to have transient infections can be further partitioned according to their antibody status 1 year after the initial finding of HBsAg (Table 2). Those patients who still had no antibody to HBsAg had y values that were relatively uniformly distributed across the four categories of Table 2, and only five of the ten patients had been correctly predicted to have a transient infection. However, 13 of 15 individuals who had antibody to HBsAg were correctly classified as transiently infected, with the majority having y values less than −0.79. Little is known about the dynamics of the metalloproteins in healthy individuals. Therefore, the observation that an increase in serum ferritin before HBV infection can be used to predict the response to the infection provides a new direction for research.

Birgegard (*7*) proposed that increased serum ferritin during unspecified acute viral infections was the result of augmented synthesis rather than release from damaged cells. The model that we derived suggests that the propensity for augmented synthesis, as indicated by initial serum ferritin levels, can increase cellular susceptibility to persistent HBV infection. Taken together, these two proposals indicate a positive feedback mechanism that results in an increased intracellular ferritin concentration. The augmented synthesis need not be abnormal. Even though the use of normal values in hemodialysis patients is questionable, the mean serum ferritin levels we analyzed were within the normal range.

The attributable risk of persistent infection with HBV to primary hepatocellular carcinoma (PHC) is nearly 100 percent (*8*). We had speculated earlier that transformed liver cells depend on surrounding cells to provide iron for their growth (*3*). This hypothesis was supported by four lines of evidence: (i) cancer cells have much less iron storage than nonmalignant cells, as judged by staining (*9*); (ii) HBV is actively replicated in the normal liver cells of patients with PHC, but not in the transformed cells (*10*); (iii) removal of iron decreases the growth of cells in culture (*11*); and (iv) carriers of HBV have increased serum iron (*1–3*).

Primary hepatocellular carcinoma occurs in areas of the world, like West Africa, where diets are deficient in iron. Sufficient iron to support the rapid growth of tumor cells, when there is no obvious interruption of normal cell function, could be accounted for by the latent period of 20 years or more between HBV infection and PHC. With persistent HBV infection in cells with increased ferritin concentrations, a cellular source for extra iron storage is possible. Carriers would have a heterogeneous distribution of intracellular iron in normal cells, with the largest concentrations in cells that are actively replicating HBV.

Blake *et al.* (*12*) suggested a mechanism by which iron can damage a cell. When moderate amounts of iron are bound to apoferritin, little free iron is available within the cell and its immediate surroundings. When excess iron results in saturation, the ferritin-free iron

Table 1. Data collected 2 months before HBsAg was found in the serum of patients undergoing hemodialysis treatment. F, log serum ferritin (serum ferritin was measured in nanograms per milliliter); ALT, alanylaminotransferase, measured in Karmen units (*14*); S, arbitrary value of 1 for males and 2 for females. Results are given as means ± standard deviations. Only the difference in F values for females was significant at the .05 level. The classification of the infection as persistent or transient was made 1 year after HBsAg was first detected.

Classification	N	Serum iron (μg/dl)	Transferrin (mg/dl)	F	ALT	S
			Males			
Persistent	23	85.0 ± 32.7	309 ± 86.9	1.98 ± 0.62	8.83 ± 3.97	1
Transient	8	87.3 ± 31.4	313 ± 110	1.60 ± 0.37	10.9 ± 6.45	1
			Females			
Persistent	6	132 ± 63.6	310 ± 95.6	2.42 ± 0.57	10.7 ± 10.3	2
Transient	17	98.6 ± 46.4	313 ± 98.1	1.64 ± 0.51	7.24 ± 6.48	2
			All patients			
Persistent	29	94.7 ± 44.0	303 ± 87.0	2.07 ± 0.63	9.21 ± 5.66	1.21
Transient	25	95.0 ± 41.8	313 ± 99.7	1.62 ± 0.46	8.40 ± 6.54	1.68

Table 2. Patients grouped by values of y from the prediction equation $y = 0.412 - 1.931\ S + 1.257\ F$. The classification of the infection as persistent or transient was made 1 year after HBsAg was first detected. The presence of antibody was determined in the 25 patients who proved to have transient infections.

Item	Number of patients in each category			
	$y < -0.79$	$-0.79 \le y < 0.00$	$0.00 \le y \le 0.68$	$y > 0.68$
Classification				
Persistent	2	3	10	14
Transient	14	4	5	2
Antibody				
Absent	3	2	4	1
Present	11	2	1	1

becomes available. Iron catalyzes the metabolism of oxygen radicals (O_2^-) to hydroxyl radicals (OH^-) and subsequent lipid peroxidation. These can damage cell and organelle membranes and lead to inflammatory damage of tissues. Vierucci *et al.* (*13*) showed that HBV proteins stimulate the production of oxygen radicals by phagocytic cells without being phagocytized. Also, the presence of the iron can attract inflammatory cells that have receptors for iron-binding proteins. These processes can contribute to the gradual destruction of the cells replicating virus and containing saturated ferritin and may thus increase the selective advantage of the cells that are not replicating HBV or that do not have saturated ferritin.

Persistent HBV infection may result in increased iron stores and be beneficial to individuals with iron-deficient diets. Therefore, early detection or prevention of PHC based on a monitoring of iron balance in carriers might serve to supplement the use of α-fetoprotein and other diagnostic aids.

EDWARD D. LUSTBADER
HIE-WON L. HANN
BARUCH S. BLUMBERG
Institute for Cancer Research, Fox Chase Cancer Center, Philadelphia, Pennsylvania 19111

Reference and Notes

1. A. I. Sutnick, B. S. Blumberg, E. D. Lustbader, *Ann. Intern. Med.* **81**, 855 (1974).
2. C. Felton *et al.*, *Proc. Natl. Acad. Sci. U.S.A.*, **76**, 2438 (1979).
3. B. S. Blumberg, E. D. Lustbader, P. L. Whitford, *ibid.* **78**, 3222 (1981).
4. An individual whose serum is positive for HBsAg for more than 6 months has a high probability of becoming a persistent carrier [W. T. London, J. S. Drew, E. D. Lustbader, B. G. Werner, *Kidney Int.* **12**, 51 (1977)].
5. E. D. Lipschitz, J. D. Cook, C. A. Finch, *N. Engl. J. Med.* **290**, 1213 (1974).
6. HBsAg was determined by Abbott radioimmunoassay, antibody to HBsAg by passive hemagglutination, serum iron by colorimetry (Hyland Ferro-chek II kits), serum ferritin by radioimmunoassay (RAINEN kits. New England Nuclear), and transferrin by single radial immunodiffusion (Behring Diagnostics). Logarithms of serum ferritin values were analyzed because of the skewed distribution of the original data. All positive results for HBsAg were confirmed by other laboratory procedures.
7. G. Birgegard, *Clin. Sci.* **59**, 385 (1980).
8. R. P. Beasley, L.-Y. Hwang, C.-C. Lin, C.-S. Chien, *Lancet* **1981-II**, 1129 (1981).
9. H. A. Edmondson and P. E. Steiner, *Cancer* 7, 462 (1954).
10. C. Brechot, M. Hadchouel, J. Scotto, F. Degos, P. Charney, C. Trepo, P. Tiollars, *Lancet* **1981-II**, 765 (1981).
11. J. A. Fernandez-Pol, *Biochem. Biophys. Res. Commun.* **78**, 136 (1977).
12. D. R. Blake, N. D. Hall, P. A. Bacon, P. A. Dieppe, B. Haliweli, J. M. C. Gutteridge, *Lancet* **1981-II**, 1142 (1981).
13. A. Vierucci, M. DeMartino, E. Graziani, M. E. Rossi, W. T. London, B. S. Blumberg, *Pediatr. Res.*, in press.
14. A. Karmen, F. Wroblewski, J. J. Ladue, *J. Clin. Invest.* **34**, 126 (1955).
15. Supported by NIH grants CA-06551, RR-05539, CA-22780, and CA-06927 and by an appropriation from the Commonwealth of Pennsylvania.
11 January 1983.

Iron-Binding Proteins and Risk of Cancer in Taiwan [1,2]

Richard G. Stevens, [3,4,5] **R. Palmer Beasley,** [6] *and* **Baruch S. Blumberg** [3]

ABSTRACT—The relationship of serum ferritin and transferrin levels to risk of cancer was examined in a population of 21,513 Chinese male government workers in Taiwan who have been followed prospectively since 1975. On the basis of a previous study in the Solomon Islands, increased ferritin and decreased transferrin levels were predicted for those men who developed cancer. The results were consistent with the prediction. The mean serum ferritin was higher at the start of the study in 192 men who had died of cancer or who had developed primary hepatocellular carcinoma (PHC) as of July 1983, as compared to their controls. The mean serum transferrin level was lower in men who had died of cancers other than PHC. The estimate of relative risk of cancer death for a man with 200 ng ferritin/ml and 200 mg transferrin/dl, as compared to a man with levels of 20 ng/ml and 400 mg/dl, respectively, is 2.9. These serum iron-binding protein levels are at the extremes of the "normal" range. Men who subsequently died of cancer had lower hemoglobin, lower hematocrit, lower albumin, and higher globulin levels at the start of the study than did the controls. These results are consistent with the hypothesis that increased iron stores increase the risk of cancer. However, direct assessment of iron stores prior to disease was not possible, and the same constellation of findings may be consistent with other explanations.—JNCI 1986; 76:605–610.

Iron is a primary requirement for microorganisms (*1, 2*), and an analogy has been made between a cancer cell and a microorganism in regard to iron acquisition in the host environment (*3*). Iron may, therefore, be a limiting nutrient for the successful growth and replication of bacteria and malignant cells in the human body (*1, 3*). On this basis, Blumberg et al. (*4*) stated the hypothesis that those chronic carriers of HBV who develop PHC may have higher iron stores than carriers who do not develop PHC.

Several population studies suggest that this need for iron at the cellular level can be seen as an increased risk of infection among individuals with high iron stores (*5–7*). In addition, we found a significant association of increased serum ferritin and decreased serum transferrin with subsequent general mortality over a 10-year time span in the Solomon Islands (*8*). Specific causes of death were unavailable in that study, and we had predicted the result on the basis of the hypothesis that increased iron stores increase the risk of infection and cancer and, therefore, of death.

Serum ferritin is thought to be directly related to body iron stores (*9–11*). Jacobs and Worwood (*12*) state, "In healthy adults the concentration of ferritin in serum is directly related to the available storage iron in the body." There is evidence that the serum transferrin level is inversely related to body iron stores (*1*). Morton and Tavill (*13*) have demonstrated a specific effect of iron on the production of transferrin by the rat liver: the higher the iron, the lower the production of transferrin.

In this paper, we present the results of a study conducted in Taiwan that was designed to test the hypothesis that increased ferritin and decreased transferrin prior to the development of disease are associated with subsequent risk of cancer. The results of the study are consistent with these hypotheses and with the hypothesis that increased iron stores increase the risk of cancer. Since ferritin and transferrin levels may be altered by factors other than iron stores, there may be additional explanations for the observed findings. Several studies are being planned to examine these possibilities in other populations.

METHODS

Study population.—Beasley and colleagues in Taiwan have conducted since 1975 a prospective study of the relationship between PHC and persistent infection with HBV (*14*). They included 21,513 male Chinese government workers whose health care is provided free of charge by the GECC in Taipei. Of these, 3,454 were HBV carriers. The men were enrolled during the period November 3, 1975, to June 30, 1978. At enrollment, a serum sample was taken from each man and tested for surface antigen of HBV (HBsAg). This serum sample has been stored at −20°C since it was drawn. The primary objective of the study was to determine the risk of PHC for carriers of HBsAg, as compared to the risk for noncarriers. Therefore, incident cases of PHC have been recorded continuously since the start of the study. In addition, mortality from any cause also has been re-

ABBREVIATIONS USED: df = degrees of freedom; GECC = Government Employees' Central Clinic; HBsAg = surface antigen of HBV; HBV = hepatitis B virus; PHC = primary hepatocellular carcinoma.

[1] Received June 6, 1985; accepted November 21, 1985.

[2] Supported in part by Public Health Service (PHS) grants CA-06551, CA-06927, CA-22780, and CA-25327-03 from the Division of Extramural Activities, National Cancer Institute (NCI); by PHS grant RR-05539 from the Division of Research Resources, National Institutes of Health; and by PHS contract Y01CP-60502 from the Division of Cancer Etiology, NCI. The study also was supported by an appropriation from the Commonwealth of Pennsylvania and by award NSC 69B-0412-02 from the China Foundation and National Science Council of Taiwan.

[3] Division of Clinical Research, Fox Chase Cancer Center, 7701 Burholme Ave., Philadelphia, PA 19111.

[4] *Present address:* Battelle, Pacific Northwest Laboratories, P. O. Box 999, Richland, WA 99352.

[5] *Address reprint requests to* Dr. Stevens *at* Division of Clinical Research, Fox Chase Cancer Center.

[6] University of Washington Medical Research Unit, American University Medical Center, 155 Li Nong St., Section 2, Shih Pai, Taiwan, Republic of China.

corded as it occurred. Beasley et al. found that nearly all cases of PHC developed in men who were carriers of HBV.

In the present study of iron-binding proteins and cancer risk, for two reasons we have considered PHC incident cases and deaths separately from deaths due to all other cancers: 1) The GECC study was designed expressly to examine the relation of HBV to PHC. 2) The liver is the principal storage site for iron and an important source for the production of transferrin (*15*) and ferritin (*16*), as well as being the organ in which HBV is produced and concentrated; PHC, even long before clinical detection, therefore, might have profound effects on iron metabolism and iron-binding protein levels.

Control selection.—The cases were all 239 men in the GECC study population who had died of any cancer or who were alive with PHC as of July 1983. All but 3 of the 94 men who developed PHC were carriers of HBV, and these 3 showed evidence of earlier infection. Initially, controls for each of the 239 cases were selected from all men who were alive and free of PHC at the date the case failed (died of cancer or developed PHC). Age and HBsAg are both associated with iron stores and with risk of cancer conditional on iron stores. Therefore, we chose as controls men of the same age (within 2 yr) and same HBsAg serum status as those of the cases. These criteria yielded several hundred possible controls for each case. We chose as actual controls the 2 men whose enrollment dates into the study were closest to that of the case. The term "risk set" is used to refer to the case and his two controls. There were two reasons for this method of control selection: 1) The sera from the case and his controls were stored for about the same length of time. 2) This method controls for any temporal trend in iron-binding protein levels in the population (although no such trend is known). We had no reason to suspect that entry date into the study was associated with subsequent risk of cancer.

When possible, the serum sample, taken at the time a subject was enrolled into the GECC study, was tested. If the enrollment serum sample had been depleted, we tested the serum sample taken at the first annual follow-up examination of the subject. Sufficient enrollment serum was available for 127 of the cases; sufficient serum from the first follow-up examination was available in an additional 65 cases. For these 192 cases, ferritin and transferrin levels were measured in the case and at least in 1 of his 2 controls.

Statistical methods.—The GECC study is prospective with 21,513 men as subjects. We did not test all sera for iron-binding proteins. Therefore, a "synthetic case-control" study design, based on the Cox proportional hazards model, was used (*17*). There were two defined outcomes: 1) death from cancer excluding PHC and 2) death from, or development of, PHC.

Weighted differences in individual variables (e.g., ferritin, transferrin, hemoglobin) between the case and the mean values for his controls are presented in the tables. The weights were chosen so as to account for the fact that some risk sets included 2 controls, whereas others included only 1 (*18*). These weights were determined by the null hypothesis of the Cox model and were derived from the fact that the score statistic weights risk sets with 3 members by 1.33 relative to risk sets with 2 members.

A score test based on the proportional hazards model (*18*) was used to detect significant differences between cases and controls in individual variables. For multivariable models, improvements in the fit to the data were detected by the use of a likelihood ratio test. Both the likelihood ratio statistic and the score statistic asymptotically have chi-square distributions.

It was important to determine whether the parameter estimates and the significance of the model were affected by a small number of risk sets; i.e., could outlying data points account for any observed significant differences. Diagnostic statistics, therefore, were applied to determine the contribution of each risk set to the score statistic. This methodology is given in Lustbader et al. (*19*) who provided a computer program for the calculations.

Medical chart abstraction and serum assays.—For each case and control whose GECC medical chart was available, the following variables were abstracted for a date as close as possible to the recruitment date of the subject: hemoglobin, hematocrit, height, weight, albumin, and total globulin. About one-third of the case charts had been removed from the GECC, placed in permanent storage, and could not be obtained.

Serum ferritin and serum transferrin levels were determined at the University of Washington Medical Research Unit in Taipei, Taiwan. Ferritin was measured by radioimmunoassay (RIANEN assay kit; New England Nuclear Corp., North Billerica, MA). Transferrin was measured by radial immunodiffusion (Behring Diagnostics, La Jolla, CA). This assay depends on measuring the diameter of a precipitan ring. To standardize measurements, a device was used that held the technician's eye at the same location over each plate. All protein determinations were made with the use of coded sera, without knowledge of the identity of the individual. The manufacturer of the ferritin kit reports a coefficient of variation for the test of between 0.1 and 7.9%, depending on the level in the serum, and the manufacturer of the transferrin test reports a coefficient of variation of 3–8.5%.

There is daily variation in serum ferritin level (*20*). In this study, we have assumed that the time of day the blood was originally drawn was not associated with levels of iron-binding proteins or with subsequent risk of developing cancer. In addition, proteins such as ferritin and transferrin may deteriorate with storage. We matched on age of serum for this reason. Such variability would serve to reduce the power to detect a real difference should it exist.

RESULTS

The causes of death for each of the 192 cases are shown in table 1 (31 men with PHC were still alive).

TABLE 1.—*No. of cancer cases as of July 1983 and mean age in 1975 of Chinese male government workers in Taiwan*

Type of cancer	No. of cases[a]	Mean age[b]
Buccal cavity and pharynx	16	50.1
Esophagus	7	50.2
Stomach	24	51.1
Colon	8	48.2
Rectum	2	56.5
PHC	31[c]	50.9
	39[d]	51.0
Gallbladder	1	57.0
Pancreas	8	49.1
Nasal cavity	1	41.0
Trachea, bronchus, and lung	31	53.8
Bone	1	40.0
Connective tissue	1	45.0
Prostate	1	50.0
Bladder	2	48.0
Other urinary cancers	3	55.0
Brain	1	57.0
Lymphosarcoma	4	50.7
Other neoplasms of lymphoid tissue	2	58.5
Myeloid leukemia	2	53.5
Other leukemia	1	42.0
Myelofibrosis	1	40.0
Nervous system	4	45.5
Unspecified	1	50.0
Total	192	

[a] No. of cases represents No. of cancer deaths unless otherwise specified.
[b] Overall mean age was 51.1 yr.
[c] Subjects living with PHC.
[d] Subjects with PHC who had died.

The most common causes of death in the case group were PHC (39), and cancers of the trachea, bronchus, and lung (31), stomach (24), and buccal cavity and pharynx (16). The mean age in 1975 of all the cases was 51.1.

Univariate Analyses

The weighted differences in ferritin and transferrin levels between each case and the mean levels for his controls are shown in table 2. Mean ferritin level in the 122 cases of death from cancer (excluding PHC) was 149.1 ng/ml (range 7-666); the mean level in 230 controls was 142.1 ng/ml (range 5-627). The weighted mean difference was 6.4 ng/ml. The mean transferrin level in the cancer death cases was 283.9 mg/dl (range 95-638); the mean level in the controls was 310.8 mg/dl (range 85-829). The weighted mean difference was −23.8 mg/dl. The mean ferritin difference and the mean transferrin difference are in the direction predicted; however, only the difference in transferrin levels is statistically significant ($P<.05$) in this univariate analysis.

Among the 70 PHC cases, the mean ferritin level was 121.4 ng/ml (range 8-381), and the mean transferrin level was 286.2 mg/dl (range 129-579). The mean ferritin level in their controls was 99.5 ng/ml (range 5-342), and the mean transferrin level was 268.9 mg/dl (range 74-579). The mean ferritin difference was 21.5 ng/ml and statistically significant. The mean transferrin difference of 18.4 mg/dl is not in the direction predicted; however, the result is not significant.

Subjects used to calculate differences in table 2 included all controls for whom protein levels had been determined. Some controls had ferritin measured but not transferrin, inasmuch as the ferritin test depleted the serum sample. A total of 358 control ferritin values and 336 control transferrin values were available. A restriction to only the 143 risk sets in which both ferritin and transferrin were determined for both controls yielded a significant transferrin difference of −25.4 in 97 cancer death risk sets (excluding PHC) and a nonsignificant ferritin difference of 5.7. Among 46 PHC risk sets, the ferritin difference of 37.3 was significant, whereas the transferrin difference of 25.3 was not. Thus these results are in accord with the results presented in table 2. The mean age of those in the complete 143 risk sets was 51 years, 40% of whom were HBV carriers. The mean age in the incomplete 49 risk sets also was 51 years; however, 78% were HBV carriers. The serum of carriers has been the subject of continuing research in this population; therefore, their sera are more likely to have been depleted.

It is apparent from table 2 that the PHC controls had lower ferritin levels than did the other cancer controls. In both PHC cases and controls, the range of ferritin values was much lower than in other cancer cases and controls, as specified above. Because the distribution of ferritin values is highly skewed, the mean is sensitive to a few high values. The median value for PHC controls was 92 ng/ml, whereas the median value for the other cancer controls was 107 ng/ml. This difference is less than the difference in the mean values. Similarly, the median in PHC cases was 112, whereas in other cancer cases it was 119. The matching of cases and controls in this study is very important, and the only comparisons

TABLE 2.—*Mean serum ferritin and transferrin levels for cases and controls in the GECC Taiwan study*

Cause of death	Ferritin level: Cases No.	Cases ng/ml	Controls No.	Controls ng/ml	Weighted mean difference[a]	Transferrin level: Cases No.	Cases mg/dl	Controls No.	Controls mg/dl	Weighted mean difference[a]
PHC	70	121.4	128	99.5	21.5[b]	70	286.2	116	268.9	18.4
Other cancers	122	149.1	230	142.1	6.4	122	283.9	220	310.8	−23.8[b]

[a] Weighted average of the difference in levels between a case and his controls. A score test was used to detect significant differences.
[b] $P<.05$.

that should be made are between the case and the control members of his risk set. Distribution of transferrin also is skewed, although not nearly so much as for ferritin.

Case-Control Analysis of Cohort Study

There were two outcomes in this study: 1) death from, or development of PHC and 2) death from all cancers excluding PHC. Therefore, a model with separate coefficients for ferritin levels in PHC, ferritin levels in cancer other than PHC, transferrin levels in PHC, and transferrin levels in cancer other than PHC was fit to the data. This is a cause-specific analysis with four terms: ferritin in cancer (all cancer deaths other than from PHC), ferritin level in PHC, transferrin level in cancer (all cancer deaths other than from PHC), and transferrin level in PHC. This model makes a significant improvement over the null model with all coefficients equal to zero (chi-square = 14.3 on 4 df).

The two coefficients associated with ferritin were not found to be significantly different from each other (chi-square = 2.8, 1 df). In addition, the transferrin-in-PHC coefficient did not significantly improve the fit (likelihood ratio test yielded a chi-square of 2.5, 1 df). Thus the model ferritin (one term for both outcomes) and transferrin-in-cancer was used to estimate coefficients and relative risks. This model made a significant improvement over the null model, with all coefficients zero (chi-square = 9.0, 2 df). It can be represented as follows:

$$H_j(a;ft) = H_{0j}\,(a)\exp[f \times B_f + I \times t \times B_t],$$

where $H_j(a;ft)$ = hazard at age a for outcome j (PHC or other cancer), given ferritin of f and transferrin of t; $H_{0j}(a)$ = base-line hazard; B_f and B_t = coefficients associated with ferritin and transferrin; and I = indicator equal to 0 if PHC and equal to 1 if other cancer.

Relative risk estimates are presented in table 3. Based on the parameter estimates and standard errors of the parameters obtained by fitting the model (ferritin and transferrin in cancer) to all 192 risk sets, the risk of PHC for a man with 200 ng serum ferritin/ml, relative to a man with 20 ng serum ferritin/ml, is 1.41. The risk of cancer other than PHC in a man with 200 ng serum ferritin/ml and 200 mg transferrin/dl, relative to a man with 20 ng ferritin/ml and 400 mg transferrin/dl, is 2.88. These values for serum ferritin and serum transferrin are at the extremes of the normal range of 20-200 ng ferritin/ml and 200-400 mg transferrin/dl (*21*).

Age

The 192 risk sets were divided into 78, in which the case was less than 50 years old in 1975, and 114, in which the case was 50 or more years old. The model did not make a significant improvement over the null hypothesis in the 78 risk sets under age 50. However, the model was highly significant when fit to the 114 risk sets over age 50 (chi-square = 15.0, 2 df). The resulting relative risk estimates were 1.7 for PHC and 6.4 for other cancers.

TABLE 3.—*Relative risk estimates based on the model ferritin and transferrin in cancer*[a]

Risk[b]	PHC[c]	Other cancers[d]
Relative	1.41	2.88
Lower bound (95% CL)	1.00	1.18
Upper bound (95% CL)	1.99	7.03

[a] Values are at the extremes of the normal range.
[b] CL = confidence limit.
[c] Risk of PHC in subjects with ferritin levels of 200 ng/ml vs. those with 20 ng/ml.
[d] Risk of other cancers in subjects with ferritin levels of 200 ng/ml and transferrin levels of 200 mg/dl vs. those with ferritin levels of 20 ng/ml and transferrin levels of 400 mg/dl.

Time on Study

Death from cancer, or development of PHC, occurred anywhere from 3 to 79 months after the blood sample was taken. Existing but undiagnosed disease could have accounted for some of the differences in the iron-binding protein levels we observed. Therefore, we analyzed 131 risk sets in which there was at least 2 years between the date at which the serum sample was collected and the date of death from cancer or development of PHC. The results were consistent with the analysis of the total sample. From this we conclude that undiagnosed cancer, which may have existed at the time at which the serum sample was taken, probably does not account for the observed overall result.

In looking at the PHC data separately, we found the mean difference in ferritin, between the 9 cases and deaths that occurred within 1 year of the date the serum sample was taken and the mean of their controls, to be 60 ng/ml, whereas the mean difference was 16.4 ng/ml among the 47 cases that occurred more than 2 years later.

Other Covariables

Hemoglobin, hematocrit, height, weight, serum albumin, and serum globulin measurements at the time of, or time closest to, enrollment were obtained from the charts (table 4). Hemoglobin and hematocrit were significantly lower in cases than in controls. The cases were shorter in stature (not significantly) and heavier and had lower albumin than controls. Globulin was higher in cases (not significant).

Diagnostic Statistics

The purpose of these statistics is to determine the contribution of each data point to the significance of the model. There were 6 risk sets that changed by more than 1 the score statistic associated with the model. All the extreme risk sets served to reduce the fit of the model, and if these 6 risk sets were removed, the score statistic would be 19.7. The relative risk estimates would be 1.7 for PHC and 6.9 for other cancers. Thus if these 6 risk sets had not been included in the study, it would have appeared that the ferritin and transferrin levels had much larger effects on risk of cancer.

TABLE 4.—*Physical variables among cases and controls in Taiwan study*

Specification	Hemoglobin, g/dl (n = 128)	Hematocrit, % (n = 125)	Height, cm (n = 130)	Weight, kg (n = 134)	Albumin, g/dl (n = 73)	Globulin, g/dl (n = 72)
Cases[a]	14.2	42.9	159.6	69.7	4.2	2.6
Difference in risk sets[b]	−0.3	−0.8	−2.2	3.0	−0.2	0.1
Score test	4.76[c]	4.9[c]	1.13	4.87[c]	8.93[c]	2.27

[a] Mean value.
[b] Weighted mean difference between a case and his controls.
[c] $P<.05$.

DISCUSSION

Overall, the results of this study are consistent with the observations of general mortality in the Solomon Islands and the predictions made from those findings (*8*). Death from cancer (and development of PHC) in Chinese male government workers in Taiwan is associated with elevated ferritin before diagnosis of the disease, and death from cancers other than PHC is associated with a low transferrin level. These findings are consistent with the hypothesis that risk of cancer is increased with increased iron stores. However, the same constellation of findings may be consistent with other explanations. In particular, we were unable to directly assess iron stores prior to disease in cases and controls, but we could assess secondary effects thought to be associated with iron stores. Our findings, therefore, are only direct tests of those covariables examined (e.g., ferritin and transferrin) and not of iron stores. However, inasmuch as this is a possible and interesting explanation, we will consider some of the consequences of differences in iron stores.

There are at least two possible mechanisms by which increased iron might increase the risk of cancer: 1) Iron may be a limiting nutrient to the growth and development of a transformed cell (*3*) and 2) iron can catalyze the production of oxygen radicals that may be proximate carcinogens (*22, 23*).

Bergeron et al. (*24*) injected a leukemia cell line into 2 groups of mice. One group received iron-dextran injections and the other iron-free dextran. Both groups were fed nonpurified Rodent Laboratory Chow from Ralston Purina. The group that received iron died sooner and had a greater tumor cell burden (peritoneal harvest prior to death). The authors concluded that their results were consistent with the hypothesis that iron may be a limiting nutrient to the growth of a cancer cell.

Current research on diet and cancer has focused on the possible protective effects of such nutrients as tocopherol, carotenoids, selenium, and ascorbic acid (*25*). While each undoubtedly affects many aspects of cellular metabolism, they share the fact that they are all antioxidants. Thus iron may catalyze the production of proximate carcinogens (oxygen radicals), whereas these antioxidants may destroy them.

Age

We found that ferritin and transferrin were significantly associated with cancer only in those over age 50 years old. If these observations reflect iron stores, then they would suggest that iron restriction is more important as a defense against cancer in older people.

Ferritin and PHC

Ferritin may reach very high levels in patients with PHC. Hence the elevated levels of ferritin detected in this study could reflect early subclinical cancer of the liver present in the apparently normal subjects who were carriers of HBV. In particular, the mean difference in risk sets, from whom serum was drawn less than 1 year prior to diagnosis, was very high (60 ng/ml), as compared to the difference of 16 ng/ml seen in those risk sets who had at least 2 years between the drawing of serum and diagnosis.

There are probably different tissue-specific, differentiation stage-specific, and tumor-specific ferritins produced during the course of the development of PHC. Some of these may be identifiable by the use of monoclonal antibodies against the different ferritins. Current research is directed toward that end.

Hemoglobin, Hematocrit, and Albumin

Table 4 is based on data only for about two-thirds of the study subjects and, therefore, must be interpreted with caution. The men in our study who developed cancer had a lower hemoglobin and lower hematocrit at the start of the study than did their controls (table 4). In our study of general mortality in the Solomon Islands (*8*), we also found, over a 10-year period, lower hemoglobin in those who died, as compared to controls who remained alive (hematocrit was not studied). Chronic infection often is associated with an anemia (referred to as the "anemia of chronic disorders") characterized by mildly depressed hemoglobin (7-11 g/dl), low hematocrit, low serum iron, low serum transferrin, slightly elevated ferritin (150 ng/ml), and increased tissue iron stores (*26*). As we suggested in our report on the Solomon Islands study, the lower hemoglobin of our cases could have been the result of lifelong increased susceptibility to infection resulting from chronically increased iron stores. The lower hematocrit observed in the Taiwan study also is consistent with this possibility as is the increased globulin (table 4).

The cases in our study had significantly lower albumin than did the controls. Konopka and Neilands (*27*) found that human serum albumin reduces the ability of certain bacterial siderophores to bind iron. These authors concluded that, "These results indicate that serum albumin may act synergistically with other factors in the serum, such as transferrin, to limit iron supply and in this way restrict the growth of invading microorganisms." If albumin can affect iron acquisition by transformed cells in a similar manner, then low albumin might be associated with increased risk of cancer.

CONCLUSIONS

The mean values of serum ferritin and transferrin in the cases and controls were within the normal range. However, all normal people are not at equal long-term risk of disease. The present study provides evidence for a difference in risk of cancer given small but significant differences in levels of ferritin and transferrin that are present before clinically apparent disease. This is consistent with the hypothesis that increased iron stores increase risk of cancer, although alternative explanations may be possible.

Ferritin, transferrin, and total-body iron stores are related to hematopoiesis, cell proliferation (including cancer cells), and survival. The studies in the Solomon Islands and in Taiwan have shown some of the interconnections of these factors. Future studies using data collected over time in individuals destined to develop cancer are directed toward a greater understanding of this process.

These studies suggest several testable hypotheses: 1) Differences in ferritin, transferrin, hemoglobin, albumin, and globulin seen in the Taiwan study also may be found in prospective cancer studies in other locations. Several such studies are now being planned. 2) Individuals who develop cancer will have had a history of more frequent infections than will individuals who have not developed the disease, if increased iron stores increase risk for both infections and cancer. 3) A decreased risk of cancer associated with low iron stores would be relatively more important with advancing age. 4) High dietary iron intake increases risk of cancer.

REFERENCES

(*1*) Weinberg ED. Iron and infection. Microbiol Rev 1978; 42:45-66.

(*2*) Lewin R. How microorganisms transport iron. Science 1984; 225:401-402.

(*3*) Weinberg ED. Iron withholding: A defense against infection and neoplasia. Physiol Rev 1984; 64:65-102.

(*4*) Blumberg BS, Lustbader ED, Whitford PL. Changes in serum iron levels due to infection with hepatitis B virus. Proc Natl Acad Sci USA 1981; 78:3222-3224.

(*5*) Murray MJ, Murray AB, Murray CJ. An ecological interdependence of diet and disease? A study of infection in one tribe consuming two different diets. Am J Clin Nutr 1980; 33: 697-701.

(*6*) McFarlane H, Reddy S, Adcock KJ, et al. Immunity, transferrin, and survival in Kwashiorkor. Br Med J 1970; 4:268-270.

(*7*) Masawe AE, Muindi JM, Swai GB. Infections in iron deficiency and other types of anemia in the tropics. Lancet 1974; 2:314-317.

(*8*) Stevens RG, Kuvibidila S, Kapps M, et al. Iron-binding proteins, hepatitis B virus, and mortality in the Solomon Islands. Am J Epidemiol 1983; 118:550-561.

(*9*) Cook JD, Lipschitz DA, Miles LE, et al. Serum ferritin as a measure of iron stores in normal subjects. Am J Clin Nutr 1974; 27:681-687.

(*10*) Forman DT, Vye MV. Immunoradiometric serum ferritin concentration compared with stainable bone-marrow iron as indices to iron stores. Clin Chem 1980; 26:145-147.

(*11*) Finch CA, Huebers H. Perspectives in iron metabolism. N Engl J Med 1982; 306:1520-1528.

(*12*) Jacobs A, Worwood M. Ferritin in serum. N Engl J Med 1975; 292:951-956.

(*13*) Morton AG, Tavill AS. The role of iron in the regulation of hepatic transferrin synthesis. Br J Haematol 1977; 36:383-393.

(*14*) Beasley RP, Lin C-C, Hwang L-Y, et al. Hepatocellular carcinoma and hepatitis B virus. Lancet 1981; 2:1129-1133.

(*15*) Jacobs A, Worwood M. Normal iron metabolism. In: Powell LW, ed. Metals and the liver. Vol 1. New York: Dekker, 1978.

(*16*) Halliday JW, Powell LW. Ferritin metabolism and the liver. In: Powell LW, ed. Metals and the liver. Vol 1. New York: Dekker, 1978.

(*17*) Breslow NE, Lubin JH, Marek LP, et al. Multiplicative models and cohort analysis. JASA 1983; 78:1-12.

(*18*) Kalbfleisch JD, Prentice RL. The statistical analysis of failure time data. New York: Wiley, 1980:75 and 168.

(*19*) Lustbader ED, Moolgavkar SH, Venzon DJ. Tests of the null hypothesis in case-control studies. Biometrics 1984; 40:1017-1024.

(*20*) Pilon VA, Howanitz PJ, Howanitz JH, et al. Day-to-day variation in serum ferritin concentration in healthy subjects. Clin Chem 1981; 27:78-82.

(*21*) Beck WS, ed. Hematology. Cambridge, MA: MIT Press, 1977.

(*22*) Halliwell B, Gutteridge JM. *Review article:* Oxygen toxicity, oxygen radicals, transition metals and disease. Biochem J 1984; 219:1-14.

(*23*) Ames BN. Dietary carcinogens and anticarcinogens. Science 1983; 221:1256-1264.

(*24*) Bergeron RJ, Streiff RR, Elliot GT. Influence of iron on in vivo proliferation and lethality of L1210 cells. J Nutr 1985; 115:369-374.

(*25*) Willett WC, MacMahon B. Diet and cancer—an overview. N Engl J Med 1984; 310:633-638, 697-703.

(*26*) Erslav AJ. Erythrocyte disorders—anemias related to unknown or multiple mechanisms. In: Williams WJ, Beutler E, Erslav AJ, et al., eds. Hematology, 3d ed. New York: McGraw-Hill, 1983.

(*27*) Konopka K, Neilands JB. Effect of serum albumin on siderophore-mediated utilization of transferrin iron. Biochemistry 1984; 23:2122-2127.

Int. J. Cancer: **56**, 364–369 (1994)

Publication of the International Union Against Cancer
Publication de l'Union Internationale Contre le Cancer

MODERATE ELEVATION OF BODY IRON LEVEL AND INCREASED RISK OF CANCER OCCURRENCE AND DEATH

Richard G. STEVENS[1,6], Barry I. GRAUBARD[2], Marc S. MICOZZI[3], Kazuo NERIISHI[4] and Baruch S. BLUMBERG[5]

[1]*Pacific Northwest Laboratory, Life Sciences Center, Richland, WA, 99352;* [2]*National Cancer Institute, Division of Cancer Etiology, Biometry Branch, Bethesda, MD, 20892;* [3]*National Museum of Health and Medicine, Armed Forces Institute of Pathology, Washington, DC, 20306, USA;* [4]*Radiation Effects Research Foundation, Hiroshima, Japan; and* [5]*Fox Chase Cancer Center, Philadelphia, PA, USA, and Balliol College, Oxford, UK.*

The purpose of the study was to address the hypothesis that elevated body iron increases the risk of cancer occurrence and death, and to determine the dose response. Subjects were 3,287 men and 5,269 women participating in the first National Health and Nutrition Examination Survey who had a transferrin saturation determination at enrollment (1971–1975), who remained alive and cancer-free for at least 4 years, and who were followed to 1988 for cancer outcome. Among 379 men who developed cancer over the study period, the mean transferrin saturation at enrollment was 32.1% whereas among 2,908 who remained cancer-free it was 30.7%; the difference for mortality was 32.3% among 233 deaths *vs.* 30.8% among 3,054 men not dying of cancer. The mean differences among women were not significant. The mean differences in TIBC and serum iron among men were consistent with the findings for transferrin saturation, and all 3 differences were stable over time when examined by years since blood test. Men and women were divided into 5 groups on the basis of baseline transferrin saturation: 0 to 30%, 30–40%, 40–50%, 50–60%, and 60% and higher. Nineteen percent of men had a baseline transferrin saturation above 40% (the last 3 groups), whereas only 10 percent of women had transferrin saturation above 40%. For men and women combined, risk of cancer occurrence in each group relative to the first was 1.0, 0.95, 1.16, 1.38 and 1.81; for mortality the relative risks were 1.0, 0.96, 1.22, 1.29 and 1.73. There is evidence, in this cohort, of elevated cancer risk in those with moderately elevated iron level. This pattern was seen in women as well as in men.

Excess body iron may increase the risk of cancer by at least 2 possible mechanisms (reviewed in Stevens and Kalkwarf, 1990). First, iron can catalyze the production of oxygen radicals (Halliwell and Gutteridge, 1990), which may be carcinogenic by increasing oxidative stress or by damaging DNA directly (Loeb *et al.*, 1988). Second, iron may be a limiting nutrient to the growth and replication of a cancer cell in the human body (Weinberg, 1984), and thus, high body-iron level may increase the chances that a cancer cell will survive and grow to become a clinically detectable neoplasm. The issue is important because the western diet includes high amounts of available iron (Cook, 1990), and this diet has been spreading throughout the world. It is also important because hereditary hemochromatosis may be far more common than previously suspected (Edwards *et al.*, 1988). The definitions of "high body-iron level" that are relevant to the 2 proposed mechanisms whereby iron might increase cancer risk are (1) high intracellular iron content, particularly that iron associated with DNA, with ferritin, and with non-specific iron complexes (*e.g.*, Fe^{+3}ADP, Floyd, 1983), and (2) high iron content of blood and interstitial fluid in a form that can be utilized by cancer cells (*e.g.*, high transferrin saturation); indeed, cancer cells often have increased transferrin receptors (Sutherland *et al.*, 1981).

In 1988, we reported on the association of iron status at enrollment (1971–1975) into the NHANES I cohort and subsequent cancer incidence through follow-up to 1981–4 (Stevens *et al.*, 1988). The purpose of this paper is 2-fold: (1) to report on cancer mortality in the cohort over the period 1971–1975 to 1987–1988, and (2) to provide detail on dose-response effects both for incidence and for mortality with a focus on moderate increases in iron level.

METHODS

Study population

The first National Health and Nutrition Examination Survey (NHANES I) was begun in 1971, with the enrollment of 14,407 subjects aged 25 to 74 selected from the non-institutionalized population of the United States (Madans *et al.*, 1986). On each subject a medical examination was performed, anthropometric measurements were taken, a questionnaire (which included diet) was administered and blood was taken for biochemical analyses. Subjects have been traced and re-interviewed twice, in 1981–1984 and in 1987–1988. The analytic cohort described in our first report (Stevens *et al.*, 1988) consisted of 3,287 men and 5,269 women who had been identified during the period 1971-75 as part of the sample chosen for NHANES I; all had a transferrin saturation determination available for analysis and had remained alive and cancer-free for at least 4 years after this blood determination. In our first report, there had been 232 cases of cancer in men and 203 in women found in the follow-up of the cohort to 1984. In the present report, the same analytic cohort has been followed for an additional 4 years. Among the men there have now been 379 cancer cases and 233 deaths from cancer; among the women, there have been 340 cancer cases and 174 cancer deaths, *i.e.*, an additional 147 cases in men and 137 in women. Cancer mortality was not previously analyzed.

Analytic strategy

Analyses have been conducted with 3 objectives: hypothesis testing, dose-response characterization, and assessment of internal consistency. The hypothesis tested was that the mean transferrin saturation was higher among subjects who developed, or died of, cancer over the follow-up period than among subjects who did not. A general linear-regression model was used in which transferrin saturation was treated as the dependent variable, with age as an independent variable and an indicator for case status (SAS, 1985). In all analyses, smoking was coded as a nominal explanatory variable with 4 categories: current smoker, former smoker, never a smoker, and unknown. Similar hypotheses tested were that serum iron was higher, and TIBC and serum albumin lower, among cases than among controls.

For dose-response characterization, subjects were grouped as before (Stevens *et al.*, 1988) on the basis of baseline transferrin saturation into quartiles (for men these were 0–22.8%, 22.9–29.1%, 29.2–36.7%, and 36.8% and higher). In addition, new analyses have been conducted on the 5 categories 0–30%, 30–40%, 40–50%, 50–60%, and 60+%, in order to gain more insight into possible dose-response effects. Risk in

[6]To whom correspondence and reprint requests should be addressed.

Received: July 20, 1993 and in revised form September 14, 1993.

TABLE I - MEAN BIOCHEMICAL VALUES AND MEAN ESTIMATED DAILY DIETARY IRON INTAKE FOR SUBJECTS IN WHOM CANCER DEVELOPED AND THOSE WHO REMAINED CANCER-FREE[1]

	Transferrin saturation (%)		Iron (µg/dl)		TIBC (µg/dl)		Albumin (g/l)		HgB (g/dl)		Dietary iron (mg/day)	
Male incidence												
2,908[2]	30.7	0.015[4]	106.2	0.04	350.9	0.035	44.3	0.011	15.5	0.99	14.1	0.51
379[3]	32.1		109.7		346.0		43.9		15.5		13.9	
Female incidence												
4,929	27.3	0.09	98.8	0.11	371.1	0.46	43.4	0.11	13.8	0.4	9.8	0.16
340	28.2		101.4		370.8		43.2		13.9		9.4	
Male mortality												
3,054	30.8	0.025	106.2	0.03	350.6	0.13	44.3	0.025	15.5	0.33	14.1	0.09
233	32.3		110.9		346.9		43.9		15.6		13.4	
Female mortality												
5,095	27.4	—	99.0	—	371.0	—	43.4	0.03	13.8	0.15	9.7	0.74
174	27.3		98.7		374.6		43.0		13.7		9.8	

[1]Adjusted for age and smoking.–[2]Number of subjects cancer-free to end of follow-up.–[3]Number of subjects who developed cancer.–[4]*p* value for comparison of means.

each category was calculated relative to the lowest category by the Mantel-Haenszel procedure adapted to cohort studies using the computer software MOX (Gilbert and Buchanan, 1984). Analyses were stratified on age (one-year intervals), smoking (see above), and sex where appropriate. The cluster sampling used in NHANES I can potentially inflate variances of estimators (Korn and Graubard, 1991). In order to address this issue, a modified version of the Mantel-Haenszel test was used to examine the dose-response relationship of transferrin saturation with cancer incidence. The modification consisted of changing the denominator of the usual score test to an estimator of the variance of the numerator under cluster sampling. This denominator was computed by replicating the numerator over 36 balanced half-samples of the data which represented the 36-strata design used for variance estimation in NHANES I (Korn and Graubard, 1991).

Internal consistency was examined by comparing means of biochemical measures (transferrin saturation, TIBC, etc.) according to the length of time from baseline blood test divided into the intervals 4 to 6 years, 6 to 8 years, 8 to 10 years, 10 to 12 years, 12 to 14 years, and 14+ years. For example, in the 6-to-8-year interval, the mean among men who were diagnosed with cancer at least 6 years but less than 8 years after blood test is compared with the mean among all men who were followed at least 6 years and who were not diagnosed with cancer for at least 8 years post-blood test. This is a conservative approach, since the non-cancer group in each follow-up interval comparison includes men who developed cancer sometime after the end of the interval.

TABLE II - MEAN BIOCHEMICAL VALUES FOR MEN IN WHOM CANCER DEVELOPED AND THOSE WHO REMAINED CANCER-FREE, BROKEN DOWN BY FOLLOW-UP INTERVAL[1]

	Transferrin saturation (%)		Iron (µg/dl)		TIBC (µg/dl)		Albumin (g/l)	
4 to <6								
3,220[2]	30.9	0.19[4]	106.6	0.48	350.4	0.11	44.3	0.07
67[3]	32.1		106.8		343.3		43.7	
6 to <8								
3,021	30.8	0.07	106.5	0.09	351.2	0.07	44.4	0.0003
91	32.6		111.7		343.6		43.3	
8 to <10								
2,808	30.9	0.25	107.0	0.40	351.1	0.35	44.5	0.25
74	31.8		108.0		348.9		44.2	
10 to <12								
2,609	30.9	0.10	106.9	0.12	351.5	0.05	44.6	0.35
55	32.9		112.6		341.4		44.5	
12 to <14								
2,388	30.9	0.28	106.9	0.14	351.9	—	44.7	—
70	31.6		111.6		355.2		45.1	
14+								
1,572	30.6	—	106.6	—	353.4	—	44.8	0.09
22	26.7		95.7		357.8		43.9	
Total								
2,908	30.7	0.015	106.2	0.04	350.9	0.035	44.3	0.011
379	32.1		109.7		346.0		43.9	

[1]4 to less than 6 years, 6 to less than 8 years, etc.; adjusted for age and smoking.–[2]Number of subjects cancer-free to end of specified follow-up interval.–[3]Number of subjects who developed cancer during specified follow-up interval.–[4]Nominal *p* value for comparison of means.

RESULTS

Table I shows the mean hemoglobin, transferrin saturation, serum iron, TIBC, albumin, and estimated dietary iron intake in subjects who developed cancer during the follow-up period and among those who did not. Among men, cancer cases had significantly higher transferrin saturation and serum iron, and significantly lower TIBC and serum albumin than men who did not develop cancer, as originally reported (Stevens *et al.*, 1988). Men who died of cancer over the study period also had significantly higher transferrin saturation and serum iron, and significantly lower serum albumin than those who did not die of cancer. Among women, only serum albumin was significantly lower among those who died of cancer.

In order to assess the internal consistency of the data, differences between cancer cases and cancer-free study subjects were calculated for intervals of time since the blood was drawn and the tests were performed. In Table II, the difference in mean transferrin saturation between cancer cases occurring between 4 and 6 years after blood was drawn and subjects who remained free of cancer for at least 6 years was 1.2%. For the interval 6 to 8 years, the difference in means was 1.8%; for 8 to 10 years it was 0.9%; for 10 to 12 years it was 2.0%; for 12 to 14 years it was 0.7%; and for 14+ years it was −3.9%. The differences in mean serum iron for each followup interval are also all positive except for the last, 14+ years; for TIBC, the differences are all negative except for the last two intervals. All but one of the differences in albumin means is negative.

When analyzed separately, among the new group of 147 cancer cases in men, there were not significant differences with cancer-free men in transferrin saturation, TIBC, serum iron, or serum albumin. These new cases, however, showed a pattern of differences depending on time since blood test as can be seen in Table II. All 22 men in the 14+ category were in the group diagnosed since 1984, and they show opposite relationships with measures of body iron than the other cases in the Table. Sixty-nine of the 70 men diagnosed between 12 and 14 years after blood test were in the new group of cases, as were 39 of the 55 men in the 10-to-12-year group. These men showed relationships in the same direction as the men analyzed previously (Stevens *et al.*, 1988). The new cases increased the total number of cases and substantially improved our ability to examine dose-response relationships.

Table III presents differences in means for cancer deaths against those who did not die of cancer by follow-up interval. Again, there is consistency across follow-up intervals, suggesting that the overall significant findings are not unduly affected by a small part of the data in a particular follow-up interval.

The transferrin saturation quartile ranges for men were 0–22.8%, 22.9–29.1%, 29.2–36.7%, and 36.8% and higher. Previously (Stevens *et al.*, 1988) we reported that although there was no dose response apparent in women across their quartiles of transferrin saturation, there was an estimate of relative risk of 1.3 among women grouped on the basis of the highest quartile cutoff among men, *i.e.*, among women with baseline transferrin saturation greater than 36.7%. On the basis of the male quartile cutpoints, the risks in each quartile of transferrin saturation in men and women combined relative to the lowest quartile were 1.0, 1.06, 1.00, and 1.29 for incidence (p for trend, 0.02); they were 1.0, 1.00, 0.94, and 1.28 for mortality (p for trend, 0.05). The trend is driven by cases in the top quartile, those men and women with a baseline transferrin saturation greater than 36.7%.

TABLE III – MEAN BIOCHEMICAL VALUES FOR MEN WHO DIED OF CANCER AND THOSE WHO DID NOT, BROKEN DOWN BY FOLLOW-UP INTERVAL[1]

	Transferrin saturation (%)		Iron (μg/dl)		TIBC (μg/dl)		Albumin (g/l)	
4 to <7								
3,229[2]	30.9	0.28[4]	106.5	0.34	350.3	—	44.3	0.19
58[3]	31.7		108.6		350.3		43.9	
7 to <10								
3,161	30.7	0.002	106.2	0.005	350.7	0.11	44.3	0.003
68	34.9		117.7		343.5		43.3	
10 to <13								
3,107	30.7	—	106.2	—	351.0	0.15	44.3	—
54	30.1		104.8		344.6		44.5	
13+								
3,054	30.7	0.23	106.2	0.17	351.1	—	44.3	0.2
53	31.8		110.8		351.2		44.0	
Total								
3,054	30.8	0.025	106.2	0.03	350.6	0.13	44.3	0.025
233	32.3		110.9		346.9		43.9	

[1]4 to less than 7 years, 7 to less than 10 years, etc.; adjusted for age and smoking.–[2]Number of subjects cancer-free to end of specified follow-up interval.–[3]Number of subjects who died of cancer during specified follow-up interval.–[4]Nominal p value for comparison of means.

In order to determine what specific ranges of transferrin saturation were most strongly associated with cancer risk, men and women were divided into 5 categories of unequal size. The baseline was chosen to be 0 to 30% because 30% is approximately the overall mean and close to the median. The next 4 categories include all study subjects for whom baseline transferrin saturation was above average and fell into the ranges 30.1–40 or 40.1–50 or 50.1–60 or greater than 60. Of the total 3,287 men, 52% were in the baseline category, and 29%, 13%, 4%, and 2% were in each of the other 4 categories in ascending order. Of the total 5,269 women, 66% were in the baseline category, and 23%, 7%, 2%, and 1% were in each of the other 4 categories. Dose response for these categories is described in Tables IV and V. In Table IV, it is apparent that both for incidence and for mortality in men and in women there is an increase in risk with increasing transferrin saturation, with the greatest increased risk in the 60+% group (the effect of the cluster sampling in NHANES I was examined in the male incidence data and found not to alter the p-value at all). In Table V, men and women are combined. The trend of increasing risk with increasing transferrin saturation is significant both for incidence and for mortality, and 90% confidence intervals for incidence relative risks in the 50–60%, and 60+% groups are significantly different from 1.0. Combining the 40–50% and the 50–60% groups into a group 40–60% yields an estimated incidence relative risk of 1.21 with 90% confidence limits of 1.02 to 1.44. For mortality the estimate is 1.22 (0.97 to 1.54).

Transferrin saturation of approximately 60% has been used to screen for hemochromatosis homozygotes (Edwards *et al.*, 1988), whether symptomatic or not. It is also a level used by Gordeuk *et al.* (1992) to define "iron overload" in Africans for another iron overload genetic predisposition not linked to the hemochromatosis locus. Therefore, those NHANES I study subjects in the 60+% category may include many individuals at genetic risk of iron overload. There is evidence in this data set, however, of elevations of cancer occurrence and of death in subjects with baseline transferrin saturations in the range 40 to

TABLE IV – RELATIVE RISK ESTIMATES DERIVED FROM THE MANTEL-HAENZEL PROCEDURE FOR TRANSFERRIN-SATURATION RANGES IN MEN AND IN WOMEN[1]

	Transferrin-saturation range (%)					p value for trend
	0–30	30.1–40	40.1–50	50.1–60	60+	
Male incidence	1.00	1.04 (0.85, 1.27)	1.17 (0.92, 1.49)	1.20 (0.77, 1.86)	1.69 (1.02, 2.77)	0.03
Female incidence	1.00	0.85 (0.68, 1.05)	1.14 (0.82, 1.58)	1.75 (1.05, 2.95)	2.09 (1.05, 4.18)	0.10
Male mortality	1.00	1.14 (0.89, 1.46)	1.33 (0.98, 1.82)	1.31 (0.76, 2.27)	1.67 (0.90, 3.10)	0.03
Female mortality	1.00	0.76 (0.56, 1.04)	1.01 (0.63, 1.60)	1.25 (0.54, 2.89)	1.90 (0.72, 5.02)	0.47

[1]Values in parentheses are 90% CI; stratified on age and smoking.

TABLE V – RELATIVE RISK ESTIMATES DERIVED FROM THE MANTEL-HAENZEL PROCEDURE FOR 5 TRANSFERRIN-SATURATION RANGES IN MEN AND WOMEN COMBINED[1]

	Transferrin-saturation range (%)					p value for trend
	0–30	30.1–40	40.1–50	50.1–60	60+	
Incidence	1.00	0.95 (0.82, 1.10)	1.16 (0.96, 1.41)	1.38 (1.00, 1.90)	1.81 (1.21, 2.71)	0.012
Mortality	1.00	0.96 (0.79, 1.16)	1.22 (0.97, 1.54)	1.29 (0.84, 2.00)	1.73 (1.03, 2.91)	0.049

[1]Values in parentheses are 90% CI; stratified on sex, age and smoking.

60% as well, a level which may include many heterozygotes for hemochromatosis (Kravitz *et al.*, 1979). Hemochromatosis homozygotes have a greatly increased risk of liver cancer (Bradbear *et al.*, 1985). There were only 6 cases among men and 6 cases among women in this study; mean transferrin saturation in these cases was 37.4 for men and 24.9 for women.

In Table VI, the means of the biochemical variables are shown by selected cancer types. In our previous report (Stevens *et al.*, 1988), cancers of esophagus, colon, lung and bladder appeared most strongly related to body iron. With the new follow-up data presented here, the strength of the association with cancers of colon and lung is less strong.

DISCUSSION

These analyses relied on serum iron, TIBC, and, in particular, transferrin saturation to assess body-iron level. Is transferrin saturation an accurate measure of body-iron level? First, transferrin saturation may simply reflect "inflammation" resulting from early disease. However, Yip and Dallman (1988) showed that transferrin saturation declined with increasing erythrocyte sedimentation rate (ESR increases in inflammation). This means that underlying inflammation from early-stage disease would have worked against increased transferrin saturation in those subsequently to develop cancer, and would have reduced the ability of our analyses to detect a positive association of cancer risk with transferrin saturation, should it exist.

The less transferrin saturation reflects body-iron level the less reliable is the inference about body-iron level that can be drawn from our report. In general, estimates of risk based on an inaccurate exposure metric (perhaps transferrin saturation as a measure of body iron level) will be closer to unity than the true relative risk associated with true exposure (*e.g.*, high body-iron level). Serum ferritin is generally considered a good indicator of body-iron level in otherwise healthy individuals (Lipschitz *et al.*, 1974). In studying the effects of repeated phlebotomies to induce mild iron deficiency in 14 normal volunteers on serum measures of iron level, Skikne *et al.* (1990) found that transferrin saturation fell from a mean of 23% at baseline to 11% during iron "depletion" to 8% during "anemia", and stated "Within this range, the transferrin saturation fell from 11% to approximately 8%, and the serum ferritin declined from 11 to approximately 6 μg/l. These changes justify the prior use of these measurements for estimating mild iron deficiency". In a study of iron overload in Africans, Gordeuk *et al.* (1992) reported a correlation of 0.70 between serum transferrin saturation and serum ferritin, and found that transferrin saturation better distinguished iron overload than did serum ferritin.

Another source of uncertainty in these analyses is the day-to-day variation in the iron-status indicators used. Borel *et al.* (1991) reported that there is considerable such variability, and they concluded that 3 to 10 repeat measurements on study subjects would much more accurately reflect body-iron status. Since we had only one blood determination of serum iron and TIBC (transferrin saturation was calculated from these), this day-to-day variability introduces further misclassification of exposure, and reduction toward unity of relative risk estimates. Finally, given that follow-up for these analyses continued for up to 16 years after blood test, further misclassification of iron status was introduced by lack of more recent iron-status blood determinations.

Unless transferrin saturation is actually associated with some unknown characteristic that increases cancer risk, the results presented here are a conservative estimate of the effect of body-iron level on risk of cancer.

There is growing evidence from animal experimentation that excess iron can increase cancer cell growth (Bergeron *et al.*, 1985) and increase tumor yield by chemical carcinogens (Siegers *et al.*, 1988; Smith *et al.*, 1989; Thompson *et al.*, 1991), also that iron restriction can reduce cancer induction (Thompson *et al.*, 1991; Hann *et al.*, 1988). Human evidence is limited, and generally supports the hypothesis of increased cancer risk from increased body iron (Stevens, 1990). Another approach to testing the hypothesis that high body-iron level increases cancer risk is to determine whether blood donors are at reduced risk of cancer (Weinberg and Weinberg, 1986). Merk *et al.* (1990) found significantly reduced risk of cancer (all sites combined) in blood donors in Sweden. The authors discuss several possible explanations for their findings, including a "healthy donor" effect in which blood donors are a self-selected group with a generally healthy life style. The authors also point out, however, that their results are consistent with a protective effect of blood donation by virtue of lowering body iron level. This effect may have been particularly apparent in Sweden because the general population is heavily supplemented with iron both voluntarily and involuntarily (Olsson *et al.*, 1978; Reizenstein, 1990).

Transferrin saturation was lower in stomach-cancer cases than among non-cancer subjects, although not significantly so (Table VI). Akiba *et al.* (1991) reported that serum ferritin was significantly lower in cases of stomach cancer than among controls in Japan. Nomura *et al.* (1992) confirmed these findings in a study in Hawaii. Both groups speculated that long-term precursors of stomach cancer could lead to poor iron absorption and could result in bleeding, both of which could lead to low body-iron level. In a letter response to our original report on the NHANES, Takkunen *et al.* (1989) reported that serum transferrin saturation was not related to cancer death in Finland. However, stomach cancer is a leading cause of cancer death in Finnish men, and after it is analyzed separately there is some support for increased risk of some cancer sites in those with high transferrin saturation or low TIBC in their study population (Knekt *et al.*, 1994). They also report a strong inverse relationship of body-iron level and risk

TABLE VI – MEAN VALUES FOR BIOCHEMICAL VARIABLES ACCORDING TO THE SITE OF CANCER IN MEN[1]

	Number	HgB (g/dl)	Transferrin saturation (%)	TIBC (μg/dl)	Iron (μg/dl)	Albumin (g/l)	Dietary iron (mg/day)
No cancer	2,908	15.5	30.7	350.9	106.2	44.3	14.2
Esophagus	9	14.6[2]	43.7[2]	319.2[2]	137.0[2]	43.8	12.2
Stomach	10	16.3[2]	26.0	361.7	93.9	44.0	11.7
Colon	25	15.8	33.0	354.1	112.2	45.5[2]	18.0[2]
Rectum	14	15.7	34.7	361.6	122.1[2]	44.1	14.0
Pancreas	15	15.0	31.0	352.5	106.8	43.2	14.6
Lung	75	15.7	31.9	351.4	111.0	43.9	12.6[2]
Prostate	94	15.2[2]	31.5	344.6	108.7	43.6[2]	13.7
Bladder	17	15.6	38.1[2]	333.4	120.8[2]	42.7[2]	15.6
Other urinary	12	15.2	22.5[2]	342.3	76.6[2]	43.5	12.3
Other	108	15.6	32.2	342.2	109.3	44.1	14.4

[1]Adjusted for age and smoking.–[2]$p < 0.05$, 2-tailed test.

of stomach cancer, consistent with the other studies cited above (Akiba *et al.*, 1991; Nomura *et al.*, 1992).

The association of low serum albumin with increased risk of cancer has now been reported in other large studies (Phillips *et al.*, 1989). This relationship is striking and as yet ill understood, although some possible explanations have been offered (Stevens and Blumberg, 1990, 1992).

Our studies have focused on a possible role for iron in the etiology of cancer. Body iron may also play an important role in prognosis after cancer diagnosis (De Sousa *et al.*, 1989), and manipulation of iron or iron-binding proteins may provide new avenues for cancer therapy (Cazzola *et al.*, 1990; De Sousa and Brock, 1989). These possibilities are based primarily on the idea that iron is a nutritional requirement for cancer cells. The possible mechanism whereby iron might increase cancer risk by generation of oxygen radicals also implies that iron may increase susceptibility to radiation-induced cancer (Stevens and Kalkwarf, 1990). Indeed, Nelson and Stevens (1992) have shown that a non-toxic level of ferritin-iron added to the growth medium of Chinese hamster ovary cells greatly increases killing of them by X rays.

Neriishi *et al.* (1991) reported that Japanese atomic-bomb survivors who reported epilation within 60 days of the bombings were at 2.5-fold higher (5 or more years later) risk of leukemia than those who did not report epilation, after adjusting for estimated radiation dose. They cite this as evidence that susceptibility to radiation-induced cancer may depend, in part, on the characteristics of the individual. One such characteristic may be levels of anti-oxidant enzymes, and another may be body-iron level. In addition, since rapid cell proliferation increases susceptibility to damage from oxygen radicals and radiation, agents which increase proliferation could increase risk of neoplastic diseases such as leukemia, and excess iron may further stimulate cellular proliferation.

An increased risk of heart disease (Sullivan, 1981; Salonen *et al.*, 1992) and of cancer (Stevens, 1990) from high body-iron level might share the common mechanism of increased oxidative stress, and cell proliferation may be a key to the different manifestations in different tissues of oxidative damage to biomolecules (Stevens and Neriishi, 1992). In slowly dividing endothelium and myocardium, DNA mutations may not be as important as accumulated cellular damage leading to tissue dysfunction. In rapidly dividing epithelium, however, cellular damage may not be so important because of the constant introduction of new healthy cells, whereas a DNA mutation has the opportunity to become fixed to a transformed phenotype in at least one cell of the tissue by virtue of cell turnover.

Finally, it is important to emphasize that the results presented here do not weaken or invalidate the considerable evidence of adverse health effects of severe iron deficiency. These data do, however, call into question the wisdom of wholesale addition of iron to food. If elevated body iron does increase cancer risk, then the lack of association of estimated dietary iron intake with cancer may be due to lack of iron absorption from the diet, or an unreliable estimate. The former is possible, whereas the latter is probable (James *et al.*, 1981).

If moderate or severe iron excess does pose a cancer risk, then periodic blood donation by those who are not anemic may be of long-term benefit to the donor as well as to society.

ACKNOWLEDGEMENTS

We thank the late Dr. David Byar for helping to put our research team together. We are grateful to Mr. J. Buchanan, of Pacific Northwest Laboratory, for valuable computer assistance. R.G.S. is supported by the Office of Epidemiology and Health Surveillance, US Department of Energy, under contract DE-AC06-76RLO 1830; K.N. is supported equally by the Japanese Ministry of Health and Welfare and the US National Academy of Sciences, through a contract with the Department of Energy; B.S.B. is supported by USPHS grants CA-40737, RR-05895 and CA-06927 from the National Institutes of Health and by an appropriation from the Commonwealth of Pennsylvania.

REFERENCES

AKIBA, S., NERIISHI, K., BLOT, W.J., KABUTO, M., STEVENS, R.G., KATO, H. and LAND, C.E., Serum ferritin and stomach cancer risk among Japanese. *Cancer,* **67,** 1707–1712 (1991).

BERGERON, R.J., STREIFF, R.R. and ELLIOTT, G.T., Influence of iron on *in vivo* proliferation and lethality of L1210 cells. *J. Nutr.,* **115,** 369–374 (1985).

BOREL, M.J., SMITH, S.M., DERR, J. and BEARD, J.L., Day-to-day variation in iron-status indices in healthy men and women. *Amer. J. Clin. Nutr.,* **54,** 729–735 (1991).

BRADBEAR, R.A., BAIN, C., SISKIND, V., SCHOFIELD, F.D., WEBB, S., AXELSEN, E.M., HALLIDAY, J.W., BASSETT, M.L. and POWELL, L.W., Cohort study of internal malignancy in genetic hemochromatosis and other chronic nonalcoholic liver diseases. *J. nat. Cancer Inst.,* **75,** 81–84 (1985).

CAZZOLA, M., BERGAMASCHI, G., DEZZA, L. and AROSIO, P., Manipulations of cellular iron metabolism for modulating normal and malignant cell proliferation: achievements and prospects. *Blood,* **75,** 1903–1919 (1990).

COOK, J.D., Adaptation in iron metabolism. *Amer. J. clin. Nutr.,* **51,** 301–308 (1990).

DE SOUSA, M. and BROCK, J.H., *Iron in immunity, Cancer and inflammation.* John Wiley, Chichester (1989).

DE SOUSA, M., POTAZNIK, D. and GROSHEN, S.C.B., Body iron stores and the risk of cancer (letter). *New Engl. J. Med.,* **320,** 1014 (1989).

EDWARDS, C.Q., GRIFFEN, L.M., GOLDGAR, D., DRUMMOND, C., SKOLNICK, M.H. and KUSHNER, J.P., Prevalence of hemochromatosis among 11,065 presumably healthy blood donors. *New Engl. J. Med.,* **318,** 1355–1362 (1988).

FLOYD, R.A., Direct demonstration that ferrous ion complexes of di- and triphosphate nucleotides catalyze hydroxyl free-radical formation from hydrogen peroxide. *Arch. Biochem. Biophys.,* **225,** 263–270 (1983).

GILBERT, E.S. and BUCHANAN, J.A., An alternative approach to analyzing occupational mortality data. *J. occup. Med.,* **26,** 822–828 (1984).

GORDEUK, V., MUKIIBI, J., HASSTEDT, S.J., SAMOWITZ, W., CORWIN, C.Q., WEST, G., NDAMBIRE, S., EMMANUAL, J., NKANZA, N., CHAPANDUKA, Z., RANDALL, M., BOONE, P., ROMANO, P., MARTELL, R.W., YAMASHITA, T., EFFLER, P. and BRITTENHAM, G., Iron overload in Africa. *New Engl. J. Med.,* **326,** 95–100 (1992).

HALLIWELL, B. and GUTTERIDGE, J.M.C., Role of free radicals and catalytic metal ions in human disease: an overview. *Methods Enzymol.,* **186,** 1–85 (1990).

HANN, H.L., STAHLHUT, M.W. and BLUMBERG, B.S., Iron nutrition and tumor growth: decreased tumor growth in iron-deficient mice. *Cancer Res.,* **48,** 4168–4170 (1988).

JAMES, W.P., BINGHAM, S.A. and COLE, T.J., Epidemiological assessment of dietary intake. *Nutr. Cancer,* **2,** 203–212 (1981).

KNEKT, P., REUNANEN, A., TAKKUNEN, H., AROMAA, A. and HAKULINEN, T., Body iron stores and the risk of cancer. *Int. J. Cancer,* **56,** 379–382 (1994).

KORN, E.L. and GRAUBARD, B.I., Epidemiological studies utilizing surveys: accounting for the sampling design. *Amer. J. public Health,* **81,** 1166–1173 (1991).

KRAVITZ, K., SKOLNICK, M., CANNINGS, C., CARMELLI, D., BATY, B., AMOS, B., JOHNSON, A., MENDELL, N., EDWARDS, C. and CARTWRIGHT, G., Genetic linkage between hereditary hemochromatosis and HLA. *Amer. J. hum. Genet.,* **31,** 601–619 (1979).

LIPSCHITZ, D.A., COOK, J.D. and FINCH, C.A., A clinical evaluation of serum ferritin as an index of iron stores. *New Engl. J. Med.,* **290,** 1213–1216 (1974).

LOEB, L.A., JAMES, E.A., WALTERSDORPH, A.M. and KLEBANOFF, S.J., Mutagenesis by autoxidation of iron with isolated DNA. *Proc. nat. Acad. Sci. (Wash.),* **85,** 3918–3922 (1988).

MADANS, J.H., KLEINMAN, J.C., COX, C.S., BARBANO, H.E., FELDMAN, J.J., COHEN, B., FINUCANE, F.F. and CORONI-HUNTLEY, J., 10 Years after NHANES I: report of initial follow-up, 1982–84. *Public Health Rep.*, **101**, 465–473 (1986).

MERK, K., MATTSSON, B., MATTSSON, A., HOLM, G., GULLBRING, B. and BJÖRKHOLM, M., The incidence of cancer among blood donors. *Int. J. Epidemiol.*, **19**, 505–509 (1990).

NELSON, J.M. and STEVENS, R.G., Ferritin-iron increases killing of Chinese hamster ovary cells by X-irradiation. *Cell Proliferation*, **25**, 579–585 (1992).

NERIISHI, K., STRAM, D.O., VAETH, M., MIZUNO, S. and AKIBA, S., The observed relationship between occurrence of acute radiation effects and leukemia mortality among A-bomb survivors. *Radiat Res.*, **125**, 206–213 (1991).

NOMURA, A., CHYOU, P.-H. and STEMMERMANN, G.N., The association of serum ferritin levels with the risk of stomach cancer. *Cancer Epidemiol. Biomarkers Prevention*, **1**, 547–550 (1992).

OLSSON, K.S., HEEDMAN, P.A. and STAUGARD, F., Preclinical hemochromatosis in a population on a high-iron-fortified diet. *J. Amer. med. Ass.*, **239**, 1999–2000 (1978).

PHILLIPS, A., SHAPER, A.G. and WHINCUP, P.H., Association between serum albumin and mortality from cardiovascular disease, cancer, and other causes. *Lancet*, **II**, 1434–1436 (1989).

REIZENSTEIN, P., Inducers and scavengers of free radicals in food. *Med. Oncol. Tumor Pharmacother.*, **7**, 67–68 (1990).

SALONEN, J.T., NYYSSÖNEN, K., KORPELA, H., TUOMILEHTO, J., SEPPÄNEN, R. and SALONEN, R., High stored-iron levels are associated with excess risk of myocardial infarction in eastern Finnish men. *Circulation*, **86**, 803–811 (1992).

SAS INSTITUTE, INC., *SUGI user's guide: statistics*. Version 5. Rev. ed., SAS Institute, Cary, NC (1985).

SIEGERS, C.P., BUMANN, D., BARETTON, G. and YOUNES, M., Dietary iron enhances the tumor rate in dimethylhydrazine-induced colon carcinogenesis in mice. *Cancer Lett.*, **41**, 251–256 (1988).

SKIKNE, B.S., FLOWERS, C.H. and COOK, J.D., Serum transferrin receptor: a quantitative measure of tissue iron deficiency. *Blood*, **75**, 1870–1876 (1990).

SMITH, A.G., CABRAL, J.R.P., CARTHEW, P., FRANCIS, J.E. and MANSON, M.M., Carcinogenicity of iron in conjunction with a chlorinated environmental chemical, hexachlorobenzene, in C57BL/10ScSn mice. *Int. J. Cancer*, **43**, 492–496 (1989).

STEVENS, R.G., Iron and the risk of cancer. *Med. Oncol. Tumor Pharmacother.*, **7**, 177–181 (1990).

STEVENS, R.G. and BLUMBERG, B.S., Serum albumin and mortality. *Lancet*, **335**, 351 (1990).

STEVENS, R.G. and BLUMBERG, B.S., Serum albumin and cancer risk. *In:* M.S. Micozzi and T.E. Moon (eds.). *Macronutrients: investigating their role in cancer*, pp. 283–292, Marcel Dekker, New York (1992).

STEVENS, R.G., JONES, D.Y., MICOZZI, M.S. and TAYLOR, P.R., Body iron stores and the risk of cancer. *New Engl. J. Med.*, **319**, 1047–1052 (1988).

STEVENS, R.G. and KALKWARF, D.R., Iron, radiation, and cancer. *Environ. Hlth Perspec.*, **87**, 291–300 (1990).

STEVENS, R.G. and NERIISHI, K., Iron and oxidative damage in human cancer. *In:* L. Spatz and A.D. Bloom (eds.), *Biological consequences of oxidative stress: implications for cardiovascular disease and carcinogenesis*, pp. 138–161, Oxford University Press, New York (1992).

SULLIVAN, J.L., Iron and the sex difference in heart disease risk. *Lancet*, **1**, 1293–1294 (1981).

SUTHERLAND, R., DELIA, D., SCHNEIDER, C., NEWMAN, R., KEMSHEAD, H. and GREAVES, M., Ubiquitous cell-surface glycoprotein on tumor cells is proliferation-associated receptor for transferrin. *Proc. nat. Acad. Sci. (Wash.)*, **78**, 4515–4519 (1981).

TAKKUNEN, H., REUNANEN, A., KNEKT, P. and AROMAA, A., Body iron stores and the risk of cancer. *New Engl. J. Med.* (letter), **320**, 1013–1014 (1989).

THOMPSON, H.J., KENNEDY, K., WITT, M. and JUZEFYK, J., Effect of dietary iron deficiency or excess on the induction of mammary carcinogenesis by 1-methyl-1-nitrosourea. *Carcinogenesis*, **12**, 111–114 (1991).

WEINBERG, E.D., Iron withholding: a defense against infection and neoplasia. *Physiol. Rev.*, **64**, 65–102 (1984).

WEINBERG, R.J. and WEINBERG, E.D., Blood-letting, iron homeostasis, and human health. *Med. Hypotheses*, **21**, 441–443 (1986).

YIP, R. and DALLMAN, P.R., The roles of inflammation and iron deficiency as causes of anemia *Amer. J. clin. Nutr.*, **48**, 48, 1295–1300 (1988).

Family Studies of a Human Serum Isoantigen System (Australia Antigen)

B. S. BLUMBERG,[1] LIISA MELARTIN,[1] R. A. GUINTO,[2] AND BARBARA WERNER[1]

[1]*The Institute for Cancer Research, Philadelphia, Pennsylvania 19111.*
[2]*Cebu Skin Clinic and the Leonard Wood Memorial-Eversley Childs Sanitarium Leprosy Research Laboratory, Cebu City, The Philippines.*

RECENTLY, we described what appeared to be a "new" serum protein (Blumberg, 1964; Blumberg *et al.*, 1965). It had been detected using isoprecipitins found in the sera of hemophilia patients who had received large numbers of transfusions. These isoprecipitins reacted, in double diffusion agar gel experiments, with an antigen in the serum and plasma of some normal and diseased individuals. Because the antigen was originally found in the serum of an Australian aborigine, the protein was referred to as "Australia antigen."

Preliminary biochemical and immunological studies indicate that Australia antigen contains lipid with electrophoretic mobility and other characteristics similar to those of low density (or beta) lipoprotein. It can, however, be readily distinguished from other lipoproteins by differences in its flotation characteristics, staining properties, and immunological reactions (Alter and Blumberg, 1966) and from the lipoprotein isoantigen system previously reported from this laboratory (Blumberg *et al.*, 1962).

Australia antigen was also found to have an unusual population distribution, which probably accounts for the failure to identify this serum protein earlier. The isoprecipitin did not react with sera from normal U. S. white and Negro populations and only rarely (1/1000) with sera from the several northern European populations studied. It was more common (approximately 1%) in Mediterranean populations and even more so in sera collected in the East (approximately 6%: Blumberg *et al.*, 1965, and Table 2). It was also found in relatively high frequencies in patients with leukemia, mongolism (Down's syndrome), and thalassemia, but not in patients with solid tumors or a variety of other diseases (Blumberg and Alter, 1965; Blumberg, 1966).

The original populations studied included some sera collected from families. In one of these (Fig. 4 of Blumberg *et al.*, 1965), an extensive pedigree from a Samaritan family of Israel, two sibs were found to have the trait. Studies in other populations (i.e., Marshall Islanders and Cashinahua Indians from Peru) also suggested a familial aggregation, and it became apparent that a systematic family study should be done. Because of the odd distribution of the trait, that is, its extreme rarity in U. S. populations, extensive family studies could not be

Received May 16, 1966.

Supported by U. S. Public Health Service research grants CA 06551 and CA 08069 from the National Institutes of Health.

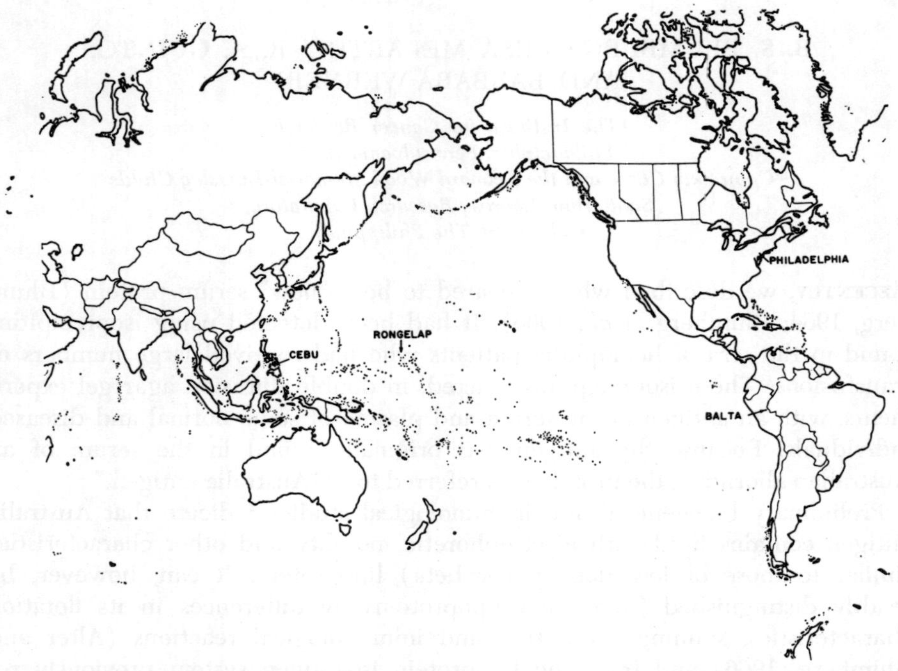

FIG. 1. Locations where families were studied. Cebu, Philippines; Rongelap, Marshall Islands District of the U. S. Trust Territories of the Pacific Islands; Balta, Peru (adjacent to the Brazilian border); Philadelphia, Pennsylvania.

readily undertaken locally. Fortunately, at this time, 978 sera collected from leprosy and nonleprosy patients in Cebu, The Philippines, were made available to us through the courtesy of Drs. Lechat, Binford, Guinto, Cohen, and others. (These have also been used in a study of the relation between leprosy and Australia antigen, the results of which are described elsewhere: Blumberg and Melartin, 1966; Blumberg *et al.*, 1966.) Sixty-four Filipinos were found to have Australia antigen. Arrangements were made to collect blood specimens from the families of some of these reactors as well as from the families of individuals who did not have Australia antigen.

The purpose of this paper is to report the population and family data collected for the purpose of determining if there is a family clustering of Australia antigen and of testing the genetic hypothesis that Australia antigen is inherited as a simple autosomal trait. The populations from which most of the families were obtained are briefly described, and the results of the experiments on the reproducibility of the immunologic experiments and the comparability of methods of blood collection are given. Information on the persistance of the trait and the sex, age, and geographical distribution of Australia antigen are also presented. Following this, the family data are analyzed in order to determine whether there is familial clustering. Finally, the segregation data are used to test a genetic hypothesis.

MATERIALS AND METHODS

Populations Studied

The family studies were done in four widely separated populations (Fig. 1). The Cashinahua, a small group of Indians classified linguistically as Panoan, live along the Curanja River of southeastern Peru and in nearby parts of Brazil. They have been studied extensively by Kenneth Kensinger of the Department of Anthropology of the University of Pennsylvania (Kensinger, 1965). Mr. Kensinger has lived with these Indians for five years between 1955 and the present. He supplied the pedigree information, which he believes to be accurate, but it is subject to the general uncertainties of family studies in primitive groups.

Rongelap is an atoll in the Marshall Islands of the U. S. Trust Territories of the Pacific Islands, whose inhabitants are classified as Micronesians. In 1954, they were subjected to accidental fallout from a nuclear weapon exploded on nearby Bikini Atoll. They were evacuated immediately but returned to their island several years later. Since the exposure, they have been examined yearly by a medical expedition directed by Dr. Robert A. Conard of the Brookhaven National Laboratory. Blood specimens have been collected yearly, and many of these were made available through the kindness of Dr. Conard. The populations and the studies which have been performed are described in the reports of the expeditions, the latest of which is Conard *et al.* (1965). Pedigree information was originally obtained by Dr. H. Hechter of San Francisco, amplified by one of the authors (B.S.B.) and finally confirmed and considerably expanded by Dr. Bryon Bender of the Trust Territory government.

Cebu is an island of the Visayan group, lying south of Manila. Its principal city, Cebu City, is the second largest in the Philippines. All of the blood specimens collected for the family study were from Visayan-speaking people from Cebu or nearby islands. Most of the families studied were obtained through patients with leprosy resident in the Eversley Childs Sanitarium or from outpatients at the Cebu Skin Clinic. In addition, blood from several families and mother-child combinations was obtained from families living in the "negative barrio," that is, a village adjacent to the Sanitarium in which discharged patients and their immediate families reside. The other families and mother-child combinations were obtained primarily among outpatients at the Cebu Skin Clinic who were attending for minor dermatologic conditions.

Through the co-operation of physicians in the Philadelphia area, sera were collected from patients with leukemia and other diseases. Some of these were found to have Australia antigen, and family sera were collected when possible. These include families 19 (thalassemia) and 20 (chronic myelogenous leukemia) of Table 5.

The Samaritan family (18) is shown in Fig. 4 of Blumberg *et al.* (1965). We are indebted to Dr. Batsheva Bonne for the studies on this family. Several other populations are referred to in Table 2. These will be described in greater detail in a forthcoming paper on population studies of Australia antigen.

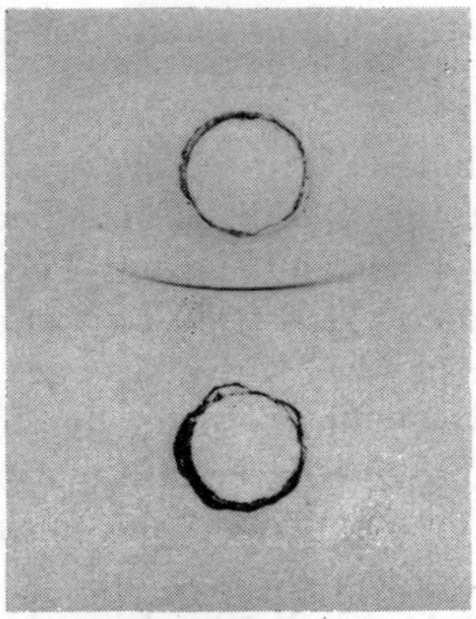

FIG. 2. Typical Ouchterlony experiment. The serum containing Australia antigen is in the upper well. The antiserum (from a patient with hemophilia) is in the lower well. The precipitin line has been stained with Azo carmine.

Methods

Blood was collected by venepuncture using vacutainer tubes (Becton, Dickinson Co.). In most cases, these contained no anticoagulant, but in several they contained Acid-Citrate-Dextrose mixture (ACD) or ethylenediaminetetraacetic acid (EDTA). For most of the family studies (Table 5, families 1–21), the red blood cells and serum (or plasma) were separated, usually within 24 hours of collection, and shipped at −20°C to the laboratory in Philadelphia. In Cebu (families 21–36), the studies were done in the field. The unfrozen sera or plasmas were usually tested within six hours of collection, none later than 12 hours after collection.

Antisera

All family sera were tested using a human and a rabbit antiserum. The human antisera are from patients L.B. and R.P. (Alter and Blumberg, 1966, Table 1), both of whom had classical hemophilia and had received hundreds of transfusions, primarily of fresh frozen plasma. The serum of patient L.B. was used unconcentrated while that of R.P. was concentrated 2½ times by ultrafiltration. The rabbit antiserum (no. 4) was produced by immunizing a rabbit with the whole serum of an Australia antigen-positive individual and absorbing with the sera of normal persons (Melartin and Blumberg, 1966).

The micro-Ouchterlony technique was used for most of the double diffusion experiments and is described elsewhere (Blumberg and Riddell, 1963).

The chi-square method was used for comparison of 2 × 2 and $n \times 2$ tables. If the number in any of the cells was smaller than 5, the exact method of Fisher was used. The genetic analyses of the families were done by the methods suggested by Smith (1956), and the mother-child studies by a method similar to that described by Maynard Smith *et al.* (1965).

RESULTS

Reproducibility of the Double Diffusion Test

A typical Ouchterlony double diffusion experiment is shown in Fig. 2. In order to test the reproducibility of the precipitin and staining methods used to determine the phenotypes, 15 samples known to contain Australia antigen (five Rongelap, five leukemia, and five Down's syndrome subjects) and 15 samples from the same populations which were known to be negative for Australia antigen were each divided into nine parts and tested on nine sequential days. All of the sera originally classified as negative remained negative on subsequent testing. All those originally classified as positive were subsequently so classified, with the exception of a single leukemia serum. This specimen had originally given a very faint reaction, and in five of the nine repeat tests gave a negative or questionable reaction. From this, it is concluded that the method of testing was consistent unless the reactions were very weak; in general, weak reactions were rare.

In order to determine the best method of collection, blood was withdrawn from ten patients with mongolism, five with and five without Australia antigen. Equal portions of each were placed in three sets of four different test tubes; the first test tube did not contain anticoagulant, the second contained ACD solution, the third EDTA, and the fourth heparin. One set was allowed to stand 24 hours, the second four days, and a third 14 days (all at 4° C) before the red blood cells were separated. The sera were then tested with a human and a rabbit antiserum. Excellent reactions were obtained with the specimens collected by each of the different methods after 24 hours of storage at 4°C. After four days of storage, more than 80% of the reactions originally positive were still positive. When the specimens were stored for two weeks, a similar result was obtained. In no case did a specimen classified as negative subsequently show a positive reaction. As noted above, the sera or plasmas either were tested immediately after collection or were separated from the formed elements within 24 hours of collection and stored at −20°C until tested.

Persistence of the Trait

In testing the inheritance of a blood trait, it is useful to know whether the supposed phenotype is constant during the lifetime of the individual or, if not, the nature of the inconsistency. If the trait varies considerably, it would be difficult to use specimens collected at one time in evaluating a genetic hypothesis. If, in general, the trait is consistent over a significant length of time, it can be assumed that the conclusions made from a single observation may be extrapolated to other (although, perhaps, not all) periods of a lifetime. In order to test persistence of the trait, it would be necessary to ascertain individuals with the trait and test them periodically over the course of several months and years. Fortunately, some serial samples were available from the Rongelap population.

Samples from a total of 300 Micronesians from Rongelap were collected during the period from 1958 to 1965. Of these, serial samples (blood collected over an interval of two or more years) were available from 250 individuals. Of the latter, 237 were consistently negative, i.e., if negative at one point in time they were negative when subsequently tested. Eleven were consistently positive, and two were inconsistent (Fig. 3). Of the latter, the inconsistency was apparent in only one of multiple collections.

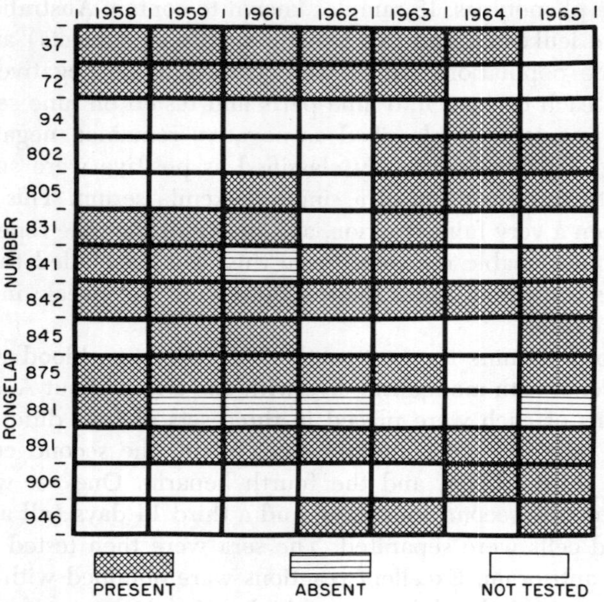

FIG. 3. Studies on the persistence of Australia antigen in blood samples collected serially from apparently normal individuals on Rongelap. The subject numbers are shown on the left. The reactions of sera collected in the years indicated at the top are given in the body of the figure.

In addition to the consistency of reaction data available from the apparently normal population of Rongelap, analogous studies were conducted on Australia antigen-positive sera from patients with leukemia and thalassemia (Table 1). These were consistent in every case.* From this, it is concluded that, in general, an individual's Australia antigen phenotype is consistent and that phenotyping undertaken at one point in time can be used to evaluate the genetic hypothesis.

Distribution of the Trait in Different Populations

A preliminary report on the distribution of Australia antigen in normal populations was given in the original paper describing the protein (Blumberg *et al.*, 1965). Additional normal populations have now been studied and are shown in Table 2. As noted below, the frequency of Australia antigen is age and sex dependent. The age and sex composition of these populations are not the same. Hence, the frequencies given are of value only for gross comparisons between the populations. The trait is absent in U. S. whites and Negroes but common in several East Asian and Pacific populations and in the Cashinahua. It is less common in Mediterranean populations.

*Inconsistent phenotypes may, however, be seen. Dr. L. Old, Cornell University Medical School, has kindly informed us that he has observed a leukemia patient who was positive when tested at one time but negative when tested several months later.

TABLE 1. PERSISTENCE OF AUSTRALIA ANTIGEN IN PATIENTS.

Patient	Diagnosis	Date	Reaction
S. B.	Chronic lymphocytic leukemia	10/2/63	+
		10/14/63	+
P. S.	Acute myelogenous leukemia	12/3/63	+
		3/25/64	+
		11/25/64	+
C. F.	Chronic reticuloendotheliosis	10/12/64	+
		10/25/64	+
		12/14/64	+
		6/20/65	+
C. de S.	Chronic lymphocytic leukemia	10/9/64	+
		11/2/64	+
		12/18/64	+
		12/28/64	+
		1/8/65	+
M B.	Thalassemia major	6/10/60	+
		1/16/62	+
		2/2/63	+
		8/6/63	+
		12/27/63	+
T. B.	Thalassemia major	3/11/60	+
		10/18/60	+

TABLE 2. AUSTRALIA ANTIGEN IN NONHOSPITALIZED POPULATIONS. Additional details on these populations will be given elsewhere.

Population	Location	Number tested	Number positive	Per cent positive
Americas				
Eskimos	Alaska (U.S.A.)	394	1	0.3
Indians, Athabascan	Alaska (U.S.A.)	204	0	0
Indians, Cashinahua	Peru	89	18	20.2
Indians, Haida	Canada	338	0	0
Indians, Mexico	Mexico	252	1	0.4
Indians, Quechua	Peru	102	0	0
Indians, Sioux	South Dakota (U.S.A.)	130	0	0
Negroes	Georgia, Maryland (U.S.A.)	607	0	0
Whites	Georgia, Maryland (U.S.A.)	896	0	0
Africa				
Pare	Tanzania	120	1	0.8
Tristan Da Cunha	Tristan da Cunha	42	0	0
Ghanese	Ghana	95	9	9.5
Asia				
Japanese	Osaka, Japan	301	1	0.3
Chinese	U.S.A., Taiwan	100	0	0
Taiwanese	Taiwan	23	3	13.0
Israelis	Israel	340	4	1.2
Jordanese	Jordan	40	0	0
Filipinos	Cebu, P. I.	764	37	4.8
Filipinos	Manila, P. I.	197	9	4.6
Vietnamese	Vietnam	128	8	6.3
Europe				
Finns	Finland	924	1	0.1
Lapps & Finnlapps	Finland	127	0	0
Italians	Ferara	101	0	0
Greeks	Greece	857	15	1.8
Portuguese	Hawaii	44	1	2.2
Oceania				
Aborigines	Australia	233	12	5.1
Maoris	New Zealand	4	1	
Melanesians	New Guinea	166	6	3.6
Micronesians	Marshall Islands	474	34	7.2
Polynesians	Bora Bora	119	3	2.5
Polynesians	Hawaii	43	1	2.3
Total		8,254	166	

TABLE 3. THE DISTRIBUTION OF AUSTRALIA ANTIGEN BY SEX IN FOUR POPULATIONS.

The Cebu populations include only individuals without leprosy since there is an increased frequency of lepromatous leprosy in males and an increased frequency of Australia antigen in lepromatous leprosy male patients (Blumberg and Melartin, 1966; Blumberg *et al.*, 1966). The male-female difference is statistically significant for the Cebu populations ($\chi^2 = 4.4$, $.05 > P > .025$) but not for the others.

	Total number	Number positive	Per cent positive
Marshall Islands			
Male	243	19	7.8
Female	226	14	6.2
Total	496	33	6.7
Cebu			
Male	430	27	6.3
Female	334	10	3.0
Total	764	37	4.8
Manila			
Male	138	6	4.3
Female	59	3	5.1
Total	197	9	4.6
Cashinahua			
Male	45	10	22.2
Female	44	6	13.6
Total	89	16	18.0

Sex Distribution

There is a higher frequency of male reactors in each of the populations with the exception of that from Manila (Table 3). The difference is statistically significant for the nonleprous Cebu population ($\chi^2 = 4.4$, $.05 > P > .025$) but is insignificant in all the others. There is no evidence of sex linkage in the family data (see below).

Age Distribution

Age distribution has been studied in several populations (Table 4). There is a decreased frequency of Australia antigen in the older age groups. If the several populations are divided into 0 to 39 and 40+ age groups, the differences between the two age groups are significant in the larger populations (Marshall Island, $\chi^2 = 4.15$, $0.05 > P > 0.02$; Cebu, $\chi^2 = 6.47$, $0.02 > P > 0.01$). In two smaller populations, however, the differences are not significant when the P values are determined using Fisher's exact method.

Family Studies

The original data on family clustering emerged from the testing of samples collected from several populations for other purposes and stored in what is now the blood collection of the Division of Clinical Research of the Institute for Cancer Research. The collection of sera from Rongelap included nearly everyone resident on that island and contained many family groups. The collections from the Cashinahua and the Samaritans included a very large percentage of these populations including family groups. After each of these populations had been tested for Australia antigen, the results were arranged

TABLE 4. THE DISTRIBUTION OF AUSTRALIA ANTIGEN BY AGE IN FOUR GENERATIONS. The Cebu population includes leprosy and nonleprosy individuals.

Age	Marshall Islands			Cebu			Manila			Cashinahua		
	No.	No. Pos.	% Pos.	No.	No. Pos.	% Pos.	No.	No. Pos.	% Pos.	No.	No. Pos.	% Pos.
0–19	196	16	9.5	348	30	8.6	22	2	9.1	27	12	44.4
20–39	148	12	8.1	948	62	6.5	132	6	4.5	49	3	6.1
40–59	85	4	4.7	378	16	4.2	36	1	2.8	13	1	7.7
60+	67	1	1.5	128	3	2.3	6	0	0	0	0	—
Total	496	33	6.7	1,802	111	6.2	196	9	4.6	89	16	18.0

TABLE 5. FAMILIES SEGREGATING FOR AUSTRALIA ANTIGEN.

Presence or absence of Australia antigen is indicated by + and 0, respectively. The sex of the sibs is shown (M-male, F-female), but the birth order is not indicated. Families 1–11 are Cashinahua Indians from Peru; 13–17 are Micronesians from Rongelap. No. 18 is a Samaritan family (see Blumberg *et al.*, 1965); 19–20 are families of patients with thalassemia and chronic lymphocytic anemia, respectively; 21–36 are Filipinos from Cebu. The three families in which both parental phenotypes are unknown (20, 25, 32) were not used in the segregation calculations.

Family Number	Father	Mother	Offspring							
1	0		+M	0M						
2	0	0	+F							
3	0	0	+M							
4	0		+M							
5	0	0	+F							
6		0	+M							
7		0	+M							
8		0	+M	+M	0M	0M	0M	0F		
9	0	0	+M							
10	0	0	+M	+F	0M	0F				
11	0	0	+F	0F						
12	+	0	+M	0M	0M	0M	0M			
13	0	0	+F	0M	0F					
14	0	+	+F	+M	+M					
15	+	0	0F	0F						
16		0	+F	0M	0F	0F	0M	0F	0F	
17	0		+M							
18	0	0	+M	+M	0M	0M	0M			
19	0	0	+M	+F						
20			+M	0M	0M					
21	0	0	+M	0M	0M	0F	0F			
22	0		+F	0M	0F					
23	0	0	+M	+F	0M	0M	0F	0F		
24		+	+M	+M	+M	0F				
25			+M	0F						
26		0	+F	+F	0F	0M				
27	0	0	+M	0F	0M	0M	0M	0F		
28	+	0	+M	+M	0F					
29	+	0	0F	0F	0M	0M				
30	+		+M	0F						
31	0		+M	0M	0M	0F				
32			+F	0F						
33	0	0	+F	+F	+F	+M	0F	0M	0M	0M
34	+	0	0M							
35	+	0	+M	0M	0F					
36		0	+M	0F	0F					

according to the pedigrees, and a pattern of segregation suggesting simple autosomal recessive inheritance was seen (Table 5, families 1–18). In addition to these two populations, two families were studied through patients with various diseases and they also are listed in Table 5 (19 and 20). In order to determine if family clustering occurred, a formal study was undertaken in Cebu. In 1964, 978 sera were studied in conjunction with the genetics and leprosy project of the Leonard Wood Memorial-Eversley Childs Sanitarium Leprosy Research Laboratory, and 64 of these were found to have Australia antigen. Blood specimens were collected from 13 families identified through positives and 40 families through negatives (Table 6). (For the purposes of this study, a family is defined as two or more individuals from the same immediate family.) Included among these were several families identified

TABLE 6. AUSTRALIA ANTIGEN IN THE FAMILIES OF PERSONS FROM CEBU IDENTIFIED THROUGH AUSTRALIA ANTIGEN-POSITIVE AND AUSTRALIA ANTIGEN-NEGATIVE INDEX CASES.

The probability for this distribution is $P = 3.32 \times 10^{-4}$ by Fisher's exact method.

Families studied	With positive index case	With negative index case
Number of families	13	40
Index cases included:		
Family members	78	153
Positive Australia antigen	24	2
Index cases subtracted:		
Family members	65	113
Positive Australia antigen	11	2
Per cent positive	16.92	1.76

through propositi selected during the 1965 field trip. There were 78 members of the families found through the positive reactors, of whom 24 were positive, and 153 members of the families found through nonreactors, of whom two were positive. In each case, the index cases were subtracted from the total number. The frequency of positive reactors in the families found through reactors is significantly higher than the frequency of reactors in the family members found through nonreactors. This difference is highly significant when the probability is determined using Fisher's exact method ($P = 3.32 \times 10^{-4}$). These results are consistent with a familial clustering of the trait in the Cebu population.*

On the basis of the observed segregation in the families (Table 5), a simple genetic hypothesis was tested using the method of Smith (1956). The hypothesis states that individuals homozygous for a gene tentatively designated Au^1 (i.e., genotype Au^1/Au^1) have detectable Australia antigen using the Ouchterlony double diffusion method [phenotype Au(1)]. Individuals homozygous for the alternate gene (Au/Au) and heterozygotes (Au^1/Au) do not have detectable Australia antigen [phenotype Au(0)].

Putative heterozygotes cannot be detected using the Preer tube technique, which is several times more sensitive than the Ouchterlony technique used in this study (A. Jones, unpublished observations). This implies that the Au^1 gene is recessive in that, using the available techniques, it is not detectable when present in single dose.

In Smith's method, families with at least one Au(1) offspring are used to compare the observed frequency of reactors in the offspring with the frequency expected on the basis of the stated hypothesis. Corrections for family size are included and the two matings, positive × negative [Au(1) × Au(0)] and negative × negative [Au(0) × Au(0)], are considered separately. The

*Many of the families studied on Cebu included patients with leprosy. We have found an increased frequency of Australia antigen in patients with lepromatous leprosy, particularly in young males (Blumberg and Melartin, 1966). It is known that leprosy may cluster in families, and the familial clustering of Australia antigen conceivably could be due to this. However, of the 14 positive individuals found in the Cebu families (exclusive of the propositi) only three had lepromatous leprosy.

TABLE 7. SEGREGATION OF AU(1) IN FAMILIES WITH AT LEAST ONE AU(1) (POSITIVE) CHILD.

Computations by the method of Smith (1956) for Cebu families (Table 5, 21–36) and for all families (all families on Table 5, excluding 20, 25, 32). In each case, the observed number (R_1, R_2) is very close to the expected (E_1, E_2), and the difference between them is always less than the variance (V_1, V_2).

Number of children in family	Number of families	Observed number of recessive children	Expected number of recessive children	Variance
***Cebu families:* mating type, positive × negative**				
c	m_c		$m_c a_c$	$m_c b_c$
2	1	1	1.333	0.222
3	2	3	3.428	0.980
4	1	3	2.133	0.782
Total	4	$R_1 = 7$	$E_1 = 6.894$	$V_1 = 1.984$
***Cebu families:* mating type, negative × negative**				
c	M_c		$M_c A_c$	$M_c B_c$
3	2	2	2.594	0.526
4	2	3	2.926	0.840
5	1	1	1.639	0.592
6	2	3	3.650	1.552
8	1	4	2.223	1.172
Total	8	$R_2 = 13$	$E_2 = 13.032$	$V_2 = 4.682$
***All families:* mating type, positive × negative**				
c	m_c		$m_c a_c$	$m_c b_c$
1	1	1	1.000	0.000
2	1	1	1.333	0.222
3	3	6	5.142	1.470
4	1	3	2.133	0.782
5	1	1	2.581	1.082
Total	7	$R_1 = 12$	$E_1 = 12.189$	$V_1 = 3.556$
***All families:* mating type, negative × negative**				
c	M_c		$M_c A_c$	$M_c B_c$
1	8	8	8.000	0.000
2	3	4	3.429	0.366
3	3	3	3.891	0.789
4	3	5	4.389	1.260
5	2	3	3.278	1.184
6	3	5	5.475	2.328
7	1	1	2.020	0.970
8	1	4	2.223	1.172
Total	24	$R_2 = 33$	$E_2 = 32.705$	$V_2 = 8.069$

results (Table 7) on the Cebu data alone as well as on all the families are consistent with the genetic hypothesis.

In addition to the family studies on Cebu, blood specimens were casually collected from residents of the "negative barrio." These included mother-child pairs. Mothers and children visiting a well-baby pediatric clinic in Cebu City were also sampled. The combined data were tested using the mother-child method. The observed results fit well with those expected on a simple autosomal recessive hypothesis (Table 8).

There was a single family ascertained through a patient with sickle-thalassemia disease in which both parents and one grandparent had Australia antigen and both of the children in the third generation also had the antigen. This complicated case will be described in greater detail elsewhere.

TABLE 8. DISTRIBUTION OF PHENOTYPES IN MOTHER-CHILD PAIRS FROM CEBU.

The gene frequencies were determined from the distribution of Au(1) in the mothers: Au^1 = .25; Au = 0.75. From the Hardy-Weinberg equation Au^1/Au^1 = 0.63, Au^1/Au = .375, Au/Au = .563.

Genotype	Mothers	Children Au^1/Au^1 Obs.	Children Au^1/Au^1 Exp.	Children Au^1/Au or Au/Au Obs.	Children Au^1/Au or Au/Au Exp.	Total Observed
Au^1/Au^1	2	1	0.5	1	1.5	2
Au^1/Au Au/Au	29	2	2.0	38	38.0	40
Total	31	3	2.5	39	39.5	42

DISCUSSION

We would now like to consider some possible sources of error in the analysis. With the exception of the Cebu group, the populations are heterogeneous and the families were not selected randomly. This does not, however, affect the methods used for the family analysis. The method of ascertainment of the mother-child pairs probably approximates randomness.

In eleven of the families, the phenotype of only one of the parents is known, and it is assumed that the untested parent is negative. If, in Cashinahua families 4, 6 and 7, the parent with unknown phenotype had been positive, there would have been no effect on the final results.

The effect of incorrect classification of the other parents in the Cashinahua, Rongelap and Cebu populations can be estimated. The frequency of Australia antigen positives in the three populations combined is approximately 9%. Of the eight unknown parents, it would be unlikely that more than two were positive. If the two most critical matings (nos. 16 and 31, Table 5) were removed from the negative × negative mating table and placed in the positive × negative table (Table 7), the resultant combined P values would be less than .3 and greater than .2, which is still consistent with the genetic hypothesis. It is obvious that if the father of family 24 were positive, the single negative offspring would be inconsistent with the recessive hypothesis. In the Cebu population, the likelihood of finding a mating pair both of whom are Australia antigen positive is 36/10,000. Given one parent positive, the likelihood of the second being positive is 6/100.

If the calculations are done using only those families in which the phenotypes of both parents are known, then P is less than .3 and greater than .2, which is still consistent with the hypothesis. In three of the families (Table 5, nos. 20, 25, and 32), neither of the parents was tested, and these families were not included in the calculations.

A very large number of families are required to establish the stated genetic hypothesis, and even when segregation data appear to fit a simple hypothesis, as they do in this case, some nongenetic explanation or another genetic mechanism may still be the correct one (see, for example, Lilienfield, 1959). Pending additional studies, we will accept the stated genetic mechanism as a working hypothesis. This does not rule out the possibility that Australia antigen found in leukemia, mongolism, and other diseases, or in other populations, may be due to nongenetic causes. For example, a virus or other infective agent could lead to the presence of the Australia antigen, perhaps due to a direct effect on the genome.

The term Australia antigen was given to this protein when it was first discovered and its chemical composition was unknown. The precedent of using geographical terms for inherited proteins is well established (i.e., hemoglobin Melbourne, transferrin Atalanti, transferrin Montreal, etc.). We now have additional chemical and immunological information on the protein, but it is suggested that the geographical term should be retained for the present.

Studies now in progress in our laboratory suggest that there may be more than one Australia antigen specificity. This is analogous to our previous findings in the first serum protein isoprecipitin system, the beta lipoproteins. In these, multiple specificities, presumably with complex genetic basis, have been described (Blumberg *et al.*, 1964). It is, of course, not known whether these specificities are inherited or what their relationship is to the phenotype discussed in the present paper.

The decrease with age in the frequency of Australia antigen should not have a marked effect on the family distribution, since the data used involve primarily a single generation. If this age dependence is confirmed in more complete studies, it suggests a measurable selection against individuals with the Au(1) phenotype. It is of interest that Australia antigen has been found to be associated with a variety of severe illnesses including leukemia, hemophilia, mongolism, thalassemia, and lepromatous leprosy (Blumberg and Alter, 1965; Blumberg, 1966; and Blumberg, Gerstley, and Hungerford, 1966). It is also possible that the decrease in frequency with age may be due to a decrease in concentration of Australia antigen in "positive" individuals over the course of the years. The data which we have on the persistence of the trait (see above) suggest that this is not an important factor in the eight years or less over which data are available. It will take additional time to determine if this is an explanation of the observed frequency decrease. Additional studies will be needed to assess the significance of the apparently increased frequency of Au(1) in males.

The striking differences in the distribution of the trait within different populations are consistent with its being a polymorphism involving widely different selective forces in different environments. No firm conclusions can be drawn from the present knowledge of distributions, since many areas of the world still have not been reported. The trait does appear, however, to cluster in tropical populations, in particular, in Malaysian, Micronesian, Polynesian, and related groups.

SUMMARY

Australia antigen shows a family clustering in the populations tested. The segregation data are consistent with simple autosomal inheritance, but additional data are required to establish firmly the genetic hypothesis. The trait appears to be persistent over several years; the frequency is higher in younger age groups; there is possibly an increased frequency in males in several populations; and the frequency varies widely in different populations.

ACKNOWLEDGMENT

We are indebted to many individuals and organizations for the valuable blood specimens used in this study, many of whom are listed in the text. We are particularly indebted to the staffs of the Eversley Childs Sanitarium, The Leonard Wood Memorial-Eversley Childs Sanitarium Leprosy Research Laboratory, and the Cebu Skin Clinic, and to many other physicians and scientists on Cebu who generously contributed to our work. Included among these are: Dr. Tranquilino Fajardo, Jr., Dr. Roland V. Cellona, Dr. Antonio C. Jovellanos, Dr. Liwanag Roman, Dr. Purificacion Espinosa, Dr. Nestor Canoy, Dr. Claude V. Reich, Dr. David A. Power, and Mother Recrude.

We thank Dr. Chapman H. Binford, Medical Director of the Leonard Wood Memorial, for his guidance and help with this project. We are indebted to Dr. R. A. Conard for the Rongelap specimens.

REFERENCES

Alter, H. J., and Blumberg, B. S. 1966. Studies on a "new" human isoprecipitin system (Australia antigen). *Blood* 27: 297–309.

Blumberg, B. S. 1964. Polymorphisms of serum proteins and the development of isoprecipitins in transfused patients. *Bull. N. Y. Acad. Med.* 40: 377–386.

Blumberg, B. S. 1966. An inherited serum isoantigen in leukemia and Down's syndrome. *J. Clin. Invest.* 45: 988.

Blumberg, B. S., and Alter, H. J. 1965. Precipitating antibodies against a serum protein ("Australia antigen") in the serum of transfused hemophilia patients. *J. Clin. Invest.* 44: 1029.

Blumberg, B. S., Alter, H. J., Riddell, N., and Erlandson, M. 1964. Multiple antigenic specificities of serum lipoproteins detected with sera of transfused patients. *Vox Sang.* 9: 128–145.

Blumberg, B. S., Alter, H. J., and Visnich, S. 1965. A "new" antigen in leukemia sera. *J. Amer. Med. Assoc.* 191: 541–546.

Blumberg, B. S., Dray, S., and Robinson, J. C. 1962. Antigen polymorphism of a low density beta-lipoprotein. Allotypy in human serum. *Nature* (Lond.) 194: 656–658.

Blumberg, B. S., Gerstley, B., and Hungerford, D. A. 1966. A serum isoantigen [Australia antigen, Au(1)] in Down's syndrome. (In preparation.)

Blumberg, B. S., and Melartin, L. 1966. Conjectures on inherited susceptibility to lepromatous leprosy. *Internat. J. Leprosy* 34: 60–64.

Blumberg, B. S., Melartin, L., Guinto, R. A., and Lechat, M. 1966. Studies on leprosy and Australia antigen. (In preparation.)

Blumberg, B. S., and Riddell, N. M. 1963. Inherited antigenic differences in human serum beta-lipoproteins. A second antiserum. *J. Clin. Invest.* 42: 867–875.

Conard, R. A., Meyer, L. M., Sutow, W. W., Lowrey, A., Cannon, B., Moloney, W. C., Watne, A. C., Carter, R. E., Hicking, A., Hammerstrom, R., Bender, B., Lanwi, I., Riklon, E., and Anjain, J. 1965. *Medical Survey of the People of Rongelap and Utirik Islands Nine and Ten Years After Exposure to Fallout Radiation (March, 1963, and March, 1964).* Upton, New York: Brookhaven National Laboratory, 167 pp.

Kensinger, K. 1965. The Cashinahua of southeastern Peru. *Expedition* 7: 5–9.

Lilienfeld, A. M. 1959. A methodological problem in testing a recessive genetic hypothesis in human disease. *Amer. J. Public Health* 49: 199–204.

Maynard Smith, S., Penrose, L. S., and Smith, C. A. B. 1965. *Mathematical Tables for Research Workers in Human Genetics.* Boston: Littie. Brown and Co.

Melartin, L., and Blumberg, B. S. 1966. Production of antibody against "Australia antigen" in rabbits. *Nature* (Lond.) 210: 1340–1341.

Smith, C. A. B. 1956. A test for segregation ratios in family data. *Ann. Hum. Genet.* (Lond) 20: 257.

Reprinted from the Archives of Internal Medicine
August 1972, Volume 130

Sex Distribution of Australia Antigen

Baruch S. Blumberg, MD; Alton I. Sutnick, MD;
W. Thomas London, MD; and Liisa Melartin, MD, Philadelphia

The sex distribution of Australia antigen was determined in ten disease groups and 13 presumed normal populations tested in our laboratory. In 22 of these 23 independent studies, the frequency of Australia antigen was higher in males than in females ($P < .001$). The increased frequency in males was most marked in the lower age groups (less than 20 years). The overall ratio of Australia antigen in males to females is 1.58.

Men are not as hardy as women, and in nearly all environments men die at an earlier age. They also have a higher susceptibility than females to infection with a variety of organisms. Washburn et al estimated the male-female ratio for cases of neonatal septicemia and meningitis by collecting data from the literature (1930 to 1965) and from the case records of the Johns Hopkins Hospital, Baltimore (1896 to 1965).[1] They concluded that, with the exception of organisms involved in prenatal infection, there was a higher frequency of infection in males than in females. Further, the difference was most marked in infants and even more striking in the post antibiotic age than before the widespread use of these drugs. They concluded from the latter that females were more amenable to therapy than males. Childs[2] proposed a genetic hypothesis to explain the sex difference. He theorized that a gene involved with the synthesis of immunoglobulins is present on the X chromosome. In order to effect dosage compensation (ie, compensate for the presence of two X chromosomes in females and only one in the male), Lyon and others have proposed that there is random inactivation of the X chromosomes in the cells of females; this allows for heterozygosity in females which does not occur in males. Hence, if resistance to the effects of infection is present on the X chromosome, the female, because of her greater variability, could more effectively resist disease. Schlegel and Bellanti[3] have suggested that a deficiency of glucose-6-phosphate dehydrogenase, which is X linked, may be one of the factors associated with susceptibility to infection, but this has been contested by Rodey et al.[4]

Childs has also pointed out that in diseases characterized by an increased production of antibody (the so-called auto-immune or iso-immune diseases) there is a higher frequency of females than males involved (eg, disseminated lupus erythematosus, primary biliary cirrhosis, rheumatoid arthritis, thyroiditis).

Australia antigen was discovered in 1963,[5] the association of Australia antigen with hepatitis was first found in 1966,[6] and since then the association has been confirmed by many investigators. Further, the hypothesis that Australia antigen is a hepatitis virus has been supported by a variety of findings including transmission to

Received for publication Aug 25, 1971; accepted Nov 2.
From the Institute for Cancer Research, Fox Chase, Philadelphia.
Reprint requests to Institute for Cancer Research, 7701 Burholme Ave, Fox Chase, Philadelphia 19111 (Dr. Blumberg).

Table 1.—Sex Ratio of Australia Antigen in Patients With Various Diseases and Normal Nonhospitalized Populations*

	Male			Female				
	No.	No. With Au(1)	% With Au(1)	No.	No. With Au(1)	% With Au(1)	Ratio	*P* Value
Disease								
Acute viral hepatitis[19]	112	40	35.7	94	18	19.1	1.87	.013
Chronic hepatitis[22]	12	3	25	19	0	0	†	.095
Down's syndrome[6,18,20]	212	67	31.6	92	17	18.5	1.71	.027
Other mentally retarded[18]	145	5	3.4	43	1	2.3	1.48	.900
Leukemia[6,23]	173	22	12.7	124	11	8.9	1.43	.394
Lepromatous leprosy (Cebu)[24]	396	45	11.4	188	10	5.3	2.14	.029
Lepromatous leprosy (India)[25]	451	32	7.1	101	2	2.0	3.58	.088
Tuberculoid leprosy (Cebu)[24]	343	19	5.5	262	7	2.7	2.07	.128
Tuberculoid leprosy (India)[25]	281	8	2.8	103	0	0	†	.184
Long-term dialysis (renal disease)[26]	5	5	100	4	3	75	1.33	.906
$\chi^2 = 42.3270$		$P = .0025$						
Population								
Marshall Islands[17]	243	19	7.8	226	14	6.2	1.26	.612
Cebu, Phillippines[24]	678	47	6.9	609	17	2.8	2.48	.001
Cashinahua[17]	45	10	22	44	6	14	1.63	.436
Manila[17]	138	6	4.3	59	3	5.1	.855	.884
Polynesia[20]	597	8	1.3	550	3	.54	2.46	.282
New Hebrides[20]	597	61	10.2	394	19	4.8	2.12	.003
New Caledonia[20]	518	13	2.5	444	7	1.6	1.59	.433
India (South)[20]	116	5	4.3	135	0	0	†	.047
British Solomon Islands[20]	127	7	5.5	160	4	2.5	2.20	.312
New Guinea[20]	532	16	3.0	287	5	1.7	1.73	.389
Micronesia[20]	330	13	3.9	292	6	2.1	1.92	.259
Surinam	310	7	2.3	315	5	1.6	1.45	.729
Bougainville[27]	931	112	12.0	906	97	10.7	1.12	.412
$\chi^2 = 47.6840$		$P = .0058$						

* All data are from our laboratory and have been published in the indicated references. Data without references have not been previously published.

† Au(1) only detected in males.

Table 2.—Sex Ratio of Australia Antigen in Patients With Various Diseases and in Nonhospitalized Populations, by Age

		Male			Female				
Group	Age (Yr)	No. With Au(1)	% With Au(1)	No.	No. With Au(1)	% With Au(1)	No.	Ratio	Value
Lepromatous leprosy (Cebu)	0-19	85	24	28.24	44	3	6.82	4.14	.009
	20+	482	45	9.34	190	9	4.74	1.97	.069
Nonlepromatous leprosy (Cebu)	0-19	183	20	10.93	184	6	3.26	3.35	.008
	20+	485	27	5.57	425	11	2.59	2.15	.038
Tuberculoid leprosy (Cebu)	0-19	58	5	8.62	59	3	5.08	1.70	.696
	20+	285	14	4.91	203	4	1.97	2.49	.145
Bougainville	0-19	416	64	15.38	390	45	11.54	1.33	.136
	20+	368	30	8.15	344	32	9.30	.876	.771
Polynesia	0-19	406	5	1.23	414	2	0.48	2.55	.432
	20+	191	3	1.57	135	1	0.74	2.12	.873
New Hebrides	0-19	370	42	11.35	255	12	4.71	2.41	.006
	20+	227	19	8.37	139	7	5.04	1.66	.320
New Caledonia	0-19	415	9	2.17	388	5	1.29	1.68	.466
	20+	93	4	4.30	56	2	3.57	1.20	.044
Down's syndrome	0-19	67	32	47.76	9	2	22.22	2.15	.276
	20+	132	32	24.24	36	5	13.89	1.75	.271

humans and animals,[7-9] by its virus-like appearance,[10,11] identification in the nuclei of affected individuals,[12,13] and in other ways. These findings have been reviewed (see for example[14-16]) and will not be elaborated further here. The use of the Australia antigen test (the test for Australia antigen in the peripheral blood) has permitted the investigation of the sex distribution in several diseases and populations. An increased frequency of Australia antigen in males was reported in our earlier publications,[17-20] and additional data have now been assembled which confirm this.

Materials and Methods

The immunodiffusion method using the reinforcement pattern and staining with azo carmine was used for the detection of Australia antigen.[21] Human and rabbit antisera (anti-Au[1]) as described in previous publications were used in the studies.

Disease Groups.—The method of collection and description of these groups studied are given in the original publications referenced in Table 1. They are also described briefly here.

Acute and Chronic Active Viral Hepatitis.[19,22]—Physicians in the Philadelphia area were asked to submit sera of their patients diagnosed or suspected of having viral hepatitis. In most cases only a single sample was available from each patient. Serum values from patients with chronic active hepatitis were obtained from well documented cases at the Mayo Clinic.

Down's Syndrome (Mongolism).—The majority of the Down's syndrome patients housed in two large institutions in the Philadelphia area are included.[6,18,20] They have been tested several times over the course of two to five years.

Leukemia.—These are transfused leukemia patients from the National Institutes of Health Clinical Center, Bethesda, Md, and from various hospitals in Philadelphia.[23]

Leprosy.—The lepromatous leprosy cases include nearly all the patients with this diagnosis in a large leprosarium on the island of Cebu in the Philippines. The tuberculoid leprosy patients were nearly all outpatients and represent about one third of the patients treated in the clinic at the time the study was performed.[24] In the India sample[25] nearly all the patients in residence at a large leprosarium in Chingleput near Madras are included. The tuberculoid patients include a small percentage of those treated in three outpatient facilities associated with the leprosarium.

Long-Term Renal Dialysis.—These are all the patients treated in a renal dialysis unit in Philadelphia.[26]

Nonhospitalized "Normal" Populations.—The sample from Bougainville, Territory of New Guinea, consists of approximately 80% of all the inhabitants over 5 years of age residing in 18 villages in south central Bougainville and are described in references (Table 1).[27,28] The Marshall Island sample includes nearly all the adult populations of Rongelap atoll and partial collections from Majuro and Utirik atolls. The Cebu population was the nonleprosy control for the Cebu leprosy study.[23] In the remaining studies, the samples were collected for quite different reasons and should be described as casual collections.

In eight of the populations, there were sufficient numbers and information to divide the group into young (0 to 19 years of age) and old (20+/years of age) subgroups for frequency analyses (Table 2).

Results

Australia antigen is more common in males than in females in all of the disease groups and in all of the populations except one (Table 1). The data were statistically analyzed in several ways:

1. A two-by-two contingency table was prepared for each disease and population group and the chi-square and probability value determined for each. There was a significant difference at the .01 level (or less) in four of the populations and two of the disease groups.

2. Since these are each independent series of observations, the P value for (a) all of the normal population groups, (b) all of the disease groups, and (c) the population and disease groups combined was determined using the formula

$$\chi^2_{(2n)} = -2 \sum_{i=1}^{n} \log_e (Pi)$$

where the degrees of freedom is equal to the total number of P values combined. In this calculation P = .0025 for the disease group, P = .0058 for the normal population group, and P = .0001 for the combined group.

3. The sign test was used to compare groups a, b, and c just mentioned. In this, a positive sign was assigned to the population or disease group in which the males had a higher frequency than the females and a negative sign for the opposite. If there were no differences between males and females, then there would be about an equal number of positive and negative signs. But this was not true; the respective values for a, b, and c are +10,–0; +12,–1; and +22, –1; and the probability that these are due to chance are .001, .003, and <.001, respectively.

From all this we conclude that there is a significantly higher frequency of Australia antigen in males than in females.

In the populations in which age breakdown was possible (Table 2), the male-female ratio is higher in the younger (0 to 19 years of age) than the older (20+ years of age) group in every case but one. The overall P value by the sign test is .035.

Comment

This combined study confirms our earlier observations[17-20] that in general there is a higher frequency of Australia antigen in males than in females. A review of sex frequency data in studies from other laboratories (A. Vierucci, MD, written communication, July 1970)[29-36] provides further confirmation and emphasizes the universality of this relationship (Table 3). There could be some trivial explanations for the higher frequency in males; for example, in some of the populations it could be argued that male activities and exposures make them more likely than females to become infected. However, it is difficult to see how this would operate in all of the populations studied. For example, there is no major difference in activity between the institutionalized

Table 3.—Sex Ratio of Australia Antigen in Patients With Various Diseases and Normal, Nonhospitalized Populations Reported by Other Laboratories

	Male			Female				
	No.	No. With Au(1)	% With Au(1)	No.	No. With Au(1)	% With Au(1)	Ratio	*P* Value
Disease								
Chronic hepatitis*	22	4	18.2	13	2	15.4	1.18	.801
Down's syndrome[29]	21	3	14.3	9	0	0	†	.595
Down's syndrome[30]	70	31	44.3	64	16	25.0	1.77	.031
Other mentally retarded[30]	70	15	21.4	64	8	12.5	1.71	.245
Renal transplant patients[31]	33	13	39.4	34	1	2.9	13.4	.00076
Dialysis patients[31]	48	15	31.3	40	5	12.5	2.50	.067
Various diseases[32]	~6,000	27	~.45	~6,000	14	~0.23	1.93	.061
Acute hepatitis[33]	16	7	43.8	11	4	36.4	1.20	.99
		$\chi^2 = 36.5549$		$P = .00242$				
Population								
Sardinia[34]	672	57	8.4	451	32	7.2	1.20	.0056
Japan[35]	4,222	52	1.23	1,024	2	0.20	6.31	.0091
Japan[36]	3,830	89	2.32	1,409	16	1.14	2.05	.465
		$\chi^2 = 21.3004$		$= .00162$				

* A. Vierucci, MD, written communication.
† Au(1) only detected in males.

male and female patients with Down's syndrome or lepromatous leprosy patients. Further, some unknown bias favoring the observed higher frequency in males could have been generated in some of the populations, but again it is hard to envision how this could happen in virtually all of the populations studied. These findings are consistent with the general observations of a higher frequency of infection in males than in females.

As noted in the introduction, Australia antigen is closely associated with viral hepatitis and may be the infectious agent or part of such an agent which causes a large proportion of the cases of hepatitis in man. Australia antigen is also associated with a group of chronic diseases ("Australia affinity group") characterized by an impairment of immune function.[14] These include Down's syndrome, the lymphocytic leukemias, Hodgkins disease, lepromatous leprosy (but not tuberculoid leprosy), and patients with chronic renal disease treated by repeated hemodialysis. In addition there are vast numbers of people living primarily in the tropics and Southeast Asia who are apparently asymptomatic carriers of Australia antigen. Family studies have been concluded in two of these populations[17,27]; on the basis of these it appears that susceptibility to persistent infection in these groups is inherited as a simple autosomal recessive trait controlled by a gene designated Au^1. This has been supported by an independent study of Sardinians living in Turin.[34]

There are several observations relating to hepatitis and to the diseases of the Australia affinity group which may bear on the sex distribution. The frequency of lepromatous leprosy is greater in males than in females. The causative organism of leprosy (Hansen's bacillus, *Mycobacterium leprae*) often localizes in the testes in patients with lepromatous leprosy whereas involvement of the ovaries is rare. Lepromatous males may become sterile.[37] Similarly, male patients with Down's snydrome are sterile while female patients are not.[38] Males also predominate in number over females in leukemia.[39] The testes are often invaded by a pathological process in male patients with acute leukemia while involvement of the ovary is less common.[40] We found evidence of Australia antigen in the testes of one Down's syndrome patient by an immunofluorescence technique.[13] Australia antigen has been identified in the testes of a male cancer patient who had Australia antigen in his blood prior to his death.[13]

In addition to the genetic hypothesis proposed by Childs,[2] these observations suggest a variety of testable hypotheses including the following: (1) Antibody to Australia antigen forms less frequently in males than in females; this may also be related to complex formation. (2) Cellular immune responses are less active in males than in females. (3) Lymphocytes of males are not as responsive to immunogenic stimuli as lymphocytes of females. Further, the age data suggest that these differences may change with increasing age. These hypotheses will be tested in forthcoming studies.

This work was supported in part by Public Health Service grants CA-06551, CA-08069, CA-06927, and RR-05539; by funds from the World Health Organization; and by an appropriation from the Commonwealth of Pennsylvania.

References

1. Washburn TC, Medearis DN Jr, Childs B: Sex differences in susceptibility to infections. *Pediatrics* 35:57-64, 1965.
2. Childs B: Genetic origins of some sex differences among human beings. *Pediatrics* 35:798-812, 1965.
3. Schlegel RJ, Bellanti JA: Increased susceptibility of males to infection. *Lancet* 2:826-827, 1969.
4. Rodey GE, Jacob HS, Holmes B, et al: Leucocyte G.-6-P.D. levels and bactericidal activity. *Lancet* 1:355-356, 1970.
5. Blumberg BS: Polymorphisms of serum proteins and the development of isoprecipitins in transfused patients. *Bull NY Acad Med* 40:377-386, 1964.
6. Blumberg BS, Gerstley BJS, Hungerford DA, et al: A serum antigen (Australia antigen) in Down's syndrome, leukemia and hepatitis. *Ann Intern Med* 66:924-931, 1967.
7. Okochi K, Murakami S: Observations on Australia antigen in Japanese. *Vox Sang* 15:374-385, 1968.
8. Gocke DJ, Greenberg HB, Kavey NB: Correlation of Australia antigen with posttransfusion hepatitis. *JAMA* 212:877-879, 1970.
9. London WT, Millman I, Sutnick AI, et al: Transmission, replication and passage of Australia antigen in African green monkeys (Vervets). *Clin Res* 18:536, 1970.
10. Bayer ME, Blumberg BS, Werner B: Particles associated with Australia antigen in the sera of patients with leukemia, Down's syndrome and hepatitis. *Nature* 218:1057-1059, 1968.
11. Almeida JD, Zuckerman AJ, Taylor PE, et al: Immune electron microscopy of the Australia-SH (serum hepatitis) antigen. *Microbios* 2:117-123, 1969.
12. Millman I, Zavatone V, Gerstley BJS, et al: Australia antigen in the nuclei of liver cells of patients with viral hepatitis by the fluorescent antibody technique. *Nature* 222:181-184, 1969.
13. Coyne VE, Millman I, Cerda J, et al: The localization of Australia antigen by immunofluorescence. *J Exp Med* 131:307-320, 1970.
14. Blumberg BS, London WT, Sutnick AI: Australia antigen as a hepatitis virus: Variation in host response. *Amer J Med* 48:1-8, 1970.
15. Blumberg BS, Sutnick AI, London WT, et al: Current concepts: Australia antigen and hepatitis. *New Eng J Med* 283:349-354, 1970.
16. Sutnick AI, London WT, Millman I, et al: Viral hepatitis: Revised concepts as a result of the study of Australia antigen. *Med Clin N Amer* 54:805-817, 1970.
17. Blumberg BS, Melartin L, Guinto RA, et al: Family studies of a human serum isoantigen system (Australia antigen). *Amer J Hum Genet* 18:594-608, 1966.
18. Sutnick AI, London WT, Gerstley BJS, et al: Anicteric hepatitis associated with Australia antigen: Occurrence in patients with Down's syndrome. *JAMA* 205:670-674, 1968.
19. London WT, Sutnick AI, Blumberg BS: Australia antigen and acute viral hepatitis. *Ann Intern Med* 70:55-59, 1969.
20. Blumberg BS, Gerstley BJS, Sutnick AI, et al: Australia antigen, hepatitis virus and Down's syndrome. *Ann NY Acad Sci* 171:486-499, 1970.
21. Blumberg BS, Bryne RJ, Chanock RM, et al: Viral hepatitis and tests for the Australia (hepatitis-associated) antigen and antibody. *Bull WHO* 42:957-992, 1970.
22. Gitnick GL, Gleich GJ, Schoenfield LJ, et al: Australia antigen in chronic active liver disease with cirrhosis. *Lancet* 2:285-288, 1969.
23. Sutnick AI, London WT, Blumberg BS, et al: Australia antigen (a hepatitis-associated antigen) in leukemia. *J Nat Cancer Inst* 44:1241-1249, 1970.
24. Blumberg BS, Melartin L, Guinto R, et al: Lepromatous leprosy and Australia antigen with comments on the genetics of leprosy. *J Chronic Dis* 23:507-516, 1971.
25. Blumberg BS, Melartin L: Australia antigen and lepromatous leprosy: Studies in South India and elsewhere. *Int J Leprosy* 38:60-67, 1970.
26. London WT, DiFiglia M, Sutnick AI, et al: An epidemic of hepatitis in a chronic hemodialysis unit: Australia antigen and difference in host response. *New Eng J Med* 281:571-578, 1969.
27. Blumberg BS, Friedlaender JS, Woodside A, et al: Hepatitis and Australia antigen: Autosomal recessive inheritance of susceptibility to infection in humans. *Proc Nat Acad Sci* 62:1108-1115, 1969.
28. Friedlaender JS: *Microevolution in south central Bougainville Island, Territory of New Guinea*, thesis. Harvard University, Boston, 1968.
29. Melartin L, Panelius M: Occurrence of Australia antigen in Finnish Mongolism patients. *Ann Med Exp Fenn* 45:157-158, 1967.
30. Szmuness W, Pick R, Prince AM: The serum hepatitis virus specific antigen (SH): A preliminary report of epidemiologic studies in an institution for the mentally retarded. *Amer J Epidem* 92:51-61, 1970.
31. Leski M, Grivaux C, Courouce-Pauty AM: Australia antigen in hemodialysis and renal transplantation units. *Vox Sang* 19:359-368, 1970.
32. Lous P, Skinhoj P, Olesen H: HAA-Antigen determination in 12,000 patients in a general Copenhagen hospital. *Vox Sang* 19:379-383, 1970.
33. Cossart Y, Taylor PE, Vahrman J, et al: Australia-SH antigen in hepatitis patients in London. *Brit Med J* 3:755-756, 1969.
34. Cepellini R, Bedarida G, Carbonara AO, et al: High frequency and family clustering of Au antigen in some Italian populations, in *Atti Convegni Farmitalla, Antigene Australia ed Epatite Virale*. Torino, Italy, Minerva Medica, 1970, pp 53-74.
35. Okochi K, Murakami S: Observations on Australia antigen in Japanese. *Vox Sang* 15:374-385, 1968.
36. Okochi K, Mayumi M, Haguino Y, et al: Evaluation of frequency of Australia antigen in blood donors of Tokyo by means of immune adherence hemagglutination technique. *Vox Sang* 19:332-337, 1970.
37. Cochrane RG, Davey TF: *Leprosy in Theory and Practice*. Bristol, England, John Wright & Sons Ltd, 1964, pp 659.
38. Benda CE: *Down's Syndrome*. New York, Grune & Stratton Inc, 1969, p 44.
39. Wintrobe MM: *Clinical Hematology*. Philadelphia, Lea & Febiger, 1967, pp 986-987.
40. Sullivan MP: *Complications in the Treatment for Acute Leukemia in Neoplasia in Childhood*. Chicago, Year Book Medical Publishers, 1969.

Printed and Published in the United States of America

Reprinted from *Science*, August 25, 1978, Vol. 201, pp. 687–692

Hepatitis B Virus and Sex Ratio of Offspring

The response to hepatitis B virus infection in parents is related to the sex ratio of their children.

Jean S. Drew, W. Thomas London, Edward D. Lustbader
Jana E. Hesser, Baruch S. Blumberg

On a genetic basis, one would expect the ratio of males to females at birth to be one. In humans, this is not the case. In most human populations there is an excess of males born, and the secondary sex ratio (the number of males per 100 females at birth) falls within a range of 104 to 107 (*1*). It is not known whether this phenomenon is due to a high sex ratio at conception, or to a high mortality of female zygotes in the earliest stage of pregnancy (that is, before pregnancy is recognized). Observations on spontaneous abortions, stillbirths, and postnatal mortality show that male mortality exceeds female by anywhere from 40 to 145 percent (*2*). Hence, by the time reproductive age is reached the sex ratio is closer to 100 (*3, 4*). It is probably this ratio that is stabilized by natural selection (*3–5*).

There are geographic and racial differences in secondary sex ratio. For example, black populations have consistently lower sex ratios at birth than nonblack populations (*1, 6*). However, significant deviations from the range of 104 to 107 are infrequent in countries with adequate ascertainment (*1, 7*). Temporal variations in sex ratio have been observed in relation to environmental factors, such as wars (*8*) and seasons (*9*); but sustained changes in secondary sex ratio probably occur too slowly to detect (*3–5*).

It has been suggested that a cultural preference for offspring of one sex could affect sex ratio. This has been seen in regard to postnatal sex ratios in societies in which a preference for one sex results in infanticide of the other (*10*). Birth control alone cannot affect secondary sex ratio, however, because the probability of a male at each birth is independent of the sex of previous births. In contrast, if couples within a population vary in their inherent or biological probability of having a male, sex preferences, expressed in terms of effective birth control, could affect the secondary sex ratio (*4, 11*).

Intrapopulation variations in sex ratio have been attributed to genetic and environmental as well as racial differences (*12, 13*). Sanghvi (*12*) found an association of the ABO blood groups of mothers and their children with secondary sex ratio. In both Bombay and New York, the sex ratio of "O" infants born to "O" mothers was significantly higher than that of "A" infants born to "A" mothers. Jackson *et al.* (*13*) found a relation between the Xg^a blood group system and sex ratio. Sex ratio was significantly higher among offspring of Xg^a+ father × Xg^a- mother matings than among those of all other Xg^a combinations. They suggested that sensitization of Xg^a- mothers by Xg^a+ female fetuses might lead to a preferential loss of such embryos. In a recent analysis of data from the United States, Erickson examined the simultaneous effects of mother's age, father's age, mother's race, birth order, legitimacy, and paternal education on sex ratio. The only significant effect detected was a negative association betwen birth order and secondary sex ratio (*14*).

A relation between sex ratio of offspring and an infectious agent has been observed in some species of *Drosophila*, in which maternal infection with small spirochetes, presumably treponemata, is associated with a reduction in male (specifically single-X chromosome) offspring (*15, 16*). This phenomenon appears to be the consequence of a disturbance in the development of male zygotes, resulting in 50 percent mortality (*15–17*).

A relation between sex ratio of offspring and an infectious agent in a human population was reported in 1975. Hesser *et al.* (*18, 19*) found a significant association between sex ratio of offspring and the responses of parents to infection with hepatitis B virus (HBV) in a Greek population. In this article, we report our investigation of the relation between HBV and sex ratio, and suggest a possible explanation for this phenomenon.

Host Responses to HBV Infection

Hepatitis B surface antigen (Australia antigen; HBsAg) is in the outer coat of the Dane particle, the presumed virion of HBV. Three types of particles (spherical, rod-shaped, and Dane particles), all with HBsAg on their surfaces, may be found in the blood of patients infected with HBV. When infected, an individual may respond by developing acute hepatitis, which is demonstrated by the transient appearance of HBsAg in his blood, elevated serum transaminase, and sometimes jaundice. Infection is followed by the development of antibody against HBsAg (anti-HBs) and the resolution of the viremia and clinical symptoms. In some cases, however, people become chronically infected with HBV, with little evidence of liver damage and few clinical symptoms. Many of these people do not develop anti-HBs, and they may be HBsAg(+) for years (*20*). Such people are termed "chronic carriers" of HBV. They may infect other uninfected persons (by transfusion, for example) and apparently are at increased risk of developing certain chronic liver diseases such as chronic aggressive and chronic persistent hepatitis, postnecrotic cirrhosis, and primary hepatocellular carcinoma (*21*).

HBV and Sex Ratio at Birth

HBV can be transmitted from mothers to offspring in the perinatal period, that is, in utero, during, or soon after birth. Infection early in life with HBV appears to increase the likelihood of becoming a chronic carrier of the virus (*22*), and vertical transmission of HBV from mother to child has been implicated as a major mechanism for the establishment of carriers in at least one population (*23*). Transmission from father to child has not

The authors are investigators in the Division of Clinical Research at The Institute for Cancer Research, The Fox Chase Cancer Center, Philadelphia, Pennsylvania 19111. Dr. London is also Associate Professor of Medicine and Dr. Blumberg is also University Professor of Medicine and Anthropology at the University of Pennsylvania, Philadelphia.

0036-8075/78/0825-0687$01.50/0

Table 1. Number of male and female live births according to parents' response to HBV (*55*).

Parents' response to HBV	Couples (No.)	Live births* Male	Live births* Female	Sex† ratio
Either parent HBsAg(+):anti-HBs(−)	33	60 (1.82 ± .22)	24 (0.73 ± .11)	250.0 (161.1, 429.1)
Both parents HBsAg(−):anti-HBs(−)	29	51 (1.76 ± .24)	35 (1.21 ± .23)	145.7 (95.7, 230.0)
Both parents HBsAg(−):either parent anti-HBs(+)	154	241 (1.56 ± .10)	222 (1.44 ± .10)	108.6 (90.5, 130.9)
Neither parent HBsAg(+):anti-HBs(−) and either parent HBsAg(+):anti-HBs(+)	16	25 (1.56 ± .39)	27 (1.69 ± .42)	92.6 (52.7, 161.1)

*The numbers in parentheses indicate the number of live births per couple (mean ± S.E.). †The numbers in parentheses indicate the 95 percent confidence limits in the sex ratio.

been studied extensively but appears to be less common (*24*).

An association between hepatitis infection and sex ratio at birth was previously suggested by Robertson and Sheard, who found a relation between an outbreak of hepatitis (of unknown etiology) and a subsequent decrease in male live births (*25*). Mazzur and Watson described a high proportion of males among siblings of HBsAg carriers in Melanesia (*26*). In 1973, we studied differences in sex ratio associated with HBsAg in parents from the Greek village of Plati in southern Macedonia (*27*). When either parent was HBsAg(+) there was a significantly higher sex ratio of live births than when both parents were HBsAg(−) (*18, 19, 28*).

The original analysis of the Plati data considered only the presence of HBsAg. We subsequently hypothesized that anti-HBs would be associated with a sex ratio alteration opposite in direction to that associated with HBsAg. In this article, we present an analysis of the Plati data, including the data on anti-HBs as well as other factors thought to affect secondary sex ratio such as parental age, number of years of marriage, and order of birth and pregnancy. The conclusions from the Plati data are compared to observations from other studies showing an interaction of HBV with sex, and a hypothesis explaining the interaction is proposed.

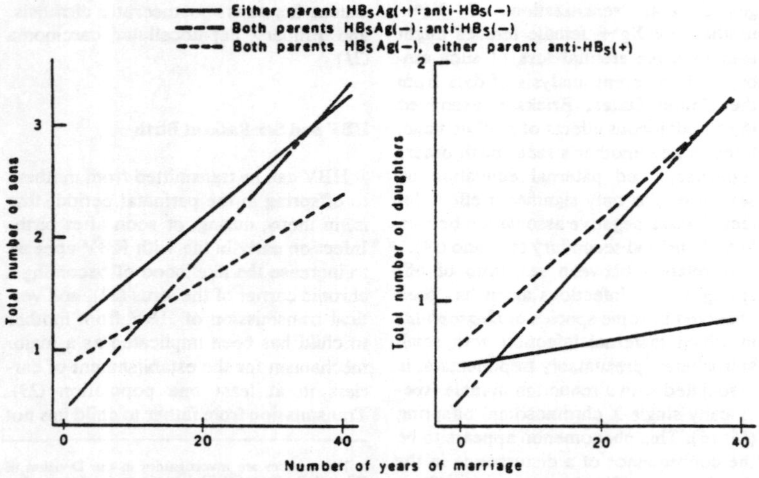

Fig. 1. Least-squares fits for the regressions of sons and daughters as a function of the number of years of marriage. By analysis of covariance (*56*), the number of sons per years of marriage is significantly higher ($P < .02$) and number of daughters significantly lower ($P < .02$) for HBsAg(+) as compared to anti-HBs(+) couples.

Results from Plati

Serums were obtained from 326 wives and 248 husbands, representing a total of 390 families. Reproductive histories, including the number and sex of live births and the number of abortions, were obtained from all the women and from those men whose wives were not available for the study. The sex ratio of the 1558 people (including parents and children) ascertained in this community was 111. This may seem high, but it is not unusual for Greece and is not significantly different from sex ratios encountered in most human populations. The normal range of sex ratios (from 104 to 107) falls within the 95 percent confidence limits of the ratio in Plati (from 100.4 to 122.7) (*6*).

Complete information, with respect to HBV serology and sex of all live births, was available from 232 couples. These couples were categorized according to immune responses to HBV, as defined by the presence or absence of HBsAg and anti-HBs in the serum of each parent (*29*). The sex ratio of offspring of 33 couples in which either parent was HBsAg(+):anti-HBs(−) was compared to that of 183 couples in which both parents were HBsAg(−). Among the 183 couples, the offspring of 29 couples in which both parents were anti-HBs(−) were compared to those of 154 in which either parent was anti-HBs(+). There were 16 couples in which neither parent was HBsAg(+):anti-HBs(−), but at least one parent was HBsAg(+):anti-HBs(+). These 16 couples are included in Table 1 but were excluded from further analysis (*30*).

Table 1 shows that sex ratio of live births was highest in families in which at least one parent was HBsAg(+):anti-HBs(−) [hereafter called HBsAg(+) couples] (*31*), intermediate in families in which both parents were HBsAg(−):anti-HBs(−), and, as predicted, lower when both parents were HBsAg(−) and at least one was anti-HBs(+) [hereafter called anti-HBs(+) couples]. A sex ratio of 145.7 may seem extreme for the HBsAg(−):anti-HBs(−) "control" population, but the 95 percent confidence limits include the normal range of 104 to 107 (Table 1). The corresponding limits for the sex ratio of children of HBsAg(+) couples do not include the normal range. The sex ratio among the 16 HBsAg(+):anti-HBs(+) families is similar to that of the anti-HBs(+) couples and suggests that the effect of anti-HBs overshadows that of HBsAg when both are present in one parent (*32*). From the analysis of these results (see Table 1), we concluded that the hypothesis that anti-HBs would be associated with a sex ratio alteration in the opposite direction from that observed with HBsAg was confirmed.

The relation between response to

HBV in parents and sex ratio of offspring was further characterized by comparing the mean numbers of sons and daughters born per family to each category of parents. Table 1 shows that HBsAg(+) couples had similar numbers of sons but significantly fewer daughters per family as compared to anti-HBs(+) couples (P = .0007) (29).

Some sex ratio effects might be explained by other factors such as parental ages, number of years married, birth order, and fetal loss. Age is of particular importance because differences in the frequencies of HBsAg and anti-HBs by age are well documented (33). We tested the hypothesis that these factors could account for our observations in several ways. First, we compared the sex ratios and mean numbers of sons and daughters per family for only couples married more than 15 years. In this way we could examine presumably "completed" families and therefore control for effects of parental ages and birth order. This subgroup included 18 HBsAg(+), 17 HBsAg(−):anti-HBs(−), and 97 anti-HBs(+) couples (for a total of 132 families, more than 60 percent of the total sample). Again, HBsAg(+) couples had a significantly higher sex ratio (P = .002), the same number of sons, and fewer daughters (P = .002) than anti-HBs(+) couples. We then compared by analysis of covariance the least-squares fits for the regressions of sons and daughters as a function of number of years of marriage. HBsAg(+) couples had significantly more sons (P < .02) and fewer daughters (P < .02) per number of years of marriage than did anti-HBs(+) couples (Fig. 1) (34). The fact that in HBsAg(+) couples total number of sons per family was similar, but that there were more sons per number of years married than in anti-HBs(+) couples, suggests that the HBsAg(+) couples had sons earlier in marriage than the anti-HBs(+) couples.

Figure 2 presents the proportions of males born to HBsAg(+) compared to the number born to anti-HBs(+) couples; these data take into account paternal and maternal ages and pregnancy and birth order (35). The HBsAg(+) couples had consistently higher proportions of male live births than anti-HBs(+) couples, regardless of the effects of the other four factors. These overall differences, as compared by the Mantel-Haenszel method for stratified analysis (36), were significant (see legend to Fig. 2).

We considered two additional questions. (i) Although parental ages and pregnancy and birth orders individually did not account for the relation between response to HBV in parents and sex ratio of offspring, could these factors account for the phenomenon if considered simultaneously? (ii) What are the relative contributions of parental response to HBV and other reproductive factors to the sex ratio of offspring?

To address these questions, we used a weighted multiple linear regression analysis for estimating sex ratio (37). To simplify interpretation, we modeled an adaptation of the sex ratio statistic (number of males divided by the number of females) on a logarithmic scale so that positive values of the log of the sex ratio indicated more males and negative values indicated more females. We considered seven factors: HBsAg and anti-HBs in mothers and fathers, total number of pregnancies, fetal loss, and number of years of marriage (38). An initial analysis suggested that the presence of HBsAg in parents favors males and that anti-HBs in mothers favors females. However, anti-HBs in fathers, total number of pregnancies, number of years married, and fetal loss did not have important effects on sex ratio. We then decided to consider together the three HBV-related factors—HBsAg in fathers, HBsAg in mothers, and anti-HBs in mothers—as one "HBV response" variable. Anti-HBs in fathers was disregarded since its effect on sex ratio was negligible. This HBV response variable had three possible values: "−1" when both parents were HBsAg(−) and the mother was anti-HBs(+), "0" when both parents were HBsAg(−) and the mother was anti-HBs(−), and "+1" when either parent was HBsAg(+) [and anti-HBs(−)].

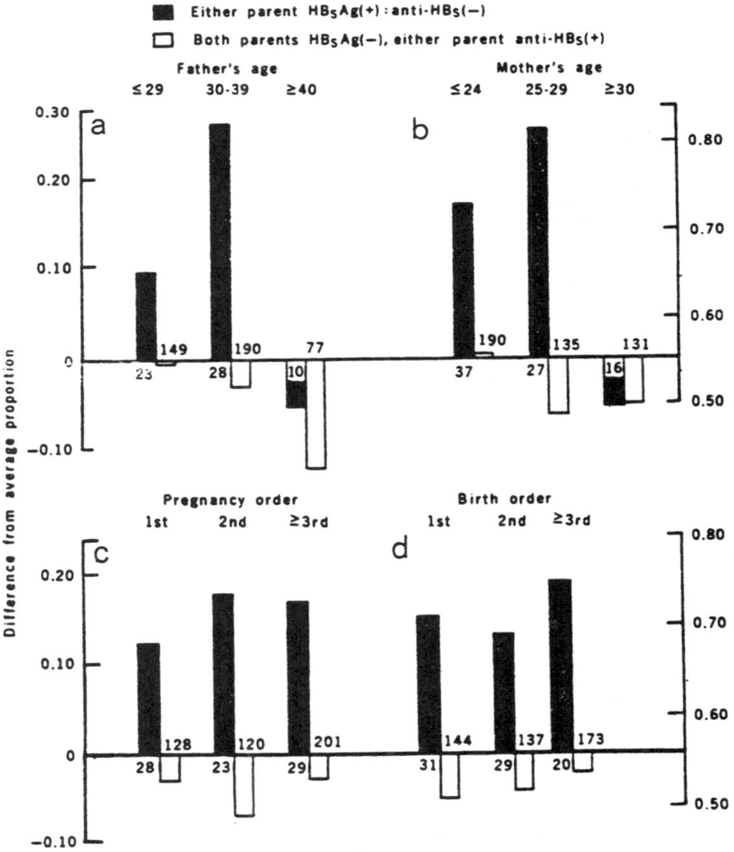

Fig. 2. The proportion of male live births according to parents' response to HBV, paternal (a) and maternal (b) ages, and pregnancy (c) and birth (d) orders. Total numbers of children are indicated at the baseline. The baseline is the proportion of males among all 633 children. The absolute proportions of males are indicated by the crossbars in reference to the scale at the right. The differences of each category from the baseline proportion are outlined by the enclosed areas in reference to the scale at the left. The Z and P values (two-tailed) calculated according to the method of Mantel and Haenszel (36) are as follows: (a) Z = 2.759, P = .006; (b) Z = 3.088, P = .002; (c) Z = 3.187, P = .001; and (d) Z = 3.207, P = .001.

This combined variable is significantly related to sex ratio (χ^2 = 5.20). In addition, fetal loss, although not significant as an individual factor, or in relation to anti-HBs(+) or HBsAg(−):anti-HBs(−) mothers, appeared to be important when considered in reference to HBsAg(+) mothers. The sex ratio of children born to HBsAg(+) mothers who had experienced abortions or stillbirths (or both) was noticeably lower than that of children of HBsAg(+) mothers who had not experienced fetal loss. Thus, a "combined HBsAg in mothers:fetal loss" variable took the values "−1" when the mother was HBsAg(+) with fetal loss, "0" when the mother was HBsAg(−), and "+1" when the mother was HBsAg(+) with no fetal loss. This combined variable was also important in predicting sex ratio (χ^2 = 2.61). When only the two combined variables are considered, the analysis predicts sex ratios for five possible categories of parents. Table 2 shows these five categories with the total number of sons and daughters, the observed sex ratio, and the sex ratio predicted by a two-variable model with the following equation: predicted sex ratio = exp [0.319 + 0.296 × HBV response + 0.553 × fetal loss].

By observation, the observed and predicted sex ratios are nearly equal. However, the model is descriptive and requires further testing with another set of data (*39*). Fetal loss in HBsAg(+) mothers has only marginal statistical significance, is based on a relatively small sample size, and is indicative of the tentative status of this model. Nonetheless, the results of this analysis reject the hypothesis that the relation between parents' responses to HBV and secondary sex ratio can be accounted for by other factors thought to affect sex ratio such as birth order, parental ages, and number of years of marriage. In addition, the analysis suggests some new hypotheses. (i) Anti-HBs has an effect on sex ratio only when it is present in mothers with HBsAg(−) husbands but not when it is present in fathers only. (ii) Fetal loss is important only in HBsAg(+) mothers and, when it occurs, is accompanied by a decrease in the generally high sex ratio of offspring of other HBsAg(+) couples. This suggests that, since HBsAg(+) mothers without fetal loss have a very high proportion of sons, those with fetal loss experience a preferential loss of males.

The regression analysis also suggested that the three original categories of HBV response be changed to the following: (i) either parent HBsAg(+):anti-HBs(−), (ii) both parents HBsAg(−):mother anti-HBs(−), and (iii) both parents HBsAg(−):mother anti-HBs(+). Figure 3 shows that these new category definitions yield clearer distinctions between the least-squares fits for the regressions of sons and daughters as compared with the number of years of marriage than did the original definitions (Fig. 1).

Table 2. Sex ratio in Plati, observed and predicted by a two-variable model including HBV response and fetal loss.

Parents' category	Sons	Daughters	Sex ratio Observed	Sex ratio Predicted
Both parents HBsAg(−), mother anti-HBs(−)	112	86	1.302	1.376
Both parents HBsAg(−), mother anti-HBs(+)	146	140	1.042	1.023
Father HBsAg(+)	13	7	1.857	1.850
Mother HBsAg(+) without fetal loss	17	5	3.400	3.216
Mother HBsAg(+) with fetal loss	10	9	1.110	1.064

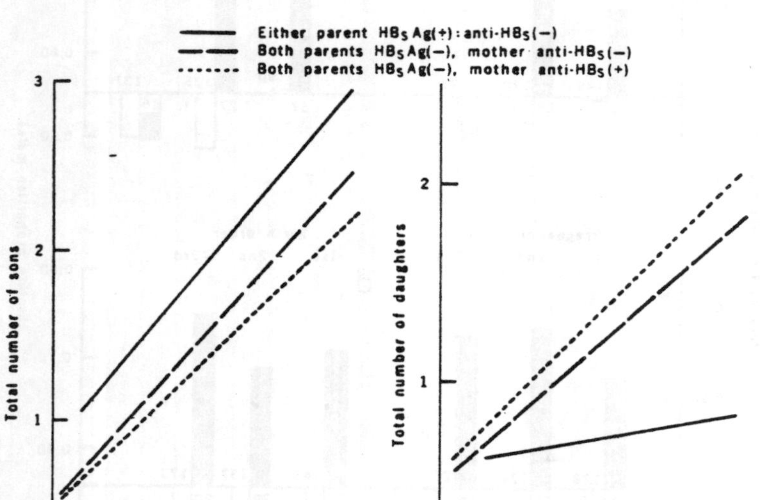

Fig. 3. Least-squares fits for the regressions of sons and daughters as a function of the number of years of marriage; revised category definitions were used. By analysis of covariance (*56*), the number of sons per years of marriage is significantly higher ($P < .02$) and number of daughters significantly lower ($P < .02$) for HBsAg(+) as compared to HBsAg(−):mother anti-HBs(+) couples.

Other HBV-Sex Interactions

In considering possible explanations for the association between parents' response to HBV infection and sex ratio of offspring, we examined two other instances in which HBV seems to interact with sex.

1) There is a sex difference in response to HBV infection. In most populations where the prevalence of chronic carriers is high, HBsAg is detected from 0.2 to 5 times more often in males than in females (*33, 40*). Similarly, both postnecrotic cirrhosis and primary hepatocellular carcinoma are more common in males (*21, 41–43*) and, although lupoid chronic [HBsAg(−)] hepatitis is more common in females, HBsAg(+) chronic hepatitis is more common in males (*44–46*). It has been argued that such differences are due mainly to an increased exposure of males to HBV infection (*47*). However, we have recently demonstrated that, among patients with end-stage renal disease treated with chronic hemodialysis, sex differences in HBsAg prevalence are due to a sex difference in response to HBV. We controlled the exposure variable by studying the responses to HBV infection, by sex, among 77 patients with renal disease, all

of whom became infected with HBV while receiving treatment at a single hemodialysis clinic (*48*). Among these patients, males had a 68 percent chance of remaining HBsAg(+) persistently once infected with HBV, whereas females had only a 33 percent chance. Conversely, females had a 55 percent chance of developing anti-HBs once infected with the virus, whereas males had only a 30 percent chance. These differences in response were significant and were not related to sex differences in age, underlying renal disease, or incidence of infection.

2) We found that the response of renal graft recipients to HBV infection before transplantation and the sex of their kidney donors was associated with the duration of graft survival. We studied the survival of 87 grafts in 79 patients whose responses to HBV infection were assessed before transplantation (*49*). Kidneys from HLA nonidentical male donors which were transplanted into male or female patients who had anti-HBs prior to transplantation had very short survival. Eight of nine such grafts were rejected within 4 months of transplantation. Survival of grafts from male donors was significantly longer in both uninfected patients and chronic carriers of HBsAg (median survival for both groups was more than 22 months). Among the few patients who received grafts from female donors, there were no significant differences in graft survival between anti-HBs(+), uninfected, and HBsAg(+) recipients (*50*).

How Can HBV Interact with Sex?

These two observations, together with the results of the Plati sex ratio study, led us to the hypothesis that there is cross-reactivity between HBsAg and a male-associated antigen (*51*). If HBsAg cross-reacts with a male-associated antigen, males would be more likely to recognize HBsAg as "self" and therefore would remain HBsAg(+) persistently. Females, however, would be more likely to recognize HBsAg as "foreign" and produce anti-HBs. This would result in the observed predominance of males among chronic carriers of HBV in dialysis patients and other populations. In kidney transplant patients, tolerance to HBsAg in kidney graft recipients would result in tolerance (that is, longer survival) of male tissues, whereas anti-HBs in a recipient would react with male antigens on renal allografts, resulting in early rejection of grafts from male donors.

Similarly, we can speculate that tolerance to HBsAg [reflected in the maintenance of HBsAg(+):anti-HBs(−) status] in pregnant women would result in lack of sensitization against male tissues developing within them, and therefore good survival of male fetuses. Anti-HBs in women, however, could react with male antigens and perhaps hinder fertilization by sperm bearing a Y chromosome or increase the probability of spontaneous abortion of male fetuses. HBsAg(+) males would have HBsAg in their semen which, conceivably, could protect Y-bearing sperm from the effects of anti-HBs or other "anti-male" antibodies in their spouses' reproductive tracts (*52*). The effect of fetal loss in HBsAg(+) women could be explained by unhindered replication of HBV in male fetuses, and this could result in a greater loss of male than female embryos.

The cross-reactivity hypothesis cannot explain all the observations from the Greek study. If the effects of HBsAg and anti-HBs were the only factors modifying sex ratio, HBsAg(+) couples would have more sons than anti-HBs(+) couples (*53*). However, the observation that, even among parents married more than 15 years, HBsAg(+) couples had the same number of sons but fewer daughters per family as compared to anti-HBs(+) couples suggests that human behavior is affected by hepatitis B infection. In this Greek community most younger women preferred small families and practiced some form of birth control. Also, in some parts of Greece a preference for sons (as observed in many other societies) has been reported (*6*). It is possible that parents chose to limit family size only after bearing a desired number of sons. If this were so, couples where both parents were HBsAg(−) and the mother was anti-HBs(+) might give birth to several daughters before achieving the desired number of sons while HBsAg(+) couples would have the desired number of sons early in marriage and, as a result, restrict the number of subsequent births and therefore the total number of daughters born. The sex ratio of offspring of HBsAg(−) couples with anti-HBs(−) mothers would be unaffected by this preference since these couples express neither of the factors (HBsAg, anti-HBs) affecting sex ratio. If these explanations are valid, then in this community biological and social factors interact to influence the secondary sex ratio.

Conclusions

We have observed an association between the types of response to hepatitis B infection in parents and the sex ratio of their offspring. In either parent HBsAg was associated with a high sex ratio of live births, while anti-HBs(+) in mothers was associated with a low sex ratio. Fetal loss in HBsAg(+) mothers was also related to a lower sex ratio. Other variables thought to affect sex ratio, such as paternal and maternal ages (or number of years of marriage) and total number of pregnancies, were not significantly related when considered in the context of HBV response and fetal loss.

Evidence from studies of sex differences in response to infection with HBV, and survival of transplanted kidneys in recipients with different responses to HBV, as well as the study of sex ratio reported in this article led to the hypothesis that there is cross-reactivity between HBsAg and a male-associated antigen. This hypothesis requires both experimental and epidemiological testing.

Very few biological factors that affect sex ratio have been identified. Studies of the kind reported here should be conducted on other populations (*54*), and the effects of additional infectious agents and biological factors on sex ratio should be evaluated.

References and Notes

1. P. M. Visaria, *Eugen. Q.* **14**, 132 (1967).
2. A. Scheinfeld, *Sci. Am.* **198**, 22 (Feb. 1958).
3. W. F. Bodmer and A. W. F. Edwards, *Ann. Hum. Genet.* **24**, 239 (1960).
4. C. Stern, *Principles of Human Genetics* (Freeman, San Francisco, ed. 3, 1973), pp. 529–551. In many countries, including the United States, there is still an excess of males at reproductive age. The sex ratio in a population at this time depends on specific mortality figures.
5. The selective advantage of a particular sex ratio at birth, therefore, depends upon the stabilization of this later sex ratio.
6. J. E. Hesser, thesis, University of Pennsylvania (1974). These differences cannot be explained by biases in reporting or by differences in fetal mortality (*1*).
7. Among the areas with populations reported to have sex ratios that deviate from the norm are Chile (103.0), Hong Kong (109.1), Aden (117.0), Korea (116.0), Gambia (116.2), Greece (113.0), and the Republic of South Africa (people of Asian origin only) (101.2) (*1*, *6*).
8. B. MacMahon and T. F. Pugh, *Am. J. Hum. Genet.* **6**, 284 (1954).
9. W. R. Lyster, *Hum. Biol.* **40**, 63 (1968); *ibid.* **42**, 670 (1970).
10. N. A. Chagnon, *Yanomamo: The Fierce People* (Holt, Rinehart & Winston, New York, 1968).
11. L. A. Goodman, *Ann. Hum. Genet.* **25**, 75 (1961).
12. L. D. Sanghvi, *Nature (London)* **168**, 1077 (1951).
13. C. E. Jackson, J. D. Mann, W. J. Schull, *ibid.* **222**, 445 (1969).
14. J. D. Erickson, *Ann. Hum. Genet.* **40**, 205 (1976).
15. D. F. Poulson and B. Sakaguchi, *Science* **133**, 1489 (1961).
16. B. Sakaguchi and D. F. Poulson, *Genetics* **48**, 841 (1963).
17. The expression of the "sex ratio" condition also appears to depend on the genotypes of both the host and the infectious agent (*16*).
18. J. E. Hesser, J. Economidou, B. S. Blumberg, *Hum. Biol.* **47**, 415 (1975).
19. J. E. Hesser, B. S. Blumberg, J. S. Drew, *ibid.* **48**, 73 (1976).
20. W. T. London, J. S. Drew, E. D. Lustbader, B. G. Werner, B. S. Blumberg, *Kidney Int.* **12**, 51 (1977).
21. B. S. Blumberg, B. Larouze, W. T. London, B.

Werner, J. E. Hisser, I. Millman, G. Saimot, M. Payet, *Am J. Pathol.* **81**, 669 (1975).
22. C. Bruyns, G. Urbain-Vansanten, C. Planard, C. DeVos-Cloeteus, J. Urbain, *Proc. Natl. Acad. Sci. U.S.A.* **73**, 2462 (1976).
23. K. Nishioka, *in Proceedings of a Symposium on Viral Hepatitis, 1975* (Slack, Thorofare, N.J., 1975), p. 322.
24. S. Mazzur, B. S. Blumberg, J. S. Friedlaender, *Nature (London)* **247**, 41 (1974); B. S. Blumberg, *Bull. Acad. Med. Toronto* **45**, 45 (1972).
25. J. S. Robertson and A. V. Sheard, *Lancet* **1973-I**, 532 (1973).
26. S. Mazzur and T. M. Watson, *Nature (London)* **250**, 60 (1974).
27. This study was conducted by J. E. Hesser from our laboratory in collaboration with our Greek colleagues, I. Economidou, S. Hadziyannis, and others.
28. A similar association has been observed in French families [P. Cazal, J.-M. Lemaire, M. Robinet-Levy, *Rev. Fr. Transfus. Immunohematol.* **19**, 577 (1976)].
29. Serums were tested for HBsAg by radioimmunoassay using a modification of the method of Ling and Overby, immunodiffusion, reaction electrophoresis, and passive hemagglutination inhibition by the Hepanosticon (Organon) method. Anti-HBs was determined by passive hemagglutination. Sex ratios were compared by the two-by-two chi-squared test, numbers of sons and daughters by the Mann-Whitney U statistic, and linear least-squares fits by analysis of covariance [C. M. Ling and L. R. Overby, *J. Immunol.* **109**, 834 (1972)].
30. There are some differences between the couples reported by Hesser *et al.* (*18*, *19*) and those analyzed in this study. We excluded the following: three couples in which the anti-HBs status of the HBsAg(+) parent was uncertain (five sons, one daughter), and eight couples in which both the father and mother were HBsAg(−) but the anti-HBs information was incomplete (13 sons, 4 daughters). We added the following: one couple in which the father was HBsAg(+):anti-HBs(−) and the mother was not tested (two sons, no daughters), eight childless couples in which both parents were HBsAg(−), and 11 couples in which both parents were HBsAg(−) and one parent was anti-HBs(+) (29 sons, 22 daughters). This last group had been excluded by Hesser *et al.* (*18*, *19*) for lack of data other than sex of offspring. Because of incomplete information, 147 families were excluded by both Hesser *et al.* and by us.
31. The HBsAg(+) category includes couples in which one parent was HBsAg(+):anti-HBs(−) regardless of the test results of the other parent.
32. There are different antigenic subtypes of HBsAg, which may occur simultaneously in one population but rarely in one individual. The presence of HBsAg and anti-HBs in one person probably represents antigen of one subtype and antibody to another.
33. B. S. Blumberg, L. Melartin, R. A. Guinto, B. Werner, *Am. J. Hum. Genet.* **18**, 594 (1966).
34. We have made similar observations on an additional population from Mali, West Africa. The fathers of these families were not studied, however.
35. The data on parental ages and pregnancy and birth orders were available for ~ 90 percent of the 633 children reported.
36. N. Mantel and W. Haenszel, *J. Natl. Cancer Inst.* **22**, 719 (1959).
37. J. E. Grizzle, C. F. Steinmer, G. G. Koch, *Biometrics* **25**, 489 (1969).
38. This analysis involved a smaller sample (~ 86 percent of the original 633 children) than the previous analyses since families with incomplete information in any variable were excluded from analysis. In families with HBsAg in one parent and anti-HBs in the other, the anti-HBs was counted as being neither present nor absent (value = 0), since the effect of HBsAg in one parent masked any effect of anti-HBs in the other. A plot of the number of pregnancies suggested a breakpoint which contrasted two or fewer pregnancies with three or more. "Years of marriage" had a correlation of greater than 0.9 with both father's age and mother's age, and therefore, could serve as a surrogate for either variable.
39. A formal goodness-of-fit test is not possible since the definitions of the "HBV response" variable and the "fetal loss" variable were suggested by the data.
40. B. S. Blumberg, L. Melartin, M. LeChat, R. A. Guinto, *Lancet* **1967-II**, 1973 (1967); J. P. Soulier, A. M. Courouce-Pauty, D. Benamom-Djiane, *Vox Sang.* **19**, 345 (1970); E. Terrier, M. Simonneau, B. Jaulmes, *ibid.*, p. 352; K. Okochi, M. Mayumi, Y. Haguino, N. Saito, *ibid.*, p. 332; P. Lous, P. Skinhoj, H. Olsen, *ibid.*, p. 379; M. Leski, C. Grivaux, A. M. Courouce-Pauty, *ibid.*, p. 359; M. Goodman, R. L. Wainwright, H. F. Weir, J. C. Gall, *Pediatrics* **48**, 907 (1971); A. Klinman, N. R. Reid, C. Lilly, J. Morrison, *N. Engl. J. Med.* **285**, 783 (1971); W. Szmuness, A. M. Prince, G. F. Etling, R. Pick, *J. Infect. Dis.* **126**, 498 (1972); B. S. Blumberg, A. I. Sutnick, W. T. London, L. Melartin, *Arch. Intern. Med.* **130**, 227 (1972); W. Szmuness, A. Prince, B. Brotman, R. Hirsch, *J. Infect. Dis.* **127**, 17 (1973); L. J. Farrow, S. G. Lamb, N. F. Coghill, R. L. Lindon, J. Preece, A. J. Zuckerman, J. S. Stewart, *Br. Med. J.* 3, 83 (1974); W. Szmuness, R. L. Hirsch, A. M. Prince, R. M. Levin, E. J. Harley, H. Ikram, *J. Infect. Dis.* **131**, 111 (1975).
41. M. J. Tong, S-C. Sun, B. T. Schaeffer, N-K. Chang, K-J. Lo, *Ann. Intern. Med.* **75**, 687 (1971).
42. C. L. Vogel, P. P. Anthony, F. Sadikali, L. F. Barker, M. R. Peterson, *J. Nat. Cancer Inst.* **48**, 1583 (1972).
43. B. Larouze, W. T. London, G. Saimot, B. G. Werner, E. D. Lustbader, M. Payet, B. S. Blumberg, *Lancet* **1976-II**, 534 (1976). Male-to-female ratios for primary hepatocellular carcinoma are 8:1 in Taiwan (*41*) and 4:1 in Uganda (*42*).
44. N. R. Shulman, *Am. J. Med.* **49**, 669 (1970); B. H. Bulkley, W. D. Heizer, S. E. Goldfinger, K. J. Isselbacher, *Lancet* **1970-II**, 1323 (1970).
45. S. Sherlock, *Resident and Staff Physician* (July 1977), p. 42.
46. Approximately 75 percent of lupoid chronic hepatitis occurs in women (*45*), while 80 percent or more HBsAg(+) chronic hepatitis occurs in men (unpublished data from Korea).
47. S. Mazzur and N. Jones, *J. Infect. Dis.* **133**, 331 (1976).
48. W. T. London and J. S. Drew, *Proc. Natl. Acad. Sci. U.S.A.* **74**, 2561 (1977). Patients were monitored monthly for the presence of HBsAg and anti-HBs in their serums for long periods of time.
49. W. T. London, J. S. Drew, B. S. Blumberg, R. A. Grossman, P. J. Lyons, *N. Engl. J. Med.* **296**, 241 (1977); J. S. Drew, W. T. London, E. D. Lustbader, B. S. Blumberg, *Birth Defects Orig. Art. Ser.*, in press. In these data, no recipients had preformed antibody to HLA prior to transplantation.
50. Most of these grafts were from cadaver donors, but the association of duration of graft survival with responses to HBV infection was independent of whether the donors were alive and related or whether the donors were cadavers. Although the early rejection of cadaver kidneys by anti-HBs(+) recipients has been observed by others, the specific association with grafts from male donors has not yet been confirmed.
51. All three studies involve populations in which the major HBsAg antigenic subtype is **ay** rather than **ad**. There is some indication from studies in the Solomon Islands (*17*) of a relation between the **ad** subtype and an alteration of secondary sex ratio in the opposite direction. In this way the genotypes of both the host and the infectious agent would be important in determining the expression of this sex ratio effect, as seen in the *Drosophila* "sex ratio" condition.
52. J. Heathcote, C. H. Cameron, D. S. Dane, *Lancet* **1974-I**, 71 (1974).
53. This appeared to be the case in the population from Mali, West Africa.
54. Preliminary results of sex ratio studies in Africa, Papua New Guinea, and Greenland are similar to those from the Plati study; HBsAg is associated with higher sex ratios, while anti-HBs is associated with lower ones.
55. The sex ratio is the number of males per 100 females.
56. B. J. Winer, *Statistical Principles in Experimental Design* (McGraw-Hill, New York, 1962), pp. 578–594.
57. Supported by NIH grants CA-06551, RR-05539, and CA-06927 and by an appropriation from the Commonwealth of Pennsylvania. We thank Dr. Ioanna Economidou, director of the Hellenic Red Cross Blood Transfusion Center, Dr. Stephanos Hadziyannis, and the Hellenic Red Cross for their work in gathering and recording the Greek data, and the members of the Division of Clinical Research for their counsel and technical assistance.

Hepatitis B Virus and Sex Ratio on Kar Kar Island

By Jean Drew,[1] W. Thomas London,[1] Baruch S. Blumberg[1] and Susan Serjeantson[2]

ABSTRACT

Hypotheses formulated from a previous study in Greece concerning the relationship between responses to hepatitis B virus (HBV) infection in parents and the secondary sex ratio of their offspring were tested in the population of Kar Kar Island, Papua New Guinea. The results were similar to those observed in the Greek village population. Sex ratios were highest (115) when either parent was a chronic carrier of HBV (HBsAg(+)) and lowest (60) when both parents were HBsAg(−) and the mother had antibody to HBsAg (anti-HBs(+)). Couples in which both parents were antigen and antibody negative had offspring with sex ratios intermediate (84) between the antigen positive and mother antibody positive couples. As in Greece, the sex ratio effects were due mainly to differences in the number of daughters, not sons, in each family; either parent antigen positive couples had the fewest (2.3) and mother anti-HBs(+), father HBsAg(−) couples had the most (3.9) daughters per couple. As a consequence, the latter couples tended to have larger total family sizes. Possible mechanisms to explain these observations are discussed.

Various epidemiological studies have demonstrated an interaction between responses to hepatitis B virus (HBV) in parents and the sex ratio at birth of their offspring (secondary sex ratio). An association between hepatitis infection and secondary sex ratio (no. of sons/no. of daughters × 100) has been observed previously in populations from Brigg, Great Britain (Robertson and Sheard, 1973), Languedoc, France (Cazal, Lemaire, and Robinet-Levy, 1976), Plati, Greece (Hesser, 1974; Hesser, Economidou, and Blumberg, 1975; Drew et al. 1978) and the Solomon Islands (Hesser, Blumberg, and Drew, 1976). In the Greek population, study of the HBV-sex ratio relation (Drew et al. 1978; Hesser, Economidou, and Blumberg, 1975) revealed the following: a) The sex ratio of live births from matings in which either the mother or father was positive for the surface antigen of HBV (HBsAg) was significantly higher than the sex ratio from matings in which neither parent had HBsAg; b) Antibody to HBsAg (anti-HBs) in parents was associated with a sex ratio alteration in the opposite direction from that observed with HBsAg; the children of matings in which either parent was antibody positive and

[1]The Insitute for Cancer Research, The Fox Chase Cancer Center, Philadelphia, Pennsylvania 19111.

[2]John Curtin School of Medical Research, Box 4, G.P.O., Canberra, A.C.T., Australia.

Human Biology, February 1982, Vol. 54 No. 1, pp. 123-135

the island, was selected for more extensive study. In a census conducted by IBP personnel in 1969 the four villages had a total population of 1286. Every resident of the village complex received a medical examination, and a sample of blood was collected for genetic, biochemical and serological studies. On presentation for medical examination, subjects were asked once again the number of living and dead children. The present investigation includes subjects from the Kaul complex and four additional villages: Dorogodan, Langlang, Orora and Mangar. During the IBP studies, family relationships were not established and, in some cases, the total number of live births reported in the demographic and medical surveys were disparate, due to some inevitable loss of accuracy with the interview technique. Therefore, one of us (SS) revisited Kar Kar Island, before the tests of hepatitis B serology were done, to determine marital and pedigree relationships and to establish more accurately any previously unrecorded adoptions in the eight villages included in the present study. Although written records were not available to confirm all of the information obtained by interview, the comparisons made in this paper are within the population and should not be biased by the methods of data collection. Furthermore, polygamy is not practiced on Kar Kar and marriages are stable (Hornabrook, 1974).

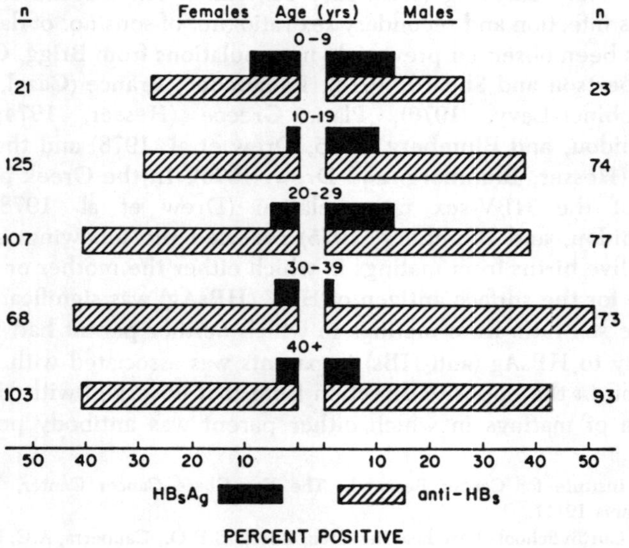

FIG. 1. Frequencies of HB_sA_g and anti-HB_s by age and sex in the Kar Kar population.

Male and Female Live Births According to Parents' Response to Hepatitis B Virus. Data from Kar Kar Island, Papua New Guinea

	A Couples	B Sons		C Daughters		D Children	E Sex Ratio Sons/Daughters × 100
	n	n/couple ± s.e.	(n)	n/couple ± s.e.	(n)	n/couple	
I. Either parent antigen positive (HBsAg(+))	24	2.6 ± .46	(63)	2.3 ± .41	(54)	4.9	115
II. Both parents antigen negative (HBsAg(−))	69	2.4 ± .22	(163)	3.0 ± .27	(206)	5.4	79
A. Both parents antibody negative (anti-HBs(−))	21	2.1 ± .41	(43)	2.4 ± .54	(51)	4.5	84
B. Either or both parents antibody positive (anti-HBs(+))	48	2.5 ± .26	(120)	3.2 ± .31	(155)	5.7	77

(Table 2, row V). Matings in which the father had anti-HBs resulted in fewer daughters per couple (2.9) and a sex ratio closer to that found in the offspring of antigen and antibody negative matings (Table 2, row III). Curiously, antibody in both parents was associated with equal numbers of male and female births.

It was possible that differences in parental ages could have accounted for the observed sex ratio differences, since the frequencies of HBsAg and anti-HBs vary with age (Figure 1). To address this matter, we examined sex ratio of live births of "completed" families only. We defined "completed" families as those in which the mother was at least 45 years old and presumably had terminated her reproductive activity. Table 3 suggests that, as we saw in Plati (Drew et al. 1978), the association of antigen with higher sex ratios and fewer daughters and that of antibody with lower sex ratios and more daughters is apparent even in this very small subgroup of children of older parents ($p = 0.002$; one-tailed).

Reliable information was not available on mothers' age at the time of the birth of each of their children or on birth order. By examining the age of the mothers at the time the data were collected and the number of sons and daughters born to each woman, it appeared that the older the woman the more sons she had, regardless of mating type. The number of daughters showed a similar increase with age of mothers among antigen negative couples, but not among antigen positive couples. To test the hypothesis that the number of daughters born to antigen positive couples increased at a different rate than those born to antigen negative or antibody positive couples, we estimated the linear least squares fits for the regressions of the numbers of sons or daughters vs. mothers' age for each of the three mating types. Analysis of covariance (not shown, available on request) revealed that the antigen positive couples had significantly fewer daughters than the antibody positive couples (see Table 1 for definitions of mating types). Also, antibody positive and antigen and antibody negative couples had a steadily increasing number of daughters with increasing maternal age, whereas antigen positive couples had a lesser increase in daughters. The number of sons among antigen positive couples increased with maternal age at about the same rate as among antigen negative couples.

Although the reliability of the number of miscarriages reported per woman is uncertain, we examined the data to see whether women who reported fetal loss had different sex ratios of offspring than those who did not. We did not observe significant differences within each mating type or for the entire population of women with reproductive information.

One of the purposes of the Plati study was to test the hypothesis that differences in fertility or family size are associated with the response of parents to HBV. In Plati the average family size was small (less than three children per couple), and significant differences were not observed. There was a suggestion of an increase in total number of live births among the anti-HBs(+) couples (3.0) compared with the HBsAg(+) couples (2.6) as a result of a decreased number of daughters and an equal number of sons born to the antigen positive parents. In Kar Kar, the family sizes are larger and the differences, which are in the same direction as those seen in Plati, are also larger (Table 1, column D). The maximum family size is observed when both parents are antigen negative, the father is antibody negative, and the mother is anti-HBs(+) (6.2 children/couple) (Table 2). This is significantly greater than the average of 4.9 children born to antigen positive couples. The regressions of the number of sons and daughters on maternal age for each mating type suggest that by age 45, each HBsAg(+) couple would have on the average 1.3 fewer daughters than either couples in which both parents are HBsAg(−), anti-HBs(−) or couples in which both parents are HBsAg(−) and either parent is anti-HBs(+). The actual data (Table 3) on completed families shows that HBsAg(+) couples had on the average 1.5 daughters compared with 3.5 for antigen and antibody negative couples and 4.5 daughters for antibody positive couples.

Discussion

The data presented here suggest that the relation of response to HBV in parents to sex ratio of offspring observed in Plati, Greece is not an isolated phenomenon. On Kar Kar Island, as in Plati, HBsAg in parents was associated with a high sex ratio of offspring while anti-HBs was associated with a significantly lower one. In Plati and Kar Kar, this difference is due primarily to a difference in the mean number of daughters rather than the mean number of sons per family. In neither Kar Kar nor Plati could these observations be accounted for by differences in parental ages.

We have suggested previously that the various observations of an interaction between HBV and sex could be caused by cross-reactivity between HBsAg and a male-associated antigen (Drew et al. 1978; Drew, London, Lustbader, and Blumberg, 1978). In antigen positive mothers, immunological tolerance to HBsAg (as reflected by the maintenance of HBsAg positivity) would be accompanied by tolerance to male embryonic

HBV in a cell would lead to release of many virus particles, stimulus of the immune system, destruction of HBV infected cells by T-lymphocytes and production of anti-HBs. Slow replication would result in prolonged release of small numbers of viral particles, gradual accumulation of particles in the circulation, little or no stimulus of the immune system to produce anti-HBs and, hence, persistent HBs antigenemia with little or no liver damage. The fact that females are more likely than males to develop anti-HBs (London and Drew, 1977; Szmuness, Harley, Ikram, and Stevens, 1978) therefore suggests that, in general, HBV replication may be faster in female than in male cells.

There is no direct information indicating that human embryos can be infected with HBV *in utero*. There is circumstantial evidence, however. High titer anti-HBs given at birth to babies born to mothers carrying infectious hepatitis B virions alters the response of these infants to infection but does not prevent infection (Stevens et al. 1981). Since hyperimmune globulin is effective in preventing infection if given at the time of exposure to HBV (Grady et al. 1976), the failure to prevent transmission to newborns suggests that infection may have occurred *in utero*. Secondly, a virus closely related to human HBV occurs in Pekin ducks (Mason et al. 1980). This virus, called duck hepatitis B virus (DHBV), is definitely transmitted vertically. It produces infection under commercial breeding conditions in duck embryos as early as five days after laying (O'Connell and London, unpublished data). If human early embryonic cells can be infected with HBV and if such infected cells die or are unable to differentiate normally, then *in utero* infection would lead to more rapid replication and spread of HBV in female than male embryos. This would result in reduced viability of female embryos and a decrease in the number of female offspring born to HBsAg(+) parents. Anti-HBs in a pregnant woman's serum and in the secretions of her reproductive tract could protect her fetus from being infected by HBV, or related agents, in the semen of an HBsAg(+) mate. If female embryos were at greater risk of a lethal HBV infection, protection by anti-HBs would result in a greater increase in the viability of female than male fetuses. This speculative hypothesis warrants experimental and epidemiological testing.

Biological factors responsible for sex ratio variations in human populations have generally been obscure. The present study describes a second population in which responses to HBV infection in parents have been significantly related to sex ratio of offspring. In addition, in this population there is a significant and large effect on fertility (carriers of hepatitis B virus have fewer children than parents with antibody). Further study of

London, W. T., J. S. Drew, E. D. Lustbader, B. G. Werner and B. S. Blumberg 1977 Host responses to hepatitis B infection in patients in a chronic hemodialysis unit. Kidney Internat. **12:** 51-58.

Mason, W. S., G. Seal and J. Summers 1980 Virus of Pekin ducks with structural and biological relatedness to human hepatitis B virus. J. Virol. **36:** 829-836.

Ogra, P. L. 1973 Immunologic aspects of hepatitis associated antigen and antibody in human body fluids. J. Immunol. **110:** 1197-1205.

Robertson, J. S. and A. V. Sheard 1973 Altered sex ratio after an outbreak of hepatitis. Lancet, **1:** 532-534.

Stanhope, J. M. and R. W. Hornabrook 1974 Fertility patterns of two New Guinea populations: Kar Kar and Lufa. J. Biosoc. Sci. **6:** 439-452.

Stevens, C. E., R. P. Beasley, W. Szmuness, C. C. Lin, L. Y. Hwang, T. S. Sun, F. J. Hsieh and K. Y. Wang 1981 Efficacy of hepatitis B immune globulin in prevention of prenatally transmitted hepatitis B: Results of a second clinical trial in Taiwan. Presented at the 1981 International Symposium on Viral Hepatitis, April 1, 1981, New York, New York.

Szmuness, W., E. J. Harley, H. Ikram and C. Stevens 1978 Sociodemographic aspects of the epidemiology of hepatitis B. *In:* Viral Hepatitis, G. N. Vyas, S. N. Cohen and R. Schmid (eds.). Franklin Institute Press, Philadelphia, pp. 297-320.

Vyas, G. N. and N. R. Shulman 1970 Hemagglutination assay for antigen and antibody associated with viral hepatitis. Science, **170:** 332-333.

Walsh, R. J. 1974 Geographical, historical and social background of the peoples studied in the I.B.P. Phil. Transac. Royal Soc. Lond. **268:** 223-228.

Reprinted from the PROCEEDINGS OF THE NATIONAL ACADEMY OF SCIENCES
Vol. 62, No. 4, pp. 1108–1115. April, 1969.

HEPATITIS AND AUSTRALIA ANTIGEN: AUTOSOMAL RECESSIVE INHERITANCE OF SUSCEPTIBILITY TO INFECTION IN HUMANS*

BY B. S. BLUMBERG, J. S. FRIEDLAENDER, ANITA WOODSIDE, A. I. SUTNICK, AND W. T. LONDON

INSTITUTE FOR CANCER RESEARCH, FOX CHASE, PHILADELPHIA, AND DEPARTMENT OF ANTHROPOLOGY, HARVARD UNIVERSITY

Communicated by Thomas F. Anderson, December 30, 1968

Abstract.—Examples of inherited susceptibility to infection controlled by genes segregating at one or a small number of loci have been identified in lower animals. In this study we report data on what appears to be a similar situation in humans. "Australia antigen" is an antigen found in the sera of patients with acute and chronic hepatitis, and it may actually be a form of virus. It is very common in many tropical areas, and people in these areas having the antigen appear to be hepatitis carriers. The antigen is detected by immunodiffusion in agar gel (Ouchterlony method). Individuals with the antigen are designated Au(1) and those without it Au(0). Family studies involving 1797 different individuals residing on the island of Bougainville are consistent with the hypothesis that susceptibility to chronic infection with the "antigen" is controlled by an autosomal recessive gene (Au^1). This confirms the conclusions previously arrived at from similar (but less extensive) studies on the island of Cebu. Individuals with this inherited susceptibility do not ordinarily have overt manifestations of hepatitis.

The sign test was used to determine family clustering. Segregation analysis was performed by the method of C. A. B. Smith. In the 41 Au(0) × Au(0) matings, 53.8 recessives were expected in the offspring and 56 were seen (0.7 $> p >$ 0.5). In the Au(1) × Au(0) matings, 40.1 recessives were expected and 42 were seen (0.7 $> p >$ 0.5). The Au(0) × Au(0) matings were also analyzed by the method of Li and Mantel, in which the recessive ratio of 0.25 is expected by the genetic hypothesis. The values observed were 0.2527 for the Bougainville study and 0.2461 for the Cebu study.

Inherited resistance and susceptibility to infection has been described in several animal species. There is a single gene-controlled factor which prevents infection of mice by a group of related viruses including yellow fever, West Nile fever, Japanese B encephalitis, and others but does not prevent infection with other viruses.[1] Gowen[2] has described inherited resistance to bacterial infection in mice and has emphasized the specific nature of the resistance which develops. There is evidence for inherited susceptibility to several forms of virus-induced neoplasms in animals. These include polyoma virus in mice,[3] in which two or three independent genes may be involved in the determination of resistance;[4] mammary tumor in mice;[5] leukemia in mice, where the inherited factor appears to be related to the histocompatibility locus;[6] Rous sarcoma;[7] fowl leucoses;[8] and others. In humans, several inherited diseases carry with them an increased susceptibility to infection; for example, patients with the inherited disease sickle-cell anemia are unusually susceptible to *Salmonella* infections, with osteomyiletis as a frequent complication, while the heterozygotes appear to have an increased resistance to falciparum malaria. Patients with inherited sex-linked agammaglobulinemia are highly susceptible to several bacterial infections. There is also evidence for differences in infectious disease susceptibility between different population groups (i.e., increased tuberculosis infection in American Indians and increased infection with coccidioidomycosis in Filipinos and American Negroes) and some of this difference in susceptibility may be inherited. These have been reviewed recently.[9–11]

In this paper we will describe family studies of a serum factor closely associated with acute and chronic viral hepatitis. The segregation of the trait in the families is consistent with simple recessive autosomal inheritance, and the data can be interpreted as an example of increased susceptibility to a chronic virus infection in humans, apparently controlled by genes segregating at a single autosomal locus.

In 1966 we reported on the family clustering and segregation of a serum antigen present in high frequency in certain tropical populations.[12] Because it had been first identified in an Australian aborigine, the antigen was termed "Australia antigen." In the family study we utilized data collected on the Visayan island of Cebu in the Philippines. In the 53 families studied, the antigen was more commonly found among relatives of individuals who had the antigen (17%) than among the relatives of those who did not have it (2%). The hypothesis of simple autosomal recessive inheritance was tested in the families in which the trait was present. (Inspection of pedigrees ruled out simple autosomal dominant and sex-linked inheritance.) Using the segregation analysis method of C. A. B. Smith[13] (Table 3), we found a very close fit to the numbers predicted by the hypothesis; the data provided strong support for the hypothesis of simple autosomal recessive inheritance.

This initial evidence for the genetic determination of the presence of the antigen was interesting since we know that Australia antigen also has features of an infectious agent associated with hepatitis.[14–17] This has been reviewed[16, 17] and will be summarized here. Since 1961 we have systematically examined the sera of transfused patients for the presence of precipitating isoantibodies against constituents of human sera.[18] With these antisera, the Ag system (inherited antigenic specificities on the low-density lipoproteins)[19] was discovered. We subsequently found that patients who received large numbers of transfusions (particularly those with hemophilia) quite often develop specific antibodies against Australia antigen.[20–22] Antisera may also be produced by immunizing rabbits with the serum of a patient with Australian antigen and absorbing with a serum which does not contain the antigen.[23]

The antigen occurs in high frequency in acute hepatitis,[14–18] but in most cases it is transient (days or weeks). The association of hepatitis with Australia antigen has been confirmed by Okochi and Murakami,[22] Prince[24, 25] (the SH antigen of Prince is indistinguishable from Australia antigen), and Vierucci,[26] with reference sera from our laboratory. It also occurs chronically (for months or years) in Down's syndrome patients,[14, 27] in three forms of leukemia,[14, 21] and in apparently normal people in vast areas of Asia and Oceania as well as elsewhere in the tropics.[12, 20] In these groups of patients it appears to be associated with chronic anicteric hepatitis.[22, 28, 29] The liver abnormality may be very slight, particularly in the case of the apparently normal tropical populations referred to.

As noted, Australia antigen is more common in the general populations in some tropical countries (6–25 %) and in Japan (1%) than it is in the United States (0.1%). The studies by Okochi and Murakami[22] in Japan show that Australia antigen may be transmitted by transfusion. Patients transfused with the blood of donors containing Australia antigen may develop the antigen in their own blood, along with hepatitis, or they may develop antibodies to Australia antigen with no evidence of disease. Hence, the detection of the antigen in donor bloods is useful in screening for hepatitis carriers. It has also proved very useful in detecting occult cases of hepatitis in patients with renal disease receiving chronic hemodialysis.[30] The "Au test" is now being used for the diagnosis of viral hepatitis and for the detection of hepatitis carriers.

Studies on the Down's syndrome patients demonstrate that both environmental (infectious) and host (congenital) factors are operating to determine the presence of the antigen.[17, 29] In large institutions, Australia antigen is very common in Down's syndrome patients but extremely rare in other mentally retarded children and in normal individuals. That is, there is a host factor in the Down's patients which makes them more susceptible to chronic infection with Australia antigen. Australia antigen was not found in Down's patients who live at home and attend day schools. Down's patients who are in-patients at small private institutions where sanitation and care are generally better have extremely low frequencies (3%) of the antigen. This is compatible with the explanation that some environmental factor, presumably an infection, is also operating and that the crowded conditions of large institutions lead to much higher frequencies of the antigen.

Australia antigen can be isolated by centrifugation in high-density sugar gradients. Isolated fractions examined under the electron microscope contain large numbers of particles of approximately 200 Å diameter;[31] this is about the size postulated for hepatitis virus. Although the appearance of the particles is consistent with that of a virus, this, of course, does not represent firm evidence that these particles are in fact virus.

All these data taken together (and briefly reviewed here) have led to the hypothesis that Australia antigen is or is associated with an infectious agent that occurs in viral hepatitis. The theoretical consequences of this will be discussed elsewhere.[17]

In view of these findings, it became extremely important to retest the genetic hypothesis by further family studies. Collections of sera on the island of Bougainville, Trust Territory of New Guinea, have made this possible. This paper contains a report of these studies.

During 1966–1967 one of us (J. F.) collected blood from individuals in 18 villages on Bougainville. In each village he attempted to collect blood from as large a number of individuals as possible. About 85% of the people over two years old were included in the collection. The majority of these sera were made available for Australia antigen testing. These populations are described in detail elsewhere.[32] A total of 1797 sera were tested, including 617 family groupings. ABO and Rh blood groups and serum haptoglobin types were available in the individuals whose serum had been tested. If the paternity of any of the offspring in a family was questioned by the use of these markers, then the entire family was removed from the calculations and data from these individuals do not appear in the tables given herein. The presence or absence of the antigen in the plasma was determined by immunodiffusion in agar gel by a micro-Ouchterlony method.[33] In this, the antiserum against Australia antigen is placed in the center well of the Ouchterlony pattern and the sera to be tested in the peripheral wells. The presence of the antigen is indicated by the appearance of a precipitin line which stains red with azo carmine. Human antiserum from a hemophilia patient[21] and an antiserum made by the immunization of a rabbit[23] with a serum containing Au(1) were used. The sera to be tested were scored as Au(1) if a precipitin band was seen under these experimental conditions and as Au(0) if a precipitin band was not seen. After the sera were scored for the presence or absence of antigen, results were recorded on the pedigree charts. In this manner the individuals testing the sera were not aware of the family relationship and the scoring was objective.

Results and Discussion.—The frequency of Au(1) in the 18 villages varied from 2.9 to 23.9 per cent. In order to determine if there is family clustering for the trait, IBM punch cards were prepared for each of the 1797 sera, given sequential numbers, and sorted in the IBM 1620 with a random numbers program. The cards were then examined in turn until the card of an individual who was scored as Au(1) was found. The next card of an individual of the same sex and within five years of the same age, but who was scored as Au(0), was selected as a control. The cards of the immediate family (mother, father, daughter, son, brother, sister, grandmother, grandfather, grandson, granddaughter) of the individual scored Au(1) were then examined and the frequency of Au(1) was determined. The same was done for the family of the control (Au(0)) individual. Twenty-four pairs were studied in this manner.

The significance of the results was tested by the Sign test.[34] If the frequency of Australia antigen was higher in the family of the Au(1) individual than in that of the control (Au(0)), it was scored as positive; if the reverse was true, as negative. If there was no clustering in the families of individuals with Australia antigen, then an equal number of positives and negatives would be expected; but this was not seen (Table 1). Twenty-one of the twenty-four pairs studied were positive, two were negative, and one was the same; and this is highly unlikely to be due to chance ($p < 0.001$). This difference was not due to the differences in Au(1) frequencies in the villages from which the case and controls were selected.

Inspection of the pedigrees indicated that if the trait were inherited, it could not be inherited as a simple autosomal dominant trait or as a sex-linked trait. A simple autosomal, recessive hypothesis, the same as that tested in the Cebu material,[12] was also tested with the present material. This hypothesis states

TABLE 1. *Clustering of Au(1) in families as shown by the Sign test.*

1	2	3	4	5	6	7	8	9	10
Au(1)		Au(0)			Au(1)		Au(0)		
Family members	Au(1), %	Family members	Au(1), %	Sign	Family members	Au(1), %	Family members	Au(1), %	Sign
14	14.7	7	14.7	±	8	37.5	18	16.6	+
3	66.6	11	0	+	33	9.1	24	8.3	+
13	15.4	14	0	+	12	25.0	14	14.7	+
9	22.2	8	0	+	10	10.0	19	31.6	−
12	50.0	10	10.0	+	17	29.4	29	0	+
17	41.1	22	13.5	+	21	9.5	3	0	+
10	20.0	11	0	+	21	9.5	13	0	+
5	40.0	8	37.5	+	33	21.2	13	0	+
21	10.5	8	0	+	27	25.8	16	12.5	+
14	50.0	8	0	+	9	44.4	15	0	+
11	27.2	34	20.6	+	9	11.1	13	0	+
15	6.7	10	30.0	−	17	11.7	13	0	+

The number of family members and the per cent with Au(1) in the family of Au(1) individuals are given in columns 1, 2, 6, and 7. The same data for individuals who are Au(0) are given in columns 3, 4, 8, and 9. When the percentage of Au(1) was greater in the families of the Au(1) individuals than it was in the families of the Au(0) individuals. the doublet was scored as positive: if the opposite was the case, as negative. There were 21 positives and 2 negatives: it is highly unlikely that this is due to chance ($p << 0.001$).

that in individuals homozygous for a gene designated Au^1 (i.e., genotype Au^1/Au^1) Australia antigen would be detectable by the Ouchterlony method (phenotype Au(1)). In individuals homozygous for the alternate gene (Au/Au) and heterozygotes (Au^1/Au), Australia antigen (phenotype Au(0)) would not be detectable.

Two methods of segregation analysis were used to test the genetic hypothesis. In Smith's method,[13] the Au(0) × Au(0) and Au(1) × Au(0) matings are considered separately and corrections are made for family size.

For the Au(1) × Au(0) matings the calculations of Smith's Table 3 are followed, and the results are shown here in the upper part of Table 2. If the genetic hypothesis is correct, then 40.1 recessives are expected, and 42 are seen. The variance (as defined by Smith) is 11.0, the $\chi^2 = 0.341$ and $0.7 > p > 0.5$. This is a good fit to the expected. Included in this calculation are ten families in each of which one parent was positive and the other untested. It was assumed in each case that the untested parent was Au(0). The probability of a positive by positive mating is about 1/100 in this population.

The calculations for the Au(0) × Au(0) matings, which follow those of Smith's Table 6, are given here in the lower part of Table 2. If the genetic hypothesis is correct, then 53.8 recessives (i.e., Au(1)) are expected, and 56 are seen, a very close fit ($\chi^2 = 0.412, 0.7 > p > 0.5$). Included in this calculation are eight families with at least one positive offspring in which the phenotype of one parent was Au(0) and in which the other parent was not tested. It was assumed in each case that the untested parent was Au(0). In this population the chance of an untested parent's being Au(1) is about 1 in 10; hence it is unlikely that more than one has been misclassified. If families in which neither parent was tested are also included, then the values for expected and observed recessive children are 77.8 and 82 with $\chi^2 = 1.044$. This is also a good fit. The results are summarized in Table 3.

TABLE 2. *Segregation of Au(1) in families with at least one Au(1) child.*

Number of children in family	Number of families	Recessives Observed	Recessives Expected	Variance
		Mating type: Au(1) × Au(0)		
c	m_c		$m_c a_c$	$m_c b_c$
1	6	6	6.000	0
2	5	5	6.665	1.110
3	4	8	6.856	1.960
4	6	17	12.798	4.692
5	3	6	7.743	3.246
Totals	24	42	40.062	11.008
	$\chi^2 = 0.341$	$0.7 > p > 0.5$		
		Mating type: Au(0) × Au(0)		
c	M_c		$M_c A_c$	$M_c B_c$
1	4	4	4.000	0
2	13	15	14.859	1.586
3	13	16	16.861	3.419
4	4	8	5.852	1.680
5	5	9	8.195	2.960
6	1	2	1.825	0.776
8	1	2	2.223	1.172
Totals	41	56	53.815	11.593
	$\chi^2 = 0.412$	$0.7 > p > 0.5$		

Computations by the method of Smith[13] as per Tables 3 and 6 of his paper. There was one family not included in the calculations in which the mother was positive and all her seven children were positive but serum was not obtained from the father. We are attempting to test this parent.

TABLE 3. *Summary of calculations by Smith's[13] method for Cebu and Bougainville families, showing variance and χ^2 (1 degree of freedom).*

Location	Type of mating	No. of families	Recessives Observed	Recessives Expected	Variance	χ^2
Cebu	Au(1) × Au(0)	7	12	12.189	3.556	0.010
Cebu	Au(0) × Au(0)	24	33	32.705	8.069	0.010
Bougainville	Au(1) × Au(0)	24	42	40.062	11.008	0.341
Bougainville	Au(0) × Au(0)	41	56	53.815	11.593	0.412

The segregation analysis method recently introduced by Li and Mantel[35] was also used to test the genetic hypothesis in the Au(0) × Au(0) families. In this, the total number of children t who are the issue of Au(0) × Au(0) matings, the number of recessives r (i.e., Au(1)) among these, and the number of children j who are the only recessive offspring in the family are determined. The proportion

$$p' = \frac{r - j}{t - j}$$

is 0.250 for an "ideal" recessive trait. Li and Mantel[35] give the rationale for this calculation and show that it compares favorably with other methods of segregation analyses. The results of these calculations are shown in Table 4, in which they are compared to the calculations for the Cebu data. For the Cebu data the value of p' is 0.2461 (±0.0585), a close fit, and for Bougainville the value is 0.2527 ± 0.0452, also a nice fit. Both of these results compare favorably to the values used as an example by Li and Mantel.

TABLE 4. *Summary of calculations by Li and Mantel's*[24] *method for Cebu and Bougainville Au(0) × Au(0) families.*

	t	j	r	p'	Standard error
Cebu	85	20	36	0.2461	0.0585
Bougainville	124	33	56	0.2527	0.0561

These segregation analyses taken together with the previous Cebu study strongly support the hypothesis that the trait follows simple Mendelian segregation in at least these two populations.

If the distributions in the families were due to simple infection with no genetic effect, then it would be expected that the frequency of Australia antigen in the genetically unrelated spouses of Au(1) individuals would be about the same as the frequency in the genetically related children. There were 33 families in which both parents were tested and at least one was Au(1). There were 27 Au(1) among the 97 offspring, but none of the mates were positive, and this difference is significant ($0.01 > p > 0.001$).

Although there may be some nongenetic explanation for this distribution, as suggested in our previous paper,[12] the close correspondence of the observed to the expected numbers of recessives in two quite different communities (Cebu and Bougainville) adds weight to the validity of the genetic hypothesis.

The data summarized in the introduction indicate that Australia antigen may also behave as an infectious agent. These findings are compatible with the explanation that there is an inherited susceptibility to chronic infection with Australia antigen mediated by the Au^1 gene. In populations where the Au^1 gene is relatively common and where the infectious agent is also common, then all those individuals susceptible to the agent would become infected and the segregation in the families would appear to follow a pattern of autosomal recessive inheritance.

By analogy with other examples of inherited susceptibility to infection, there are probably factors such as age, sex, state of nutrition, etc. which also affect susceptibility and resistance. Under appropriate circumstances an individual without the inherited susceptibility factors could become infected. Most patients in the United States who contract acute viral hepatitis (usually post-transfusion hepatitis) may have the antigen transiently, but it is only rarely chronic, since they are not genetically susceptible to the chronic infection. In these patients, the presence of the antigen is usually associated with evidence of clinical hepatitis and/or highly elevated serum glutamic pyruvate transaminase levels. In individuals with chronic Australia antigen, such as the large numbers of apparently normal individuals in Asia and Oceania, the effect on the liver is much less striking and may be reflected only in minor elevations of serum glutamic pyruvate transaminase, or in no changes at all. A symbiotic accommodation of the organism with the host could have developed in these individuals over the course of many generations, and even though they become infected more readily, they do not develop severe symptoms of the disease. This may represent a protective value of the Au^1 gene and could be associated with the development of what appears to be a genetic polymorphism in this population.[36]

The data presented here are compatible with the hypothesis that there is an inherited susceptibility in humans to chronic infection with a virus of hepatitis. The susceptibility, in this analysis, appears to be controlled at a single autosomal locus, and individuals homozygous for a recessive gene termed Au^1 are more likely to be chronically infected. The infection is not accompanied by any apparent illness, although people with Australia antigen are probably hepatitis carriers.

* Supported in part by USPHS grants CA-06551 and CA-08069 from the National Cancer Institute, by a grant from the World Health Organization, and by an appropriation from the Commonwealth of Pennsylvania. Dr. Friedlaender was the recipient of a N.S.F. graduate fellowship and N.S.F. Dissertation Improvement Award.

[1] Sabin, A. B., in Hooker, D., and C. C. Hare, *Genetics and Inheritance of Integrated Neurological and Psychiatric patterns* (Baltimore: Williams and Wilkens, 1954).

[2] Gowen, J. W., *Ann. N.Y. Acad. Sci.*, **91,** 689 (1961).

[3] Chang, S., and W. H. Hildemann, *J. Natl. Cancer Inst.*, **33,** 303 (1964).

[4] Tahkola, M., *J. Natl. Cancer Inst.*, **35,** 595 (1965).

[5] Heston, W. E., and G. Vlahakis, in *Carcinogenesis: a Broad Critique* (Baltimore: Williams and Wilkens, 1967), p. 347.

[6] Tennant, J. R., and G. D. Snell, *Natl. Cancer Inst. Monograph*, **22,** 61 (1966).

[7] Waters, N. F., and B. R. Burmeister, *J. Natl. Cancer Inst.*, **27,** 655 (1961).

[8] Gross, L., in *Oncogenic Viruses* (New York: Pergamon Press, 1961), p. 76.

[9] Motulsky, A. G., *Human Biol.*, **32,** 28 (1960).

[10] Cox, R., and C. Mac Leod, in Burdette, W. J., *Methodology in Human Genetics* (San Francisco: Holden-Day, Inc., 1962), p. 156.

[11] Blumberg, B. S., *Arch. Environ. Health*, **3,** 612 (1961).

[12] Blumberg, B. S., Liisa Melartin, R. A. Guinto, and Barbara Werner, *Am. J. Human. Genet.*, **18,** 594 (1966).

[13] Smith, C. A. B., *Ann. Human Genet.*, **20,** 257 (1956).

[14] Blumberg, B. S., Betty Jane S. Gerstley, D. A. Hungerford, W. T. London, and A. I. Sutnick, *Ann. Int. Med.*, **66,** 924 (1967).

[15] Sutnick, A. I., W. T. London, and B. S. Blumberg, *J. Clin. Invest.*, **46,** 1122 (1967).

[16] Blumberg, B. S., *Tokyo J. Med. Sci.*, **76,** 1 (1968).

[17] Blumberg, B. S., A. I. Sutnick, and W. T. London, *Bull. N.Y. Acad. Med.*, **44,** 1566 (1968).

[18] Allison, A. C., and B. S. Blumberg, *Lancet*, **1,** 634 (1961).

[19] Blumberg, B. S., S. Dray, and J. C. Robinson, *Nature*, **194,** 656 (1962).

[20] Blumberg, B. S., *Bull. N.Y. Acad. Med.*, **40,** 377 (1964).

[21] Blumberg, B. S., H. J. Alter, and S. Visnich, *J. Am. Med. Assoc.*, **191** (1965).

[22] Okochi, K., and S. Murakami, *Vox Sanguinis*, in press.

[23] Melartin, Liisa, and B. S. Blumberg, *Nature*, **210,** 1340 (1966).

[24] Prince, A. M., these PROCEEDINGS, **60,** 814 (1968).

[25] Prince, A. M., *Lancet*, **3,** 462 (1968).

[26] Vierucci, A., manuscript in preparation.

[27] Blumberg, B. S., *J. Clin. Invest.*, **45,** 988 (1966).

[28] Sutnick, A. I., W. T. London, and B. S. Blumberg, *J. Clin. Invest.*, **46,** 1122 (1967).

[29] Sutnick, A. I., W. T. London, B. S. Blumberg, and M. D. Cronlund, *J. Am. Med. Assoc.*, **205,** 670 (1968).

[30] London, W. T., Marion Di Figlia, A. I. Sutnick, J. Ziegenfuss, and B. S. Blumberg, *Clin. Res.*, **16,** 547 (1968).

[31] Bayer, M. E., B. S. Blumberg, and Barbara Werner, *Nature*, **218,** 1057 (1968).

[32] Friedlaender, J. S., "*Microevolution in south central Bougainville Island, Territory of New Guinea,*" Ph.D. thesis, Harvard University (1968).

[33] Blumberg, B. S., and Nancy M. Riddell, *J. Clin. Invest.*, **42,** 867 (1963).

[34] Dixon, W. J., and F. J. Massey, Jr., *Introduction to Statistical Analysis* (New York: McGraw-Hill, 1957), p. 448.

[35] Li, C. C., and N. Mantel, *Am. J. Human Genet.*, **20,** 61 (1968).

[36] Ford, E. B., *Ecological Genetics* (London: Methuen and Co., 1964), p. 84.

The Natural History of Australia Antigen

B. S. Blumberg

The Institute for Cancer Research, Fox Chase, Philadelphia, Pennsylvania, U.S.A.

Since our discovery of Australia antigen (Au) (*4*) and its relation to hepatitis (*5*) much of the work in this field has been focused on this disease. In this paper the relation of Au to human biology (particularly genetics and maternal effects) and to the possible cause (or causes) of cancer will be emphasized. Au is, in some situations, associated with leukemia, hepatoma, and possibly other cancers, and further knowledge of the nature of Au may help us to understand the nature of agents which may cause these and other cancers.

In the earliest papers on Au it was suggested that it might be a virus. In 1965 (*4*) we said, " . . . the Au is related to the virus which has been suggested as the cause of leukemia," and the following year (*6*) (in reference to hepatitis) " . . . a virus or other infective agent could lead to the presence of Au, perhaps due to a direct effect on the genome."

When the association with hepatitis was reported, we immediately set about testing the hypothesis that Au was the virus which causes " viral hepatitis," and during the next 4 years accumulated a body of evidence, which for the most part supported this view (*7*). There were, however, some features of Au which were not typical of a virus although none of these actually ruled out the hypothesis. At the same time we continued to test the notion stimulated by our earlier work on the lipoprotein "Ag" system (*1*) that Au had the properties of an inherited serum constituent, a protein polymorphism. None of the accumulated data has ruled out this hypothesis either. At first consideration it might appear that these two hypotheses are mutually exclusive; but there is, in fact, no reason to assume that this is so. Hence, since neither of the hypotheses had been ruled out they were combined to give a third hypothesis (definition) which is, in effect, a restatement of the properties of the agent. Au is associated with an infectious agent which causes (some types of) hepatitis in man and has the properties of a serum protein polymorphism.

An extension of this view states that there are additional agents, related to other diseases, which have properties similar to Au. In order to designate this group of agents we have suggested the term " Icron," an achronym on the name of the Institute for Cancer Research. The statement of this hypothesis does not imply that the preceding ones are ruled out; a considerable amount of energy and effort is still being spent testing the virus hypothesis. For example, it is possible that Au is not the infectious agent causing hepatitis, but that there is another causative agent (perhaps an extremely small one) which has not yet been discovered but which has many of the properties of Au. However, testing the " Icron " hypothesis has the advantage of encouraging studies of characteristics not included within the rubric of a conventional virus.

The Nature of Australia Antigen

On the basis of studies, initially from our laboratory and subsequently others, a picture is beginning to emerge of the nature of Au. Some of these features are consistent with the definition of a virus, others appear to be atypical, but none rule out the possibility that Au acts as an infectious agent causing (some) hepatitis in man (*14*). In this section, we will attempt to summarize these results in the form of a general description of the properties of the agent.

Infectious agents are always associated with the disease they cause, although the reverse is not necessarily true, *i.e.*, infectious agents may be associated with diseases they do not cause, and in addition the agent may be found in apparently normal individuals who do not have the disease (or diseases) caused by the agent. Au is found in higher frequency in patients with acute viral hepatitis (both " infectious " and " serum ") and in some form of chronic hepatitis. It is also found in the blood of many individuals who are not apparently ill and who appear to be asymptomatic carriers. The frequency of carriers varies from very low values (about 0.1% in North American and North European communities) up to about 20% in some Oceanic, Asian, and tropical communities. In this respect its distribution is similar to that of a serum protein polymorphism whose frequency may vary from population to population, presumably depending on local ecological and selective factors. There is a striking difference in host reaction to " infection " with Au, which has been discussed in detail elsewhere (*8*, *14*). The same agent which has had no noticeable effect on some individuals, the carriers, will cause acute, chronic, and sometimes fatal illness in others. One man's meat is another's poison; " normal flora " in a carrier is a pathogen in another person. This dichotomy between carriers and affected individuals holds true for many, conceivably most, infectious agents.

Australia Affinity Group

Au is also associated with a group of other disease, that is, it occurs in higher frequency in these diseases than in controls. These diseases, termed the "Australia affinity " group are characterized by impairments in their immune mechanism. They include the lymphocytic leukemias, Down's syndrome, lepromatous leprosy (as opposed to tuberculoid leprosy) chronic renal disease and hepatoma.

In some of these (for example, Down's syndrome) the presence of Au is accompanied, in general, by chronic anicteric hepatitis. It is possible that Au is related to the etiology of some of these diseases (*i.e.*, hepatoma). Another possibility is that individuals who are susceptible to persistent infections with the agents which are responsible for these illnesses are also susceptible to persistent infection with Au; that is, these agents have an affinity to each other in that they have an affinity to a common susceptibility factor (this susceptibility factor may be inherited (see below)). If this is true they may also have other features in common and, as a consequence, knowledge about Au may help to identify and understand the mode of action of other agents related to the "Australia affinity" diseases.

For example, there is an increased frequency of Au in patients with lymphocytic leukemia who have received transfusions of donor blood, some of which contains Au. The frequency is higher in the transfused leukemia patients than in other patients not in the "Australia affinity group" who have also received transfusions. The leukemia patients, when exposed are more likely to develop a persistent infection with Au. The agent responsible for lymphocytic leukemia may have other characteristics of Au, and these characteristics can be used as a basis for formulating hypotheses about leukemia and its etiologic agent, *i.e.*, the sex distribution of lymphocytic leukemia is similar to that of Au (M>F); family distribution is similar, age distribution similar, *etc.*, and these hypotheses can be tested. The same predictions relating to sex, age, genetics, maternal effect which have been formulated from the experience with Au can also be made for hepatoma and other members of the affinity group (*7*).

Several years ago we isolated Au and examined it under the electron microscope (*3*), and this has now been done by many investigators. Its appearance is compatible with that of a small virus; it is 200Å in diameter and has fine structure detail. Larger units (400Å) may also be seen, but whether these are antigenically related to Au is still unclear. An unusual feature, in respect of the viral hypothesis, is that there is doubt as to the existence of nucleic acid in these particles. Joswiak *et al.* (*20*) using a starch block method for isolation, reported that Au contained about 5% RNA. This was subsequently confirmed by Millman *et al.* (*28*), but they were unable to show (in a chimpanzee) the incorporation of tritiated uridine into the Au. This raises the possibility that the detected RNA may derive from the host and not the infectious agent. Hence, it is possible that the material identified as Au does not contain nucleic acid or, if it does, the amount present is very small. This may mean that another particle containing the appropriate amount of nucleic acid and responsible for reproductivity and infection does exist; so far this has not been found. It also may mean that Au is the infectious particle and the means of its reproduction and infectivity are not now clear.

The ability to grow in tissue culture is a characteristic of many viruses. We have shown that liver tissue taken from a patient with hepatitis who has Au in his blood and liver appears to produce Au in the cells and tissue fluid following many passages of the tissue. Zuckerman has shown that Au may be grown in fetal liver tissue. However, these experiments appear to be difficult to do and have not, to my knowledge, been confirmed. This again scores against, but does not rule out the viral hypothesis. Some viruses, apparently, have not been grown on tissue culture, or, can be grown only with difficulty.

Cellular Localization of Australia Antigen

Infectious agents which cause disease might be expected to be found within the cells which are affected in the illness; Au fulfills this criterion. In the initial studies using fluorescent anti-Au we found material with the antigenic specificity of Au in the liver cells of patients with Au in their blood and/or with hepatitis (*29*). This has now been confirmed many times (see *e.g.*, Nowoslawski *et al.* (*33*) and Hadziyannis *et al.* (*18*)). The fluorescence has been reported to be present in the nucleus in some reports, in the cytoplasm in others. These differences could be due to differences in technique, the kinds of disease studied or perhaps the times after infection at which the cells were sampled. The presence of Au in the nucleus has been confirmed by electron microscope studies using ferritin labeled anti-Au (*19*). These findings are compatible with the notion that the agent can spend some of its time in the nucleus, some in the cytoplasm; that it can don nuclear or cellular coats and presumably that it can interact with the cellular genetic material of the host.

Transmission of Australia Antigen to Nonhuman Primates

An important criterion for judging an agent to be or not to be infectious is its transmission to an experimental animal. In order to find a suitable animal for such studies we initially surveyed many animal species for the natural occurrence of Au and found it only in certain primate species (*9*). London was then able to transmit isolated and partially purified Au to infant African green monkeys and passage the material twice (*22*). The amount of antigen present in the final monkey was greater than would have been expected by simple dilution. The possibility of transmission to primates was in part confirmed some years later by London and his colleagues (*23*). (This second London is not the same person who reported the original primate studies.) They were able to transmit to rhesus monkeys a material which resulted in the production of antibodies to Au. The primate studies, therefore, imply that Au can be transmitted to experimental animals in the manner expected of an infectious agent.

All the studies taken together do not rule out the possibility that Au is an infectious agent with many similarities to a virus, but with some atypical features.

Australia Antigen as a Blood Polymorphism

During the period in which we were testing the infectious agent hypothesis we continued to study Au as a serum protein polymorphism. I use the term blood protein polymorphism in the sense introduced by Ford, E. B.. It means that there are two (or more) inherited forms of a protein; and the frequency of the least common form is so large that it could not be maintained in the population by recurrent mutation alone. Examples of polymorphism are the sickle cell hemoglobin trait, the red blood cell antigens, serum protein polymorphisms such as haptoglobins, gamma globulins (Gm, InV) low density lipoproteins (Ag) and others. Implied in the definition of polymorphism is the concept that there is differential selection in respect of the different phenotypes, and this may be related to differential mortality and fertility.

Australia Antigen Viewed as a Serum Protein

In our initial isolations of Au (*2*), we were able to show that, although it was clearly distinct from serum low density lipoprotein, it had many of its characteristics. Later Millman *et al.* (*30, 31*) isolated Au from blood and purified it to the extent that it was not possible to detect any serum proteins using immunologic methods. When this apparently purified material is treated with Tween 80 it partially disassociates into soluble components. The products released in this manner include what appear to be serum proteins including gamma globulin (IgG), (both heavy and light chain) complement, β-lipoprotein, albumin and possibly others. These findings have not been confirmed, although studies by Pert and his colleagues appear to be consistent with this result. Further, Sukeno *et al.* using optical rotary dispersion and circular dichroism methods found that Au has characteristics that are more similar to human lipoproteins than to a normal virus (*37*). (As we have seen above it is not clear if Au contains RNA). Hence, Au apparently is made up of serum components which have antigenic characteristics related to those of the host, many of which may be occult under most circumstances, but may be revealed in others.

Amount of Australia Antigen in the Blood

During the course of hepatitis the amount of Au present in the blood may vary greatly. In acute viral hepatitis the antigen level may be high before there is any other evidence of disease and then decrease to lower levels as the symptoms of the disease develop. In chronic anicteric hepatitis the levels may vary over the course of the weeks, months and years that the disease persists. This is shown in Fig. 1, which incidentally is of historic interest (*38*). It shows the course of the illness, in terms of Au and SGPT levels, in one of the first patients (J. B.) in whom the association between Au and hepatitis was detected (in 1966). The amount of Au present is not directly related to the severity of the hepatitis, as expressed by the levels of SGPT.

This disparity between amounts of Au detected in the blood, and the severity of illness may also be seen in Table 1, taken from the paper by Mazzur *et al.* (*26*). The first column designates the population or disease group studied. These include patients with Down's syndrome with anicteric hepatitis associated with Au, the apparently normal Lau population of Malaita in the Solomon Islands, patients with leprosy, patients with acute hepatitis and apparently normal U. S. blood donors who are asymptomatic carriers of Au. The second column indicates the median

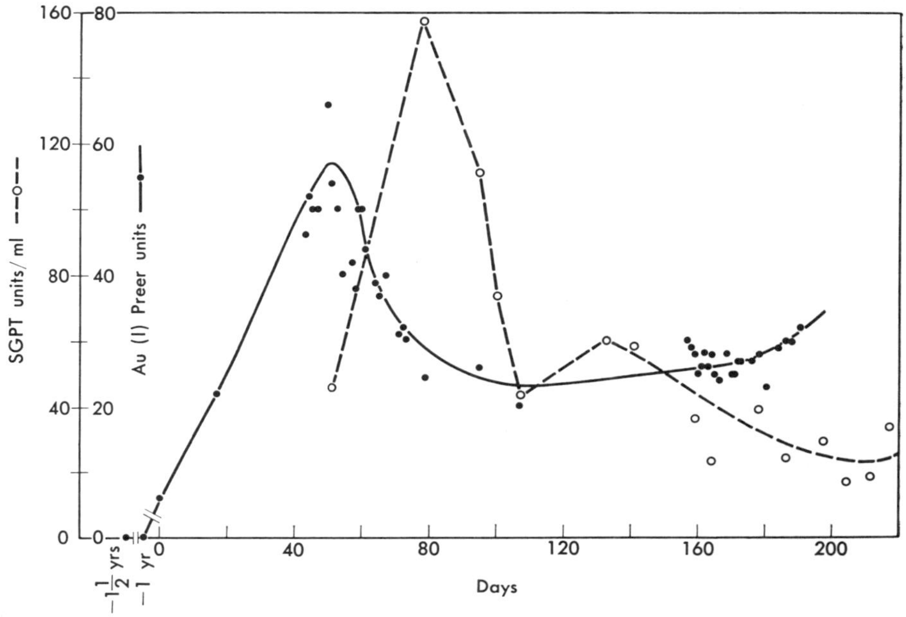

FIG. 1. The levels of Au and SGPT in a patient (J. B.) with Down's syndrome and chronic anicteric hepatitis. The levels of Au vary considerably with time and do not have a simple relation to the activity of the disease, measured by SGPT levels. (Adopted from Sutnick *et al.* (*38*))

TABLE 1. Comparison of Au Levels in Several Populations (Complement-fixation test)

	Median	Down's syndrome	Lau	Leprosy	Hepatitis
Down's syndrome	1 : 512	—			
Lau	1 : 256	0.151	—		
Leprosy	1 : 128	0.010	0.326	—	
Hepatitis	1 : 128	0.005	0.048	0.378	—
Blood donors	1 : 32	0.000003	0.00003	0.002	0.014

For explanation see text. (Mazzur *et al.* (*26*))

levels of Au in the group (measured by complement fixation) and the other columns the p values (Mann-Whitney test) for the comparisons between the groups. The highest levels are found in individuals with a nearly occult form of hepatitis (Down's syndrome) and in apparently normal people; whereas the Au level in the acutely ill hepatitis patients is nearly the lowest. This is somewhat unusual for an infectious agent; in general it would be expected that the amount of the infectious agent in the blood would be roughly related to the severity of the illness. It may be interesting to view the varying blood levels of Au as a response of the host to the infection, that is the infectious etiologic agent and the host response to it involve the same materials or they have some measurable characteristic(s) in common.

Genetics of Australia Antigen

We have repeatedly emphasized the striking differences in the response of putative hosts exposed to infection with Au (*8*), and very early in the work proposed the hypothesis that persistent presence of Au (as in the asymptomatic carriers) is inherited as a simple autosomal recessive trait (*4*). We have published several studies which have supported this model (*10*), and it has been independently confirmed (The data has recently been summarized (*11*)). These data are consistent with the view that there is a gene (designated Au^1) which, when present in double dose (genotype Au^1/Au^1), renders the bearer more susceptible to persistent infection with the infectious agent associated with Au, if the bearer is appropriately exposed to infection. Many of these individuals will be asymptomatic; they are included among the large numbers of carriers that can occur in high frequency in parts of the world. In these populations the susceptibility gene actually represents an advantage to the host in that the agent assumes the role of normal flora. Hosts with the alternate genotypes, however, may respond by the development of acute hepatitis, by the formation of antibody, or one may conjecture by the development of " complex disease " if both antigen and antibody coexist.

The genetic concepts may also be applied to the "Australia affinity " group of diseases, discussed above. According to this view, individuals of the Au^1/Au^1 genotype are also susceptible to persistent infection with the postulated organism responsible for some of the affinity diseases, for example, lymphocytic leukemia. In this case, persistent infection would not be benign (as it may be with Au) but associated with clinical disease. The combination of the appropriate genotype(s) (modified by factors such as age and sex) plus infection with the appropriate agent would result in disease. For the leukemia agent infection of individuals with the alternate genotypes (Au^1/Au, Au/Au) would result in an acute infection and some kind of relatively benign or conceivably asymptomatic condition.

The existence of genetic control implies that there are inherited differences between people which lead to variation in the consequences of infection. We are now examining the proposition that inherited antigenic differences in the hosts which can interact with the antigenic specificities of the Au agent account for these differences. This interaction could take place at one or more locations, *i.e.*, tissue cell, formed elements of the blood, the serum proteins. That is, if person A is infected with Au particles which contain constituents that are antigenically very similar to his own then he will have little immunological response (he will recognize these as " self ") and will tend to develop a persistent infection with Au. On the other hand, if the constituents of the agent are antigenically different from his, he will develop an immune response to the agent (anti-Au and possibly other antibodies) and will have a transient infection. One can also picture an intermediate situation in which some of the host/agent specificities are shared, and some are not. The result could then be intermediate; some degree of infection will occur, but antibodies will also develop and this could lead to antigen/antibody complexes and, possibly, another form of disease. During the course of infection in person A, new Au (1) particles would be synthesized which contain the proteins and lipids of A. He in turn can infect person B and the same alternatives present themselves. If the relevant proteins of B are antigenically similar to the proteins contained in A's Au then B will develop a persistent infection; if different he will develop a transient infection or develop antibody. These considerations lead to interesting kinds of epidemiologic studies. The spread of disease is dependent not only on the antigenic makeup of the agent (*i.e.*, the " Icron ") and on the genetically determined makeup of the host, but also on the genetics of the human or other hosts from which the infection came.

Relation of Australia Antigen to Gamma Globulin Polymorphism

During the past year we have started to investigate the specific serum proteins which may be involved in this interaction of antigenic specificities and polymorphisms and the first among these is gamma globulin (Gm). This protein appears to be one of the constituents of Au (*31*). Patients with thalassemia who receive large numbers of transfusions will often develop antibodies against IgG (anti-Gm antibodies) as well as antibodies against Au. In an extensive study of Italian thalassemia patients (*12*, *39*) these two antibodies coexist in the same patient more often than expected by chance (Table 2). This would happen if there were an interaction between the Gm polymorphism and the Au polymorphism. Gm heterozygotes would be more likely to have a specificity in common with an infecting organism and therefore less likely, on average, to form Au antibody. In the thalassemia patients the frequency of occurrence of any of the Gm specificities is greater in the patients who were persistent Au carriers than in those who have developed antibody; a finding consistent with the Gm heterozygote hypothesis.

Another study which appears to confirm this association has been published by

TABLE 2. Anti-Gm and Anti-Au in Thalassemia Patients (Adopted from London *et al.* (*25*))

	Anti-Gm		
	Present	Absent	% present
Anti-Au	22	8	73.3
Au	35	21	46.1
Neither	5	12	29.4

"Anti-Au " indicates the presence of antibody, "Au " the presence of antigen and " neither " that neither of these were found. The differences between anti-Au and Au; and between anti-Au and neither are both statistically significant ($p = 0.02$; $p = 0.009$).

TABLE 3. Anti-Gm and Anti-Au in Drug Addicts

	Anti-Gm		
	Present	Absent	% present
Anti-Au present	8	10	44.4
Anti-Au absent	15	70	17.6

The association between anti-Au and anti-Gm is significant ($p = 0.02$).

TABLE 4. Comparison of the Frequency of Au to the Ratio (R) of Gm Heterozygotes to Gm Homozygotes in Several Communities in Bougainville and Malaita, Solomon Islands (Adopted from Blumberg *et al.*(*13*))

Population	Au		R = hetero/homo		
	N	% pos.	N	Gm	InV
Nasioi	169	1.8	161	0.074	—
Kwaio	379	4.7	451	0.121	—
Baegu	445	16.6	147	0.518	0.55
Aita	334	19.8	307	1.040	—
Lau	374	24.9	143	0.740	0.72

The Gm frequencies published by Steinberg *et al.* (*36*) were used and the definitions of heterozygote and homozygote were derived from the " phenogroups " determined in this study.

Nathenson, Millman, and their colleagues (*32*). A group of drug addicts (who used improperly cleaned needles) had high frequencies of both anti-Au and anti-Gm; and there is a significant co-occurrence of these antibodies in the same person (Table 3).

If this finding on the relation of Au and Gm heterozygotes, and anti-Au and Gm homozygotes can be supported in additional studies it represents an important connection between these two polymorphisms. It gives one genetic basis for the development of anti-Au (*i.e.*, it is related to the genetics of Gm) and further suggests that the presence of anti-Gm in many individnals is due to infection with Au and similar organisms.

Another prediction can be made from the association of Gm heterozygotes with persistent infection and Gm homozygotes with the development of antibody. The frequency of persistent Au would be directly, and perhaps linearly, related to the Gm heterozygote/homozygote ratio in communities in which Au is common. This was tested in several locations on the islands of Bougainville and Malaita in the Solomon Islands (*13*) (Table 4). In this study the prediction is fulfilled and there is a (nearly) linear relation between the heterozygote/homozygote ratio for Gm and InV and the frequency of persistent Au (A similar close relation was not found using Gm data from an earlier study in the Solomon Islands. However, in this study, Gm 21 was not distinguished and as a consequence many individuals scored as homozygotes would have been scored as heterozygotes if Gm 21 had been determined).

If this finding can be confirmed in other populations then it implies that the Gm frequencies are, in some regions, related to the frequency of occurrence of an infectious agent, Au, and/or the gene which determines the frequency of its persistent infection.

Modifications of the Genetic Hypothesis—Sex and Age

In the initial genetic studies there was a good fit between expected and observed ratios. Despite this, in the very first study it was clear that effects other than genetic had a bearing on the presence of persistent Au (*6*). As is often the case, deviations from a simple hypothesis may be the most interesting aspects of a phenomenon.

There is a striking sex difference in the frequency of Au. In every disease group we studied there is a higher frequency of Au in males than in females.

TABLE 5. Sex Ratio (Male/Female) of Au in a) Patients with Various Diseases and b) Apparently Normal Nonhospitalized Populations (Immunodiffusion method)

	Male			Female			Ratio	p
	N	Au(1)	%Au(1)	N	Au(1)	%Au(1)		
a) Disease								
Acute viral hepatitis	112	40	35.7	94	18	19.1	1.87	0.013
Chronic hepatitis	12	3	25	19	0	0	—[a]	0.095
Down's syndrome	212	67	31.6	92	17	18.5	1.71	0.027
Other mentally retarded	145	5	3.4	43	1	2.3	1.48	0.900
Leukemia	173	22	12.7	124	11	8.9	1.43	0.394
Lepromatous leprosy (Cebu)	396	45	11.4	188	10	5.3	2.14	0.029
Lepromatous leprosy (India)	451	32	7.1	101	2	2.0	3.58	0.088
Tuberculoid leprosy (Cebu)	343	19	5.5	262	7	2.7	2.07	0.128
Tuberculoid leprosy (India)	281	8	2.8	103	0	0	—[a]	0.184
Chronic dialysis (renal disease)	5	5	100	4	3	75	1.33	0.906
$\chi^2 = 42.3270$ $p = 0.0025$								
b) Population								
Marshall Island	243	19	7.8	226	14	6.2	1.26	0.612
Cebu, Philippines	678	47	6.9	609	17	2.8	2.48	0.001
Cashinahua	45	10	22.2	44	6	13.6	1.63	0.436
Manila	138	6	4.3	59	3	5.1	0.855	0.884
Polynesia	597	8	1.3	550	3	0.54	2.46	0.282
New Hebrides	597	61	10.2	394	19	4.8	2.12	0.003
New Caledonia	518	13	2.5	444	7	1.6	1.59	0.433
India (South)	116	5	4.3	135	0	0	—[a]	0.047
British Solomon Island	127	7	5.5	160	4	2.5	2.20	0.312
New Guniea	532	16	3.0	287	5	1.7	1.73	0.389
Micronesia	330	13	3.9	292	6	2.1	1.92	0.259
Surinam	310	7	2.3	315	5	1.6	1.45	0.729
Bougainville	931	112	12.0	906	97	10.7	1.12	0.412
$\chi^2 = 47.6840$ $p = 0.0058$								

[a] Au(1) only detected in males.

This is also (nearly) true for all the carrier populations tested (Table 5), and this has been confirmed in studies from other laboratories. There are some interesting observations on sex which may be related to this finding. In the affinity diseases (*i.e.*, lymphocytic leukemias, Hodgkin's disease, lepromatous leprosy, Down's syndrome) there is a similar predominance of males. Further, the Au agent appears to involve the testes. We have identified Au (by immunofluorescence) in an autopsy specimen of a human testis (*16*). Leukemia involvement of the testes is apparently more common than of the ovaries. Male Down's syndrome patients appear to be sterile, while females are not. Recently my colleague, Dr. London, has studied a 22-year-old mentally retarded boy who was a chronic carrier of Au and subsequently developed a seminoma, a malignant tumor of the testes. An association of Au with testicular tumors should be searched for.

There is a higher frequency of Au in younger males than older, except that the frequency appears to be lower in very young people. In the populations in which we have been able to obtain extensive data there is a peak in frequency in

TABLE 6. Mortality Rate in Thalassemic Patients Observed over 5-Year-Period

Patients	Number	Dead	%	Median age at death (yr.)	Age of death range (yr.)
With Au(1)	18	4	22.2	5.0	(4–6)
With anti-Au(1)	35	4	11.4	10.5	(8–16)
Without Au(1) or anti-Au(1) [Au(0)]	116	11	9.5	7.0	(1–12)

	p (Mann-Whitney)
Au(1) *vs.* anti-Au(1)	0.014
Au(1) *vs.* Au(0)	0.114
Anti-Au(1) *vs.* Au(0)	0.086

the 6–9-year-age group. This change with age is more common in some populations than others. In the genetic study of Sardinians by Ceppellini and his colleagues (*15*), it was necessary to make a correction for penetrance in the older individual in order to have an adequate fit of the observed segregation from that predicted from the genetic model, whereas this was not necessary in the Cebu and Bougainville studies.

What is the significance of this decrease in frequency with age? It could signify that there is a selection against individuals with Au who therefore die before reaching older age. There is no direct evidence for this. Some data is available from the thalassemia study of Vierucci and his colleagues (Table 6). (These findings are, of course, not directly applicable to the asymptomatic individuals identified in the population studies since they apply to severely ill patients receiving large numbers of transfusions.) There is a significantly increased frequency of early deaths among the individuals who earlier developed Au in their blood as compared to those who developed anti-Au ($p=0.0035$, Mann-Whitney).

It is also possible that the amount of Au in the blood decreases with age until it reaches levels undetectable by immunodiffusion or techniques of similar sensitivity. In the population of Lau (on the island Malaita) (*13*) we found a decrease in frequency of Au whith age using the immunodiffusion method. However, using the more sensitive radioimmunoassay (RIA) method the striking decrease with age does not occur. This could indicate that small amounts of Au are present in the older people which, although not detectable by insensitive methods, can be found by the much more sensitive RIA.

Shulman, his colleagues, and others have shown that the presence of anticomplementary (AC) activity in populations in which Au is common may signify the formation of Australia-antigen-antibody complexes, although there may be many other reasons for the existence of AC activity. In the Lau population there is an increase of AC activity with age, and this could mean that the Au is more often present in the form of complex in later life and is not, therefore detected by immunodiffusion. This is rather thin evidence, but, nevertheless, sufficient to make the hypothesis that the Au, anti Au complex occurs more commonly in older than younger people, which is consistent with what is known for other complexes (*i.e.*, rheumatoid factor).

Maternal Effect

One of the potentially most interesting studies on Au concerns the maternal effects since this bears a similarity to the transmission of certain tumors and leukemias in lower animals. In the estimation of the genetic segregation ratios, the matings in which one parent is positive (*i.e.*, has persistent Au) are considered separately from the negative by negative matings. In the initial calculations using these matings the families in which the mother was positive and the father negative were considered together with matings in which the father was positive and the mother negative, and these fit the segregation ratios well. In Table 7 these matings are considered separately. When the mother is positive and the father negative many more of the offspring have Au than when the opposite is true (*i.e.*, father positive and mother negative). This is statistically significant in several of the populations and highly significant in the total study.

In another study Mazzur and Blumberg (*27*) used data on the complex specificities of Au to investigate the family distribution of Au. In our initial genetic studies (see above) it was first shown that there is a family aggregation of Au. Family aggregation is characteristic of an inherited trait, but it is also characteristic of an infectious one. In the genetic studies it was shown that the family data were compatible with a simple genetic model. In addition to this it would be useful to attempt to rule out simple infectious transmission as the cause of the family aggregation. If infectious transmission were the cause of the family aggregation then it would be expected that all members of the family group would have the same "strain" of the organism, *i.e.*, Au with the same specificities. It was possible to test this hypothesis using the families which had been used for the Bougainville genetic study. In 11 of the 32 families tested there were mixed subtypes. This finding appears to rule against the simple family infection explanation.

When the data were examined by the sex of the positive parent in matings in

TABLE 7. Maternal Effect

Location	Father positive				Mother positive				p
	Number of families	Children			Number of families	Children			
		N	Au(1)	Au(0)		N	Au(1)	Au(0)	
Cebu, *etc.* (*6*)	7	20	5	15	2	7	6	1	0.87×10^{-2}
Bougainville (*10*)	28	68	8	61	36	86	52	34	0.17×10^{-9}
Turin (*15*)	7	26	9	17	2	7	5	2	0.09
Lau, Malaita (*13*)	6	17	5	12	9	29	16	13	0.08
Baegu, Malaita (*13*)	2	3	1	2	11	32	14	18	0.61
Total	50	135	28	107	59	161	93	68	0.55×10^{-10}

Number of children with Australia antigen [Au(1)] and without Australia antigen [Au(0)] in matings in which at least one parent is Au(1). The matings in which father is Au(1) and mother Au(0) are shown under "Father positive" and the alternative mating under "Mother positive." In every population the Au(1) mother has a larger number of Au(1) children than the Au(0) mother. (Adopted from Blumberg *et al.* (*11*)) which one of the parents was positive an interesting pattern was seen. In families in which the mother was positive the children (nearly) all had the same specificities as the parent, while this was not true when the father was positive or neither parent was positive. This suggests (as does the data in Table 7) that transmission can occur directly from the mother to the offspring either *in utero*, early in life, or later. It would not appear to rule out the genetic hypothesis, but rather suggests that both forms of vertical transmission may occur, *i.e.*, 1) persistent infection (or transmission) controlled by a gene segregation as a simple Mendelian trait and 2) maternal transmission.

There is now considerable information on other aspects of transmission from mothers to their offspring. London *et al.* (*24*) originally found that a mother who had Au at birth did not transmit the antigen to her offspring immediately, that is, it was not found in the cord blood or in blood examined a few days immediately after birth. It is known that if a mother has hepatitis at the time of birth it is highly likely that her offspring will develop persistent Au within weeks or months. In a detailed study of pregnant mothers who were asymptomatic carriers of Au, Okochi and his co-workers (*35*) demonstrated, using a sensitive detection method, that Au was found in the child's blood as early as one month after delivery. A particularly important finding, again from the Japanese workers (*34*), was the demonstration of a maternal influence in families in which there were clustering of hepatoma, chronic liver disease and Au carriers.

We are continuing to look for other biological effects in mothers related to the presence of Au. Two recent observations may be of particular importance in that they may have detected an effect on fertility related to the presence of Au; and this could have a bearing on differential selection in relation to the Au polymorphism concept.

Kukowski *et al.* (*21*) in a study in northern India found that the gestation time of mothers with Au is increased and that they are younger compared to mothers who do not have antigen in their blood (Table 8). It will now be necessary to test this hypothesis in other populations. If supported, it indicates that there is a measurable effect of Au on the biology of parturition.

In another study of (mostly) completed families in Bougainville, we found that

TABLE 8. Comparison of Pregnancy and Progeny Characteristics of Mothers with and without Au

Group I	Group II	Criterion	N_1	N_2	U	Z	p
Au(1)	Au(0)	Gestation time	7	198	395.0	−1.9651	0.025 longer
		Mother's age	7	788	1,301.5	−2.4158	0.0079 younger
		Mother's gravida	7	737	1,911.5	−1.225	0.110
		Baby length	6	693	1,996.5	−0.171	0.864
		Baby head circ.	6	679	1,363	−1.445	0.148
		Baby weight	6	697	1,772	−0.642	0.521

N_1 and N_2 are, respectively, the number of Au(1) and Au(0) mothers. U is the Mann-Whitney Rank Sum statistic, and Z is the expression of this statistic as a standard normal deviation. The significance level, p, derives from the value of Z. The gestation time is significantly larger in the Au mothers and the age significantly younger than the mothers without Au.

in matings in which the mothers had Au and the father did not, there were significantly fewer children than in the matings in which the father had Au, and the mother did not. If this observation is supported in further studies, it could indicate that there is an overall decrease in progeny when the mother is a carrier.

Modes of Transmission of Australia Antigen—Similarity of Tumor Viruses in Lower Animals

It appears, therefore, that there may be (at least) 3 general modes of transmission of Au. It can be transmitted " horizontally " from person to person as might a " conventional " infectious agent. This is seen, for example, in the transmission of hepatitis and/or Au by blood containing Au; and this forms the basis for the widespread use of the "Au test " for the detection of hepatitis carriers (*7*, *25*). Other forms of horizontal transmission are also probable, for example, fecal-oral, by blood-biting insects, *etc*. The agent may also be spread by 2 forms of " vertical " transmission. If the genetic studies described above continue to be sustained then it implies that the ability to become persistently infected is controlled (at least in part) as a Mendelian trait. The data are also consistent with the notion that the agent could be transmitted with the genetic material in a lysogenic form (*8*), but there is no direct evidence for this.

The data also indicate that there may be a maternal effect. It cannot be deduced at what time during the maternal-child interaction the infection occurs: in the egg, *in utero*, at the moment of birth or at some time afterwards. Further direct studies will be needed to determine this.

Several of the tumor agents responsible for cancers in lower animals have these kinds of vertical and horizontal transmission. Particularly interesting examples are provided by the Avian RNA tumor viruses (*17*). There are additional interesting similarities of Au to RNA tumor viruses of lower animals, and these are now being investigated.

SUMMARY AND CONCLUSION

Au has many properties of an infectious agent, but also some that are atypical. On the basis of available data relating to the biological properties of Au, it is possible to suggest models which can be used for the formulation of hypotheses and experiments. One such model is suggested here.

The Au agent is pictured as having antigenic characteristics which may or may not be shared by putative hosts. The degree of matching will determine the results of infection, *i.e.*, persistent infection, acute infection, development of antibody, formation of complex, combinations of these, *etc*. The antigenic specificities of the host which may or may not match with the Au agent are, at least in part, genetically determined; some of these inherited host antigenic specificities may be reflected in serum protein polymorphisms, for example the immunoglobulins (Gm). There is little direct information on the action of Au on the cells of the host after infection takes place. Presumably it can enter the cellular mechanism of the host and cause the host to produce additional quantities of Au. The presence of Au in the cells may also have an effect on the cellular immunity of the host.

Au can be transmitted from individual to individual by horizontal transmission (simple infection). The nature and pathogenesis of the resulting infection may depend on the route of infection, *i.e.*, through the blood, the gastrointestinal tract, *etc.* It may also be transmitted vertically, in that there appears to be an inherited susceptibility to persistent infection. In addition there is another form of vertical transmission determined by a maternal effect. This may be manifest in the mother's egg cell although the present data do not rule out the operation of the maternal effect later in the life of the child. Factors in addition to genetic ones, such as sex and age, also have a measurable effect on the transmission of the agent.

On the basis of association studies Au appears to be related to a variety of diseases in addition to hepatitis. It is possible that in some cases this may reflect a pathogenetic role (for example, in hepatoma) but in others may mean that the agents which cause the disease (say, for example, lymphocytic leukemia) have properties in common with Au.

These considerations do not, of course, rule out the possibility that an agent more similar to a virus and related to Au, may eventually be identified.

This work was supported by U.S.P.H.S. Grants CA-06551, CA-06927 and RR-05539 from the National Institutes of Health, and by an appropriation from the Commonwealth of Pennsylvania.

REFERENCES

1. Allison, A. C. and Blumberg, B. S. An Isoprecipitation Reaction Distinguishing Human Serum Protein Types. Lancet, *1*: 634–637, 1961.
2. Alter, H. J. and Blumberg, B. S. Studies on a " New " Human Isoprecipitin System (Australia Antigen). Blood, *27* (3): 297–309, 1966.
3. Bayer, M. E., Blumberg, B. S., and Werner, B. Particles Associated with Australia Antigen in the Sera of Patients with Leukemia, Down's Syndrome and Hepatitis. Nature, *218*: 1057–1059, 1968.
4. Blumberg, B. S., Alter, H. J., and Visnich, S. A " New " Antigen in Leukemia Sera. J. Amer. Med. Assoc., *191*: 541–546, 1965.
5. Blumberg, B. S., Gerstley, B. J. S., Hungerford, D. A., London, W. T., and Sutnick, A. I. A Serum Antigen (Australia Antigen) in Down's Syndrome, Leukemia and Hepatitis. Ann. Int. Med., *66*: 924–931, 1967.
6. Blumberg, B. S., Melartin, L., Guinto, R. A., and Werner, B. Family Studies of a Human Serum Isoantigen System (Australia Antigen). Amer. J. Human Genet., *18*: 594–608, 1966.
7. Blumberg, B. S., Sutnick, A. I., London, W. T., and Millman, I. Australia Antigen. *In*; Call, E. A. and Moslofi, F. K. (eds.), The Liver. International Academy of Pathology, Monograph No. 13, pp. 269–285, Williams and Wilkins Co., Baltimore, Maryland, 1973.
8. Blumberg, B. S., London, W. T., and Sutnick, A. I. Australia Antigen as a Hepatitis Virus. Variation in Host Response. Amer. J. Med., *48*: 1–8, 1970.
9. Blumberg, B. S., London, W. T., Sutnick, A. I., and Millman, I. Australia Antigen,

Hepatitis and Susceptibility to Leukemia. *In*; Dutcher, R. M. (ed.), Comparative Leukemia Research 1969, Bibl. Haemat., No. 36, pp. 659–677, Karger, Basel/München/Paris/New York, 1970.

10. Blumberg, B. S., Friedlaender, J. S., Woodside, A., Sutnick, A. I., and London, W. T. Hepatitis and Australia Antigen. Autosomal Recessive Inheritance of Susceptibility to Infection in Humans. Proc. Natl. Acad. Sci. U. S., *62*: 1108–1115, 1969.
11. Blumberg, B. S. Australia Antigen: The History of Its Discovery with Comments on Genetic and Family Aspects. *In*; Vyas, G. N., Perkins, H. A., and Schmid, R. (eds.), Viral Hepatitis and Blood Transfusion, pp. 63–83, Grune & Stratton, New York, 1972.
12. Blumberg, B. S., Vierucci, A., London, W. T., and Sutnick, A. I. Association of Antibodies to Australia Antigen with Anti-Gm Antibodies in Italian Patients with Thalassemia. Nature, *236*: 28–30, 1972.
13. Blumberg, B. S., Hesser, J., Mazzur, S., Hertzog, K., Millman, I., Bloom, J., and Damon, A. Australia Antigen in the Solomon Islands. In preparation.
14. Blumberg, B. S., Sutnick, A. I., and London, W. T. Australia Antigen and Hepatitis. J. Amer. Med. Assoc., *207*: 1895–1896, 1969.
15. Ceppellini, R., Bedarida, G., Carbonara, A. O., Trinchieri, G., and Fillipi, G. High Frequency and Family Clustering of Au Antigen in Some Italian Populations. Atti Convegni Farmitalia "Antigene Australia ed Epatite Virale," pp. 53–74, Minerva Medica, Torino, 1970.
16. Coyne, V., Millman, I., Blumberg, B. S., Cerda, J., London, W. T., Gerstley, B. J. S., and Sutnick, A. I. The Localization of Australia Antigen by Immunofluorescence. J. Exp. Med., *131*: 307–320, 1970.
17. Emmelot, P. and Bentvelzen, P. (eds.). RNA Viruses and Host Genome in Oncogenesis, North Holland Publishing Co., Amsterdam, 1972.
18. Hadziyannis, S., Moussouros, A., Vissoulis, C., and Afroudakis, A. Cytoplasmic Localisation of Australia Antigen in the Liver. Lancet, *1*: 976–979, 1972.
19. Huang, Shao-Nan, Millman, I., O'Connell, A., Aronoff, A., Gault, H., and Blumberg, B. S. Virus-like Particles in Australia Antigen-associated Hepatitis. Amer. J. Pathol., *67*: 453–462, 1972.
20. Joswiak, W., Kosceilak, J., Madalinski, K., Brzosko, W. J., Nowoslawski, A., and Kloczewiak, M. RNA of Australia Antigen. Nature (N. B.), *229*: 92–94, 1971.
21. Kukowski, K., London, W. T., Sutnick, A. I., Kahn, M., and Blumberg, B. S. Comparison of Progeny of Mothers with and without Australia Antigen. Human Biol., *44*: 489–500, 1972.
22. London, W. T. Transmission of Australia Antigen to Man and Non-human Primates. Proc. Natl. Acad. Sci. U.S., *66*: 235, 1970.
23. London, W. T., Alter, H. J., Lander, J., and Purcell, R. H. Serial Transmission in Rhesus Monkeys of an Agent Related to Hepatitis-associated Antigen. J. Infect. Dis., *125*: 382–389, 1972.
24. London, W. T., DiFiglia, M., and Rodgers, J. Failure of Transplacental Transmission of Australia Antigen. Lancet, *2*: 900, 1969.
25. London, W. T., Sutnick, A. I., McKenna, P. J., and Blumberg, B. S. Australia Antigen and Posttransfusion Hepatitis. Transfusion, *8*: 318, 1968.
26. Mazzur, S. R., Vanstory, L., London, W. T., Sutnick, A. I., and Blumberg, B. S. Serological Studies on Populations Associated with Australia Antigen. Res. Commun. Chem. Pathol. Pharm., *3*: 435–454, 1972.

27. Mazzur, S. and Blumberg, B. S. Silent Maternal Transmission of Australia Antigen. In preparation.
28. Millman, I. In prepration.
29. Millman, I., Zavatone, V., Gerstley, B. J. S., and Blumberg, B. S. Australia Antigen in the Nuclei of Liver Cells of Patients with Viral Hepatitis by the Fluorescent Antibody Technique. Nature, *222*: 181–184, 1969.
30. Millman, I., Loeb, L. A., Bayer, M. E., and Blumberg, B. S. Australia Antigen (a Hepatitis-associated Antigen): Purification and Physical Properties. J. Exp. Med., *131*: 1190–1199, 1970.
31. Millman, I., Huhtanen, H., Merino, F., Bayer, M. E., and Blumberg, B. S. Australia Antigen: Physical and Chemical Properties. Res. Commun. Chem. Pathol. Pharm., *2*: 667–686, 1971.
32. Nathenson, C., Cohen, M. I., Millman, I., and Blumberg, B. S. Association of Antibodies to Gm and Antibodies to Australia Antigen in Adolescent Drug Addicts. In preparation.
33. Nowoslawski, A., Brzosko, W. J., Madalinski, K., and Krawczynski, K. Cellular Localisation of Australia Antigen in the Liver of Patients with Lymphoproliferative Disorders. Lancet, *1*: 494–497, 1970.
34. Obayashi, A., Okochi, K., and Mayumi, M. Familial Clustering of Asymptomatic Carriers of Australia Antigen and Patients with Chronic Liver Disease or Primary Liver Cancer. Gastroenterology, *62*: 618–625, 1972.
35. Nishioka, K., Mayumi, M., Okochi, K., Okada, K., and Hirayama, T. Natural History of Australia Antigen and Hepatocellular Carcinoma. *In*; Nakahara, W. (ed.), Analytic and Experimental Epidemiology of Cancer, pp. 137-146, University of Tokyo Press, Tokyo, 1973.
36. Steinberg, A. G., Damon, A., and Bloom, I. Gamma Globulin Allotypes of Melanesians from Malaita and Bougainville, Solomon Islands. Amer. J. Phys. Anthrop., *36*: 77–84, 1972.
37. Sukeno, N., Shirachi, R., Shiraishi, H., and Ishida, N. Conformational Studies of Australia Antigen by Optical Rotatory Dispersion and Circular Dichroism. J. Virol., *10*: 157–158, 1972.
38. Sutnick, A. I., London, W. T., Millman, I., Coyne, V. E., and Blumberg, B. S. Viral Hepatitis. Revised Concepts as a Result of the Study of Australia Antigen. Med. Clin. N. Amer., *54*: 805–817, 1970.
39. Vierucci, A., London, W. T., Blumberg, B. S., Sutnick, A. I., and Ragazzini, F. Australia Antigen and Antibody in Transfused Children with Thalassemia. Arch. Dis. Childhood, *47*: 760–765, 1972.

AMERICAN JOURNAL OF PHYSICAL ANTHROPOLOGY 243–254 (1984)

Sex-Related Differences in Transmission of Hepatitis B Infection in a Melanesian Population

ALEXANDRA LANGENDORFER, WILLIAM DAVENPORT, W. THOMAS LONDON, BARUCH S. BLUMBERG, AND SCOTT MAZZUR
Institute for Cancer Research, Fox Chase Cancer Center, Philadelphia, Pennsylvania 19111

KEY WORDS Hepatitis B virus, HBeAg, HBsAg

ABSTRACT Factors involved in the spread of hepatitis B virus (HBV) in the largely Melanesian population (N = 909) of Graciosa Bay, Ndeni, are examined. Based upon cultural information from Ndeni and ways in which HBV is spread in other populations, certain practices and interactions, predicted as effective routes of HBV infection, are analyzed. Most significant are father-to-son transmission and older-brother-to-younger-sibling transmission, indicating that males are instrumental in the transmission of HBV on Ndeni. Other possible routes of HBV transmission that are not shown to be significant on Ndeni are significant modes of disease spread in other cultures. Cross-cultural differences are discussed with regard to behavior, age of infection, and persistence of HBeAg (an antigen linked to HBV infection). Particularly with regard to maternal transmission, it is clear that the average age of infection combined with the degree of HBeAg persistence limits the number of infective mothers in this population. Cultural practices, however, may explain the disproportion of male carriers over 30 implicating a largely extinct set of culturally sanctioned practices involving very early exposure to HBV in boys and their age-related inability to make antibody to the virus.

This paper examines the spread of hepatitis B virus (HBV) in a largely Melanesian population living along the shores of Graciosa Bay. Graciosa Bay is located on the South Pacific island of Ndeni in the Solomon Islands (formerly Santa Cruz of the British Solomon Islands Protectorate). The data are based on census material and sera collected during a series of field trips conducted from 1973 through 1977 by the late Scott Mazzur.[1] The study uses cultural information and knowledge about the possible routes of HBV transmission to predict routes of infection within families and other subgroups of the population. These predictions are then tested using serological and census-related data, and the apparent routes of HBV infection are determined. As the spread of infection is dependent upon certain behaviors, the results of this analysis also generate new hypotheses about behavior.

Hepatitis B is caused by a relatively small virus (42 nm) containing a distinctive and partially double-stranded circular DNA, an associated polymerase, and several distinguishable protein components. The DNA polymerase incorporates nucleotides into the incomplete single-strand portion prior to viral replicaton. Both the DNA and its polymerase are contained within a protein core known as the core antigen or HBcAg [Hoofnagle et al., 1974]. The outer surface of the virion is comprised of a second protein termed the surface antigen HBsAg, or initially, Australia antigen [Blumberg et al., 1967]. A third antigen, HBeAg, is typically found in conjunction with the DNA and its polymerase [Magnius et al., 1975] and is probably a component of the core.

Received September 15, 1983; revised January 13, 1984; accepted January 23, 1984.

[1]Scott Mazzur died in 1979. In accordance with her wishes, the Division of Clinical Research of the Institute for Cancer Research maintains the data base and records of her expeditions to Graciosa Bay.

Specific antibodies produced by the host are often present in the serum after infection with HBV. These antibodies are associated with the major proteins of the virus, HBsAg, HBcAg, and HBeAg. Anti-HBs, the antibody to HBsAg, protects the host against subsequent HBV infection [Blumberg, 1977].

There are a variety of responses to HBV infection.

1) Acute hepatitis followed by complete recovery and serologically marked by the appearance of HBsAg and anti-HBc initially, and succeeded by the appearance of anti-HBs and the disappearance of the two earlier markers of infection.

2) Acute hepatitis with 5–10% of cases developing the chronic carrier state and symptoms. Serologically, anti-HBs and anti-HBc persist.

3) Chronic hepatitis with symptoms of chronic liver disease. HBsAg and anti-HBs persist.

4) Asymptomatic carrier. Initially there are no symptoms. Serologically, HBsAg and anti-HBc persist. HBeAg may also be present and this appears to be well correlated with viral production and the ability of one human to infect another [Magnius et al., 1975]. These individuals are at increased risk of developing chronic liver disease and primary hepatocellular carcinoma.

5) Asymptomatic sustained antibody production. Serologically, no trace of HBsAg but persistence of anti-HBs [Blumberg, 1977].

Factors that help determine which response a given host will exhibit are probably both genetically related and dependent upon the age, sex, and health of the host [Blumberg and Hesser, 1975].

The probability of being infected with HBV is an equally important issue. Infection with HBV is due to direct contact with either an infected individual's blood or secretions. Possible infection routes include injections with unsanitary needles, tattooing, sexual intercourse [Szmuness et al., 1975], sharing of chewed material [Leichtner et al., 1981], birth [Papaevangelou et al., 1974], and direct contact with menstrual blood [Mazzur, 1973].

The Graciosa Bay Villagers of Ndeni share a high risk for both HBV infection and chronic carrier status with the other inhabitants of the Solomon Islands [Blumberg et al., 1974]. The Graciosa Bay Villagers, though, are primarily Melanesian with a small amount of Polynesian admixture. They live in 14 continuous villages consisting primarily of leaf hut dwellings. Though several dialects and some distinct language differences exist, the villages are culturally homogeneous, and intermarriage between villages frequently occurs. Villagers are generally in good health despite the prevalence of malaria, respiratory illnesses, and limited native knowledge of infection.

Certain behaviors that could facilitate the spread of HBV are prevalent in the interactions of individuals on Ndeni based on age, sex, and family position. All children are breast-fed and, generally, mothers feed only their own children who are under 2 years of age. Between the ages of 2 and 5 years, children are weaned with premasticated food. While mothers are primarily involved in weaning children, other members of the family often participate. In particular, fathers, proud of young sons, are generally seen carrying their sons around the village.

Hepatitis B virus spread facilitated by homosexual intercourse is also a factor on Ndeni. Homosexual relations, as a form of sexual gratification, are culturally sanctioned, and sexual relations among both men and adolescent boys are prevalent [Davenport, 1964].

In the past, older men often engaged in intercourse with very young boys. Then, the men's house was the foremost hut in the village. It was the place where men slept and adolescent boys ventured to have their first heterosexual encounters with the village concubines. Premarital sexual intercourse was strictly forbidden except in the men's house, where both married and marriageable village women were banned. Almost complete public segregation of the sexes was also practiced and adultery was uncommon. As a result, homosexual relations, while not preferred over heterosexual encounters, were a major source of sexual gratification in adolescents and widowers [Davenport, 1964].

In the last 30 years, western missionary moralizing and government regulation has destroyed the village concubine system and reduced the men's house to a vestige of its former role. Today, men's houses are makeshift dwellings where adolescent boys and unmarried young men sleep and engage in sexual relations. Ironically, the missionary-inspired breakdown of the men's house has been accompanied by the loosening of premarital sexual segregation so that today adolescent heterosexual promiscuity is commonplace [Davenport, 1965].

In addition to homosexual experiences, husbands engage in sexual intercourse with wives, and heterosexual contact may spread infection as it does in other societies [Szmuness et al., 1975]. Also, in keeping with tradition, boys spend much of their time with their fathers or with age peer groups. Here the sharing and chewing of betel nut between men and their sons, as well as the contact between young boys in play, may facilitate the spread of HBV. Girls, on the other hand, spend much of their time working in the fields with their mothers and little time chewing betel nut or playing with a group.

The high prevalence of HBV infection on Ndeni, the homogeneity of the villages, and the knowledge of virus transmission behaviors (i.e., the mode of weaning, betel nut chewing, sexual practices, and sex differences in child rearing) allowed us to make the following hypotheses prior to the analysis of HBV transmission on Ndeni: 1. Carrier (HBsAg(+)) mothers have significantly more infected offspring aged 0 to 2 years than noncarrier mothers. 2. Infants of anti-HBs(+) mothers show an antibody response due to acquired immunity. 3. Carrier mothers and/or carrier fathers have significantly more infected children aged 3 to 5 years than noncarrier mothers and/or fathers. 4. Older carrier children have significantly more infected sibs aged 0 to 2 years and 3 to 5 years than older noncarrier children have. 5. Due to sexual segregation practices, there is a stronger relationship between a father's carrier status and infection in his sons than between a father's status and infection in his daughters. 6. Men's house cohabitants are at greater risk of infection and carrier status than siblings or cohabitants who live at home. 7. Carrier husbands have significantly more infected wives than do noncarrier husbands.

MATERIALS AND METHODS

From 1973 through 1977 a census of the Bay Villages was compiled and updated by Scott Mazzur. All individuals who were serologically tested were listed in the census and all parents, spouses, and sibs were recorded for use in intrafamily infectivity studies. In addition, the residence of each individual was noted so that cohabiting relationships could also be studied with regard to HBV. One of us, William Davenport, completed a census of this same island population in the 1950s and this earlier census was used in both verifying and enriching the information in the later census. Subsequent visits to Ndeni by Davenport have enabled an assessment of the changes in the traditional life style so that an accurate and current description of village life on Graciosa Bay was obtained.

In 1974, 771 individuals were tested for HBsAg and anti-HBs and the frequency of these viral markers was reported [Mazzur and Jones, 1977a]. In 1977 a more comprehensive sera collection was undertaken, and 909 individuals, roughly 84% of the population,[2] were sampled. In addition to HBsAg and anti-HBs, anti-HBc, HBeAg and anti-HBe were tested.

The serological analyses in this paper rely exclusively upon the 1977 sample with but one exception. In the analysis of acquired immunity, some mothers were only tested in 1974. As a result, the 1974 anti-HBs status of these mothers is analyzed and the results are discussed separately.

During both blood collections, vacutainer samples were obtained from adults while capillary tubes were used to obtain samples from infants and young children. HBsAg testing by a commercial reverse passive hemagglutination assay (RPHA) system (Auscell, Abbott) and anti-HBs testing by use of lyophilized HBsAg-coated red blood cells (Abbott) were done in the field. All initial results were confirmed in the field by repeating the test after neutralizing the sample with the appropriate antigen or antibody.

In 1977, samples of sufficient volume were sent to Dr. N. Nath, American Red Cross, Bethesda, Maryland and tested for anti-HBc activity by radioimmunoassay (CORAB, Abbott). Field identified HBsAg positive samples were retested by radioimmunoassay (Ausria II, Abbott). Those with sufficient volume were tested for HBeAg and anti-HBe by rheophoresis (Austect, Abbott) and/or by radioimmunoassay by Dr. H.A. Field, Center for Disease Control, Phoenix, Arizona. Field identified anti-HBs samples were retested by hemagglutination (Electronucleonics).

To describe this serological data, our report uses the terms HBsAg status and HBV outcome. HBsAg status describes an individual as a carrier (HBsAg(+)) or a noncarrier (HBsAg(−)). The HBV outcome describes

[2]The official 1970 government de facto census lists the population of the Graciosa Bay Villages as 1,080 (Western Pacific High Commission, British Solomon Islands Protectorate, 1970). At any given time, an additional 2–3% of the population is engaged in work outside the villages.

three types of responses to contact with a hepatitis B infectious carrier: 1) not infected; 2) infected but not a carrier; and 3) carrier. Those who are not infected tested negative for all the antigen and antibody responses specific for HBV that are described above. Those who are carriers tested positive for the surface antigen. Those who are described as infected but not carriers tested positive for some markers of HBV infection but not HBsAg or HBeAg.[3] Those individuals had produced anti-HBs and/or anti-HBc when infected with HBV.

In analyzing the serological data within the context of the census material all of those sampled are included. As a result, families with more members appear more frequently in the tables than families with few members. For example, in analyzing the association between a mother's HBsAg status and a 0- to 5-year-old child's outcome, the outcome of each child within the given age group is matched to the mother's HBsAg status. In this way, the status of women who have two children in this age group is listed twice, while the status of women with one child in this group is listed only once. As exposure and response to HBV among sibs vary, this approach is comprehensive. Furthermore, as larger families contribute more individuals to the population, it seems appropriate to score each individual rather than each family.

RESULTS AND DISCUSSION

Maternal transmission

Mothers' and/or fathers' HBsAg status was not associated with the HBV outcome of their children aged 0 to 2 years (see Table 1). Of 53 children ages 0 to 2, 48 were uninfected. Of the five children who were infected, four were HBsAg(+) and only one child was anti-HBs(+). All five of these infected children were the offspring of HBsAg(−) mothers, but two of these children had HBsAg(+) fathers.

The lack of detectable maternal transmission from mothers to children less than 2 years of age suggested that on Ndeni HBV was not spread from mother to child during pregnancy and birth or during the many months of breast feeding and related child care following birth. These findings are consistent with Mazzur and Jones' [1977b] review of the 1974 data from Ndeni that compared the HBsAg status of all mothers with their children regardless of age. Maternal transmission, however, is an effective route of HBV infection in Africa [Maupas et al., 1981], Asia [Stevens et al., 1975; Okada et al., 1975] and other parts of the South Pacific [Dickie et al., 1982].

In places where maternal transmission occurs, perinatal infection is one way HBV is spread. In Taiwan, Stevens et al. [1975] reported an extremely high rate of maternal transmission. Of 158 babies born to asymptomatic HBsAg(+) women, nearly 40% (63 of 158) tested HBsAg(+) by 6–12 months of follow-up. To test a specific route of maternal transmission, Beasley and Stevens [1975] compared breast fed and bottle fed babies of HBsAg(+) mothers in Taipei and found no apparent evidence for transmission via breast feeding. Thus, Stevens et al. [1975] suggested that transmission probably occurred during labor and delivery.

A specific delivery practice does not seem to preclude maternal transmission. In the United States and in Europe transmission from asymptomatic HBsAg(+) mothers to their children does not occur even though delivery practices are similar to those in Taiwan. On Ndeni, where hospital care is not available, maternal transmission could not account for any infection among children under 2 years of age.

It is evident that HBsAg(+) women differ in the ability to transmit HBV to their infants. Schweitzer et al. [1973], in a 3-month follow-up of 21 children born to HBsAg(+) asymptomatic mothers, found only one HBsAg(+) child. Similar results were reported in studies from Europe (Skinhøj et al. in Denmark [1972] and Papaevangelou in Greece [1973], Thailand (Punyagupta et al. [1973], and the South Pacific (this study). Yet in Japan [Okada et al., 1975], the rate of vertical transmission from asymptomatic mothers to infants is similar to that reported by Stevens et al. [1975] for Taiwan. Similarly, in the United States Schweitzer et al. [1972] reported relatively high levels of transmission when mothers tested HBsAg(+) with acute hepatitis. In their study, seven of 26 women had HBsAg(+) children within 3 months of delivery.

HBeAg and maternal transmission

HBsAg alone is not enough of a predictor of maternal vertical transmission. Other factors involving the response of the host to

[3] HBeAg testing was only done when individuals tested HBsAg(+).

TABLE 1. HBsAg status of parents and HBV outcome in children aged 0–2 years and 3–5 years

HBsAg		Age of	Outcomes in children			
Father	Mother	children	Carrier	Not infected	Infected, not carrier[1]	Totals
Negative	Negative	0–2 yrs	2	36	1	39
		3–5	2	38	9	49
Negative	Positive	0–2	0	2	0	2
		3–5	0	6	0	6
Positive	Negative	0–2	2	10	0	12
		3–5	3	13	0	16
Positive	Positive	0–2	0	0	0	0
		3–5	0	1	0	1

[1]Anti-HBs(+) and/or anti-HBc(+).

TABLE 2. Prevalence of HHBeAg and anti-HBe in Graciosa Bay HBsAg(+) carriers

		HBeAg(+)		Anti-HBe(+)	
	N	#	%	#	%
Unmarried females	11	3	27	1	9
Married females	17	1	6	4	24
All females	28	4	16	5	18
Unmarried males	24	6	25	3	12
Married males	24	1	4	4	17
All males	48	7	12	7	12
Total	76	11	14	12	16

HBV infection, particularly during pregnancy and delivery, need to be considered. An important factor linked to HBV infectivity is HBeAg. Presence of HBeAg and absence of anti-HBe has been shown to facilitate horizontal HBV transmission [Shikata et al., 1977].

HBeAg was detected in 11 (14%) of 76 HBsAg(+) individuals tested. The level of anti-HBe was similar (16%). The distribution of HBeAg and anti-HBe was similar for both the 48 HBsAg(+) males and the 28 HBsAg(+) females tested (see Table 2).

Among the 17 married HBsAg(+) women who were tested, only one was HBeAg(+) while four (24%) were anti-HBe(+) (see Table 2).

The small percentage of HBeAg(+) mothers among the relatively small percentage (9–14%) of HBsAg(+) women of childbearing age (see Table 3) limits widespread transmission of HBV in our study population. Only two of the 53 children under 2 years of age had HBsAg(+) mothers (see Table 1).

In contrast, HBeAg is found in asymptomatic carriers at relatively high frequencies in Taiwan where maternal transmission is documented. In one study 20 of 62 carrier Chinese mothers (32%) were HBeAg(+) while only 2% were anti-HBe(+) [Beasley et al., 1977]. In this same study, 13 of the 42 HBeAg(−) mothers (31%) had HBsAg(+) offspring, while 17 of the 20 HBeAg(+) mothers (85%) had carrier children (see Tables 4, 5). In Europe, where vertical transmission of HBsAg is low, HBeAg is found in virtually none of the asymptomatic carriers tested, while anti-HBeAg is found at levels well over 50% [Eleftheriou et al., 1975; Magnius et al., 1975] (see Table 4).

Persistence of HBeAg through childbearing years

Among HBsAg(+) village children, however, the proportion of HBeAg(+) carriers is much higher than among older individuals of marriageable age. A comparison of married carriers (generally all those over 20) with

TABLE 3. Hepatitis B virus marker distribution by age and sex in the 1977 sample of the Graciosa Bay Village population

Age of group	Sex	Number and (%) Carrier HBsAg(+)	Not infected	Infected, not carrier[1]	Totals
0–2 yrs	M	1 (3)	33 (97)	0 (0)	34
	F	0 (0)	14 (93)	1 (7)	15
	u	3	1	0	4
3–5 yrs	M	5 (14)	24 (69)	6 (17)	35
	F	2 (8)	21 (81)	3 (11)	26
	u	0	13	2	15
6–10 yrs	M	12 (17)	25 (35)	35 (48)	72
	F	2 (3)	39 (61)	23 (36)	64
11–15 yrs	M	6 (8)	23 (30)	47 (62)	76
	F	4 (9)	19 (43)	21 (48)	44
16–20 yrs	M	2 (8)	7 (27)	17 (65)	26
	F	3 (14)	6 (27)	13 (59)	22
21–30 yrs	M	5 (11)	7 (16)	32 (73)	44
	F	6 (10)	19 (30)	38 (60)	63
31–40 yrs	M	10 (25)	5 (12)	26 (63)	41
	F	7 (13)	9 (16)	39 (71)	55
41–50 yrs	M	7 (22)	3 (10)	21 (68)	31
	F	4 (16)	5 (19)	17 (65)	26
> 50 yrs	M	5 (24)	0 (0)	16 (76)	21
	F	0 (0)	2 (13)	14 (87)	16
Totals		84 (11)	275 (38)	371 (51)	730

[1]Anti-HBs(+) and/or anti-HBc(+).

unmarried carriers (generally those under 20) showed a four- to five-fold decrease in the proportion of HBeAg(+) individuals in the older, married HBsAg(+) carrier group (Table 2). In studying Japanese HBsAg carriers, Ohbayashi et al. [1976] also noted a decrease in the frequency of HBeAg with age (Table 4).

However, Ohbayashi et al. [1976] reported frequencies of HBeAg for both adults and children that were much higher than those found on Ndeni. Among the HBsAg(+) Japanese studied, the percentage of HBeAg(+) individuals under 20 was well over 40% (see Table 4). In comparison, the percentacge of those HBeAg(+), among unmarried carriers on Ndeni, was well under 30% (see Table 2). Similarly, while the proportion of HBeAg(+) Japanese carriers does decrease from 41% at ages 11 to 20 to 17% at ages 21 to 40, the proportion of infectious HBsAg(+) Japanese adults remains at a level that is three times as great as the frequency of infectious HBsAg(+) adults on Ndeni (see Tables 2 and 4). In another Asian population, a Korean population with chronic active liver disease, Kim et al., [in preparation] reported even higher levels of HBeAg persistence (>than 20%) among HBsAg(+) adults (see Table 4).

The persistence of HBeAg at levels greater than 15% among populations of Asian adult carriers [Beasley et al., 1977; Okada et al., 1976; Ohbayashi et al., 1976; Kim et al., in preparation] helps maintain a large number of infectious adult carriers in these populations (Table 4). The level of HBeAg persistence found in Asia, but not on Ndeni, is probably responsible for the significant amount of maternal, neonatal, HBV transmission reported in Asia but not apparent on Ndeni. In turn, population differences in the persistence of HBeAg may be directly related to factors such as the age of onset of infection.

Age at onset of infection related to outcome

At a time when mother and child contact is so intimate, it is fortunate that there are so few HBeAg(+) mothers in the Graciosa

TABLE 4. Prevalence of HBeAg and anti-HBe among HBsAg(+) individuals in various populations[1]

Location	Population description	N	% HBeAg(+)	% Anti-HBe(+)	Reference
Taiwan	Asymptomatic mothers	62	32%	2%	Beasley et al. [1977]
Japan	Asymptomatic mothers	23	43%	30%	Okada et al. [1976]
	Asymptomatic mothers	49	28%	30%	Shiraki, Yoshihara,and Sakuri [1980]
	Asymptomatic individuals				Ohbayashi et al. [1976]
	Ages 1–10	13	77%	15%	
	11–20	22	41%	23%	
	21–40	46	17%	28%	
	41–60	21	14%	38%	
	61 or older	13	15%	54%	
Korea	Patients with chronic liver				Kim et al. [in preparation]
	disease Ages 20–29	14	21.4%	42.9%	
	30–39	70	32.9%	42.8%	
	40–49	113	50.4%	25.7%	
	50–59	77	37.7%	36.4%	
	> 60	39	20.5%	48.7%	
Sweden	Blood donors				Magnius et al. [1975]
	Asymptomatic	18	0%	56%	
	individuals	21	43%	0%	
	Individuals with abnormal liver function				
England	Patients at London				Eleftheriou et al. [1975]
	Hospital	99	0%	14%	
	Acute hepatitis	79	44%	3%	
	Chronic active hepatitis	16	13%	44%	
	Chronic persistent	5	0%	80%	
	hepatitis	21	0%	73%	
	Primary hepatic carcinoma				
	Asymptomatic carriers				

[1]Sampling and testing methods also vary in studies by population.

TABLE 5. HBeAg and anti-HBe and maternal transmission of HBsAg in Japan[1]

	Infants			
Population description and maternal status	N	# HBV infected	# HBsAg(+)	# anti-HBs(+)
Asymptomatic HBsAg(+) carrier mothers Japan				
HBeAg	14	13 (93%)	12 (86%)	0
Anti-HBe(+)	31	9 (29%)	2 (6%)	5 (16%)
Both HBeAg and anti-HBe(−)	4	2 (50%)	1 (25%)	0
[Shiraki, et al., 1980]				
HBeAg(+)	10	10 (100%)	10 (100%)	0
Anti-HBe(+)	7	0	0	0
[Okada et al., 1976]				

[1]Sampling and testing methods also vary in studies by population.

Bay Villages because, once infected, children aged 0 to 2 are more likely to become carriers than older children. A significant difference in response (P = .04) was noticed when viral markers were compared in older children ages 3 to 5. A decrease in HBsAg(+) responses to infection was apparent: four of five infected 0–2-year-olds were HBsAg(+) (80%) while seven of 18 infected 3–5-year-olds (39%) were HBsAg(+), and only 14 of 72 infected 6–10-year-olds (19%) were carriers (Table 3).

Very few of the children aged 0 to 2 were tested for anti-HBc because the serum samples were generally of insufficient volume. Comparisons between the 0–2-year-olds and the 3–5-year-olds using only HBsAg and anti-HBs data revealed the same significant difference in outcome between age groups as reported above.

This same pattern is seen in Senegal, where 34% of children under 2 years of age were infected with HBV and half of these were

HBsAg(+); but by age 13, the prevalence of HBsAg among those infected is greatly reduced, and while 91% are infected only 13% of these are HBsAg(+) [Maupas et al., 1981]. Studies done in America in institutional settings have also shown a relationship between early exposure to HBV and increased likelihood of becoming a chronic carrier [Szmuness et al., 1970]. Similar results from Malo Island concerning newborns led Dickie et al. [1982] to conclude that perinatal exposure to HBV is likely to result in a failure to produce antibody and a lifelong HBsAg carrier status.

Primary hepatocellular carcinoma on Ndeni

Chronic HBV infection has been associated with liver disease including post hepatitic cirrhosis and primary hepatocellular carcinoma (PHC). There is also evidence that the onset, and possibly the pathogenesis, of HBV-related chronic liver disease may depend in part on the age at which chronic infection begins [Blumberg et al., 1975].

A review of medical records at a hospital on Kira Kira, a nearby island, revealed the death of a 5-year-old village girl from primary hepatocellular carcinoma. The clustering of HBsAg(+) carriers within her nuclear family makes early exposure to HBV infection a likely component in this child's terminal illness. In 1977, a year or more after the girl's death, her mother tested HBsAg(+), HBeAg(−), and anti-HBe(−). The lack of detectable HBeAg in the mother in 1977 may not be a good indicator of the mother's infectivity 5 to 10 years earlier when the girl and three surviving siblings were born. Two of the three surviving siblings are HBsAg(+) but were not tested for HBeAg and anti-HBe. The other sib was HBsAg(−), anti-HBs(−), and anti-HBc(+) at age 7. The presence of only anti-HBc indicates exposure to HBV years before; and in a child of 7, this means early exposure to the virus. The carrier status of the two other young sibs are likely outcomes to early HBV infection. The father was HBsAg(−), anti-HBs(−), and anti-HBc(−). The absence of HBsAg in the father and its presence in the mother coupled with the markers exhibited by the children indicate maternal transmission as the likely pathway for infection in this family. It is conceivable that perinatal transmission of HBV resulted in immunologic tolerance in three of the four children, and that in one of these three young HBsAg(+) carriers, early HBV infection promoted the development of primary hepatocellular carcinoma. It is likely that as better medical services become offered on Ndeni, more individuals will be identified with PHC and other HBV-related liver diseases.

Lack of acquired maternal antibody

Six of the 30 children less than 1 year old had mothers who were anti-HBs(+) in 1977. Though four of these six children were only 6 months old, none showed an antibody response. A 7-month-old child and another under 1 year also showed no trace of acquired immunity. Two more children aged 8 months and just under 1 year had mothers with detectable anti-HBs in 1974. Neither of these children had a measurable anti-HBs response.

The lack of detectable acquired maternal antibody in children 6 months old or greater is consistent with reports from Senegal [Maupas et al., 1981]. While 30% of newborn Senegalese babies are protected by passive immunity, only 4% maintain anti-HBs titers during the second 6 months of life. In fact, the loss of acquired immunity, infection, and seroconversion to carrier status was reported by Dickie et al. [1981] in a child only 9 months of age in a population from the New Hebrides. A review of mother-child serological responses to HBV in this New Hebrides population led Dickie et al. [1981] to believe that acquired antibody did little to prevent eventual HBsAg(+) outcomes and, in fact, that it may promote such an outcome. Our data do not support the latter conclusion. None of the children aged 0 to 2 years and born to antibody-positive mothers were HBsAg carriers, but all four HBsAg carriers had mothers with no detectable viral markers of infection. Thus, among the Graciosa Bay Villagers of Ndeni, acquired antibody was not measurable in those 6 months of age or older and its effect was not shown to be detrimental in very young children.

Paternal Transmission and Sexual Segregation

Neither mothers' nor fathers' HBsAg status had a significant effect on the HBV outcome of their children aged 0 to 3 years and 3 to 5 years (see Table 1). However, when HBV outcomes of boys aged 0 to 5 years and 6 to 10 years were compared with fathers' HBsAg status, a highly significant association between fathers' HBsAg status and sons'

TABLE 6. HBsAg status of parents by sex and HBV outcome of children aged 0–5 and 6–10, by sex

HBsAg status of parent	Children		HBV outcome of children			
	Age	Sex	Carrier	Not infected	Infected, not carrier[1]	Totals
Father: negative	0–5 yrs	M	1	44	7	52
		F	3	36	5	44
positive		M	5	11	1	17
		F	0	13	1	14
Mother: negative		M	6	50	8	68
		F	3	45	6	54
positive		M	0	5	0	5
		F	0	4	0	4
Father: negative	6–10 years	M	6	15	22	43
		F	1	25	17	43
positive		M	3	5	7	15
		F	0	7	3	10
Mother: negative		M	7	19	24	50
		F	1	30	20	51
positive		M	2	1	5	8
		F	0	2	0	2

[1]Anti-HBs(+) and/or anti-HBc(+).

HBV outcome at ages 0 to 5 was found. Carrier fathers were more likely to have carrier sons (P = .002). Fathers had no measurable effect on sons aged 6 to 10 and daughters aged 0 to 5 and 6 to 10 (see Table 6).

Mothers' HBsAg status had no apparent association with the HBV outcome of children regardless of sex (Table 6). In comparing children of HBsAg(+) parents, there were relatively fewer children in the HBsAg(+) mothers group, indicating again that the small percentage of HBsAg(+) mothers limits the role of mothers in transmission of HBV to offspring. In contrast, the greater percentage of HBsAg(+) men (see Table 3) helps to facilitate the spread of HBV from father to offspring. Due to practices of sexual segregation, fathers involved in the weaning of sons, but not daughters, transmit HBV to sons and not to daughters.

Horizontal transmission by male siblings

Children 0 to 2 years old with one or more older HBsAg(+) siblings and children 0 to 2 without carrier siblings did not vary significantly in outcome. A comparison of children 3 to 5, having at least one older HBsAg(+) sibling, with children in the same age group without older carrier siblings, showed a significant difference in outcome (P = .03). At this age, children with older carrier sibs had a greater chance of becoming carriers. In analyzing the data on older siblings by sex, it was found that the HBV status of older brothers had a significant effect on the sibs ages 3 to 5 (p = .01) (see Table 7). The HBV status of older sisters did not have any measurable effect (p = .71), but only two children ages 3 to 5 had an HBsAg(+) older sister.

Strictly defined behavioral differences between the sexes enhance contacts between boys that could lead to infection. Sexual segregation rules dictate that boys eat prepared meals together and from a common bowl (30 years ago all village males ate together this way). Girls, though, are isolated and the number of contacts in food sharing is limited to female family members. This type of segregation of the sexes may help perpetuate the significant difference (.05 < P < .1) in the frequency of infection between preadolescent males and females aged 6 to 10 (see Table 3). The isolation of females also limits the effect that an infectious female has in spreading HBV. Further field observations could test these hypotheses.

Horizontal transmission of HBV between children is a significant mode of HBV infection in other societies as well. Horizontal transmission among American children in Boston as a means of infection was also shown by Leichtner et al. [1981]. In the Bos-

TABLE 7. HBsAg status of brothers and sisters compared with HBV outcome of younger sibs aged 3–5 years

HBsAg status	Outcomes in children aged 3–5 years: Carrier	Not infected	Infected, not carrier[1]	Totals
All sisters negative	4	37	9	50
At least one sister positive	0	2	0	2
All brothers negative	3	40	7	50
At least one brother positive	3	3	2	8

[1]Anti-HBs(+) and/or anti-HBc(+).

ton study, shared chewing gum was the probable vehicle of infection. On Ndeni, the sharing of food, particularly premasticated food with those who have trouble chewing, appears to be one mode of spreading infection.

As both mothers and brothers share food with young family members, it is interesting that carrier brothers, and not carrier mothers, spread HBV to children aged 3 to 5. While behavior may be the main factor affecting these results, age-related differences in the frequency of infectious carriers may also be an important factor. The frequency of carrier boys is similar to the frequency of carrier women of marriageable age. Twenty-three of 183 boys (13%) aged 3 to 15 tested HBsAg(+), and 17 of 144 women (12%) aged 21 to 50 tested HBsAg(+) (see Table 3). Married women (those generally older than 20) have a much lower proportion of infectious HBeAg(+) carriers than unmarried carrier boys (see Table 2). The four-fold difference in infectious carriers may facilitate the spread of HBV from older brother to younger sib but not from mother to child.

Homosexual practices and transmission

Among teenage boys who cohabit, the outcome of carrier is associated with having an HBsAg(+) sib. While carriers represent 11% (Table 3) of the total population of villagers, 50% of the carrier boys in cohabitation units have carrier brothers at home. A comparison with sisters' HBsAg status revealed that 67% of cohabiting boys who are carriers have carrier sisters. Thus, on Ndeni there was no significant difference in the proportion of those either infected with HBV or positive for HBsAg when young cohabiting men were compared with sibs that had no history of cohabitation. Perhaps the limited number of sexual contacts in a Graciosa Bay Village (average population < 100) accounts for the apparent ineffective spread of HBV via homosexual contact.

The similarity in HBsAg marker prevalence between boys who cohabit and their sibs suggests an additional explanation for the lack of apparent HBV infection through homosexual contact. Most males on Ndeni are infected as children long before sexual cohabitation begins. This explanation is supported by the family clustering of HBV among sibs as reported by Mazzur and Jones [1977b] using data collected in 1974 from this population.

This pattern strikingly contrasts the spread of HBV through homosexual practices in New York City. In the study of New York City homosexuals [Szmuness et al., 1975], 37–51% of male homosexuals exhibited a serological marker of HBV infection. This level of infection is at least five times greater than that in the general population (7%). Sexual activity among homosexuals was correlated with infection, and the greater the number of sexual contacts, the greater the infection rate.

Heterosexual practices and transmission

There was no apparent association between carrier husbands and infection in their wives (P = .35). Similarly, the spread of HBV through heterosexual contacts, as seen in New York City [Szmuness et al., 1975], was not apparent on Ndeni. In New York City, evidence of HBV infection was found in 26% to 28% of the spouses of HBsAg(+) individuals, while spouses of noncarriers had only a 10% to 11% rate of HBV marker prevalence.

The relatively low general prevalence of infection in New York City decreases the chances of childhood infection through family contacts. As a result, both homosexual and heterosexual contact become significant factors in the spread of HBV. Conversely, children on Ndeni are at high risk for HBV infection, and by age 15 most individuals have been exposed to HBV. As a result, sexual contacts on Ndeni are not as significant a factor in HBV infection as they are in New York City.

Sex-related differences in HBsAg frequency

One interesting feature of HBV marker distribution on Ndeni is the striking increase in HBsAg(+) men over 30 years old. While 11% of 21–30-year-old men are HBsAg(+), 22% or more of men over 30 are HBsAg(+). Among those older than 30, the prevalence of HBsAg among males is significantly higher than among females [Mazzur et al., 1981].

This higher frequency in older males may be a cohort effect and may not be seen in 30 years when the boys in our study become men. The increased prevalence of carriers among older men could be related to the more traditional life-style that once existed. Post-World War II government rules, while met with resistance, did destroy the most important traditional men's institution, the men's house. This men's house of the past, unlike today's vestige, was a place where all village men gathered, sometimes bringing young sons. Food sharing and homosexual encounters among very young boys and adults were probably responsible for an early and more widespread exposure to HBV among boys. Girls, then as today, would be largely unaffected by men's house activities. As early exposure to HBV seems more likely to result in an HBsAg(+) response, the men's house of old may have institutionalized the exposure of very young males to HBV and the subsequent development of the carrier state.

This study has focused on behavior to elucidate routes of HBV infection in a specific population. As behavior is mediated by culture, populations with different cultures have different practices. When these practices involve the spread of HBV infection, it is likely that effective routes of disease spread will differ between cultures. These differences are evident when factors such as age, sex, and family interaction are compared with regard to HBV marker prevalence. This study has determined routes that are effective in the spread of hepatitis B in Graciosa Bay Village society. Other routes of infection exist, and in other populations, some of these are the routes that spread HBV most efficiently. It is only by analyzing HBV in many populations that we can determine the dynamics of its infection.

ACKNOWLEDGMENTS

This work was supported by USPHS grants CA-06551, RR-05539, CA-06927 and CA-22780 from the National Institutes of Health and by an appropriation from the Commonwealth of Pennsylvania.

LITERATURE CITED

Beasley, RP, Trepo, C, and Stevens, CE (1977) The e antigen and vertical transmission of hepatitis B surface antigen. Am. J. Epidemiol. *105:*94–98.

Beasley, RP, and Stevens, CE (1975) Evidence against breast feeding as a mechanism for vertical transmission of hepatitis B. Lancet *2:*740–741.

Blumberg, BS (1977) Australia antigen and the biology of hepatitis B. Science *197:*17–25.

Blumberg, BS,and Hesser, J (1975) 10. Anthropology and infectious disease. In A Damon (ed): Physiological Anthropology. Oxford: Oxford University Press.

Blumberg, BS, Larouze, B, London WT, Werner, B, Hesser, JE, Millman, I, Saimot, G, and Payet, M (1975) The relation of infection with hepatitis B antigen to primary hepatic carcinoma. Am. J. Pathol. *81:*669–682.

Blumberg, BS, Mazzur, S, Hertzog, K, Millman, I, Bloom, J, and Damon, A (1974) Australia antigen in the Solomon Islands. Human Biology *46:*239–262.

Blumberg, BS, Gerstley, BJS, Hungerford, DA, London, WT, and Sutnick, AI (1967) A serum antigen (Australia antigen) in Down's syndrome, leukemia and hepatitis. Ann. Intern. Med. *66:*9924–931.

Davenport, W (1965) Sexual patterns and their regulation in a society of the Southwest Pacific. In Sex and Behavior. Frank Beach (ed) New York: Wiley, pp 164–207.

Davenport, W (1964) Social structure of Santa Cruz Island. In WH Goodenough (ed): Exploration in Cultural Anthropology. New York: McGraw-Hill Book Co., pp 57–93.

Dickie, ER, Knight, RM, and Merten, C (1982) Ethnographic observations on child care and the distribution of hepatitis B virus in the nuclear family. Med. Anthropol. *6:*21–36.

Dickie, ER, London, WT, Merten, C, and Blumberg, B (1981) Hepatitis B virus infection in infants of anti-HBs(+) mothers. Lancet *1:*1000.

Eleftheriou, N, Thomas, HC, Heathcote, J and Sherlock, S (1975) Incidence and clinical significance of e antigen and antibody in acute and chronic liver disease. Lancet *2:*1171–1173.

Hoofnagle, JH, Gerety, RJ, Nily, et al. (1974) Antibody to hepatitis B core antigen: A sensitive indicator of hepatitis B virus replication. N Engl J Med *290:*1336–40.

Leichtner, AM, Leclair, J, Goldmann, DA, Schumacher, BS, Gewolb, IH, and Katz, AJ (1981) Horizontal nonparenteral spread of hepatitis B among children. Ann. Intern. Med. *94:*346–349.

Magnius, LO, Lindholm, A, Lundin, P, Iverson, S. (1975) A new antigen-antibody system. Clinical significance in long-term carriers of hepatitis B surface antigen. JAMA *231:*356–359.

Maupas, P, Barin, F, Chiron, JP, Coursaget, P, Goudeau, A, Perrin, J, Denis, F, and Diop Mar, I (1981) Efficacy of hepatitis B vaccine in prevention of early hepatitis B surface antigen carrier state in children: Controlled trial in an endemic area (Senegal). Lancet *1:*289–292.

Mazzur, S (1973) Menstrual blood as a vehicle of Australia antigen transmission. Lancet *1:*749.

Mazzur, S, Bastiaans, MJS, and Nath, N (1981) Hepatitis B virus (HBV) infection among children and adults in the Solomon Islands. Am. J. Epidemiol. *113:*510–519.

Mazzur, S, and Jones, N (1977a) Distribution and persistence of hepatitis B surface antigen and antibody in a Melanesian population. Am. J. Epidemiol. *105:*107–111.

Mazzur, S, and Jones, N (1977b) Limited family clustering of hepatitis B surface antigen in a Melanesian population. Am. J. Epidemiol. *105:*113–117.

Ohbayashi, A, Matsuo, Y, Mozai, T, Imai, M, and Mayumi, M (1976) Decreasing frequency of e antigen with age in serum of symptom-free carriers of HB antigen. Lancet *2:*577–578.

Okada, K, Yamada, T, Miyakawa, Y, and Mayumi, M (1975) Hepatitis B surface antigen in the serum of infants after delivery from asymptomatic carrier mothers. J. Pediatr. *87:*360–363.

Okada, K, Kamiyama, I, Inomata, M, Imai, M, Miyakawa, Y, and Mayumi, M (1976) e Antigen and anti-e in the serum of asymptomatic carrier mothers as indicators of "+" and "−" transmission of hepatitis B virus to their infants. N. Engl. J. Med. *294:*746–749.

Papaevangelou, GJ (1973) Hepatitis B in infants. N. Engl. J. Med. *288:*972.

Papaevangelou, G, Hoofnagle, J, and Kremastinou, J (1974) Transplacental transmission of hepatitis B virus by symptom-free chronic carrier mothers. Lancet *2:*746–748.

Punyagupta, S, Olson, LC, Harinasuta, U, Akarawong, K and Varawidhya, W (1973) The epidemiology of hepatitis B antigen in a high prevalence area. Am. J. Epidemiol. 97:349–354.

Schweitzer, IL, Mosley, JW, Ashcavai, M, Edwards, VM, and Overby, LB (1973) Factors influencing neonatal infection by hepatitis B virus. Gastroenterology *65:*277–283.

Schweitzer, IL, Wing, A, McPeak, C, Spears, RL (1972) Hepatitis and hepatitis-associated antigen in 56 mother-infant pairs. JAMA *220:*1092–1095.

Shikata, T, Karasawa, T, Abe, K, Uzawa, T, Suzuki, H, Oda, T, Imai, M, Mayumi, M, and Moritsugu, Y (1977) Hepatitis B e antigen and infectivity of hepatitis B virus. J. Infect. Dis. *136:*571–576.

Shiraki, K, Yoshihara, N, Sakurai, M, Eto, T, Kawana, T (1980) Acute hepatitis B in infants born to carrier mothers with the antibody to hepatitis B e antigen. J. Pediatr. *97:*768–770.

Skinhøj, P, Olesen, H, Cohn, J, Mikkelsen, M (1972) Hepatitis-associated antigen in pregnant women. Acta Pathol. Microbiol. Scand. [B] *80:*362–365.

Stevens, CE, Beasley, RP, Tsui, J, and Lee, W-C (1975) Vertical transmission of hepatitis B antigen in Taiwan. N. Engl. J. Med. *292:*771–774.

Szmuness, W, Much, MI, Prince, AM, Hoofnagle, JH, Cherubin, CE, Harley, EJ, Block, GH (1975) On the role of sexual behavior in the spread of hepatitis B infection. Ann. Intern. Med. *83:*489–495.

Szmuness, W, Pick, R, and Prince, AM (1970) The serum hepatitis virus specific antigen (SH): A preliminary report in an institution for the mentally retarded. Am. J. Epidemiol. *92:*51–61.

Western Pacific High Commission, British Solomon Islands Protectorate (1970) Report on the Census of the Population of the British Solomon Islands Protectorate, 1970. Honiara, N.D. (p. 129, Table 2. Graciosa Bay Population = Uta + Leupe + Pala wards).

© 1990 Elsevier Science Publishers B.V. (Biomedical Division)
Hepatitis B: A sexually transmitted disease in heterosexuals.
P. Piot and F.E. Andre, editors

3

Sex-related aspects of hepatitis B infection and its consequences

BARUCH S. BLUMBERG

Balliol College, Oxford 0X1 3BJ, England, UK, and Fox Chase Cancer Center, Philadelphia, USA

Introduction

Soon after the discovery of the hepatitis B virus (HBV), there were occasional observations within families from which it was inferred that heterosexual transmission of HBV was a common event. If one member of a connubial pair was a carrier of HBV, his or her partner had a much higher probability of having anti-HBs than an individual randomly chosen from the population. It was also found that the probability that the mate of a HBV carrier would become a carrier was not much greater than expectation. This was consistent with the explanation that, whereas there is a high risk of infection in a sexual partner of a carrier, there is an additional requirement (possibly genetic) to increase the probability that the partner would develop the carrier state (1). With increasing experience it became more apparent that sexual activity may be one of the most common forms of HBV transmission, particularly in regions of moderate or low endemicity of HBV. Appropriate public health and health education steps are now being taken to decrease the probability of transmission through this route.

Sexual transmission of hepatitis B and vaccination

It is likely that, overall, HBV is transmitted sexually more commonly than HIV. In the case of HBV a safe and effective vaccine is available; unfortunately this is not so for HIV. An obvious solution to sexual transmission of HBV would be universal vaccination which included adolescents and adults. This is unlikely to occur before infant vaccination programs are more advanced, although in some countries these programs will develop sooner. Individuals can undertake vaccination independent of public health programs, and in many cases will probably do so.

Sex-related aspects of hepatitis B infection

Consequences of HBV infection such as acute hepatitis, fulminant hepatitis, chronic persistent or active hepatitis, liver cirrhosis and hepatocellular carcinoma have been well studied and are generally known. Some aspects of the consequences of HBV infection have not, however, been the subject of intensive research, but nevertheless are of considerable biological and possibly Public Health importance. They are related to sex but not only in the literal sense.

4

Nearly everything about hepatitis B infection and host responses to it are sex related. When infected, males are much more likely to become carriers and females to develop anti-HBs (2). Males are likely to remain carriers longer than females. In most populations the ratio of male to female carriers is about two or greater (see for example ref. 3). Following treatment with interferon, females are more likely to respond than males. It has been inferred that male carriers are more likely to develop primary hepatocellular carcinoma (PHC) than female carriers and the ratio of males to females among patients with primary carcinoma can be 5 or higher in most populations.

For obvious reasons the female carrier parent is more likely to transmit HBV to a child than the male carrier parent (4). The mother may transmit HBV *in utero* or at the time of birth. After delivery mothers in most societies have more social interaction with their children than do fathers, and this increases the probability of maternal transmission. We have compared the transmission of HBV within families in village communities in Senegal. Transmission was most likely to occur from mothers to children, next from sibs to sibs, and transmission from fathers to children was not greatly different from random transmission from outside the family (5). There was a different pattern in a traditionally male homosexual community in the Pacific where there were correlations indicating that male to male transmission was a common transmission mode (6).

HBV infection and sex ratio

One of the most curious sex differences for HBV was uncovered during a family study conducted in a community in Macedonia in Northern Greece in 1973 (7). We had been interested in family studies since our earliest investigations of the genetics of the carrier state (8). Sera were collected from most of the inhabitants of the town and tested for HBsAg and anti-HBs. The results were analyzed by family groupings according to the response of parents following exposure to HBV. We found that, if one of the parents was a carrier (families with both parent carriers were uncommon) then the sex ratio (defined as the number of male offspring divided by the number of female offspring times 100) was higher than in families in which the parents were not carriers. In a subsequent analysis we found the sex ratio was highest in HBsAg(+) carriers, lowest in families in which the parents (in particular the mother) was anti-HBs and intermediate in families in which the parents were uninfected (9). In this population both the carrier and anti-HBs families had about the same number of boys, but had fewer girls. The carrier families therefore have fewer children and fewer girls than the anti-HBs; hence, there is an effect both on sex ratio and family size.

We have postulated several biological and behavioral explanations for these findings. The observations could be explained by the relatively decreased viability of a female compared to a male conceptus infected with or exposed to HBV. There could also be a behavioral component. In societies where male children are valued more than females, there is often a desire to have a certain number of male offspring. In the carrier families, there is a high probability of producing boys; from our data, about 2 appears to be the favoured number. Hence the carrier parents stop producing children after the desired number is achieved. However, the anti-HBs families are more likely to have girls, and they continue to have children until they have the desired number of boys, ending up with a larger family with more girls than the carrier parents.

Because of the curious nature of these findings we have attempted to validate them in other populations, including Papua New Guinea (10), The Philippines (11) and Greenland (12). The sex-ratio difference observed in Greece was confirmed in the other populations. There has been some confirmation from other laboratories but the problem has, for the most part, remained unstudied.

Iron, ferritin, HBV infection and sex

We have completed a large number of studies on serum iron, ferritin and transferrin in relation to HBV infection and its outcome. These have been extended to population studies of mortality in general, and morbidity and mortality from cancer, but these are not relevant to the present paper and will not be discussed further (13). Carriers of HBV have higher levels of iron than individuals who are uninfected (14-16). In a study in Senegal the difference was found to be most striking in females. In the tropical regions where HBV is common, iron is often deficient in the diet and hemoglobin levels are, on average, lower than in temperate regions. One could conjecture that the HBV carrier state may be relatively advantageous in that given the same level of dietary iron intake, the carriers would retain more iron. This would be particularly valuable in females who, between menstruation and menopause, loose blood regularly and require more iron than men. In previous times, when life expectancy was lower than at present, the HBV carrier state was probably less of a disadvantage. Carriers would usually die from other causes before the effects of chronic HBV infection could shorten life as a consequence of chronic liver disease and primary hepatocellular carcinoma.

We have also investigated the relation between serum ferritin levels and the probability of becoming a hepatitis B carrier following exposure to HBV (17). Previously, several factors have been shown to increase the probability of becoming a carrier. These include: (1) age, the young are more likely to become carriers than older people, (2) sex, males are more likely to become carriers than females, and (3) some sort of genetic predisposition. There is now a growing body of data that HLA types and immune states of the host may be related to the probability of carrier state development.

Ferritin levels may be another factor. The study of ferritin was conducted in a renal dialysis unit in Philadelphia beginning in the early 1970's. At that time, blood screening for HBV was not done and HBV infection was a frequent consequence of renal dialysis and the blood transfusion associated with it. We had stored serum samples from renal dialysis patients before, at about the time of, and after infection with HBV. We tested the hypotheses that individuals who had higher ferritin levels before infection were more likely to become carriers than those with lower ferritin levels and that those with the lower ferritin levels were more likely to develop anti-HBs. The data supported the hypothesis and we concluded that in this population, increased ferritin levels (which might be an indicator of total body iron stores) increased the probability that an infected individual would develop the carrier state.

In later studies (18) we found that, in patients with chronic hepatitis B infection, those who went on to develop PHC were more likely to have had elevated ferritin and decreased transferrin levels than the patients who did not develop PHC within the observation period. A variety of studies in mice have confirmed these original impressions and have been extended to the relation of increased ferritin and decreased albumin levels to the more rapid development of cancers in general.

6

Transmission of HBV by bedbugs and other insects

There is another body of information which has an unusual relation to sexual activity. There has been a considerable interest in the possibility of insect transmission of HBV. Several studies on mosquitoes have shown that there is a high field infection rate (of the order of 1:200) and that detectable virus may remain in the mosquitoes for several weeks after a blood meal (19–22).

In Senegal we collected bed bugs (*Cimex hemipterus*) from beds whose main occupants were carriers of HBV. More than sixty percent of the insects had detectable HBsAg (23, 24). After a single feed of HBsAg(+) blood, HBsAg could be detected in the body of the insect for up to 6 weeks and in the faeces for 2 weeks after the cessation of a blood meal. The infected faeces accumulates as dust in the bed and, theoretically could enter the blood stream of another occupant through ulcerous or other lesions on the skin. Hence, virus could be transmitted to another occupant of the bed, i.e. a child or sexual partner. In the latter case, it represents an unusual form of 'venereal' transmission. There was no evidence that replication occurred either in the bed bugs or mosquitos.

An international controlled study is now in progress in the Gambia to determine if systematic insect control can decrease HBV infection. When complete this could provide important information of potential public health importance. A final item concerning HBV and sex. In the experiment with bed bugs, it was found that the female imbibes more blood than does the male (25).

Conclusion

Sexual transmission is undoubtedly an important mechanism for the perpetuation of HBV in human populations. An understanding of the complex relation of HBV to human sex differences will reveal a great deal of important information of biological value, but also help in developing appropriate mechanisms of control.

References

1. Blumberg BS, Melartin L, Guinto RA, and Werner B, (1966) Am J Human Genet, 18: 594.
2. London WT and Drew JS (1977) Proc Natl Acad Sci (U.S.A.) 74: 2561.
3. Blumberg BS, Sutnick AI, London WT and Melartin L (1972) Arch Intern Med 130: 227.
4. Larouzé B, London WT, Saimot G, Werner BB, Lustbader ED, Payet M and Blumberg BS (1976) Lancet 2: 534.
5. Marinier E, Barrois V, Larouzé B, London WT, Cofer A, Diakhate L and Blumberg BS (1984) In: Immunita ed Infezione Nelle Malattie del Fegato e del Tratto Gastroenterico del Bambino, Ed. A. Vierucci. Masson Italia Editori S.p.A., Milano, pp. 379.
6. Langendorfer A, Davenport E, London WT, Blumberg BS and Mazzur S (1984) Am J Phys Anthropol 64: 243.
7. Hesser JE, Economidou J and Blumberg BS (1975) Human Biol. 47: 415.
8. Blumberg BS, Melartin L, Guinto RA, and Werner B (1966) Am J Human Genet, 18: 594.
9. Drew JS, London WT, Lustbader ED, Hesser JE and Blumberg BS (1978) Science 201: 687.
10. Drew J, London WT, Blumberg BS and Serjeantson S (1982) Human Biology 54: 123.

7

11. London WT, Stevens RG, Shofer FS, Drew JS, Brunhofer JE and Blumberg BS. (1982) In: Viral Hepatitis, 1981 International Symposium. Eds W Szmuness, H. J. Alter and J. E. Maynard Franklin Institute Press, Philadelphia, p. 195.
12. Drew JS, Blumberg BS and Robert-Lamblin, J (1986) Human Biology 58: 115.
13. Blumberg BS (1986) In: Viruses, Immunity and Immunodeficiency. Ed A. Szentivanyi and H. Friedman. Plenum Publishing Corp, New York, p. 81.
14. Sutnick AI, Blumberg BS and Lustbader ED (1974) Ann. Internal Med. 81: 855.
15. Felton C, Lustbader BD, Merten C and Blumberg, BS (1979) Proc. Natl. Acad. Sci. USA 76: 2438.
16. Blumberg BS, Lustbader ED and Whitford, PL (1981) Proc Natl Acad Sci USA 78: 3222.
17. Lustbader ED, Hann HL and Blumberg, BS (1983) Science 220: 423.
18. Hann HL, Kim CY, London WT and Blumberg BS (1989) Int J. Cancer 43: 376.
19. Wills W, Saimot G, Brochard C, Blumberg BS, London WT, Dechene R and Millman I (1976) Am J Trop Med Hyg 25: 186.
20. Blumberg BS, Wills W, Millman I and London WT (1973) Res Commun Chem Path Pharm 6: 719.
21. Metselaar D, Blumberg BS, Millman I, Parker AM and Bagshawe AF (1973) Lancet 2: 758.
22. Wills W, Larouze B, London WT, Millman I, Werner BG, Ogston W, Pourtaghva M, Diallo S and Blumberg BS (1977) Lancet 2: 217.
23. Ogston CW, Wittenstein FS, London WT and Millman I (1979) J Infect Dis 140: 411.
24. Ogston CW and London WT (1980) Trans Roy Soc Trop Med Hyg 74: 823.
25. Yanovski AD and Ogston CW (1982) J Med Entomol 19: 45.

Australia Antigen and Viral Hepatitis in Drug Abusers

Alton I. Sutnick, MD;
James J. Cerda, MD;
Philip P. Toskes, MD;
W. Thomas London, MD; and
Baruch S. Blumberg, MD, PhD,
Philadelphia

Much evidence supports the hypothesis that Australia antigen (Au[1]) causes viral hepatitis. We found Au(1) in 30 of 43 drug abusers with hepatitis (69.8%), a similar frequency to posttransfusion hepatitis (56.3%), but different from infectious hepatitis (30.1%). There were no distinctive histologic features in liver biopsies from drug abusers. Two of nine drug abusers with Au(1) followed serially, developed persistent Au(1); two of 22 other Au(1) hepatitis patients had persistent Au(1). Australia antigen was found in nine of 215 "normal" addicts (4.2%), significantly different from the general population (two of 2,412), but not different from 261 nonaddicts exposed to hepatitis by multiple blood transfusions (3.8%). Acute hepatitis in drug abusers is likely mainly viral, not toxic in origin.

There is now a great deal of evidence to support the hypothesis that Australia antigen (Au[1]) is a causal agent of viral hepatitis.[1-4] Persistent serum Au(1) has been found in patients with subacute and chronic active hepatitis,[5-7] and in up to 20% of apparently normal people in some tropical populations.[8,9] Persistent Au(1) also occurs in leukemia,[8,10,11] Down's syndrome,[8,12] lepromatous leprosy,[9] and chronic renal disease,[13] a group of disorders in which immunologic abnormalities have been found.

Because of the widespread illicit use of drugs and the high incidence of hepatitis among drug abusers, we have investigated the question of the presence of Au(1) in drug abusers. This study was undertaken to determine the frequency of Au(1) in drug abusers with hepatitis, to gain an insight into the etiology of hepatitis in these patients (ie, toxic vs infectious), and to assess the relationship of drug abuse to chronicity of hepatitis and persistence of Au(1).

Received for publication June 2, 1970; accepted Nov 24.

From the Institute for Cancer Research, Fox Chase, Philadelphia; the Gastroenterology Section, University of Pennsylvania Medical Division, Veterans Administration Hospital; and the Department of Medicine, University of Pennsylvania School of Medicine, Philadelphia. Dr. Cerda is recipient of a Clinical Investigator Award from the Veterans Administration.

Reprint requests to Institute for Cancer Research, 7701 Burholme Ave, Fox Chase, Philadelphia 19111 (Dr. Sutnick).

Materials and Methods

The subjects of this study are 258 habitual users of addicting drugs (ie, opiates, amphetamines, barbiturates). Users of marihuana alone were not included. Forty-three had acute hepatitis at the time they were studied, and 215 were asymptomatic and presumably healthy. There were 231 men and 27 women. The age range was from 16 to 58 years with a median of 23. The clinical diagnoses of the patients with acute hepatitis were supported by elevated serum glutamic pyruvic transaminase (SGPT) levels and abnormalities in other liver chemistry tests. Nineteen of these patients had liver biopsies performed which without exception supported the diagnosis of acute hepatitis. These patients were compared with 103 other patients with viral hepatitis who were not known drug abusers. The latter group was obtained since the series of hepatitis cases reported in our previous study,[14] and blood was generally drawn soon after the onset of symptoms. Australia antigen commonly appears only transiently early in the course of hepatitis[1,6]; this is reflected by a higher frequency of Au(1) in the more recent series. This group includes 32 patients with posttransfusion hepatitis, and 71 patients without a history of transfusions or injections, whose condition was diagnosed as infectious hepatitis. To gauge the effect of drugs on the persistence of Au(1), nine drug users with hepatitis and Au(1) in their blood were followed up for two months or longer and compared with 22 other hepatitis patients with Au(1) who do not use drugs and who were followed up for the same length of time.

A group of 261 patients at Jefferson Medical College Hospital who had received multiple blood transfusions were used as a control group for the drug addicts without clinical hepatitis. Patients with leukemia and chronic renal disease were excluded, since they are known to have a high frequency of persistent Au(1).[8,10,11,13]

Australia antigen was determined by immunodiffusion in agar gel by a micro-Ouchterlony technique.[15] Determinations of SGPT were performed by the method of Wroblewski and LaDue.[16] Liver biopsy specimens were obtained with a Menghini needle, and the sections were stained with hematoxylin-eosin and special stains. Eleven drug abusers with hepatitis and Au(1) in their blood were matched by sex and age with 11 other patients with acute hepatitis. The histologic characteristics of their liver biopsies were compared. Statistical comparisons were made by the chi-square test or the sign test.

Results

Nine of the 215 asymptomatic drug abusers (4.2%) had Au(1) in their blood. This was significantly different from the frequency previously reported in the general population in the United States (0.1%)[1,6,12,14,17] (X^2 = 70.1667, $P <$ 0.0001). However, it was not significantly different from the frequency in the 261 multiply transfused patients who were not drug abusers (3.8%) (X^2 = 0.0015) (Table 1). Australia antigen was found in the sera of 30 of the 43 drug abusers with hepatitis (69.8%) (Table 2). This is close to the frequency found in the 32 patients with posttransfusion hepatitis (56.3%), and is significantly

Table 1.—Frequency of Australia Antigen in Asymptomatic Drug Abusers

	No.	Au(1)	%	P	
Drug abusers	215	9	4.2	$\chi^2 = 0.0015$ NS*	$\chi^2 = 70.1667$ $P < 0.0001$
Multiply transfused patients	261	10	3.8		
General US population	2,412	2	0.1		

* Not significant.

Table 2.—Frequency of Au(1) in Hepatitis in Drug Abusers Compared With Posttransfusion and Infectious* Hepatitis

	No.	Au(1)	%	P	
Hepatitis in drug abusers	43	30	69.8	$\chi^2 = 0.9272$ NS†	$\chi^2 = 14.7105$ $P < 0.0002$
Posttransfusion hepatitis	32	18	56.3		
Infectious hepatitis	71	22	30.1		

* No history of transfusions or injections over previous six months.
† Not significant.

Table 3.—Persistence of Au(1) Following Acute Hepatitis

	No.	Persistent Au(1)	%	P
Hepatitis in drug abusers	9	2	22.2	$\chi^2 = 0.1598$ NS*
Other hepatitis patients	22	2	9.1	

* Not significant.

different from the frequency of Au(1) in the 71 patients with infectious hepatitis (30.1%) ($P < 0.0002$). Because of the unreliability of these patients, there was some difficulty in obtaining blood specimens over a long period of time. However, nine patients with Au(1) had serial blood specimens drawn for periods from six weeks to one year. Two of these nine had Au(1) in their last serum samples, one 63 days after the first blood sample tested, and the other nearly one year afterward. Of the 22 hepatitis patients with Au(1) who were not drug abusers and who were followed for a similar period, two (9.1%) also had persistent antigen. This difference was not significant (Table 3). Coded slides of the liver biopsy of the drug abusers with hepatitis and the hepatitis patients who did not use drugs were examined by two pathologists independently, and the findings were graded. They were compared by means of the sign test, and no important differences were found.

One patient appeared to have been infected with two organisms as a consequence of drug abuse. The patient is a 19-year-old boy who had taken drugs both orally and parenterally, including mescaline, marihuana, amphetamine, dextroamphetamine, methamphetamine, chlordiazepoxide hydrochloride (Librium), and secobarbital. He was admitted to the hospital with severe acute hepatitis. His SGPT level rose to 2,400 Karmen units, and his serum bilirubin level to 15 mg/100 ml. His serum was positive for Au(1) and he had a marked rise in heterophil antibody titer of the type seen in infectious mononucleosis. The antibody titer against EB virus rose during his hospitalization. There were marked changes of acute hepatitis with early fibrotic changes on liver biopsy. Fluorescent granules were seen in the nuclei of his liver cells after staining with fluorescein labeled anti-Au(1) antibody.[18] Au(1) subsequently disappeared from his blood and his heterophil antibody titer diminished to within the normal range.

Comment

This study has demonstrated a high frequency of Au(1) in drug abusers with hepatitis in the Philadelphia community. This frequency is not significantly different from that found in patients who develop hepatitis following transfusions as reported in this and earlier studies[6,14,17]; it is significantly higher than that in hepatitis patients who had no transfusions or injections. These results are in agreement with a recent report by Cherubin et al in New York.[19]

There are now a variety of published data of a clinical, immunological, epidemiological, and electron microscopic nature favoring the hypothesis that Au(1) is a virus, or part of a virus, which causes human hepatitis (see reference 4 for summary). Australia antigen is not associated with toxic hepatitis.[4,6] The high frequency of Au(1) in drug abusers with hepatitis thus favors a viral rather than a toxic etiology of hepatitis in these patients, as has been suggested by others.[20,21] The significant difference from the frequency in infectious hepatitis and the similarity to that in patients with posttransfusion hepatitis suggests that the mode of transmission is the same as in the latter. However, two drug abusers with Au(1) and hepatitis gave histories of solely oral drug intake, no transfusions or injections, but contact with other drug abusers who contracted hepatitis from needles.

There is a possibility that more than one agent may be transmitted simultaneously with the same contaminated equipment. This was suggested by a young man who developed severe acute hepatitis with Au(1) in his blood accompanied by an elevation of heterophil antibody titer and a rising EB virus antibody titer, which indicates the presence of infectious mononucleosis as well. It was considered likely that both Au(1) and the virus of infectious mononucleosis were transmitted by injection at approximately the same time. After a severe illness, his Au(1) disappeared, and the heterophil antibody titer returned to normal range.

Serial blood samples were obtained over a period of two months or longer from nine of the drug abusers with hepatitis who initially had Au(1) in their serum. Two of these patients (22.2%) had persistent Au(1) in the last blood specimens 63 days and one year, respectively, following onset of the disease. This was not significantly different from the 9.1% frequency of persistence in 22 hepatitis patients who did not use addicting drugs.

We also demonstrated a high frequency of Au(1) in drug abusers without hepatitis compared with the general population. However, when these were compared with nonaddicts who had been exposed to multiple blood transfusions, there was no significant difference. It does not appear that the abuse of drugs provokes persistence of Au(1) any more than that seen in nonusers of drugs; certainly, if it does, it is not striking. In other studies,[5] persistent Au(1) has been shown to be associated with chronic hepatitis, and persistence of viral activity has been implicated in this condition. By analogy, it seems likely that those drug abusers with hepatitis in whom Au(1) persists are also destined to develop chronic hepatitis; it thus appears that the chronic disease in these patients is virus-induced and not drug-induced as has been suggested by others.[20,21] The possibility of some influence of drugs has not been entirely excluded however.

Of practical importance is the known tendency of drug users to become professional blood donors to support their habit.[22] An increased risk of posttransfusion hepatitis has been documented following transfusions of blood obtained from commercial sources,[23-26] and at times can be associated with addict donors.[27,28] Furthermore, a study of Au(1) in 5,927 blood donors[29] has shown that certain high-risk donor groups may be identified by the "Au test."[30] The current study further emphasizes the hazard of the use of sources of blood obtained from poorly screened donors.

This investigation was supported by Public Health Service grants CA-08069, CA-06551, CA-06927, FR-05539, 5-MO1-RR-40-09 and 5-TO1-AM-5462-05 and by an appropriation from the Commonwealth of Pennsylvania. Donald J. Ottenberg, MD, and Jean C. Keyser, Eagleville Hospital and Rehabilitation Center, Eagleville, Pa, obtained blood specimens from asymptomatic drug addicts; H. Fittingoff, MD; G. Schless, MD; C. Goepp, MD; J. Medoff, MD; W. Snape, MD; and H. P. Potter, MD, supplied serum from some of their patients and permitted reviews of patients' charts; and T. J. McKenna, MD, obtained the sera from multiply transfused patients at Jefferson Medical College Hospital. Werner Henle, MD, and Gertrude Henle, MD, performed the EB virus antibody studies.

References

1. Blumberg BS, Sutnick AI, London WT: Hepatitis and leukemia: Their relation to Australia antigen. *Bull NY Acad Med* **14**:1566-1586, 1968.
2. Blumberg BS, London WT, Sutnick AI: Australia antigen as a hepatitis virus: Variation in host response. *Amer J Med* **48**:1-8, 1970.
3. London WT, Millman I, Sutnick AI, et al: Australia antigen and viral hepatitis. *Rev Franc Etud Clin Biol* **14**:961-963, 1970.
4. Sutnick AI, London WT, Millman I, et al: Viral hepatitis: Revised concepts as a result of the study of Australia antigen. *Med Clin N Amer* **54**:805-817, 1970.
5. Gitnick GL, Gleich GJ, Schoenfield LJ, et al: Australia antigen in chronic active liver disease and cirrhosis. *Lancet* **2**:285-288, 1969.
6. Wright R, McCollum RW, Klatskin G: Australia antigen in acute and chronic liver disease. *Lancet* **2**:117-121, 1969.
7. Vierucci A, Scalise G, Bianchini AM, et al: L'antigene Australia: III. Presenza in pazienti con epatite cronica. *Pediat Int* **18**:1-8, 1968.
8. Blumberg BS, Gerstley BJS, Hungerford DA, et al: A serum antigen (Australia antigen) in Down's syndrome, leukemia and hepatitis. *Ann Intern Med* **66**:924-931, 1967.
9. Blumberg BS, Melartin L, Lechat M, et al: Association between lepromatous leprosy and Australia antigen. *Lancet* **2**:173-176, 1967.
10. Blumberg BS, Alter HJ, Visnich S: A "new" antigen in leukemia sera. *JAMA* **191**:541-546, 1965.
11. Sutnick AI, London WT, Blumberg BS, et al: Australia antigen (a hepatitis associated antigen) in leukemia. *J Nat Cancer Inst* **44**:1241-1249, 1970.
12. Sutnick AI, London WT, Gerstley BJS, et al: Anicteric hepatitis associated with Australia antigen: Occurrence in patients with Down's syndrome. *JAMA* **205**:670-674, 1968.
13. London WT, DiFiglia M, Sutnick AI, et al: An epidemic of hepatitis in a chronic hemodialysis unit: Australia antigen and differences in host response. *New Eng J Med* **281**:571-578, 1969.
14. London WT, Sutnick AI, Blumberg BS: Australia antigen and acute viral hepatitis. *Ann Intern Med* **70**:55-59, 1969.
15. Blumberg BS, Riddell NM: Inherited antigenic differences in human serum beta-lipoproteins: A second antiserum. *J Clin Invest* **42**:867-875, 1963.
16. Wroblewski F, LaDue JS: Serum glutamic pyruvic transaminase in cardiac and hepatic disease. *Proc Soc Exp Biol Med* **91**:569-571, 1956.
17. Gocke DJ, Kavey NB: Hepatitis antigen: Correlation with disease and infectivity of blood donors. *Lancet* **1**:1055-1059, 1969.
18. Millman I, Zavatone V, Gerstley BJS, et al: Australia antigen in the nuclei of liver cells of patients with viral hepatitis detected by the fluorescent antibody technique. *Nature* **222**:181-184, 1969.
19. Cherubin CE, Hargrove RL, Prince AM: The serum hepatitis related antigen (SH) in illicit drug users. *Amer J Epidem* **91**:510-517, 1970.
20. Barrett PVD, Boyle JD: "Hippie hepatitis": The possible role of methamphetamine in chronic active hepatitis. *Gastroenterology* **54**:1219, 1968.
21. Holmes AW, Rosenblate H, Eisenstein R, et al: Addict hepatitis: Toxic or viral? *J Clin Invest* **48**:39a, 1969.
22. Dismukes WE, Karchmer AW, Johnson RF, et al: Viral hepatitis associated with illicit parenteral use of drugs. *JAMA* **206**:1048-1052, 1968.
23. Allen JG, Dawson D, Sayman WA, et al: Blood transfusions and serum hepatitis: Use of monochloracetate as an antibacterial agent in plasma. *Ann Surg* **150**:455-467, 1959.
24. Kunin CM: Serum hepatitis from whole blood: Incidence and relation to source of blood. *Amer J Med Sci* **237**:293-303, 1959.
25. Grady GF, Chalmers TC: Risk of post-transfusion viral hepatitis. *New Eng J Med* **271**:337-342, 1964.
26. Walsh JH, Purcell RH, Morrow AG, et al: Post-transfusion hepatits after open-heart operations: Incidence after the administration of blood from commercial and volunteer donor populations. *JAMA* **211**:261-265, 1970.
27. Cohen SN, Dougherty WJ: Transfusion hepatitis arising from addict blood donors. *JAMA* **203**:427-429, 1968.
28. Potter HP Jr, Cohen NN, Norris RF: Chronic hepatic dysfunction in heroin addicts: Possible relation to carrier state of viral hepatitis. *JAMA* **174**:2049-2051, 1960.
29. Goeser E, London T, Sutnick A, et al: Post-transfusion hepatitis and frequency of donor blood Australia antigen in population subgroups. *Clin Res* **18**:380, 1970.
30. London WT, DiFiglia M, Sutnick AI, et al: Australia antigen associated hepatitis epidemic in a hemodialysis unit: The Au test. *Clin Res* **16**:567, 1968.

Reprinted from *Transfusion*
Vol. 14, No. 1, Jan/Feb. 1974

Hepatitis Carriers Among Soldiers Who Have Returned from Vietnam. Australia Antigen Studies

B. S. Blumberg, W. T. London, A. I. Sutnick, F. R. Camp, Jr., A. J. Luzzio and N. F. Conte

From the Department of Medicine, University of Pennsylvania, the Institute for Cancer Research, Philadelphia, Pennsylvania and the U. S. Army Medical Research Laboratory, Ft. Knox, Kentucky

Australia antigen (HBAg) or antibody (HBAb) was detected by immunodiffusion in 17 of 803 Vietnam returnees, but in none of 846 recruits. It is estimated that about 58,000 hepatitis carriers will have been added to the population by servicemen returning from Vietnam. A past history of hepatitis appeared to be no more common among those with Australia antigen (HBAg) or antibody (HBAb) than in those in whom they were not found. It is recommended that all servicemen returning from Southeast Asia be tested to detect asymptomatic hepatitis carriers.

Australia antigen (HBAg) is an antigen associated with an infectious agent which can cause hepatitis in man.[1, 2] The detection of Australia antigen can be useful in the diagnosis of acute and chronic hepatitis and to detect hepatitis carriers. The latter observation is the basis of the present use of the Australia antigen test in many blood banks and laboratories to decrease the incidence of posttransfusion hepatitis.

In our initial study of Australia antigen, we reported a high frequency of it in the indigenous populations of Vietnam,[3, 4] and a frequency of about 6 per cent was reported in subsequent studies.[5] This made it probable that US soldiers living under field conditions in Vietnam would have a much greater exposure to Australia antigen (and hepatitis) than similar troops stationed in the continental United States or in Europe, where the frequency of Au in the general population is about 0.1 per cent. Many individuals are (relatively) asymptomatic carriers of persistent Australia antigen. Several studies support the hypothesis that this persistent carrier state is genetically determined and controlled by a gene designated *Au*[1, 5, 6, 8] The frequency of the gene determining this trait varies from population to population and is very common in portions of the tropics and Southeast Asia, such as Vietnam. An implication of this hypothesis is that individuals with genotypes different than the one which determines the carrier state would be likely to contract acute hepatitis when exposed to the agent.

In 1968, we began to test soldiers who had returned to Philadelphia from Vietnam. Hepatitis was common among these individuals and Australia antigen was common in the patients with hepatitis. It was known that hepatitis was a major cause of morbidity in Vietnam (Table 1). These two findings made it likely that the hepatitis carrier rate and hence the frequency of Australia antigen among apparently asymptomatic Vietnam returnees would be greater than in US soldiers who had not been to Vietnam. We tested this hypothesis in a study of Armed Forces personnel located at Fort Knox, Kentucky.

Received for publication March 10, 1973; accepted September 4, 1973.

This investigation was supported by USPHS grants CA-06551, CA-08069, CA-06927, and RR-05539 from the National Institutes of Health, and by an appropriation from the Commonwealth of Pennsylvania.

TABLE 1. *The Incidence Rates of Viral Hepatitis in US Armed Forces for the Census Years (CY) 1965 to 1970.**

CY	World-wide†	RVN†	USAREUR†	CONUS†
1965	1.00	5.70	0.35	0.50
1966	1.28	3.99	0.47	0.62
1967	2.41	7.01	0.25	0.97
1968	3.37	8.59	0.23	1.83
1969	3.02	6.41	0.35	1.98
1970	4.04	7.24	0.39	3.34

* Data obtained from the Analysis Reports, and Publication Division, Office of The Surgeon General, Washington, DC 20314.

† Worldwide; RVN = Republic of Vietnam; USAREUR = Europe; CONUS = Continental United States. Rates per 1,000 average strength per year.

Materials and Methods

It was decided to study a group of recruits prior to service and to test a different group of soldiers recently returned from Vietnam. We tested the null hypothesis that there was no significant difference in HbAg, HBAb, and both in these two groups.

At the US Army Reception Center in Fort Knox, Kentucky, blood was collected from 846 recruits who had not been to Vietnam and 803 members of the 194th Armored Brigade; the latter were collected within one to 18 months (average about eight months) of their return from Vietnam. These blood specimens were used for the determination of red blood cell groups, the remaining sera were coded and sent to Philadelphia for Australia antigen and antibody testing.

Australia antigen and antibody were detected by agar gel diffusion.[7] After the tests were completed the code was revealed and the distribution of antigen and antibody determined.

The probability values for Tables 2 and 3 were determined by Fisher's exact method.

The estimates of Table 4 were determined as follows. The "most likely" entries are the products of the expected number of Vietnam returnees (2.6 million) and the fraction of cases having the given characteristics. The "minimum estimates" were obtained as 90 per cent lower one-sided confidence bounds based on the normal approximation to the binomial distribution. The lower bounds on the estimated proportions were multiplied by the expected number of returnees to obtain the tabulated values.

Results

The results are shown in Table 2. Six individuals had HBAg and 11 had HBAb; all were Vietnam returnees. The probability that this distribution could be due to chance is 0.132×10^{-1} for antigen, 0.352×10^{-3} for antibody, and 0.440×10^{-5} for both combined (Fisher's exact method). Hence, the null hypothesis was rejected for antigen, antibody, and both; that is, it is highly likely that service in Vietnam leads to a significant increase in Australia antigen and hepatitis carriers, and in individuals with anti-Au.

Of the 17 individuals who had either antigen or antibody, only one gave a history of hepatitis or jaundice (Table 3). He had been hospitalized for hepatitis in Vietnam about four months prior to the collection of his serum. Of the 786 soldiers in which neither antigen or antibody was detected, 28 gave a history of hepatitis or jaundice; these groups are not significantly different from each other. From this, we can conclude that the persistent antigen state, or detectable antibody, is not, in this study, associated with clinical hepatitis.

In order to determine if a known hepatitis exposure could account for the presence of antigen or antibody, the 17 individuals found to have either antigen or antibody were asked again if they had had hepatitis, or if they had had transfusions. Replies were received from

TABLE 2. *Frequency of Au(1), Anti-Au(1), and Either in Recruits and Soldiers Returned from Vietnam.*

Study Group	Total	Au (1) N	Au (1) %	Anti-Au (1) N	Anti-Au (1) %	Either N	Either %
Recruits	846	0	0	0	0	0	0
Returnees	803	6	0.75	11	1.37	17	2.12
p value		0.132×10^{-2}		0.352×10^{-3}		0.446×10^{-5}	

12, including all six of the individuals with antigen, and six of those with antibody. Only one (a person with antigen) gave a history of hepatitis, and none had had transfusions. Similar inquiries were sent to 17 soldiers randomly selected from the 194th Armored Brigade who had returned from Vietnam and 17 recruits from the US Army Reception Center who had not been to Vietnam. None gave a history of blood transfusion or hepatitis. From this, we concluded that the presence of antigen or antibody in these soldiers was not associated with a history of hepatitis or of transfusion.

Discussion

From these results, it is possible to conclude that service in the Army, specifically in Vietnam, significantly increases the frequency of Australia antigen and of hepatitis carriers. We did not investigate whether service in other locations also increased the frequency of carriers. However, the reported incidence of hepatitis among US Army personnel serving in Vietnam (Table 1) suggests that the acquired carrier rate would also be highest among these men.

It is possible to estimate the total number of individuals among those returning from Vietnam who will develop persistent Australia antigen and antibody and become carriers. Confidence limits were established for each of these estimates and the results are shown in Table 4. These estimates are based on the assumption that the exposure rate will be about the same during the entire period of service in Vietnam and that about 2.6 × 10^6 individuals will return from there. Because of the small sample size, the variance is large; this can be decreased by the study of larger numbers of individuals. Such a study is now in progress.

The estimate of the number of hepatitis carriers is based on the assumption that the detection of Australian antigen by the immunodiffusion method identifies about one-third of possible carriers. It can be recalculated when better data are available.

TABLE 3. *Responses to the Question "Have You Had Hepatitis or Jaundice?"*

Au(1) or Anti-Au(1)		Hepatitis or Jaundice: Yes	No
	Yes	1	16
	No	28	758

p = .468

N = 803 returnees from Vietnam. The data are arranged in a two by two contingency table to distinguish between soldiers who did and did not have Au (1) or anti-Au (1).

From these calculations, it can be said that the Vietnam experience will have added about 58,200 carriers to the US population. Each of these individuals can potentially spread hepatitis to people to whom they donate blood or with whom they come into contact.

The results can also be interpreted to mean that the individuals who became carriers usually do not have a history of hepatitis. According to the genetic hypothesis we have presented elsewhere, many of the carriers are individuals homozygous for the gene Au^1 which, according to this view, controls the development of the chronic

TABLE 4. *Most Likely and Minimum Estimates (90% Confidence)**

Characteristic	Most Likely Estimate	Minimum Estimate
Au (1)	19,400*	9,522
Anti-Au (1)	35,616	21,966
Au (1) or anti-Au (1)	55,016†	38,138
Hepatitis carriers‡	58,200	40,896

* Estimates are of the number of individuals 1) with persistent Au (1), 2) with anti-Au (1), 3) with either, 4) who are hepatitis carriers, among the estimated 2.6 × 10^6 individuals who have or will have returned from Vietnam.

† Includes an estimated 3,100 individuals who would have Au (1) prior to their residence in Vietnam. This is based on the estimate 1/1000 Au (1) carriers in the general population.

‡ Calculated on the assumption that the immunodiffusion test detects about one-third of the hepatitis carriers.

antigenemic state. These individuals are (generally) asymptomatic while those with alternate phenotypes would be more likely to develop acute hepatitis when exposed. The latter are represented by the returnees listed in Table 3 who have a history of hepatitis but did not develop persistent antigen.

It is recommended that servicemen returning from Southeast Asia, and possibly other areas, should be tested for the presence of Australia antigen and antibody and should be informed of the results. Those with antigen are not eligible to donate blood.

One out of the 11 antisera was sufficiently potent to be used as a reagent for antigen testing. We estimate that about 10 per cent of sera containing antibodies detected by immunodiffusion are useful as reagents. If only a fraction of these would donate or sell their blood, sufficient antisera would be available for national and international laboratory use as Australia antigen test reagents. This is based on the assumption that the antisera detected in this program have as wide a specificity as antisera from transfused patients.

Acknowledgment

We are indebted to Dr. Samuel Litwin, who developed the statistical programs used in this paper.

References

1. Blumberg, B. S., B. J. S. Gerstley, D. A. Hungerford, W. T. London, and A. I. Sutnick: A serum antigen (Australia antigen) in Down's syndrome, leukemia and hepatitis. Ann. Intern. Med. **66:** 924, 1967.
2. ———, A. I. Sutnick, W. T. London, and I. Millman: Current concepts. Australia antigen and hepatitis. N. Engl. J. Med. **283:** 349, 1970.
3. ———: Polymorphisms of the serum proteins and the development of isoprecipitins in transfused patients. Bull. N. Y. Acad. Med. **40:** 377, 1964.
4. ———, H. J. Alter, and S. Visnich: A "new" antigen in leukemia sera. JAMA **191:** 541, 1965.
5. ———, L. Melartin, R. A. Guinto, and B. Werner: Family studies of a human serum isoantigen system (Australia antigen). Am. J. Hum. Genet. **18:** 594, 1966.
6. ———, J. S. Friedlaender, A. Woodside, A. I. Sutnick, and W. T. London: Hepatitis and Australia antigen: autosomal recessive inheritance of susceptibility to infection in humans. Proc. Nat. Acad. Sci. USA **62:** 1108, 1969.
7. ———, R. J. Byrne, R. M. Chanock, W. C. Cockburn, Y. Kono, J. Koza, R. W. McCollum, D. Menache, K. Penttinen, R. H. Purcell, P. E. Taylor. F. G. Wewalka, and A. J. Zuckerman: Viral hepatitis and tests for the Australia (hepatitis-associated) antigen and antibody. Bull. WHO **42:** 957, 1970.
8. Cepellini, R.: Considerazeoni sulla distribuzione dell antigene Australia nells popolazione italiana. Proc. Int. Sympos. Australia Antigen and Viral Hepatitis, Milan, 1970.

Proc. Natl. Acad. Sci. USA
Vol. 84, pp. 274–278, January 1987
Medical Sciences

Effects of an extract from *Phyllanthus niruri* on hepatitis B and woodchuck hepatitis viruses: *In vitro* and *in vivo* studies

(antiviral agent/*Marmota monax*/DNA polymerase/hepatitis B surface antigen/woodchuck hepatitis surface antigen)

P. S. VENKATESWARAN*, I. MILLMAN, AND B. S. BLUMBERG

Fox Chase Cancer Center, Philadelphia, PA 19111

Contributed by B. S. Blumberg, September 18, 1986

ABSTRACT An aqueous extract of the plant *Phyllanthus niruri* inhibits endogenous DNA polymerase of hepatitis B virus and binds to the surface antigen of hepatitis B virus *in vitro*. The extract also inhibits woodchuck hepatitis virus (WHV) DNA polymerase and binds to the surface antigen of WHV *in vitro*. The extract, nontoxic to mice, was tested for antiviral activity in woodchucks (*Marmota monax*). In a trial using six long-term WHV-carrier woodchucks, five treated animals showed a faster decrease in woodchuck hepatitis virus surface antigen titer compared to one untreated control. In animals recently infected with WHV, the extract was effective when administered i.p. in three out of four animals in reducing and within 3–6 weeks eliminating both the surface antigen titer and DNA polymerase activity in serum. The treatment was discontinued after 10 weeks, and the treated animals have remained free of detectable markers of WHV for more than 45 weeks. In contrast, three untreated controls remained positive for both markers for WHV. One of the controls died after 8 weeks; the other two controls have remained positive for WHV markers for more than 45 weeks. In a third trial with long-term carriers, test animals treated subcutaneously with the extract for 12 weeks did not respond; but on switching the mode of administration to i.p., two out of the five animals showed a significant decrease in woodchuck hepatitis virus surface antigen titer compared to controls.

Chronic carriers of the hepatitis B virus (HBV) may remain asymptomatic for long periods, but many are at high risk of eventually developing post-hepatitic cirrhosis and primary hepatocellular carcinoma. Carriers are often infected within the first few years of life, but symptoms of chronic liver disease and primary hepatocellular carcinoma may not be perceived until the third, fourth, or later decades; pathogenesis, even though relentless, is slow (1, 2).

Materials of animal, bacterial, and plant origin (3) have been described that appeared to interfere with the binding of the HBV surface antigen (HBsAg) to the HBsAg antibody (anti-HBs). Subsequently, about 1200 species of plant were tested, and about one-third were found to inhibit anti-HBs–HBsAg binding. To obtain more specificity and increase the probability of obtaining an effective therapeutic agent, in addition to the inhibition of HBsAg–anti-HBs binding, we examined plant extracts *in vitro* to determine if they inhibited the endogenous DNA polymerase (DNAp) of HBV, which is necessary for its replication. The first plant tested was *Phyllanthus niruri*, which has been and is used widely (4) in southern India and elsewhere for the treatment of jaundice. The treatment of HBV carriers has not been recognized in traditional indigenous medical systems. The inhibition of anti-HBs–HBsAg binding by *P. niruri in vitro* has been reported by Thyagarajan *et al.* (5).

To assess the effects of *P. niruri* on the replication of HBV-like viruses *in vivo*, we used the woodchuck (*Marmota monax*) as an animal model. The carrier state in woodchucks and humans is similar. Liver diseases including primary hepatocellular carcinoma induced by woodchuck hepatitis virus (WHV) in woodchucks are very similar to those induced by HBV in humans. WHV is similar to HBV (6, 7) with substantial immunological cross-reactivity (8) and significant homology of DNA (9). The endogenous DNAp of both viruses exhibited optimal activities in the same range of pH, $MgCl_2$ concentration, and showed similar sensitivity to inhibitors like phosphonoformic acid and arabinofuranosyl nucleotides (10).

In this paper we report that *P. niruri* has profound effects *in vitro* on HBsAg, on woodchuck hepatitis virus surface antigen (WHsAg), and on the DNAp of both viruses and *in vivo* on the replication of WHV and on liver histopathology. In some controlled studies, it appeared to eliminate WHV from carriers.

MATERIALS AND METHODS

Preparation of the Aqueous Extract of *P. niruri*. Dried whole plant (40 g) was pulverized in a Waring blender and mixed with 200 ml of water. The mixture was shaken periodically (60°C) for 2 hr and filtered through nylon mesh. The filtrate was centrifuged at 8000 rpm for 1 hr in a Beckman JA10 rotor at 20°C. The supernatant was filtered through a 0.45-μm filter (Millipore) for *in vivo* studies.

Assay for HBsAg or WHsAg Binding Activity. Serial dilutions of the aqueous extract of *P. niruri* were mixed with an equal volume of sera positive for HBsAg or WHsAg, and the mixture was incubated for 1 hr at 20°C. The mixture was assayed directly for HBsAg or cross-reacting WHsAg using Ausria II ELISA kits (Abbott). Binding activity was expressed as the decrease (in percent) in the absorption of the test sample compared to that of the control composed of 1:1 (vol/vol) mixture of surface antigen positive serum and PBS. (PBS = 0.01 M sodium phosphate/0.85% NaCl, pH 7.2.)

Assay of WHsAg Titers in Serum. Serum titers of WHsAg were determined by assaying serial dilutions of serum with the Ausria II kit using the value obtained for sera of uninfected woodchucks as controls.

Inhibition of Endogenous Viral DNA Polymerase Activity. Suspensions of the virus were added to a reaction mixture containing the nucleotides required for DNA synthesis by DNAp. The formation of DNA was determined by gel electrophoresis. Serial dilutions of the extracts of the plant were added to determine their inhibitory ability.

The publication costs of this article were defrayed in part by page charge payment. This article must therefore be hereby marked "*advertisement*" in accordance with 18 U.S.C. §1734 solely to indicate this fact.

Abbreviations: DNAp, DNA polymerase; HBV, hepatitis B virus; WHV, woodchuck hepatitis virus; WHsAg, woodchuck hepatitis surface antigen; HBsAg, hepatitis B surface antigen; anti-HBs, antibody to HBsAg.

*To whom reprint requests should be addressed.

Table 1. Effect of *P. niruri* extract on the binding of HBs to HBsAg or WHsAg

P. niruri extract, mg/ml	% inhibition of anti-HBs binding	
	WHsAg	HBsAg
5	61	63
2.5	45	35
1.25	25	18
0.63	11	9
0.5	13	0
0.31	13	0
0.1	5	0

Serum (50 μl) containing HBV (or WHV) particles was layered over a 10–20% sucrose gradient and centrifuged at 45,000 rpm for 3 hr in a SW 55 Ti rotor in a Beckman ultracentrifuge. Pelleted virus was separated from the supernatant and was resuspended in 5 μl of 0.05 M Tris·HCl, pH 8.0. Equal volume of serial dilutions of the extract of *P. niruri* was mixed with the virus suspension. Then nucleotide triphosphates dATP, dGTP, dCTP, and [^{32}P]dTTP were added in the presence of 0.05 M Tris·HCl, pH 8.0, containing 10 μM $MgCl_2$, 0.15 M NaCl, 1 mM dithiothreitol, and 0.1% Nonidet P-40, and the mixture was incubated at 37°C for 2 hr. For control, the virus suspension was mixed with 0.05 M Tris·HCl buffer, pH 8.0, instead of the potential inhibitor. The reaction was stopped by the addition of 15 μl of Pronase (0.5 mg/ml) in 0.1% $NaDodSO_4$ containing 0.01 M EDTA in Tris·HCl, pH 7.4; and the mixture was electrophoresed on 1.5% agarose gel using bromophenol blue as the tracking dye. The DNA formed was detected by autoradiography.

A quantitative determination of the inhibition was obtained by substituting [^{3}H]dTTP and [^{3}H]dGTP for [^{32}P]dTTP as radiolabel according to the method reported by Hantz *et al.* (10) with modifications. In this experiment the reaction was terminated by the addition of 2.5 ml of 5% (wt/vol) trichloroacetic acid [containing 2% (wt/vol) pyrophosphate]. Then 40 μl of 2.5% bovine serum albumin and 100 μl of 5% (wt/vol) calf thymus DNA were added as carriers, and the mixture was filtered through a glass fiber filter (Whatman). The filter was washed thrice with 5% (wt/vol) trichloroacetic acid containing pyrophosphate, thrice with 95% (vol/vol) ethanol, dried under a heat lamp, and the radioactivity remaining on the filter was measured in a scintillation counter (Packard Instrument, Downers Grove, IL).

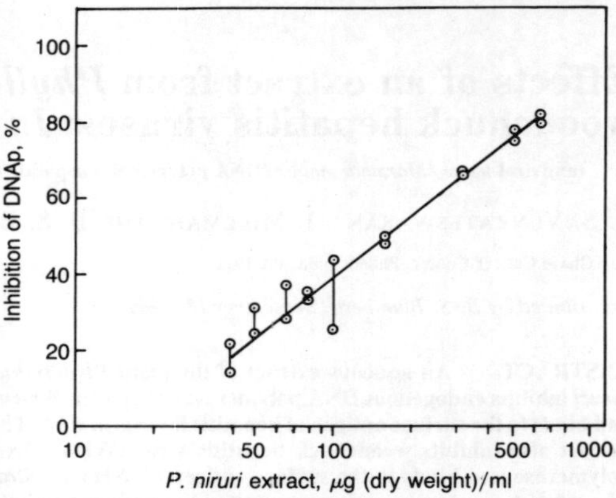

FIG. 1. Inhibition of DNAp of WHV by aqueous extracts of *P. niruri.* The data are presented as μg (dry weight) per ml of assay mixture.

To determine the WHV DNAp activity in sera of woodchucks used for experiments *in vivo*, a virus pellet, centrifuged from 50 μl of serum on sucrose density gradient, was used directly without addition of any diluent buffer as described above.

RESULTS

The Effect of *P. niruri* Extract on WHsAg and HBsAg *in Vitro*. *P. niruri* extract inhibits the reaction of HBsAg with anti-HBs and of WHsAg with anti-HBs (Table 1). The inhibition was concentration dependent for HBsAg and WHsAg.

Inhibition of WHV DNAp by the Extract *in Vitro*. Serial dilutions were assayed for their ability to inhibit WHV DNAp activity. Inhibition (Fig. 1) was directly proportional to the concentration of the extract up to 600 μg/ml at which the inhibition was 82%.

Toxicity Tests in Mice. Forty outbred albino female mice were used to determine the toxicity of the extract in accordance with a National Institutes of Health recommended assay to determine acute toxicity. The mice were divided into eight

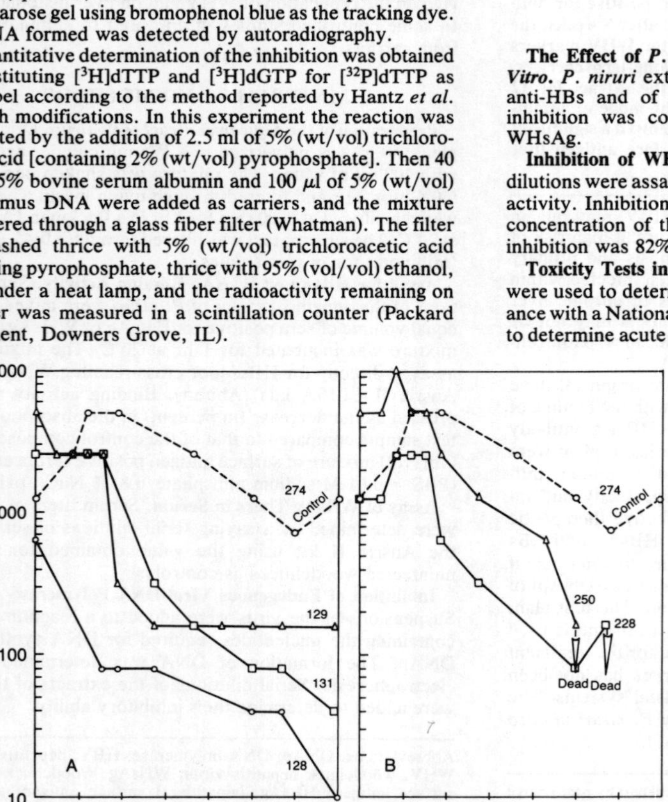

FIG. 2. Effect of aqueous extracts of *P. niruri* given i.p. on long-term chronic-carrier woodchucks (solid lines) compared to a single control carrier (dashed lines). For clarity, the results have been arbitrarily separated in two groups, *A* and *B*. The numbers of the animals are given next to the appropriate curve.

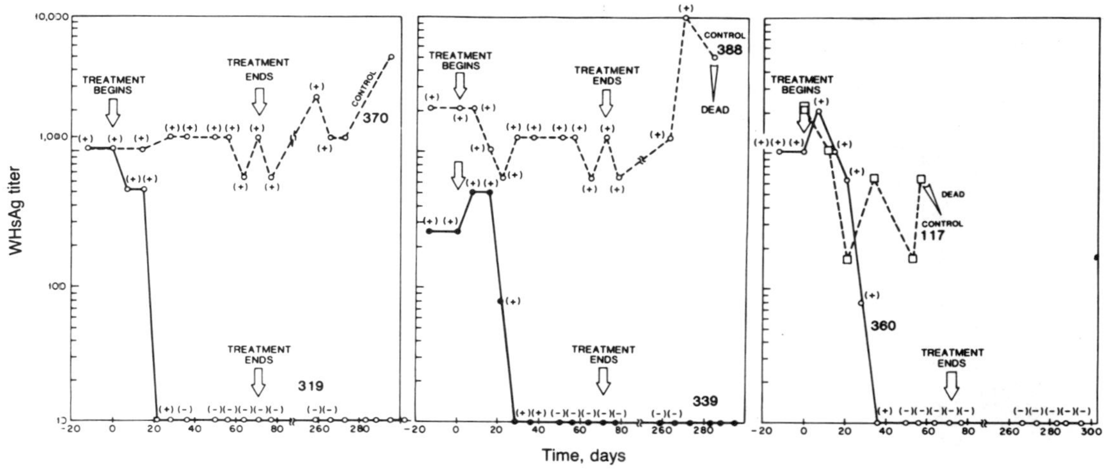

FIG. 3. Effect of aqueous extracts of *P. niruri* given i.p. on recently infected woodchucks. Solid lines indicate treated animals, and dashed lines indicate the controls. Arrows indicate the beginning and end of treatment. The DNAp activity [positive (+) or negative (−)] is indicated in parentheses.

sets of five mice and were weighed. Four sets designated "test" were given i.p. aqueous extract of *P. niruri* [0.1 ml, 1.8 mg (dry weight) per mouse], while the other four sets, designated "control," were given i.p. PBS, 0.1 ml per mouse. All sets were weighed after 3 and 7 days. There was no loss in weight among the treated or control mice after 3 days. There was a net gain in weight among all the mice after 7 days. One of the treated mice died on the 3rd day of internal injury unrelated to the test.

Long-Term Chronic-Carrier Woodchucks. An initial study was conducted on woodchuck carriers whose sera showed the presence of both the surface antigen and WHV DNAp activity at the time these animals were trapped. The length of time these animals were carriers of WHV was unknown, but it was probably more than several months. Of the six carrier woodchucks that were available, five animals, 128, 129, 131, 228, and 250, were treated; and one, 274, was used as control. Test animals were given 0.5 ml of the extract [9 mg (dry weight)] i.p. once a week, while the control animal was given the same volume of buffered saline i.p. once a week. The animals were bled periodically, and the titer of WHsAg was determined (Fig. 2). A comparison of the slope of the curves indicated that the extract appeared to reduce the titer of WHsAg in long-term chronic carriers.

Woodchucks Recently Infected with WHV. The second study was conducted on animals that were negative for WHV at the time of capture but became positive while in the enclosures in our woodchuck colony. By the time of the experiment, these animals had been infected for at least 1 month. The aqueous extract of *P. niruri* [0.5 ml, 9 mg (dry weight) per woodchuck] was administered i.p. twice a week to four woodchucks, 319, 339, 360, and 376A; the other three, 370, 388, and 117, received 0.5 ml of PBS twice a week. They were followed by weekly bleeding and assay for WHsAg and WHV DNAp. Treatment was terminated after 72 days, but the weekly bleedings were continued for over 300 days. Liver biopsy was performed on day 80, except for control woodchuck 117, which was autopsied when it died on day 57.

In woodchuck 319, WHsAg started dropping soon after the start of treatment, becoming undetectable about 21 days later (Fig. 3). WHV DNAp activity stayed positive for about 1 week after the surface antigen titer became undetectable, but subsequently it also became undetectable. Although the treatment with extract was terminated after 72 days, there were no detectable levels of the surface antigen or DNAp

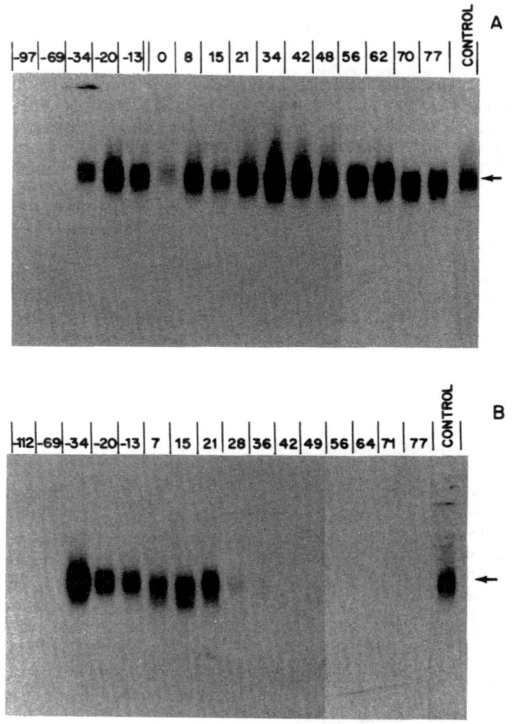

FIG. 4. Autoradiography of DNAp activity on the day indicated (−, before treatment) in the sera of a control animal, 370 (*A*) and a treated animal, 339 (*B*). These are typical of the other two control and treated animals (see Fig. 3). Arrow indicates the band of 3.3-kilobase, closed, circular DNA of WHV.

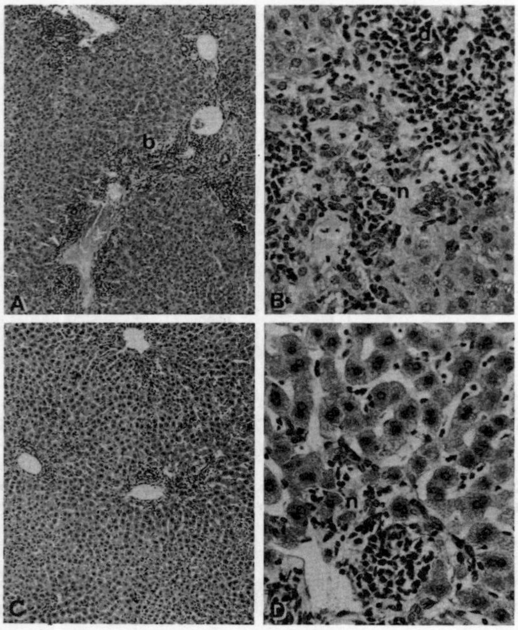

FIG. 5. Comparison of livers from the untreated woodchuck 370 (*A* and *B*) and the treated woodchuck 339 (*C* and *D*) that were infected with WHV and obtained at biopsy 8 days after the termination of treatment. All preparations were sectioned in paraffin at 5 μm and stained with hematoxylin and eosin. (*A*) Typical pattern of chronic progressive viral hepatitis with granulomatous inflammation centered chiefly around portal triads with frequent "bridging" intervening spaces between them (b). Note blurring of intralobular cords as compared with *C* (treated), owing to inflammatory swelling of individual hepatocytes. (×75.) (*B*) A periportal lesion, with an agglomeration of lymphocytes, plasmacytes, histiocytes, fibroblasts, and necrotic hepatocytes (n); proliferating biliary ductules (d) are conspicuous, which, along with fibrosis, indicate a progression to cirrhosis. (×300.) (*C*) Minimal inflammatory foci are barely visible in the portal triads of this treated animal. Hepatocytic swelling and loss of crispness of cords is seen to the right of center. (×75.) Overall, these effects are much less than in the untreated liver. (*D*) Small periportal granuloma; hepatocyte and cords are well preserved, apart from rare necrosis (n). (×300.)

activity in 319 up to 300 days after the start of treatment (i.e., 228 days after termination of treatment). The control animal 370, on the other hand, did not show a drop in either WHsAg titer or DNAp activity up to 300 days.

Treated animals 339 and 360 showed a similar drop in WHsAg between 21 and 35 days after the start of treatment, followed 7–14 days later by a drop in WHV DNAp; the markers stayed undetectable up to 300 days. Control animals 117 and 388, however, showed high levels of the WHsAg and DNAp during the same period. Autoradiography of the product of DNAp reaction in the serial bleedings of control woodchuck 370 and *P. niruri* extract-treated woodchuck 339 is given in Fig. 4. The presence of a band at 3.3 kilobases (arrow) indicated WHV DNAp activity up to 21 and 28 days. These became undetectable thereafter. DNAp of control animal 370 did not change.

Woodchuck 376A, one of the four treated animals, became bacteremic early in the experiment. Chloramphenicol was administered, but the animal succumbed to the infection. It did not respond to the extract.

The histopathology of the livers of control woodchuck 370 and treated animal 339 are shown in Fig. 5. Data on the liver biopsy performed 8 days after the termination of treatment are given in Table 2. (Liver biopsies before treatment were not available.) The three untreated controls, 370, 388, and 117, showed extensive portal infiltration and focal necrosis; all three were diagnosed as chronic active hepatitis. The livers of the treated animals 319, 339, and 360, on the other hand, showed marginal or negative portal infiltration and focal necrosis. The diagnosis of woodchuck 319 was early mild viral hepatitis, of 339 was early active hepatitis, and of 360 was minimal portal hepatitis.

Table 2. Data on liver biopsies performed 8 days after termination of treatment in experiment involving woodchucks recently infected with WHV

	Pathology		
	Portal infiltrate	Focal necrosis	Diagnosis
Control animal			
370	+++	++	Chronic, active hepatitis
388	++	++	Chronic, active hepatitis
117	+++	++	Chronic, active hepatitis
P. niruri extract-treated animal			
319	±	±	Mild viral hepatitis
339	+	+	Active hepatitis
360	±	0	Minimal, portal hepatitis

0, None. ±, Marginal. +, Minimal positive. ++, Positive. +++, Extensive.

Subcutaneous Administration of the Extract in Long-Term Carrier Woodchucks. Five of eight long-term WHV-carrier woodchucks (327, 437, 456, 488, and 492) were administered 0.5 ml [9 mg (dry weight)] of extract subcutaneously twice a week. The remaining three (318, 429, and 471) were given 0.5 ml of PBS subcutaneously twice a week. The animals were bled weekly, and the titer of WHsAg and WHV DNAp was monitored. During 3 months of treatment, there was no appreciable change in either marker in the treated or control animals.

We concluded that subcutaneous administration was ineffective. We hypothesized that either the active principle was not absorbed by this route, that antibodies were developed against the extract, or some other mechanism rendered it ineffective. After 90 days the mode of administration was changed to the intraperitoneal route that had apparently been successful in the previous studies. Two of the treated animals showed a drop in WHsAg about 60 days after switching to i.p. administration (Fig. 6). One control and two treated animals died due to bacteremic infections. None of the control animals showed any significant change in WHsAg.

DISCUSSION

London and Blumberg (1) have proposed a model to explain the observations on the relation between primary hepatocellular carcinoma and HBV. It postulates the existence of fully differentiated liver cells which, when infected, allow complete replication of HBV. (They are designated S cells; i.e., susceptible to replication.) Less-differentiated liver cells (common in the fetus and newborn but less so in the adult liver), when infected, do not allow replication, although penetration of the virus and integration of virus DNA may occur. (They are designated R cells; i.e., resistant to repli-

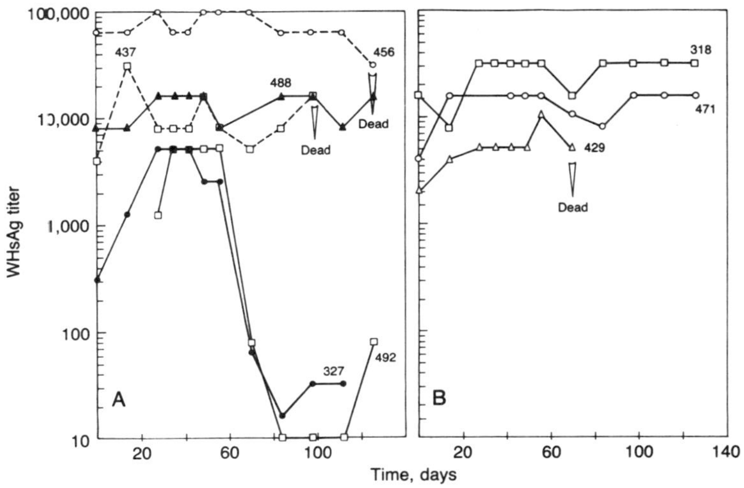

FIG. 6. Effects of *P. niruri* extracts given i.p. to animals that did not respond to subcutaneous administration of the same extract. (*A*) Treated animals; (*B*) controls.

cation.) The proliferating HBV in the differentiated cells leads to cell death, in part caused by the immune response of the host to its S cells altered by HBV. The less-differentiated R cells are not damaged by the virus, since the virus does not replicate. The R cells divide and multiply in response to the death of the S cells. With increased division of R cells, chromosomes are more liable to disruption, deletion, rearrangement, or mutation by the resident HBV (or, possibly, another agent). This could result in a favored clone that divides rapidly and is eventually perceived as a cancer.

The course of these events could be stopped if virus were eliminated from the carrier. If this is not possible, then decreasing the virus load or inhibiting its entry into liver cells could slow the death of liver cells so that perceptible disease would not be expected to occur until the carrier had lived out a life span. We have referred to this as "prevention by delay" (2).

In an attempt to achieve this goal, we looked for agents that would affect the virus or its entry into liver cells. The aqueous extract of *P. niruri* inhibits HBV DNAp and WHV DNAp and interferes with the binding of anti-HBs to HBsAg and WHsAg apparently because of its ability to bind the surface antigen.

The extract of *P. niruri* was tested in three independent *in vivo* experiments in woodchuck carriers. In the first, with long-term carriers, a significant drop in the titer of WHsAg was observed in the extract-treated animals compared to the control. In the second, using short-term carriers, the WHsAg and WHV DNAp in three of four treated animals became undetectable and stayed that way even after the treatment was terminated, while the levels of these markers stayed high in the controls. These results indicated a possible break in the carrier state directly attributable to the treatment with *P. niruri* extract. The conclusion from the third experiment was that the extract was not effective when administered subcutaneously. However, on switching the mode of administration to i.p., two out of three extract-treated animals showed a drop in WHsAg titer. The advantages of intraperitoneal chemotherapy for liver ailments have been reported (11) and may be applicable to WHV and HBV infections since replication of both these viruses takes place in the liver. Although this experiment was faulted because of its *post hoc* design, the results were in the direction predicted from the two earlier experiments.

Our preliminary results indicate that there are one or more active materials in *P. niruri* that inhibit the replication of WHV *in vivo* and decrease the pathological effects of WHV on woodchuck liver. By inference, the substance should affect HBV infection in humans similarly. Using a variety of techniques (including HPLC), we have identified fractions from *P. niruri* containing the DNAp inhibitory activity and the surface antigen binding activity.

We are indebted to Philip Custer for the histopathological studies, Howard Blatt for the biopsies, and Terry Halbherr and Renee Grob for technical assistance. The research was supported in part by grants from the Kaiser Family Foundation, the Mary L. Smith Charitable Lead Trust, and the Commonwealth of Pennsylvania.

1. London, W. T. & Blumberg, B. S. (1982) *Hepatology (Baltimore)* **2,** 10S–14S.
2. Blumberg, B. S. & London, W. T. (1982) *Cancer* **50,** 2657–2665.
3. Millman, I. & McMichael, J. C. (1978) *Infect. Immun.* **21,** 879–885.
4. Chopra, R. N., Neyer, S. L. & Chopra, I. C. (1956) *Glossary of Indian Medicinal Plants* (CSIR, India), p. 191.
5. Thyagarajan, S. P., Tiruneelakantan, K., Subramanian, S. P. & Sundaravelu, T. (1982) *Ind. J. Med. Res.* **76,** 124–130.
6. Summers, J., Smolec, J. M. & Snyder, R. (1978) *Proc. Natl. Acad. Sci. USA* **75,** 4533–4537.
7. Summers, J., Smolec, J. M., Werner, B. G., Kelly, T. J., Tyler, G. V. & Snyder, R. L. (1980) in *Viruses in Naturally Occurring Cancers*, eds. Essex, M., Todaro, G. & zur Hausen, H. (Cold Spring Harbor Laboratory, Cold Spring Harbor, NY), Vol. 7, pp. 459–470.
8. Millman, I., Southam, L., Halbherr, T., Simmons, H. & Kang, C. M. (1984) *Hepatology* **4,** 817–823.
9. Galibert, F., Chen, T. N. & Mandart, E. (1981) *Proc. Natl. Acad. Sci. USA* **78,** 5315–5319.
10. Hantz, O., Ooka, T., Vitvitski, L., Pichoud, C. & Trepo, C. (1984) *Antimicrob. Agents Chemother.* **25,** 242–246.
11. Speyer, J. L. (1985) in *Liver Cancer*, eds. Bettino, I. C., Onfell, R. W. & Muggia, F. M. (Nijhoff, Boston), pp. 225–235.

Reprinted from THE LANCET, *October 1, 1988, pp. 764–766*

EFFECT OF PHYLLANTHUS AMARUS* ON CHRONIC CARRIERS OF HEPATITIS B VIRUS

S. P. THYAGARAJAN[1] S. SUBRAMANIAN[1]
T. THIRUNALASUNDARI[1] P. S. VENKATESWARAN[2]
B. S. BLUMBERG[2]

Department of Microbiology, Post Graduate Institute of Basic Medical Sciences, University of Madras, India,[1] and Fox Chase Cancer Center, Philadelphia, Pennsylvania, USA[2]

Summary In a preliminary study, carriers of hepatitis B virus were treated with a preparation of the plant *Phyllanthus amarus* for 30 days. 22 of 37 (59%) treated patients had lost hepatitis B surface antigen when tested 15–20 days after the end of the treatment compared with only 1 of 23 (4%) placebo-treated controls. Some subjects have been followed for up to 9 months. In no case has the surface antigen returned. Clinical observation revealed few or no toxic effects. The encouraging results of this preliminary study recommend continued evaluation of this plant and the active principles isolated from it.

Introduction

PLANTS of the genus *Phyllanthus* have been used widely by traditional medical practitioners for the treatment of jaundice and other diseases. Their use was described in Indian Ayurvedic literature more than 2000 years ago. *Phyllanthus* species are also used in China, the Philippines, Cuba, Nigeria, Guam, East and West Africa, the Caribbean, Central America, and South America.

Extracts of *Phyllanthus amarus*[1,2] and *P tenellus* inhibit the reaction between HBsAg (the surface antigen of the hepatitis B virus [HBV]) and the antibody to it (anti-HBs). HBV contains an endogenous DNA polymerase which is required for its replication. *Phyllanthus* extracts inhibit the endogenous DNA polymerase of HBV virus, woodchuck hepatitis virus,[2] and duck hepatitis B virus (unpublished). The inhibition is quantitatively specific for the DNA polymerases of these HBV-like viruses; much larger quantities are required to inhibit the DNA polymerases of bacteria and mammals (unpublished).

Five studies of the effect of extracts of *Phyllanthus* species on woodchuck (*Marmota monax*) carriers of woodchuck

*The plant tested in this study is known in India as *Phyllanthus niruri* and has been so described in previous reports (Thyagarajan 1982, Venkateswaran 1987, Thyagarajan 1986). It has now been identified as *P amarus* (Grady Webster, University of California, Davis, CA; Personal communication).

2

hepatitis virus have been completed[2] (and published). The extract-treated animals were much more likely than placebo-treated controls to lose the virus, and the virus titres fell in more of the extract-treated than placebo-treated animals.

Acute (1 month) and chronic (3 months) toxicity studies in mice have been reported.[2,3] There was no clinical or histopathological evidence of toxicity. The traditional medical uses indicate that this plant is non-toxic, although, to our knowledge, there have been no previous systematic studies of toxicity in man or animals. Various methods of administration are described in the traditional systems. In most of these methods the whole plant, rather than individual components, is used. Our eventual aim is to identify, purify, and synthesise an active agent from *Phyllanthus* which can be used in the treatment of HBV infection. We are concentrating on the fraction of the extract that inhibits the HBV DNA polymerase.

It is important to establish first whether the whole plant is effective in abolishing the carrier state. It is also possible that there are several active substances, or that several activities must act in conjunction for effective therapy or to prevent toxicity. Studies with the whole plant would contribute to an understanding of these questions.

Phyllanthus is used widely in India and elsewhere but, to our knowledge, there has been no controlled trial of its effectiveness in carriers. In this study, we have tested the hypothesis (stated and implied previously[1,2]) that treatment with *P amarus* will lead to the loss of the HBV carrier state.[3] Additional explanatory variables, particularly HBV serological responses, have also been measured to differentiate between responders and non-responders to the treatment.[4]

Subjects and Methods

Carriers of HBV in three categories were identified. These included 28 symptom-free carriers detected among voluntary blood donors at the Institute of Child Health and the Hospital for Children, Madras; 14 carriers with symptoms, who were outpatients with the diagnosis of chronic liver diseases at the Department of Gastroenterology, Government General Hospital, Madras; and 18 patients from the Institute of Child Health and Hospital for Children with glomerular nephritis, who were also chronic carriers of HBV.

Before the trial, the carriers were followed for 6 months with repeated (4–6) serological tests for HBsAg and anti-HBc IgM. Only those who remained carriers for 6 months were included in the study. The glomerular nephritis patients were followed for 1 year before inclusion. Carriers were stratified by sex and age (within 3 years) and assigned randomly to treatment and placebo groups. Treatment and follow-up were carried out with the subjects as outpatients. The treated patients were given 200 mg of the dried, powdered, sterilised plant in a capsule three times a day for 30 days. Placebo-treated patients were given an equal amount of sterile lactose in a capsule. The subjects were instructed to continue their regular diet and activities. Of the 78 cases (40 treated, 38 placebo) randomised, 18 (3 treated, 15 placebo) failed to complete the course of capsules or did not return after the treatment. The population available for follow-up therefore consisted of 37 treated and 23 placebo subjects. All the clinical studies were done in Madras with plants collected locally.

Full-grown plants (30–45 cm) were selected. The whole plants, including roots, were quickly washed with cold water. A technique of intermittent sterilisation was used. The washed plants were maintained at 45°C for 30 min daily for 3 successive days. The roots were removed and the remainder of the plant powdered in an electric blender. The powder was dispensed under sterile conditions in 200 mg doses into gelatin capsules which had been presterilised with ethylene oxide. Samples of the dried powder were tested for aerobic and anerobic bacteria and fungi by incubation in the appropriate media (blood agar plate, thioglycollate broth, Sabouraud's glucose agar). Randomly selected capsules from four batches of capsules were tested similarly on a regular basis. There was no evidence of contamination in any of the plant substance used for treatment. One batch of powder showed five colonies of *Bacillus subtilis,* and this batch was not used. Lactose was used as a placebo. It was presented in gelatin capsules identical to those used for the plant powder in the same doses for the same time. Ninety capsules of plant or placebo were placed in a coded container. The doctors and nurses administering the substances were told to give a coded container from one or the other group alternately to carriers as they were admitted to the study. They were unaware who was receiving which treatment. The code was maintained at the Department of Microbiology which is 10–15 km from the hospitals where the capsules were given. The initial serological tests were carried out before the subjects were randomised. All the serological tests were done at the Department of Microbiology.

60 subjects returned for the first follow-up v sit 15–20 days after the end of treatment. They were then asked to return each month. All 60 returned for the first 2 months. All but 1 treated and 4 placebo subjects returned for the third monthly visit and follow-up continued for some subjects for up to 9 months. At the first visit after treatment, each subject was examined clinically and a clinical questionnaire was completed with the help of the examining physician. The questionnaire was designed to determine whether any symptoms or signs occurred during treatment, including those that could be ascribed to side-effects or toxicity of the plant preparation. The information obtained included age, sex, and a history of jaundice, blood transfusions, surgery, and injections. A general medical history was also obtained. Symptoms which might be attributed to side-effects of the treatment included: fatigue, malaise, fever, chills, urticaria, anorexia, nausea, abdominal pain, diarrhoea, lymphadenitis, myalgia, arthralgia, headache, dizziness, disturbed sleep, paraesthesia, respiratory illness, and skin rash.

On each subsequent visit, the subjects were asked if they had had any symptoms or findings which might be attributable to the treatment. Blood samples were collected by venepuncture before treatment and at each follow-up visit.

Laboratory Methods

The serum was tested in Madras for HBsAg by enzyme-linked immunosorbent assay (ELISA; 'Auszyme II'), hepatitis B e antigen (HBeAg) by ELISA (HBe EIA), and IgM antibody against the core

4

TABLE I—COMPARISON OF SUBJECTS WHO RETURNED FOR FOLLOW-UP AND THOSE WHO DEFAULTED

—	Number (%)							
	Male	Age (yr)			HBsAg(+) only	HBsAg(+), HBeAg(+), anti-HBc IgM(–)	HBsAg(+), HBeAg(+), anti-HBc IgM(+)	HBsAg(+), HBeAg(–), anti-HBc IgM(+)
		< 10	11–30	> 30				
Returned (n = 60)	40 *(67)*	20 *(33)*	20 *(33)*	20 *(33)*	25 *(42)*	16 *(27)*	13 *(22)*	6 *(10)*
Did not return (n = 18)	14 *(78)*	4 *(22)*	12 *(67)*	2 *(11)*	5 *(28)*	6 *(33)*	3 *(33)*	6 *(22)*

5

antigen (anti-HBc IgM) by 'Corzyme-M' (Abbott Laboratories, North Chicago, Illinois, USA).

To exclude the possibility of systematic error in the determinations of HBsAg, the major outcome variable in this study, the serum samples that remained after the initial testing were also tested for HBsAg in Philadelphia. The western blot technique of McMichael et al,[5] which requires only 3 μl serum, was used because only small amounts of serum remained for the confirmatory test. 42 samples were tested by this method. There was a very high correlation between the ELISA and western blot methods; all but 7 results were concordant. Of these, 4 showed a "trace" or "1+" score in the western blot compared with zero in the ELISA. The results of the ELISA method were used in the final analysis. To determine whether there were any differences in clinical chemistry between the treated and placebo groups, a series of studies were done on the blood in the pretreatment collection. Bilirubin,[6] serum alanine and aspartate aminotransferases,[7] and alkaline phosphatase[8] were measured for the symptomatic carrier group and the glomerular nephritis group but not for the symptom-free carriers. Renal function and other studies were carried out for the glomerular nephritis group only. These included blood urea,[9] serum creatinine,[10] serum cholesterol,[11] serum total protein, albumin, and globulin,[12] 24 h urine protein,[13] and blood glucose.[14]

The results of the studies on the placebo and treated groups were compared by the Mann-Whitney statistic. The only significant difference was a higher level of bilirubin in the plant-treated group. From this we infer that there were no significant clinical differences between the groups.

Results

The demographic and serological characteristics of the subjects lost to follow-up and those who where followed for at least 2 months after the end of treatment are shown in table I. There were no significant differences between the groups. There was a significant difference between the placebo and treatment groups in the numbers defaulting (15 *vs* 3), but we are unable to explain why this occurred. According to the study design, the physicians and subjects were unaware of who was receiving the treatment or placebo, and there is no reason to believe that this rule was disregarded.

Overall, 22 (59%) of the 37 treated subjects compared with only 1 (4%) of the 23 placebo-treated patients had lost HBsAg at the first follow-up visit, a highly significant difference ($p < 0.0001$, Fisher's exact test).

All subjects in the treated group, except for 1 symptom-free carrier, were followed for at least 3 months. HBsAg did not return in any of those who had lost HBsAg at the first follow-up visit. Follow-up has continued for up to 9 months in some subjects (table II), and there have been no cases of antigen reappearing.

In the placebo group only 1 of those followed up (a symptom-free carrier) had lost the antigen by the first follow-up visit, and she remained negative at the 3rd month follow-up (her last visit).

6

TABLE II—NUMBER OF SUBJECTS AND LENGTH OF FOLLOW-UP

Follow-up (mo)	No of subjects	
	Treated	Placebo
1	37	23
2	37	19
3	36	19
4	28	10
5	19	4
6	10	0
7	6	0
8	3	0
9	1	0

Overall, males and females appeared to respond equally, but in the symptomatic carrier group there was a suggestion of a poorer response by males. The numbers are too small to allow conclusions to be drawn on possible sex differences in response to treatment.

13 of the 14 plant-treated carriers (93%) who were positive for HBsAg but negative for HBeAg and anti-HBc IgM cleared the carrier state. Of the 9 treated carriers who were positive for HBsAg and HBeAg but negative for anti-HBc IgM, only 4 of 9 (44%) were cleared. Of the 8 who were positive for all three markers, only 1 (13%) was cleared. 4 of the 6 (67%) treated carriers positive for HBsAg and anti-HBc IgM but negative for HBeAg lost the carrier state. Overall, HBsAg-positive carriers with HBeAg were less likely to respond to treatment than those without HBeAg (5 of 17 [29%] *vs* 17 of 20 [85%]; $p<0{\cdot}001$). The single placebo-treated carrier who lost HBsAg had been positive for HBsAg only and was symptom-free.

The toxicity questionnaire described above was used during the follow-up visits. None of the treated or placebo groups seen at the first or subsequent follow-up visits had symptoms or signs attributable to toxic side-effects of the plant.

A rash developed in 1 treated subject, a boy aged 2·5 years, after 10 days of treatment. The treatment was discontinued. The child was found to be infected with roundworm, and his physician believed that the rash was a consequence of the infestation and not the *Phyllanthus* treatment. He was not included in the analysis.

Discussion

Our hypothesis has been supported by this preliminary study of human HBV carriers. Treated carriers were much more likely to lose the carrier state (as determined by the HBsAg ELISA) than those given placebo (59% *vs* 4%). Those that lost HBsAg remained clear for several months after the end of treatment. Follow-up on the carriers is continuing. Carriers with HBeAg were less likely to respond. The presence of anti-HBc IgM also appeared to reduce further the probability of clearance. Anti-HBc IgM

7

is thought to indicate recent infection, and presumably more active replication is taking place during this period. A smaller percentage of the carriers with evidence of viral replication (HBeAg) responded to treatment. This trial does not provide sufficient information on mode of action to explain this observation. It may be due to an inadequate dose and this can be tested in subsequent trials. Only 1 placebo subject cleared the virus.

We should consider the limitations of this preliminary clinical study. For reasons that are unclear, a significantly greater number of placebo subjects were lost to follow-up. On the basis of the experience with followed placebo subjects we would expect to see 1 subject among the placebo group not followed lose HBsAg. Although the inconsistent follow-up represents a potential flaw in the study, it seems unlikely that it could account for the highly significant difference between the treated and placebo groups.

The clinical and laboratory data collected in this preliminary study were limited by the facilities available in the outpatient services. However, the main outcome variables (HBsAg, anti-HBc IgM, HBeAg) were tested by sensitive and standard methods. It was not feasible to carry out extensive laboratory studies which might have detected side-effects of the medication. However, clinically, there were no obvious detrimental effects. This observation is supported by the lack of evidence of toxicity in the animal studies.[1-3]

These results must be regarded as tentative because of the flaws in design and execution already discussed, and the inherent uncertainty of trials of plant preparations, for example, possible inconsistency of dosage. They will, however, allow for a more rigorous design of future studies in India and elsewhere of the whole plant, and serve as a benchmark for eventual clinical trials of isolated and characterised active agents.

We thank Dr N. Madanagopalan (Department of Digestive Health and Diseases, Government Peripheral Hospital, Annanagar, Madras) Dr Vimala Ramalingam (Blood Bank), and Dr B. R. Nammalwar (Department of Nephrology, Institute of Child Health and Hospital for Children) for their collaboration with the clinical studies.

The work at Madras was supported by ICMR grants in IRIS no: 8301410. The work in Philadelphia was partly supported by USPHS grant CA-06927 from the National Institutes of Health and by an appropriation from the Commonwealth of Pennsylvania.

Correspondence should be addressed to S. P. T., Department of Microbiology, Postgraduate Institute of Basic Medical Sciences, Taramani, Madras 600 113, India.

REFERENCES

1. Thyagarajan SP, Thiruneelakantan K, Subramanian S, Sundaravelu T. *In vitro* inactivation of HBsAg by *Elipta alba* hassk and *Phyllanthus niruri* Linn. *Ind J Med Res* 1982; **76:** 124-30.
2. Venkateswaran PS, Millman I, Blumberg BS. Effects of an extract from *Phyllanthus niruri* on hepatitis B and woodchuck hepatitis B viruses: *In vitro* and *in vivo* studies. *Proc Natl Acad Sci USA* 1987; **84:** 274–78.

8

3. Jayaram S, Thyagarajan SP, Panchanadam M, Subramanian S. Anti-hepatitis B properties of *Phyllanthus niruri* Linn and Elipta alba hassk: *In vitro* and *in vivo* safety studies. *Biomedicine* 1987; **7:** 9–16.
4. Thyagarajan SP, Thirunalasundari T, Subramanian S, Nammalwar BR, Prabha V. Effect of a medicinal plant on hepatitis B surface antigen carriage: a double blind clinical trial (a research report). *Virus Inform Exchange Newsletter* (western Australia) 1987; **4:** 9.
5. McMichael JC, Greisiger LM, Millman I. The use of nitrocellulose blotting for the study of hepatitis B surface antigen electrophoresed in agarose gels. *J Immunol Method* 1981; **45:** 79–94.
6. Malloy HT, Evelyn KA. The determination of bilirubin with the photoelectric colorimeter. *J Biol Chem* 1937; **119:** 481–90.
7. Reitman S, Frankel S. A colorimetric method for the determination of serum glutamic axalacetic and glutomic pyruvic transaminases. *Am J Clin Pathol* 1957; **28:** 56–63.
8. King J. A note on the colorimetric assay of coenzyme I and II dependent enzymes. *J Med Lab Techn* 1960; **17:** 89–92.
9. Natelson S. Microtechniques for the routine laboratory. Springfield: C. C. Thomas, 1957: 381–91.
10. Owen JA, Iggo B, Scandrett FJ, Stewart CP. Determination of creatinine in plasma or serum, and in urine; a critical examination. *Biochem J* 1954; **58:** 426–37.
11. Zak B. Simple rapid microtechnic for serum total cholesterol. *Am J Clin Pathol* 1957; **27:** 583–88.
12. Reinhold JC. In: Reiner M, ed. Standard methods of clinical chemistry. New York: Academic Press, 1953: 88.
13. Varley H. Practical clinical biochemistry. New York: Interscience Books, 1967: 349–93.
14. Marks V. An improved glucose oxidase method for determining blood, cerebrospinal fluid, and urine glucose levels. *Clin Chim Acta* 1959; **4:** 395–400.

Printed in Great Britain by Robquest Ashford Kent

Reprinted from *Advances in New Crops*
Eds. J. Janick and J. E. Simon
Timber Press, Inc. Portland, OR pp. 518–521, 1990

Phyllantus Species: Sources of New Antiviral Compounds

David W. Unander, P. S. Venkateswaran, Irving Millman, Herbert H. Bryan, and Baruch S. Blumberg

INTRODUCTION

Phyllanthus has more than 700 species in at least 10 sub-genera (Holm-Nielsen 1979, Webster 1956–58). A survey of 300 ethnobotanical references of *Phyllanthus* spp. arranged taxonomically suggested some uses were clustered by subgenus. A summary of the number of species per sub-genus used traditionally in various places to treat possible symptoms of liver disease (such as jaundice) is presented in Table 1. Among the herbs of subgenus *Phyllanthus* used medicinally, we found that aqueous extracts of *P. amarus* Schum. & Thom., *P. debilis* Klein, *P. fraternus* Webster, *P. niruri* and *P. urinaria* L. all inhibited viral DNA polymerase (DNAp) of hepadnaviruses *in vitro.* (Hepadnaviruses include human hepatitis B virus and several animal hepatitis viruses.) Additionally, extracts of less common species in the same sub-genus, such as *P. mimicus* Webster or *P. odontadenius* Muell. Arg., also had such anti-DNAp activity. Extracts of species from sub-genera not reported as being used for symptoms of liver disease generally had much weaker inhibitory activity (data unpublished).

The ultimate goal of our research is to develop a therapy for carriers of hepatitis B virus (HBV) which would reduce the incidence of liver cancer. Since its identification (Blumberg et al. 1967, Blumberg 1977), the virus has been found to be associated with cirrhosis, chronic liver disease and primary liver cancer as well as acute serum hepatitis. More than 200 million people worldwide are estimated to be carriers, many of them asymptomatically (Ghendon 1987). A series of studies have shown that persistent HBV infection is associated with a greatly elevated risk of developing liver cancer (Beasley et al. 1981, Dodd and North 1987). Although a safe and effective vaccine exists, and is being vigorously promoted, there is no effective therapy for carriers.

Aqueous extracts of *P. amarus* collected near Madras, India, (initially published under the name *P. niruri* L.) inhibited viral DNAp *in vitro*. In addition, they eliminated detectable virus from the sera of woodchucks (*Marmota monax*) acutely or chronically infected with the woodchuck hepatitis virus (WHV) (Venkateswaran et al. 1987). Endogenous DNAp is packaged with viral DNA in hepadnaviruses like HBV or the closely related WHV. In a small clinical trial in India, administration of powdered *P. amarus* eliminated detectable antigen from the sera of 59% of the treated human carriers as compared to 4% of controls (carriers receiving a placebo) who cleared the virus naturally (Thyagarajan et al. 1988). Studies are in progress to characterize the compound(s) responsible for these effects.

Table 1. *Phyllanthus* subgenera, number of representative species, regions where found, number of species used as liver disease remedies and references.[z]

Subgenus	Approx. no. of spp.	Location: North America	South America	West Indies	Africa	Asia	Australia	Oceania	No. spp. used as liver disease remedies
Isocladus	70	X	X	X	X	X	X	X	3[y]
Kirganella	50	X	X	X	X	X	X	X	1[x]
Phyllanthus	140	X	X	X	X	X	X	X	9[w]
Gomphidium	155		X				X	X	0
Conami	20	X	X	X	X				0
Eriococcus	30					X			0
Phyllanthodendron	10					X			0
Botryanthus	15		X	X					0
Xylophylla	60		X	X					0

[z]Summarized from a series of papers in preparation by D. W. Unander and G. L. Webster examining 300 ethnobotanical references from a taxonomic perspective and incorporating recent taxonomic changes by G. L. Webster.
[y]Kangsu Medical Institute 1975, Kirtikar and Basu 1975, Stehlé 1986.
[x]Al-Kindi ca. 850, Antarkar et al. 1980, His Majesty's Government of Nepal 1982, Kirtikar and Basu 1975, Oliver 1950, Said 1969.
[w]Amadeo 1988, Bunyapraphatsara 1987, Cruz 1965, Freise 1934, Gooding et al. 1965, His Majesty's Government of Nepal 1984, Kangu Medical Institute 1975, Kirtikar and Basu 1975, MacRae et al. 1988, Morton 1981, National Institute of Health 1977, Oliver 1960, Petelot 1954, Puri 1970, Singh 1986, Stehlé and Stehle 1962, Watt 1892.

SOURCES OF ANTI-DNAp VARIABILITY

In Experiment 1, plants of *P. urinaria* and *P. debilis* were grown in the greenhouse in a 2 × 2 factorial design comparing soils of differing fertility (fertilized regularly with a 0.04N-0.06P-0.07K solution vs. not fertilized) and different moisture and structure conditions (a commercial peat mix regularly watered vs. sand infrequently watered). Six month old plants were harvested, lyophilized and ground. Aqueous extracts were made using 5 g ground plant in 20 ml water at 60° C for two hours, mesh-filtered, centrifuged, and the supernatant passed through a 0.45 μm micropore filter (Nalgene 125, Nalge Co., Rochester, NY). Dried whole and ground plants and extracts were stored at −15°C. Extracts were made of two random samples of ground plant per treatment combination. Activity was expressed as μg of dry matter per ml of extract producing 50% inhibition of WHV DNAp, after Venkateswaran et al. (1987), and analyzed after testing for homogeneity of variance.

Various morphological variables (e.g., plant height, weight, amount of branching), were significantly affected by differing soil conditions (data not shown). The only effects on inhibition of viral DNAp, however, were for gross soil fertility on *P. urinaria* (84 and 38 μg/ml for fertilized and unfertilized, respectively) and, in the pooled analysis, species (139 and 61 μg/ml for *P. debilis* and *P. urinaria*, respectively). Regression analyses showed significant effects on *P. urinaria* of available N or P or percent saturation of K. Genetic differences in viral DNAp inhibition were greater than any effects from soil conditions.

In Experiment 2, plants of *P. amarus* were grown in the greenhouse in a 2 × 3 factorial design comparing soil moisture and structure conditions as in Experiment 1, and three levels of soil pH, and in a 2 × 2 factorial comparing soil moisture and structure conditions and high vs. low Ca levels. *P. amarus* was commonly found in calcarious sites in Puerto Rico but much less commonly in other soils, so a preference for high pH, high Ca or both was hypothesized, and the effect of these variables on activity of plant extracts against WHV DNAp was tested. Media of average pH 5.5, 7.5 and 10.5 were made using aluminum sulphate, calcium sulphate and hydrated lime, respectively. The treatments in sand had mean pH values about one log unit greater than those in the peat mix (i.e., 11.5 vs. 10.5). To test the effect of Ca apart from pH, pulverized limestone was added or not added to the sand or peat mix. In this test, the pots with and without Ca had pH means of 8.0 and 7.0, respectively. Water samples taken from the bottom of pots during watering were used to follow pH. A more acidic treatment was also attempted, but *P. amarus* did not survive more than two weeks at about pH 4.5. In this experiment individual plants served as replicates with three replicates in the first test and two in the second. Extracts, assays and analyses were done as described.

Soil conditions significantly affected plant growth. Optimum growth seemed to be in soil slightly above neutral and with high Ca, as expected. For the effects of plant extracts on viral DNAp, soil pH as a main effect was non-significant. Using data from both tests, the linear regression of DNAp inhibition on soil pH showed an effect at $P = 0.09$. The R^2 value (0.11), however, suggested relatively little variability in the inhibition of viral DNAp was explained by pH. Soil moisture or structure significantly affected DNAp inhibition of plant extracts in the pH test (158 and 263 μg/ml for the sand and peat, respectively). The interaction of moisture and pH had an effect at $P = 0.10$. Means over pH treatments ranged from 174–246 μg/ml, with greatest activity at higher pH; over all six treatment combinations, means ranged from 85–332 μg/ml (greatest and least activity at high pH, sand and low pH, peat, respectively). Plants from high pH in sand were significantly stunted relative to other treatments, with average dry weight per plant of 1.7 g vs. an average for the other five treatments of 6.6 g. Plants at high soil pH in peat had mean inhibition values (262 μg/ml) similar to those at lower pH, illustrating the interaction observed between pH level and peat vs. sand. No regressions of DNAp inhibition on specific soil macronutrient data were significant. In the test examining high Ca vs. no Ca and sand vs. peat, neither main effect was significant.

In Experiment 3, genetic differences among accessions were tested. Plants of *P. urinaria* from seed collected in India, the Ivory Coast, Hawaii, Puerto Rico, Trinidad, Venezuela or Vietnam were grown randomized on one greenhouse bench using the same growing regime. Based on an analysis of extracts derived from two plants/seed source, there were highly significant differences among seed sources in inhibition, with means ranging from 70 to 532 μg/ml. This suggests that genetic variability for inhibition of DNAp exists in *P. urinaria.* Plants of *P. amarus* from seed collected from two sites in India and three sites in Puerto Rico were grown in a similar experiment with three plants/seed source but seed source was non-significant at $P = 0.25$, with means for viral DNAp inhibition ranging from 192 to 409 μg/ml. A third test using the same procedure examined different accessions of *P. amarus* from sites in Florida, India and Puerto Rico. No significant differences were found; means ranged from 122 to 163 μg/ml. These latter two tests were respectively grown in the winter and the summer.

CULTIVATION

Although common in the tropics, the *Phyllanthus* species of interest are generally small and often scattered in distribution. The only reference found which discussed cultivation (Kangsu Medical Institute) recommended fertile, well-drained soil for growing *P. urinaria.* To produce sufficient quantities for large scale extraction, a system was developed at the University of Florida Tropical Research and Education Center at Homestead, using black plastic mulch and trickle irrigation. Since plants with the strongest inhibition of viral DNAp in our studies had consistently come from the warmest environments, a tropical site was desired. Southern Florida was chosen as being both tropical and efficient for shipment of harvested plant. The soil at Homestead, which is a crushed oolitic limestone, was also considered well-suited to *P. amarus.* Wild plants of *P. amarus* were abundant in fallow fields around Homestead following periods of rain.

The first plot served as an experiment to establish *P. amarus* as a row crop and to test two fertility levels. Plots were fertilized with 560 or 1120 kg/ha 6N-5.2P-10K before mulching in a randomized block design with three replicates. Seed of *P. amarus* were sown using 1.5% cellulose gel (N-Gel, Hercules Co.) at 0.15 g seeds/liter (100 seedweight = 0.015 g). Approximately 50 cc of gel/hill were applied. Hills were holes set 30 cm apart in the mulch. Hills were in four rows spaced 30 cm apart on raised beds 1.7 m wide for a planting density of approximately 7,500 hills/ha.

Six months after planting, whole plants weighed an average of 7.5 g (lyophilized dry weight). Aqueous extracts were made as described. Extracts were made of shoots and roots by fertility treatment and replicate. The average DNAp inhibitory activity over all samples was 142 μg/ml. The two levels of fertility did not significantly affect viral DNAp inhibition. Roots were significantly more active than shoots (66 and 151 μg/ml respectively), but roots only accounted for about 10% of the total dry weight. This range of activity was consistent with earlier results obtained with plants collected in the wild (Venkateswaran et al.). Cultivation, irrigation and fertilization thus did not reduce DNAp inhibitory activity over the far more laborious collection from the wild. Stand establishment, measured as hills with any plants six months after planting, was only about 70% in this first plot. Variables affecting seed germination and longevity have not yet been well-characterized. Total potential production was estimated to be at least 40 kg/ha (dry weight) but should be substantially greater once stand establishment can be improved. At the rates used by Thyagarajan et al., 1.8 kg of dried whole plant would have provided a course of treatment for 100 persons.

CONCLUSIONS

P. amarus and closely related species appear to contain activity against the endogenous DNAp of hepadnaviruses. This may be the basis for traditional uses of these species against disease symptoms such as jaundice, which, retrospectively could, in at least some cases, have been caused by hepatitis B virus. This DNAp inhibitory activity was relatively unaffected by soil conditions, but differed among accessions, suggesting that genetic variability does exist for this trait. Plants proved easy to grow and biological activity of the cultivated material was equivalent to material collected from the wild.

REFERENCES

Al-Kindi. Circa 850. *The medical forumlary or* aqrabadhin *of Al-Kindi: translated with a study of its materia medica by Martin Levey.* 1966. Univ. of Wisconsin Press. Madison, WI.

Amadeo, A. J. 1888. The botany and vegetable materia medica of the island of Porto-Rico. *Pharmaceut. J.* Apr. 28, 1888, p. 906.

Antarkar, D. S., A. B. Vaidya, J. C. Doshi, A. V. Athavale, K. S. Vinchoo, M. R. Natekar, P. S. Tathed, V. Ramesh and N. Kale. 1980. A double-blind clinical trial of Arogya-Warhani—an Ayuvedic drug—in acute viral hepatitis. *Indian J. Med. Res.* 72:588–593.

Beasley, R. P., L-Y Hwang, L-C Lin, C-S Chien. 1981. Hepatocellular carcinoma and hepatitis B virus. *Lancet* 1981, ii:1129–1133.

Blumberg, B. S., B. J. S. Gerstley, D. A. Hungerford, W. T. London and A. I. Sutnick. 1967. A serum antigen (Australia antigen) in Down's syndrome, leukemia and hepatitis. *Ann. Internal Med.* 66:924–931.

Blumberg, B. S. 1977. Australia antigen and the biology of hepatitis B. *Science* 197:17–25.

Bunyapraphatsara, N. 1987. Personal communication: uses of *Phyllanthus* species in traditional Thai medicine. Translated from nine books in Thai. Medicinal Plant Information Center, Mahidol Univ., Bangkok.

Cruz, G. L. 1965. *Livro verde das plantas medicinais e industriais do Brasil.* Velloso S. A., Belo Horizonte.

Dodd, R. Y. and N. North. 1987. Increased risk for lethal forms of liver disease among HBsAg-positive blood donors in the United States. *J. Virol. Methods* 17:81–94.

Freise, F. W. 1934. *Plantas medicinales Brasileiras.* Instit. Agron. do Estado, São Paulo. p. 442–443.

Ghendon, Y. 1987. Perinatal transmission of hepatitis B virus in high incidence countries. *J. Virol. Methods* 17:69–79.

Gooding, E. G. B., et al. 1965. *Flora of Barbados.* Ministry of Overseas Development, Barbados.
His Majesty's Government of Nepal, Ministry of Forest and Soil Conservation, Dept. of Medicinal Plants. 1982. *Medicinal plants of Nepal.* Third edition. His Majesty's Government Press, Kathmandu.
His Majesty's Government of Nepal, Ministry of Forest and Soil Conservation, Dept. of Medicinal Plants. 1984. *Medicinal plants of Nepal.* (Supplement volume). His Majesty's Government Press, Kathmandu.
Holm-Nielsen, L. B. 1979. Comments on the distribution and evolution of the genus *Phyllanthus* (Euphorbiaceae), p. 277–290. In: K. Larsen and L. B. Holm-Nielsen (eds.). *Tropical botany.* Academic Press, New York.
Kangsu Medical Institute. 1975. *Encyclopedia of Chinese Medicine.* 3 vols. Shanghai Publisher of Science and Technology. Chinese; English translations by M. P. Wong, Fox Chase Cancer Center, Philadelphia, PA.
Kirtikar, K. R. and B. D. Basu. 1975. *Indian medicinal plants.* 2nd ed. 4 vols. Jayyed Press, New Delhi.
MacRae, W. D., J. B. Hudson and G. H. N. Towers. 1988. Studies on the pharmacological activity of Amazonian Euphorbiaceae. *J. Ethnopharmacol.* 22:143–172.
Morton, J. F. 1981. *Atlas of medicinal plants of Middle America: Bahamas to Yucatan.* Charles C. Thomas Publ., Springfield, IL.
National Institute of Health. 1977. *A barefoot doctor's manual: the American translation of the official Chinese paramedical manual.* (for Hunan province.) DHEW Publication No. (NIH) 75-695. Reprinted by Running Press, Philadelphia 19103.
Oliver, B. 1960. *Medicinal plants in Nigeria.* Nigerian College of Arts, Science and Technology, Ibadan.
Paul, S. R. 1977. Medicinal plants of Netarhat, Bihar (India). *Quart. J. Crude Drug Res.* 15:79–97.
Petelot, A. 1954. *Plantes medicinales du Cambodge, du Laos et du Vietnam.* Tome III (Amarantacées a Selaginellacées).
Plotkin, M., V. Randrianasob, L. Sussman and N. Marshall. n.d. *Ethnobotany in Madagascar.* World Wildlife Fund, Washington, DC 20009.
Puri, H. S. 1970. Indian medicinal plants used in elixirs and tonics. *Quart. J. Crude Drug Res.* 10:1555–1566.
Said, H. M. 1969. *Hamdrad pharmacopoeia of Eastern medicine.* Hamdard Natl. Fdtn., Karachi, Pakistan.
Singh, Y. N. 1986. Traditional medicine in Fiji: some herbal folk cures used by Fiji Indians. *J. Ethnopharmacol.* 15:57–88.
Stehlé, H. and M. Stehlé. 1962. *Flore medicinale Illustrée.* Vol IX. *Flore agronomique des Antilles Francaises.* Anibal Lautric, Pointe-à-Pitre, Guadeloupe.
Thyagarajan, S. P., S. Subramanian, T. Thirunalasundari, P. S. Venkateswaran and B. S. Blumberg. Effect of *Phyllanthus amàrus* on chronic carriers of hepatitis B virus. *Lancet* 1988, ii:764–766.
Venkateswaran, P. S., I. Millman and B. S. Blumberg. 1987. Effects of an extract from *Phyllanthus niruri* on hepatitis B virus and woodchuck hepatitis viruses: *In vitro* and *in vivo* studies. *Proc. Nat. Acad. Sci.* (USA) 84:274–278.
Watt, G. 1892. *A dictionary of the economic products of India. Vol VI.* W. H. Allen & Co., London.
Webster, G. L. 1956–58. A monographic study of the West Indian species of *Phyllanthus. J. Arnold Arbor.* 37:91–122, 217–268, 340–359; 38:51–80, 170–198, 295–373;39:49–100, 111–212.

Proc. Natl. Acad. Sci. USA
Vol. 91, pp. 2235–2239, March 1994
Microbiology

Secretion of human hepatitis B virus is inhibited by the imino sugar *N*-butyldeoxynojirimycin

(antivirals/glycosylation)

TIMOTHY M. BLOCK*†, XUANYONG LU‡, FRANCES M. PLATT§, GRAHAM R. FOSTER¶, WOLFRAM H. GERLICH‡, BARUCH S. BLUMBERG§, AND RAYMOND A. DWEK§

*Jefferson Cancer Institute, Jefferson Medical College, Philadelphia, PA 19107; §Glycobiology Institute, University of Oxford, Oxford OX1 3QU, United Kingdom; ‡Institute of Medical Virology, Justus Liebig Universitat, Giessen, 6300, Germany; and ¶Department of Medicine, Imperial College of Science, Technology and Medicine, St. Mary's Hospital, London W2 1PG, United Kingdom

Contributed by Baruch S. Blumberg, October 18, 1993

ABSTRACT **The imino sugar *N*-butyldeoxynojirimycin (NBDNJ) is a potent inhibitor of the oligosaccharide-trimming enzyme α-glucosidase I. Hepatitis B virus (HBV) contains three surface proteins (HBs proteins) of different sizes that are singly or doubly N-glycosylated and are essential for the formation of infectious virus. Therefore, the replication and secretion of HBV in the human hepatoma cell line HepG2 were studied in the presence of NBDNJ. In the stably HBV-transfected HepG 2.2.15 cells and in HBV-infected HepG2 cells, NBDNJ suppressed secretion of HBV particles and caused intracellular retention of HBV DNA. The secretion of subviral particles was less affected. These data suggest that inhibitors of oligosaccharide trimming may be useful for antiviral therapy of hepatitis B and for the study of the intracellular transport of the viral glycoproteins.**

Hepatitis B virus (HBV) is a causative agent of acute and chronic liver disease (1). Although effective vaccination is available, there are still >300 million people worldwide chronically infected with the virus (2). For them, the vaccine has no therapeutic value. Between 25% and 40% of those who are chronically infected with HBV develop serious liver disease (2). It is therefore important to find effective anti-HBV therapies. Interferon α has been used with promising results in some patients (3, 4). In addition, recent reports suggest that the virus-encoded DNA polymerase, which functions as a reverse transcriptase, is an attractive target (5–8). Other virus-mediated processes have not been targeted for antiviral intervention. Effective antiviral therapy for HBV is likely to involve multiple strategies, including agents that influence the host immune system as well as those that interfere with different steps in the life cycle of the virus. It was, therefore, of interest to explore the possibility that other, non-polymerase-mediated steps in the virus life cycle are vulnerable to intervention.

The HBV envelope contains three co-carboxyl-terminal proteins (HBs proteins), termed large (LHBs), middle (MHBs), and small (SHBs) proteins (see Fig. 1). These proteins result from the alternate translation initiation of a single open reading frame (9). All three HBs proteins occur with complex-type N-linked oligosaccharides at amino acid 146 of the S domain (Fig. 1; ref. 10). MHBs (but never LHBs) also occurs with hybrid-type oligosaccharides within the pre-S2 domain (ref. 10; X.L. and W.H.G., unpublished data). During natural infection with HBV, the liver produces a large excess of HBs proteins that are secreted as either filamentous or spherical subviral particles of 20 nm in diameter (9, 10, 30). HBs spheres are most abundant and contain 5–10 times less LHBs than do HBs filaments and HBV particles. MHBs is a minor component of all three types of particles (10).

The morphogenesis of HBV is complex. Preassembled viral core particles are believed to attach to the cytosolic sides of viral envelope (surface) proteins, which have inserted into the endoplasmic reticulum (ER) membrane (9). After acquiring envelopes, virions bud to the lumen of the ER, from which they are transported through the Golgi apparatus into the extracellular fluid (11). Immature glycoproteins contain three terminal glucose residues on the N-linked oligosaccharides. The removal of terminal glucose residues is thought to play an important role in the migration of immature glycoproteins from the ER to the Golgi (12). The imino sugar *N*-butyldeoxynojirimycin (NBDNJ) is a potent inhibitor of α-glucosidase I, a cellular enzyme that removes terminal glucose residues from nascent oligosaccharides and has been found to suppress formation of cytotoxic human immunodeficiency virus *in vitro* (13, 14). Since HBV secretion requires LHBs and SHBs, both of which bear N-linked oligosaccharides, the effect of NBDNJ upon virus synthesis was examined.

METHODS

Cells and Media. HepG2 cells were purchased from the European Collection of Animal Cell Cultures (Porton Down, U.K.). HepG2 2.15 (2.2.15; ref. 15) cells were obtained from George Acs (Mt. Sinai Medical College, New York). All tissue cultures were maintained in 5% CO_2 in RPMI 1640 (GIBCO) medium, supplemented with 10% heat-inactivated fetal calf serum (Techgen, London), 50 units of penicillin and streptomycin per ml, and 1 mM glutamine (GIBCO). For 2.2.15 cells, 200 μg of G418 per ml (GIBCO) was added to the medium, as in ref. 15. Cell viability was measured by flow cytometry using a FACScan cytometer (Becton Dickinson) after incubation with propidium iodide, as in ref. 16.

Infection of HepG2 Cells. HBV was purified from human serum or from the medium of cultured cells by sedimentation to densities between 40% and 60% sucrose (wt/wt) following ultracentrifugation (17). Virions were dialyzed in 0.02 M potassium phosphate buffer (pH 7.4), concentrated, treated with V8 protease overnight at 37°C, and centrifuged for 8 hr through a 20% sucrose/0.01 M Tris, pH 7.4/0.14 M NaCl/0.005 M EDTA (TNE) cushion. The pellets were resuspended in growth medium and used to inoculate HepG2 cells.

Imino Sugar Compounds. The synthesis of NBDNJ has been described (18) and was provided by Searle/Monsanto as compound SC-48334.

Detection of Viral DNA. Medium from ≈5 × 10^6 cells was precipitated with polyethylene glycol (PEG) 8000 (Sigma), after clarification, as in ref. 15, resuspended in 0.5 ml of

The publication costs of this article were defrayed in part by page charge payment. This article must therefore be hereby marked "*advertisement*" in accordance with 18 U.S.C. §1734 solely to indicate this fact.

Abbreviations: HBV, hepatitis B virus; NBDNJ, *N*-butyldeoxynojirimycin; ER, endoplasmic reticulum; mAb, monoclonal antibody.
†To whom reprint requests should be addressed.

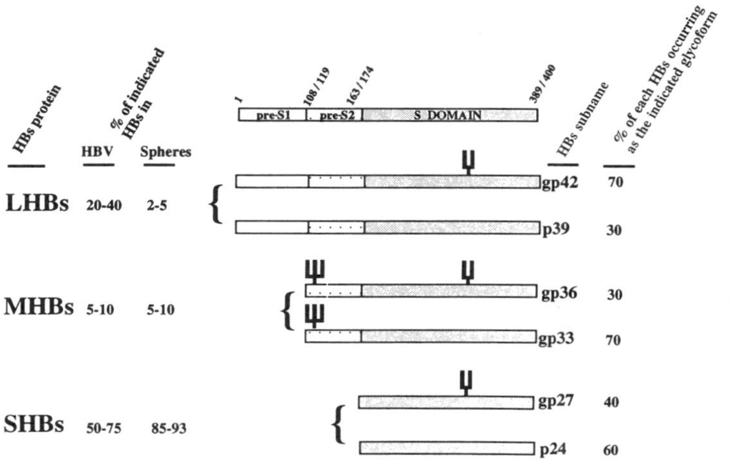

FIG. 1. Genetic map of HBV envelope proteins. The uppermost box is a linear map of the HBs genes, with the pre-S1, pre-S2, and S domains indicated. Numbers above this map indicate the boundaries of the domains, expressed as amino acid number, which vary with the genotype. Most genotypes have the longer pre-S1 domain, but genotype D expressing SHBs/ayw has an 11-amino acid deletion at the amino terminus of LHBs. Other numbers are as indicated, with values from ref. 10.

phosphate-buffered saline (PBS), and sedimented through a cushion of 20% sucrose in PBS for 5 hr at 50,000 rpm in a Beckman T100. 3 rotor (≈75,000 × g). DNA was prepared from the pellets as in ref. 15. For Southern blots (19), DNA was resolved by electrophoresis through 1.2% agarose, transferred to H+ bond (Amersham) filter paper (19), and hybridized with radioactive ^{32}P (Amersham) HBV probe [made by the random priming method described by the kit manufacturer (Amersham)] using pHBV as template (see ref. 20). Progeny virus from HepG2 cells infected with serum-derived protease-treated HBV was immunoprecipitated from 1 ml of medium 1 and 5 days after inoculation using anti-pre-S2 monoclonal antibody (mAb) Q19/10 (21). HBV DNA in the immunoprecipitate was amplified by a PCR using primers from nucleotides 2815 and 190 (using the *Eco*RI site as nucleotide 3221/1). This PCR assay was standardized with a European reference sample for HBV particles (Eurohep Group). The PCR was shown to distinguish 2.5- to 3.0-fold differences of HBV DNA within the range from 10^2 to 10^5 HBV particles per ml. Details will be described elsewhere (X.L. and W.H.G., unpublished data). DNA was prepared from cell lysates by standard methods (19).

Detection of HBV Proteins by ELISA. mAbs are described in refs. 10, 21, and 22. Samples were incubated in microtiter wells coated (overnight at 4°C) with mAbs specific for the LHBs, MHBs, or SHBs epitopes and blocked with 1% bovine serum albumin in PBS. After incubation with virus samples for 1 hr (37°C), plates were washed four times with PBS/0.1% Tween 20. Bound antigen was detected by incubation with peroxidase-conjugated goat anti-HBs antibody (Behring) followed by development in *ortho*-phenylenediamine (0.33 mg/ml of PBS/peroxide solution. Optical densities were read in a Behring plate reader. Experiments on purified virus and total medium were conducted over a series of sample dilutions to ensure proper quantification.

Western Blots. Samples were dissolved in loading buffer, resolved by electrophoresis through 13.5% SDS/polyacrylamide gels (SDS/PAGE), transferred to polyvinylidene difluoride (Millipore) membranes, and blocked with 5% powdered milk, as in ref. 23. After incubation with primary antibody overnight at room temperature in 1% bovine serum albumin in TNE and second antibody (peroxidase-conjugated goat anti-mouse IgG serum) for 1 hr in TNE at room temperature, membranes were developed in peroxide/diaminobenzidine (Sigma) PBS, as described in the manufacturer's instructions.

RESULTS

NBDNJ Reduces the Amount of Virion-Associated HBV DNA Released into the Medium by 2.2.15 Cells. 2.2.15 cells are derived from the HepG2 line and chronically secrete infectious HBV as well as subviral particles (HBs filaments and spheres) into the culture medium (15, 24). To determine the effect of NBDNJ upon the production and secretion of HBV, 2.2.15 cells were maintained in medium containing a range of NBDNJ concentrations for 6 days, with one change of medium on the third day. After 6 days in the compound, DNA was isolated from virus (*Methods*). Viral DNA was detected by hybridization of Southern blots to radioactively labeled HBV DNA probes, as shown in Fig. 2*A*. Virus-specific DNA, migrating with the expected mobility of relaxed circles and linear genome-length DNA (24), was detected in the samples derived from untreated culture medium (lanes 1 and 2). There is a clear dose-dependent decrease in virus-specific DNA obtained from the medium of cells treated with NBDNJ (lanes 3–5). The autoradiograph shown in Fig. 2*A* was quantified by densitometry (data not shown). Densitometry revealed that NBDNJ at 500 μg/ml (2.28 mM) and 1000 μg/ml resulted in decreases of 90% and 99%, respectively. These decreases were not due to

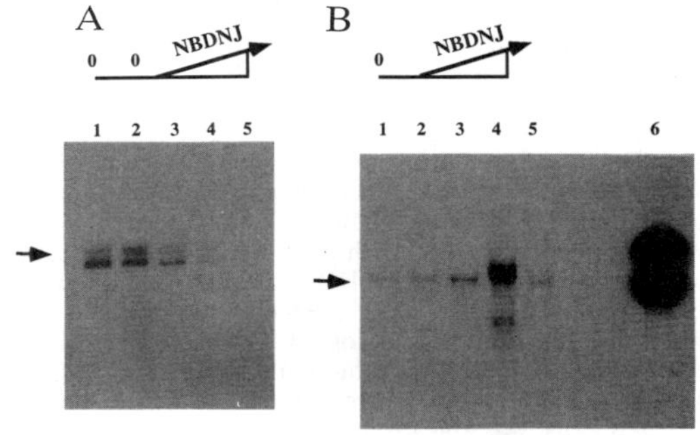

FIG. 2. HBV DNA in the medium and cells of cultures treated with NBDNJ. (*A*) Autoradiogram of a Southern blot of DNA recovered from the medium of 2.2.15 cells maintained in medium without NBDNJ (lanes 1 and 2) and in the following NBDNJ concentrations: 200 μg/ml (lane 3), 500 μg/ml (lane 4), and 1000 μg/ml (lane 5). (*B*) Autoradiogram of a Southern blot of total intracellular DNA, digested to completion with *Eco*RI, from cells maintained in the absence of NBDNJ (lane 1) and in the presence of the following NBDNJ concentrations: 200 μg/ml (lane 2), 500 μg/ml (lane 3), and 1000 μg/ml (lane 4). Lane 5, *Eco*RI-digested DNA isolated from virions prepared from untreated 2.2.15 cells; lane 6, plasmid DNA as a hybridization control. The arrows indicate the expected mobility for relaxed circular HBV genomes (*A*) and linearized 3.2-kb genomes (*B*).

toxicity of NBDNJ, since 90% of the cells maintained in NBDNJ at 1000 μg/ml for 6 days were as viable as untreated controls, as determined by fluorescence-activated cell sorting analysis of propidium iodide-stained cells and [^{35}S]methionine incorporation into proteins (data not shown).

2.2.15 Cells Treated with NBDNJ Contain Elevated Levels of Intracellular HBV DNA. The NBDNJ-mediated decrease in virion-associated HBV DNA in the medium could have been due to a decrease in viral DNA synthesis. Alternatively, it could have been due to a postsynthetic event such as virion assembly, transport, or egress from the cell. To distinguish between these possibilities, the amount of intracellular HBV-specific DNA in untreated and NBDNJ-treated 2.2.15 cells was compared. Total cellular DNA was prepared from treated and untreated cells and digested to completion with *Eco*RI to linearize viral genomes. Near-equal microgram amounts of digestion products (as determined by ethidium bromide staining and hybridization to a cell-specific probe; data not shown) were resolved by electrophoresis, Southern blotted, and hybridized with the HBV-specific probe (Fig. 2*B*). Unit-length HBV genomes migrating as 3.2-kb bands are detected in DNA derived from untreated 2.2.15 cells (lane 1) and virions isolated from control culture medium (lane 5). Lanes 2, 3, and 4 contain DNA derived from cells treated with NBDNJ at 200, 500, and 1000 μg/ml, respectively. There is a clear dose-dependent increase in the amount of HBV DNA present in NBDNJ-treated cells as compared with untreated cells. Densitometry of this autoradiogram (data not shown) suggests that cells treated with 200, 500, and 1000 μg/ml exhibit 1.7-, 3.0-, and 5.1-fold increases in HBV copy abundance, respectively, when adjusted for loading variation, using hybridization to a cellular major histocompatibility complex class III gene, G1, as a loading control (data not shown).

HepG2 Cells Infected with HBV and Treated with NBDNJ Release Less Progeny Virus. 2.2.15 cells are a useful system to study HBV production in a stably transfected environment. However, HBV pregenome synthesis, in these cells, may occur from integrated viral DNA templates and not covalently closed circular viral DNA templates, as is thought to occur during natural infection (24). Moreover, these cells produce naked core particles as well as various subgenomic viral DNA products that are released into the medium (ref. 24; T.M.B., unpublished data).

Therefore, HepG2 cells were infected with protease-modified HBV (*Methods*; X.L. and W.H.G., unpublished data). The next day, culture medium was replaced with either control medium or medium containing various concentrations of NBDNJ. Five days after infection, progeny virions were immunoprecipitated with mAb specific for the central portion of the pre-S2 domain. HBV-specific DNA sequences were amplified by PCR using HBV-specific primers. Products of the reaction were resolved by agarose gel electrophoresis and imaged after ethidium bromide staining (Fig. 3*A*). Note that only V8 protease-treated virus, and not untreated virus, was able to cause the secretion of HBV in the medium of HepG2 cells 5 days after infection, and no progeny was detected in the culture medium at 1 day after infection, suggesting that there was no detectable residual HBV from the inoculum (data not shown). The 519-bp products (Fig. 3*A*, arrow) were quantified by densitometry analysis and the plot is shown in Fig. 3*C*. Although PCR may underestimate the differences between initial DNA concentrations in samples and the slight increase in PCR product in samples treated with NBDNJ at 200 μg/ml was not reproducible, it is evident that media from cells infected with protease-processed virus and posttreated with NBDNJ at 700 μg/ml (3.2 mM) contain an order of magnitude less viral DNA than do untreated samples. These results show that the NBDNJ-mediated decrease

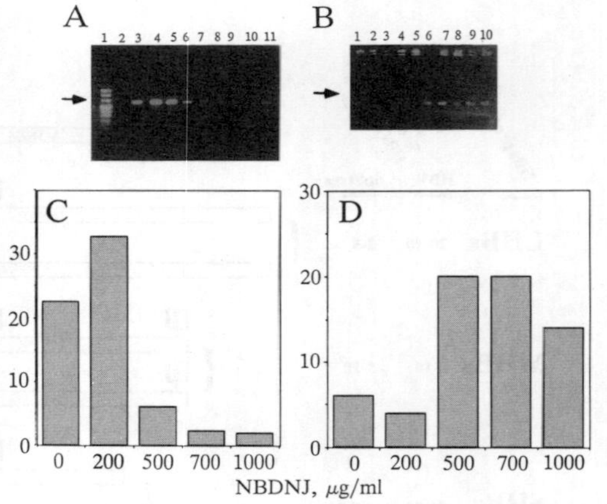

FIG. 3. HBV DNA in the cells and medium of HepG2 cells infected with HBV and treated with NBDNJ. (*A* and *C*) DNA from virions, immunoprecipitated with mAb to MHBs, was amplified by PCR and resolved by agarose gel electrophoresis. Lane 1, molecular mass markers; lane 2, blank; lane 3, medium from cells that received no NBDNJ; lanes 4 and 5, medium from cells that received NBDNJ at 200 μg/ml; lanes 6 and 7, 500 μg/ml; lanes 8 and 9, 700 μg/ml; lanes 10 and 11, 1000 μg/ml. These bands were imaged by densitometry and the areas under the peaks are shown in *C*. Note: the average of the two samples of each NBDNJ concentration was plotted. (*B* and *D*) DNA from the intracellular compartment of the cultures infected (in *A*) with HBV and treated with NBDNJ was amplified by PCR. The lanes contain amplified DNA as follows: lanes 1 and 2, no NBDNJ; lanes 3 and 4, NBDNJ at 200 μg/ml; lanes 5 and 6, 500 μg/ml; lanes 7 and 8, 700 μg/ml; lanes 9 and 10, 1000 μg/ml. (*D*) Histogram of the area under the averaged peaks of the densitometric tracing of the gel in *B* expressed in arbitrary units.

in HBV released into the medium is not peculiar to the 2.2.15-transfected cell system.

HepG2 Cells Infected with HBV and Treated with NBDNJ Contain Increased Amounts of Intracellular HBV DNA. Since the culture medium of HBV-infected HepG2 cells treated with NBDNJ was similar to that of 2.2.15 cells, with respect to the reduced amount of virion-associated DNA, it was of interest to know if there was a concomitant increase in viral DNA within the treated cells. Total cellular DNA was prepared from samples corresponding to those presented in Fig. 3 *A* and *C*. Intracellular HBV-specific DNA was amplified using PCR. Products of the reaction were resolved by agarose gel electrophoresis, and the ethidium bromide-stained gel (Fig. 3*B*) was analyzed by densitometry (Fig. 3*D*). Clearly, HepG2 cells infected with HBV and posttreated with NBDNJ accumulate greater amounts of viral DNA than do untreated cells.

The Culture Media of NBDNJ-Treated and Untreated 2.2.15 Cells Contain Similar Amounts of HBV Envelope Antigen. 2.2.15 cells secrete virions as well as subviral particles (refs. 17 and 24; T.M.B., unpublished data). The NBDNJ-mediated reduction of virion-associated DNA in the culture medium could be a reflection of a generalized decrease in the secretion of all HBs containing particles. Alternatively, there may have been a selective diminution of the rare virion particle with a relative sparing of the other form(s), which is in vast excess. To distinguish between these possibilities, the amount and nature of envelope antigens in the culture medium were determined by ELISA and Western analysis. ELISA of SHBs, MHBs, and LHBs antigens present in clarified culture medium is shown in Fig. 4 *A–C*. The results show that there is no significant effect upon the total amount of SHBs (S) and LHBs (pre-S1) antigens in the medium. There is a modest,

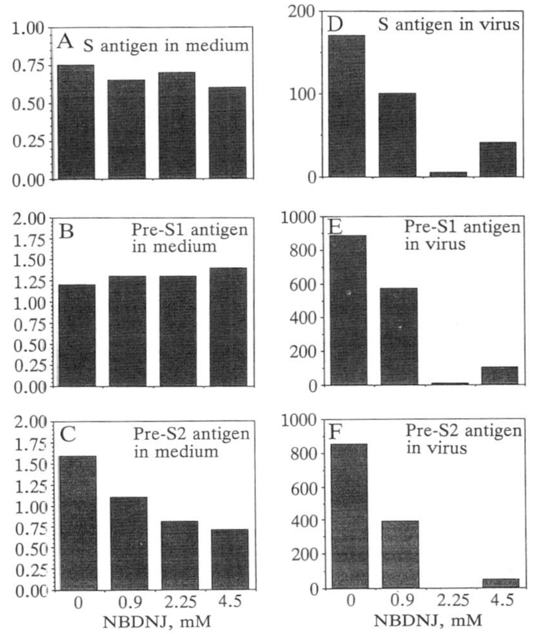

FIG. 4. HBV antigens present in the unfractionated culture medium and partially purified virion preparations. Equal volumes of the indicated samples were tested for HBV envelope antigens using the ELISA method. Samples of unfractionated medium (*A–C*) or partially purified virus (*D–F*) from 3 day cultures were tested for HBV SHBs ("S"; *A* and *D*), LHBs ("Pre-S1"; *B* and *E*), or MHBs ("Pre-S2"; *C* and *F*) epitopes. Cultures were maintained in the indicated concentrations of NBDNJ (for comparison, 2.28 mM approximately equals 500 μg/ml). The *y* axis shows the colorimetric signal of the ELISA reaction, in arbitrary OD units, read by the plate reader.

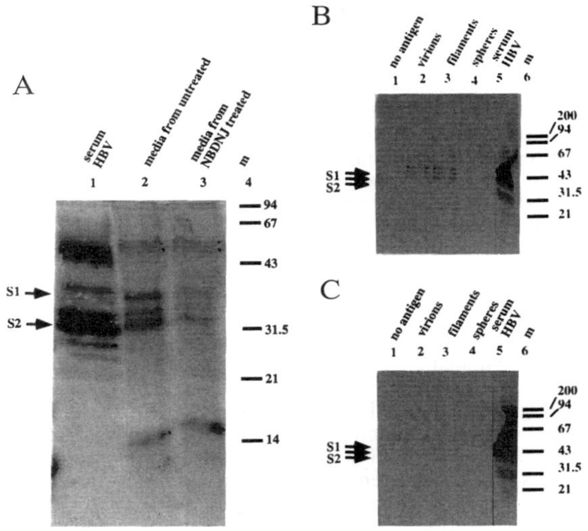

FIG. 5. Western blot analysis of LHBs and MHBs in the medium of NBDNJ-treated and untreated cultures. PEG precipitates of culture medium (*A*) or partly purified virions, HBs filaments, and spheres from untreated (*B*) and NBDNJ-treated (*C*) cultures were resolved by SDS/PAGE (13.5% acrylamide), transferred to Immobilon paper, and incubated with pre-S1- and pre-S2-specific mAb. One microgram of HBV genotype D, purified from human serum, was used as a control. Molecular mass markers (lanes m) are indicated in kDa. Arrows show LHBs (S1) and MHBs (S2) polypeptides.

dose-dependent decrease in the total amount of MHBs (pre-S2) antigen in the medium. This decrease is ≈2.5-fold at the highest NBDNJ concentration (4.5 mM, or ≈1000 μg/ml). These results were confirmed by Western blot analysis. Fig. 5*A* shows Western blots of medium from control cultures and those treated with NBDNJ at 1000 μg/ml. Media from NBDNJ-treated (Fig. 5*A*, lane 3) and untreated cultures (Fig. 5*A*, lane 2) contain similar amounts of MHBs (S2) and LHBs (S1) antigens.

Therefore, NBDNJ does not cause a generalized reduction in the amount of SHBs and LHBs antigens in the culture medium. There is, however, a 2-fold decrease in the amount of MHBs antigen in the medium of 2.2.15 cells treated with NBDNJ at 1000 μg/ml as determined by ELISA.

The Culture Medium of NBDNJ-Treated 2.2.15 Cells Contains Reduced Amounts of HBV Envelope Antigens Sedimenting as Intact Virions. To determine the amount of HBs present in intact virions, medium from the indicated cultures was fractionated through sucrose gradients by ultracentrifugation. Secreted virions derived from treated and untreated cultures and sedimenting to 40–46% sucrose were concentrated and tested for HBs proteins by ELISA. Fig. 4 *D–F* shows the results. All forms of HBs were easily detectable in samples containing virions prepared from the medium of untreated 2.2.15 cells. On the other hand, there were virtually no detectable HBs proteins in samples prepared from the medium of cells treated with 2.25 and 4.5 mM NBDNJ.

The ELISA results were confirmed by Western blot analysis of fractions from the sucrose gradient containing either intact virus, filaments, or spheres [Fig. 5 *B* (from untreated cultures) and *C* (from NBDNJ-treated cultures)]. Equal volumes from fractions of the sucrose gradient were resolved by SDS gel electrophoresis, transferred to membranes, incubated with antibody specific for LHBs (pre-S1), imaged, and then further incubated with antibody specific for MHBs (pre-S2) epitopes. Partially purified HBV derived from human serum is presented in lanes 5 (Fig. 5 *B* and *C*) as a control. Fractions distal to the virion-containing fractions (near the bottom of the gradient) were resolved in lane 1 to show specificity of the antibody. Lanes 2, 3, and 4 (Fig. 5 *B* and *C*) contain the intact virus-, HBs filament-, and sphere-containing fractions, respectively, as defined by sedimentation in sucrose and the presence (for virions) and absence (for subviral particles) of viral DNA (data not shown). There is a decrease in the amount of all HBs proteins present in virions, filaments, and spheres prepared from the medium of NBDNJ-treated cultures (compare lanes 2, 3, and 4 in Fig. 5 *B* and *C*). The images shown in Fig. 5 *B* and *C* were quantified by densitometry. Densitometry analysis (not shown) after the double antibody staining revealed that the decrease of LHBs in NBDNJ-treated virion samples, compared to untreated samples, was ≈4-fold. The decrease in MHBs in the same lanes (Fig. 5*C*, lane 2, compared with Fig. 5*B*, lane 2) was 12-fold. It is noted that the gradient used to separate the various forms of HBs protein results in fractions "enriched" for different forms—that is, the virion-containing fractions are likely to also contain filaments and spheres. This may cause an underestimation of the effect of NBDNJ upon the release of virions, relative to spheres and filaments, as judged by immunological analysis. Nevertheless, the results of the Western blot analysis are consistent with those of the ELISA, in that medium from NBDNJ-treated cultures contains similar amounts of total HBs antigens (not shown) but greatly reduced antigenic material in virion fractions.

DISCUSSION

This study explores the possibility of interfering with the life cycle of HBV by influencing host glycosylation processing. The objective was to cause viral envelope antigens to retain immature oligosaccharides, rather than prevent glycosylation altogether. It was reasoned that glycoproteins bearing defective, or immature, oligosaccharides may behave very differently from those completely lacking glycosylation.

NBDNJ is a sugar analogue that is a potent inhibitor of the ER α-glucosidase I, which removes the terminal glucose from immature glycans. Inhibition of this enzyme causes oligosaccharides to remain unprocessed and carry terminal glucose residues. The removal of terminal glucose residues is believed to be important in the movement of some glycoproteins from the ER to various cellular locations (16, 25). Mammalian cells can compensate for inhibition of α-glucosidase I activity, to some extent, by the use of a Golgi endomannosidase pathway (26). This alternative pathway may account for the relatively low cytotoxicity of NBDNJ, which does not inhibit endomannosidase, in mammals and in cultured cells (tolerated dose 50 in cultured human mononuclear cells was >5 mM; M. Bryant, personal communication).

This report shows that treatment of cells expressing HBV gene products and releasing progeny virions into the medium, with nontoxic concentrations of NBDNJ, results in a >10-fold decrease in the secretion of virions and virion-associated DNA with a concomitant 3- to 5-fold increase in the amount of intracellular viral DNA. This effect was not peculiar to 2.2.15 cells, since HepG2 cells infected with HBV and treated with NBDNJ behaved similarly. Analysis of 2.2.15 cells showed that although NBDNJ caused only a 25–50% reduction in the overall secretion of HBs proteins into the medium, the amount of virions was greatly reduced by >90–95%.

Given the known inhibitory effect of NBDNJ upon host α-glucosidase I (18), it is possible that the production of HBV envelope antigens bearing immature, unprocessed, N-linked oligosaccharides leads to a failure of virion secretion by the cell. Since, in the absence of envelopment, HBV genomes are not secreted (27), it is hypothesized that, in NBDNJ-treated cells, virions bearing defective envelopes accumulate within a subcellular compartment. Compared with subviral particles, virions are enriched for LHBs. This protein is known to be retained in the ER (9). Its retention can only be overcome by an excess of SHBs or MHBs within the same particle. The data in this report suggest that LHBs with unprocessed oligosaccharides may be even more inefficiently transported from the ER. HBs spheres contain relatively little LHBs and are thus not retained in the ER, to any great extent. MHBs is more completely glycosylated than LHBs or SHBs. Perhaps the presence of unprocessed LHBs and MHBs on virions leads to a transport failure.

Other explanations for the observed effects are possible. In the duck hepadnavirus system, it has been shown that the large envelope antigen influences the accumulation of covalently closed circular viral DNA (28). Perhaps HBV envelope antigens with unprocessed glycan are somehow defective in this function. The presence of immature glycans on virion envelope antigens has not been shown in this study and it is possible that NBDNJ negatively influences a host function involved in the transport of HBV from the cell. Alternatively, NBDNJ may be exerting an influence upon the cell rather than upon the α-glucosidases. In this regard, the effect upon HBV secretion reached a plateau at NBDNJ concentrations approaching 2.26 mM. In other systems, complete inhibition of glucosidase I was reached at <1 mM (13, 29). The reason that higher concentrations of compound are needed in the HepG2 cell system is unclear. However, HepG2 cells were resistant to concentrations of NBDNJ (2000 μg/ml or 9.2 mM) that are growth inhibitory to other cultures (such as Huh7 and Vero; G.R.F. and T.M.B., unpublished data). Perhaps HepG2 cells either export the drug from the intracellular compartment or have unusually high levels of glucosidase I.

This paper serves to raise the possibility that inhibitors of host glucosidases may be valuable as tools to explore the role of immature glycans upon HBV envelope protein trafficking within the cell and as potential HBV therapeutic agents.

We thank Terry Butters (University of Oxford) for many helpful discussions and Gabriele Reinkensmeir (University of Oxford) for technical assistance. This work was supported, in part, by the Searle/Monsanto Company, The American Cancer Society/International Union Against Cancer, The Hepatitis B Foundation, and Deutsche Forschungsgemeinschaft SFB 249.

1. Ayoola, E. A., Balayan, M. S., Deinhardt, F., Gust, I., Kureshi, A. W., Maynard, J. E., Nayak, N. C., Brodley, D. W., Ferguson, M., Melnick, J., Purcell, R. H. & Zuckerman, A. J. (1988) *Bull. W.H.O.* **66**, 443–455.
2. Eder, G., McDonel, J. L. & Dorner, F. (1986) in *Progress in Liver Diseases*, eds. Popper, H. & Schaffner, F. (Grune & Stratton, Orlando, FL), Vol. 8, pp. 367–394.
3. Hoofnagle, J. & Jones, E. (1989) *Semin. Liver Dis.* **9**, 231–233.
4. Perrillo, R. P. (1989) *Semin. Liver Dis.* **9**, 240–248.
5. Doong, S.-L., Ching-Tsai, T., Schinazi, R., Liotta, D. & Cheng, Y. C. (1991) *Proc. Natl. Acad. Sci. USA* **88**, 8495–8499.
6. Lee, B., Lus, W., Suzuki, S., Robins, M. & Tyrell, D. J. (1989) *Antimicrob. Agents Chemother.* **33**, 336–339.
7. Price, P. M., Banerjee, R. & Acs, G. (1989) *Proc. Natl. Acad. Sci. USA* **86**, 8541–8544.
8. Venkateswaran, P. S., Millman, I. & Blumberg, B. S. (1987) *Proc. Natl. Acad. Sci. USA* **84**, 274–278.
9. Ganem, D. (1991) in *Hepadnaviruses*, eds. Mason, W. S. & Seeger, C. (Springer, Berlin), pp. 61–84.
10. Heerman, K.-H. & Gerlich, W. H. (1991) in *Molecular Biology of the Hepatitis B Virus*, ed. McLachlan, A. (CRC, Boca Raton, FL), pp. 109–143.
11. Patzer, E. J., Nakamura, G. R., Simonsen, G. C., Levinson, A. L. & Brands, R. (1986) *J. Virol.* **58**, 884–892.
12. Datema, S., Olofsson, P. & Romero, P. A. (1987) *Pharmacol. Ther.* **33**, 221–286.
13. Karpas, A., Fleet, G. W. J., Dwek, R. A., Petursson, S., Namgoon, S., Ramsden, N. G., Jacob, G. S. & Rademacher, T. W. (1988) *Proc. Natl. Acad. Sci. USA* **85**, 9229–9233.
14. Walker, B. D., Kowalski, M., Goh, W. C., Kozarsky, K., Krieger, M., Rosen, C., Rohrschneider, L., Haseltine, W. A. & Sodrowski, J. (1987) *Proc. Natl. Acad. Sci. USA* **84**, 8120–8124.
15. Sells, M. A., Chen, M. L. & Acs, G. (1987) *Proc. Natl. Acad. Sci. USA* **84**, 1005–1009.
16. Platt, F., Karlsson, G. B. & Jacob, G. S. (1992) *Eur. J. Biochem.* **208**, 187–193.
17. Seifer, M., Heerman, K. H. & Gerlich, W. H. (1990) *Virology* **179**, 300–311.
18. Fleet, G. W., Karpas, A., Dwek, R. A, Fellows, L. E., Tyms, A. S., Petersson, S., Namgoong, S. K., Ramsden, N. G., Smith, P. W., Son, J. C., Wilson, F., Witty, D. R., Jacob, G. S. & Rademacher, T. W. (1988) *FEBS Lett.* **237**, 128–132.
19. Maniatis, T., Fritsch, E. F. & Sambrook, J. (1982) *Molecular Cloning: A Laboratory Manual* (Cold Spring Harbor Lab. Press, Plainview, NY).
20. Foster, G. R., Ackrill, A. M., Goldin, R. D., Kerr, I. M., Thomas, H. C. & Stark, G. R. (1991) *Proc. Natl. Acad. Sci. USA* **88**, 2888–2892.
21. Heermann, K. H., Kruse, F., Seifer, M. & Gerlich, W. H. (1987) *Intervirology* **28**, 14–21.
22. Heermann, K. H., Waldeck, F. & Gerlich, W. H. (1988) in *Viral Hepatitis and Liver Disease*, ed. Zuckerman, A. J. (Liss, New York), pp. 697–701.
23. Gültekin, H. & Heermann, K. H. (1989) *Anal. Biochem.* **172**, 320–329.
24. Sells, M. A., Zelent, A., Shvartsman, M. & Acs, G. (1988) *J. Virol.* **62**, 2836–2844.
25. Lodish, H., Kong, A. (1984) *J. Cell Biol.* **98**, 1720–1729.
26. Lubas, W. A. & Spiro, R. G. (1988) *J. Biol. Chem.* **263**, 3990–3998.
27. Bruss, V. & Ganem, D. (1991) *Proc. Natl. Acad. Sci. USA* **88**, 1059–1063.
28. Summers, J., Smith, P., Huang, M. & Yu, M. (1991) *J. Virol.* **65**, 1310–1317.
29. Karlsson, G. B., Butters, T. D., Dwek, R. A. & Platt, F. M. (1993) *J. Biol. Chem.* **268**, 570–576.
30. Bayer, M. E., Blumberg, B. S. & Werner, B. (1968) *Nature (London)* **218**, 1057–1059.

Journal of Acquired Immune Deficiency Syndromes and Human Retrovirology
15:387–390

Decreased Survival of HTLV-I Carriers in Leprosy Patients From the Democratic Republic of the Congo: A Historical Prospective Study

*Michel F. Lechat, †Daniel I. Shrager, ‡Etienne Declercq, *Françoise Bertrand, §William A. Blattner, and †Baruch S. Blumberg

**Universite Catholique de Louvain, Brussels, Belgium; †Fox Chase Cancer Center, Philadelphia, Pennsylvania, U.S.A.; ‡Institute of Hygiene and Epidemiology, Brussels, Belgium; and §National Cancer Institute, Bethesda, Maryland, U.S.A.*

Summary: In this historical prospective study using sera stored for 22 years, we investigated the effect of HTLV-I infection on survival in a population of leprosy patients in the Democratic Republic of the Congo (formerly Zaire). We also determined the distribution of HTLV-I by subpopulation, age, and gender. Stored sera taken from a population of leprosy patients and controls in 1969 were tested for HTLV-I. Follow-up survival data on these patients were obtained in 1991. The sera collected in 1969 from 520 individuals was used to determine the prevalence of HTLV-I. Included in this number were 328 patients resident in the sanatorium. Survival and other data were available for 327 of these. A multivariate survival analysis using a logistic regression model was performed to evaluate the influence of HTLV-I status, age, type of leprosy, gender, duration of hospitalization, and ethnic group on survival. The overall prevalence of HTLV-I among the 520 individuals in the prevalence study was 34%, with 37.4% in the leprosy group and 25.2% in the control group ($p < 0.01$). Multivariate analysis using logistic regression showed that females of the Mongo and Ngombe ethnic group taken together were significantly more likely to be infected than the other groups (OR = 3.67, 95% CI: 2.14 to 6.30). A comparison of the death rates directly standardized for age and sex showed that the rate was significantly higher for HTLV-I positive (5.5/100 person-years of observation) compared with HTLV-I negative (3.6/100 person-years of observation). A survival analysis using the Cox model showed a risk ratio of 1.4 (CI: 1.04 to 1.89) for those infected with HTLV-I. An increase in the death rate was associated with HTLV-I infection in leprosy inpatients. The decreased survival associated with HTLV-I infection may result from an increased susceptibility to a variety of diseases. **Key Words**: HTLV-I—Leprosy—Democratic Republic of the Congo—Survival.

Human T-cell lymphotropic virus type I (HTLV-I), a retrovirus, is endemic throughout certain regions of the world. HTLV-I, first isolated by Poiesz et al. (1), is common in Japan, particularly in Kyushu (prevalence ≤15%) (2), elsewhere in Asia, where the frequency is usually lower (although areas of high frequency such as North-eastern Iran have been reported), the Pacific Islands, and Australia. It is common in the Caribbean (prevalence ≤10%) particularly in populations of African origin and occurs at lower levels in autochthonous populations of the Americas, with some pockets of high frequency. European and North American cases are ordinarily found in persons from the endemic regions. The prevalence in blood donors in the United States is 0.01%. In Africa, there are areas of high prevalence. The Equator Province of the Democratic Republic of the Congo (DRC, for-

Address correspondence and reprint requests to Baruch S. Blumberg, M.D., Ph.D., Fox Chase Cancer Center, 7701 Burholme Avenue, Philadelphia, PA 19111, U.S.A.
Manuscript received July 26, 1996; accepted March 26, 1997

merly Zaire) has one of the highest prevalences (about 15%), and high frequencies of infection are found in Cameroon and Chad (9% to 12%) (3,4).

It is likely that there are tens of millions of carriers in the world, although it is difficult to arrive at an accurate estimate with available regional data. It is important to have some reckoning of the disease load imposed on carriers compared with those who are not infected to evaluate public health programs and determine the cost benefit ratios of prevention programs.

Several diseases are associated with HTLV-I. Adult T-cell leukemia (ATL) is relatively uncommon; a small percentage of carriers, after a long incubation period, do develop the disease. HTLV-I is associated with myelopathy/tropical spastic paralysis (HAM/TSP), a chronic and slowly progressive paraparesis accompanied by impotence and other dysfunctions. The annual attack rate can vary from 0.03 per 1000 in Japan to 0.13 to 0.25 per 1000 in the DRC. There may be a relatively short incubation period, and cases occurring a few months after infection by transfusion have been reported.

There are other diseases for which an association has been reported including polymyositis, arthritis, uveitis, Sjögren's syndrome, retinal vasculitis, and various pulmonary diseases. Diseases associated with immune deficiency have also been reported, including Norwegian scabies, infectious dermatitis, and Kaposi's sarcoma. It has been estimated that about 5% of HTLV-I carriers go on to develop these diseases, particularly ATL and TSP (4).

We report a significantly increased mortality among carriers of HTLV-I compared with noncarriers in a population of leprosy patients from Equator Province, DRC. An increased mortality in Japanese carriers who were monitored for a mean of 5.3 years has been reported by Iwata et al. (5). If these observations are supported in other groups, particularly HTLV-I carriers in the general population, it indicates that infection with the virus is a public health concern.

In previous publications, we have studied hepatitis B virus prevalence (6) and the prevalence and nucleic acid sequences of HTLV-I and HTLV-II in the population considered in this study. HIV was not detected (7).

METHODS

Population

In 1969, 936 blood samples were collected at the Iyonda Leprosy Sanatorium, Mbandaka, Equator Province, DRC, from leprosy patients and control subjects to test the hypothesis of an association between leprosy and hepatitis B virus (6). These specimens were stored at –25°C and were thawed occasionally for other uses. A sample of 520 specimens was taken in 1991 from the initial collection. Selection was based on the amount of serum remaining in the containers, but it was otherwise random. It included 377 leprosy patients (186 with multibacillary and 191 with paucibacillary disease), of whom 49 lived away from the sanatorium, and 143 controls without leprosy, of whom 21 lived within the sanatorium and 122 lived in the vicinity (Table 1). Information on gender, age, ethnic affiliation, and type of leprosy was available only for the inpatients. These specimens were tested for HTLV-I and were used in the prevalence study.

In 1991, information was obtained from the nursing staff in Iyonda on the survival of 327 of the 328 inpatients whose sera had been tested for HTLV-I. Information on the cause of death was obtained, but it was usually not specific, except for accidents. This population was used for the survival study. The preservation of the stored sera allowed a 22-year historical prospective study to be undertaken in a relatively short time.

Serologic Tests for HTLV-I

Plasmas were screened for antibodies using commercially available the whole-virus HTLV-I enzyme-linked immunosorbent assay (ELISA; Cellular Products, Buffalo, NY, U.S.A.), according to the manufacturer's instructions. Positive samples were retested in duplicate. Repeatedly reactive samples were confirmed using Western blot assays (DuPont, Wilmington, DE, U.S.A.), which contain whole viral lysate and recombinant HTLV-I (8).

Statistical Methods

A multivariate analysis using a logistic regression model was performed to evaluate the influence of age, type of leprosy, gender, duration of hospitalization and ethnic group on the probability of HTLV-I infection. Death rates for HTLV-I positive and HTLV-I negative leprosy inpatients were calculated after standardizing (direct method of standardization) for age, sex, and type of leprosy. The Cox proportional hazards regression model (9) was used to test the influence on mortality of factors possibly associated with higher probability of death: HTLV-I status, age, sex, and type of leprosy. The statistical package SPSS was

TABLE 1. *Prevalence of HTLV-1 according to disease status, type of leprosy, and residence*

Population	Number sampled	Number positive	Percent positive
Leprosy patients	377	141	37.4
Inpatients	328	130	39.6
Outpatients	49	11	22.4
Multibacillary	186	74	39.8
Inpatients	181	72	39.8
Outpatients	5	2	40.0
Paucibacillary	191	67	35.1
Inpatients	147	58	39.5
Outpatients	44	9	20.5
Controls (without leprosy)	143	36	25.2
Controls in sanatorium	21	11	52.4
Controls in villages	122	25	20.5
Resident of sanitarium	349	141	40.4
Nonresident of sanitarium	171	36	21.1
Total	520	177	34.0

TABLE 2. *Prevalence of HTLV-1 according to age and sex, inpatients only*

Age (1969)	Males *N*	Males +	Males %	Females *N*	Females +	Females %
0–29	13	5	38.5	6	3	50.0
30–39	34	12	35.3	17	5	29.4
40–49	45	10	22.2	46	24	52.2
50–59	66	21	31.8	42	20	47.6
60+	34	12	35.3	25	18	72.0
Total	192	60	31.3	136	70	51.5

used for the logistic regression model (10) and SAS for the Cox regression model (11).

RESULTS

Prevalence Study

The prevalence of HTLV-I in the entire 1991 sample, arranged according to disease status, type of leprosy, and residence, is shown in Table 1. The overall prevalence of HTLV-I in the 520 individuals in the prevalence study was 34%, with 37.4% among the inpatients and 25.2% among the controls, a significant difference ($p < 0.01$).

The results for inpatients according to age and sex is shown in Table 2. No regular trend in age-specific prevalences was observed, but there was a higher prevalence among females than males in the 40 and older age groups. The patients 60 and older had a higher prevalence (50.8%, CI: 34.3 to 72.7) than the overall prevalence for the combined younger age groups (37.2%, CI: 31.4 to 43.3).

To determine the influences of other interdependent variables (e.g., type of leprosy, duration of hospitalization, ethnic group), a multivariate analysis using logistic regression was performed. The patients were categorized in two ethnic groupings: Mongo and Ngombe, the local and most frequent groups (204 persons) and others (124 persons). The prevalence rates for these groups were 47.1% and 27.4%, respectively. Females of the Mongo or Ngombe ethnic groups are significantly more often infected (OR = 36.7; 95% CI: 2.14 to 6.30) than the other groups in our study.

Survival Study and Cause of Death

The survival study was conducted on the 327 of 328 patients for whom follow-up data were available: the entire 22-year study period for the known survivors (102) or until the time of death for those who died (186). For the 39 patients who left the institution but whose subsequent fate is unknown, the end point for the follow-up was the year of discharge. The total number of person-years of observation was 4654.

The death rates of carriers compared with noncarriers directly standardized for age and sex was significantly higher ($p < 0.01$) for HTLV-I–positive persons (5.5 per 100 person-years of observation) compared with HTLV-I–negative persons (3.6 per 100 person-years). The Cox's model of multivariate analysis for survival showed that death rates depended on age (as expected), sex (females had a lower risk ratio), and HTLV-I status (carriers had a higher risk ratio). There was no effect of type of leprosy on death rates. Risk ratios were calculated: for HTLV-I, 1.4 (95% CI: 1.04 to 1.89); for age, 1.05 (95% CI: 1.04 to 1.07), and for sex (female), 0.7 (95% CI: 0.52 to 0.96) (Table 3).

The presumed cause of death had been recorded for 160 patients. There were 3 violent deaths (i.e., suicide, accidents). The reported causes of the other deaths are not specific, referring to symptoms or undocumented illnesses (21 categories in total). These included heart disease (28), cachexia (26), senility (24), cancer (10), stroke (10), amyloidosis (10), kidney disease (10), liver cirrhosis (8), pulmonary tuberculosis (5), nutritional deficiencies, (4), pneumonia (3), and leukemia (2). There was no difference with respect to HTLV-I carrier status in these cause-of-death categories.

DISCUSSION

The prevalence of HTLV-I in Equator Province, DRC, was reported to be high in recent years (4) and was also high in 1969. From these prevalence figures in 1969, it can be inferred that HTLV-I was not introduced recently, as appears to be the case for HIV (7), and that it has been present at high levels for at least 2 decades. The overall frequency is one of the highest reported from any population. The prevalence of HTLV-I was significantly higher among leprosy patients than controls. There was, however, a higher prevalence among the controls living in the sanatorium (52.4%) than those living in the villages (20.5%), and the prevalence was higher among inpatients (39.6%) than outpatients (22.4%). The higher frequency of infection among leprosy patients compared

TABLE 3. *Cox's model of multivariate analysis for survival*

Variable	Relative risk	95% Confidence interval	Greater survival
HTLV-I status	1.4	(1.04–1.89)	Neg > Pos
Age	1.05	(1.04–1.07)	Young > Old
Gender	0.7	(0.52–0.96)	Females > Males
Leprosy type		No difference	

with controls could be accounted for by the higher prevalence in general in the sanatorium. It is possible that the depressed immune function, particularly T-cell–mediated immunity, of leprosy patients could account for their apparent increased prevalence of HTLV-I.

Older females are more likely to be positive for HTLV-I (2,12,13). This has been ascribed to the observation that sexual transmission of the virus from males to females is much more common than the reverse (2). As females age, their cumulative sexual exposure increases their risk of HTLV-I infection, but the increased risk would be less for males. The major mechanism of childhood transmission appears to be breast-feeding (14), which is likely to be about equal for male and female children.

The leprosy inpatient population was considered appropriate for the survival study because stored sera were available, because there is a high prevalence of HTLV-I in the population, and because follow-up for 22 years was feasible.

There is a significantly higher mortality rate for carriers than for those who are not infected with HTLV-I. The risk figures in this study are similar to those reported in the Japanese study (5). Whatever their infection status, males have a higher mortality rate than females. It raises the possibility that HTLV-I is associated with increased risk for deadly diseases that are more common than ATL or TSP. Because the persistent HTLV-I infection may increase susceptibility to infectious diseases (15,16), it is important to compare disease-specific death rates for these diseases according to HTLV-I status (5). In our study, we did not find any differences in relation to the recorded cause of death, but with few exceptions, the listed causes of death were not specific.

It is important to undertake HTLV-I survival studies, historical prospective studies such as this one or conventional prospective studies, with nonhospitalized and patient populations from different regions. Historical prospective studies can be done relatively simply and inexpensively if the appropriate stored sera are available. Investigators who hold these collections may want to consider such studies. We have stored sera from epidemiologic studies of hepatitis B virus completed during the past 30 years from many locations and hope to undertake similar survival investigations in the future.

Acknowledgments: We are indebted to Dr. Mputu Luenga Boyau, Medecin-Directeur, Bureau National de la Lepre, Ministere de la Sante Publique, Kinshasa, DRC, for obtaining authorization to publish our results. We are also indebted to Dr. Keith Fleshman and Mr. Roger Moris for the 1969 sera collection. The collaboration of the late Sister Amanda Lemeire, head nurse of the Iyonda Sanatorium since 1953, is particularly acknowledged for her relentless efforts in assembling the survival data. This work was supported by USPHS grant CA-06927 from the National Institutes of Health and by an appropriation from the Commonwealth of Pennsylvania.

REFERENCES

1. Poiesz BJ, Ruscetti FH, Gazdar AF, et al. Detection and isolation of type C retrovirus particles from fresh and cultured lymphocytes of a patient with cutaneous T-cell lymphoma. *Proc Natl Acad Sci USA* 1990;77:7415–9.
2. Kajiyama W, Kashiwagi S, Ikematsu H, Hayashi J, Nomura H, Okachi K. Intrafamilial transmission of adult T cell leukemia virus. *J Infect Dis* 1986;154:851–7.
3. Goubau P. HTLV—studies in the comparative virology of non-HIV human retroviruses. Doctoral thesis, Katholieke Universiteit Leuven, 1993.
4. Delaporte E, Peeters M, Durnad JP, et al. Seroepidemiological survey for HTLV-I infection among randomized populations of western central African countries. *J AIDS* 1990;2:410–3.
5. Iwata K, Ito S, Saito H, et al. Mortality among inhabitants of an HTLV-I endemic area in Japan. *Jpn J Cancer Res* 1991;85:231–7.
6. Lechat MF, Melartin Prehn L, Blumberg BS, Moris R. Australian antigen in Zaire. Studies on leprosy. *Ann Soc Belge Med Trop* 1973;53:173–8.
7. Dube DK, Dube S, Erensoy S, et al. Serological and nucleic acid analyses for HIV and HTLV infection on archival human plasma samples from Zaire. *Virology* 1994;202:379–89.
8. Lillehoj EP, Alexander SS, Duule CJ, et al. Development and evaluation of a human T-cell leukemia virus type I serological confirmatory assay incorporating a recombinant envelope polypeptide. *J Clin Microbiol* 1990;28:2653–8.
9. Cox R. Regression models and life tables. *J R Stat Soc B* 1972; 34:187–202.
10. Marija J. Norusis/SPSS Inc, SPSS/PC—Advanced Statistics 4 for the IBM PC/XT/AT and PS/2. Chicago, IL: SPSS, 1990.
11. SAS Institute Inc, SAS Technical Report P-229, SAS/STAT Software: Changes and Enhancements, Release 6.07. Cary, NC: SAS Institute, 1992:620.
12. Blattner WA, Saxinger C, Riedel D, et al. A study of HTLV-I and its associated risk factors in Trinidad and Tobago. *J AIDS* 1990; 3:1102–8.
13. Lavanchy D, Bovet P, Hollanda J, Shamlaye CF, Burczak JD, Lee H. High seroprevalence of HTLV-I in the Seychelles. *Lancet* 1991; 337:248–9.
14. Hino S, Doi H. Mechanisms of HTLV-I transmissions. In: Roman GC, Vernant JC, Osarne M, eds. *HTLV-I and the nervous system: neurology and neurobiology,* vol 51. New York: Alan R. Liss, 1989:495–501.
15. Veyssier-Belot C, Couderc LJ, Desgranges C, et al. Kaposi's sarcoma and HTLV-I infection. *Lancet* 1990;336:575.
16. Mollison LC, Lo STH, Marning G. HTLV-I and scabies in Australian aborigines. *Lancet* 1993;341:1281–2.

Reprinted from JOURNAL OF THE HISTORY OF MEDICINE AND ALLIED SCIENCES
1958, Volume XIII, Number 3

Bernard Connor (1666-1698)

And His Contribution to the Pathology of Ankylosing Spondylitis

BARUCH S. BLUMBERG* and JEAN L. BLUMBERG

INTRODUCTION

BERNARD Connor's early contribution to the pathology of ankylosing spondylitis has been recognized for some time.[1,2,3,23,27,33,34,37,44] However, other than a short but excellent section in the *Dictionary of National Biography*,[43] and a review of some aspects of his contributions with some biographical data,[29] there has been little study of his life.

He was a youth of considerable charm and attraction who had an exciting though brief career. He was acquainted with many interesting medical personalities of the seventeenth century and was extensively travelled. The present short account contains a biography of his life with some comments on the people he met and the places he visited and a description of his contribution to pathology.

BIOGRAPHY

Bernard Connor or O'Connor was born in the year 1666** in the county of Kerry, Ireland. Because of the destruction of large numbers of records of this time in Ireland, it has been difficult to obtain genealogical data. There is some evidence, however, that Connor may have been a brother of Thomas O'Connor, a prominent Dublin physician and potential claimant to the Chieftainship of the O'Connor Kerry about whom information is available.[33] Bernard Connor is referred to in the manuscript, *Kerry Clans and Confiscations*, written in 1854 by the Reverend Charles James O'Connor Kerry,[28] but his exact relationships continue obscure. From his will, still preserved in the Principal Probate Registry, Somerset House,[19] it is known that he had more than one sibling (" . . . the largest diamond ring to be sold and the value given to my younger sister . . .") and that a cousin, Charles Connor, was living in London at that time.

* Balliol College, Oxford, England; Fellow of the Arthritis and Rheumatism Foundation.

** In his funeral oration[24] delivered in 1698, it is stated that he died in his thirty-third year. The entry in the Medical School at Rheims[35] gives his year of birth as 1666.

He was educated privately at home but was unable to secure higher education in Ireland or England. Therefore, in about 1686 "... il sortit de son pays à l'âge de vingt ans et vint en France pour y étudier en médicine et chercher les moyens de s'avancer."[31]

He attended the schools at Paris, Montpellier, and Rheims and received his M.D. from Rheims in 1693. The date 1691 given in the *Dictionary of National Biography*[43] is probably incorrect. In the archives of the medical school of the University of Rheims there is a manuscript by Raussin, Dean of the Faculty of Medicine in 1756,[35] which contains a listing of the medical graduates with the date of their theses. Listed under the O'Connors is "O'Connor Bernard, Irlandis, 18 Septembre 1693, né dans le comte de Kerry en 1666. . . ." There is no O'Connor listed for 1691. The date used in the *Dictionary of National Biography* may have been derived from a paper in the British Museum, published in Paris, which is either the thesis or a publication gleaned from it.[6] This bears the date 1691 in Latin numerals but a knife cut has removed further writing from this line (Fig. 1.). It is this paper which contains the first description of the spondylitic skeleton of which more will be said later.

There were a surprisingly large number of Irishmen at Rheims at this time. Of the 3,322 medical doctorates granted between 1550-1793 no less than 654 were granted to Irishmen. Among the ten O'Connors listed there were two, a "Joaness Bernardus" (1777) and "Bernardus" (1787) both of county Kerry, who were probably related to the Bernard O'Connor with whom we are concerned.[26]

It was around this time that O'Connor decided to delete the "O" from his name. In the Paris paper[6] the "O" has been removed on the title-page by a knife cut. In a publication of the following year[7] the same alteration has been made. This is seen in the copy in the British Museum as well as the copy which the author presented to the library of the Royal Society. In a book published in 1695,[8] the name "Connor" was printed in the first instance with no alterations on the title-page. One may surmise that sometime between 1694 and 1695 he decided to alter his name and effected these changes on the title-pages of his previous publications. His reasons for doing this are obscure. Perhaps it was related to his desire to practice in England or to his political activities which will be discussed below.

LETTRE
E'CRITE
A
MONSIEUR LE CHEVALIER
GUILLAUME DE WALDEGRAVE,
PREMIER MEDECIN
DE SA MAJESTE' BRITANNIQUE.

CONTENANT UNE DISSERTATION PHYSIQUE ſur la continuité de pluſieurs os, à l'occaſion d'une fabrique ſurprenante d'un tronc de Squelette humain, où les vertebres, les côtes, l'os Sacrum, & les os des Iles, qui naturellement ſont diſtincts & ſeparez, ne font qu'un ſeul os continu & inſeparable.

Par Mr. BERNARD CONNOR, *Doctеur en Medecine, & Anatomiſte.*

A PARIS,

Chez JEAN CUSSON, ruë ſaint Jaques, à l'Image de ſaint Jean Baptiſte. *Avec Privilege du Roi.*

M. DC. XCI

FIG. 1. The title-page of the article containing the first description of the skeleton with ankylosing spondylitis. The "O" has been removed by a knife cut from in front of Connor. An alteration has also been made on the date line. (Photograph courtesy of the British Museum.)

During his seven years' sojourn in France (1686-93) he received recognition from the medical profession and even from the French Court itself. In his thesis[6] he notes that he had students in anatomy, and several of his publications bear the following description under his name:[7,8]

> ". . . doctore Medico & Anatomico è Regia Medicorum Camerae Regiae Parisiensis Societate."

It was through his court connections that he made the acquaintance of the two sons of the High Chancellor of Poland who were visiting the country. There were strong connections between the Polish and French kingdoms at this period. The wife of the Polish monarch was French and he received a subsidy from Louis XIV.

In company with the two Polish nobles, Connor travelled to Italy, going as far south as Naples, perhaps visiting Sicily as well.[47] He attended the lectures of the anatomists and physiologists Malpighi, Bellini, Redi, and others. In later years he lectured on their discoveries. In Naples he investigated a crypt where the dead were said to rise to life and in 1694 published his observations.[7] Within the covers of this same book is a description of an eruption of Vesuvius. Since no great eruption is recorded during the years 1693-94, much of his information must have been hearsay.

He continued on to Venice which was at that time a rich commercial city, the main entrepôt between the East and southern Europe. Here, Connor encountered William Legge, Lord Dartmouth, who had fallen ill and was not responding to the ministrations of the local physicians. Connor tells us that he successfully treated him[16] and conducted him to Padua to convalesce under the care of his companion, the Earl of Kildare.

From Italy he proceeded with his Polish friends through the Tyrol, Bavaria, Austria, and down the Danube by ship to Vienna. He rested at the court of the Emperor for some time and then continued through Moravia and Silesia to Cracow, the capitol of Poland. He was joined there by a large company of voyagers and they travelled to Warsaw where the King was holding court.[16]

Poland was at that time a wild, remote country only rarely visited by Western Europeans. It was a monarchy of a unique sort; the nobles elected a king who reigned until his death but did not pass on the sceptre to his heir. The large majority of the inhabitants were peasants and included groups of Cossacks and other nomadic or semi-nomadic people. The incumbent at the time of Connor's visit, John Sobieski, had gained prominence as a consequence of a brilliant military career. In 1667 at the head of an army which he had created at his own expense, Sobieski defeated the combined Turkish, Tartar, and Cossack forces. He again defeated their ancient enemy, the Turk, in 1673 and in the following year, when the Polish throne became vacant, he was elected by the nobles to fill it. His greatest achievement was the rescue of Vienna from the Turks in 1683 which effectively halted the further spread of Islam into Christendom.[22,36]

When Connor arrived in Poland sometime in 1694,* the king was already in poor health. Connor had received an introduction to the Venetian minister in Warsaw, Hiernonomo Alberto de Conti, through the courtesy of Conti's wife, Lady Margaret Paston, who was the sister of William, Earl of Yarmouth. While at the court, Connor's medical reputation was enhanced by an astute diagnosis. The sister of Sobieski fell ill, and the King's private physician, the Italian, Jonas, diagnosed "the ague." Connor opined that she had an abscess of the liver. When she died a short time later his view was confirmed at autopsy. He was soon appointed physician to the King and attended him in the illness which in 1696 finally led to his death.

In *The History of Poland*[16] Connor states, "When I came first into this Nation, my chief Design was to converse with Physicians and other Naturalists, to improve my knowledge in the practice of Physick and in Natural History. . . ." However, he found that there was little to learn in medicine and instead he busied himself studying the history and customs of the country.

In addition to formal historical material, he included in the *History* many bizarre observations. He gives an account of the fire and serpent worship that existed in Lithuania before its conversion to Christianity in the fourteenth century and which still persisted in the remote sections of the country. He relates his observations (and includes an illustration) of children who had been found in the woods by bears and had been raised and suckled by them. He also had much to say about the Cossacks and the Tartars who inhabited part of the country. They were horsemen

* In his book, *The history of Poland*[16] Connor states that he had not ". . . been in it [Poland] a full twelve month." He left the country on 11 November 1694.

and warriors and in every way "unacquainted with luxury, only they love aqua-vita extremely."

Late in 1694 he was entrusted with the duty of escorting the daughter of the King, Princess Teresa Cunigunda, to Brussels where she was to marry the Elector of Bavaria. They departed on 11 November with a large company and proceeded through Posnan, Frankfort, and Berlin to Wesel-on-the-Rhine where they were met by the royal suitor. He was so taken with his intended bride that he insisted that the marriage ceremony be celebrated forthwith. This was done. A short time later the party continued on to Brussels, arriving on 12 January 1695. During this long journey, Connor continued to question his Polish companions about their country and much of the information thus gathered was included in his book.

After remaining in Brussels for a short time, he left the Princess in the care of the physician of the Elector, Dr. Pistorini, and proceeded via Holland to London in February 1695.

It may be presumed that he had by this time been received into the Established Church of England for, as a consequence of the Test Acts of 1673, he would not otherwise have been permitted to attend the universities or to practise in London. He took up residence in Bow Street, Covent Gardens, which at that time was a fashionable district of the city. Strype, writing in 1720,[46] states that, "The street is open and large, with very good houses, well inhabited, and restored unto by gentry lodgings: as are most of the other streets in this parish."

Among Connor's neighbors was John Radcliffe who lived on this street from 1687 to 1714. Radcliffe (1650-1714) was one of the most celebrated and successful practitioners of all time, and much has been written about him. He was frequently called to see the British royal family and on one occasion received 1200 guineas from King William after curing a sick nobleman. When Radcliffe was being consoled by his friends on the loss of 5000 guineas in a business venture, he said, "I would only have to go up 250 pairs of stairs to make myself whole again."[43] He had been appointed physician to Princess Anne of Denmark in 1686 but later offended her by neglecting to visit her when sent for, saying that her distemper was nothing but the vapours. He had always supported scientific investigation during his lifetime, and on his death much of his considerable estate went towards the establishment of a hospital and library at Oxford which still bear his name.

Connor dedicated a letter in his book, *The History of Poland*[16] to Radcliffe and in a letter which he wrote to Hans Sloane on 28 May 1695[11] he requested that Sloane inform Radcliffe about the dedication.

Some information on this phase of Connor's life is gained from two letters which he wrote to Sloane who was then Secretary of the Royal Society.[11,17] The circumstances which led to their preservation is interesting.

Sloane had a brilliant career in science and medicine. In 1687 he had accompanied the Duke of Albermarle to the island of Jamaica in the West Indies when that nobleman became the governor there. In addition to his duties as physician to the governor, he made a collection of 800 specimens of plants. After his return to England he became a Fellow, Secretary (1693-1712), and later (1727-41) President of the Royal Society. While occupying these offices he had an extensive correspondence with the physicians and scientists of this period. His collections and correspondence were carefully preserved and on his death in 1753 were placed in the hands of public trustees. The collection (along with several others) became the basis of the British Museum of Natural History, and the correspondence, still preserved in the library of the Museum, has become source material of the utmost importance in the medical history of the period from the time of Charles II to that of George II.[43] The first of the letters which Connor wrote is shown in Figure 2.

In one of these Connor informs Sloane about his French article on the spondylitic skeleton. We may presume that it was this notification which led Sloane to effect the translation and publication of the article in the *Philosophical Transactions* of 1695.[9]

Connor did not remain in London long but proceeded to Oxford where he spent the winter and spring of 1695. There are no records in the archives of the University to authenticate his visit but this is a poorly documented period in the University's history.* His name does not appear on any of the college rolls, from which we can be sure that he did not receive a degree. He must have been in Oxford on 28 May 1695 for one of the letters to Sloane was dated from Oxford on that day.

He remained in Oxford long enough to publish a small treatise in Latin.[8] This was a series of four separate essays bound in the

* The available records were searched through the courtesy of I. G. Philip, M.A., Deputy Keeper of the Oxford University Archives.

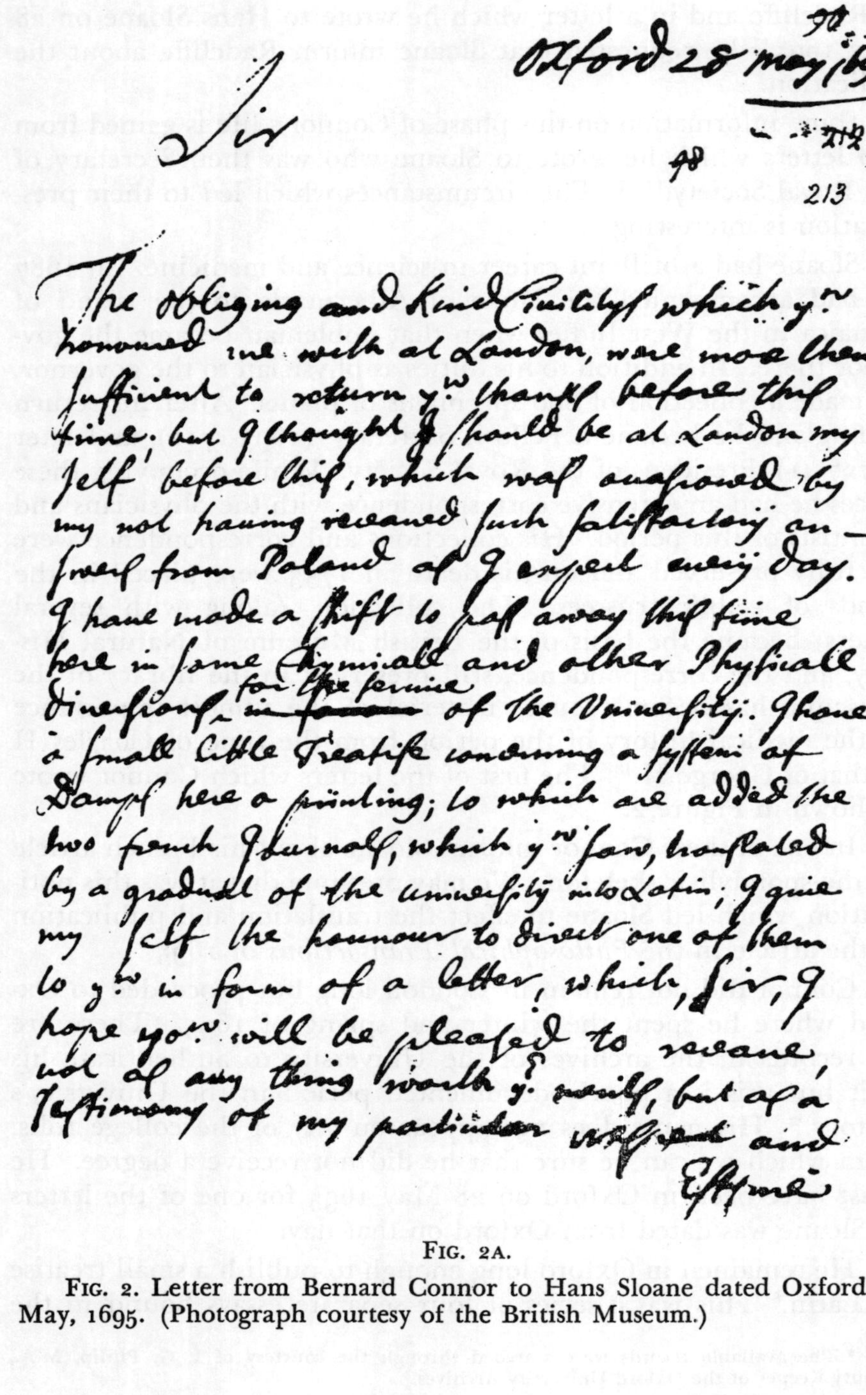

Oxford 28 may 1695

Sir

The Obliging and Kind Civilitys which yr. honoured me with at London, were more then sufficient to return yr. thanks before this time. but I thought I should be at London my self before this, which was occasioned by my not having received such satisfactory answers from Poland as I expect every day. I have made a shift to pass away this time here in some Chymicall and other Physicall Diversions for the service of the University. I have a small Little Treatise concerning a system of Damps here a printing; to which are added the two french Journals, which yr. saw, translated by a graduat of the University into Latin; I gave my self the honour to direct one of them to yr. in form of a letter, which, Sir, I hope you will be pleased to receive not as any thing worth yr. notice, but as a Testimony of my particular respect and Esteme

FIG. 2A.

FIG. 2. Letter from Bernard Connor to Hans Sloane dated Oxford, 28 May, 1695. (Photograph courtesy of the British Museum.)

same cover with a separate title-page for each essay. The first title-page, carrying the etching of the Sheldonian Theatre where the Oxford University Press was then located, is shown in Figure 3.

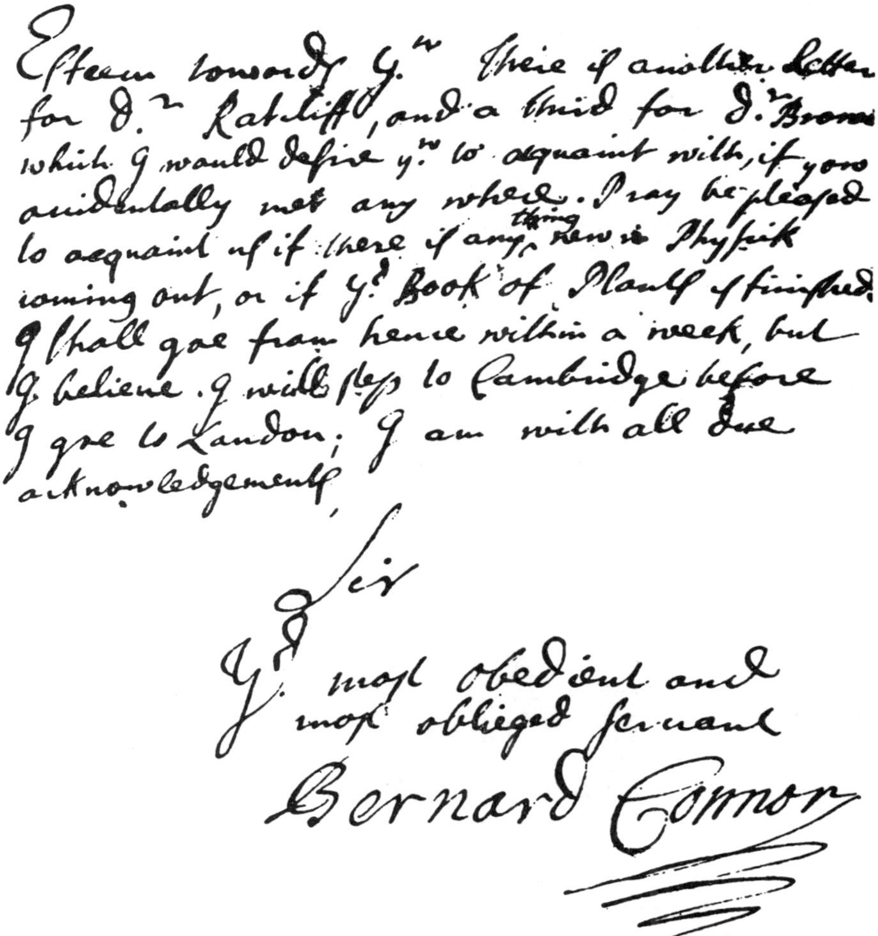
Esteem towards Y.r There is another Letter
for D.r Ratcliff, and a third for D.r Brown
which I would desire y.u to acquaint with, if you
accidentally met any where. Pray be pleased
to acquaint us if there is any thing new in Physick
coming out, or if Y.r Book of Plants is finished.
I shall goe from hence within a week, but
I believe I will pass to Cambridge before
I goe to London; I am with all due
acknowledgements

Sir

Y.r most obedient and
most obliged servant
Bernard Connor

FIG. 2B.

His lectures at Oxford were well received. An unknown "Gentleman in Cambridge" wrote to him on 15 October 1695[10] and requested that he come to lecture to the students there, which Connor finally did in the summer of the following year.[16] Again, however, there is no record of his stay in Cambridge in the archives of the University.[45]

358 *Journal of the History of Medicine:* JULY, 1958

DISSERTATIONES MEDICO-PHYSICÆ.

De Antris Lethiferis.
De Montis Veſuvii Incendio.
De Stupendo Oſſium Coalitu.
De Immani Hypogaſtrii Sarcomate.

A D. *BERNARDO CONNOR* M. D.
Sereniſſimi Poloniæ Regis Medico è Regia
Cameræ *Pariſienſis* Societate.

OXONII,
E THEATRO SHELDONIANO,
Sumptibus *Henrici Clements* Bibliopolæ *An. Dom.* 1695.

FIG. 3. Title-page of *Dissertationes Medico-Physicae*. The pathological description of the ankylosing spondylitis is given in the third essay. (Photograph courtesy of the New York Academy of Medicine.)

It appears that Connor did not concern himself with scientific questions alone, if we are to believe the testimony of a contemporary resident of Oxford. The last two decades of the seventeenth century were an unsettled and unhappy time in the British Isles. The relatively bloodless revolution of 1688-89 had led to the exile of James I and with him the group that favored the French and Catholics. William III, the Stadtholder of the Dutch state who had been called to head the new kingdom, was supported by the Anglicans, Dissenters, and the anti-French group. In 1692 William's position had been strengthened by his victory over the Irish and French forces under the returned James at the Battle of the Boyne. Throughout the country an immoderate suspicion of Catholics, Irishmen, and French supporters led to many unjust and hasty acts. These sentiments were particularly strong in Oxford, which in 1688 had seen the Fellows of Magdalen College turned out and replaced by the followers of James.[4] By 1695 some degree of moderation was returning to the country but it was not sufficient to remove from public suspicion a person such as Connor who was at the same time an Irishman and a Catholic (although he had presumably been converted to the Established Church on his return to England), and had been educated in France.

During this period Sir William Trumbull (1639-1716) was the secretary of state, a privy councillor, and a member of Parliament for Oxford University. His correspondence had been preserved at his ancestral home, Easthampstead Park, Berkshire, and in 1924 was published by the Historical Manuscripts Commission as the *Downshire Reports.* Among the letters received by Trumbull were two concerning Connor.[25]

Dr. Robert Gorge of Oxford, writing to Trumbull in July 1695, states that he had heard of Connor, and although Connor professed to be a Polander (which we have no evidence he ever attempted to do) he was in fact an Irishman and that he had come to Oxford not to instruct in Physick but to excite the young men to rebellion against the throne and to stir up Jacobite sympathizers.

We know that Connor changed his name in 1694 but it is hard to believe that he was attempting to hide his Irish birth by the single device of removing the "o." In a second letter Gorge states that he had received a letter in code which purported to prove that Connor was a French spy. He said that he received this information from Dr. Wallis, at that time a Fellow of the Royal Society. If this were true, it is surprizing that the fact did not prevent Connor's enrollment as a Fellow of the Society the following November.

It is difficult to evaluate these accusations. One is left with the impression, however, that Connor was more anxious to advance his medical and scientific career than to be embroiled in partisan politics.

Gorge stated that Connor had "gone off privately but whither is unknown."[25] From the date line of Connor's letters we know that he returned to his Bow Street residence to carry on his medical practice and experiments. He continued to lecture in London on his "animal oeconomy" and also on the anatomical and physiological knowledge he had acquired in France and Italy.

He had by this time become sufficiently prominent to be considered for a Fellowship in the Royal Society of London. His election is duly noted in the *Journal Book* of the Royal Society for 27 November 1695,[34] and Connor signed the first *Journal Book* of the Society under that date.[40] His only other mention in the records of the society is on 23 February 1697 when it is noted that he had presented a copy of *The History of Poland* to the Society and that it was given to Mr. Waller, the secretary.[41] A long and detailed review, presumably in the hand of Mr. Waller, is preserved in the classified papers of the Society.[39] Other than the letter in the *Philosophical Transactions,* to which reference has already been made, he apparently made no contribution to the proceedings of the Society. Connor presented several other books to the library of the Society[7,8,15] and these contain the notation, possibly in the hand of Connor, that they are the gift of the author.

Connor was further honored in 1696 by election as a licentiate of the Royal College of Physicians of London. This is noted in an entry from the Register of the College for 16 April 1696.[38] It is noteworthy that he had no compunction in taking the oath of loyalty to the Crown as was required of all licentiates. The only other entry under Connor's name is one recording the imprimatur of the College on his *Evangelium medici.*[15]

Ever since his arrival in England, Connor had written and spoken about his plan for the "Animal oeconomy," the eventual object of which was "To explain the miracles relating to human bodies by the principles of Physick." These he had outlined in a series of published open letters[10, 12–14] in *The History of Poland* and elsewhere in his publications. In this day of specialization into one narrow segment of human knowledge, one cannot but be impressed by the great size of the task that Connor had set himself.

> It is plain to me [he wrote] that a man void of all prejudice, and who considers that all the operations of bodies are performed by natural causes without miracles may be easily convinced that the causes of the disease and the true use of applications to cure them, can be rendered very intelligible so that the vulgar axiom that there is no certainty in Physick will be found most eroneous.[10]

Connor felt that some explanation should be made for the miracles of revealed religion in order that sceptics and deists might be confronted with these proofs of divine actions. He protested that he was not being irreligious in these activities but indeed that they would strengthen the faith of the unbelievers. He proposed to do this by first understanding the structure of the human body, "which I have taken apart [in] my anatomy and resolved into essential elements or minute particles by chemistry." Following this, he would investigate the "Laws of Motion," the suspension of which led to miracles. Having established this information as a consequence of experiments he planned to do, "since experiments are the only foundation we can surely build upon in Physick,"[13] he could then investigate the various miracles of revealed religion. These included the creation of Adam, the Virgin birth, the resurrection, and others. Needless to say, this led to a good deal of censure in ecclesiastical circles despite Connor's professed interest in religion. He was publicly criticised by the Archbishop of Canterbury and Connor published a letter addressed to the Archbishop defending himself against this charge of heterodoxy.[12]

Notwithstanding this censure and his rather tenuous political situation, in 1697 he published in London a Latin treatise, *Evangelium medici*[15] which embodied his radical ideas. It was an immediate sensation and had another London printing later that year, a third in Amsterdam in 1699,[18] and a fourth in Jena in 1724.[20]

Connor was now at the height of his brilliant career. He had arrived in England a stranger with serious political and religious handicaps and in a short space of three years had published several books,* had lectured at Oxford, Cambridge, and London, had created an intellectual stir sufficiently great to receive a public censure from the Archbishop of Canterbury.

At this time, when he was "In the vigor of his youth . . . when the world was in expectation of great things from him, and when possibly improvement in Natural knowledge, reputation in his profession and advancement in his fortune filled and employed his thoughts,"[24] he suddenly died.

A good deal of information about his last days is available. In his will,[19] written shortly before his death, he designated that his cousin and executor, Charles Connor, should have a sermon preached for him by Dr. William Hayley of St. Giles-in-the-Fields, Connor's parish church, and that the address be printed.

In his sermon (later used as biographical material in *Wilford's Memorials*[47]), Hayley tells that Connor had a febrile illness, sometimes accompanied by delirium, and that he was ill for several days before dying. (No further information is given which would help in identifying the illness.) After questioning Connor about the intention of his controversial writings and being assured that he was a penitent member of the Church, Hayley had pronounced the last rites. Connor was later visited by an Irish Catholic priest who did the same for him. Hayley concluded by drawing a lesson on the necessity for being prepared for death at any time. The sermon was delivered on 30 October 1698 and Connor was interred at St. Giles-in-the-Fields three days later. The name of Dr. Bernard Connor is entered in the Registry of Deaths of the church for November 3d.**

It is not unlikely that Connor would have gone on to still greater accomplishments in his profession had he been spared a few more years. He was a man of great energy and ambition. Although some of his efforts appeared to have been directed towards gaining recognition from his colleagues and the public, he was genuinely interested in studying disease, using the new techniques of scientific experiment and rational thought that were becoming current in the seventeenth century.

* According to McGee,[30] Connor published no less than 17 medical volumes in addition to *The History of Poland* during his stay in England. However, not all of these have been traced.

** The Death Register was examined through the kindness of Dr. G. C. Taylor, the present Rector of St. Giles-in-the-Fields.

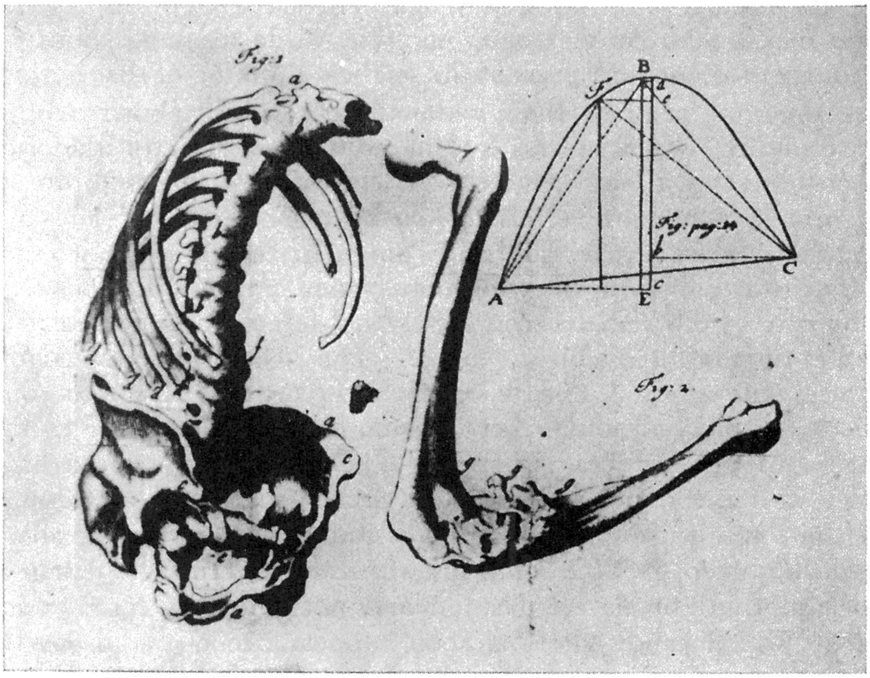

FIG. 4. Illustration from the article in the *Philosophical Transactions* published in 1695. Picture 1. is the pathological spine. Picture 2. is an ankylosed knee (from another skeleton) which might have been caused by rheumatoid arthritis or infectious arthritis. (Photograph courtesy of Dr. Charles Ragan, Presbyterian Hospital, New York City.)

Connor's Description of the Pathology of Ankylosing Spondylitis

The history of ankylosing spondylitis has recently been reviewed.[33] Although there has been some description of skeletons with abnormalities which might have been produced by ankylosing spondylitis prior to the seventeenth century, it is probable that Connor's is the first adequate account of what was undoubtedly an example of this disease. The description is contained in no less than three separate papers, in as many languages,[6,8,9] the first being the French publication of 1693.[6] The accuracy of his description makes it of contemporary interest, while his suggestions on etiology reveal some aspects of the medical thinking of that period.

Sometime prior to 1693, while Connor was demonstrating in anatomy in France, he was told of an unusual skeleton that had been found in a local cemetery. It consisted of the os ilium, os sacrum, five lumbar vertebrae, ten thoracic vertebrae, and five right and three left ribs, all firmly united at "the joinings" so that they made one uniform continuous bone. The joints between ribs and vertebrae were completely obliterated and the normal juncture was impossible to detect. The "Cartilagineous edge," that is, the annulus fibrosis of the intervertebral disc was completely ossified, although there was no evidence of joining in the center of the vertebrae where the nucleus pulposis would have been in life. Connor tested this by cutting the vertebrae through at the commissures and "found this uniting did not enter above two lines deep."[9] This conforms with the modern view that it is the annulus fibrosis calcification and not the longitudinal ligaments which causes the "bamboo" appearance of the spine on x-ray.[5] In many cases, however, the longitudinal ligaments are involved as well.

The marked curvature of the spine, characteristic of the disease, is well seen in the diagram which accompanied the article in the *Philosophical Transactions*.[9] (Fig. 4). Connor further commented on the unnatural arrangement of the direction of the ribs and their possible effects on respiration. Although there is no evidence that Connor ever saw a case when alive, he hazarded some opinions on the clinical state of the patient. He remarked that the body of the person when alive must have been immovable, that he could neither stretch, rise up, or lie down, or turn upon his side, having only the head, feet, and hands movable. Since the ribs were fixed, it would be impossible for the patient to expand and contract his chest in the usual manner. He had previously seen that the motion of the chest was not necessary for respiration as in "hysteric fits, etc.", and observed that the diaphragm could still be completely movable. This auxiliary means of respiration could supply the patient with sufficient air if he were to utilize short, rapid breathing. Recent studies have shown that many of these patients do not have disabling cardiorespiratory symptoms and that although their air volume is considerably reduced, they can use the air available to them as well as normal subjects.[21]

The latter part of the article is concerned with some speculations on the etiology of this condition. Although his remarks seem naïve and quaint to modern ears, it should be recalled that even now we have very little more information on the etiology of this perplexing illness.

Connor dismissed the possibility that the unification occurred post-burial as a consequence of unusual soil conditions or was acquired during adult life (although his reason for rejecting the latter is not made clear). He suggested that the deformity originated *in utero* as a consequence of pressure from an abscess tumor in the womb or elsewhere which led to an abnormal development of the spine with the other changes secondary to this.

In concluding, the author raises the question of why some people should become bony in some regions of the body and others not and says that some explanation of this fact might be necessary before any understanding of the pathogenesis is achieved. This problem still remains to be solved.

REFERENCES

1. BLUMBERG, B. S. Bernard Connor, F.R.S. (1666-1698) and his contribution to the pathology of ankylosing spondylitis. Abstracts of papers, 8th International Congress of the History of Science, Florence-Milan, 3-9 September 1956.
2. BLUMBERG, B. AND RAGAN, C. The natural history of rheumatoid spondylitis. *Medicine*, 1956, *35*, 1-31.
3. BUCKLEY, C. W. Spondylitis deformans. *Brit. med. J.*, 1931, *1*, 1108-12.
4. CLARK, G. *The later Stuarts 1660-1714*. Oxford, Clarendon Press, 1955. 479 pp.
5. COLLINS, D. H. *The pathology of articular and spinal diseases*. London, E. Arnold, 1949. 331 pp.
6. CONNOR, B. *Lettre écrite à Monsieur le chevalier Guillaume de Waldegrave, premier médecin de sa Majesté Britannique*. Paris, 1693 (1691 ?). 17 pp. (B.M.)
7. CONNOR, B. *Ζωοθαvασιρν θαvμαστον seu mirabilis viventium interitus in charonea Neopolitana crypta*. Cologne, 1694. 70 pp. (Bodleian and Royal Society.)
8. CONNOR, B. *Dissertationes medico-physicae*. Oxford, Sheldonian Theatre, 1695. 206 pp. (N. Y. Acad. Med. and Royal Soc.)
9. CONNOR, B. An extract of a letter to Sir Charles Walgrave, published in French at Paris: Giving an account of an extraordinary humane sceleton, whose vertebrae

of the back, the ribs and several bones down to the os sacrum, were all firmly united into one solid bone, without joynting or cartilage. *Philos. Trans.*, 1695, *19*, 21-27.

10. [CONNOR, B.] *A letter from a gentleman in Cambridge to Dr. Connor of London, concerning the method used in his physical and anatomical lectures at Oxford.* 1695. 8 pp. (B. M.)
11. CONNOR, B. A letter to Hans Sloan dated Oxford, 28 May 1695. Sloane MS 4036, The British Museum.
12. CONNOR, B. *A copy of letter sent to his Grace *** from Dr. Connor . . . concerning his Medicina arcana de mystico corporis humani statu; or a Latin treatise, in which he designs to explain the miracles relating to human bodies, by the principles of physick.* London, 1696. *s. sh.* fol. (B. M.)
13. CONNOR, B. A letter to his worthy friend, D.B.M. from Dr. Connor, member of the College of Physicians and Royal Society, concerning his Medicina arcana de mystico corporis humani statu. London, 1696. *s. sh. fol.* (B. M.)
14. CONNOR, B. *A letter to James Tyrrel, Esq. containing a further explanation and vindication of the plan of animal oeconomy.* London, 1695. 8 pp. (B. M.)
15. CONNOR, B. *Evangelium medici; seu medicina mystica; de suspensis naturae legibus, sine de miraculis.* London, S. Briscoe, 1697. 246 pp. (Royal Soc.)
16. CONNOR, B. *The history of Poland in several letters to persons of quality.* London, Brown, 1698. 2 vols. 352 pp., 356 pp. (Bodleian.)
17. CONNOR, B. A letter to Hans Sloan dated 17 August 1698. Sloane MS 4037, The British Museum.
18. CONNOR, B. *Evangelium medici.* Amsterdam, Walters, 1699. 193 pp. (B. M., with some handwritten notes on the flyleaf.)
19. CONNOR, B. The will of Bernard Connor, probated 14 April 1699. The Principal Probate Registry, Somerset House, London.
20. CONNOR, B. *Evangelium medici.* Jena, 1724. (B. M.)
21. D'SILVA, J. L., FREEDLAND, D. E., AND KAZZNTZIS, G. The performance of patients with ankylosing spondylitis in the maximum ventilatory capacity test. *Thorax*, 1953, *8*, 303-08.
22. DYBOSKI, R. *Outlines of Polish history.* London, Allen and Unwin, 1931. 285 pp.
23. GOLDING, F. C. Spondylitis ankylopoietica. *Brit. J. Surg.*, 1936, *23*, 484-500.
24. HAYLEY, W. *A sermon preached in the parish church of St. Giles in-the-Fields at the funeral of Bernard Connor, M.D. who departed this life October 30, 1698. With a short account of his life and death.* London, 1699. Bodleian Library, Oxford Sermons 9 (12).
25. HISTORICAL MANUSCRIPTS COMMISSION. Report on the manuscripts of the Marquess of Downshire preserved at Easthampstead Park, Berks. Papers of Sir William Trumbull. Vol. 1, part 2. London, H.M.S.O., 1924. 1025 pp.
26. JACQUINET, A. *Le centre universitaire médical de Reims (1550-1950) son oeuvre, son rayonnement.* Largentière (Ardeche) 1950. 29 pp.
27. JONES, L. *Arthritis deformans.* New York, Wood, 1949, p. 278.
28. KERRY, C. G. O'C. Kerry clans and confiscations. MSS. 1854. Quoted by B. O'Connell of The National Library of Ireland.
29. KNOTT, J. A forgotten Irish medical exile and scientific pioneer of the 17th century (1666-1698). *J. med. Sci., Dublin*, 1907, *57*, 131.
30. MCGEE, T. D. *Irish writers of the 17th century.* Dublin, Duffy, 1846. 252 pp.
31. MORERI, L. *Le grand dictionnaire historique.* Paris, Libraires Associés, 1759, vol. 4, p. 45.
32. O'CONELL, B. The National Library of Ireland. Personal correspondence.
33. O'CONELL, D. Ankylosing spondylitis. The literature up to the close of the nineteenth century. *Ann. rheum. Dis.*, 1956, *15*, 119-23.
34. POLLEY, H. F. A study of 1035 cases of rheumatoid spondylitis. Thesis, Graduate School, University of Minnesota, March 1945.
35. RAUSSIN, LOUIS-JEROME. MSS quoted by A. Jacquinet in personal correspondence.
36. REDDAWAY, W. F., PERESON, J. H., HALECKI, O., AND DYBOSKI, R. *The Cambridge history of Poland.* Cambridge, 1950, p. 554.
37. ROLLESTON, H. The history of chronic rheumatism. In: *A survey of chronic rheumatic diseases.* R. G. Gordon, Ed. London, Oxford Univ. Press, 1938, p. 3.

38. ROYAL COLLEGE OF PHYSICIANS OF LONDON. The register. 1695-1710. Entry for 16 April 1696.
39. ROYAL SOCIETY. Classified papers. Accounts of books. Vol. 22 (1), 1660-1740.
40. ROYAL SOCIETY. Journal book of the Royal Society. Vol. 8, 1690-1696. Entry for 27 November 1695, p. 318.
41. ROYAL SOCIETY. Journal book of the Royal Society. Vol. 9, 1696-1702. Entry for 23 February 1697, p. 81.
42. ROYAL SOCIETY. *The signatures in the first Journal Book and the Charter Book of the Royal Society.* London, Oxford University Press, 1950.
43. STEPHEN, L. and LEE, S. *Dictionary of national biography.* London, Smith, Elder and Co., 1908.
44. VAN SWAAY, H. *Spondylitis ankylopoetica. Een pathogenetische studie.* Thesis, Leiden, 1950, p. 182.
45. VENN, J. A., Keeper of the Archives, University of Cambridge, Personal correspondence.
46. WHEATLEY, H. B., and CUNNINGHAM, P. *London past and present.* London, 1891. 3 vols.
47. WILFORD, J. *Memorials and characters, together with the lives of divers eminent and worthy persons.* London, 1741, p. 345-7.

Creativity

Baruch S. Blumberg

A sermon given at Balliol College Chapel on Rogation Sunday
5 May 1991

Rogation Sunday traditionally includes events related to agriculture in it's celebration. I would like to focus my comments on the related themes of growth, faith, creation and the considerations of the divine that these themes imply. This might appear to encompass a broad spectrum of items for a short sermon, but I promise not to present each of them in great detail.

The first tractate of the Jewish Talmud that was written in Babylon over a period ranging from a few centuries before to a few centuries after the birth of Christ deals with the many agricultural practices presented in the Hebrew Testament. The first book of this tractate is devoted to prayer . The question was raised as to why a book of agriculture should include, indeed start, with a discussion of prayer. The rabbis answered that agriculture is an extraordinary testament of faith. Planting a crop or a tree implies faith that the planter will be there to reap when the crop matures ; that he, she or their descendants will eat the fruit of the tree or utilize it's lumber at an older age or in subsequent generations.

This faith was very much in my mind during the last vacation, which we spent in part on our farm in Maryland. The land banking program of the U.S.Department of Agriculture encourages farmers to put crop land into forest. Over the course of two, seasons we have planted 2400 American Walnut trees - Jugulans nigra - and my new grandson, Stephano Baruch Blumberg and his two grandfathers planted one of these trees together.

I have been a scientist all of my adult life,and have found that scripture and it's interpretations have supported my enthusiasm to see order in the world and to try to understand natural phenomenon. The first book of the Hebrew Testament , Bereshith , Beginnings, or Genesis starts with the account of the creation of the cosmos. Well into the process, starting at Genesis II,19 God makes one of his first demands of newly created man. I will read a translation of this section.

" Lord God brought out of the earth all the animals of the field and all the birds of the air (Note: he did not bring forth fish or amphibian) and he brought them to man to see what he would call them, and everything that man called the living things, that was to be there name. "

He then continues:

" And the man gave names to all the cattle (I believe the translation might actually mean all domesticated animals) and to the birds of the air and to all the beasts of the field..."

the birds of the air and to all the beasts of the field..."

What was the meaning of this command from God to his newly created being ? He required that man set up a system of classification. Naming implies that you can distinguish one individual or group of individuals from another and that sufficient singularities are identified to give it a separate name.

Naming implies power and control and in many cases the notion of possession. Parents name there children and signal there image of what he or she will be by doing so. If man names the species he actually , in an odd way, creates them. Those of us who have been involved in the traditional methods of plant classification know very well that it is the specialist in the particular genus or family to which a plant belongs who decide which species it is and indeed if a new species or genus is to be created. Species, as we know them are made by the systematist, and not directly by God.

The human systematist does not operate without constraint. He must conform to the characteristics that are present in nature. If arbitrary classifications are made to satisfy the ego, and the desire for control by the scientist, they will ultimately be corrected by other scientists who impose the demands made on a science that attempts to interpret nature as it is and not as a product of a theoretical model devised by humans which disregards the real world.

The human role in species creation and destruction was illustrated recently when the international committee responsible for the naming of bacteria and establishing the characteristics which distinguish one species from another decided to review the existing catalog utilizing advances made by recent research. There was a major loss of species from the book, a destruction probably greater than wrought by any recent contaminations of the environment.

Let us return again to this theme of creation and the closely allied concept of creativity. The scholar I. Epstein in his book " Judaism " published in 1968 comments on this in relation to the first sections of Bereshith- Genesis- to which we have already referred.

" The story begins with the creation of the world by God and the formation of man in his image. But this creative process did not cease when the world and man had been made , for that which had been created had to be developed and fostered. This task God entrusted to man. Made in the image of God, man must conform to character of God and because God creates, man must also create. He must work and cooperate with God in maintaining and developing the work He has committed to man's care. "

This raises the question of co-creativity- humans creating along with God- a very heady message indeed and one, I understand that is a frequent subject of discussion in religious circles. Does the scientist discover a natural object or phenomenon that has always been there, or does he create it ? Does an individuals intercession in discovery convey some unique characteristic of the discoverer/inventor or would any other scientist have revealed the exact same phenomenon?

The answer to this is probably, sometimes yes and sometimes no. For example, one could argue that Columbus's discovery of America (granted that many others may have done so before him) would be the exact equivalent of any other person's discovery. Suppose, however, that the Japanese had been the first discoverers and settlers. The country would actually be very different . It would have been populated from the west and not the east. Cities would have been located in different places because of the differences in importance put on geographic sites by different cultures. The immigration policies, language, art, culture would have been different. Perhaps even this kind of discovery does have an element of creativity in it.

The scientists in my laboratory were responsible for the discovery of the hepatitis B virus an important virus that infects man and causes great deal of disease. I believe that the subsequent progress of the work was to large extent influenced, at least in it's first few years by the creative instincts and the scientific and other interests of my colleagues and I. There were several clinicians among us. Hence we were familiar with the medical problems and practical application came sooner rather than later. We were interested in population biology and preventative medicine; hence the research had a much stronger inclination in that direction and probably influenced the speed at which the discoveries had application to the public health. We were interested in genetics hence family studies were done early in the history of the program. We were not virologists and that aspect of the investigation advanced more slowly than it might have. In effect, because a particular group of scientists had been involved the outcome was different than if another group had been the discoverers. There was an element of creativity or originality in the process.

From all that I have said so far it can be argued, in fact has been, that it is a responsibility we have to God the creator to help in the program of creation and to do this by the accumulation of knowledge and an analytical and, later, synthetic understanding of the world. It can be viewed as a form of worship and serving that fulfills the discipline that has been put upon us.

There is a mythic, even a mystical character to this kind of creativity/discovery. The first and predominant caution is to beware the sin of <u>hubris</u> the notion that we can understand all and therefor control all. The history of science is filled with the lesson that the knowledge available to humans is always incomplete; every solution to a problem raises additional problems, often not comprehended or even comprehensible before the discovery had been made. Therefore application of discovery, if it does occur, must always proceed with the sure knowledge that no matter how advantageous the application may seem and actually be, there is always lurking the possibility of unfortunate and unwholesome outcomes that the scientists should prepare for.

A great deal of the anti-scientistic sentiment that is so common in our age arises from these unhappy, even disastrous outcomes of discovery . This may lead to the demand that science should be severely limited. It is important to remember that not doing something to solve a problem is a form of action . If we are compelled to seek out knowledge and to comprehend the world, if that is a mission of humankind, then hazards exist and, if possible, they should be anticipated. The best way that this can be accomplished is by striving to make our knowledge of a subject as broad as possible in order to be best prepared to deal with unexpected and negative outcomes when they emerge.

The story by Mary Wolstencraft Shelly, the great pioneer of science fiction of the Monster created by Victor Frankenstein is a caution to scientist. There was recently a wonderful dramatization of the story presented in Philadelphia with imaginative sets modeled after 1930's science fiction movies and many dramatic staging devices. Victor's Monster was compared to Adam, and there creators contrasted. In a final confrontation between the monster and his inventor (at the North Pole, a place of great mystery to the scientists and intellectuals of the period) the Monster admits to the horrible acts he has committed and then accuses Victor of abandonment. He created him but did not love him, and the awful fate that befell the Monster, his victims and mankind was a consequence of this awful neglect. Later that night I awoke from a nightmare of fear, fear that there could be unfortunate consequences from the medical practices that had resulted from our work and the guilt that could accrue if the responsibilities that accompany discovery were not sustained.

To conclude, I believe that we have a responsibility to comprehend what we can of the world and to explain it in an ordered and comprehensible manner; that we have to learn to deal with the consequences of discovery and creativity, tasks that have been assigned to us in fulfilling God's plan. A colleague of mine, a dedicated Christian made this point extraordinarily well in a recent meeting in the Vatican when he said that it was his God given gift of scientific skill that motivated him towards discovery and the application of his findings. In fulfilling this imperative we must avoid the dangers of overconfidence, the conceit of perfection of knowledge, and the sin of pride.

A Sermon

Baruch S. Blumberg

Balliol College Chapel

12 June 1994

Science is a creature of the western world. Although descriptions of nature and perhaps a pre-science was present in the classical world, what we think of as modern scientific inquiry did not originate until several centuries later. It has been suggested that the Holy Testament, with its creation myth emboldened humans to develop a scientific explanation of the world and universe. For the Bible told of a world created and organized in a reasonable manner, by a Creater who had a plan for the universe. If the world is organised, then it is possible to understand it. The alternate view is that the <u>is</u> no plan, that there isn't a reason for all that is in nature and therefore it cannot be understood by any rational method. Prediction of events would be fruitless, since general laws would not apply.

If the world is ordered and <u>can</u> be understood, and if predictions of future events <u>can</u> be made on the basis of generalisations derived from observation and theory, then humans can have some control of their destinies. That is, they can hope to change the world, and change it for the better. It is this that I wish to speak of.

There is a Hebrew expression, <u>Tikun olom</u>, which can be roughly translated as "repair of the world." But, it has a broader meaning, something along the lines of setting wrongs right, making things better, reacting to error; and further that it is not an abstract ideal, but an obligation. Our College has, traditionally, tried to impart this meaning to our students, to prepare them, not only for their own development but for what they can do for others.

Is this goal being realised? In some cases I can attest that it is. When our students come to me before they leave it is not at all uncommon to hear that they are planning to enter a field of work that will prepare them to be of service, sometimes in a very direct way, by working abroad or at home with those who have been hard stricken by a harsh world; or in their everyday life, to prepare to volunteer for good works. I leave the College with the assurance that our tradition of service remains, perhaps not as strong as it has been in the past, but it's there.

What does distress me, however, is the strong measure of cynicism which appears to pervade this and other countries. I use the term to mean a sneering disbelief in sincerity and integrity, that is, that no person means what they say, all declared motives are false, that, in effect, nothing can be accomplished, because the stated goals are false. The media is often criticised for excessive cynicism, but the media reflects the society with which it is communicating, and it is unfair to blame the messenger for the news. A classic example of this cynicism, was published recently in an article describing President Clinton's visit to Oxford. Those who attended the ceremony would, in general, agree that it was a pleasant day, and an appropriate way to honour a foreign head-of-state who had attended the University. But the writer of the article painted everyone as either false, slightly ridiculous, or very ridiculous. The President, his wife, the Secret Service folk, the protesters, the motivations of the University itself. How can anyone aspire to make things better if we start with a suspicion of the motives of all we encouter?

It is often a single, seemingly minor incident that serves to characterise an attitude. As many of you will know, we award the Glen/Bowlby grants to students for "adventurous travel". I attended the reporting session of one of our students who had been funded for an exciting, and sometimes perilous trip to the Himalayas. I was introduced to a friend of his, not from Balliol, who said, "The reason the College grants the travel fund is to encourage students after graduation to become benefactors of the College", that is, that the motivation for the awards was for the financial gain the College would obtain at some distant time in the future. In fact, that isn't our goal and his statement is a distortion. The College wants to provide special opportunities for the summer breaks, to encourage students to attempt projects beyond their prior experience, and to launch them on an imaginative direction that opens them to a new world. I'm happy to say that the recipient himself didn't agree with the calloused view of his friend. But, I was troubled by this immediate cynical response.

What is required is a sense of hope, the notion that there is an exciting future for the young as they enter the world after their education. Science is particularly fortunate in this regard. There is hardly a scientific field that is not overflowing with promise. Attending any of the many seminars held in Oxford leaves one full of ideas of what can be accomplished both in basic science and in its applications. In politics, one can argue that there is so much dissatisfaction with the current state of affairs that there is immense scope for improvement, room for new political leaders, better equipped, more capable to take over for the new century. In international affairs, there is certainly plenty that has gone wrong that is ripe for repair. The daily news is filled with tragedies of a global scale looking for new solutions. And for British youth there are some particularly exciting and proximate problems. In Europe itself, there is a vision of a peaceful and economically successful continent. How can this be realised? British students with a tradition of stable, democratic government and a mastery of

English, a language which has become a requirement for any international dealings, are to my mind, particularly well placed to take a role in these developments. And looming beyond are other possibilities. The Commonwealth is an existing organisation; one of the first acts following the recent election in South Africa was to rejoin the Commonwealth, and several other African countries are now being considered for entry. One hundred years ago, many of our graduates spent their careers in the far reaches of the then Empire, a life in many cases of fulfilment and excitement. Is it time again to see what opportunities may lie in the life of the Commonwealth? A sermon may not be the appropriate place to continue this list of career possibilities, but the intent is to raise the very pertinent notion of Hope.

In the last few weeks we have witnessed moving ceremonies related to the landing on the Normandy beaches and the commencement of the campaigns which led to the successful termination of the war. The Second World War was the defining event for the people of my generation. It represented a combined effort to achieve a common goal for which many were prepared to sacrifice all, and tragically, many did. We still have great problems to solve, victories to win and the same spirit of endeavour. What is required is faith, that there is a future, hope that we will take part in it, and the motivation to "repair the world", tikun olom.

A LETTER FROM OXFORD

BARUCH S. BLUMBERG,* M.D., Ph.D., *Senior Advisor to the President of the FCCC*

SCIENCE IN CONTEMPORARY MOVIES

Does popular culture propel societal behavior, or just reflect it? Are violence and crime increasing as a reaction to the constant mayhem displayed on television, the movies and the other media, or do the popular arts reproduce the concerns that engross and interest the public? Is the media vision of science, particularly that shown in blockbuster movies, a generator of distrust, or does it reflect the public unease with the scientific endeavor and its products or those that it is thought it can produce?

Consider three recent movies. The Fugitive is a chase adventure with spectacular scenes of train crashes, a leap into the tailrace of a mighty dam, and brutal murders. And who are the villains behind it all'? A pharmaceutical company and an avaricious clinical investigator bent on gaining FDA approval for a therapeutic agent with unfortunate side effects that they are attemting to conceal. Lorenzo's Oil deals with a rare neurological disease without a cure. The attempts of parents to find non-conventional treatments are opposed by self-seeking scientists lusting for recognition and glory. Jurassic Park — with its vast public relations campaign and spectacular special effects, perhaps the most influential of the three — provides a massive attack not only on the products of science but the very process by which it achieves its goals.

In a previous essay (The Application of Science, (FCCC Scientific Report, 1990-91)), I discussed "the Frankenstein Model" for public perceptions of science, and I will not elaborate directly on this theme now. But I would like to deal with several aspects of Jurassic Park (JP) which appear to have a growing impact on how the public regards science and its practitioners and which have become particularly pertinent since the development of molecular biology.

JP is one of the most successful movies and novels of recent times. Millions of copies of the novel have been printed, and the movie has attracted more viewers in the early stages of its release than any in recent history. Tens and perhaps hundreds of millions of people have been exposed to this flick (or will be during the next few years); it has had an impact on more people than nearly any other popular exhibition of science. The novel and movie differ in detail from each other, but I shall refer to both without attempting to distinguish differences.

John Hammond, a wealthy entrepreneur without scientific training, borrows a large amount of capital to establish a theme park, located on a remote island off the coast of Costa Rica, featuring genetically cloned, fully developed dinosaurs. The Park scientists obtain DNA from blood-sucking insects which had fed on dinosaurs, and were subsequently fossilized in amber. The DNA obtained was incomplete, and they patched it with conserved DNA from

* *Present address* — Master of Balliol College, Oxford University, Oxford OX1 3BJ, England.

modern amphibia. They then used fertilized reptilian ova replacing the DNA from the ova with that of the putative dinosaur DNA. The "conceptus" was incubated in an artificial egg and the investigators waited to see what would emerge. The molecular biology was accomplished by Henry Wu, a Stanford graduate, who is the metaphor for what could be called the reductionist approach to biology.

Wu's antithesis is Ian Malcolm, a mathematician or, more explicitly, a "chaotician" from Texas. In both the movie and the novel, he produces explanations, at a cocktail party level, of the significance of chaos theory in complex biological systems. Chaos theory connotes an underlying model for a natural phenomenon that can be described by mathematical equations, but the model cannot be used effectively for prediction of outcome. Small, even infinitesimal, changes in the original parameters can have major — and unpredictable — effects on the end result. To illustrate his point, he uses the metaphor of the butterfly which "...flaps its wings in Peking, (sic) and weather in New York is different." That is, even extraordinarily small initial changes, which may not even be detectable, will have an unpredictable effect on the phenomenon that is being observed.

Portrayed through Henry Wu and Ian Malcolm, the presentations of the models for, on the one hand, reductionism, and the other, overwhelming biological complexity, are extreme, but that is the choice of Michael Crichton, a medical doctor, the author of both the novel and the screenplay. Henry Wu believes that all the scientist needs to know is the DNA code, and everything else pertaining to a complex organism is then understood. To put it differently, you can grind up a mouse, extract the DNA, throw away the rest of the animal, and still learn all you need to know about the animal. Although this is an extreme statement of the case, it bears a resemblance to the claims of contemporary molecular biology where knowledge of a phenomenon is explained only in terms of the DNA code, without study or consideration of the multiple forces that impinge upon it from the products of other genes, from the environment, or from the complex interaction of time and substance. The witness of the novel is that this approach is doomed to failure.

The mathematician Malcolm has an opposite and also absolutist view. He reasons that all systems, even those seem simple, are subject to initial parametric changes that cloud outcome. A pessimist (he dies in the novel but is spared in the movie), he predicts that outcomes usually will be bad, a quantitative restatement of Murphy's Law, namely, that anything that can go wrong, will. When called in as a consultant by John Hammond, he tells him that because of the complexity, the project will not succeed.

What is the message that the public might carry away from this story? That scientists have a presuming creative arrogance and that they attempt to remake the world, if not in their image, then in an image they conceive. In an interesting exchange, Wu tells Hammond that he wants to introduce a new edition of the dinosaurs (numbered Version 4.4, in the same manner as software program updates) in order to make them conform to the public's, or at least Wu's perception, of how dinosaurs should look and behave. He wants to be creative — not only duplicate the acts of nature. He can do better than the previous creator. In Wu's concoction, dinosaurs are made lysine-deficient so that they cannot subsist on a natural diet and would not survive (or so they thought) if they escaped the island. And to get the

Park ready for business as soon as possible, Wu adds hormonal control genes to accelerate the growth of their dinosaurs. As Hammond emphasizes, he needs to see an income as soon as possible to satisfy his investors. Wu also wants the animals to move more slowly because, in his view, the public thinks of dinosaurs as ponderous beasts based on the interpretations of the early paleontologists, and he wants to cater to this perception.

Henry Wu bemoans his loss of control of the project as the task of animal creation is completed and the Park goes on an operational basis, compounding his sin of arrogance. One would hope that his motivation is to exert responsible and moral direction, but the character created by Crichton seems more interested in personal power. This negative image of the presuming creator-scientist is bad enough, but the message of his opposite number, the chaotician Malcolm is, if anything, worse. He tells us, and with an overwhelming certainty, that all attempts to solve complex biological problems will fail because of the unpredictable, non-linear and chaotic character of the system. What has this to do with the scientist's image of himself as a problem solver, his great justification for public support? Is Crichton trying to tell us not to be scientifically creative — not, for example, to discover new therapies — because it will always lead to a sticky end? This notion is reinforced by the message of The Fugitive where pharmaceutical scientists are seen as not only unable to produce new drugs, but with a desire only for personal gain, irrespective of the damage the treatment might cause. And, in Lorenzo's Oil, the establishment scientists object to the unconventional drug, not because it is ineffective, but because it would diminish their claims to recognition and reward.

In JP, the inevitability of evolution and adaptation is posited as another obstacle to scientific control. The animals have all been designed as females to ensure that they cannot reproduce and escape the island. However, Wu has used frog DNA to patch in the bits lost from the dinosaur DNA extracted from the amber, and this included a gene for gender reversal. He is not a biologist and doesn't realize that under conditions of population decrease and in the face of gender imbalance, animals can change sex in order to maintain their numbers. The attempt to prevent the dinosaurs escaping the island through the introduction of the lysine deficiency gene was also thwarted. When the animals reach the mainland, they develop feeding behaviors to selectively seek out and eat lysine-rich foods, presumably until evolution takes hold and selects out the deficiency gene. The message: don't fool around with Mother Nature.

Altogether, not a happy picture of the scientist with creative arrogance on one hand and built-in failure on the other. This bleak image would discourage anyone from entering a field in which defeat is assured from the start. The only scientist who appears to come off with any glory is Alan Grant, the paleontologist who is cast in the role of scientist as hero and problem solver, who rescues the children and also seems to win the favor of the glamorous paleobotanist, Ellie Sattler. He is reminiscent of the archeologist-hero, *Indiana Jones,* who can get out of any scrape, the man of action in the cowboy hero mold so well understood in the corpus of traditional American cinema. Perhaps the favorable image of Grant and Indiana Jones is the result not only of their bravery, but of their professions. They seek to observe and catalog nature, in contrast to Wu, the molecular biologist afflicted with hubris who wants to interfere with nature and be a creator.

How much of an impact does this negative image of science have? Is it counterbalanced by the generally favorable views of science conveyed by the regular writers in the columns of the science sections of the newspapers and the television programs depicting scientific accomplishment? Still, even in these friendly columns there is an extraordinary interest in actual and supposed scientific fraud and deceit. How then are we as scientists to resolve this problem which can have so serious an impact on the ability of science to accomplish its goals of increasing knowledge of the natural world and providing applications that ultimately convey more good than harm?

I hope that none of my readers was expecting an answer to these conundrums. There are, however, some obvious observations. If we can agree that scientists are not motivated only by power and control, as implied by the movies we've discussed, then there must be a more effective way to convey to the world that scientists are driven by a concern for knowledge for its own sake and that even when financial gain is a primary goal, professional responsibility, integrity and honesty will insure that the best interests of the public will prevail. Now is the time for aspiring novelists and screenwriters to make their move to rescue the good name of science.

PUBLICATIONS

BLUMBERG, B.S. Ethics and cancer therapy, *Biomed. & Pharmacother*, 47: 393–395, 1993.

LEGA, L., A. VIERUCCI, B.S. BLUMBERG, G. SARACCO, M. RIZZETTO, M. ZHU, and M. A. FEITELSON. Hepatitis B x antigen and polymerase antibodies in the serum of hepatitis B carriers with or without delta virus infection. Effects of interferon treatment. *J. Hepatol*, 14: 286–293, 1992.

STEVENS, R.G. and B.S. BLUMBERG, Serum albumin and the risk of cancer. *In* Macronutrients: Investigating Their Role in Cancer, edited by M.S. Micozzi, T.E. Moon, Marcell Dekker, Inc., New York, NY, pp. 283–292, 1992.

UNANDER, D.W., G.L. WEBSTER, and B.S. BLUMBERG. Usage and bioassays in Phyllanthus (Euphorbiaceae): a compilation III. The subgenera Eriococcus, Conami, Gomplidium, Botryanthus, Xylophylla and Phyllanthodendron, and a complete list of the species cited in the three-part series. *J. Ethnopharmacol.* 36: 103–112, 1992.

Reflections on a two-wheel way of life

Picture: ROB JUDGES

Blumberg by the Oxford Canal: 'Recreational cycling is not the everyday task of getting from place to place but adventure'

Dr Baruch Blumberg has just come to the end of his five-year term as Master of Balliol College, Oxford. Here he describes the joys of his main extra-curricular activity

I DON'T remember learning to ride a bicycle; it must have come as a gift. Perhaps this was an intimation of the mystical relationship cyclists often have with the activity; it is not only a form of locomotion but a way of life. Miles Kington, one of the most astute cycle commentators, argues that the bicycle is not of earthly origin but was created on another planet and sent here to accommodate and irritate humans. I agree.

I remember the first occasion when, as a teenager, the romantic possibilities of self-propelled locomotion came to me. It was the first long trip I took away from home, all the way from our neighbourhood of Far Rockaway in New York City to a remote suburb normally reached only by car or train.

Although it was only a few miles, it seemed we had been beyond the Pillars of Hercules as we hurried back in the gathering darkness to the safety of home. I realised a bicycle could carry you to places only previously imagined.

As a student at Oxford in the 1950s, I rode a decrepit £3 wonder that grudgingly carried me from place to place, vanquished at traffic-light sprints by schoolboys and pregnant north Oxford housewives. Then fate intervened. My old wreck, parked in the lee of the ancient wall of Exeter, was crushed when the wall collapsed. This is still a site of family pilgrimage but the outcome wasn't all bad.

I bought, second-hand, a beautiful bespoke road bike from the studio of a noted British bicycle builder. Exposed to excellence, I discovered the joy of riding a well-made machine, skilfully and elegantly crafted.

Years later, when I decided once again to embark upon serious cycling, I attempted to purchase a mountain bike at a "hip" hi-tech cycle store in Philadelphia. I was waved off, told to apply for an appointment when I could come properly clothed and psychologically set to be interviewed by a "cycle consultant". In due course he put me on a Specialised Stump Jumper Sport, an outstanding machine which has brought joy equalled only by that gained from a happy marriage, children, grandchildren and the Nobel prize.

But on the more down-to-earth subject of the pleasures of cycling round Oxford, first is the convenience. I live in the centre of the city and can get anywhere I require in less than 20 minutes. No hassle with traffic, no parking problems, only the ever-present risk of automobile injury.

I've always enjoyed recreational cycling. This is not the everyday task of getting from place to place but adventure. It is visiting a village whose name has always intrigued you but you have never seen, aiming for sites marked on the Ordnance map with weird gothic type or with the mysterious icons that raise a wanderer's interest. It intensified when at 64 I gave up 25 years of road-running. I reckoned I had only a limited number of miles left in my lower joints, and I wanted to save these joints for my later years. Because Oxford is built on several narrow, raised gravel beds between the rivers that intersect the city, and the flood plains are extensive and preserved, it's possible to be in the countryside within 10 minutes of leaving the city centre. And there are many delightful places to see.

The Thames and the Oxford Canal have towpaths which traverse the city and are easily accessible. On the canal you can go all the way to Birmingham through lovely traffic-free countryside. The towpath isn't great for cycling but it's possible to negotiate it with some effort and a masochistic amount of carrying.

The Thames is designated as a national trail, from London to the remote headwaters near Lechlade. Parts of it are not continuous and the rough towpath route wanders inland from time to time. The Thames locks, frequently staffed by avid gardeners, vie with each other for the most beautiful grounds and for most of the year they are a delight. The handsome canal locks, by contrast, are not staffed and have a serene charm — with silent fisherman, fleets of swans, ducks, and geese drifting and flying by, and the occasional walker.

I've also gone further afield. An hour or so with the cycle in the back of the car and you are in another congenial part of Oxfordshire or the surrounding counties. The Ordnance Survey publishes a series of round-trip cycle tours of 30–40 miles, mapped in great detail with encyclopaedic information on grades and difficulty of cycling.

The Ridgeway is an excellent example. The section above Didcot and Harwell is lovely. I've often gone there in the middle of winter and gazed out over the cooling towers at the Didcot power station, the containment vessels at Harwell, the long grassy horse rides that run parallel to the track, and the Neolithic remains that characterise the Ridgeway. They are all juxtaposed on this linear path through time and place, and the magic wheels can carry you through.

A word about image. In the city, of course, the dress is that of the ordinary citizen — plus a safety helmet and a reflector and/or bright clothing. But as it has been known to rain in these latitudes, quick on-and-off waterproofs are useful. However, when on an extended trip in the *cycliste* mode, there is ample scope for a clothing statement. For many years I favoured a random-access approach to kit-selection but now I'm inclined to a nattier bicycle style with purpose-designed shorts, shirts and gloves. I'm also considering coloured shades.

Preparation for winter cycling is akin to suiting up for a space walk. Layering is my message; on occasion I have had 22 items of thin clothing lightly but effectively shielding my vulnerable body. Silk undershirt, polypropylene jersey, wool pullover, windproof jacket and a waterproof covering. Equivalent pieces covered the lower parts, plus additional items for head and hands.

Cycling in Oxford has its special disadvantages. Cycle theft has been elevated here to a folk art. On top of that is the occasional conflict between the perceived needs of motorists and, more particularly, pedestrians. Cyclists may see the car and lorry as great monsters imperilling life and limb; pedestrians, in turn, may view the bicycle as a natural enemy. The remedy? Stick to traffic and safety rules, and be polite. However, if the burden of anxiety is too great, think again. The bicycle may be a blessing from another world, but it may not be for you.

Reprinted from ANNALS OF INTERNAL MEDICINE Vol. 102; No. 3 March 1985
Reprinted in USA

The Daedalus Effect: Changes in Ethical Questions Relating to Hepatitis B Virus

BARUCH S. BLUMBERG, M.D.; and RENÉE C. FOX, Ph.D.; Philadelphia, Pennsylvania

The Daedalus myth is a metaphor for aspects of the scientific process. When a problem is solved it often raises others, and when these in turn are solved they generate additional questions. Although perfect solutions may not be possible, major improvements can be made in many cases. Research on the hepatitis B virus is an example. The ability to detect carriers of hepatitis B virus contributed to the control of post-transfusion hepatitis but raised social and ethical problems inherent to the identification of carriers in the community. Partial solutions to this problem resulted from the ability to distinguish infectious from noninfectious carriers by the use of the hepatitis B e antigen (HBeAg) test and the development of an effective vaccine against hepatitis B virus. These solutions will undoubtedly lead to other problems and their solutions, which will in turn lead to other ethical and medical questions.

THE DAEDALUS MYTH can be viewed as a metaphor for the problem-solving and problem-creating aspects of scientific life (1). Daedalus, a legendary figure of the classical world, was a craftsman, inventor, architect, and artist. He was also a generator and solver of problems. Every time a question was answered it raised several more, and these in turn led to other questions; his search was endless. Daedalus repeatedly learned that even the most ingenious solutions were neither flawless nor definitive. His actions had both positive and negative consequences for the persons affected. Some of these consequences were anticipated and others were not.

Daedalus must have been aware of the imperfect nature of his solutions, but he was captivated by the problems presented to him, encouraged by the progress that he made, and challenged by the new questions his solutions raised. He was also realistic and prudent and recognized limits beyond which he must not venture. Daedalus attempted to convey his perspectives on problem-solving to his son Icarus. But in keeping with the limitations and imperfections of which he was so keenly aware, he did not succeed in saving his son.

The medical scientist will recognize a familiar pattern in the career of Daedalus. When a hypothesis is tested it generates others, and when tested these yield still more. Scientific research has an infinite quality; the more we know, the more we know about what we do not know, and this unknown must in turn be understood. Each time a medical treatment, diagnostic technique, or public health measure is introduced, no matter how effective it may be, it nearly always raises other problems. All drugs have undesirable side effects; diagnostic procedures are rarely, if ever, entirely specific or infinitely sensitive; and public health measures often have medical, social, and even political consequences that are difficult to foresee.

The metaphor, however, is not pessimistic. It may be a paradox, but knowledge of what we do not know contributes to understanding and enlarges the imaginative possibilities of an investigation. There may be an infinite amount that can be known about a medical problem but it is amazing how little need be known in order to provide effective, although not perfect treatments. Penicillin, for example, provided remarkable treatment long before there was any understanding of its mode of action. The avitaminoses were treated and prevented decades before there was any knowledge of vitamins, or even before they were named.

The solution of problems generates other problems; however, they are often of less consequence than the problems that were resolved. Drugs may have serious side effects, but the problems they solve, such as the cure of infectious disease by penicillin, are far more beneficial than the side effects that they cause. Anyone who remembers the era before antibiotic agents can testify to the immense improvement in patient care that resulted from the introduction of penicillin and the drugs that followed. Although perfection in medical care may not be possible, major improvement is, and the history of medicine in the past 50 years contains many examples of such advances.

The Daedalus metaphor is not only relevant to biomedical problem solving; it is also applicable to ethical questions associated with medicine. Medical research can raise ethical problems, but may also contribute to their resolution.

Daedalus

Daedalus, the legendary builder, designer, and inventor, murdered his nephew Perdix for reasons that are ambiguous. Daedalus fled Athens, went to Crete, and was employed at the court of King Minos. Known for his ability to make clever images of animals, Daedalus was employed by Pasaphae, the Queen of Crete and wife of Minos, when she became hopelessly enamoured of a great white bull presented to her and her husband as a gift by the god Poseidon (the mythic father of Minos). Daedalus designed and built a life-size, hollow cow. The queen positioned herself in the model, which was placed in a seductive situation in a field, and eventually she was impregnated by the white bull. Her problem, the need for

▶ From the Institute of Cancer Research, Fox Chase Cancer Center, and the Department of Sociology, University of Pennsylvania; Philadelphia, Pennsylvania.

the bull, was solved, but it created another: the offspring of this unfortunate union was the half-beast, half-human Minotaur. This awful animal terrorized the citizens; the noise it created was like thunder, when it stamped its feet the earth shook, and it was a nuisance and danger in other ways as well.

Daedalus designed and built the labyrinth in which the Minotaur could be kept away from others. But the Minotaur required the sacrifice of young men and women. Each year, youths were sent from Athens to be offered up to his insatiable appetite. When Theseus, the son of the king of Athens, was sent with the sacrificial youths, Daedalus told him and Ariadne (the daughter of Minos and Pasaphae, thus half-sister of the Minotaur) of a secret door by which they could enter the labyrinth to slay the Minotaur. He also gave Ariadne a magic spool of thread that would unwind and show them how to escape the labyrinth after they had slain the beast. Theseus and Ariadne slew the Minotaur and fled Crete, with many adventures along the way. Ariadne was eventually abandoned by Theseus on the island of Naxos.

Hence, another problem was solved but others developed. Minos imprisoned Daedalus and his son Icarus in the labyrinth for their part in the death of the Minotaur and the escape of the killers; the mazemaker was trapped in his own maze—an ironic metaphor within a metaphor. But the imprisonment had its solution. Daedalus invented wings and fashioned them from feathers and wax. He and his son escaped their jailers but, in the ensuing flight, against paternal advice, Icarus flew too high. The heat of the sun melted the wax, the wings came away, and Icarus fell to his death in the sea that now bears his name. Daedalus flew on to the Greek colony of Cumae, south of Naples (or, in some versions of the myth, to Sicily), and additional adventures and travels followed.

The sun and its heat, the wings and their wax, the sea into which Icarus plummeted, his attitudes and his death, all became part of the continuing cycle of questions and answers, problems and solutions that are inherent to creative and responsible science and to human questing.

Bioethics and Medical Research

Bioethical problems cannot be fully understood without a thorough grasp of the scientific and technical phenomena that surround them. It is necessary to understand the scientific content of a problem in order to formulate the ethical problem. However, it is not the scientific factors alone that cause the ethical dilemma. There is always an interplay of belief, value, and social factors in ethical problems and their designation as such. Scientists and physicians need to understand this process in order to more knowledgeably and responsibly serve their patients and the general public. Advances in biological and medical knowledge do not only lead to ethical problems; they can also alleviate and even solve these problems. Continuing research often mitigates a previously difficult issue or changes it into one for which a solution can more readily be found.

Science is at once a problem-solving and problem-creating process. Each time a question is answered (a hypothesis effectively tested), it raises additional questions that in turn are answered and produce another generation of questions. The process may be infinite.

Another pattern in contemporary medical science ought to be considered in arriving at decisions as to when and how a scientific finding should be applied. Research is often adopted in clinical practice before it is actually completed. When the possibilities of a scientific advance are perceived, they are frequently applied to clinical practice as if their value had already been tested. Or, to put it in a more formal statement, a distinction is not made between the hypothesis when it is first stated ("Based on laboratory observation the hypothesis is made that drug Alpha is beneficial for cardiac arrhythmias"), and the hypothesis after it has been adequately tested and supported ("We have already tested Alpha in a series of controlled trials and find it to be better than the previously available therapy"). In the early phase of research there often appears to be greater awareness and more frequent and enthusiastic reporting on the problem-solving ("beneficial") consequences of the discovery than on its problem-creating ("side effects") consequences. It is a common experience to read the glowing first accounts of a newly introduced procedure or therapy proclaiming its great effectiveness and nearly trouble-free character. Only later one learns that the procedure is not as effective as first thought and carries with it significant toxic or undesirable consequences.

These patterns are cultural as well as cognitive. The ritualized optimism that is involved and the questing hope that a "perfect" medical drug, procedure, or device can be found that has only good consequences and no bad ones are part of Western values and beliefs, especially in their American form (2). It is the first-hand experience of many, however, that a problem-free "magic bullet" form of scientific knowledge does not exist; perhaps it never will. Although commitment to the principle of constant improvement through new problems and their solutions is essential to science, it is unrealistic and in some respects unethical for scientific progress to be tied to a utopian and hubris-ridden notion of perfection. The practice of medicine cannot become perfect, but it can be made much better.

Bioethical Problems Related to the Detection of Hepatitis Carriers

The test for what was then known as the Australia antigen was reported in 1964 (3). By 1966, there was substantial evidence that Australia antigen was on the surface of hepatitis B virus. In 1967 (4), screening of blood donors for occult carriers of hepatitis virus was routine in Philadelphia, and by 1970 it was widely applied in the United States and other countries (4, 5).

In 1968, reports were heard from persons, usually health care personnel, who had been identified as carriers either as a result of a blood donor or hospital staff testing program. For example, a hospital nurse identified as a carrier was told that she would be fired because of the possibility that she might infect people by personal contact. An applicant for a hospital position was deemed ineligible for employment because he was a carrier. A homosexual man was informed that he was a carrier, but given no specific instructions on how he should conduct himself socially.

Along with the many incidents of this sort, policy questions about hepatitis B virus carriers were raised by various institutions. Military medical authorities in a foreign country where hepatitis B virus carriers were common asked if they should screen applicants for admission to medical school and disqualify those found to be carriers. The same question was raised with respect to admission to officers' training school and graduation from nursing school: should these applicants and graduates be screened for the carrier state to determine if they should be allowed to enter, or undertake the practice of their chosen professions? (We recommended that screening should not be done.) Still another difficult policy problem arose in relation to the adoption of Southeast Asian refugee children. In the mid-1970s, many children from Indochina (primarily orphans, but in some cases children who still had parents) were placed for adoption. Because hepatitis B virus carriers were known to be prevalent in Vietnam, the advisability of screening these children for the virus became an issue. Should the results of this screening test determine who would be accepted for adoption and who would not? (United States health officials decided that carrier testing should not be done.) These carrier-related questions surfaced in increasing numbers as the use of the hepatitis B virus test became more widespread.

What appeared to be emerging was the possibility that a new class of stigmatized persons and groups—hepatitis B virus carriers—was being created by the introduction of a single laboratory test (6). These were persons who did not have any recognizable external characteristics. Their carrier state was "occult," only discernible by use of a blood test that had been introduced. Many carriers already had been, and even more would be, identified as a result of the donor and other blood-testing programs that had been launched. (Tens of millions of donor blood samples are collected each year, and in due course most of these would be tested.)

It was known that some carriers could transmit hepatitis readily by blood transfusion. From this knowledge, but without quantitative data on actual transmission, it was inferred that carriers could also convey hepatitis through social interaction. This assumption began to take hold despite the fact that it was apparent that most carriers were not very infectious. It had been estimated that there were about 700 000 carriers throughout the United States. If they were very infectious—for example, to the degree that people infected with smallpox are—there would have been far more hepatitis in this country and elsewhere than there is known to be. Furthermore, several preliminary studies of health care workers known to be carriers showed that they had not transmitted hepatitis B virus to their patients. The scientific evidence strongly suggested that, although some carriers might be infectious, many were much less so if at all, and that the danger to public health was probably not immediate or enormous. Nevertheless, persons who had been identified as hepatitis B virus carriers were being medically and socially marked in potentially disadvantageous ways. They were having employment and career difficulties as well as interpersonal and family embarrassment and stress as a result of their disclosed carrier status.

Consideration of these problems made it clear that an evaluation of a general screening program should be done before any such program was organized. (This program refers to the screening of persons other than blood donors. Blood donor screening had been evaluated, found to be justified, and accepted.) What was called for was a judicious, well-informed set of decisions about screening that would potentiate its public health benefits, and protect individual hepatitis carriers from undue economic, psychic, and social harm. It became apparent that there were not sufficient data on which to make sound decisions that would appropriately balance collective needs and individual rights. It was not at all clear what the rules for a screening program ought to be. Should screening be compulsory? Should the activities of carriers be restricted and, if so, in what way? How could infectious carriers be distinguished from those who were not infectious? What instructions should be given to identified carriers? What kind of protection could and should be offered to those with whom carriers came in contact? The history of previous medical applications strongly suggested that these issues should be addressed before the screening procedure acquired the routine familiarity and authority of an established practice; once a procedure has been instituted, it becomes increasingly hard to question or freshly evaluate it.

There were also broader issues that had been raised and required consideration. Most of the infectious diseases that people contract are transmitted either directly or indirectly from other people, in many cases from carriers. However, screening for these agents is difficult and not done routinely on large segments of the general population. For example, *Staphylococcus aureus* carried on skin surfaces may be spread from person to person and has caused large and calamitous epidemics in hospital nurseries. Nevertheless, routine screening is not done for this bacteria except after infection has been found. *Salmonella* species may be spread by food handlers and cause serious epidemics of diarrheal disease, but routine screening is not done because of its expense and difficulty. Should hepatitis carriers be targeted for screening simply because the test is easy to do and widely available? Beyond this, how much should biological knowledge be allowed to influence and control our social relationships? To what extent should medical and public health practices be allowed to affect our social behavior—particularly in the face of the kind of inadequate information about hepatitis B virus carriers and the infectious risk they constitute that existed at this point?

For various reasons, hepatitis screening for blood donors was accepted quickly, whereas screening for many other infectious agents has not been accepted. There was an unambiguous, long-standing need to screen blood donors for hepatitis. Post-transfusion hepatitis was a real and significant problem that had been recognized for years and any solution was bound to be accepted quickly. The tests for hepatitis B virus (particularly the radioimmunoassay that was introduced relatively early in the program) were sensitive and specific. There was a large commercial interest in these tests. Test reagents for hepatitis B virus to the value of tens of millions of dollars are sold yearly, aided by extensive advertising and skilled promotion. Further, several law suits had been brought against hospitals, blood banks, and physicians by defendants who developed post-transfusion hepatitis and claimed that the institution and health care workers were liable because they had not used the screening test for hepatitis B virus. In addition, blood has a powerful symbolic meaning in our culture. It is associated with life and vigor, lineage and kinship, and sacrifice and the sacred in ways that are likely to confer special positive significance on technical procedures that guarantee its "purity."

The question of a general screening program (the testing of persons who were not blood donors) had to be viewed very differently. We, and others in the field, took the position that there had not yet been enough research on hepatitis B virus carriers to justify general screening programs, except for those that were part of a research protocol (7). Adequate research had been done to support the screening of blood donors, and here the balance was clearly in favor of the public health. But the research was not yet sufficient to know how to use the results of a general screening program. Continued research to resolve the social and ethical as well as scientific and medical aspects of these problems was recommended. Concrete projects for research, several of which were already in progress, were also advocated (7).

Advances in Research and Their Impact on Ethical Questions

Since our original publication on these issues, there have been scientific and technical advances that have changed the ethical issues in hepatitis B virus carrier screening, and altered our views in the process. In 1972, Magnius and Espmark (8) reported their finding of the hepatitis B e antigen (HBeAg). This antigen appears to be a part of the core of hepatitis B virus, and its presence in a carrier indicates that significant amounts of whole infectious virus are present in the blood (9, 10). Several studies have shown that carriers of hepatitis B surface antigen (HBsAg) who also have HBeAg are much more likely to be infectious than carriers without HBeAg, as seen in transmissions from a carrier mother to her offspring, and in horizontal transmission from one person to another. Furthermore, HBsAg carriers who also have antibody to hepatitis B e antigen (anti-HBe) in their blood are much less likely to transmit the agent than those with HBeAg or those without any sign of HBe antigen or antibody. By separating the carrier group into those who are potentially infectious and those unlikely to transmit hepatitis, the problems of screening were narrowed and focused.

In 1969, Millman and Blumberg introduced a unique method for the production of a vaccine to prevent infection with hepatitis B virus (11). The vaccine has been tested (12), accepted by government agencies, and is now widely used by individuals and for public health projects in this country and abroad. If the use of the vaccine continues to be as safe and effective as these original trials indicate, it will be possible to protect persons with whom carriers come in contact. In time, in some regions of the world where hepatitis prevalence is high, all or nearly all of the population will either have natural protection against hepatitis B virus (they will have developed the antibody to hepatitis B surface antigen [anti-HBs] after natural infection with the virus), or they will have been vaccinated. When this happens, the public health impact of the carriers will be greatly minimized in these areas. As a consequence, the chief ethical problem of the carrier will have been eliminated.

In areas with low intensity of hepatitis B virus infection, the vaccine will probably only be used in high-risk populations, including health care personnel, travelers, military personnel, blood handlers, homosexual men (who have a high risk of infection with hepatitis B virus), drug abusers, family members of carriers, and certain other groups. In some of these high-exposure populations, the frequency of naturally occurring anti-HBs may be common (15% to 50%). Because the cost of vaccine is high, it would be prudent in some areas to screen these populations for the presence of anti-HBs (and possibly HBsAg), because they would not profit from vaccination. Under these changed circumstances, the screening of general populations would be warranted. The initial concerns about general population surveys would be set aside, because measures of known value could be taken as a consequence of the survey.

There are other situations in which screening surveys may be warranted. Current policy encourages the "mainstreaming" of mentally retarded children by placing them in small, home-like settings in the general community, and in regular schools with other children. There is a relatively high frequency of carriers among children and young adults with Down's syndrome, who make up a sizeable portion of deinstitutionalized, mentally retarded people. Research should continue to determine the risk that these children may impose on their classmates, and on the feasibility of vaccine protection. Similarly, adopted children from communities with high frequencies of carriers (for example, Southeast Asians) may also be screened after adoption has occurred. Vaccine could then be recommended to family members if the adopted child is found to be a carrier and have HBeAg. In selected cases, connubial partners may request screening tests in order to decide if an uninfected partner should be vaccinated.

Nevertheless, these screening procedures will continue to raise ethical problems. Although there is now a more obvious public health value that would derive from screening in certain populations, the identification of hepatitis carriers still involves some degree of individual and group stigmatization and is a potential disadvantage.

A recent example illustrates the difficult problems that may develop. A limited screening program of East Asian immigrants living in the United States was part of a vaccination project. A young student about to attend a private, residential school, was found to be a carrier. This raised the question of whether the school should be informed, and, if so, in what way? Should the student's roommate be vaccinated? Should other students be vaccinated? Who should pay for these vaccinations? Would knowledge of the fact that this student was a carrier adversely affect the way he was viewed by other students? What repercussions might this have for attitudes toward the student's ethnic group? This last possibility was of special concern to the immigrants community who had come to the United States from an area of Asia where hepatitis B virus is endemic. Not only was the community worried about the prejudice and discrimination that the student might experience, but also about the "impure" image of their ethnic group that might emerge if a significant number of carriers were identified in their community. Would Americans come to regard their whole group as "marked" by hepatitis, and, consequently, as a public threat? What implications would this have for the respect and status they would be accorded in American society, and for their educational, occupational, and economic opportunities for advancement and success? In this particular case, such questions (some of which were explicitly stated, others of which were implied) became so acutely problematic that government health authorities were consulted. In the end, the issues surrounding the case were only partly resolved.

Other advances that would further alter the ethics of screening are likely to occur. Research is now being directed toward an understanding of how to either eliminate hepatitis B virus from carriers, or to effect virus replication so that the amount of virus produced and excreted is so low that the carrier is no longer an infectious hazard (13). If such measures become possible, they would probably also decrease the likelihood of the carrier developing chronic liver disease or primary cancer of the liver. Under these conditions, identified carriers could be offered a quasi-therapeutic procedure. (It is not exactly therapy because carriers often do not perceive any illness.) This would provide a powerful new medical and moral rationale for general hepatitis B virus carrier screening that might further offset the negative personal, social, and cultural side effects associated with it.

Summary

We have used the term "Daedalus effect" to identify and describe the never-perfect nature of medical research and practice: the way in which each solution to a problem generates new problems. We have shown that this effect can relate to bioethical as well as biomedical problems.

It may be important for scientists and practitioners to recognize this phenomenon. It can help them to distinguish between the imperfections that are a consequence of the "Daedalus effect," and those that emanate from inadequate scientific thought or poor experimental design. Comprehension of this effect may help to mitigate both the hubris that results from the expectation that perfect solutions are possible, and the discouragement that comes from not achieving all desired ends.

Although it may not be possible to achieve perfection, it is within the capacity of research to make things much better than they currently are. For the contemporary medical investigator and practitioner this is a goal, as it was for Daedalus.

ACKNOWLEDGMENTS: The authors thank Dr. Mary Tiles, Reading, United Kingdom, for her insightful discussions on the Daedalus model, and Mr. Mark J.T. Thompson of the British Broadcasting Co., London, for his critical review of the paper.

Grant support: in part by United States Public Health Service grants CA-06651, RR-05539, and CA-06927 from the National Institutes of Health, and by an appropriation from the Commonwealth of Pennsylvania.

▶ Requests for reprints should be addressed to: Baruch S. Blumberg, M.D.; Fox Chase Cancer Center, Institute for Cancer Research, 7701 Burholme Avenue; Philadelphia, PA 19111.

References

1. MALVILLE K. A feather for Daedalus. In: *Explorations in Science and Myth.* Cummings. 1975; Menlo Park, CA.
2. FOX RC, SWAZEY JP, CAMERON EM. Social and ethical problems in the treatment of end-stage renal disease patients. In: NARINS RG, ed. *Controversies in Nephrology and Hypertension.* New York: Churchill Livingstone Inc.; 1984:45-70.
3. BLUMBERG BS. Australia antigen and the biology of hepatitis B. *Science.* 1977;197:17-25.
4. BLUMBERG BS, GERSTLEY BJS, HUNGERFORD DA, LONDON WT, SUTNICK AI. A serum antigen (Australia antigen) in Down's syndrome, leukemia and hepatitis. *Ann Intern Med.* 1967;66:924-31.
5. BLUMBERG BS. The Australia antigen story. In: MILLMAN I, EISENSTEIN TK, BLUMBERG BS, eds. *Hepatitis B: The Virus, the Disease and the Vaccine.* New York: Plenum Publishing Corp; 1984:5-31.
6. PERKINS HS, JONSEN AR. Conflicting duties to patients: the case of a sexually active hepatitis B carrier. *Ann Intern Med.* 1981;94:523-30.
7. BLUMBERG BS. Bioethical questions related to hepatitis B antigen. *Am J Clin Pathol.* 1976;65:843-53.
8. MAGNIUS LO, ESPMARK JA. New specificities in Australia antigen positive sera distinct from the LeBouvier determinants. *J Immunol.* 1972;109:1017-21.
9. NORDENFELT E, ANDERN-SANDBERG M. Dane particle-associated DNA polymerase and e antigen relation to chronic hepatitis among carriers of hepatitis B surface antigen. *J Infect Dis.* 1976;134:85-9.
10. HINDMAN SH, GRAVELLE CR, MURPHY BL, BRADLEY DW, BUDGE WR, MAYNARD JE. "e" antigen, Dane particles and serum DNA polymerase activity in HBsAg carriers. *Ann Int Med.* 1976;85:458-60.
11. MILLMAN I. The development of the hepatitis B vaccine. In: MILLMAN I, EISENSTEIN TK, BLUMBERG BS, eds. *Hepatitis B: The Virus, the Disease and the Vaccine.* New York: Plenum Publishing Corp; 1984:137-47.
12. SZMUNESS W, STEVENS CE, HARLEY EJ, et al. Hepatitis B vaccine: demonstration of efficacy in a controlled clinical trial in a high-risk population in the United States. *N Engl J Med.* 1980;303:833-41.
13. BLUMBERG BS, LONDON WT. Hepatitis B virus and the prevention of primary cancer of the liver. *J Natl Cancer Inst.* 1984. (In press).

Scientific and artistic process

Baruch S. Blumberg

Presented at the Awards Ceremony

Pennsylvania Academy of the Fine Arts

Broad and Cherry Streets

Philadelphia, Pennsylvania

14 May, 1980

Students, faculty and officers of the Pennsylvania Academy of the Fine Arts and their guests.

It is always bad to start an address with an excuse, but I am afraid that I must do so. I know very little about the practice of art--and most of that from listening and watching my wife who has today completed four years of her training at the Pennsylvania Academy of the Fine Arts. I am, however, interested in scientific process; that is, how scientists go about getting new ideas and how they test and use these ideas. In talking to artists, mostly students and teachers at PAFA and my friend and neighbor Sam Maitin, I have noticed similarities in the process used by both scientists and artists. In my short address today I would like to present some ideas on these similarities in the hope that they will strike a resonant chord in one or more artists and provide them with a new way to look at some of their activities. In science, as in art, we try to find new insights of natural phenomena: of objects around us, new colors, new angles. Looking at nature in many ways helps us to understand it more readily and feel more at home with it.

Hypothesis Formation

There is a general belief that in science one can develop absolute proofs, that a statement can be made for which there is no possibility of refutation. This may hold for some general statements, but it can rarely be said about hypotheses and theories in many branches of science, for example, biology and medicine. Contemporary views of scientific process, which have been very influenced by the British philosopher Sir Karl Popper, hold that it is never possible to prove a hypothesis. One can only attempt to disprove it; and, if after a consistent and valiant effort to do so it is not disproven, it can be treated as if it has some of the properties of truth but with the knowledge that the next test may disprove it.

A hypothesis is a statement created by the scientist which gives his opinion about a specific condition in nature. Where does the hypothesis come from? It usually is based upon observations that the scientist has made; that is, it is generated from real world experience. It can also be derived from the observations of others, recorded in the scientific literature, or it can be theoretical; that is, induced from an "established" law about how the world is put together. I'll have more to say later about how new hypotheses are made, an extremely important step in the creative process.

The next step is to collect data to test the hypothesis, or to be more exact, to attempt to reject it. A hypothesis that cannot under any conceivable circumstances be rejected is not very valuable in advancing understanding. When a sufficient amount of data are assembled, a decision is made as to whether the hypothesis has been rejected or supported. Independent of the fate of this hypothesis, the new data can generate new hypotheses; that is, new questions or ideas. These, in turn, are subjected to testing and the process is repeated indefinitely. As the scientist answers questions, he will actually generate more unanswered questions, and more answers generate even more questions so that he ends up with more answered questions, but also more unanswered ones.

There are some analogies for the artistic process. Art is constantly evolving and one of the most difficult questions an artist has to answer is "When is a work of art completed?" There are stories of painters going to the exhibit hall after their work has been hung to make changes on a painting they considered not quite finished. Printmakers prepare a series of proofs in an effort to help decide when their image is nearing completion. One of the difficulties in arriving at a decision arises because it is perceived that when an experiment or a painting is deemed complete, it is the equivalent of saying that you know everything there is to know about it; this is a form of hubris, of overwhelming pride, both in the scientist and the artists who, with even moderate experience, must know that they cannot learn everything about a subject in nature--even a subject which is apparently simple. However, despite these difficulties, it is necessary to complete a painting, or an experiment, at some time.

Oddly, there isn't any rule which indicates how much proof is necessry before other scientists will accept a hypothesis. Rather, it's some undefinable combination of amount, quality, the method in which the results are presented and other difficult-to-define features. Similarly, the decision as to when a work of art is complete seems to reside primarily in the artist itself; but this usually requires confirmation by someone else (colleagues, critics, museums, purchasers of the art) establishing a major forum for confrontation.

New ideas often arise when new combinations of events are seen, when the scientist puts together what he calls variables that have not been combined before. These are suggested by continually enlarging the scope of experience in order to see new combinations. There is, I believe, a similar process among artists who combine ideas, often abstract, or combine objects in new variations, and the search for such combinations is part of the constant growing process of the artist.

The system of repeated testing of hypotheses demonstrates the historical character of scientific process. A hypothesis is the child of a preceding hypothesis and its character, in turn, determines the hypotheses that follow it. Although the rejection of a hypothesis may be compared to a small death, it is not a total death since it gives birth to a whole family of new hypotheses.

This sense of evolution and historical continuity is also a part of art. The work of an artist is often shown in historical perspective; works are arranged in chronological order to show how one idea evolved into another. Artists are often considered in the light of artistic ancestors by whom they were informed at different periods of their careers. There is a similarity between this change and growth in the artist and in the repeated process of hypothesis development and rejection.

There are parallels in naming scientific phenomena on the one hand, and works of art on the other. Very often, with the discovery of a new material or a new phenomenon, a name may be assigned which implies that the character or the function of the material is known, even though there may only be very preliminary data about it. Naming something implies control and power, and the name originally assigned by the scientist can determine the direction of the research. This can have the effect of actually preventing research in the directions not specified in the name.

Painters are often reluctant to give names to their works--or they give them innocuous and non-informative names (i.e. Landscape No. 2). An explanation that has been offered to me is that if a painting is named, it directs attention to a particular view. This could have the effect of dampening the imagination when the title provides the viewer a wholly adequate interpretation. Hence, names can actually limit the possibilities of a work of art in the same manner in which premature naming of a scientific phenomonen may restrict its possibilities.

A recent important development in the study of the scientific process has been the notion that science advances by revolutionary steps. Kuhn, a philosopher now in Boston, described the process. There is a model, or what he refers to as a paradigm, which guides the interpretation of the data of a particular branch of science. This paradigm has usually been learned by a scientist during his school and apprenticeship training period, and it is the view of the establishment on the state of things. Most practitioners of science serve this model and add to the bricks of information which build and reinforce the edifice of the paradigm. The model may become increasingly unsatisfactory as more and more data accumulate that it cannot readily explain. That is, there are a growing number of exceptions to the model. It is a feature of the human mind that people are not prepared to abandon an explanatory model, no matter how faulty, unless there is another which can replace it. In due course, a new paradigm which more efficiently incorporates all the new facts is developed by a scientist who is prepared to discard the old paradigm; that is, break with tradition. What follows is a period of what Kuhn has referred to as "crisis" when there is a great deal of excitement and discord as the loyal followers of the old paradigm reluctantly abandon it and accept the new model. This is often very difficult to do since there is a considerable emotional investment in the old view. Finally, the new paradigm is accepted and it now becomes the establishment position until it, in turn, is replaced by a newcomer.

There are similarities to this in art. The "academic" models are learned in school and at the feet of one's teachers. In due course, these are found not to satisfy all the needs for expression. The establishment method is then attacked by those who are constituted to act in that way and new forms of expression instituted. These, in turn, become the accepted, normative model until they too are attacked and replaced by others. Of course, it is the role of the academies to attempt to maintain the establishment position; for if they did not, there would be nothing for the young to rebel against and this would do damage to the whole system.

During the past four years I've had an opportunity to learn something about the nature of this Academy. It is an unusual institution and embodies unique Philadelphia characteristics. It has a profound identification with its past and a recognition of its important role in maintaining high standards among the artists of our city and country, and I am particularly pleased that it has these qualities.

I offer my congratulations to those who have finished their programs and my very best wishes for a satisfying and successful career.

AN ADDRESS

UNION COLLEGE BICENTENNIAL CELEBRATION. FOUNDATION DAY

February 25, 1995, Schenectady, New York

PERCEPTIONS OF SCIENCE

Baruch S. Blumberg

Honorary Chancellor, Union College, 1995

DRAFT FEB. 23, 1995

Revision Dec. 18, 1996

Bicentennials are special events, and not only because they occur once in 200 years. In 1976, I survived the mother of all bicentennials, that is the Independence Day celebrations of 1976. My home city of Philadelphia managed to put on a stirring event, despite the tragic epidemic of Legionnaires disease which effectively halted widespread public participation after August of that year. My recollections have been dimmed by the years, but I will try to recall the essence of the events.

The major celebrations started at the last stroke of midnight on the last day of 1975 when, during a howling winter gale, the Liberty Bell was transferred from its site in Independence Hall to a purpose-built pavilion a block away. Mounted on a caisson drawn by the hard hatted construction workers who had completed

-2-

the new building, accompanied by a ceremonial guard in their splendid period uniforms, and hailed by the assembled throng of alcoholically stimulated Philadelphians, the Bell was installed in its new home. The drama was heightened by the elaborate security generated by threats of a radical political group to blow up the Bell during the celebrations. Members of the heavily armed Philadelphia Police Swat Team guarded nearby rooftops, the police cavalry were out in strength and helicopters hovered overhead, searching for possible miscreants, their high intensity lights illuminating the frenzied celebrations. It was a good start for a splendid year of commemoration and thinking about our national future.

Now let us consider our own Union College bicentennial. In 1795, the country was still new, fresh, and experimental. We had started on one of the great political adventures of all time, and it caught the imagination of the world. The foundation of our College was stimulated by the perception, strong in the history of the United States, that we would prosper by maintaining an educated citizenry who could take advantage of the new learning generated by the cultural and scientific renaissance of the 18th century that had fired many of our Founding Fathers. It is remarkable how many of the Founders thought of themselves as scientists. Many were members of the American Philosophical Society. Founded in Philadelphia, it was the premier scientific organization of the Colonial and Federal period. Benjamin Franklin, had he not achieved fame in politics, would have been remembered as a major contributor to the science of electricity and for his many inventions. Thomas Jefferson was, for many years,

President of the Philosophical Society and made many contributions to architecture, invention, anthropological science, and other fields. George Washington started his career as a land surveyor and geographer. They, and many others, had a vision of the possibilities of rational inquiry and application; the laws for the protection of invention by means of patents were incorporated in the Constitution. The importance of government in technical and scientific education and application has been recognized through the years by, for example, the creation of the Land Grant College system and, for many years, the generous Federal funding of basic scientific research.

I plan today to talk about science on the brink of the next millennium and the beginning of our next one hundred years. I know this may raise apprehensions of a heavy and boring speech, but, don't despair. As Mark Twain said about the music of Wagner, "It's not as bad as it sounds." My discourse will be brief and anecdotal. Incidentally, it is always something of a problem for a speaker to know how long he or she should talk at an event such as this. A very helpful response to this question was given to a distinguished scholar when he asked his host how long he should speak. The response was that he could talk as long as he wished, but they were leaving after 10 minutes.

Discussions by scientists of the future of their field usually have a triumphilist character, and with good reason. The particle physicists believe they will soon produce a validated model for the entire family of fundamental particles, though some doubt is beginning to creep in, and provide us with a theory of everything that will tell us all we need to know about the nature of matter. The

-4-

biologists, who are now able to read the code of the genes and manipulate them, will, by the year 2005, have a complete map of the human chromosome and possibly the location of the 80,000 or so genes that are hypothesized to control the production of the proteins of which humans are made. But, don't think that this is the end of all needed knowledge in biology. We will still have to know what these genes produce and how the product of the genes function, what is added on to the proteins after they are translated from the genes, and, even more demanding, how this mass of products interact to produce the phenomenon that make up human physiology, pathology, and behavior. Science at its frontiers is also pursuing that most difficult of problems; how to deal with complexity. The "chaoticians" are telling us that we may be able to model, describe, and predict extraordinarily complex problems such as turbulence, the emergence of life, and the mechanisms of the brain and mind. And similar examples of bright tomorrows could be gleaned from many areas of science.

Despite these promises of brave new worlds, there is a sense of unease. The support of higher education where a great deal of basic research is carried out is decreasing. The concern about lower taxes could seriously deplete the resources available to basic scientific research and private industry won't, and can't be expected to, provide the level of support that governments traditionally have.

But there are other problems that govern the public perception of science. There are inflated expectations of what can be accomplished. The scientific process is one of the best methods we have to solve problems, but the areas in

-5-

which it can be used are limited. They have to be subject to experimentation and the data has to be reproducible. Topics which raise emotion and bias are difficult to deal with because it becomes impossible to maintain objectivity in evaluating data; and this can lead to distorted conclusions.

There is also the notion that science is a certain guide in a world of uncertainties. Science, however, is not built on certainties, but on probabilities. Every scientific hypothesis, even those firmly established by years of experiment and observation, is subject to overthrow when new data warrants its rejection and replacement by a better explanation of nature. Hence, science is really made up of a complex of probabilities and uncertainties. It is to religion, philosophy, and Mother that one turns to find absolutes.

Further, knowledge is always incomplete. Every time a hypothesis is tested, new data are generated and these new data can lead to new hypothesis, that is, new questions. We are faced with the paradox that the more we know, the more we know about what we don't know. Answers generate questions, and the business of science is to continue on this path of inquiry, indefinitely, to substantiate our knowledge and at the same time, create new mysteries for the imaginative inquiries of the future.

Despite the uncertainties and incompleteness one can be consoled by the knowledge that complex problems have been solved even though we may know only a portion of what can be known. Next time you fly off in a 747, you may want to reflect on this observation.

-6-

There is also, in many quarters, a fear of science that could be called the *Frankenstein* phenomenon. The novel, by Mary Shelley, was an entry in a contest among members of a party including the Shelleys, the Byrons, and their friends during a weather-beset vacation in the Swiss mountains in the early 19th Century. It has been continuously in print since its initial publication and has been the subject of numerous movies and plays. It has been a major influence on the perceptions of science. It reflects the fear of human creativity and of the potentially destructive nature of the inventive imagination.

It is remarkable how pervasive this theme is in the popular media, for example, the cinema. Scientists in the movies often have a negative image or are portrayed as silly and crazy; *Dr. Strangelove*, the whacko scientist of *Back to the Future*, the evil drug company officials and scientists of *The Fugitive,* all fit this unwholesome model. Perhaps the most malicious example in the recent cinema are the scientists in the widely popular *Jurassic Park*. The molecular biologist who has created dinosaurs from DNA extracted from the blood imbibed by a Jurassic insect from its animal neighbors, attempts to better the efforts of the original Creator by inserting genes to produce an animal more to the liking of the scientist and his employers. His efforts are doomed, as explained by the mathematician employed by the investors in the Park, because of the extraordinary complexity of natural systems and the inevitability of failure resulting from the profound effects of even small changes in the parameters which describe the early sequence of events.

These perceptions aren't restricted to the movies. Cormac McCarthy, in his extraordinary novel *Blood Meridian*, describes the peregrinations of a band of scalp hunters roaming the border regions of Texas, New and Old Mexico, Arizona, and California in the early 19th century. Judge Holden emerges as the Nietzhean leader of the group, a man of exceptional evil and brutality. But he is depicted as a scientist, a collector of plants, animals, insects, rocks, who has a crazed desire to control all that he sees by classifying and wrenching them into his power. This is, again, an expression of the fear of the scientific endeavor, even the fear of classification itself.

I don't believe that there is time to defend the role of the scientist. It's important to remember that *Frankenstein* and *Jurassic Park* are fiction. The Monster and the dinosaurs never happened; further, most of us in science never see the evil portrayed in these pieces of fiction. And, despite these grim perceptions, the public is interested in the products of research and recognizes the potential value, not only of the things, but of the explanations that science can provide. It is the responsibility of the scientist to explain to the public the nature of their endeavor; to attempt to allay some of these fears that could hold back interest and support.

We know that the education provided at Union and at our sister institutions in the next millennium will continue to advance culture, learning and science and fulfill the hopes of the Founders of our Nation and our College.

Proc. Natl. Acad. Sci. USA
Vol. 83, pp. 2233–2237, April 1986
Microbiology

A newly identified hepatitis B type virus in tree squirrels

(chronic carrier/hepadnavirus)

MARK A. FEITELSON*, IRVING MILLMAN, THERESA HALBHERR, HEIDI SIMMONS, AND BARUCH S. BLUMBERG

Fox Chase Cancer Center, Philadelphia, PA 19111

Contributed by Baruch S. Blumberg, December 11, 1985

ABSTRACT **Virus-associated particles have been isolated from the livers of three common gray tree squirrels (*Sciurus carolinensis pennsylvanicus*) that have histological evidence of hepatitis. Two of these livers were also positive by orcein staining, suggesting the presence of surface antigen in the cytoplasm of hepatocytes. Fractionation of these particles by CsCl density equilibrium gradient centrifugation and assay of the fractions for surface antigen, core antigen, and DNA polymerase activities demonstrate the presence of all three at an approximate density peak of 1.27. Electron microscopic examination of purified virus preparations showed spherical particles with a mean diameter of 25 nm. Initial characterization of the DNA polymerase product by gel electrophoresis showed a single DNase I sensitive band, migrating slightly faster than the woodchuck hepatitis virus DNA polymerase product. The presence of apparently cross-reacting antibodies was demonstrated by purified hepatitis B surface and/or core antigens binding to some squirrel sera in solid phase assays. Infected tree squirrels appear to lack detectable antigen in their sera. These results suggest that the tree squirrels studied are chronic carriers of a hepatitis B type virus. The host–virus interaction described herein may be useful in understanding the chronic carrier state associated with hepatitis B in man.**

Elucidation of hepatitis B virus (HBV) molecular biology relevant to the asymptomatic carrier state and disease, including chronic hepatitis and hepatocellular carcinoma, has been hampered by its narrow host range of infection and its inability to replicate in a tissue culture system. The discovery of hepatitis B type viruses in woodchucks (1), ground squirrels (2), and ducks (3) has provided important systems to address questions concerning host–virus interactions (4), including hepatocellular carcinoma (1, 4), and has recently made significant contributions toward better understanding the viral replication scheme (5). The continued development of these systems and the search for new ones will be important in learning more about their biology and that of HBV. In this study virus-associated particles have been found in liver extracts obtained from locally captured tree squirrels. Characterization of such particles demonstrates that they are distinct from, but similar to, HBV and related viruses. It is proposed that this tree squirrel hepatitis virus (THBV) is another member of the group including HBV, woodchuck hepatitis virus, ground squirrel hepatitis virus, and duck hepatitis B virus.

MATERIALS AND METHODS

Animal Studies. Ninety-four tree squirrels (*Sciurus carolinensis pennsylvanicus*) were collected with humane traps in areas adjacent to the Fox Chase Cancer Center (Philadelphia and Montgomery Counties, PA), around Reading, PA, and in selected sites within several neighboring states.

Isolation of Tree Squirrel Antigen. The liver was obtained from each tree squirrel immediately after capture and frozen at −70°C until use. Approximately 200 mg of each liver was homogenized in 20 mM Tris·HCl, pH 7.5 (grinding buffer), at 0°C (6). The mixture was sonicated (twice for 30 sec) and centrifuged at 10,000 rpm at 4°C for 30 min. The pellet was resuspended in fresh grinding buffer, sonicated, and centrifuged again as described above. The supernatants were pooled, layered over 20% (wt/vol) sucrose in grinding buffer, and centrifuged at 20,000 rpm at 4°C for 18 hr in a Beckman Ti 60 rotor. Pellets were resuspended in grinding buffer and centrifuged once more at 50,000 rpm at 4°C for 4 hr in a Ti 60 rotor. These pellets were further analyzed by CsCl density equilibrium centrifugation by resuspension in phosphate-buffered saline (PBS, 0.015 M sodium phosphate, 0.15 M sodium chloride), pH 7.3, followed by addition of enough CsCl to make the density 1.2090 g/ml. Centrifugation was carried out at 40,000 rpm, 10°C, for 72 hr in a Beckman SW 55 rotor. Alternatively, pellets were taken up in 0.01 M Tris·HCl, pH 7.5, 0.15 M NaCl (Tris/saline) and chromatographed over a Sepharose 4B column (3 × 60 cm, LKB) equilibrated with the same buffer.

Antigen and Antibody Assays. Crude liver extracts (antigen samples prior to CsCl density equilibrium centrifugation or Sepharose 4B chromatography), sera, column, or gradient fractions were tested for cross-reactive hepatitis B surface antigen (HBsAg) by the Ausria II or Auszyme assay kits (Abbott). For hepatitis B core antigen (HBcAg) assay, samples were spotted onto nitrocellulose (BA85; Schleicher & Schuell) and developed with ^{125}I-labeled anti-HBc from the Corab assay kit (Abbott). Samples were also tested for core reactivity by direct assay in the Corab kit.

Tree squirrel sera were screened for cross-reacting anti-HBs or anti-HBc by using several different assays. For anti-HBs detection, the commercially available Ausab (Abbott) or passive hemagglutination assay kits (Hemasure-Ab, Electro-Nucleonics, Bethesda, MD) were used. Alternatively, squirrel sera were serially diluted in PBS, pH 7.3, containing 10% (vol/vol) fetal bovine serum (PBS/FBS) and used to coat Immulon 2 Removawells (Dynatech, Alexandria, VA). Following overnight adsorption, the wells were washed six times in PBS/FBS, and 1 μg of purified HBsAg (7, 8) was added per well. Following overnight binding, the wells were washed and then incubated for 1 hr at 37°C with goat anti-HBs conjugated to horseradish peroxidase obtained directly from the Auszyme II kit (Abbott). After washing, binding was detected colorimetrically by using *o*-phenylenediamine. Anti-HBc screening was carried out using the Corab kit (Abbott) or by counterimmunoelectrophoresis. After electrophoresis, the slides were fixed and stained in

The publication costs of this article were defrayed in part by page charge payment. This article must therefore be hereby marked "*advertisement*" in accordance with 18 U.S.C. §1734 solely to indicate this fact.

Abbreviations: HBV, hepatitis B virus; DNAp, DNA polymerase; THBV, tree squirrel hepatitis B virus; HBsAg, hepatitis B surface antigen; HBcAg, hepatitis B core antigen; anti-HBs, anti-hepatitis B surface antigen; anti-HBc, anti-hepatitis B core antigen.
*To whom reprint requests should be addressed.

Coomassie blue to detect antigen–antibody precipitation lines. Alternatively, purified HBcAg particles (9) were added to the Immulon wells in place of HBsAg as described above. Sheep anti-HBc (a kind gift from Wolfram Gerlich) was added, followed by incubation at 37°C for 1 hr. After washing, horseradish peroxidase-conjugated rabbit anti-sheep Ig (Cappel, West Chester, PA) diluted 1:100 in PBS/FBS was added, and binding was detected by using *o*-phenylenediamine.

DNA Polymerase Reaction and Analysis of Labeled Products. Tree squirrel antigen pelleted from a crude lysate, from Sepharose 4B peaks, or from CsCl gradient fractions was radiolabeled for 1 hr by the endogenous DNA polymerase (DNAp) reaction as described (10, 11) and then digested for 30 min at 37°C with pronase at a final concentration of 0.5 mg/ml in 10 mM Tris·HCl, pH 7.4, 10 mM EDTA, and 0.1% $NaDodSO_4$. The digest was electrophoresed in 1.5% agarose gels using buffer and electrophoretic conditions described (3). The gels then were dried under vacuum and exposed to x-ray film for screen-intensified autoradiography.

DNase I Digestion. Some agarose gels containing the tree squirrel antigen-associated DNAp reaction product were run in the presence of 0.1 μg of ethidium bromide/ml (Calbiochem) next to *Hin*dIII-digested phage λ DNA markers (Bethesda Research Laboratories). A trough was cut just in front of the 2-kilobase λ DNA fragment in the lane containing the DNAp product, and the radioactive band recovered by electroelution. The material was ethanol precipitated in the presence of 20 μg of carrier salmon sperm DNA and washed twice, and then half was treated with DNase I (Sigma; ribonuclease free) in 0.05 M Tris·HCl (pH 7.2), 0.01 M $MgSO_4$, 0.1 mM dithiothreitol, and 50 μg of bovine serum albumin/ml (fraction V, Sigma). The product before and after DNase I treatment was then analyzed on agarose gels as described above.

Electron Microscopy. Semi-purified antigen resulting from CsCl density equilibrium centrifugation or Sepharose 4B column chromatography was layered on continuous CsCl gradients (density 1.15–1.30) and centrifuged at 10°C for 2.5 hr in an SW 40 rotor at 34,000 rpm. The gradients were then fractionated and assayed for crossreactive surface and core antigen containing fractions as well as for DNAp activity. One hundred microliter fractions positive by all three criteria were diluted 1:10 in 0.005 M Tris·HCl/0.0005 M EDTA (pH 7.5) (TE) and centrifuged overnight at 10°C in a type 25 rotor at 20,000 rpm. After carefully decanting the tubes, the pellets were resuspended in 10 μl of TE and transferred to nitrocellulose–carbon coated 300 mesh copper grids for electron microscopy. Grids were negatively stained with saturated, aqueous uranyl acetate and viewed on a Philips 420 electron microscope.

RESULTS

Assay of Tree Squirrel Sera and Livers for Markers of an HBV-Type Infection. Ninety-four tree squirrels *(Sciurus carolinensis pennsylvanicus)* were trapped, their livers were removed, and each was assayed for histological evidence of hepatitis and possible virus markers cross-reacting with HBsAg or HBcAg. Sera were collected from 53 of the 94 animals and assayed for cross-reacting HBsAg, anti-HBs, and anti-HBc using commercially available assays. None of the sera were positive for HBsAg by Ausria II, for anti-HBs by passive hemagglutination and by Ausab, or for anti-HBc by counterimmunoelectrophoresis. These results suggest that at least with reagents suitable for the detection of HBV and its corresponding antibodies, no serological evidence for infection was found. However, when these tests were reconstructed so that the binding of purified HBsAg or HBcAg to increasing dilutions of squirrel sera was determined, many

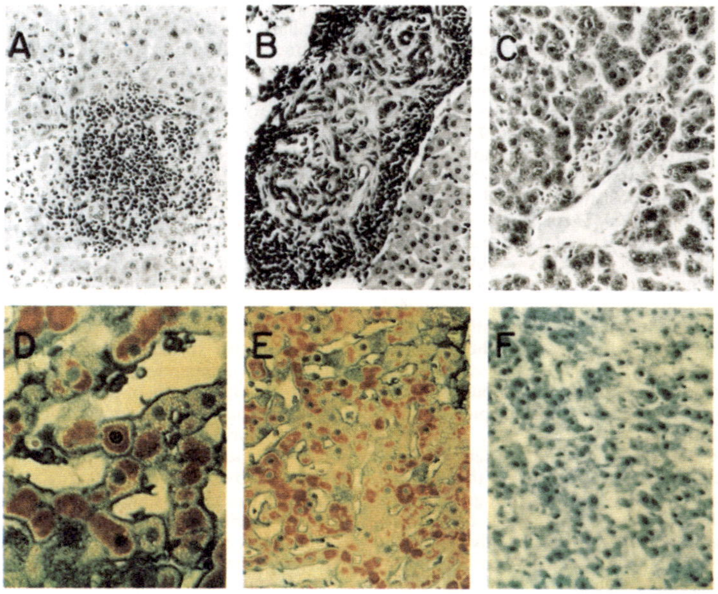

FIG. 1. Liver sections from two infected tree squirrels, 308957 (*A*) and 308960 (*B*), showing representative areas of inflammation. (×80.) (*C*) Similar section from a squirrel with no signs of hepatitis. (×80.) Liver extracts from this animal were negative in the DNAp and surface antigen assays. The serum was also negative for the presence of apparently cross-reacting anti-HBs and anti-HBc. Sections were stained with hematoxylin/eosin. (*D*) Liver section from squirrel 308960 stained with orcein and counterstained with fast green. (×150.) (*E*) Same liver as in *D* at lower magnification. (×80.) (*F*) Orcein staining of a liver section from a tree squirrel negative for hepatitis and markers of viral infection. (×80.) The orcein staining is striking in the liver cell cytoplasm of the infected animals.

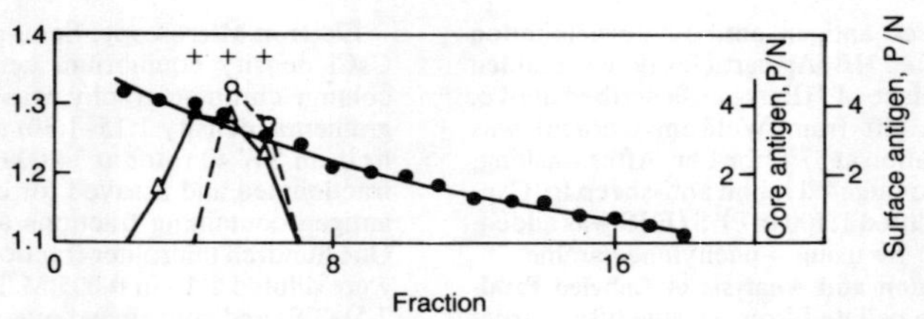

FIG. 2. CsCl density equilibrium gradient of liver homogenate from squirrel 308957 demonstrating the copurification of surface antigen, core antigen, and endogenous DNAp activities in a single peak. Twenty microliters of each fraction was tested for surface antigen reactivity in the Ausria II assay (○). For the core assay, 5 μl of each fraction was spotted onto nitrocellulose. The blot was soaked overnight in complete Towbin buffer (13) and finally incubated with ^{125}I-labeled anti-HBc from the Corab kit. The core antigen binding curve (△) was generated from densitometer area readings of each spot on the autoradiogram. Only ratios that were ≥2 compared to the average background values (in other regions of the gradient) were scored as positive and plotted. These results were confirmed by the fact that equal aliquots of the appropriate gradient fractions inhibited detection of anti-HBc in the Corab kit. To test for endogenous polymerase activity, tree squirrel antigen in 25 μl of each gradient fraction was centrifuged overnight in a type 25 rotor at 4°C, and the pellets were incubated with [α-^{32}P]ATP in the endogenous polymerase reaction as described (11). Fractions were digested with pronase, and the products were assayed by agarose gel electrophoresis (3). DNAp-positive fractions, 4–6, are designated by +. CsCl densities (●) were determined by refractive index measurements of gradient fractions.

tree squirrels demonstrated the presence of apparently cross-reacting anti-HBs and anti-HBc, suggesting a high frequency of exposure to an HBV-type agent. The nature of these responses will be considered in greater detail below.

Independent examination of hematoxylin/eosin stained liver sections from each of the 94 animals showed marked scattered focal necrosis, periportal infiltration, pericholangitis, and/or bile duct proliferation consistent with possible viral hepatitis in 14 animals, two of which are shown in Fig. 1. A small piece of liver tissue (0.1–0.2 g) from each of these animals as well as from each of the 80 normal appearing animals was homogenized and centrifuged at low speed to remove cell debris and at higher speed to pellet material for the DNAp assay. Three of the 14 livers that demonstrated pathology by hematoxylin/eosin staining were also positive in the DNAp assay and were used for further characterization (squirrels 308957, 308960, and 309017). Similarly prepared crude extracts from all livers were tested for HBsAg by the Ausria II and Auszyme assays. Only the three livers that were DNAp positive were also positive for cross-reacting surface antigen. Material from liver 308960 had a positive to negative (P/N) ratio of 2.42; material from 309017 had a ratio of 4.46; and material from liver 308957 had a ratio of 2.59 by Ausria II. A result was scored as positive if the P/N ratio was greater than 2.1. Respective positive values (P/N) in the Auszyme assay were 15 for 308960, 25 for 309017, and 21 for 308957. The immunological specificity of these reactions was demonstrated by the ability of anti-HBs to block these surface antigen tests when mixed with the crude extracts prior to assay. Identical analyses of the 80 livers that showed no sign of pathology and of the other 11 livers with evidence of pathology were negative in these assays. These results suggest the presence of HBsAg cross-reactive and DNAp positive material in crude liver extracts from 3 out of the 14 different tree squirrels showing a histological picture consistent with viral hepatitis.

The presence of tree squirrel antigen in liver sections was also assayed using orcein, a histochemical stain that detects HBsAg in HBV-infected hepatocytes (12). In a group of 32 tree squirrel livers selected for orcein staining, two of the three livers containing virus were strongly positive (Fig. 1). As with HBsAg, hepatocyte staining was strictly cytoplasmic. Livers lacking detectable tree squirrel antigen by the criteria described above were uniformly negative. These results are consistent with the presence of material similar to HBsAg exclusively in the cytoplasm of infected hepatocytes in virus-containing animals.

Purification and Partial Characterization of Tree Squirrel Antigen Particles. CsCl density equilibrium centrifugation of crude extracts from squirrels 308957 or 309017 was carried out to isolate the reacting fractions and ascertain whether the markers expected for HBV and related viruses copurified. Gradient fractions from liver 308957 assayed for surface antigen by Ausria II demonstrated a peak P/N ratio of 3.83 at density 1.271 g/ml (Fig. 2). Fraction samples spotted on nitrocellulose and blotted with ^{125}I-labeled anti-HBc from the Corab kit demonstrated binding in the HBsAg positive fractions (Fig. 2). Fractions pelleted overnight and assayed for DNAp demonstrated such activity only within the HBsAg and HBcAg positive peak (Fig. 2). The same results were obtained from an independent analysis of tree squirrel antigen from liver 309017. Further characterization of the DNAp product by agarose gel electrophoresis consistently showed a small band equivalent to 2.0–2.4 kilobases of linear DNA marker (Fig. 3*A*). Since this band migrated faster than similarly labeled DNA of HBV and related viruses, a crude extract from another DNAp positive liver was processed differently to see if the same DNAp product was still present. Purification of crude extracts from squirrel 308960 by chromatography on Sepharose 4B yielded five discrete peaks by absorbancy (unpublished data). Assay of each pooled peak for DNAp activity showed that such activity resided exclusively in the second eluted peak and that further analysis of the product resulted in exactly the same result as that obtained from CsCl fractions (Fig. 3*B*). The DNase I sensitivity of the DNAp product following pronase digestion also suggests that the radiolabeled band is due to the presence of DNA (Fig. 3*C*). These results are consistent with the presence of a nucleic acid species that becomes endogenously radiolabeled in the DNAp assay and copurifies with the crossreactive surface and core antigen containing fractions using two independent approaches.

Tree squirrel antigen partially purified by density equilibrium ultracentrifugation or column chromatography was further purified by rate zonal sedimentation in preformed CsCl gradients. Fractions positive for DNAp activity, and crossreactive surface and core reactivities were used to prepare samples for electron microscopy. Fraction aliquots were centrifuged overnight, and pellets were resuspended. Negatively stained samples were examined (Fig. 4). Particles having a mean diameter of 25 nm, with a size range between 21 nm and 35 nm, were observed. Such particles were not observed in DNAp negative fractions derived from isolates of infected livers, nor from parallel analyses of four uninfected livers. These results are consistent with the conclusion that these particles are virus associated, although it is not clear whether they represent virions or subviral particles. Their size and appearance, however, suggest that they are similar

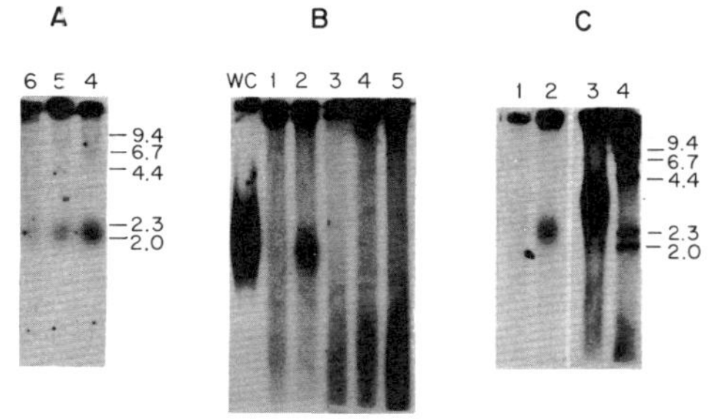

FIG. 3. Analysis of the DNAp reaction product from livers of different infected squirrels by using two independent methods of purification and sensitivity of the product to DNase I. (*A*) DNAp-positive fractions (lanes 4–6) from a CsCl density equilibrium gradient of tree squirrel antigen from squirrel 308957. Only fractions 4–6 from this gradient, as shown in Fig. 2, were positive in the DNAp reaction and were further analyzed by agarose gel electrophoresis. The numbers to the right are molecular size markers derived from *Hin*dIII-digested λ DNA and are expressed as kilobase pairs. (*B*) Sepharose 4B absorbancy peaks (lanes 1–5) analyzed for the tree squirrel antigen 308960 DNAp product. Fractions in each absorbancy peak were pooled and reduced by a factor of approximately 10 in volume. Ten percent of the concentrated materials (by volume) was pelleted and finally assayed for DNAp activity. The polymerase product (lane WC) from woodchuck hepatitis virus liver-derived core particles used as a positive control and for comparison of mobilities to that of the tree squirrel antigen DNAp product. (*C*) DNase I sensitivity of the tree squirrel antigen DNAp reaction product. Lane 1, gel purified tree squirrel antigen DNAp product was treated with DNase I and analyzed by agarose gel electrophoresis. Lane 2, same product, untreated. Lane 3, HBV-DNAp product labeled under the same conditions as the tree squirrel antigen sample. Lane 4, *Hin*dIII-digested λ DNA markers terminally labeled by polynucleotide kinase (a gift from A. O'Connell). The numbers in the right-hand column are in kilobase pairs.

to the small spherical HBsAg or HBcAg particles often found in animals infected with HBV, ground squirrel hepatitis virus, or woodchuck hepatitis virus.

Serological Profiles Associated with Tree Squirrel Antigen Infection. The binding of purified HBsAg and/or HBcAg to different dilutions of tree squirrel sera in solid-phase assays from 26 of the 53 sera tested is consistent with the presence of cross-reacting antibodies and suggests that exposure to the tree squirrel antigen is widespread. Analysis of anti-HBs and anti-HBc in each serum resulted in three different types of response to infection (Fig. 5). Twelve sera demonstrated a pattern of reactivity consistent with there being a strong anti-HBc response and little or no anti-HBs, as exemplified in Fig. 5*A*. In 28 sera, neither anti-HBc nor anti-HBs responses were detectable, as shown in Fig. 5*B*. In the remaining 13 sera, both activities were present, as presented, in Fig. 5*C*. The titers in many squirrel sera were similar to positive controls using known anti-HBs and anti-HBc containing sera (Fig. 5*D*). Further, the P/N ratios in Fig. 5*B* with tree squirrel sera were similar to those obtained using human sera with the Ausab and Corab kits known to be negative for the corresponding antibodies. The apparent presence of an anti-HBs response without anti-HBc in tree squirrel sera was not detected.

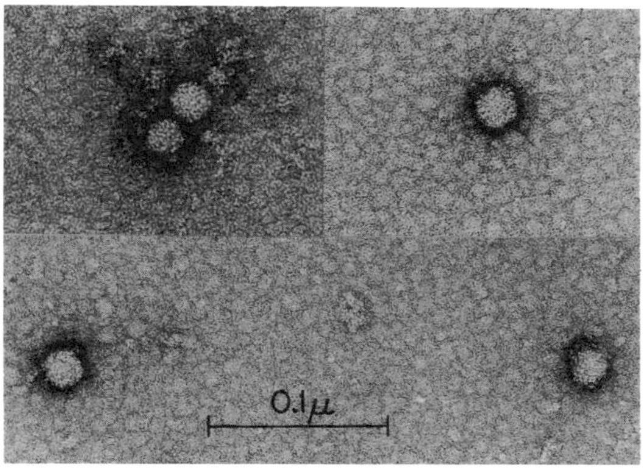

FIG. 4. Electron micrograph of purified tree squirrel antigen particles from several areas on a grid. The particles shown here were prepared from infected liver 309017.

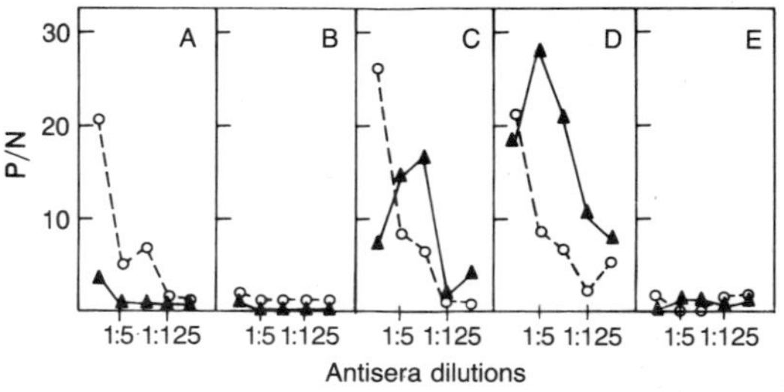

FIG. 5. Presence of apparently cross-reacting anti-HBs and/or anti-HBc in the sera of tree squirrels. Increasing dilutions of individual tree squirrel sera were used to coat plates and assay the binding of known amounts of HBsAg or HBcAg. P/N is the positive to negative test ratio in which the binding of HBsAg or HBcAg to squirrel sera is compared to the behavior of a pooled normal human serum (which is similar to that of many squirrel sera) in the same assays. (*A–C*) Apparent serological profiles of different groups of tree squirrels: mostly or entirely anti-HBc (*A*); no detectable anti-HBc or anti-HBs (*B*); and detectable anti-HBc and anti-HBs (*C*). (*D*) Positive control in which wells were coated with human carrier sera known to contain anti-HBs or anti-HBc. (*E*) Negative control in which wells were coated with normal human sera lacking anti-HBs and anti-HBc. The squirrel sera used to coat the wells in the assays presented here are as follows: 309000 (*A*); 309004 (*B*); and 309017 (*C*). ▲, HBsAg; ○, HBcAg.

Comparison of these serological responses to the appearance of hepatitis in the liver showed no correlation. Tree squirrel antigen carrying animals also showed different serological responses. Animals 308960 and 308957 did not demonstrate apparently cross-reacting anti-HBs or anti-HBc, while animal 309017 assayed strongly positive for both binding activities. All three antigen positive animals, however, demonstrated one or more features characteristic of hepatitis upon examination of their liver sections.

DISCUSSION

The discovery of histological evidence of hepatitis in 14 of 94 tree squirrel livers examined prompted the search for a possible viral etiology. In 3 of the 14 diseased livers, crude extracts were positive for DNAp and crossreactive surface antigen activities. None of the 80 livers normal by histological criteria or of 11 others with histological signs of hepatitis were positive for these viral markers. The presence of cross-reactive HBsAg and HBcAg determinants copurifying with DNAp positive particles together suggest that tree squirrel antigen is a virus similar to, but distinct from, HBV. These data suggest the presence of a new HBV type of virus that we propose to name tree squirrel hepatitis B virus (THBV).

The results of this study demonstrate the presence of surface antigen and endogenous DNAp activity in particles from THBV-infected tree squirrel livers. Particles having similar characteristics have also been isolated from HBV-infected human livers (14). In the case of infected human liver, the major particle-associated DNA forms are double-stranded relaxed circular and linear, both of which are analogous to the DNA forms predominant in serum-derived Dane particles. In this study, characterization of the DNAp reaction product of THBV resulted in a broad band 2–2.4 kilobases in size, which is smaller than the 3–3.3 kilobases full-length genomes of HBV and other related viruses. Similar results have been obtained when virus-derived ground squirrel hepatitis virus DNA was compared to HBV DNA by gel electrophoresis and Southern blotting under nonstringent conditions (15). Since these DNA molecules were detected by nonstringent hybridization prior to the DNAp reaction, a simple explanation for the relatively fast migration of THBV DNA in comparison to HBV DNA following the DNAp reaction may be incomplete filling in of the single-stranded region. The DNase I sensitivity of the DNAp reaction product from THBV is also consistent with it being DNA. Extending the endogenous polymerase reaction from 1 to 6 hr resulted in an increased incorporation of radioactivity with time (unpublished data) further suggesting that the reaction product has a partially single-stranded structure that is filled in over time, as observed in HBV and other related viruses.

The absence of detectable virus particles from the sera of infected animals could indicate either that viral replication is proceeding at a low rate, with release of a few virions into the blood, or that a vigorous antibody response is present. While the latter explanation may be operative in squirrel 309017, the lack of any detectable binding of HBsAg or HBcAg to the sera of 308960 and 308957 suggests that perhaps low levels or complete lack of viral replication, which is characteristic of the HBsAg-negative chronic carrier state in HBV infection (16), may be present in these animals. Alternatively, noncross-reacting viral antibodies may exist in these sera. If antibodies cross-reacting with purified HBsAg and/or HBcAg roughly reflect the frequency of virus exposure, then the THBV-negative carrier state may be the dominant one in infected adult tree squirrels. Among the hepatitis B type viruses, long-term duck hepatitis B virus infection is also often characterized by a lack of circulating antigen (4, 17). The inverse relationship between liver disease and serum DNA or endogenous polymerase activity observed with duck hepatitis B virus infection (17) also seems to be consistent with the observations herein. Since HBV replication generally decreases as liver disease progresses from chronic passive to chronic active hepatitis and is infrequent or absent in primary hepatocellular carcinoma (17–21), it may be possible to study this type of disease progression in a mammalian host. If further work more firmly establishes the presence of a chronic carrier state, comparison of the host–virus relationship in this system with that of infected woodchucks and ground squirrels may help elucidate some of the factors in related mammalian hosts that account for the differences in pathogenesis among them.

We would like to thank Drs. Richard Crowell and John Taylor for their critical reviews of the manuscript and helpful suggestions for improvement. We would like to acknowledge Drs. Crane and Custer for their help in screening hematoxylin/eosin-stained liver sections and Dr. M. E. Bayer and S. Shepardson for help in obtaining the electron micrographs of purified THBV. Dr. Howard Blatt arranged the trapping, transport, and care of the tree squirrels. This work was supported by Public Health Service Grants CA-06551, RR-05539, and CA-06927 from the National Institutes of Health and by an appropriation from the Commonwealth of Pennsylvania.

1. Summers, J., Smolec, J. M. & Snyder, R. (1978) *Proc. Natl. Acad. Sci. USA* **75,** 4533–4537.
2. Marion, P. L., Oshiro, L. S., Regnery, D. C., Scullard, G. H. & Robinson, W. S. (1980) *Proc. Natl. Acad. Sci. USA* **77,** 2941–2945.
3. Mason, W. S., Seal, G. & Summers, J. (1980) *J. Virol.* **36,** 829–836.
4. Marion, P. L. & Robinson, W. S. (1983) *Curr. Top. Microbiol. Immunol.* **105,** 99–121.
5. Summers, J. & Mason, W. S. (1982) *Cell* **29,** 403–415.
6. Onodera, S., Ohori, H., Yamaki, M. & Ishida, N. (1982) *J. Med. Virol.* **10,** 147–155.
7. Feitelson, M. A., Marion, P. L. & Robinson, W. S. (1981) *J. Virol.* **39,** 447–454.
8. Gerlich, W. H., Feitelson, M. A., Marion, P. L. & Robinson, W. S. (1980) *J. Virol.* **36,** 787–795.
9. Feitelson, M. A., Marion, P. L. & Robinson, W. S. (1982) *J. Virol.* **43,** 687–696.
10. Robinson, W. S. & Greenman, R. L. (1974) *J. Virol.* **13,** 1231–1236.
11. Robinson, W. S. (1975) *Am. J. Med. Sci.* **270,** 151–159.
12. Shikata, T., Uzawa, T., Yoshiwara, N., Akatsuka, T. & Yamasaki, S. (1974) *Jpn. J. Exp. Med.* **44,** 25–36.
13. Towbin, H., Staehelin, T. & Gordon, J. (1979) *Proc. Natl. Acad. Sci. USA* **76,** 4350–4354.
14. Miller, R. H. & Robinson, W. S. (1984) *Virology* **137,** 390–399.
15. Ganem, D., Greenbaum, L. & Varmus, H. E. (1982) *J. Virol.* **44,** 374–383.
16. Brechot, C., Lugassy, C., Dejean, A., Pontisso, P., Thiers, V., Berthelot, P. & Tiollais, P. (1984) in *Viral Hepatitis and Liver Disease*, eds. Vyas, G. N., Dienstag, J. L. & Hoofnagle, J. H. (Grune & Stratton, New York), pp. 395–409.
17. Omata, M., Uchiumi, K., Ito, Y., Yokosuka, O., Mori, J., Terao, K., Wei-Fa, Y., O'Connell, A. P., London, W. T. & Okuda, K. (1983) *Gastroenterology* **85,** 260–267.
18. Andres, L. L., Sawhney, V. K., Scullard, G. H., Smith, J. L., Merigan, T. C., Robinson, W. S. & Gregory, P. B. (1981) *Hepatology* **1,** 583–585.
19. Lieberman, H. M., Labrecque, D. R., Kew, M. C., Hadziyannis, S. J. & Shafritz, D. A. (1983) *Hepatology* **3,** 285–291.
20. Werner, B. G., Murphy, B. L., Maynard, J. E. & Larouze, B. (1976) *Lancet* **ii,** 696.
21. Musca, A., Cordova, C., Barnaba, V., Bonavita, M. S., Levrero, M., Zaccari, C. & Balsano, F. (1983) *Hepatogastroenterology* **30,** 3–5.

College lectures

Complex interactions of hepatitis B virus with its host and environment

The Croonian Lecture 1994

Baruch S Blumberg
MD, PhD, FRCP

Although one can say that most (80–90% or more) hepatocellular carcinoma (HCC) is caused by hepatitis B virus (HBV), hepatitis C virus (HCV) or both, such a bold statement misses the richness of the interactions of the virus with its hosts, the environment, other viruses and time. An understanding of the complexity reveals many places where intervention can take place to prevent or treat a disease. The more details of the life of the virus that are known, the more the interventional approaches that can be taken for the benefit of public and personal health. Despite the complexity, it is reassuring to know that it is not necessary to understand everything in order to effect changes. The history of medicine is filled with applications developed before all that could be known about a disease had been revealed.

Classification of the hepatitis viruses

There are five known and lettered hepatitis viruses (A, B, C, D, E), probably a sixth (?F), and undoubtedly other viruses yet to be discovered. In 1967, after we identified HBV (the first of the viruses to be discovered) there was a tacit assumption that the other viruses which were known to exist but had not yet been identified, would be similar to HBV. The symptoms and signs of the various forms of hepatitis have a large overlap and it was thought that the viruses would also be similar. However, the viruses are all in different families (Table 1). HAV is a *picornavirus,* initially classified in the genus *enterovirus* but now placed in a genus of its own, *hepatovirus.* HBV has characteristics of a *retrovirus,* but is a double- and single-stranded DNA virus rather than an RNA virus. A new family, *hepadnavirus* was created to include HBV and related animal viruses. HCV is a *flavivirus* or *pestivirus.* HDV, which can infect only those who are also infected with HBV, is akin to plant viruses (viroids, virusoids and plant satellite RNA). HEV is a *calicivirus,* again another family.

The A and E viruses are spread by the faecal-oral route, B, C and D by the parenteral route (needle stick, transfusion, etc) as well as by sexual contact, and from mother to child (Table 2). It appears that these viruses, all with a tropism for the liver, developed in different forms rather than evolving towards similarity.

Clinical aspects

Viral hepatitis is an infection of the liver by one or more of the hepatitis viruses and results in inflammation accompanied by a variety of symptoms, the most striking and pathognomonic being yellow jaundice. In the acute form, seen with HAV, HBV and HEV infection, there is a prodromal period of malaise, loss of appetite and sometimes flu-like symptoms. The urine darkens, the faeces may become light coloured, pruritus may be troublesome and fever develop. When jaundice appears, the symptoms usually abate and resolution can occur in weeks or months, but fatigue and malaise may persist for longer. The usual outcome is the development of protective antibody. Rarely, acute hepatitis may lead to fulminant hepatitis which is often deadly.

Acute infections with HBV, HCV and HDV may become chronic. However, more commonly, chronic infection with HBV will commence with an initial infection (particularly in the newborn and young) which is asymptomatic but persists for years. Many of those infected will, during the course of several decades, develop a life-shortening chronic liver disease and cirrhosis, and also HCC (which is deadly).

Epidemiology of the hepatitis viruses

Taken together, hepatitis viruses may constitute the most common viral infections of humans associated with disease states. For example, in the USA which by world standards is a relatively low endemicity area, viral hepatitis was the second most common infectious disease reported in 1990.

HAV

In high endemicity areas essentially all the population becomes infected with HAV by the time adulthood is reached. Even in relatively low endemicity areas such as the USA, 75% of adults will have been infected at some time in their lives.

This article is based on the Croonian lecture given at the Royal College of Physicians on 27 May 1994 by **Professor Baruch S Blumberg,** Master of Balliol College, Oxford.

B S Blumberg

Table 1. The hepatitis viruses

Name	Virus family	Particles, shape	Virion size (nm)	Nucleic acid	Nucleotides
HAV	Picornavirus	Naked, spherical, icosahedral	~27	Linear, single-stranded RNA Does not integrate	7,480
HBV	Hepadnavirus ('para-retro virus')	Spherical	42	Circular, partially double-stranded, small portion triple-stranded DNA, RNA Integrates	3,200
HCV	Flavivirus, pestivirus	Spherical	55	Positive strand RNA intermediate Does not integrate	9,500
HDV	Plant viroid, virusoids or plant satellite RNA (?)	Spherical, covered with HBsAg	36 (some 40)	Single-stranded circular RNA	~1,700
HEV	Calicivirus	Spherical, non-enveloped	32–34	Single-stranded, polyadenylated RNA	~7,500

HBV

HBV infection is a much more serious cause of long-term mortality and morbidity; it leads to chronic infection, which markedly increases the risk of life-shortening chronic liver disease and HCC. There are probably more than 350 million carriers in the world, about 25% of whom are likely to develop the late sequelae of the disease. Endemicity is high in sub-Saharan Africa, in Asia and Oceania and, strangely, among Eskimos. It has been estimated that there are several million deaths per year as a consequence of HBV. A dramatic example of mortality in a high endemicity country comes from Taiwan where death from primary cancer of the liver, which is mostly due to chronic HBV infection, is the second most common cause of death in middle-aged men.

HCV

The frequency of HCV is low in most European countries, roughly similar to the prevalence of HBV carriers, but it is extraordinarily high in Zaire, Egypt and elsewhere in Africa and Asia. In Japan, HCV is now the most common cause of primary cancer of the liver and, until the introduction of donor screening programmes for HCV, was the major cause of post-transfusion hepatitis.

HDV

HDV occurs only in individuals who are also infected with HBV and is common worldwide among HBV carriers. The frequency varies from about 4% in parts of Europe to over 10% in southern Europe, Africa and North America. It is often associated with deadly fulminant hepatitis and primary cancer of the liver. It has been estimated that worldwide about 15 million people are infected with HDV. In the USA about 7,500 cases of acute HDV infection occur annually and there are about 70,000 carriers. About 1,000 persons die each year as a consequence of HDV infection.

HEV

HEV can cause enormous epidemics when water supplies become contaminated with the faeces of infected persons. It is probably responsible for many of the widespread epidemics of hepatitis in India and other regions where it is common. A particularly tragic feature of these epidemics is the high mortality among pregnant women (which may average about 20%). The reason for this unusual morbidity pattern is unknown, and its unravelling may teach us about viruses and both female and fetal physiology. Epidemics occur in Asia and Africa but have also been reported from Mexico and Borneo. Only sporadic cases occur in the USA and these appear to be imported.

It is clear that the hepatitis viruses are common and widespread, and that in many regions of the world two or more of them may occur in the same individual. This makes it likely that interactions will occur frequently.

Virology and molecular biology: the vaccine

HBV was the first human pathogen to be sequenced, possibly because of its very short genome (for review, see Ref 1). Following our discovery of HBV in 1967, applications of immunodiffusion, electrodiffusion and, later, radioimmunoassay and enzyme immuno-

Table 2. Modes of transmission of the hepatitis viruses

Name	Method of transmission	Form of hepatitis
HAV	Faecal-oral, enteric, water-borne	Acute
HBV	Blood-borne, venereal, mother to child, transfusion, needle injection, parenteral	Acute, chronic, carrier, primary cancer of the liver
HCV	Blood-borne, parenteral	As HBV
HDV	Same as HBV	Chronic, carrier-associated after HBV infection
HEV	Same as HAV	Acute, high mortality in pregnant women

assay, resulted in the establishment of excellent diagnostic techniques. Screening of donor blood almost completey eliminated post-transfusion hepatitis due to HBV.

The diameter of the HBV virion is about 42 nm. The surface antigen (HBsAg) which is glycosylated and contains lipid, coats the virus and encloses the core. The core itself is encircled by the core antigen (HBcAg), which contains the DNA genome and the proteins involved in replication and other activities. In addition to the whole virion there are small spherical particles 17–25 nm in diameter and filaments of the same width but of variable length which contain only the surface antigen (Fig 1). These surface antigen particles were identified in our laboratory in 1968 and were used in the preparation of blood-derived vaccine.

The vaccine was invented before molecular techniques were available using simple epidemiological, clinical and biochemical methods. It had been observed that the presence of the antibody against HBsAg was rarely accompanied by the virus itself. A vaccine was therefore produced from the HBsAg particles which were present in enormous quantities in the blood of HBV carriers. This was a unique method for producing a vaccine which has been used safely and effectively in millions of people. Later, when recombinant methods were developed, the S gene was used to prepare a vaccine nearly identical to that produced by the blood extraction technique. The recombinant vaccine has the merit of easy and uniform production and the psychological benefit of *not* being blood-derived, but it is no more effective or safe than its naturally-derived predecessor.

The genome consists of 3,200 base pairs (Fig 2) and is double-stranded. Open reading frames (ORFs), of which there are four (S: surface antigen; C: core antigen; P: polymerase; X: function originally unknown), are defined as the bases between the first start and stop codons. The HBV genome has evolved towards minimal length; the total genome is only slightly larger

Fig 1. Schematic model of the HBV and HBs particles (reproduced with permission from Ref 2).

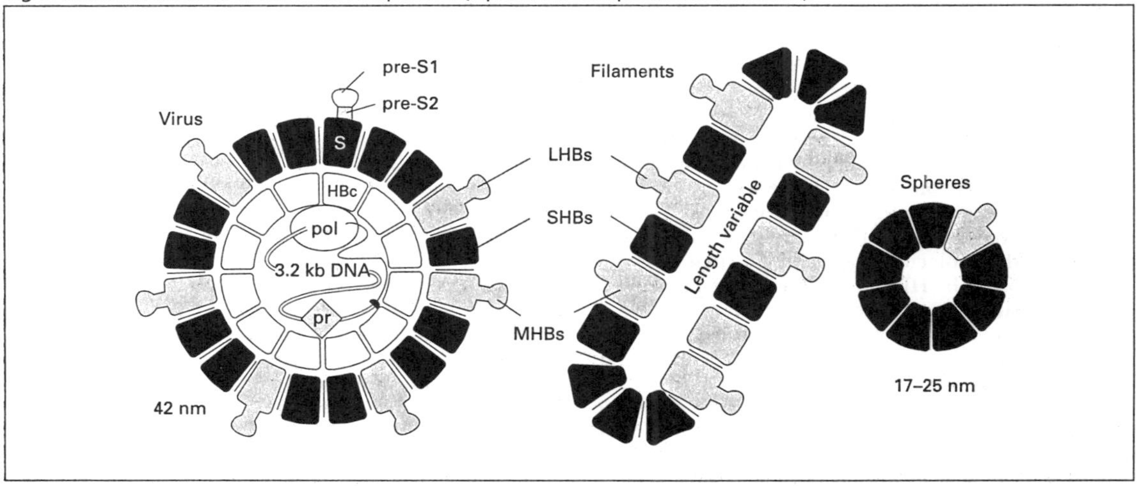

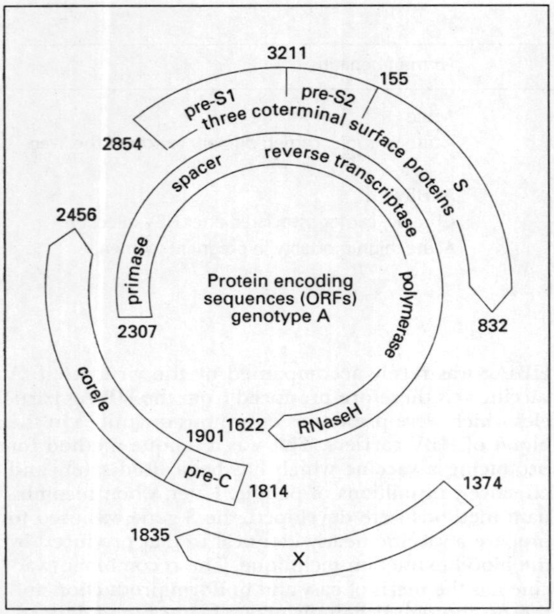

Fig 2. Hepatitis virus genome open reading frames (ORF) (reproduced with permission from Ref 2).

than the P genome, leaving a small region where there is no overlap.

The S ORF includes pre-S1, pre-S2 and S. The products of these three form a small (S product only), middle (S, pre-S2), and large HBs (S, pre-S1, pre-S2) protein. They occupy different positions on the virion and on the smaller particles which contain only HBsAg (Fig 1). S is the predominant surface antigen protein; the others are represented in different proportions in the several particles. We were the first to describe the existence of antigen subtypes on the surface antigen [3] which are manifested in the S protein; they have since been described in greater detail. All subtypes have a common antigenic portion, *a*, which renders them cross-antigenic. Therefore, a single vaccine, usually composed of a 'cocktail' of the subtypes, can be used for protection against all the subtypes with the exception of what may be escape mutants which have emerged in recent years. The sharing of this antigen portion of the molecule may reflect a further evolutionary economy by HBV, but it has proven detrimental to the virus in that it has made the manufacture of the vaccine much simpler. There are at least six genotypes of the S antigen but these do not correspond to the antigen subtypes.

The HBc and HBe proteins are the products of the C ORF. It packages its own messenger RNA (mRNA) as well as that of the viral polymerase, and also includes protein kinase, probably of host origin. This particle is encapsulated by the HBs protein before export. The ORF includes the pre-C region of 29 codons which, when translated along with the C region, produces HBe protein which can be secreted from the cells. Variants due to mutations in the pre-C sequence prevent the production of HBe but do not inhibit replication of the virus. (The possible reasons for this will be discussed in the strategies section).

The products of the P gene are not well defined but are necessary for the replication of the virus. It appears to produce a primase, necessary for priming minus strand synthesis, RNA- or DNA-dependent polymerase (reverse transcriptase), and an RNAaseH which can cleave the RNA in RNA/DNA hybrids. The polymerase protein is packaged in the core.

The X sequence is absent or displaced in at least one of the hepadnaviruses. The nature of the X codon suggests that it has been introduced only recently into the virus from the eukaryotic genome, and homologous sequences are found in primate genomes. The product of this ORF is involved in the interaction of the virus with the host cells and therefore may be involved in the evolutionary relations between HBV and its human host. In *in vitro* experiments it transactivates a variety of genes. It may also be important in the carcinogenic potential of HBV and has been shown to have a tumorigenic effect in immortalised mouse hepatocyte cultures and transgenic mice.

Life cycle of HBV

The sequence of events in the life cycle of HBV is based on experimental studies and inferences from the life cycle of other viruses (Fig 3). There is no agreement on the nature of the attachment site for HBV; in fact, there may be several, relatively weak attachments. The pre-S1 sequence 21–47 appears to attach to a membrane-bound interleukin-6, but it does not appear to be a major site. Pre-S1 may also be involved in attachment to an immunoglobulin (IgA) receptor in a process involving molecular mimicry. The glycan linked to pre-S2 may be a site of attachment since it has a weak affinity for HepG2 cells (derived from a human cancer patient), but again this does not seem to be a sufficient site. The attachment of pre-S2 to polymerised serum albumin may serve as a means for attachment, but this too is not sufficient. The transferrin binding site on cells may mediate the attachment of pre-S2, and could be related to the unusual relation of HBV replication to serum iron storage levels (see section on body iron stores). The apparently weak attachments of HBV to liver (and other) cells may be part of the evolutionary strategy designed to maintain a high level of viraemia for long periods. If affinities were strong, the viraemia would be less and the cells become overburdened with virus, decreasing the probability of chronicity because the cells would be more readily eliminated.

How does the virus get into the cell? It has been sug-

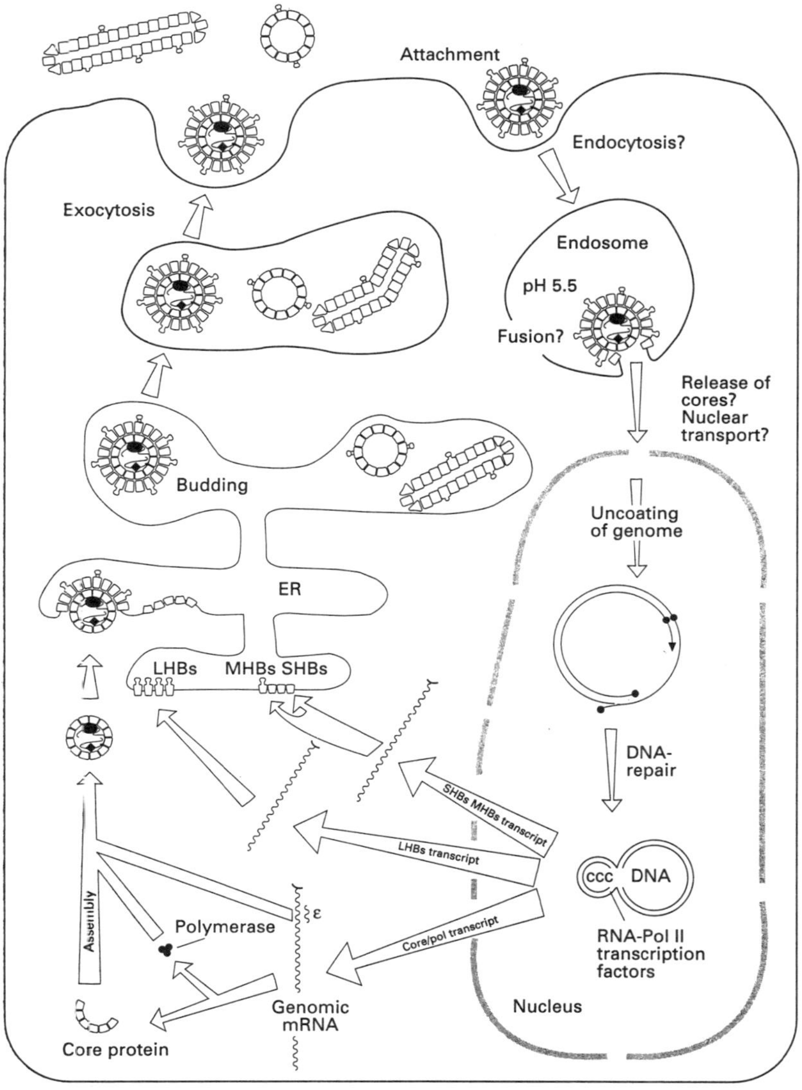

Fig 3. Schematic view of the life cycle of HBV (reproduced with permission from Ref 2).

gested that endocytosis, mediated by the transferrin receptor mechanism, may effect entry. The viral genome is then released into the nucleus and forms a double-stranded covalently closed circle. Viral polymerase helps to close the gap, though other proteins may also be utilised. The HBV DNA can integrate into the host DNA, a characteristic it shares with retroviruses. In contrast to retroviruses, though, it is not essential to the replication process and may, in fact, hinder it since the linearisation will interrupt at least one of the HBV genes. The integrated HBV DNA does not appear to replicate, but occasionally fuses with the host DNA. These fusions are found in patients with HBV-induced cancer of the liver.

It is interesting to conjecture how viruses containing portions of human genome might behave if they could replicate and be transmitted. They could bring DNA sequences from one human host into another, and integration could then occur. If this took place in human germ cells and the genes were transmitted to the next generation, it would represent a method of acquiring heritable genetic material after birth—a fascinating technique for altering the evolutionary process.

After the covalently closed DNA is formed, the episomal DNA is transcribed to RNAs which serve as templates for their respective proteins (eg polymerase, core). These proteins combine with their mRNA in the replication complex. The primase portion of the polymerase serves as a primer for the reverse transcriptase which then proceeds to produce the negative strand DNA. An RNAaseH associated with the transcription removes the RNA template with the exception of an 18-base RNA fragment. This serves as a primer for the generation of the positive strand DNA, and the structure of the virion DNA is reproduced.

The transcripts for the several viral proteins are excreted from the nucleus into the cytosol. The core protein and polymerase are assembled into the core particle which contains the replicated DNA, and enter the rough endoplasmic reticulum where they are covered with the surface protein assembled from the S antigen component transcripts. The whole virions as well as the particles containing only HBsAg are manufactured in the endothelial reticulum and then exported from the cell, presumably by exocytosis. They are then free to enter other cells or to be transmitted to other individuals by the mechanisms already discussed. As Gerlich has noted [1], this complicated process of replication can be seen as an effort by the virus to persist at reasonably high levels in the blood so that it can be transmitted sexually, from mother to child, and by other mechanisms of blood transmission from one person to the next.

The recombinant vaccine

Using recombinant techniques, several investigators isolated clones of the S reading frame and inserted them in a variety of cells (*Escherichia coli*, yeast, human, etc). Vast quantities of vaccine are now manufactured by the recombinant process. It is the first (and, so far, only) commercial recombinant vaccine and one of the financially most successful recombinant products. It is now being used in population-based vaccination programmes involving millions of individuals.

Hepatocellular carcinoma

The earliest evidence for a relationship between HBV and HCC was discovered during field and epidemiological studies in Africa and Asia. Chronic HBV infection was much more common in the HCC cases than in controls, particularly in regions where the HBV carrier rate was high. Prospective studies in Taiwan, and later in Alaska and elsewhere, showed that individuals who were carriers at the start of a study were much more likely to develop HCC than those in the same environment who were not carriers. In Taiwan, the risk was more than 200-fold (much greater than that of heavy cigarette smokers for cancer of the lung).

Integrated sequences of HBV are found in the DNA of host liver cells. Integration is also seen in HBV carriers, and several explanatory models for the role of integration in the pathogenesis of HCC have now emerged. It is likely that a number of mechanisms are operative and may provide concurrent or alternative pathways to disease.

The host's immune response results in the slow but continuous death of liver cells infected with HBV. This, in turn, promotes cell division in the remaining cells and could increase the likelihood of mutational carcinogenic events occurring over time. The integration sites do not appear to be specific. It was suggested that insertional activation of potential oncogenes might be the explanation. This is rarely seen in humans, but quite commonly in the HCC which occurs in woodchucks infected with woodchuck hepatitis virus (WHV). Experimental evidence from studies on transgenic mice indicates that unregulated expression of either the X protein, a transcriptional activator, or the large surface antigen might be carcinogenic. Integrated HBV sequences may alter host cell growth in transgenic mice by unregulated expression of proteins. Loss of heterozygosity on chromosomes 1p, 4q, 11p, 13q, and 16q in patients with HCC is consistent with the loss of 'anti-oncogenes', and so increases the risk of cancer.

The existence of many models increases the opportunities for identifying medications that could interfere with the pathogenic processes. If taken in the appropriate combinations they could improve treatment of a deadly cancer.

Adaptations

HBV is one of the simplest viruses, yet it has developed remarkable adaptive and evolutionary techniques to perpetuate its existence. It can be argued, for example, that it has adapted to the computer age. In the mid-1970s an HBV epidemic occured among laboratory and hospital staff in an institution where computer techniques had been installed for the recording and transmission of laboratory data. Data were recorded on computer cards which were despatched from laboratory to clinic with the necessary information. Some of these computer cards became soaked with the blood of an HBV carrier, and the virus was transmitted to individuals who cut their fingers on the sharp edge.

An epidemic of acute hepatitis B involving several hundred cases developed in Swedish orienteerers. Its cause and remedy were soon found: running through the brush, the runners, wearing only shorts, suffered multiple scratches on their legs and drops of blood were left on the low-lying brush. Contestants following them could be infected with HBV if a previous runner was a carrier. In addition, there were check-points where runners could wash the blood from their legs using a common bucket and/or towel, which provided mechanisms for transmission. When the sportsmen were required to wear long pants, the epidemic ceased, a simple remedy to a complicated pathogenic process.

Interactions between the viruses

Hepatitis viruses A, B, C, D, E and possibly F have evolved different strategies for survival, transmission, replication and probably pathogenesis. Interactions with human immunodeficiency virus (HIV) may be included here because co-infection often occurs and, in the past few decades, HIV may have affected the evolution of some of the hepatitis viruses.

In general, superinfection with other hepatitis viruses will decrease the replication of HBV. This is particularly true with delta virus (HDV); it suggests that an attenuated delta-like virus might be used to decrease HBV replication. In chimpanzees with chronic HCV infection, the intensity of acute hepatitis due to HBV or HAV is attenuated. Also, co-infection with HBV appears to decrease the replication of HCV.

During the great Shanghai epidemic of HAV in 1988 when thousands became infected as a consequence of contaminated shellfish, 15 of the 47 deaths were associated with chronic HBV carriage and in 11 of them there was active HBV replication. However, this effect has not been seen in regions of lower endemicity for HBV.

Interactions between HIV and HCV may have had a profound effect on both viruses. The spread of HIV started some time in the late 1970s (probably mostly due to changes in human behaviours). The spread of HCV may also be recent, possibly starting at the same time and for the same reasons. This could be a consequence of mutual 'support' between the two viruses. The transmission of HCV from mother to child is ordinarily not very effective, but if the mother is also infected with HIV, perinatal infection with HCV is more likely to occur. Infection of HCV carriers with HIV results in a more severe clinical course, probably associated with increased viral replication. This, in turn, might increase the probability of subsequent HCV transmission.

There are undoubtedly many interactions that affect the mutual evolution of these viruses, and continuing studies of all of them together (not an easy task) are likely to yield fascinating insights into general principles of viral evolution.

Molecular and immunological evolutionary strategies

A small compact genome may contribute to the hardiness of a virus. All the hepadnaviruses appear to have evolved toward minimal genome length, and the whole genome is only slightly larger than P, its largest ORF, and much overlapping has occurred. This can decrease the probability of mutation since a change in one gene will, in most cases, be accompanied by a mutation in another (usually P), which could be lethal to the virus.

Perhaps because of its limited genetic resources HBV has learnt to use the host extensively. The overall strategy appears to be the maintenance of a high level of virus in the blood and other body fluids to allow efficient transmission if even small amounts of blood and/or fluid are transmitted from one individual to the next. As noted, the absence of high-affinity receptors for HBV on hepatocytes and other cells helps to maintain a high level of circulating virus and HBs particles. A second major strategy is the development of chronic carriage infectious virus; for this, it must evade the immune system at least during the period when natural transmission is most effective.

HBV has seized on the most necessary biological functions of humans to ensure its own perpetuation, namely, sexual intercourse and childbirth. In premodern times, together with folk practices such as ritual circumcision, tattooing, skin cicatrix decoration, ceremonial blood exchange ('blood brotherhood') and others by which blood could be transferred from one individual to the next, these were the main methods of transmission. Before the availability of disposable equipment, the re-use of needles and other medical instruments has been a major mechanism for transmission. Until the development of blood screening programmes in the past decade, blood transfusion was an important method of transmission, but this has now been controlled.

The mechanisms by which the immune system is evaded are complex and, in some cases, unique. HIV uses a direct and brutal frontal attack on the cells which affect the immune responses of the body, thereby wreaking a tragic toll on those it has infected, whereas HBV is more subtle and benign. In many cases, infected individuals can remain asymptomatic carriers for many decades, thereby extending the period in which the virus can be transmitted.

Large numbers of circular and elongated particles which contain only surface antigen are produced by the infected cell. These can serve as a 'smoke screen' in the blood and body fluids which will engage the immune 'attack' systems of the host and spare the less numerous replicating whole virion. They may also block receptors on cell surfaces and maintain high levels of viraemia. Ironically, this strategy has worked to the disadvatage of the virus in that these HBsAg particles have been used to prepare a vaccine which has led to control, and could conceivably lead to elimination, of the virus.

The pathological effects of HBV are caused not by a direct destructive effect on liver cells but by the attack of the host immune system on the hepatocytes containing HBc and probably other viral antigens on their surface. To maintain chronic infection, destruction of the infected cells is mitigated by an interesting viral mechanism involving the HBc/HBe system of protein encoded by the C reading frame. If translation starts at the second start codon, thereby not translating the pre-C region, HBc is produced. It is not excreted, but remains in the cell to coat the DNA and other constituents that form the core of the virus.

If translation begins at the first start codon, both pre-C and C are translated and HBe is produced. HBe shares most of its amino acid sequences with HBc and can be excreted, because the pre-C region peptides encode a signal sequence which probably facilitates secretion through the cell membrane. HBe appears to inhibit the destruction of hepatocytes which contain HBV and exhibit HBc on their surface by interacting and 'disarming' the host immune cells, probably by making them tolerant to HBcAg which is part of the HBe protein. This prolongs the life of the cells and allows replication to continue as an unhurried and lengthy process, thus preserving the life of the host usually through the decades of its most intensive sexual activity and the child-bearing period, so that he or she will have ample time to serve as a transmitter of HBV. When HBe is absent but replication of the virus persists, which is the case in pre-C mutant viruses, a much more severe and progressive disease may ensue.

HBe may also have an important effect on transmisson to the newborn. There is a remarkably high frequency of maternal transmission (90% or more) when the mother is HBeAg-positive. The excretable HBe present in the pregnant carrier can pass through the placenta and 'tolerise' the immune cells of the fetus. When exposed to the virus some time during or after the pregnancy and delivery, the immune systems will not reject it, and a long period of carriage can ensue.

HBV appears to have developed additional mechanisms to alter the host immune responses and perpetuate the carrier state. Interferon is effective in resolving acute HBV infection, but this appears to be altered in chronic HBV infections. Peripheral blood mononuclear cells from patients with acute and chronic HBV infection produce less interferon, apparently mediated by the core region of HBV. In addition, chronically infected cells, both hepatocytes and immune cells, are only minimally activated by interferon, which may be due to the virus itself or to interference by HBV polymerase.

HBV has been successful in invoking a series of mechanisms which perpetuate the life of the infected cells, induce long-term infection with consequent advantages to the transmission of the virus, and allow the host to live relatively symptom-free, usually for many decades and often entirely encompassing the child-bearing period. After that time, the effects of viral mutation can accumulate, the flexibility of the host immune response may decrease and finally become exhausted. Severe hepatic damage ensues and the host dies. In pre-modern times this may not have had much effect on overall human survival since life expectancy was short and most carriers would have succumbed to other causes of death before the destructive phase of HBV chronic infection could take over. In any case, the main transmission goals of the virus would have been accomplished by that time, and extended chronicity would have been of only marginal advantage to the virus.

Complex interactions with host and environment

Gender

Interactions of HBV with its host and the external environment have a significant bearing on infection with HBV and the pathogenesis of liver disease and HCC.

Gender is an important variable. Men, when infected, are more likely to become carriers while women are more likely to develop antibody to the surface antigen and remain protected. Therefore, in most populations more men than women are carriers. Male carriers are at much greater risk of developing HCC than women. Ironically, while the carrier females transmit the virus to their children, their male children are more likely to become carriers.

An even more remarkable observation is that in populations with a high frequency of carriers, the nature of the parents' response to HBV infection has a bearing on the gender of their offspring. In a series of studies in Greece, Papua New Guinea, the Philippines and Greenland, carrier families (ie families in which either parent is a carrier) were likely to have a higher sex ratio (more sons than daughters) than families in which the mother had anti-HBs. This effect is large and could account for the lower than expected number of female births in areas of high carrier endemicity such as south China. If verified, it is a strange observation that a viral infection could have a role in human sex determination; and therefore in human evolution.

Aflatoxin

Aflatoxin is a carcinogenic toxin produced by fungi which infect stored grains and other foods such as ground-nuts. Epidemiological studies have shown a high prevalence of HCC in areas where aflatoxin is common. Ross *et al* reported a study from Shanghai which involved 18,244 men with a total of 35,299 years of follow-up [4]. The risk ratio for HCC in individuals who were not carriers of HBV and were aflatoxin-negative was taken as 1.0. Those who had HBV without elevated urinary aflatoxin had a risk ratio of 4.8 (CI 1.2–19.7), for those with elevated aflatoxin who were not carriers it was 1.4 (CI 0.5–7.5), while those who were both carriers and had elevated aflatoxin had a remarkable risk ratio of 60.1 (CI 6.4–561.8). Hence, aflatoxin alone may not cause HCC, but combined with HBV infection its effect is profound. If these studies are confirmed, a potent and preventable cofactor for the development of HCC will be revealed and will validate an important method of control in addition to HBV vaccination.

Genetic control

Some of our earliest studies appeared to show genetic control of the propensity for a person infected with HBV to become a carrier. Recent studies are consistent with this hypothesis. The gene, if it exists, has not been identified. One or more of the sites of HBV integration could be a possible location of such a gene(s).

There also appears to be genetic control of the carcinogenic and other effects of aflatoxin. The story, as described by McGlynn and her colleagues [5], is complex. A mutation in the p53 tumour suppressor gene has been identified in HCC in some populations, in particular those with a high prevalence of aflatoxin. There is a mutational hotspot in codon 249 of exon 7 of the p53 gene which is postulated to be related to aflatoxin exposure. Two gene loci, EPHX and GSTM1, have been identified in relation to aflatoxin detoxification. A mutant allele of EPHX and a null allele of GSTM1 decrease the aflatoxin detoxification capability. Both genes occur significantly more frequently in patients with aflatoxin, HCC and/or the p53 mutation. This observation, if confirmed, indicates a genetically controlled susceptibility to the carcinogenic effects of aflatoxin, and helps to explain why some people exposed to aflatoxin and HBV develop HCC while others do not.

Body iron stores

Plasma ferritin levels are directly correlated with body iron stores, and transferrin levels are inversely related. A study in Taiwan found that increased ferritin and decreased transferrin (ie increased body iron stores) determined at the start of a prospective study are associated with a higher risk of developing HCC or certain other cancers. In Korean patients with chronic liver disease who were carriers of HBV, those with elevated ferritin levels were significantly more likely to develop HCC than those with lower levels. This effect is not restricted to HCC. In Philadelphia, McGlynn *et al* [5] have shown that increased ferritin is also associated with shorter survival in cancer of the lung, breast and colon. Total body iron stores can be controlled by dietary intake, and it is possible that excess iron intake may be detrimental.

Additional evolutionary factors

HBV-like viruses in other species

Hepadnaviruses have been found in woodchucks (WHV) and ground squirrels (GSHV). Less well characterised viruses include the tree-squirrel hepatitis B virus (TSHBV), and a virus reported in kangaroos. Avian viruses have been reported in the domestic Peking duck (DHBV) and in herons. Viruses which appear to be similar to DHBV have been found in wild Mallard ducks in Michigan and Illinois. Several of these non-human hepadnaviruses have been sequenced and their phylogenetic relationships estimated using base pair comparisons.

It is interesting that the cauliflower mosaic virus family *phytopararetrovirus,* DNA plant viruses utilising reverse transcription during replication, are similar to human hepadnaviruses. It has been suggested that these two families, animal and plant, might be grouped into a superfamily to indicate their relationship.

An additional observation is that HDV is similar to several of the very small plant RNA viruses.

Is it possible to infer from these observations that these viruses have remote affinities with plants and plant viruses, and also that anti-viral strategies might be sought in the plant kingdom?

Human inheritance of HBV genomes: a conjecture

In common with the retroviruses, hepadnaviruses integrate into the DNA of the host genome. However, the integration is not essential for replication; on the contrary, it probably *decreases* replication and production of gene products, in that linearisation of the circular DNA splits at least one of the reading frames. Integration is common in cancer of the liver, but also in individuals who have been carriers for some years.

Does this unusual viral strategy serve a biological function? It may be conjectured that, with massive infections and in large epidemics, HBV DNA may infect and integrate into the DNA of the host germ cells. At the time of reproduction these could enter the gene pool of the population and segregate in the same manner as ordinary host genes; it would be analogous to mutation. Although germ cell integration may be relatively rare, in areas of massive infection such as East Asia and sub-Saharan Africa the absolute number of such integrations could be large. (Sections of the human genome have homologies with HBV, but little is known about how they relate, if at all, to this conjectural model.) It would represent a form of inheritance of acquired base pair sequences, and could lead to interesting models and observations on the effect of viruses on human population genetics. HBV DNA, so integrated, could be viewed as an 'engine' of evolution.

Conclusion

I have emphasised *complexity* in the natural history and pathogenesis of HBV infection. A traditional goal of the scientific process has been simplicity, unitary aetiological hypothesis and linear relationships between cause and effect. But in the natural world, interaction and involvement with the external and internal environments are the norm. To describe the world as it is, we have to attempt to embrace complexity. Furthermore, the more we learn of the constituents of this complex system, the more possibilities there are to intervene for prevention and treatment.

References

1 Gerlich W. In: Zuckerman AJ, Thomas HC (eds). *Viral hepatitis.* Edinburgh: Churchill Livingstone, 1993.

2 Zuckerman AJ, Thomas HC (eds). *Viral hepatitis.* Edinburgh: Churchill Livingstone, 1993.

3 Levene C, Blumberg BS. Additional specificities of Australia antigen and possible identification of hepatitis carriers. *Nature* 1969;**221**:195–6.

4 Ross RK, Yuan JM, Yu MC, Wogan GN, *et al.* Urinary aflatoxin birmarkers and risk of hepatocellular carcinoma. *Lancet* 1992;**239**:943–6.

5 McGlynn KA, Stevans RG, Lusbader ED, Blumberg BS, *et al.* Body iron stores and the prognosis of cancer. Presented at American Society of Preventive Oncology, Tuscon, AZ, USA, 20–23 March 1993 (as yet unpublished).

Address for correspondence: Professor Baruch S Blumberg, Fox Chase Cancer Center, 7701 Burholme Ave, Philadelphia, PA 19111, USA.

IV CORPUS ACKNOWLEDGMENTS

We wish to thank the authors and the following publishers for their permission to reproduce the articles found in this volume.

Article	*Publisher*
Blumberg B S and Ragan C. The natural history of rheumatoid spondylitis. *Medicine* **35**: 1–31, 1956.	Williams & Wilkins
Blumberg B S and Oster G. Light-scattering studies on hyaluronic acid. *Science* **120**: 432–433, 1954.	American Association for the Advancement of Science
Blumberg B S. A memorial oration, The University Church of St. Mary the Virgin, Oxford, England, November 2, 1996. Alexander George Ogston, MA (Oxon), D. Phil, FRS, FAA. January 30, 1911 — June 29, 1996.	permission not required
Alberdi F, Allison A C, Blumberg B S, Ikin E W and Mourant A E. The blood groups of the Spanish Basques. *J Roy Anthrop Inst* **87**: 217–221, 1957.	Royal Anthropological Institute, London
Blumberg B S, Ikin E W and Mourant A E. The blood groups of the pastoral Fulani of northern Nigeria and the Yoruba of western Nigeria. *Am J Phys Anthropol* **19**: 195–201, 961.	John Wiley & Sons, Inc.
Bangham A D and Blumberg B S. Distribution of electrophoretically different haemoglobins among some cattle breeds of Europe and Africa. *Nature* **181**: 1551–1552, 1958.	Macmillan Magazines Ltd
Corcoran P A, Allen F H Jr, Allison A C and Blumberg B S. Blood groups of Alaskan Eskimos and Indians. *Am J Phys Anthropol* **17**: 187–193, 1959.	John Wiley & Sons. Inc.
Blumberg B S, Bloch K J, Black R L and Dotter C. A study of the prevalence of arthritis in Alaskan Eskimos. *Arthritis and Rheumat* **4**: 325–341, 1961.	Lippincott-Raven Publishers

Melartin L and Blumberg B S. Albumin Naskapi: A new variant of serum albumin. *Science* **153**: 1664–1666, 1966.	American Association for the Advancement of Science
Schell, L M and Blumberg B S. The genetics of human serum albumin. In *Albumin Structure, Function and Uses*, Rosenoer V M, Oratz M and Rothschild M A (eds.). Pergamon Press, Oxford and New York pp. 113–141, 1977.	Pergamon Press
Lampl M and Blumberg B S. Blood polymorphisms and the origins of New World populations. In *The First Americans: Origins, Affinities, and Adaptations*, Laughlin W S and Harper A B (eds.). Gustav Fischer, New York, Stuttgart pp. 107–123, 1979.	Gustav Fischer
Franklin S G, Wolf S I, Ozdemir Y, Yuregir G T and Isbir T. Albumin Naskapi varian in North American Indians and Eti Turks. *Proc Natl Acad Sci USA* **77**: 5480–5482, 1980.	National Academy of Sciences
Allison A C and Blumberg B S. An isoprecipitation reaction distinguishing human serum protein types. *Lancet* **1**: 634–637, 1961.	The Lancet Ltd
Blumberg B S, Bernanke D and Allison A C. A human lipoprotein polymorphism. *J Clin Invest* **41**: 1936–1944, 1962.	The Rockefeller University Press
Blumberg B S. Polymorphisms of serum proteins and the development of iso-precipitins in transfused patients. *Bull N Y Acad Med* **40**: 377–386, 1964.	New York Academy of Medicine
Blumberg B S. Iso-antibodies in humans against inherited serum low density beta-lipoproteins, the Ag system. *Ann N Y Acad Sci* **103**: 1052–1057, 1963.	American Association for the Advancement of Science
Blumberg B S, Alter H J and Visnich S. A "new" antigen in leukemia sera. *J Am Med Assoc* **191**: 541–546, 1965.	American Medical Association

Blumberg B S, Gerstley B J S, Hungerford D A, London W T and Sutnick A I. A serum antigen (Australia antigen) in Down's syndrome leukemia and hepatitis. *Ann Int Med* **66**: 924–931, 1967.	American College of Physicians
London W T, Sutnick A I and Blumberg B S. Australia antigen and acute viral hepatitis. *Ann Int Med* **70**: 55–59, 1969.	American College of Physicians
Sutnick A I, London W T, Gerstley B J S, Cronlund M M and Blumberg B S. Anicteric hepatitis associated with Australia antigen. Occurrence in patients with Down's syndrome. *J Am Med Assoc* **205**: 670–674, 1968.	American Medical Association
Coyne (Zavatone) V, Millman I, Cerda J, Gerstley B J S, London W T, Sutnick A I and Blumberg B S. The localization of Australia antigen by immunofluorescence. *J Exp Med* **131**: 307–320, 1970.	The Rockefeller University Press
Senior J R, Sutnick A I, Goeser E, London W T, Dahlke M B and Blumberg B S. Reduction of post-transfusion hepatitis by exclusion of Australia antigen from donor blood in an urban public hospital. *Am J Med Sci* **267**: 171–177, 1974.	Charles B. Slack, Inc., Medical Publishers
Blumberg B S. Blood transfusion and the discovery of viruses. James Blundell Award Lecture: 1989. British Blood Transfusion Society Newsletter No. 19, January 1990, pp. 1–3, 1990.	British Blood Transfusion Society
Coller J A, Millman I, Halbherr T C and Blumberg B S. Radioimmunoprecipitation assay for Australia antigen, antibody, and antigen-antibody complexes. *Proc Soc Exp Biol Med* **138**: 249–257, 1971.	Blackwell Science Inc.
Millman I, Loeb L A, Bayer M E and Blumberg B S. Australia antigen (a hepatitis-associated antigen). Purification and physical properties. *J Exp Med* **131**: 1190–1199, 1970.	The Rockefeller University Press
Blumberg B S. The current state of the prevention of HBV infection, the carrier state and hepatocellular carcinoma. *Res Virol* **148**: 91–94, 1997.	Elsevier Science Ltd.

Blumberg B S, Larouze B, London W T, Werner B, Hesser J E, Millman I, Saimot G and Payet M. The relation of infection with the hepatitis B agent to primary hepatic carcinoma. *Am J Pathol* **81**: 669–682, 1975.	American Society for Investigative Pathology
London W T and Blumberg B S. A cellular model of the role of hepatitis B virus in the pathogenesis of primary hepatocellular carcinoma. *Hepatology* **2**: 10s–14s, 1982.	W. B. Saunders W.
Payne R J H, Nowak M A and Blumberg B S. Analysis of a cellular model to account for the natural history of infection by the hepatitis B virus and its role in the development of primary hepatocellular carcinoma. *J Theor Biol* **159**: 215–240, 1992.	Academic Press Ltd
Payne R J H, Nowak M and Blumberg B S. The dynamics of hepatitis B virus. *Proc Natl Acad Sci USA* **92**: 6542–6546, 1996.	National Academy of Science, USA
Maupas Ph, Larouze B, London W T, Werner B, Millman I, O'Connell A, Blumberg B S, Saimot G and Payet M. Antibody to hepatitis-B core antigen in patients with primary hepatic carcinoma. *Lancet* **2**: 9–11, 1975.	The Lancet Ltd
Larouze B, London W T, Saimot G, Werner B G, Lustbader E D, Payet M and Blumberg B S. Host responses to hepatitis B infection in patients with primary hepatic carcinoma and their families. A case/contol study in Senegal, West Africa. *Lancet* **2**: 534–538, 1976.	The Lancet Ltd
Larouze B, Blumberg B S, London W T, Lustbader E D, Sankale M and Payet M. Forecasting the development of primary hepatocellular carcinoma by the use of risk factors: Studies in West Africa. *J Natl Cancer Inst* **58**: 1557–1561, 1977.	United State Government Printing Office
Marinier E, Barrois V, Larouze B, London W T, Cofer A, Diakhate L and Blumberg B S. Lack of perinatal transmission of hepatitis B virus infection in Senegal, West Africa. *J Pediatrics* **106**: 843–849, 1985.	Mosby-Year Book, Inc.

Feret E, Larouze B, Diop B, Sow M, London W T and Blumberg B S. Epidemiology of hepatitis B virus infection in the rural community of Tip, Senegal. *Am J Epidemiol* **125**: 140–149, 1987.	Johns Hopkins University School of Hygiene and Public Health
Blumberg B S and London W T. Hepatitis B virus and the prevention of primary cancer of the liver. Guest Editorial. *J Natl Cancer Inst* **74**: 267–273, 1985.	United State Government Printing Office
Blumberg B S. Proc. International Conference on Prospects for Eradication of Hepatitis B Virus. *Vaccine* Vol. 8, Introduction: pS5; Conclusion pS139, 1990.	Elsevier Science Ltd.
Blumberg B S. Hepatitis B virus, iron and iron-binding proteins. In *Viruses, Immunity and Immunodeficiency*. Szentivanyi A and Friedman H (eds.). Plenum Publishing Corp. pp. 81–99, 1986.	Plenum Publishing Corp
Felton C, Lustbader E D, Merten C and Blumberg B S. Serum iron levels and response to hepatitis B virus. *Proc Natl Acad Sci USA* **76**: 2438–2441, 1979.	National Academy of Science
Lustbader E D, Hann H L and Blumberg B S. Serum ferritin as a predictor of host response to hepatitis B virus infection. *Science* **220**: 423–425, 1983.	American Association for the Advancement of Science
Stevens R G, Beasley R P, Blumberg B S. Iron-binding proteins and risk of cancer in Taiwan. *J Natl Cancer Inst* **76**: 605–610, 1986.	United States Government Printing Office
Stevens R G, Graubard B I, Micozzi M S, Nerushi K and Blumberg B S. Moderate elevation of body iron level and increased risk of cancer occurrence and death. *Int J Cancer* **56**: 364–369, 1994.	John Wiley & Sons. Inc.
Blumberg B S, Melartin L, Guinto R A and Werner B. Family studies of a human serum isoantigen system (Australia antigen). *Am J Human Genet* **18**: 594–608, 1966.	Cambridge University Press

Blumberg B S, Sutnick A I, London W T and Melartin L. Sex distribution of Australia antigen. *Arch Intern Med* **130**: 227–231, 1972.	American Medical Association
Drew J S, London W T, Lustbader E D, Hesser J E and Blumberg B S. Hepatitis B virus and sex ratio of offspring. *Science* **201**: 687–692, 1978.	American Association for the Advancement of Science
Drew J, London W T, Blumberg B S and Serjeantson S. Hepatitis B virus and sex ratio on Kar Kar Island. *Human Biol* **54**: 123–135, 1982.	Wayne State University Press
Blumberg B S, Friedlaender J S, Woodside A, Sutnick A I and London W T. Hepatitis and Australia antigen: Autosomal recessive inheritance of susceptibility to infection in humans. *Proc Nat Acad Sci* **62**: 1108–1115, 1969.	National Academy of Sciences, USA
Blumberg B S. The natural history of Australia antigen. In *Analytic and Experimental Epidemiology of Cancer. Proceedings of the 3rd International Symposium of the Princess Takamatsu Cancer Research Fund.* Nakahara W, Hirayama T, Nishioka K and Sugano H (eds.). University of Tokyo Press, pp. 109–151, 1973.	University of Tokyo Press
Langendorfer A, Davenport W, London W T, Blumberg B S and Mazzur S. Sex-related differences in transmission of hepatitis B infection in a Melanesian population. *Am J Phys Anthropol* **64**: 243–254, 1984.	John Wiley & Sons. Inc.
Blumberg B S. Sex-related aspects of hepatitis B infection and its consequences. In *Hepatitis B: A sexually transmitted disease in heterosexuals*, Piot P and Andre F E (eds.). Elsevier Science Publishers B.V. (Biomedical Division), pp. 3–7, 1990.	Elsevier Science Ltd
Sutnick A I, Cerda J J, Toskes P P, London W T and Blumberg B S. Australia antigen and viral hepatitis in drug abusers. *Arch Int Med* **127**: 939–941, 1971.	American Medical Association

Blumberg B S, London W T, Sutnick A I, Camp F R Jr, Luzzio A J and Conte N F. Hepatitis carriers among soldiers who have returned from Vietnam. Australia antigen studies. *Transfusion* **14**: 63–66, 1974.	Lippincott Raven Publishersy
Venkateswaran P S, Millman I and Blumberg B S. Effects of an extract from *Phyllanthus niruri* on hepatitis B and woodchuck hepatitis B viruses: *In vitro* and *in vivo* studies. *Proc Natl Acad Sci USA* **84**: 274–278, 1987.	National Academy of Sciences, USA
Thyagarajan S P, Subramanian S, Thirunalasundari T, Venkateswaran P S and Blumberg B S. Effect of *Phyllanthus amarus* on chronic carriers of hepatitis B virus. *Lancet* **2**: 764–766, 1988.	The Lancet Ltd
Unander D W, Venkateswaran P S, Millman I, Bryan H H and Blumberg B S. Phyllanthus species: Sources of new antiviral compounds. In *Advances in New Crops*, Janick J and Simon J E (eds.). Timber Press Inc., Portland, OR pp. 518–521, 1990.	Timber Press, Inc.
Block T M, Lu X, Platt F M, Foster G R, Gerlich W H, Blumberg B S and Dwek R A Secretion of human hepatitis B virus is inhibited by the imino sugar N-butyldeoxynojirimycin. *Proc Natl Acad Sci USA* **91**: 2235–2239, 1994.	National Academy of Sciences, USA
Lechat M F, Shrager D I, Declercq E, Bertrand F, Blattner W A and Blumberg B S. Decreased survival of HTLV-1 carriers among leprosy patients from Zaire, a historic prospective study. *JAID&HR* (In press, 1997.	Lippincott-Raven Publishers
Blumberg B S and Blumberg J L. Bernard Connor (1666–1698) and his contribution to the pathology of ankylosing spondylitis. *J Hist Med and Allied Sci* **13**: 349–366, 1958.	Oxford University Press
Blumberg B S. Creativity. A sermon given at Balliol College Chapel on Rogation Sunday. Oxford, England. May 5, 1991.	permission not required

Blumberg B S. A Sermon. Balliol College Chapel, Oxford University, Oxford, England. June 12, 1994.	permission not required
Blumberg B S. A Letter from Oxford. Science in Contemporary Movies. In Fox Chase Cancer Center Scientific Report 1992–1993. Fox Chase Cancer Center, Philadelphia, PA USA.	permission not required
Blumberg B S. Reflections on a two-wheel way of life. London Telegraph, Saturday, October 8, 1994.	The Daily Telegraph, London
Blumberg B S and Fox R C. The Daedalus effect: Changes in ethical questions relating to hepatitis B virus. *Ann Int Med* **102**: 390–394, 1985.	American College of Physicians
Blumberg B S. Scientific and artistic process. Awards Ceremony, Pennsylvania Academy of the Fine Arts, Philadelphia, PA USA, May 14, 1980.	permission not required
Blumberg B S. An Address. Perceptions of Science. Union College Bicentennial Celebration, Schenectady, New York, USA, February 25, 1995.	permission not required
Feitelson M A, Millman I, Halbherr T, Simmons H and Blumberg B S. A newly identified hepatits B type virus in tree squirrels. *Proc Natl Acad Sci USA* **83**: 2233–2237, 1986.	National Academy of Sciences, USA
Blumberg B S. Complex interactions of hepatitis B virus with its host and environment. *J Royal College Physicians London* **29**: 31–40, 1995.	Royal College of Physicians